NASA SP-4017

ASTRONAUTICS AND AERONAUTICS, 1972

Chronology of Science, Technology, and Policy

Text by
Science and Technology Division
Library of Congress

Sponsored by
NASA Historical Office

Scientific and Technical Information Office   1974
NATIONAL AERONAUTICS AND SPACE ADMINISTRATION
Washington, D.C.

Foreword

NASA's twelfth annual chronology of events in astronautics and aeronautics reflects a U.S. space program redefined to be less costly and perhaps less spectacular in the 1970s than in the previous decade—yet this volume records a continued and indeed matured response to the challenges of space, both in scientific exploration and in practical uses.

The year 1972 closed with the triumphant end of the Apollo program. The mass of data from the manned explorations of the moon and data still being collected by instruments left on the moon's surface combined with events on earth to intensify awareness of our planet, especially the fragility of its environment and the preciousness of its resources. In addition to bringing insights into how the earth evolved and into the history and nature of the moon, the flights—culminating in the new space records and investigations of *Apollo 16* and *Apollo 17* during 1972—let us see our planet "as it really is," influencing long-range planning for future activities in space, as well as in many other areas of life. The assimilation and interpretation of lunar data will occupy many scientists around the world for at least a decade to come.

Several major events underscored the shift in emphasis from lunar exploration toward study of our planet. President Nixon announced that the United States would proceed to develop the reusable space shuttle, to reduce the cost and risk of using space. NASA selected a configuration with reusable boosters and orbiter and let the initial contracts to develop a practical and economical system for both manned and unmanned missions in the 1980s. Meanwhile, all Skylab components were delivered and flight crews were readying for the 1973 missions that would feature prolonged research in astronomy, earth resources, materials processing, and life processes, all under the weightless conditions of earth orbit.

A major step forward in space applications, turning space knowledge and technology into practical benefits, was the launch of *Erts 1* to begin global observation of the earth's resources and environmental factors. Data from *Erts 1*, from Skylab, and from remote-sensing aircraft were to be available to users worldwide. Weather and communications satellites—including *Nimbus 5, Noaa 2, Anik 1* for Canada, and two Intelsats—were orbited to continue almost routine service to man from space. NASA's technology utilization program expanded the nation's technology base and attacked problems in health and medical care, air and water pollution, transportation, urban construction, and fire safety.

Although a stretched-out space program reduced plans for a future Grand Tour of all the outer planets to a focus on Jupiter and Saturn, and the NERVA nuclear rocket program was reoriented toward a small, high-energy, nuclear propulsion system for planetary missions, 1972 produced significant firsts in space science. *Pioneer 10* made the first probe of the Asteroid Belt and flew on toward its December 1973 rendezvous with Jupiter and eventual escape from the solar system. *Pioneer 11* was to follow in 1973 for a meeting with Jupiter in 1974. *Mariner 9*, launched in 1971, completed its mission during 1972, photographically mapping the entire globe of Mars and sending back information that revised our understanding of that planet. We now know that Mars is geologically active; free-flowing water may have existed in its ancient past. *Mariner 9* data laid the groundwork for 1975–1976 Viking missions to softland on Mars and search for evidence of life. *Mariner 10,* to be launched in late 1973, would make the first explorations of Mercury, flying by Venus on its way.

Oao 3 (Copernicus) opened new channels of information into the distant reaches of the universe. The Orbiting Astronomical Observatory made the first ultraviolet observations of Uranus, observed a supernova, and, at this writing, has returned compelling evidence of the existence of black holes in space. *Explorer 47* and *48* joined satellites launched in previous years to extend our knowledge of physics and astronomy.

International cooperation took unprecedented steps in 1972. A U.S.–U.S.S.R. agreement on cooperative use of outer space included plans for an Apollo-Soyuz docking mission in 1975; another provided for cooperation in science and technology. The European Space Research Organization worked on plans to develop a space laboratory to be flown on NASA's shuttle. NASA launched four satellites for other nations and four cooperative spacecraft, and the President announced a U.S. policy of launch assistance to other countries and international organizations.

NASA's major efforts in aeronautics were directed chiefly toward quieter engines; efficient, short-haul air transportation; and reduction of exhaust emissions. The supercritical wing was flight-tested on the Navy T–2C trainer and the Air Force TF–8A jet aircraft in joint advanced transport technology projects to permit savings in structural weight and increases in cruise speed without increasing power. In the NASA–USAF program to develop technology for landing manned space vehicles horizontally, the M2–F3 lifting body completed its 27th and final flight, having reached a speed of 1718 kilometers (1066 miles) per hour and a 21 800 meter (71 500 foot) altitude. The program was to continue in 1973 with flights of the X–24B in a new configuration.

Thus, designed for realistic progress at a constant budget level of $3.4 billion, NASA's 1972 program in space and in the air was nonetheless marked by the excitement of new tasks, new achievements, and the continued increase in man's knowledge in many fields.

Willis H. Shapley
Associate Deputy Administrator

February 11, 1974

Contents

	PAGE
Foreword	iii
Associate Deputy Administrator Willis H. Shapley	
Illustrations	vii
Preface	ix
January	1
February	41
March	77
April	127
May	169
June	213
July	247
August	279
September	305
October	335
November	371
December	403
Summary	447
Appendix A: Satellites, Space Probes, and Manned Space Flights, a Chronicle for 1972	457
Appendix B: Chronology of Major NASA Launches, 1972	485
Appendix C: Chronology of Manned Space Flight, 1972	489
Appendix D: Abbreviations of References	493
Index and List of Abbreviations and Acronyms	497

Illustrations

	PAGE
President Nixon and NASA Administrator Fletcher examine a model of the space shuttle	3
An instrumented B-57 aircraft tests methods to detect clear-air turbulence	44
A manned helicopter lands automatically	57
Pioneer 10 in preparation for launch toward Jupiter, with the plaque to be carried for readers in space	82
A hospital surgery team applies spacecraft sterilization techniques	92
Dr. Fletcher, Dr. Low, and Dale Myers explain the space shuttle configuration	103
Apollo 16 EVA routes at the Descartes lunar landing site	143
Astronaut Young works on the lunar surface	145
Stars in the Milky Way photographed from the moon	147
Dr. Kvenvolden displays a rock thought to hold evidence of the beginning of photosynthesis on earth	159
Dr. Jones with a wind-tunnel model of the pivot-wing aircraft	192
Signing of the U.S.-U.S.S.R. space cooperation agreement	196
Docked Apollo-Soyuz Test Project spacecraft models	196
A mobile medical gas analyzer and an ultrathin, low-volt switching circuit, from space technology	249
Erts 1 in preparation for launch	268
First *Erts 1* photos are examined by Dr. Clark, Dr. Fletcher, and Charles Mathews	269
Atlas-Centaur stages arrive at Cape Kennedy for the *Oao 3* launch	296
Flight Research Center marks its 25th anniversary	312
Astronauts Thornton, Bobko, and Crippen emerge from the Skylab medical experiments altitude test	324
The Skylab airlock module in the clean room	324
The Bell X-1 breaks the sound barrier—Oct. 14, 1947	351
The X-24B lifting body arrives at Flight Research Center	351
Potential Mars landing sites for 1976 Viking spacecraft	376

	PAGE
A laser instrument detects sulfur-dioxide air pollution	391
Apollo 17 liftoff lights the night sky	412
Astronaut Cernan checks out the Rover	414
Geologist-Astronaut Schmitt works beside a boulder on the moon	415
The crescent earth rises over the moon, seen from *Apollo 17*	417
Evans retrieves film cassettes during space walk	418
"Picture of a manned spacecraft," cartoon by Scott Long	450

Preface

Astronautics and Aeronautics for 1972, twelfth in this series of annual chronologies, gives a brief day-by-day account of key events of the year in space and in the air. It pinpoints dates, actors, new technology, decisions, programs, achievements, and early evaluations of results, with notations to sources. Some of the indirectly related technological and scientific events, some indications of the impact on society and national and international concerns, and a sampling of public and official reaction also are noted. Not intended to develop a full picture, the brief references, when used with the index and cross-references, can nevertheless outline the progress of a given subject during the year and point to origins in previous years.

Designed primarily for reference use within the Federal Government, the volume is also offered to the public. The principal sources are those immediately available in NASA, other Federal agencies, Congress, the professional societies, as well as the journals and the press, with verification from participants. Often details that could be lost with time can be captured and questions clarified while the participants are still at hand.

General editor of this volume was the Deputy Director of the NASA Historical Office, Frank W. Anderson, Jr., and the technical editor was Mrs. Carrie E. Karegeannes. The entire Historical Office participated in source selection, review, and publication. Archivist Lee D. Saegesser collected current documentation. The Science and Technology Division of the Library of Congress, under an Exchange of Funds agreement, drafted monthly segments in comment edition form; which were circulated for corrections, additions, and use. At the end of the year, the entire manuscript was reworked to include comments received and additional information that was not available when the monthly segments were prepared. At the Library Mrs. Patricia D. Davis, Mrs. Carmen Brock-Smith until October and Mrs. Nancy L. Brun from December, and Mrs. Shirley M. Singleton carried principal responsibility. Arthur G. Renstrom of the Library prepared the monthly indexes and Informatics TISCO, Inc., prepared the detailed index for the annual volume. The index, indispensable to the usefulness of a chronology, also serves as a glossary of abbreviations and acronyms.

Appendix A. "Satellites, Space Probes, and Manned Space Flights, 1972," Appendix B, "Chronology of Major NASA Launches, 1972," and Appendix C, "Chronology of Manned Space Flight, 1972," were prepared by Leonard C. Bruno of the Library of Congress. Appendix D, "Abbreviations of References," was prepared by Mrs. Davis.

Without the assistance of many individuals throughout NASA and other Federal agencies, the content of this volume would be less reliable and complete. Comments, additions, and criticisms are always welcomed by the NASA Historical Office.

Monte D. Wright
Director, NASA Historical Office

January 1972

January 1: *Apollo 15* Astronauts David R. Scott, James B. Irwin, and Alfred M. Worden participated in pregame pageant at 1972 Orange Bowl Classic football game in Miami. Pageant, "The First Decade," was dedicated to U.S. space program. (*NASA Activities*, 2/15/72, 35)

January 2: Research in U.S. was shifting from pure to practical, Thomas O'Toole wrote in *Washington Post*. "Changes that stress practicality have already been made to parts of U.S. science, but they're scarcely detectable against the $16 billion the federal government spends every year to support U.S. science and technology. One of the few programs of any size begun by the Nixon administration is a $49 million project at the National Science Foundation code-named RANN—for Research Applied to National Needs. RANN supports a host of small programs aimed at improving things like subways, cancer treatment, earthquake forecasts and sewers."

Changes within NASA required "little new money, but they serve to spotlight the emphasis the Nixon administration places on the practical goals of science. In the last six months NASA has taken on study projects from the Departments of Commerce, Interior, and Transportation—none of which involves space exploration. The space agency also has a contract from the National Science Foundation to investigate the use of solar cells (used on most unmanned spacecraft) to electrify homes and buildings. It has another contract from the Department of Housing and Urban Development to look into ways an apartment complex might recycle its sewage and water the way future astronauts will do inside their spacecraft." (*W Post*, 1/3/72, A1)

January 3–7: NASA's *Mariner 9* Mars probe (launched May 30, 1971, and inserted in Mars orbit Nov. 13, 1971) began systematic mapping of Mars. Full 360° of longitude for band of latitudes would be mapped every 20 days. Mapping had been delayed because major dust storm on planet had obscured visibility. By Jan. 7, after 55 days in Mars orbit, *Mariner 9* had responded to 20 000 ground commands and had taken 3500 pictures. Dust was clearing and mission plan had been revised to meet mission objectives to greatest possible extent. (NASA proj off)

January 3: National Science Teachers Assn. had received more than 15 000 applications for participation in NASA Skylab Student Project, NASA announced. NSTA was managing project to stimulate interest in science and technology by promoting participation of U.S. students in grades 9–12 in experiments, demonstrations, or activities to be performed by astronauts during 1973 Skylab missions. NASA would select 25 proposals on basis of compatibility with Skylab requirements. Selectees and their teachers would attend Skylab Educational Conference and award presentation at Kennedy Space Center at Skylab launch time. NASA would build required hardware in consultation with students. Regional and national selectees would be announced in April. (NASA Release 72–1)

January 3

- NASA announced issuance of requests for proposals (RFPs) for six-month studies to evaluate new propulsion systems for ensuring quiet and clean operation of short takeoff and landing (STOL) aircraft. Evaluation would include system noise and emission levels, thrust performance, size, and weight. Systems would be investigated for use with propulsive-lift concepts in which jet engines helped produce lift for takeoff and landing on short runways. Proposals were due Jan. 31. At least two companies would be selected. NASA planned to issue RFPs for quiet, clean STOL experimental engine in late 1972. (NASA Release 72-2)
- Aerospace industry outlook for 1972 was described in *Aviation Week & Space Technology* editorial: "Much of the political miasma and public criticism . . . appears to be dissipating. The cruel bite of unemployment . . . has made it politically less attractive to attack technology. The legislators who were glad to equate their anti-technology votes with a blast at skin cancer now find these votes equated with stimulating unemployment. The industry will continue to be under vigorous public and legislative scrutiny and deserves it. But it has shown the first signs of fighting back and beginning to state its own very valid case. Hopefully, this will continue during the election year in an effort to force politicians to assess technology more accurately in their appeals to the electorate." (Hotz, *Av Wk*, 1/3/72, 7)
- Science and Technology Div. of New York Public Library was closed after failing to meet $86 000 budget deficit by $9000. Library officials had said previously they expected to raise required funds and reopen by mid-January. (Andelman, *NYT*, 1/1/72, 18; NYPL PIO)
- In interview published in *Time* magazine President Nixon discussed impact of his impending visit to People's Republic of China: ". . . what really matters here is not the fact that the trip to China is announced" but results of trip. "Our people have become accustomed to the spectaculars. It is exciting. A trip to China is like going to the moon." (*Time*, 1/3/72, 14-5)

January 4: Apollo 16 lunar module (LM) pilot Charles M. Duke, Jr., was admitted to Patrick Air Force Base Hospital for treatment of bacterial pneumonia. Duke's illness was not expected to have serious impact on training preparations for Apollo 16 launch, scheduled for March 17. (MSC Release 72-02)

- Dr. Hannes O. Alfven, professor of applied physics at Univ. of California at San Diego and winner of 1970 Nobel Prize in physics, was named to receive U.S.S.R.'s 1971 Lomonsov Gold Medal for "outstanding achievements in plasma physics and astrophysics." Award was highest granted by Soviet Academy of Sciences. (UPI, *LA Times*, 1/5/72)
- Completion by U.S.S.R. of world's first fast-breeder nuclear reactor for commercial purposes was announced by *Izvestia*. Reactor, at Shevchenko on Caspian Sea, would produce 250 000 kw of nuclear power to convert salt water from Caspian Sea into fresh water. (Shabad, *NYT*, 1/15/72, 11)

January 5: President Nixon announced decision U.S. should develop space shuttle system. In statement released from San Clemente, Calif., following 45-min meeting with Dr. James C. Fletcher, NASA Administrator, and Dr. George M. Low, Deputy Administrator, President said: "I have decided today that the United States should proceed at once with the development of an entirely new type of space transportation system designed to help transform the space frontier of the 1970's into familiar territory, easily accessible for human endeavor in the 1980's and 1990's.

"This system will center on a space vehicle that can shuttle repeatedly from earth to orbit and back. It will revolutionize transportation into near space, by routinizing it. It will take the astronomical costs out of astronautics. In short, it will go a long way toward delivering the rich benefits of practical space utilization and the valuable spinoffs from space efforts into the daily lives of Americans and all people.

"The new year 1972 is a year of conclusion for America's current series of manned flights to the moon. Much is expected from the two remaining Apollo missions—in fact, their scientific results should exceed the return from all the earlier flights together. Thus they will place a fitting capstone on this vastly successful undertaking. But they also bring us to an important decision point—a point of assessing what our space horizons are as Apollo ends, and of determining where we go from here.

"In the scientific arena, the past decade of experience has taught us that spacecraft are an irreplaceable tool for learning about our near earth space environment, the moon, and the planets, besides being an important aid to our studies of the sun and stars. In utilizing space to meet needs on earth, we have seen the tremendous potential of satellites for intercontinental communications and worldwide weather forecasting. We are gaining the capability to use satellites as tools in global monitoring and management of natural resources, in agricultural

January 5: *President Nixon and* NASA *Administrator James C. Fletcher examined a model of the reusable space shuttle at San Clemente, Calif., following the President's announcement that the United States would proceed with shuttle development.*

applications, and in pollution control. We can foresee their use in guiding airliners across the oceans and in bringing televised education to wide areas of the world.

"However, all these possibilities, and countless others with direct and dramatic bearing on human betterment, can never be more than fractionally realized so long as every single trip from earth to orbit remains a matter of special effort and staggering expense. This is why commitment to the space shuttle program is the right next step for America to take, in moving out from our present beachhead in the sky to achieve a real working presence in space—because the space shuttle will give us routine access to space by sharply reducing costs in dollars and preparation time."

It was significant, President said, "that this major new national enterprise will engage the best efforts of thousands of highly skilled workers and hundreds of contractor firms over the next several years. The amazing 'technology explosion' that has swept this country in the years since we ventured into space should remind us that robust activity in the aerospace industry is healthy for every one—not just in jobs and income, but in the extension of our capabilities in every direction. The continued pre-eminence of America and American industry in the aerospace field will be an important part of the shuttle's 'payload.'"

Shuttle program would "give more people access to the liberating perspectives of space, even as it extends our ability to cope with physical challenges of earth and broadens our opportunities for international cooperation in low-cost, multi-purpose space missions."

President quoted Oliver Wendell Holmes: "We must sail sometimes with the wind and sometimes against it, but we must sail, and not drift, nor lie at anchor." President concluded, "So with man's epic voyage into space—a voyage the United States of America has led and still shall lead." (*PD*, 1/10/72, 27–8; Transcript of Fletcher, Low press conference, 1/5/72)

- Dr. James C. Fletcher, NASA Administrator, released statement from San Clemente, Calif., following President's announcement of decision to proceed with development of space shuttle. Decision was consistent with NASA FY 1972 budget and was "a most historic step in the nation's space program—it will change the nature of what man can do in space. By the end of this decade the nation will have the means of getting men and equipment to and from space routinely, on a moment's notice if necessary, and at a small fraction of today's cost. This will be done within the framework of a useful total space program of science, exploration, and applications at approximately the present overall level of the space budget."

Dr. Fletcher described shuttle as "airplane-like orbiter, about the size of a DC–9 . . . capable of carrying into orbit and back again to earth useful payloads up to 15 feet [4.6 meters] in diameter by 60 feet [18 meters] long, and weighing up to 65,000 lbs [29 500 kg]. Fuel for the orbiter's liquid-hydrogen liquid-oxygen engines will be carried in an external tank that will be jettisoned in orbit." Orbiter would be launched by unmanned booster and would operate in space for one week, during which crew could launch, service, or recover unmanned spacecraft; would perform experiments; and, in future, would resupply and restaff space modules brought to space by shuttle.

He cited reasons "space shuttle is important and is the right step in manned space flight and the U.S. space program." First, "the shuttle is the only meaningful new manned space program which can be accomplished on a modest budget. Second, the space shuttle is needed to make space operations less complex and less costly. Third, the space shuttle is needed to do useful things. Fourth, the shuttle will encourage greater international participation in space flight."

Dr. Fletcher said NASA and contractors would focus until late February on technical areas, including comparisons of pressure-fed liquid- and solid-fuel rocket-motor options for shuttle's booster stage, before NASA issued requests for proposals in the spring. "This summer we will place the space shuttle under contract and development work will start."

At press conference in San Clemente following release of shuttle statements, Dr. Fletcher and Dr. Low discussed President's decision. Dr. Fletcher said President was "particularly anxious that I stress the international aspects of this. This program will be open to all nations of the world, and it is his hope some day that foreign visitors from all over the world will be able to participate by moving to and from space in the space shuttle."

NASA anticipated that "at its peak, the direct employment on the space shuttle will be of the order of 50,000 individuals. Of course, a great many of those will be employed in this area."

Responding to question, Dr. Fletcher said decision on shuttle launch site, "which will be the same as the landing site," depended on recommendation of launch site selection group. "It is not necessarily Cape Kennedy or any other site at this time." (Text: Transcript)

- Ames Research Center, Langley Research Center, and Manned Spacecraft Center had received over 250 proposals for using space-related technology to solve problems of air pollution, water pollution, solid waste management, and clinical medicine, NASA announced. Proposals were in response to request for ideas to demonstrate possible applications of NASA-developed technology to public problems. They would be reviewed by experts and one or more projects in each category might be selected for maximum $75 000 contract. (NASA Release 72-3)

- Results of computer study made by Massachusetts Institute of Technology for International Club of Rome, organization of distinguished world social scientists, were described in *Washington Post*. Study, made with dynamic worldwide model, traced effects of population, capital investment, geographical space, natural resources, food production, and pollution upon each other. Findings had shown "crisis level" 40 or 50 years away with few remaining options. Current age might be golden age, with higher life quality than in future. Next century might find four-sided dilemma—suppression of industry by shortage of natural resources, decline of population because of pollution, limits on population by food shortage, or population collapse from war, disease, and stresses of overcrowding. Birth control might be self-defeating by bringing improved food supply and living standards, causing resurgence of population growth; poor countries might have no "realistic hope" of reaching living standards of rich countries. Disparity between developed and underdeveloped nations might be equalized as much by decline in developed nations as by improvement in underdeveloped nations; rich industrial societies might be "self-extinguishing"; and poor countries might be unwise to persist in industrialization. "They may be closer to an ultimate equilibrium with the environment than industrialized nations." (Sterling, *W Post*, 1/5/72, A16)

- Federal Pay Board rejected contract settlements that would have provided first-year increase of 12% to more than 100 000 aerospace industry workers in 1973. (Shabecoff, *NYT*, 1/6/72, 1)
- Erhard Milch, pioneer of German civil aviation who had been in charge of supply and development for wartime Luftwaffe, died of liver ailment at age 79. (AP, *W Star*, 1/30/72, B10)

January 6: Apollo 16 command module (CM) pilot Thomas K. Mattingly II described some of Apollo 16 experiments at Manned Spacecraft Center press conference. ALFMED (Apollo light-flash moving-emulsion detector) experiment would test hypothesis that light flashes seen by astronauts on previous Apollo missions were "high energy particles, nuclei of heavy particles coming from the sun with a great deal of energy penetrating and causing some response . . . in the eyeball by perhaps triggering one of the photocells in your eye. It may . . . be hitting the . . . optic nerve, or it may be going into the head or some other location and creating a stimulus which appears to be a visual clue."

MEED (microbial ecological evaluation device), a package containing organisms, would be carried to determine effect of ultraviolet (UV) radiation, solar radiations, and zero g. MEED would be opened during extravehicular activity (EVA) and exposed to sun; parts of experiment would be exposed to UV radiation and parts to different bands of solar spectrum. Organisms would be covered by superimposed filters; colony of one material could be exposed to different wavelengths by using different filters. Biostack, West German-made package of organisms in silicon biological material, would be carried to investigate effects of high-energy particles on the organisms.

Photographic tasks would include photography of solar corona and two new areas at galactic poles, Gegenschein experiment, and UV photography of earth. (Transcript)

- NASA's *Mariner 9* Mars probe photographed area of Mars 800 km (500 mi) from south pole from 3335-km (2072-mi) altitude. Photos, released to press Jan. 11, showed pits and hollows—including two large closed basins 16 km (10 mi) wide—with more detail than seen before. According to Dr. Carl E. Sagan of Cornell Univ., spots occurred in sizes from more than 160 km (100 mi) to size of Yankee Stadium in New York. Many blotches seemed to be associated with craters, indicating that they might be wind shadows where hill or other feature had protected downwind area from dust deposition. (Sullivan, *NYT*, 1/12/72, 1)
- NASA announced establishment of joint Civil Aeronautical Research and Development (CARD) Review Group by National Aeronautics and Space Council, Dept. of Transportation, and NASA to help maintain close relationships among agencies with interests and responsibilities in aeronautics. (NMI 1052.160)
- Sen. Walter F. Mondale (D-Minn.) in press release called for "full and open debate" on space shuttle development issue, "leading to the rejection of this wasteful program." It was "typical" that Nixon

Administration "can squander $6.5 billion to fly four people into orbit when it refuses to invest less than one-third that amount to provide desperately needed day care and development programs for millions of children." Space shuttle was "many times worse than SST" in "magnitude of its cost, in the folly of its concept and in its damage to the country." (Text)

- President Nixon's decision to support development of space shuttle was expected to stir "some loud opposition in Congress," *New York Times* said, "but probably not enough to sidetrack initial funding. . . ." Opposition was expected "from those who question the President's decision to back an expensive new space effort at a time when so many down-to-earth social programs have to wait in line for a share of the Federal dollar." NASA officials had expressed "confidence that they have sufficient Congressional support for an initial shuttle appropriation." (Wilford, *NYT*, 1/6/72, 14)
- *Soviet Space Programs, 1966-70* (Senate Document 92-51) was released by Senate Committee on Aeronautical and Space Sciences. The 670-page report—prepared by Library of Congress under direction of Dr. Charles S. Sheldon II, Chief of Science Policy Research Div.—noted that Soviet space program remained "a strong and growing enterprise" unhindered by budgetary strain and undimmed by deaths of three Soyuz cosmonauts in 1971. Current level of Soviet space activity exceeded that of U.S. at its peak in 1966, with perhaps 2% of Soviet gross national product devoted to space (U.S. spent about 1% of its GNP at peak). Although U.S. launch record had been declining steadily since 1966, Soviet launches had continued to climb, at least until 1970. Contrary to assertion that it was concentrating its efforts on unmanned program rather than manned program, U.S.S.R. apparently was still planning manned lunar landing and might spend $49 million on both advanced unmanned and manned lunar programs. U.S. would have committed $35 billion to total Apollo program when complete.

  In unmanned research satellite program "the level of activity currently runs ahead of the corresponding level of work at NASA." Compendium of every known Soviet launch that had achieved orbit showed that, along with scientific satellites, Cosmos program consisted of unmanned tests of spacecraft designed to carry cosmonauts, lunar and planetary probes that failed to leave earth orbit, unsuccessful Mars probes, and military research and development and observation satellites that accounted for 80% of Cosmos payloads. According to study, Cosmos series through April 1971 had included at least 16 tests of fractional orbital bombardment system (FOBS) and 22 tests of experimental spacecraft designed to inspect and destroy other nations' observation and navigation satellites.

  Soviet achievements in near future were dependent on success or failure of G booster (in Saturn V class) or similar booster. Studies of moon with Luna series spacecraft would continue, perhaps with spacecraft launched by advanced booster carrying both rovers and sample returners. Planetary program would continue, probably with flyby mission to Jupiter and landing mission to Mars. Dr. Sheldon stressed that "the Soviet program is not a sham. It may be exploited for political purposes, but it is real and it is pursued in earnest." (Text)

- Super Arcas sounding rocket, launched by NASA from Wallops Station, carried Pennsylvania State Univ. experiment to 66.8-km (41.5-mi) altitude to measure electrical conductivity in upper atmosphere with parachute-borne payload and to flight-certify payload and vehicle for later launches. Although some data were missed because of lower apogee than planned, flight was judged successful. (NASA Rpt SRL)
- National Science Foundation released *Scientists, Engineers, and Physicians from Abroad: Trends Through Fiscal Year 1970*. Immigrant scientists and engineers admitted to U.S. in FY 1970 totaled 13 300—one third more than in 1969 and two and one half times 1965 number. Increases over 1965 level had occurred under October 1965 amendments to immigration law. More than 50% of immigrant scientists and engineers in 1970 had last lived in Asia. In 1965, only 10% of smaller total had been from Asia. Total of 2900 Indian scientists and engineers in 1970 had been largest number admitted from any one country over last 20 yrs. (Text)

January 7: Launch of Apollo 16 manned lunar landing mission, scheduled for March 17, was postponed by NASA because of problems with suit fitting, lunar module (LM) battery, and command module (CM) docking-ring jettison device. Apollo 16 would be launched April 16—with LM landing on moon April 20 and lifting off moon April 23—and would return to earth April 28. (NASA Release 72-8; AP, B *Sun*, 1/8/72, A5)

- Dr. Robert C. Seamans, Jr., Secretary of the Air Force, issued statement in Washington, D.C., on President Nixon's Jan. 5 decision to proceed with development of space shuttle. Decision "initiates a program which holds great promise for scientific and technological advances in the interests of the nation and all mankind. We are also interested in its potential as a means for performing our military mission more efficiently and economically. The Air Force role in the program is to provide NASA data to help assure that the Shuttle will be of maximum utility to the DOD and we are pleased that the proposed vehicle is configured to meet potential DOD needs. We will continue our close coordination with NASA as their development program proceeds." (Text)
- Space shuttle concept endorsed by President Nixon made it "all but certain" that Kennedy Space Center would be selected for shuttle's launch site, *New York Times* said. Space experts seemed to feel that only by "converting the assembly buildings, launching pads and communications center" at KSC "would it be possible to develop a shuttle port for $300 million." (*NYT*, 1/7/72, 7)
- Newspaper editorials commented on decision to develop space shuttle.

    Kansas City Star: "The next great step in space for the United States seems assured. President Nixon has pressed the go-ahead button and apparently it will be largely a matter of adequate funding by Congress. In view of the record of space accomplishments, there is not much doubt that the planned space shuttle program will be fruitful both from the standpoint of technological success and long-range scientific benefits." (*KC Star*, 1/7/72)

    St. Louis Globe-Democrat: "There are so many potentialities for the shuttle that it is hard to envision all of them. But there is no doubt that if it is completed space travel for the average citizen may become a distinct possibility in the 1980s." (*St Louis G-D*, 1/7/72)

Christian Science Monitor said "space discovery versus human needs argument should not be seen as an either/or issue. This is so even leaving aside the jobs creation or economic return arguments." *Monitor* believed that "a space shuttle program, because it is so practically linked to fulfilling man's age-old vision of mastering the heavens, will unleash more constructive human energy than it will consume. And this excess will make meeting earthside demands the easier." (*CSM*, 1/7/72)

Atlanta Journal Constitution: "The program . . . has the almost immediate benefit of employing some 50,000 highly trained aerospace workers who have faced a dismal job picture as the Apollo program has phased out. And, again, the proposal demonstrates that imagination and vision and a willingness to dare are still part of the American way of doing things." (*Atlanta JC*, 1/7/72)

Milwaukee Journal: "The shuttle offers the opportunity to make space flight fairly routine, of taking large numbers of people and supplies in and out of space, of manning earth resource space stations continuously, of monitoring and repairing communications and resource satellites. It is a worthwhile program." (*MJ*, 1/7/72)

Los Angeles Times said military aspect of space shuttle needed clarification. "Most taxpayers welcomed the decision to end the wasteful competition between NASA and the Air Force on the shuttle and orbiting space station." NASA had not explained how shuttle's civil and military functions would be kept separate. "No American wants the Soviet Union to tip the balance of power with its own space program. But it might help avert a new confrontation if the military applications were enumerated along with a statement of whether the American government thinks shuttles are bound by the international agreement barring nuclear weapons from outer space." (*LA Times*, 1/7/72)

Seattle Times: "The history of warfare shows that victory usually goes to the one who occupies the high ground. Near-earth space is the 'high ground' of modern military planning. In approving a full go-ahead for . . . space shuttle program, President Nixon this week served notice that this nation intends to occupy the 'high ground.'" (*Seattle Times*, 1/7/72)

- Half-hour documentary film "The Worlds of von Braun" was televised by WMAL–TV in Washington, D.C. Documentary, filmed at NASA Wallops Station, included footage never before televised, from NASA, U.S.S.R., and moon. In film, Dr. Wernher von Braun, NASA Deputy Associate Administrator for Planning, discussed his careers as World War II rocket expert and with NASA in interview with WMAL–TV Public Affairs Director James Clarke. *Washington Post* TV critic Victor Cohn said of film: "Wernher von Braun is a compelling man and a true believer . . . and the joy is that of seeing any man so thoroughly absorbed in a discipline that he absorbs you, even at the distant end of a TV tube." (NASA Special Ann, 1/4/72; *W Post*, 1/7/72, B1)

- U.S.S.R.'s *Cosmos 463* (launched Dec. 6, 1971) and *Cosmos 464* (launched Dec. 10, 1971) had been launched "in quick succession" to observe India-Pakistan war and had been brought down ahead of schedule in Soviet "rush to analyze the pictures," George C. Wilson said in *Washington Post*. "The rapid-fire space shots are fresh evidence that the era of open skies has arrived" even though neither U.S. nor

U.S.S.R. openly acknowledged it. "Instead, both superpowers look down on the other from space, cameras rolling, and mobilize this new space tool for special missions in time of crisis.... Some ... specialists think that since both superpowers depend heavily on their satellites for information, there is now a form of mutual deterrence in space which will keep one side from attacking the other's satellites. However, there is widespread agreement that the United States has nothing to compare with the Soviets' satellite inspection ability." (*W Post*, 1/7/72, A16)

- Lyudmila liquid-hydrogen chamber at joint nuclear research institute in Dubna, U.S.S.R., had been completed, Tass reported. Institute scientist Aleksander Baldin had said chamber would be used with world's largest nuclear accelerator at Serpukhov to study photo-birth of particles—crushing of one proton by high-energy protons. (FBIS–Sov, 1/10/72, L1)

- *San Francisco Examiner* editorial praised U.S. Pay Board's veto of 12% wage increase for aerospace workers: "Even in its ailing state the aerospace industry has continued as a major American exporter. It has thus prevented the serious deficit in this country's balance of trade from becoming truly horrendous. But it holds no monopoly on aerospace technology. Unless it can maintain high productivity and competitive wage costs, foreign-made planes will become as common in our skies as foreign-made cars are on our highways." (*SF Exam*, 1/7/72)

January 8: Sen. William Proxmire (D-Wis.) said in press release that President Nixon's decision to proceed with space shuttle development was "a great mistake, and an outrageous distortion of budgetary priorities" and "the last thing we need at this time." It would keep NASA and aerospace industry occupied but "it will also insure a steady and very substantial drain on the federal budget. And it will insure that other needs will continue to go unmet." (Text)

- Newspapers continued editorial comment on decision to develop space shuttle.

  New York Times: "The space shuttle is far less expensive than the Apollo project, and, fortunately, is not being launched in the atmosphere of Cold War hysteria that attended the birth of the lunar landing effort. On the contrary, NASA has already stated publicly it would welcome cooperation on the space shuttle from other industrially developed nations, including the Soviet Union. The argument that the space shuttle is simply another SST boondoggle has little validity. The potential benefits are far greater, and the space shuttle presents none of the environmental problems that counted so heavily against the supersonic transport." (*NYT*, 1/8/72, 28)

  Detroit News: "It is good that the $24 billion outlay on the Apollo moon program is not to be wasted by the United States dropping out of the space program. Mission Impossible became mission accomplished because there was a national commitment to beat the Russians to the moon. The fact that the race has been won, however, should not blind Americans that it was only the first, tentative step in unlocking the mysteries of space. If we don't press on, others will." (*D News*, 1/8/72)

January 9: *Pravda* article by scientists from Soviet Academy of Sciences Institute of Space Studies described preliminary results from U.S.S.R.'s

Mars 2 and *Mars 3* probes. Spacecraft (launched May 19 and May 28, 1971) had measured atmospheric humidity and surface temperature, photometered atmosphere and surface, and photographed surface. Temperature along route from 58° south latitude, 330° longitude to 30° north latitude, 190° longitude had been determined. Highest temperature—recorded near mid-day region—was 253 K ($-20°C$; $-4°F$). On night side of planet temperature dropped to 183 K ($-90°C$; $-130°F$).

Dust storm that had covered Mars since *Mars 2* reached planet Nov. 27, 1971, was subsiding. "Photometric recordings obtained by means of the red filtre show dark areas corresponding to the Martian 'seas' that were previously absolutely undistinguishable. But the recordings made with the ultraviolet filter again showed bright clouds." Onboard instruments to record emissions in three spectral bands containing lines of hydrogen, oxygen, and argon atoms showed that hydrogen emissions stretched to more than 3400-km (2100-mi) altitude and oxygen atoms concentrated near surface up to 600–800 km (370–500 mi). (Tass, FBIS–Sov, 1/10/72, L3)

- Space shuttle's "subtle costs" were discussed in *New York Times* article by John N. Wilford: "Because NASA must initiate shuttle development within its current $3.2-billion budget framework, other programs will presumably by sacrificed." Possible "victim" was Grand Tour mission planned for late 1970s. "The choice is certain to be unpopular with much of the scientific community." Scientists had criticized NASA's emphasis on "big-money manned projects at the expense of the less expensive science-oriented unmanned missions." Shuttle development also would "restrict the funds available for other scientific projects in NASA and elsewhere in Government." (*NYT*, 1/9/72, 6)

- *New York News* editorial commented on space shuttle decision: "We would like to see the U.S. taking on, too, some of the more exciting space challenges that could be achieved with America's present know-how. But at least Mr. Nixon has furnished assurance that he will keep the U.S. in the space business. We can be thankful for that at a time when undeserved abuse is being heaped on the whole program by people who prefer to shut out the beckoning glitter of the stars so they can concentrate on the mud at their feet." (*NY News*, 1/9/72)

- Interview with Mrs. Ruth Bates Harris, NASA Director of Equal Employment Opportunity, was published by Washington *Sunday Star*. Mrs. Harris had adopted as her philosophy phrase, "We are all fellow astronauts on Spaceship Earth." She had said, "My position is to make sure the space program is for everybody. We have to get the alienated minorities into every NASA level, especially space exploration." (Trescott, W *Star*, 1/9/72, C1)

January 10: European participation in space shuttle development was discussed by NASA Associate Administrator for Manned Space Flight Dale D. Myers during press briefing at Manned Spacecraft Center. European space officials had released to European industry "fairly extensive study . . . leading to the definition of the tug that the Europeans might build for the shuttle to carry." Tug was 3rd stage of shuttle transportation system. Myers said NASA had examined three possible areas for European participation: "One would be in elements, physical pieces of the shuttle that would be managed . . . as the prime contractor manages a subcontractor. We will expect the prime contractors to work directly with the European industries in that case." Europeans would supply money for parts developed by Europe. Second area was tug. "If the tug were developed, we would expect it to be paid for by the Europeans." Third area was sortie can, "experiment-carrying device to be carried in the payload compartment," which would give them "an opportunity to enter into the use of the shuttle very extensively." European space officials would meet with NASA within two months to respond on interest in participating.

Timetable called for selection of industrial contractors in summer 1972, with subsonic and suborbital test flights beginning in 1975, manned orbital tests in 1978, and first fully operational mission by 1980. Site evaluation board would make its next report to NASA within few weeks, and site for shuttle launch and recovery would be selected by spring. One site would be selected initially, with another or others selected later according to program needs. "We don't see the shuttle as a highly limited operation. And, I think that eventually there will be more than one site." Location depended on mission. "Today . . . there is a preponderance of due-east missions out of KSC [Kennedy Space Center], and . . . different azimuths out of the West Coast. We can't go due east out of the West Coast very well . . . because of the populated areas east . . . and we can't get as efficient in polar orbits out of Kennedy . . . so there is a tendency to . . . go to the two coasts. Now whether that really ends up being that way in the shuttle system, we don't know yet." (Transcript)

- Events leading to President Nixon's endorsement of space shuttle development were reported by *Aviation Week & Space Technology:* "The struggle between NASA and OMB [Office of Management and Budget] officials reached the point of acrimony when the latter suggested design changes and demanded to know from NASA engineers the dollar savings that would result. Dr. James C. Fletcher, NASA Administrator, finally appealed directly to the office of the President with a complaint that OMB was attempting to design the U.S. space shuttle rather than merely budget it." President Nixon had removed OMB accountant from conference with NASA and demanded shuttle agreement. "The dispute culminated in Fletcher's triumphal flight to San Clemente . . . to join in the announcement that the space shuttle had been approved by the Administration." (*Av Wk*, 1/10/72, 15)

- NASA notified staff of Civil Service Commission directive canceling special salary rates for engineers, scientists, accountants, mathematicians, and others. Approximately 500 NASA employees affected would be converted to General Schedule and placed on saved pay rates or in higher pay steps with longer waiting periods for step increments, effective Feb. 6. Persons hired to affected positions after Feb. 6 would be paid General Schedule rates. (NASA Hq *WB*, 1/10/72, 2)

- European space leaders were "growing increasingly impatient" over U.S. Government's failure to resolve controversy over method for participation in international air traffic control satellite program (aerosat), *Aviation Week & Space Technology* reported. European Space Research Organization official had said European space community was concerned that U.S. would not honor agreement between Federal Aviation Administration and ESRO for joint backing of twin-satellite systems covering Atlantic and Pacific Oceans, but would back Communications Satellite Corp. proposal that aerosat be developed under international consortium. Concern had arisen "from the fact that the [U.S.] executive branch has not taken a firm stand on what its policy is to be." (*Av Wk*, 1/10/72, 20)

- *Washington Daily News* quoted Harvard Univ. nutritionist Dr. Jean Mayer, Presidential Adviser on Nutrition, as saying some scientists had let their opposition to war, poverty, and pollution lead them into unscientific distortions and exaggerations. As result, laymen were confused about nuclear testing, defoliation, antiballistic missiles, supersonic transport, and extent of hunger and malnutrition in U.S. Dr. Mayer had said credibility of all scientists would be undermined as long as a few insisted on offering opinions "with little basis in fact." (*W News*, 1/10/72, 16)
- *New York Times* editorial praised "encouraging response" of public and major corporations that would enable Science and Technology Div. of New York Public Library to reopen Jan. 17 after shutdown for lack of operating funds [see Jan. 3]. "But it is evident that the fate of the privately financed research divisions has merely been deferred—unless continuous and long-range efforts are made to put this vital facility on a sound financial footing." (*NYT*, 1/10/72, 32)

January 10–13: Third Annual Lunar Science Conference held by NASA and Lunar Science Institute at Manned Spacecraft Center was attended by more than 600 researchers from U.S. and 16 foreign countries. Foreign visitors included Soviet scientists Y. I. Belyayev, M. S. Chupakhin, and K. P. Florenskiy, who participated in conference, consulted with U.S. scientists, looked at lunar samples stored at MSC, and helped select samples to return to their colleagues in U.S.S.R. NASA provided three grams (one tenth ounce) of lunar material collected on *Apollo 14* mission as part of U.S.–U.S.S.R. agreements for exchange of samples. Samples included chip from widely studied crystalline rock, small piece of 9-kg (20-lb) boulder, and core samples soils, and polished thin sections.

Conference papers dealt with scientific results of Apollo lunar landing missions, analysis of Soviet *Luna 16* samples, and results of study of lunar surface by U.S.S.R.'s *Lunokhod 1*. *Apollo 15* Astronauts David R. Scott, James B. Irwin, and Alfred M. Worden discussed their observations of lunar surface made from lunar orbit during July 26–Aug. 7, 1971, mission. Dr. N. C. Costes, Marshall Space Flight Center representative of Apollo Soil Mechanics Investigation Science Team, discussed mechanical properties of lunar soil.

Washington Post later compared conference with First Lunar Science Conference in 1970, when "there were few facts available and many theories." At third conference, "situation seemed to have reversed." Dr. Paul W. Gast, Chief of MSC Planetary and Earth Sciences Div., said, "You're not throwing facts out into a vacuum any more."

Among facts stated at conference were: moon once had strong magnetic field, of which remnants remained to suggest moon once had metallic core up to 280 km (175 mi) thick; lunar crust was 65 km (40 mi) deep and layered; existence of extinct plutonium 244 in *Apollo 14* samples indicated plutonium had been trapped when lunar crust solidified shortly after origin of solar system; and Copernicus

Crater had been gouged 850 to 950 million yrs ago, making process one of more recent major lunar events.

Dr. James R. Arnold of Univ. of California at San Diego reported that moon's radioactive elements were concentrated in two broad plains, Ocean of Storms and Sea of Rains. "As we go away from this region . . . we drop to quite low levels of radioactivity." Scientists agreed that these radiation emissions should not prove hazardous to astronauts who might visit for short period.

Dr. Isidore Adler, Goddard Space Flight Center scientist, reported that map of lunar surface chemistry produced from *Apollo 15* x-ray data indicated highlands of moon contained more aluminum than did plains. Finding supported hypothesis that moon's early crust was rich in feldspar.

Dr. Gary V. Latham of Columbia Univ., Apollo Program chief seismic investigator, debated presence of volcanoes on moon with Cornell Univ. astronomer Dr. Thomas Gold. Dr. Latham said that while main episode of lunar vulcanism seemed to have ceased 3 billion yrs ago, "it's by no means certain that we do not have continuing vulcanism on a very minor scale on the moon." Dr. Gold continued to support theory that moon was dead, rigid body with basins of compacted dust.

Dr. Leon T. Silver of California Institute of Technology proposed theory of "parentless lead" on moon to explain presence in some lunar locations of more lead than could have been produced by decay of radioactive uranium. In moon's extreme vacuum it was possible for vaporized lead to travel great distances from its parent, uranium. Dr. Silver said other materials could be distributed over lunar surface in same way, all of which could confuse analysis of moon's age.

Dr. John A. Wood of Smithsonian Astrophysical Observatory discussed tiny piece of green glass among *Apollo 14* samples that scientists had nick-named "Genesis bean." Bean and other glass fragments were almost identical chemically to rare meteorites, Howardites. Evidence suggested green glass had originated on moon and not from meteorite impacts on earth and thus might be some of earliest material of solar system.

Consensus of conference on bulk composition of moon was summarized by Dr. Gast: feldspar-rich highland material with anorthosite as one prevalant component, basalts (solidified lava, low in radioactive elements and relatively young) in plains and basins, and KREEP (highly radioactive material rich in potassium, rare earth elements, and phosphorus).

Conference closed with discussion of geologic setting of Apollo 16 Descartes landing site and plans for mission's extravehicular activity. Soviet scientists returned to U.S.S.R. Jan. 23 with lunar samples for Soviet Academy of Sciences. (MSC Release 72-06; NASA Release 72-18; *Marshall Star*, 1/12/72, 1; Wilford, *NYT*, 1/14/72, 22; NASA Trans Sect)

January 11: Apollo 16 lunar module pilot Charles M. Duke, Jr., was released from Patrick Air Force Base Hospital where he had been treated for bacterial pneumonia since Jan. 4. (AP, B *Sun*, 1/11/72)

• Tass had released first photos of Soviet atomic energy reactor at Atomic Energy Institute near Moscow, *New York Times* reported. Tass had said photos had been taken in December 1971 to commemorate 25th anniversary of inauguration of graphite and uranium low-power reactor on Dec. 25, 1946. (*NYT*, 1/11/72, 26)

- NASA announced issuance by Manned Spacecraft Center of requests for proposals to eight aerospace firms for design, development, and testing of ground-based test unit of integrated medical and behavioral laboratory measurement system. IMBLMS would obtain basic medical and behavioral data in zero g space environment and provide clinical support for astronaut crew in long-duration space missions. Cost-plus-fixed-fee contract for 48-mo study would include construction in remote area of IMBLMS field test unit to evaluate requirements for remote health care in space. Unit might also be used to evaluate delivery of health services to remote or isolated earth communities. (NASA Release 72-7)

- *St. Louis Post-Dispatch* editorial commented on space shuttle decision: "Man's ability to perform capably in a prolonged state of weightlessness has yet to be proven and until it is the arguments for a large shuttle program must remain inconclusive. Beyond this, the recent agreements between this country and the Soviet Union to exchange biomedical and other space data and to work toward compatible docking systems would seem to make a full scale shuttle program a prime candidate for a co-operative enterprise." (*St Louis P-D*, 1/11/72)

January 12: U.S.S.R. launched *Cosmos 471* from Baykonur into orbit with 317-km (197-mi) apogee, 194-km (120.5-mi) perigee, 89.5-min period, and 64.9° inclination. Satellite reentered Jan. 25 (GSFC *SSR*, 1/31/72; *SBD*, 1/13/72, 51)

- NASA's *Mariner 9* Mars probe (launched May 30, 1971) photographed vast Mars chasm with branching canyons resembling network of dry river beds. Describing photo, which was released to press Jan. 19, Harold Masursky, of U.S. Geological Survey, said in interview that chasm probably had been caused by "faulting of the crust," a cracking and sliding process associated with earthquakes. Masursky believed line of small craters running parallel to canyon had probably been caused by venting volcanic gases and ash. Wind erosion and landslides had probably altered canyon, but landscape in photo appeared to be geologically young. (Wilford, *NYT*, 1/20/72, 25)

- Nike-Apache sounding rocket launched by NASA from Andoeya, Norway, carried National Oceanic and Atmospheric Administration experiment to 208-km (129-mi) altitude. Objectives were to obtain information on precipitating charged-particle flux associated with quiet auroral form and measure electric-field strengths in and near auroral form. Rocket and instruments functioned satisfactorily. (NASA Rpt SRL)

- Geraldine Cobb, former NASA employee, had piloted twin engine Islander aircraft for average six hours daily during 15-day search for survivors of Lockheed Electra airliner that crashed in Peruvian jungle Dec. 24, 1971, with 92 aboard, Reuters reported. Miss Cobb, who had picked up only survivor of crash and had flown her to missionary settlement, had passed all NASA tests to become astronaut, Reuters said. She had been "No. 1 choice to be the first woman to fly an Apollo mission until the space program was cut back and it became clear that no woman would be selected by NASA." (B *Sun*, 1/12/72, A2)

- Lewis Research Center Deputy Director Eugene J. Manganiello was introduced as new President of Society of Automotive Engineers at SAE banquet in Detroit. (*NASA Activities*, 2/15/72, 35)
- Marshall Space Flight Center engineers John R. Rasquin and M. F. Estes had received copies of patent issued to NASA for Rasquin-Estes invention—adaptation of magnetomotive hammer to diamond-making process—MSFC announced. Hammer, developed for Saturn V program, had been used to remove distortions from rocket tank sections misshapen by welding. In process developed by Rasquin and Estes, hammer was mounted on copper plate atop hardened steel horn. Hammer created shock wave in horn with sufficient energy to produce pressure and temperature necessary to change graphite to diamonds. (MSFC Release 72-1)
- Cost reduction suggestion by Marshall Space Flight Center scientist Dr. Thomas A. Parnell had saved NASA $57 000 in development of gamma ray telescope for scheduled balloon probe to investigate gamma ray sources, MSFC announced. Dr. Parnell had suggested substitution of scrap crystal (sodium iodide) for large, expensive crystal used to shield gamma-ray detectors from radiation sources other than desired target during probes. Scrap crystal material in tank containing oil-based liquid served as effective shield. Balloon probe was to be launched in March or April from National Center for Atmospheric Research. (MSFC Release 72-2)
- William Hines commented on space shuttle decision in *Chicago Sun-Times*: "What the President offered last week was not a completely reusable workhorse aerospace plane at all, but a scaled-down hodge-podge of obsolete, current and avant-garde technology that cannot possibly meet the stated goal of $100-a-pound payloads in 1980. The sad truth is that the original shuttle couldn't, either, once it became apparent that development costs would approximate $13 billion rather than the $5.2 billion estimated in 1969. As backers of the defunct supersonic transport finally had to admit, when you markedly alter the development cost of a flight system, you alter its economics also." (*C Sun-Times*, 1/12/72)
- Success of first 27 test flights of new Maverick missile had enabled Air Force Systems Command to end demonstration phase of Maverick flight testing ahead of schedule, AFSC reported. Maverick Program Director Col. James A. Abrahamson (USAF) had canceled 13 additional tests originally scheduled under contract with Hughes Aircraft Co. (AFSC Release 256.71)

January 12–14: Secretariat for Electronic Test Equipment—organization funded by NASA, Dept. of Defense, and Federal Aviation Administration to collect and disseminate test equipment information for Government agencies—held electronic test equipment conference at Kennedy Space Center. NASA presentations described design, development, procurement, and deployment of automatic test equipment for Apollo program. SETE was administered by New York Univ. under direction of David M. Goodman of New York Univ. School of Engineering and Science. (KSC Release 8-72; KSC PAO)

January 13: Federal Pay Board, by vote of 8 to 2, approved 8.3% first-year increase in aerospace salaries and increases in second year of three-year contracts, to bring total increases to 12%. Board had rejected 12% first-year increases Jan. 5. Although approved increase would be well over 5.5% guideline established by Pay Board, unions had indicated dissatisfaction, *Washington Post* later reported. (Rowe, *W Post*, 1/14/72, A1)

- NASA launched Nike-Tomahawk sounding rocket from Andoeya, Norway, carrying Norwegian particles and fields experiment. Rocket and instrumentation performed satisfactorily. (SR list)
- Manned Spacecraft Center announced Technicolor Inc. had received one-year $2 346 800, cost-plus-award-fee contract extension for photographic support services. Work would include science photography, still and motion picture laboratory, precision laboratory service for earth resources and lunar photography, and audiovisual support for Public Affairs Office. (MSC Release 72–10)
- Atomic scientist Dr. Edward Teller urged legislation to make public Government's scientific findings within year after they were classified. In speech before Florida Presbyterian College in St. Petersburg he said, "I am convinced we must do something about secrecy if democracy can function." (AP, *W Post*, 1/15/72, A11)

January 14: NASA's Office of Advanced Research and Technology (OART) was renamed Office of Aeronautics and Space Technology (OAST). Dr. James C. Fletcher, NASA Administrator, said NASA was steadily increasing its aeronautical activity and said name change "highlights the importance we attach to aeronautics." (NASA Release 72–10)

- Dr. Robert R. Gilruth, Manned Spacecraft Center Director, was appointed to newly created position of NASA Director of Key Personnel Development. Dr. Gilruth would be responsible for integrating NASA's management work in planning to fill key positions, identify actual and potential candidates, and guide them through appropriate work experience. Dr. Gilruth would be succeeded by Dr. Christopher C. Kraft, Jr., MSC Deputy Director. (NASA Release 72–11)
- Symposium on significant accomplishments in sciences was held at Goddard Space Flight Center. GSFC scientists presented papers covering their studies and experiments in astronomy during 1970. Dr. Norman H. MacLeod said imagery obtained by high-resolution infrared radiometer on *Nimbus 3* (launched April 14, 1969) could now be handled quantitatively and might permit estimations of the location, amount, and seasonal changes in water available. Data could be used by governments for better management of natural resources. Study of effect on earth's surface temperature and climatology of climatic modification by carbon dioxide, hydrogen, and aerosol was discussed by Dr. S. Ichtiaque Rasool. Study had indicated that "runaway greenhouse effect" from accumulation of man-made carbon dioxide alone probably was not possible. But, as aerosol dust in air increased, albedo increased "very fast, compared to the opacity in the infrared; as the albedo increases, the solar radiation decreases, and the surface temperature goes down." Temperature could decrease in next 30 yrs as much as about 3 K [5.4°F]. Several climatologists had suggested that "3K decrease in global temperature is sufficient to trigger an Ice Age!" (Transcript)
- Display of sculpture "Scott and Irwin on the Moon" by Red Grooms opened at Guggenheim Museum in New York. Diorama showing moon, *Apollo 15* lunar roving vehicle Rover and huge figure of astro-

naut had been commissioned as part of NASA–National Gallery of Art "Eyewitness to Space" program but Grooms had paid all of his own expenses. *Washington Post* art critic Paul Richard later said work managed "to capture something of the scale, something of the heroism and the hubris, of man's exploration of the moon." It should "move to some public place, the National Portrait Gallery, perhaps, after it leaves the Guggenheim on Feb. 27." (*W Post*, 1/19/72, B1)

- U.S. planned to build world's largest and most sensitive radiotelescope, *Washington Post* reported. President Nixon's FY 1973 budget request for National Science Foundation contained $3 million toward construction of $62.5-million array of 27 antennas, 35 km (22 mi) deep, that could pick up signals from outer space. Array had been designed by National Radio Astronomy Observatory, at Green Bank, W. Va., and was expected to take 6 to 10 yrs to build. (Cohn, *W Post*, 1/14/72, A3)

- Discovery of phenomenon of triplet-triplet transfer of energy between organic molecules was recorded by Committee on Inventions and Discoveries of U.S.S.R. Council of Ministers in Moscow. Discovery by Soviet scientists V. L. Yermolayev and A. N. Terenin was described in *Sotsialisticheskaya Industriya* as "new landmark in the knowledge of the laws of the universe" with "great scientific and practical significance" for photophysics, radiation chemistry, photobiology, and photochemistry. "It can be said . . . that the science of the microparticles of matter holds in its hands the entire future of a new technology. . . ." (FBIS–Sov, 1/14/72, L1)

- Edwin E. Aldrin, *Apollo 11* astronaut and second man to set foot on the moon, announced he was retiring from Air Force in summer. Col. Aldrin had been Commandant of Aerospace Research Pilots School at Edwards Air Force Base, Calif., since he retired from NASA July 1, 1971. He said he was leaving USAF because his 10 yrs with space program had hurt his chances for promotion. (*W Post*, 1/15/72, A9)

- Appointment of Dr. Eberhardt Rechtin, Principal Deputy Director of Defense for Research and Engineering and former Director of Advanced Research Projects Agency in Dept. of Defense, to new post of Assistant Secretary of Defense for Telecommunications was announced by Secretary of Defense Melvin R. Laird. New office had been established by DOD directive Jan. 11. (DOD Release 29–72)

- Newspaper editorials commented on President Nixon's decision to develop space shuttle.

    Washington Post said arguments for or against shuttle were not "error free" since "major ones rest on projections into the future which are exceedingly difficult to make and others rest on basically undemonstrable assumptions about the quest for knowledge. Part of the difficulty springs from the fact that no one can know what space-based research will discover. Is the key to the hydrogen atom and thus to unlimited energy out there . . . ? Will the world some day need to import minerals from space to sustain life here? Will man have to be in space to accomplish things such as these or can machines do them all? Above all, where does this kind of program fit in a national budget that cannot provide for doing all the things at home that ought to be done?" Such questions would make debate over space shuttle "different in character and significance from last year's debate over the SST." Standards applied to "project which involves scientific research and military considerations . . . must be somewhat different from those applied to a project, such as the SST, which involved only another way to move people from place to place." (*W Post*, 1/14/72, A22)

Houston Post: "With the last Apollo moon flight scheduled later this year and the Skylab project to terminate in 1973, the shuttle holds the key to an on-going space program. Our failure to build it would indicate our lagging interest in space research to the possible detriment of future cooperation with the Russians in this field. It is true that we have important social priorities which need attention, but our space research posture in relation to the Soviet Union, whether competitive or cooperative, also carries a high priority." (*H Post*, 1/14/72)

- L/G August Schomburg (USA, Ret.), former Commandant of Industrial College of the Armed Forces and former Commander of Ordnance Missile Command at Redstone Arsenal, died at age 63. He had served as consultant to NASA Administrator 1967–1970. (*W Post*, 1/16/72, D11)

January 15: Pioneer-F spacecraft to probe Jupiter's environmental and atmospheric characteristics arrived at Kennedy Space Center to begin final checkout before Feb. 27 launch. Pioneer-F 3rd stage had been erected Jan. 7 and overall vehicle interface test had been conducted Jan. 10. Pioneer-F would mark first use of Atlas-Centaur booster in Pioneer program and first use of Atlas-Centaur with 3rd stage—TE–M–364–4 solid-propellant engine with 66 750-newton (15 000-lb) thrust. (KSC Release 7–72; NASA Release 72–25)

- NASA launched two Booster Arcas I sounding rockets from Churchill Research Range, Canada, carrying Univ. of Houston experiments to investigate auroral zone disturbances and study auroral phenomena at altitudes previously unstudied. Data would supplement data from three Canadian Black Brant sounding rockets launched by Canadian National Research Council during substorm movement. Rockets reached 53.1- and 60-km (33- and 37.3-mi) altitudes and functioned satisfactorily, with each set of instruments providing data for one hour. (NASA Rpts SRL)

- Pan-Pacific Education and Communications Experiments Using Satellites (PEACESAT) program, using NASA's orbiting *Ats 1* satellite for long-distance educational exchange, was described in *New York Times*. System would be used to convey information instantaneously by voice, print, and pictures between classrooms, laboratories, or libraries in different countries. Key component was inexpensive, easily operated two-way broadcasting and receiving unit. Impulses beamed by transmitter from primary station at Univ. of Hawaii to satellite could be received by similar device on ground anywhere in Pacific area. University's unit had been constructed for $1200, with $130 taxi radio as nucleus. Teachers on Honolulu campus had used system to conduct orientation classes for students traveling to Hawaii from Japan, Hong Kong, and Philippines aboard passenger ship thousands of kilometers away and to conduct five-week speech course between Honolulu and Hilo campuses.

Other applications of PEACESAT might include multistation hookups for scholarly discussions among teachers and students, long-distance collaboration in medical activities, and exchange of library material. (Trumbull, *NYT*, 1/15/72)

January 15

- Col. Daniel D. McKee (USAF, Ret.), manager of Bell Aerosystems in Houston and former Assistant Director of NASA Manned Satellite Programs in NASA Hq., died in Houston at age 53 after apparent heart attack. (MSC Hist Off)

January 16–17: *Apollo 15* Astronauts David R. Scott and Alfred M. Worden, during visit to Poland, discussed *Apollo 15* mission with scientists at Univ. of Warsaw's Institute for Experimental Physics. (FBIS–Poland, 1/19/72, G5; Reuters, *C Trib*, 1/18/72)

January 17: Sen. Edward M. Kennedy (D-Mass.), in speech before National Press Club in Washington, D.C., asked, "Shall we spend our dollars on a space shuttle and an SST for the few to fly the heavens, when many here on earth have simple unmet needs like homes and schools and health?" (Text)

- Federal Aviation Administration national plan for development of new microwave instrument landing system for common use by civil and military aircraft was announced by Secretary of Transportation John A. Volpe. System covered broad area, would increase number of flight paths and airport acceptance rate, and would ease noise over surrounding communities. It would provide continuous distance information, eliminating need for marker beacons that gave limited progress information on final approach. (FAA Release 72-12)

- New satellite Hot Line between White House and Kremlin was targeted to begin operation in 1973, *Aviation Week & Space Technology* reported. President Nixon's approval had been announced Sept. 24, 1971, of U.S.–U.S.S.R. agreement on direct communication link using two satellite circuits, one established by each country. Agreement had been signed Sept. 30, 1971. (*Av Wk*, 1/17/72, 9; *A&A 1971*)

January 17–19: American Institute of Aeronautics and Astronautics (AIAA) held 10th Aerospace Sciences Meeting in San Diego, Calif. *Apollo 15* Astronaut James B. Irwin and Nobel Prize-winning physicist Dr. Harold C. Urey reviewed space activities for local high school students.

Dr. John C. Houbolt, Executive Vice President of Aeronautical Research Associates, delivered Dryden Research Lecture "Atmospheric Turbulence." Lecture had been named in honor of late Dr. Hugh L. Dryden, first NASA Deputy Administrator, in 1967. It succeeded Research Award established by AIAA in 1960. Dryden Lecture award, sponsored by General Electric Co., carried $1000 honorarium and $1500 travel allowance for repetition of lecture before selected AIAA sections. Dr. Houbolt reviewed latest data on atmospheric turbulence and its influence on aircraft design and flight.

Dr. Allen E. Puckett, Executive Vice President and General Manager of Hughes Aircraft Co., was installed as 1972 AIAA President at Jan. 19 Honors Banquet.

AIAA presented Goddard Award to Gary A. Plourde, project engineer at United Aircraft Corp. Pratt & Whitney Div., and Squadron Leader Brian Brimelow (RAF) and Howard E. Schumacher, both of Air Force Aero Propulsion Laboratory. Trio was honored as team for original research leading to first understanding of gas-turbine compressor stalls induced by turbulent inlet flow conditions.

Sylvanus Albert Reed Award was presented to Dr. Max M. Munk, retired engineer, for "enduring contributions to the science of aerodynamics."

Robert M. Losey Award was given to Dr. David Q. Wark, Senior Scientist with National Oceanic and Atmospheric Administration, "in recognition of outstanding contributions to the science of meteorology as applied to aeronautics."

Edward W. Price, Head of Naval Weapons Center Aerothermochemistry Div., received G. Edward Pendray Award for "continued outstanding contributions to the literature of solid rocket internal ballistics and combustion, particularly for his contributions on combustion instability and ignition."

Space Science Award was presented to Dr. Norman F. Ness, Chief of Goddard Space Flight Center Laboratory for Extraterrestrial Physics, for "significant contributions to the description and understanding of the interplanetary medium and the interactions of this medium with the earth and other large bodies." (AIAA Releases 1/10/72, 1/13/72; *AIAA Bull*, 1/72, 2/72, 67)

January 17–21: Fourth Annual Earth Resources Program Review was held by NASA at Manned Spacecraft Center. Scientists from foreign nations and organizations, Government, universities, and private institutions attended. Reports were presented on 1971 research and analysis of aircraft and spacecraft remote sensing data.

In Jan. 18 press briefing NASA Associate Administrator for Applications Charles W. Mathews said pollution monitoring was natural activity for global surveillance capability because atmospheric pollution was potentially global problem. Not well understood were manner in which pollutants were dispersed throughout globe or manner in which earth corrected itself. "There's a lot of work to be done in this area."

In Jan. 21 briefing Dr. John M. DeNoyer, NASA Director of Earth Observations, said program was moving from exploratory to flight phase, from learning to applying techniques. Among applications was spring Arctic ice survey, in which some 80 scientists would be on ice measuring strain rates, meteorological conditions, and ice types, supported by Convair 990 aircraft from Ames Research Center. Also microwave techniques in remote sensing could identify ice-free lanes in Great Lakes region, making possible efficient shipping much earlier in spring than at present. Soo Canal into Lake Superior, which carried more tons of freight than any other canal in world, was now open only one half of year. Weather and cloud cover made aircraft surveillance inefficient; microwave techniques from satellite altitude was more practical in this instance. Another important program would be thermal mapping by remote sensing to permit better industrial planning from knowledge of how development would actually affect environment. Land-use planning also would be greatly aided by remote sensing data processed by computerized inventory system being developed at MSC. (MSC Release 72–11; MSC PAO; Transcript)

January 18–20: Opening hearings on aeronautical research and development, Chairman Ken Hechler (D-W. Va.) of House Committee on Science and Astronautics' Subcommittee on Advanced Research and Technology announced Subcommittee name had been changed to Subcommittee on Aeronautics and Space Technology. Change followed receipt of Jan. 14 letter from Dr. James C. Fletcher, NASA Administrator, announcing change in NASA office name from OART to OAST.

On first day of hearings Dr. Fletcher testified on "steady increase" in NASA funding for aeronautics: Funding had grown from $42 million in FY 1966 to $110 million in FY 1972, "a doubling in terms of constant dollars. When you add the funding for salaries and other in-house costs related to supporting the aeronautics programs, the total . . . has grown from 1.6% of the total NASA budget in fiscal year 1966 to 7.1% in fiscal year 1972." Number of NASA personnel members working in aeronautics research and technology had grown from 2600 in 1966 to 5300. Increase had occurred while NASA staff had been reduced from 34 000 to 27 500.

Dr. Fletcher said joint Dept. of Transportation and NASA Study on Civil Aviation Research and Development (CARD) had "forced the establishment of new and more effective coordination mechanisms between NASA and DOT." Mechanisms were being used "to develop a formal CARD Policy Implementation Plan to specify clear goals, responsibilities, planned achievement milestones, and projected resource requirements for DOT and NASA jointly." CARD policy study had caused "better focus on the priority problem areas of aircraft noise . . . and terminal congestion" and directed more specific attention to role R&D might play in developing better total operating system for low-density aviation market so that civil aviation might be used more economically as tool for regional development. Focus did not compromise "ability to continue a comprehensive research and technology program in all the aeronautical disciplines to provide a technology base for the future."

Deputy Director Clarence A. Syvertson of Ames Research Center outlined general recommendations of CARD study: scope of civil-aviation R&D programs should be expanded to emphasize nontechnological factors; economists and social scientists should be assigned to NASA and DOT staffs. R&D staffs of Dept. of Defense, NASA, DOT, and Civil Aeronautics Board should be interchanged to provide broad systems background for more effective Government action in civil aviation programs. CAB should explore policy of allowing intermodal mergers of airlines for expanded use of air cargo to encourage innovative industry R&D in this area. Dept. of Commerce should refine and monitor measurable indicators of U.S. progress in civil aviation industry. National Aeronautics and Space Council should review and recommend national policies guiding civil aviation that embraced several agencies.

George W. Cherry, Deputy Administrator for Programs in OAST, reported "considerable ongoing activity abroad in designing, building and selling new, small economical aircraft for the world's low-density short-haul market. This raises the questions: Where will replacement aircraft come from? Can the U.S. industry compete with the foreign manufacturers to fill the market?" Answers required "fundamental policy decisions by the United States with regard to the desirability of providing continuing or expanded public-need air service, and also . . . to whether Government support can or should be given to the manufacturer."

U.S. needed to know more about market and economics of operating in low-density area to write specifications for aircraft and operating system. "If industry designs fall short of meeting the economic specifications, there should be enough information available to determine if Government assistance . . . could result in the operator providing viable service. If subsidy is necessary, subsidizing the nonrecurring costs of manufacture, rather than the recurring costs of operation, might make good sense."

NASA Associate Administrator for Aeronautics and Space Technology Roy P. Jackson described QUESTOL project plans for two quiet, experimental, short takeoff and landing (STOL) transport research aircraft: "The experimental airplanes will function as a versatile facility for NASA flight research programs developed in cooperation with other Government agencies and interested industry groups, including the airlines. The information provided will reduce the technical risk associated with development, by industry, of both civil and military STOL transports" and "provide a comprehensive technical foundation on which Government regulatory agencies can establish realistic criteria for certification of commercial subsonic STOL transport aircraft and for en route and terminal area operations." (Transcript)

January 19: NASA announced selection of flight crews for Skylab missions. Prime crewmen for first, 28-day flight were Charles Conrad, Jr. (commander), Dr. Joseph P. Kerwin (science pilot), and Paul J. Weitz (pilot). Backup crewmen were Russel L. Schweickart, Dr. Story Musgrave, and Bruce McCandless II.

Crewmen for second, 56-day mission were Alan L. Bean, Dr. Owen K. Garriott, and Jack R. Lousma. Third mission, for 56 days, would be flown by Gerald P. Carr, Dr. Edward G. Gibson, and William R. Pogue. Backup crew for second and third missions was Vance D. Brand, Dr. William B. Lenoir, and Dr. Don L. Lind. (NASA Release 72-12)

- Skylab crew press conference, with prime and backup crewmen, was held at Manned Spacecraft Center. Astronaut Charles Conrad, Jr., said preparations were on schedule for April 1973 launch. Contractor checkouts and tests of hardware were expected to be completed for delivery to Kennedy Space Center in July. Skylab would carry some 20 000 pieces of stowage equipment on board to provide life support for nine men for 140 days. "So it all goes up at one time, and we've got a great deal of work to do, not only to learn how to operate this vehicle but also all the experiments in it. It became apparent that we could not be 100 percent cross-trained as we had been in Apollo, so we've . . . defined some areas for each guy to become expert in . . . . That allowed us to balance out the training hours. Right now . . . we have some 2000 training hours per man defined. We've been working on the basic training for the past year." Training hardware would be available about February 1. Commander would have overall responsibility for mission and would be command and service module expert. Science pilot would be expert in all medical equipment and in Apollo Telescope Mount (ATM) and its associated hardware. Pilot would be expert in Orbital Workshop systems and electrical systems. Remaining experiments would be divided among crewmembers according to availability and choice.

Astronaut Alan L. Bean said photographic data returned from Skylab would be "far superior" to previous data because actual film would be returned to earth. To return film, crews would conduct two extravehicular activities (EVAs) on first mission, three on second, and two on third. "We're going to be moving . . . 240 pounds [109 kg] worth of film, in canisters, magazines, in and out of the cameras. We'll have to remove the used exposed film and replace it with new film. It's going to take about 3 hours."

Experiments were described by Astronaut Gerald P. Carr. They included EREP earth resources experiment containing five sensors to calibrate earth's atmosphere, corollary experiments submitted by high school students, engineering experiments to study effects of zero g on metal alloys and flammability of certain items, habitability experiments to study noise levels and atmospheric composition, and scientific experiments. Scientific experiments would include studies of effect of zero g on single human cells, studies of circadian rhythm of pocket mice and vinegar gnats, ultraviolet stellar astronomy and x-ray astronomy. Technological experiments would include crew vehicle disturbances, manual navigation sightings, and "a study of the little contamination cloud that we carry around with us . . . kind of like Pigpen in Peanuts."

In response to question on illness during missions, Astronaut Joseph P. Kerwin said NASA would provide equipment and training on board to take care of most illnesses and injuries that could be handled in a doctor's office. Science pilot would be able to provide emergency care for illnesses "up to and including quite severe. We have quite a package of drugs on board, administrable by all routes; we have tracheotomy equipment and equipment for maintaining an airway for stopping bleeding. We are attempting to develop in time for the flight intravenous fluids. We feel that, almost regardless of the severity of the injury, we will at least be able to stabilize the patient." (Transcript)

- NASA Convair 990 aircraft was being used for scientific probe of high-altitude cloud formations up to 14 000 m (45 000 ft) in attempt to identify and measure ice and water particles with remote sensing devices, Ames Research Center announced. Goddard Space Flight Center scientist Dr. Warren A. Hovis and team of scientists from NASA, Univ. of Arizona, and Arthur D. Little Co. were conducting experiment to improve short-term weather forecasting and to transfer technology to weather satellites. (NASA Release 72-14)

January 19-20: NASA Office of Space Science held semiannual Spacecraft Sterilization Seminar at Kennedy Space Center to review progress made on research tasks of NASA and academic prime investigators, inform American Institute of Biological Sciences' Planetary Quarantine Advisory Panel of this progress, and to foster interchange of recent developments in spacecraft sterilization concepts. (KSC Release 9-72; KSC PIO)

January 20: President Nixon delivered State of the Union address before joint session of Congress and sent written message to Congress. He said in speech: "In reaching the moon, we demonstrated what miracles American technology is capable of achieving. Now the time has come to move more deliberately toward making full use of that technology here on earth, of harnessing the wonders of science to the service of man." He would propose new "Federal partnership" in technological research and development "with Federal incentives to increase private research, federally supported research on projects . . . to improve our everyday lives."

In written message President said dealings with U.S.S.R. had "advanced the prospects for limiting strategic armaments. We have moved toward greater cooperation in space research and toward improving our economic relationships."

President said he would request Dept. of Defense budget increase "to preserve the sufficiency of our strategic nuclear deterrent, including an allocation of over $900 million to improve our sea-based deterrent force." He had instructed DOD "to develop a program to build additional missile launching submarines, carrying a new and far more effective missile."

Space program had been reoriented to produce increased domestic benefits: "In recent years, America has focused a large share of the technological energy on projects for defense and for space. These projects have had great value. Defense technology has helped us preserve our freedom and protect the peace. Space technology has enabled us to share unparalleled adventures and to lift our sights beyond earth's bounds. The daily life of the average man has also been improved by much of our defense and space research—for example, by work on radar, jet engines, nuclear reactors, communications and weather satellites, and computers. Defense and space projects have also enabled us to build and maintain our general technological capacity, which—as a result—can now be more readily applied to civilian purposes.

"America must continue with strong and sensible programs of research and development for defense and for space. I have felt for some time, however, that we should also be doing more to apply our scientific and technological genius directly to domestic opportunities. Toward this end, I have already increased our civilian research and development budget by more than 40 percent since 1969 and have directed the National Science Foundation to give more attention to this area . . . .

"I recently announced support for the development of a new earth orbital vehicle that promises to introduce a new era in space research." Reusable space shuttle would lower cost and risk of space operations and open new opportunities in weather forecasting, communications, monitoring natural resources, and air traffic safety. "The space shuttle is a wise national investment."

Proposals included increase of $700 million in civilian R&D, 15% increase over previous year. And technically oriented agencies like NASA and Atomic Energy Commission would work more closely with agencies with primary social mission like Dept. of Transportation. (*PD*, 1/24/72, 74-92)

- Air Force launched two unidentified satellites from Vandenberg Air Force Base by single Titan IIID booster. First satellite entered orbit with 340-km (211-mi) apogee, 150-km (93-mi) perigee, 89.3-min period, and 96.9° inclination and reentered Feb. 29. Second entered orbit with 550-km (342-mi) apogee, 470-km (292-mi) perigee, 94.8-min period, and 96.5° inclination. (Pres Rpt 73; *SBD*, 1/24/72, 100)
- NASA test pilots Stuart M. Present, age 41, and Mark C. Heath, age 37, were killed when their T-38 aircraft crashed on landing at Matagorda Island off coast of Texas. Radar contact was lost during third instrument approach to landing strip while fog bank was moving in. Pilots were checking out T-38 for use by astronauts to maintain flight proficiency. (MSC Release 72-20)

January 21: Grumman Corp. expected 1971 net loss of between $15.4 million and $18.2 million because of accounting write-down on Navy F-14A fighter jet program, *Wall Street Journal* reported. Write-down on value of inventories and work in progress on F-14A would lead to overall profit for 1972 and 1973 but Grumman Aerospace Corp. estimated loss on F-14A contract of about $65 million, before tax adjustments. (*WSJ*, 1/21/72, 7)

- Japanese space budget for FY 1973 was approved. Of 24 080 million yen ($80 million), Y19 810 million ($66 million) was for further development of N rocket and satellites by Science and Technology Agency for use by National Space Development Agency. FY 1972 space budget had been Y8630 ($28 million). (*SF*, 4/72, 121)

- *New York Times* editorial commented on results of Third Lunar Science Conference at Manned Spacecraft Center [see Jan. 10–13]: "Science fiction writers predicted long ago that the moon would be colonized; some even projected a future 'revolution' by lunar settlers against domination from the earth. Though such writers could skip over the many difficulties and high costs required by any genuine settlement attempt, there is now increasing evidence that the moon may not be quite so inhospitable as scientists once believed. The possibility that on this, too, the science fiction writers may have been prophets no longer seems as fantastic as it did before the first men walked on the moon." (*NYT*, 1/21/72, 46)

- Col. H. J. Odenthal (USAF, Ret.), pioneer airman who once assisted in rescue of Orville Wright after his aircraft had crashed, died at age 84. Col. Odenthal had helped organize Signal Corps air photographic and mapping division and had been awarded Croix de Guerre for his mapping flights in France during World War I. (Barker, *W Post*, 1/23/72, B16)

January 22–25: *Intelsat-IV F-4* comsat was launched by NASA for Communications Satellite Corp. on behalf of International Telecommunications Satellite Consortium (INTELSAT). Launched from Eastern Test Range at 7:12 pm EST by Atlas-Centaur booster, satellite entered elliptical transfer orbit with 36 523-km (22 694.3-mi) apogee, 5984-km (3718.3-mi) perigee, and 28.2° inclination. Primary objective was to place satellite in transfer orbit accurate enough for onboard propulsion systems to transfer it to planned synchronous orbit for commercial communications over Pacific.

Apogee-kick motor was fired at 7:32 pm EST Jan. 24 and *Intelsat-IV F-4* entered circular orbit with 35 787-km (22 237-mi) apogee, 25 625-km (22 136.4-mi) perigee, 23-hr 52-min period, and 0.7° inclination over Pacific at 165° east longitude. It would drift 1.1° per day to reach station at 174° east longitude in early February. Satellite's first major mission would be coverage of President Nixon's visit to People's Republic of China in February.

Intelsat-IV F-4 was third comsat in Intelsat IV series. Satellite was 238 cm (93.7 in) in diameter and 528 cm (208 in) high and weighed 1402 kg (3090 lbs) at launch. It had 12 transponders, providing 12 TV channels and 3000–9000 telephone circuits, and was capable of multiple-access and simultaneous transmissions. Expected lifetime was seven years. *Intelsat-IV F-3* had been launched Dec. 19, 1971, and was operating satisfactorily over Atlantic. (NASA proj off; GSFC *SSR*, 1/31/72; NASA Release 72–16)

January 22: Dr. James C. Fletcher, NASA Administrator, discussed NASA plans for continuing fruitful space program in interview published by *Christian Science Monitor:* "For the next few years, we may have fewer of the more spectacular explorational missions . . . . We will focus more on . . . communications satellites, weather satellites, earth resources survey satellites."

Next Applications Technology Satellite (ATS) would demonstrate direct TV broadcasting in region of Rockies and Alaska, using large satellite with big antenna that could beam programs directly to augmented receivers in individual homes. Demonstration—in cooperation with Dept. of Health, Education, and Welfare—would work through educational network rather than directly to homes, however.

Comsats were "not quite so exciting an aspect of space technology" in U.S. but, "worldwide, there has been more interest in this than in almost any other aspect of our program." Interest equaled that of ERT satellite. "Countries with vigorous mineral exploitation and agricultural development are interested in ERT satellites . . . . Countries with underdeveloped communications and education, such as India, are keenly interested in the broadcasting satellites."

In weather satellite field, new sophisticated instruments "give hope that we can get a really world weather observing network going." Infrared sensors, microwave sensors, and high-resolution optics that obtained temperature profiles throughout atmosphere by looking "through holes in cloud systems" should help "to get points all over the globe where we now get only pieces of the picture."

NASA would try processing materials in zero g during Skylab program. "If it works with men doing it, there are cheaper ways we could do it without men. It may be one of those things that will be worth doing when we can get routine access to space with the shuttle."

In response to question, Dr. Fletcher said second Skylab would be "a very good thing for the country. But I'm not sure the budget needed for both a Skylab and the shuttle would be salable. The shuttle is more important in the long run. Not to do more of the Skylab sort of thing will be bad for prestige. The Russians will be doing much more than we will. I'm sure there will be criticism from Congress and the public." But with shuttle and current budget level, "we could do virtually all those things we have talked about doing in this century. We could go back to the moon and establish crude bases . . . . We could establish a permanent space station. We just might be able to send men to Mars and get them safely home."

Apollo industrial base had already been destroyed, Dr. Fletcher said. "We could assemble a nucleus of people who know these things within NASA. But that's about all." As NASA turned toward space shuttle, "We hope and expect that many of the people who worked on Apollo will be in that program, so we won't lose all the basic space flight know-how."

In international cooperation, France had asked to participate in Skylab mission and U.S.–U.S.S.R. meetings to establish compatible space docking system had progressed. "It looks as though we can do it technically. With existing hardware, we could do a docking in something like 1974 to 1975."

Scientific community was "still debating whether it would prefer an intensive Jupiter and Saturn explorational program or some form of grand tour to other planets as well," Dr. Fletcher said. There was "no slackening of interest in Grand Tour" on NASA's part. (*CSM*, 1/22/72)

- First photos of Mars taken by U.S.S.R.'s *Mars 2* and *Mars 3* spacecraft were shown on Moscow TV. One photo showed equatorial zone with sun at low angle throwing long shadows over mountainous area. Second photo showed sharply defined, ring-shaped formation against light-colored background. Origin of formation, visible through dust clouds, was not established. Third photo showed craters of Syrtis Major. (*NYT*, 1/23/72)

- Library of Congress Congressional Research Service published *United States and Soviet Progress in Space: Summary Data Through 1971 and a Forward Look* (72–17 SP). Report—prepared by Dr. Charles S. Sheldon II, Chief of Science Policy Research Div.—updated Senate Document 92–51 released Jan. 6 and discussed future directions and significant aspects of space programs. Soviet planetary probes *Mars 2* (launched May 19, 1971) and *Mars 3* (launched May 28, 1971) were studying temperatures, density, and constituents of Martian ionosphere, atmosphere, and surface. Joint experiment with France on *Mars 3* had been studying solar wind. With Soviet start in planetary exploration during 1971, U.S. should continue to feel competition in planetary program until NASA Viking landed on Mars in 1976. NASA future plans gave prominent priority to space shuttle. While Soviet plans for reusable space vehicle had not been divulged, "it is hard to imagine their attainment of announced goals . . . without use of a shuttle." U.S.S.R. talked of putting up space station in five years. "It may be that the long awaited new large launch vehicle will find use in lifting major components for such a station. Using this vehicle, the U.S.S.R. could put up its equivalent of the U.S. Skylab any time from 1972 on." U.S.S.R. had recorded "wide variety of commitments . . . to conduct wide-ranging program" of solar system exploration and of earth applications. They had discussed possibility of Grand Tour flights and of manned planetary missions. In space cooperation, "although no firm commitment has been made and no real timetable set, the possibility has been opened that some of the surplus Saturn IB and Apollo hardware might be used to send an American crew to dock with a future Soviet space station. Then perhaps later if a remaining Saturn V is used to put up a Skylab 2 . . . a Soviet Soyuz might send a crew to dock with it. More joint planning and more funds will be required to accomplish these suggestions." (Text)

- *Apollo 15* Astronauts David R. Scott, Alfred M. Worden, and James B. Irwin arrived in Belgrade from Warsaw, Poland, for tour of Yugoslavia. They would meet with Yugoslav scientists to discuss *Apollo 15*. (AP, W *Star*, 1/23/72, A11)

- RCA Global Communications, Inc., and China National Machinery Import and Export Corp. signed $2.9-million contract under which RCA would sell People's Republic of China P.R.C.'s first permanent satellite earth station. Station, to be completed in time to cover President Nixon's visit to Shanghai Feb. 28, would initially carry TV transmission, 23 two-way voice bands, and 12 two-way teleprinted channels and would be expanded to 60 voice-grade circuits. (*W Post*, 2/16/72, A29; 3/24/72, D10)

- L/G Frank A. Bogart (USAF, Ret.), Manned Spacecraft Center Associate Director since November 1969, retired. Gen. Bogart had joined NASA in 1964 and had received numerous high awards and honors, including 1969 NASA Distinguished Service Medal for his contributions to Apollo lunar landing program. (MSC Release 72–18)
- Council of Europe meeting in Strasbourg, France, called for introduction of civil supersonic aircraft to be postponed pending full scientific study of effects on people and environment. (Reuters, *W Post*, 1/23/72, A34)

January 23: American Institute of Aeronautics and Astronautics released *International R&D Trends and Policies: An Analysis of Implications for the U.S.:* "U.S. leadership in technology and in world trade is being challenged and the nation could lose its top ranking in both areas during the seventies" because of "diminishing" Federal support of research and development when "other nations are accelerating efforts toward specific scientific and technological goals." U.S.S.R. was challenger in defense and space technology. "In world trade, the U.S. is being challenged by the nations of Western Europe and Japan, who are conducting strong programs, particularly in certain research-intensive industries such as aerospace and electronics." Report recommended that U.S. adopt "explicit technological strategy with policies which define national R&D goals and provide levels of support adequate for their attainment." (Text)

January 24: President Nixon in message transmitting FY 1973 budget to Congress said: "We have been reordering our research and development investments in defense and space. We have reassessed the space program and placed it on a firm future footing with increased attention to practical and economical applications of space and reductions in the cost of manned space flight.

". . . we have strengthened our defense research and development capability to insure that the country will not face the possibility of technological surprise or lack of the deterrent power necessary to protect our national security." Therefore budget authority requested for Dept. of Defense research, development, technology, and engineering (RDT&E) was at all-time high.

President proposed measures "to emphasize this Administration's strong belief that science and technology can make significant contributions to the quality of American life and to economic growth"; to encourage private investment in R&D, including investment by small firms with innovative ideas; to use NASA and Atomic Energy Commission talents on clean, economical energy and safe, fast transportation—"this year we shall have the agency which sent men to the moon and back begin to assist the Department of Transportation in finding better ways to send people downtown and back"—and to review economic policies which might restrict utilization and development of technical advances.

President also said human resources spending would be 45% of FY 1973 budget, while defense programs would total 32%. Outlays requested for major environmental programs in 1973 were $2.5 billion, "more than three times the 1969 level." (*PD*, 1/31/72, 104–17)

- President Nixon sent $246.3-billion FY 1973 budget request to Congress—increase of $9.6 billion over FY 1972. National total included research and development budget of $17.8 billion (increase of $1.4 billion), with $3.740 billion requested for civilian and military space R&D.

  Total request for NASA new obligational authority (NOA) of $3.379 billion (1.3% of total U.S. budget) was $83 million more than FY 1972 NOA of $3.296 billion. NASA expenditures were budgeted to increase by $11 million, against decline of $200.9 million in FY 1972. Increase to $3.192 billion reflected funds for design and development of space shuttle engine and airframe. Of budget request, $2.6 billion would go for R&D, $77.3 million for construction of facilities, and $701 million for research and program management.

  Apollo funding decrease of $472.5 million, to $128.7 million, reflected completion of Apollo program with Apollo 16 and 17 scheduled for launch during calendar year. Manned space flight operations—including $540.5 million for Skylab, $200 million for space shuttle, and $23 million for orbital systems and experiments—would increase from $682.8 million in FY 1972 to $1.094 billion in FY 1973. Space life sciences would receive $25.5 million and development, test, and mission operations would receive $305.2 million, including funds for detailed design and development of shuttle airframe and engine. Advanced missions would receive $1.5 million, to bring total for manned space flight (including Apollo) to $1.224 billion, down $61 million from FY 1972.

  Funding for NASA space science and applications programs would increase $123.7 million, from $740.4 million in FY 1972 to $864.1 million in FY 1973. Increases would go to physics and astronomy program for High Energy Astronomy Observatory ($59.6 million in FY 1973) and orbiting explorers ($32 million); to lunar and planetary exploration (up $29.7 million, to $321.2 million, with $229.5 million for Viking project, $31.6 million for Mariner-Venus/Mercury 1973, and $18.7 million for supporting research and technology advanced studies); and to launch vehicle procurement. Funding for Pioneer planetary exploration program would decrease from $12.8 million in FY 1972 to $10.2 million. Increase of $7.2 million in applications programs, to $194.7 million in FY 1973, included $48.4 million for earth resources survey, $61.2 million for ATS, and $28.3 million for Nimbus.

  Total requested for NASA aeronautics and space technology programs increased from $212.8 million in FY 1972 to $249.3 million. Aeronautical research and technology funding would increase from $110 million in FY 1972 to $163.4 million, including $27.5 million for quiet, experimental, short takeoff and landing (QUESTOL) aircraft and $3 million for vertical takeoff and landing (VTOL) research/experimental vehicle programs. Nuclear engine for rocket vehicle application (NERVA) development was terminated in favor of reoriented program to define smaller nuclear rocket system.

  Tracking and data acquisition funding would fall $4.9 million, to $259.1 million.

  Dept. of Defense FY 1973 budget of $83.5 billion was all-time high, $6.3 billion above FY 1972 budget authority, 30% of total Federal outlays for FY 1972, and 6.4% of Gross National Product—but percentage of GNP had dropped from FY 1972's 7%. DOD total included $8.6 billion for research, development, technology, and engineering, $1.1 billion increase over FY 1972. Of requested RDT&E total, military astronautics—with major programs including military comsats, first spaceborne ballistic missile early warning system, continued flight experiment programs, and applied research and technology programs—would receive $454 million (up $49.4 million). Military sciences would receive $572.7 million (up $34.9 million). Aircraft RDT&E would receive $1.948 billion (down $36.8 million) and missiles RDT&E $2.383 billion (up $437.8 million).

Major DOD increases were programmed for undersea long-range missile system (ULMS) to supplement and eventually replace Poseidon-Polaris fleet ($942.2 million, up $802 million), B-1 advanced strategic bomber ($444 million, up $74 million), F-15 air superiority fighter ($910 million, up $490 million), airborne defense warning and control system (AWACS) ($470 million, up $331 million), and Safeguard antiballistic missile program ($1.483 billion, up $366 million). Some $299 million was requested for advance procurement of components for new nuclear-powered CVAN-70 aircraft carrier. Increases were also programmed for prototype development of vertical or short takeoff and landing aircraft for Navy, STOL transports for possible C-130 replacement, Air Force SCAD strategic bomber penetration decoy, and engines for future transport and fighter aircraft.

National Oceanic and Atmospheric Administration FY 1973 budget included $39.9 million for meteorological satellite operations, $6.8 million more than FY 1971 allocation of $33.1 million. Funding would complete financing of fourth satellite in ITOS series, continue financing of fifth and sixth ITOS satellites and one Geostationary Operational Environmental Satellite (GOES), and initiate procurement of seventh ITOS.

Dept. of Transportation budget request of $8.56 billion was decrease of $91.5 million over FY 1972. DOT funding would start 10-yr, $10-billion program to improve urban mass transportation and reduce highway congestion. Federal Aviation Administration would continue efforts to reduce aircraft noise, ensure that advanced aircraft engines did not adversely affect atmosphere, and design safer and more productive airport system.

Atomic Energy Commission FY 1973 budget request of $2.56 billion, up $247 million over FY 1972, would be used for development of liquid-metal fast-breeder power reactor and for controlled thermonuclear fusion research.

National Science Foundation FY 1973 budget request of $653 million would cover basic scientific research, experimental program to test incentives to stimulate non-Federal R&D investment, program to analyze effect of R&D on national economy, continuation of research applied to national needs in advanced technology and environmental problems, program to improve research management at institutions of higher learning, and funding for astronomy to permit development of very large array of antennas for radio astronomy. (OMB, *Budget of US Govt, FY 1973;* OMB, *US Budget in Brief;* OMB, *Special Analysis;* NASA budget briefing transcript; DOD budget briefing transcript; DOT budget briefing transcript; AIAA Release 72-3; NSF Release 72-106; Kelly, W *Star,* 1/24/72, A14; *Av Wk,* 2/14/72, 14)

- NASA released briefing (held Jan. 22) on FY 1973 budget request in which Dr. James C. Fletcher, NASA Administrator, said budget would "enable the United States space program to move forward on all fronts, on a basis consistent with realistic budgetary constraints both now and in the future." Budget "fully supports President Nixon's decision to proceed with development of the space shuttle, . . . achieves this and other basic objectives in science, in applications, and in aeronautics within a fiscal '73 budget approximately equal to that of last year," and "provides programs to do the kind of work this agency should be doing."

  In past NASA had presented programs that required relatively modest outlays in first year and subsequent substantial increases. In FY 1973 budget, "We have a well-balanced program that can be supported at an essentially constant budget level over the next several years."

  "Most important aspects" of NASA's FY 1973 program were: "The decision to proceed with the space shuttle; a strong effort in space science and applications; a major increase in aeronautics; a change of scope in the outer planet exploration program; and a reorientation of our nuclear rocket program." Significant changes from FY 1972 plans were in Grand Tour and nuclear engine for rocket vehicle application (NERVA) programs. Grand Tour of outer planets would be replaced with less complex missions to Jupiter and perhaps Saturn. In place of NERVA, NASA would concentrate on defining small, high-energy nuclear propulsion system for missions to explore planets.

  During 1973 and beyond, "NASA will put even greater emphasis on turning the knowledge and technology acquired through space research to the immediate and lasting benefits of mankind. The recent establishment of the Office of Applications was an important step in that direction.

  "In aeronautics we will increase our efforts by 50 percent this year, directed mainly at the pressing domestic needs for quiet engines and quiet short-haul air transportation systems."

  Dr. Fletcher said NASA FY 1973 budget "maintains our manned space flight capability, contributes to national security, increases the opportunity for acquiring new scientific knowledge, encourages international cooperation, addresses the problems of society and the environment, and helps to assure continued American leadership in aeronautics."

  Dr. George M. Low, NASA Deputy Administrator, discussed space shuttle funding: "The shuttle, as we now visualize it"—if it had pressure-fed boosters and 4.6- by 18.3-m (15- by 60-ft), 30 000-kg (65 000-lb) capacity—"will cost $5.5 billion to develop. This includes two flight vehicles." Funding would "spread over more than 6 years, but the bulk of the funding will be expended over a six-year time period. In addition . . . there will be an investment in facilities of about $300 million. And if more orbiters and boosters are needed for flight applications, these will cost $250 and $50 million apiece. Our current estimates for the cost per flight are $7.7 million. This is less than most of the launch vehicles we're using today." (Transcript)

- Federal Communications Commission had been asked to decide on feasibility of U.S. Postal Service using satellites to transmit facsimile mail for business items, *Newsweek* reported. (*Newsweek*, 1/24/72)

January 24-25: Federal Aviation Administration held international symposium in Washington, D.C., on area navigation. Talks covered operational experiences of airlines, military, and general aviation; capabilities of specialized area navigation equipment; and R&D for future area navigation application. *Apollo 9* Astronaut James A. McDivitt, Apollo Program Manager, addressed symposium banquet. (FAA Release 72-02; FAA PIO)

January 25: Apollo 16 spacecraft had developed fuel leak in reaction control system and would be removed from launch pad for replacement of fuel tank, NASA announced. Leak had apparently been caused by overpressurization of teflon bladder by ground support equipment. NASA later said repair was not expected to delay launch, scheduled for April 16. (Reuters, *NYT*, 1/26/72, 27; *W Post*, 1/28/72, A5)

- *Cosmos 472* was launched from Plesetsk by U.S.S.R. Orbital parameters: 1536-km (954.4-mi) apogee, 193-km (119.9-mi) perigee, 102.2-min period, and 82° inclination. Satellite reentered Aug. 18. (GSFC *SSR*, 1/31/72; 8/31/72; *SF*, 6/72, 262)

- Subcommittee on NASA Oversight submitted to House Committee on Science and Astronautics *Space Shuttle—Skylab: Manned Space Flight in the 1970's*, status report prepared by Committee staff at request of Subcommittee Chairman, Rep. Olin E. Teague, from information from NASA space flight Centers and key industrial contractors.

    Conclusions on Skylab program were that program was within projected costs; bulk of equipment and systems were meeting design requirements in performance and reliability; and, despite some problems in procuring and integrating experiments, schedules were being met and no flight delays or significant cost increases were anticipated. Serious consideration was recommended for several options: possibility of flying backup Skylab B in 1974-1976 period; possible revisits to Skylab A after first three flights; and use of remaining Saturn IB launch vehicles in earth resources and applications flights after Skylab A program, during 1974-1978 period.

    Conclusions on space shuttle program were: Sufficient technology existed to undertake development of fully reusable, low-operation-cost, earth-orbital shuttle. Number of design possibilities had been studied, covering range of development costs. Regardless of design chosen, as development costs were reduced, system recurring operating costs increased. In addition to reducing cost of near-space operations by order of magnitude, shuttle offered many opportunities for increased flexibility of earth resource surveys and management, space manufacturing, and short-term rapid-response laboratory facilities. And shuttle could also be used to improve national security position. (Text)

- Tenth anniversary of start of Saturn V project at Marshall Space Flight Center. Since inauguration of Advanced Saturn development Jan. 25, 1962, Saturn V launch vehicles had been used successfully to launch eight manned space flights. Apollo 16 Saturn V booster would be 26th Saturn launch vehicle flown from Kennedy Space Center since firing of first Saturn I Oct. 27, 1961. (MSFC Release 72-3)

- Tass said Soviet test of lunar iron samples from *Luna 16* and *Apollo 11* had shown lunar iron resisted rust better than iron found on earth. (Reuters, *C Trib*, 1/26/72)

- President Nixon submitted to Senate nominations of Kenneth Rush to succeed David Packard as Deputy Secretary of Defense and of Eberhardt Rechtin to new post of Assistant Secretary of Defense for Telecommunications. Packard's resignation was effective Dec. 13, 1971. (*PD*, 1/31/72, 119-20, 152)
- Manned Spacecraft Center announced award of $407 630 to General Electric Space Div. for development of checkout system for space shuttle software. GE would develop system specifications, computer program compatible with MSC system, procedural documents, and demonstration plan; would document results; and would demonstrate system. (MSC Release 72-21)
- "Political advantages and votes" were in the 50 000 jobs of space shuttle program, *New York Times* said, "and in the lift it would give to the nation's ailing aerospace industry." Shuttle jobs over next four fiscal years would offset expected losses in jobs that would occur as Apollo and Skylab programs ended. "Thus, if President Nixon's plan is approved, private employment under Government space contracts will level out during this period at not much less than its present total of 111,000 jobs." Administration's proposed $5.5-billion shuttle investment was to be augmented by up to $2 billion before program became operational in later 1980s. "During that decade, $3-billion to $4-billion more will probably be spent on shuttle operations—one-third of them for military flights—if the current plan is followed." Annual shuttle spending would double in FY 1973 to $200 million, triple in FY 1974 to $600 million, and reach $1 billion in FY 1975. (Lyons, *NYT*, 1/25/72, 16)

January 25-27: House Committee on Science and Astronautics held 13th meeting with Panel on Science and Technology, with theme of remote sensing of earth resources. Dr. James C. Fletcher, NASA Administrator, said in theme address that key element of long-range earth survey program was development of machine capability to accept raw data from sensors in air, in space, on ground—infrared, microwave, etc. This level of machine capability did not yet exist. "What we will need are computers of a different level of sophistication that can adapt from their own experience and changing information needs as they digest continuing inputs. This work will lead, in turn, to the development of models of the natural world." Earth resources, meteorological, and environmental satellite data and data relay capabilities would force "quantum jumps necessary to create the kinds of interconnected models of natural and human action that will be essential in the not too distant future." Ability to observe and measure phenomena affecting everyday life was "one of the most important products of our space program."

Dr. H. Guyford Stever, National Science Foundation Director, said in keynote address U.S. had necessary equipment and techniques for remote sensing. "The scientific potential in geology, oceanography, hydrology, and a host of other disciplines is well known. . . . What we don't know very well is the impact of the application of remote sensing technology on society and its goals and values."

Dr. Edward E. David, Jr., Presidential Science Adviser, said that it was easier to acquire data than to interpret it and that "traditionally, unfortunately, interpretation is neglected until it is too late." Automatic processing and retrieval of images would be major resource. Machine processing of images, still in infancy, would be a most important part of earth resources experiment. (Transcript)

January 26: Apollo 15 Astronauts David R. Scott, Alfred M. Worden, and James B. Irwin were received by Vice President Boris Bakrac of the Assembly of Croatia and met with President of Yugoslav Academy of Sciences and Arts, Dr. Grga Novak, during two-day visit to Zagreb, Yugoslavia. Astronauts were touring Europe to discuss *Apollo 15* mission with scientists. (FBIS–Yugoslavia, 2/4/72, I12)

- Requests for participation in Skylab Student Project [see Jan. 3] had increased to more than 80 000 from some 15 000 in early January, NASA announced. (NASA Release 72–19)
- Washington *Evening Star* editorial supported space shuttle development: "Congress should go along with this proposed investment, knowing that if the country's highly efficient space organization is killed, it will not easily be revived. There is still much work to be done on those peaceful cosmic frontiers that can spread many benefits and much inspiration on this troubled earth." (W *Star*, 1/26/72, A17)
- Senate Committee on Aeronautical and Space Sciences favorably reported H.R. 11487, bill that would authorize sale of land at Kennedy Space Center for Chapel of the Astronauts. (*CR*, 1/26/72, D28)

January 27: High-speed interferometer (HSI) developed for NASA at Jet Propulsion Laboratory would be used by JPL engineers to record levels of chemical components of Los Angeles smog, NASA announced. California Div. of Highways had requested JPL assistance in monitoring trace constituents of city's atmosphere to help fight air pollution. Data gathered would also be given to Los Angeles Air Pollution Control District. HSI, designed for use on future NASA spacecraft, could detect five parts of carbon monoxide per billion parts of atmosphere. Instrument would monitor Santa Monica Freeway, particularly during traffic peaks. (NASA Release 72–15)

- Marshall Space Flight Center announced selection of Aerojet-General Corp., Lockheed Propulsion Co., Thiokol Chemical Co., and United Technology Center to receive separate, two-month, $150 000 study contracts to analyze possible use of solid-fuel motors with 305- and 396-cm (120- and 156-in) diameters to power space shuttle booster. (MSFC Release 72–4)
- Rep. Bob S. Bergland (D-Minn.) was elected member of House Committee on Science and Astronautics. (*CR*, 1/27/72, H367)
- NASA launched Black Brant VC sounding rocket from Wallops Station carrying Goddard Space Flight Center performance test payload. Rocket and instrumentation performed satisfactorily. (SR list)
- Federal Aviation Administration announced award of contracts totaling approximately $3 million to six companies for first phase of five-year program to develop microwave landing system (MLS) for civil and military use [see Jan. 17]. Companies were Airborne Instrument Laboratory, Bendix Corp., Texas Instruments Inc., Raytheon Co., Hazeltine Corp., and ITT Gilfillan, Inc. (FAA Release 72–18)

January 28: Preliminary description of *Apollo 15* lunar samples was presented by *Apollo 15* Preliminary Examination Team in *Science*. More than 350 individual samples weighing total of 77 kg (170 lbs) had been collected from 10 areas in Hadley region during July 26–

Aug. 7, 1971, mission. Samples were mare basalts and breccias with variety of premare igneous rocks. "The bulk chemical compositions and textures of these rocks confirm the previous conclusions that the lunar maria consist of a series of extrusive volcanic rocks that are rich in iron and poor in sodium. The breccias contain abundant clasts of anorthositic fragments along with clasts of basaltic rocks much richer in plagioclase than the mare basalts. These two rock types also occur as common components in soil samples from this site. The rocks and soils from both the front and mare region exhibit a variety of shock characteristics that can best be ascribed to ray material from the craters Aristillus or Autolycus." (*Science*, 1/28/72, 363–74)

- Dr. James C. Fletcher, NASA Administrator, in letter to *New York Times*, criticized interpretation of shuttle economics in Jan. 25 *Times* article: "NASA has a good record on cost estimating, as demonstrated by bringing in the Apollo program below the original estimates. Even if a contingency amount of $1 billion should be required to overcome unforeseen problems beyond those provided in the $5.5 billion estimates, the combined total for development and initial investment will be substantially below the $10 billion to $14 billion estimate in the Times." (*NYT*, 2/7/72)

- LeRoy E. Day, NASA Space Shuttle Program Deputy Director, described studies of shuttle's effect on payload development costs during speech before American Institute of Astronautics and Aeronautics Rocky Mountain Section in Boulder, Colo.: "A number of . . . payloads have been flown and historical design and program cost data were available. We have determined by actual preliminary design, preparation of program plans, and detailed costing analyses, that these payloads could have been redesigned to fly on the shuttle with payload development cost savings . . . of 50% of the historical baseline. The studies conclude that the cost of developing payloads (the major portion of space program costs) can be reduced primarily as a result of the shuttle's ability to retrieve payloads for reuse, refurbishment and updating."

  Dichotomy of unmanned and manned space programs should not be perpetuated, Day said. U.S. needed to continue "vigorous and meaningful manned space program" that did not "obscure the automated satellite programs." Shuttle was "the *one* mechanism that can effectively support both goals in an optimum way" within national space budget. (Text)

- President Nixon's "enthusiastic support" of space shuttle had "pumped fresh breath of life" into NASA, "buoyed the spirits of a flagging aerospace industry and lost no friends in politically vital West Coast and southern states," Robert Gillette said in *Science* article. Shuttle endorsed by President was "considerably less ambitious, and at least a billion dollars cheaper, than the shuttle NASA has fought for" but "in basic outline it remains unchanged. . . ." Shuttle "and its presidential stamp of approval stand as a tribute to NASA's deft and persistent salesmanship, a talent marked by careful acquiescence to political and economic realities and by a willingness to bleed other programs, including Apollo, to keep the shuttle alive."

  While "some budget bureau officials are said to be unhappy with the magnitude, the design, and the pace of the shuttle project as it now stands," prospects for "scuttling the shuttle are slim." Shuttle possessed few of SST's intrinsic weaknesses. Central moral issue raised by the SST was "the propriety of government subsidy for an essentially commercial enterprise." Shuttle "has no such commercial overtones, and, by any measure, it is a more truly national enterprise."

Congressional attack on supersonic transport program, while largely based on project's economics, had drawn strength from "vast and vocal public constituency aroused by alarms . . . of environmental damage." No such catalytic issue seemed likely to arise in shuttle debate. Gillette quoted unidentified Senate aide as saying, "Unless we can pull together the kind of public campaign that brought down the SST, the shuttle is going to get by this year without a thorough examination and without an adversary hearing in Congress." (*Science*, 1/28/72, 392-6)
- Nike-Apache sounding rocket was launched by NASA from White Sands Missile Range carrying Univ. of Minnesota experiment to measure composition of neutral atmosphere between 95 and 125 km (59 and 78 mi). Structural failure of undetermined origin caused vehicle to break up three seconds after launch. (NASA Rpt SRL)
- Creative science in U.S. was in transition state, Dr. William D. McElroy, National Science Foundation Director, said in *Science* editorial. "And much of the feedback associated with this fermentation is focused on NSF, which in this country is often equated with creative science and scientists." Historically, NSF had devoted large portion of its resources to "pursuit of disciplinary science—research and science education motivated solely by the intrinsic needs of a discipline or the creative needs of individual scientists. This kind of programming has been highly successful and must continue, for it is the bedrock of all scientific enterprise. However, there must also be a heightened awareness of the requirements placed on all science, and for this reason, a significant share of the total resources available to NSF in the future must be devoted to the social and technological needs of the nation. This, however, does not mean that the Foundation should be diverted from its earlier and historical purpose; in fact, this diversification should be construed as a means of strengthening that purpose." (*Science*, 1/28/72, 361)
- Manned Spacecraft Center announced issuance of two requests for proposals for space shuttle heating contracts. One request was for design study of ablative materials with low density to protect space shuttle orbiter and one was for design and construction of heating unit to produce high temperatures for testing materials used externally on shuttle. (MSC Releases 72-24, 72-25)

January 29: Vice President Spiro T. Agnew scored Senate critics of President Nixon's decision to proceed with development of space shuttle in speech before Florida Jaycees state convention in Daytona Beach: "They would, in effect, bring to a virtual halt this country's technological progress in a field which has already proved of enormous benefit to mankind and holds even greater promise for the future. Worse, they would have us abandon our hard-won leadership in an area where United States ingenuity and creativity visibly overcame the lead of our principal world competitor and proved the American free enterprise system is still Number One when the chips are down." (Text)

- U.S.S.R.'s *Luna 19* (launched Sept. 28, 1971) had orbited moon 1358 times and was operating satisfactorily, Tass announced. Spacecraft had studied moon's gravitational field, measured interplanetary magnetic field and meteor-flow density, studied characteristics of space radiation in lunar space for comparison with *Mars 2* and *Mars 3* data, and collected data on dynamics of changes in intensity of space radiation corpuscular flows. (FBIS–Sov, 1/31/72, L1)
- Paralytic patients at Southwest Research Institute in San Antonio, Tex., were using system developed at Langley Research Center to adjust their immediate environment, United Press International reported. System used paddles and switches activated by patients' eye movements and breath to control room lights, radio, TV, and other electronic devices. (*C Trib*, 1/29/72)
- Delivery to Dulles International Airport of ARTS III—computerized automated radar terminal system to improve air traffic handling in busy terminal areas—was announced by Federal Aviation Administration. System processed radar beacon signals emitted by airborne transponders giving aircraft identity and altitude and presented information on radar displays for air traffic controllers. (FAA Release 72-19)

January 30: First of four lunar eclipses to occur during 1972 was partially visible in portions of North and South America. Smithsonian Astrophysical Observatory spokesman said later that 37-min eclipse was not seen over Northeastern U.S. because of cloud cover. (*NYT*, 1/29/72, 19; 1/31/72, 3)

January 31–February 23: European Space Research Organization's 117-kg (257-lb) *Heos 2* Highly Eccentric Orbit Satellite was launched by NASA from Western Test Range by three-stage thrust-augmented Thor-Delta (DSV-3L) booster at 9:20 am PST. Orbital parameters: 244 011-km (151 621.4-mi) apogee, 416-km (258.5-mi) perigee, and 90.2° inclination.

All vehicle systems monitored during launch functioned normally. Second burn of 2nd stage and ignition and burnout of 3rd stage occurred over Indian Ocean south of African continent. By Feb. 6 all flight experiments had been turned on and were working properly and all spacecraft systems were still functioning normally Feb. 23. On Feb. 18 NASA officially judged mission successful.

Primary NASA objective was to place spacecraft into orbit that would permit successful achievement of ESRO scientific objectives and provide tracking and telemetry support. Satellite—16-sided cylinder with 70% of its outer surface covered with solar cells—carried seven experiments to investigate strength and direction of magnetic fields, energy distribution of protons and electrons, and nature of solar winds; make very-low-frequency solar observations; and detect micrometeorites.

Heos 2 was second HEOS mission. *Heos 1* had been launched Dec. 5, 1968, and was still in orbit. After orbital insertion, spacecraft was controlled from European Space Operations Center (ESOC) and tracked by European Space Tracking Stations (ESTRACK). European Space Technology Center (ESTEC) was responsible for project management and would reimburse NASA $6.6 million for booster and prelaunch and launch support services. (NASA proj off; NASA Release 72-17; GSFC SSR, 1/31/72)

January 31: Mathematica, Inc., published *Economic Analysis of the Space Shuttle System.* Report of study directed by Klaus P. Heiss and Oskar Morgenstern under NASA contract concluded development of space shuttle system was economically feasible assuming level of space activity equal to average of U.S. unmanned space program of past eight years; thrust-assisted orbiter shuttle (TAOS) with external hydrogen and oxygen tanks was economically preferred choice among current shuttle configurations; and choice of thrust assist for orbiter shuttle was still open, with pressure-fed boosters and solid-fuel rocket motors, either using parallel burn, as main economic alternatives. Third economic choice was use of series-burn boosters. (Text)
- NASA launched two sounding rockets from Wallops Station in support of winter anomaly program. Nike-Apache carried Univ. of Illinois experiment to 200-km (124.3-mi) altitude to measure D- and E-region ionospheric characteristics during winter anomaly. Super Arcas carried Pennsylvania State Univ. experiment to 82.3-km (51.1-mi) altitude to measure electron density in upper atmosphere. Rockets and instruments functioned satisfactorily. (NASA Rpts SRL)
- New map depicting geologic features of moon's visible side had been published by U.S. Geological Survey, *Chicago Tribune* reported. Map provided "unique picture of the distribution of the various kinds of rock materials." (*C Trib*, 1/31/72)
- Marshall Space Flight Center announced award of $9.8-million contract modification to McDonnell Douglas Astronautics Co. for work on backup Skylab Workshop. Modification included production completion and post-manufacturing verification of backup hardware and launch site support operations through completion of second manned mission. (MSFC Release 72-5)
- Senate received nomination of *Apollo 10* astronaut Col. Thomas P. Stafford (USAF) to be brigadier general. Col. Stafford was Director of Flight Crew Operations at Manned Spacecraft Center. (*CR*, 1/31/72, S824; *A&A 1971*)
- Unidentified senior analysts said in Washington, D.C., that People's Republic of China had deployed "handful" of ballistic missiles with estimated 2400- to 4000-km (1500- to 2500-mi) range. Missiles used storable liquid propellant that permitted underground installation in concrete and steel silos. (Beecher, *NYT*, 2/1/72, 1)
- Major thrust of Dept. of Defense FY 1973 expenditures was to improve strategic weapon capability, *Aviation Week & Space Technology* editorial said. Budget [see Jan. 24] was "the best news the aerospace industry has had in several years . . . a clear signal for action and forward thrust. It also imposes an awesome responsibility on the aerospace industry. Once again it is being called upon to produce new and demonstrably superior hardware on which it is possible to base a military power sufficient to deter major war and sufficient to win quickly any minor wars. It must produce swiftly and efficiently. Neither the aerospace industry nor the country can afford any more of the expensive development fiascos of the McNamara era." (*Av Wk*, 1/31/72, 9)

During January: Space Science Board of National Academy of Sciences and National Research Council released *Outer Planets Exploration, 1972–1985.* Analysis of new cost and engineering data for unmanned spacecraft for missions to outer planets had been made for NASA by panel chaired by Dr. Francis S. Johnson of Univ. of Texas at Dallas. Extensive study of outer solar system was "one of the major objectives of space science in this decade." Panel recommended thermoelectric outer planets spacecraft (TOPS) be developed and used in Grand Tour missions of late 1970s; Pioneer-level program be maintained for exploration of Jupiter and Saturn and their satellites every two years; Pioneer-F and -G evaluate radiation environment of Jupiter and Pioneer-H be readied for Jupiter magnetosphere mission to further evaluate planet's radiation if necessary; NASA ground-based research on planetary exploration be continued; even-segment Titan booster or its equivalent be developed for outer planet exploration; support be provided for scientific advisory committees, development of instrumentation, and studies of alternative mission strategies for Jupiter and Saturn exploration; and NASA development of methods of solar and nuclear electric propulsion be continued.

Report recommended programs for three funding levels: four TOPS Grand Tour missions in 1976–1980 period at $400-million annual budget level, compromise between Grand Tour missions and outer solar systems programs like modified Pioneer missions at $250-million-a-year level, and varied program of outer solar system exploration using modified Pioneer spacecraft at lowest funding level, approximately $400 million over one decade. (NAS–NRC–NAE *News Rpt*, 1/72, 8–9)

- Advanced technology aircraft to evolve from increased NASA, Dept. of Transportation, and Dept. of Defense use of research and experimental aircraft were described in *Astronautics & Aeronautics* by Langley Research Center Director Edgar M. Cortright. "NASA, in coordination with DOD and DOT, plans a program to design and manufacture two STOL transports. Although these aircraft probably will be smaller than anticipated production models, they will suit the needs of developing promising STOL augmented-lift concepts, control systems, and crosswind landing gear. They will be able to operate within acceptable noise limits to and from STOL strips about 2000 ft [610 m] long."

  Joint NASA-Army plan "should lead to advanced helicopter and tilt-rotor research vehicles. Although oriented primarily toward military applications, the research results will apply to commercial VTOL transports."

  Advanced subsonic transports of the 1980s "will incorporate most of the applicable advanced technology . . . with significant gains in productivity. A longer range, higher capacity SST [supersonic transport] would fill a need on the transatlantic routes and the longer routes across the Pacific. And no forecaster can discount the possibility of a hypersonic transport (HST) for the 1990s." (*A&A*, 1/72, 30–4)

February 1972

February 1: NASA released preliminary timeline for Apollo 16 manned lunar landing mission. Apollo 16 would be launched from Kennedy Space Center at 12:54 pm EST April 16—with lunar module landing on moon at 3:41 pm EST April 20 and lifting off moon at 4:39 pm EST April 23. Returning command module would splash down at 3:30 pm EST April 28. (KSC Release)

- H.R. 12824, $3.379-billion FY 1973 NASA authorization bill, was introduced by Rep. George P. Miller (D-Calif.), Chairman of House Committee on Science and Astronautics. (CR, 2/1/72, H556)
- *Apollo 15* Astronauts David R. Scott, James B. Irwin, and Alfred M. Worden visited President Nixon at White House to report on their 16-day goodwill tour to Poland and Yugoslavia. (PD, 2/7/72, 196)
- Development of automated visual sensitivity tester was announced by Ames Research Center. Simple, accurate, and easy-to-use device to map individual vision patterns was originally developed for use during long-term confinement, such as during space flight. It mapped position and extent of normal blind spot of each eye and plotted abnormal blind spots. Device was expected to be valuable in diagnosis and treatment of many disorders, including brain damage caused by tumors or injuries, optic track degeneration, glaucoma, and detached retina. It could be used in testing visual acuity, color blindness, dynamic ocular tracking, form discrimination, Optokinetic Nystagmus Reflex, and color sensitivity. (NASA Release 72-21)
- Research scientists had been invited to submit proposals for Skylab experiments studying use of weightlessness, NASA announced. Experiments would use weightlessness in space to develop improved techniques for preparing biological materials and for studying crystal growth, solidification, and other aspects of nonorganic substances. One invitation was for electrophoresis—motion of charged particles through fluid while under influence of electrical field. Use of technique on ground was hampered by effects of heat convection in fluid and sedimentation. Second invitation was for investigations of solidification effects, crystal growth, and other phenomena in weightless materials, using either small multipurpose electric furnace or system for levitating small samples of molten materials and closely observing them as they cooled and solidified. First experiments could be flown on Skylab missions beginning in 1973. (NASA Release 72-22)
- S.R. 193, joint resolution to redesignate Cape Kennedy as Cape Canaveral, was introduced by Sen. Edward J. Gurney (R-Fla.) and Sen. Lawton M. Childs, Jr. (D-Fla). Resolution would retain name John F. Kennedy Space Center for NASA facilities on Merritt Island. Sen. Gurney said it was from KSC, "and not from the Cape itself, that President Kennedy's dream was realized." Desire of Cape's residents for restoration of 400-yr-old name Canaveral and "desire of a lasting memorial to the

President" could be accomplished through adoption of resolution. "From a historical standpoint, no greater justice could be done." (*CR*, 2/1/72, S862–4)

- Manned Spacecraft Center announced contract award and issuance of requests for proposals for work on space shuttle. One-year, $99 985, firm-fixed-price contract went to Lowey/Snaith, Inc., to study methods for making space shuttle orbiter interior pleasant environment for working and living. RFPs were for design study for orbiter's orbital maneuvering system (OMS). Firm selected would receive one-year, $250 000 firm-fixed-price contract. (MSC Releases 72–29, 72–30)
- First avionics testbed aircraft for Air Force Airborne Warning and Control System (AWACS)—modified Boeing 707–320 topped by 9-m (30-ft) rotodome—was rolled out at Boeing Co. facility in Renton, Wash. AWACS would detect and track aircraft at high and low altitudes over land and water and track low-altitude aircraft over extended areas. (USAF memo for correspondents)
- Highest frequency measurement reported to date was announced by National Bureau of Standards scientists in *Applied Physics Letters*. Scientists had measured frequency of infrared light waves generated by helium-neon laser to 88 376 245 million cycles—100 times higher than any frequency recorded until 1968. (Evenson, Day, *et al.*, *Applied Physics Letters*, 2/1/72, 133–4)
- Brazilian Agriculture Minister Cirne Lima inaugurated new meteorological satellite tracking station in Brasilia, Brazil. Station was second built in Brazil; first was in Rio de Janeiro. (FBIS–Brazil, 2/3/72, D3)

February 2: *Mariner 9* news conference was held at NASA Hq. Spacecraft (launched May 31, 1971) was in 81st day in Mars orbit, had accepted 25 000 commands, and had taken about 5000 pictures. Dr. Bradford Smith of New Mexico State Univ. said wind-transported dust that had obscured early *Mariner 9* photography was more widespread and occurred more frequently than generally had been supposed; substantial quantities of surface dust were displaced each Martian year by local winds that varied with season and with specific areagraphic location; and large-scale atmospheric instabilities occurred, producing hemispheric and global dust storms which were seasonal but apparently did not occur every Martian year. Storm observed by *Mariner 9* had begun in late September 1971, reached its peak in late October, and had cleared slowly until mid-December. Planet was quite clear in February.

Mariner 9 photos included pictures of huge volcano with base 500 km (300 mi) wide, Harold Masursky of U.S. Geological Survey told press. ". . . as the storm cleared we were able to establish that the dark spots that we saw barely sticking up through at first" were "the very summit of Nix Olympica. This tall mountain with a complex crater at the top and the scalloped edges and the flat, or non-raised rims indicated to us that this was probably a multiple volcanic vent. As the storm continued clearing finally we could see all the way down to the great plains at the base of the mountain. Now this is a great volcanic pile . . . twice as big as the great volcanic pile that forms the Hawaiian Islands. The Hawaiian pile is about 225 kilometers [140 miles] across and . . . about 9 kilometers [6 miles] high. It is the largest volcanic pile on the earth. It is something like 32,000 feet [9800 meters] from the floor of the Pacific to the top of the mountain."

Photos also showed "spectacular channel" 100 km (60 mi) long with numerous side branching valleys that ended in featureless areas. Absence of vents at ends of valleys made gaseous eruption formation improbable and suggested channel might have been formed by water erosion. Highest amount of water vapor—one thousandth of that in earth's atmosphere—above Mars' south polar cap suggested that water vapor was being released from region as cap retreated. Absence of vapor over north polar cap indicated that water vapor there was condensed in carbon dioxide and was transported by storm or by general circulation from south polar cap to north polar cap. Vapor over south polar regions was slightly less than 20 precipitable micrometers.

Mariner 9 photos of Mars had provided data which conflicted with data from 1969 *Mariner 7* Mars probe. From 1969 photos of heavily cratered planet experimenters had decided that Mars was "a dead planet, very primordial . . . the product of the accumulation of the cosmic debris that fell in to form the planet." *Mariner 9* photos showed "something very, very different indeed. . . . we can see the great volcanic piles and because of the crispness of the edges and the lack of craters, we think these are geologically young . . . [and] we have a geochemically evolved planet." Comparison of 1969 and 1971 photos showed that south polar cap regions completely covered by frost in 1969 were only partially covered in 1971. During period when photographed by *Mariner 9* little separations had widened and solid carbon dioxide coverage had broken up until many dark spots showed through as the ice sublimated and revealed underlying dark rocks.

Dr. Charles A. Barth of Univ. of Colorado reported that Mars atmosphere was primarily carbon dioxide but also contained atomic hydrogen and atomic oxygen. "The atomic oxygen comes presumably from the photo-dissociation of carbon dioxide and the atomic hydrogen comes most probably from the photo-dissociation of water vapor." Amount of atomic hydrogen varied very little with altitude. Atomic hydrogen extended some 20 000 km (12 000 mi) from planet like a "great big sphere of glowing hydrogen atoms with the spacecraft running around through it. Those hydrogen atoms are put there from photo-dissociating water vapor and because of the very low gravitational field of Mars, it is relatively easy for these atoms to escape. In fact, the picture that we have, is that water comes up out of the surface of Mars, goes into the lower atmosphere . . . [and] gets deposited in the polar caps; when the polar caps thaw, it is released into the atmosphere." Atomic hydrogen was escaping into upper atmosphere at rate equal to 380 cu m (100 000 gallons) of water per day. (Transcript)

- NASA was working with Dept. of Transportation (DOT) to define characteristics of clear-air turbulence (CAT) better and to develop methods of detecting it in flight, Flight Research Center announced. Instrumented B-57 aircraft—equipped with DOT prototype radiometric sensor able to detect CAT up to 80 km (50 mi) ahead of aircraft—was being flown at altitudes up to 15 000 m (50 000 ft) in areas over western U.S. where CAT was expected. Aircraft also carried Univ. of Wyoming aerosol and ozone detector to determine relationship between presence of aerosols or ozone, or both, and atmospheric conditions that caused CAT. Since tests began one year before, B-57 had completed 13 flights.

February 2: *Flight Research Center flew a B-57 aircraft equipped with Department of Transportation prototype radiometric sensor and Univ. of Wyoming aerosol ozone detector in a NASA-DOT program to develop methods of detecting clear-air turbulence.*

When CAT was encountered, 30-min flight pattern was flown to record data in turbulent area. (FRC Release 1-72)

- Kennedy Space Center announced two contract awards. McDonnell Douglas Corp. was awarded $15 945 000, one-year contract modification for prelaunch, launch, and postlaunch operations of Skylab airlock module, Orbital Workshop, associated ground support equipment, special test devices, and facilities. Modification brought total value of contract to $60 599 698.

  Boeing Co. Field Operations and Support Div. was awarded $22 904 790, one-year contract extension for base support services at KSC and Cape Kennedy Air Force Station. Extension brought total value of contract to $42 122 790. (KSC Releases 24-72, 25-72)

- NASA launched Nike-Tomahawk sounding rocket from Poker Flat Rocket Range, Fairbanks, Alaska, carrying Rice Univ. auroral fields and particles experiment. Rocket and instrumentation performed satisfactorily. (SR list)

February 3: *Cosmos 473* was launched by U.S.S.R. from Baykonur into orbit with 356-km (221.2-mi) apogee, 176-km (109.4-mi) perigee, 89.6-min period, and 65° inclination. Satellite reentered Feb. 15. (GSFC *SSR*, 2/29/72; *SBD*, 2/7/72, 191)

- Apollo Telescope Mount (ATM) flight unit was being readied for three-month, post-manufacturing checkout at Marshall Space Flight Center. ATM would be moved in May from Quality and Reliability Assurance Laboratory to Astronautics Laboratory for vibration tests and would be delivered to Manned Spacecraft Center June 1 for thermal and vacuum tests. ATM would be launched on first Skylab mission in 1973.

  Series of preliminary Skylab experiment planning simulations was being conducted by flight controllers, flight planners, and other experts at MSC. Purpose of simulations was to exercise basic plans and procedures necessary before detailed simulations with Skylab team could begin. (MSFC Release 72-8; MSC Release 72-32)

- Senate passed and sent to House H.R. 11487, Chapel of the Astronauts bill, after amending it to restrict authority to sell land at Kennedy

Space Center for chapel to two years after act's enactment date. (CR, 2/3/72, S1097-8)

- Award of $185 000 grant for joint Dept. of Health, Education, and Welfare, NASA, and Corp. for Public Broadcasting project to beam educational TV programs to rural residents in eight Rocky Mountain states was announced by HEW. Total of $500 000 was to be spent on early childhood and career education programs to be transmitted via ATS-F satellite scheduled for May 1973 launch. (AP, NYT, 2/6/72)
- Senate confirmed nominations of Kenneth Rush as Deputy Secretary of Defense and of Eberhardt Rechtin as Assistant Secretary of Defense (Telecommunications). (CR, 2/3/73; DOD Directory, Spring 1973)

February 4: Marshall Space Flight Center announced award of one-month $1-million interim contract extension to North American Rockwell Corp. Rocketdyne Div. for space shuttle main engine. Contract was awarded pending completion of General Accounting Office review of original $500-million contract awarded July 13, 1971. Review had been requested by competing contractor, United Aircraft Corp. Pratt & Whitney Div., on Aug. 3, 1971. (MSFC Release 72-9; *A&A 1971*)

- Manned Spacecraft Center announced award of $175 183 cost-plus-fixed-fee contract to Cornell Aeronautical Laboratory, Inc., to conduct simulations of space shuttle orbiter approach and landing. Simulations would enable engineers to establish more accurately aft location of orbiter's center of gravity and systems for vehicle aerodynamic control. Contract would end July 31. (MSC Release 72-31)
- Ultrasonic wrench developed by NASA to connect tubing in Saturn boosters was available to private industry for possible commercial use, Marshall Space Flight Center announced. Wrench induced flexuous vibrations in nut being tightened and guaranteed calibrated tension. It could be used by industry on conventional nuts and bolts, which could replace expensive lock nuts and bolts that changed color as tension increased. Ultrasonic tightening also could remove normal friction forces and ensure proper tension by preselected torque dial. (MSFC Release 72-10)
- Largest NASA-owned aircraft—Lockheed C-141 StarLifter weighing 147 000 kg (325 000 lbs) with 49-m (160-ft) wing span and 43-m (140-ft) length—was delivered to Ames Research Center. Aircraft, modified to accept infrared telescope 91 cm (36 in) in diameter and two computers, would be used by NASA as international facility for infrared astronomy. Aircraft observatory would be unique when telescope was installed later in year. (ARC *Astrogram*, 3/16/72, 1)
- President Nixon sent message to Congress outlining plan for Federal Partnership in District of Columbia's observance of American Revolution Bicentennial in 1976: "... there will be a handsome new building for one of the Mall's oldest tenants, the Smithsonian Institution. This structure, which will house the National Air and Space Museum with exhibits ranging from Kitty Hawk to Hadley Rille and with a former astronaut [Michael Collins] in charge, can be ready in 1976 if the Congress will move now to approve FY 1973 construction funds for it; the plans are nearly complete." (PD, 2/7/72, 184-93)

February 6: Fire leveled eight-trailer modular complex being used by NASA as temporary office space near Astrionics Laboratory at Marshall Space Flight Center. Redstone Arsenal Fire Dept. extinguished blaze within one hour but officials said portion of building containing high

February 6

vacuum and thin film research section of hybrid microelectronics research laboratory had suffered minor damage. MSFC Director, Dr. Eberhard F. M. Rees, later appointed six-member board to investigate cause of fire that led to estimated $100 000 loss, but no injuries. (*Huntsville Times*, 2/8/72)

February 7: Experiment turn-on and checkout of *Heos 2* Highly Eccentric Orbit Satellite (launched by NASA for European Space Research Organization Jan. 31) was completed and spacecraft became fully operational. *Heos 2* carried seven experiments to study interplanetary physics and high-latitude magnetosphere. (NASA Release 72–29)

- U.S.S.R.'s *Luna 19* lunar probe (launched Sept. 28, 1971) was still relaying high-quality photos of lunar surface. Two photos showed region around Eratosfyen Crater and around Godyen and Agrippa Craters. (*SBD*, 2/7/72, 34)
- NASA launched Nike-Tomahawk sounding rocket from Kiruna, Sweden, carrying Swedish auroral aeronomy experiment. Rocket and instrumentation performed satisfactorily. (SR list)
- United Auto Workers filed suit in U.S. District Court in Washington, D.C., in attempt to recoup part of first-year wage increase denied 30 000 aerospace workers by Federal Pay Board Jan. 13. UAW contended portion of wage increase requested represented cost-of-living agreement made in 1968. (Rowe, *W Post*, 2/8/72, A7)

February 8: Apollo 16 spacecraft, returned to Vehicle Assembly Building Jan. 27 for replacement of command module fuel tank damaged during testing, was rolled out to Kennedy Space Center Launch Complex 39, Pad A. Spacecraft would be mated with Saturn V booster Feb. 11 and was to begin flight readiness test Feb. 29 in preparation for April 16 launch toward moon. (KSC Release 26–72)

- Dr. James C. Fletcher, NASA Administrator, presented NASA's FY 1973 authorization request before House Committee on Science and Astronautics as hearings opened. He emphasized importance of space program to U.S. national economy: "Scientific knowledge, scientific exploration, and the practical applications of aeronautics and space are enormously important in their own right. But perhaps most important of all is the need for the United States to have a continuously advancing technology. To meet the pressing social problems of our times requires above all a sound economy operating at a high level of employment to generate the tax revenues required at all levels of government. To maintain such an economy in a competitive world, we must increase our productivity year after year, decade after decade. The only way in the long term to keep increasing our productivity is through advancing our technology.

  "I know of no other activity which has done and can do as much to keep the United States strong in advanced technology as NASA's programs in space and aeronautics." Economic necessity for advanced technology; "the direct practical benefits of space applications and improved aircraft, including their significance for national defense; and . . . the human and future practical values of increased understanding of the earth, sun, moon, planets, and universe—these are the basic reasons for maintaining a strong national program in space and aeronautics."

Dr. Fletcher said that NASA's FY 1973 program "moves forward in that it fully supports President Nixon's decision to proceed with the development of the Space Shuttle, the keystone to the Nation's future in space; it continues our major ongoing space programs; and it provides a 50 percent step-up in our work in aeronautics. It is realistic in that it supports these objectives within a fiscal year 1973 budget approximately equal to that of last year, and under a plan that does not commit the Nation to higher total NASA budget levels in future years." (Transcript)

- *Intelsat-IV F-4*—launched by NASA Jan. 22 for Communications Satellite Corp. on behalf of International Telecommunications Satellite Consortium—was adjudged successful by NASA. Satellite had been launched into satisfactory transfer orbit and subsequently placed into synchronous orbit at 165° east longitude, from which it was drifting to station over Pacific at 174° east longitude. (NASA proj off)

- Manned Spacecraft Center announced issuance of requests for proposals for development of concepts, construction of hardware, and testing of thrust chamber for space shuttle orbit maneuver engine (OME). Two firms would be selected, each to receive one-year, $275 000, firm-fixed-price parallel contract. (MSC Release 72–33)

- President Nixon, in message to Congress outlining 1972 environmental program, said: "The time has come to increase the technological resources allocated to the challenges of meeting high-priority domestic needs." Temptation "to cast technology in the role of ecological villain must be resisted—for to do so is to deprive ourselves of a vital tool available for enhancing environmental quality. . . . The difficulties which some applications of technology have engendered might indeed be rectified by turning our backs on the 20th century, but only at a price in privation which we do not want to pay and do not have to pay." Technology "must be wisely applied so that it becomes environmentally self-corrective. This is the standard for which we must aim."

  President said he had requested in budget $23-million increase in research and development funds for reducing aircraft noise and additional $88 million for development of "broad spectrum of new technologies for producing clean energy." These would include "new or increased efforts on fusion power, solar energy, magnetohydrodynamics, industrial gas from coal, dry cooling towers for power plant waste heat, large energy storage batteries and advanced underground electric transmission lines." He also proposed establishment of $100-million Voluntary Fund for Environment by United Nations to which U.S. would contribute "fair share . . . on a matching basis over the first 5 years" to improve global environment. "We are now growing accustomed to the view of our planet as seen from space—a blue and brown disk shrouded in white patches of clouds. But we do not ponder often enough the striking lesson it teaches us about the global reach of environmental imperatives. No matter what else divides men and nations, this perspective should unite them." (*PD*, 2/14/72, 218–27)

- Adapted, ultrasensitive, fast-scanning, infrared optical equipment developed by NASA to test miniaturized electronics circuits was being used by B. F. Goodrich Co. to test automobile and aircraft tires. Goodrich equipment permitted, for first time, effective nondestructive tire testing, much less expensive than destructive testing. Real-time cathode-ray-tube picture of heat in tires was produced as tires spun in testing device. Hot spots, indicating design or construction flaws, appeared as bright areas. Sensitive IR camera read heat from 600 000 points on each tire every second, presenting view as if tire were not spinning. Equipment could be used in testing other products and in quality control for electronic circuitry, carbon dioxide gas-laser research, and void detection in honeycomb structures. (NASA Release 72-24)

- House passed H.R. 10243 to establish Office of Technology Assessment for Congress by vote of 256 to 118. Office would identify and consider impacts of technological application. (CR, 2/8/72, H865-7)

February 8-11: NASA and European Space Conference (ESC) Joint Experts Group met in Neuilly, France, to identify potential areas for European participation in U.S. post-Apollo space program. In statement issued Feb. 11 Group said NASA, "noting President Nixon's approval on Jan. 5, 1972, to proceed with development of the space shuttle," had encouraged European participation in shuttle development. NASA expected that European participation would be "within the context of a broader program which included multilateral European responsibility for development of a major element such as reusable space tugs to provide access to geosynchronous and other orbits" beyond shuttle's capability or "shuttle-borne orbital laboratories" for research by U.S. and European scientists. Group recommended calendar for actions to permit governmental decisions consistent with July 1 NASA commencement of shuttle development. (NASA Note to Editors, 2/11/72)

February 9: President Nixon sent *United States Foreign Policy for the 1970's: The Emerging Structure for Peace* to Congress. Third annual report on U.S. foreign policy analyzed world situation. Of space, President said: "As our astronauts have seen, the unity of the Earth is experienced most vividly from outer space. And conversely, seen from our planet, space itself is a frontier to mankind as a whole, not merely to individual nations. Space is, therefore, an unparalleled field for cooperation among nations.

"As we move into the second decade of space exploration, the U.S. is committed to work with others in space for the benefit of all mankind. We are taking whatever steps can reasonably and properly be taken to work with other countries in the development of their space skills.

"Specifically, we have assured the European Space Conference that its member countries may obtain our assistance in launching satellites which are for peaceful purposes and which are consistent with international obligations embodied in such agreements as the Outer Space Treaty and the arrangements for the International Telecommunications Satellite Consortium (INTELSAT). We are prepared to consider such assistance to other interested countries. In addition, we are working closely with the Europeans on the concepts and design of a reusable space transportation system.

"Over the past year, NASA has agreed with the Soviet Academy of Sciences to significant cooperation in specific space tasks, and in the exchange of information and plans concerning our respective space programs. We have exchanged samples of lunar soil. We are examining together the means to enable Soviet manned spacecraft and our own to rendezvous and dock in space. Joint expert groups have been meeting to arrange details of further collaboration in space meteorology, biology, and medicine, in the study of the natural environment, and in exploration of the moon and planets.

"In 1971, after years of negotiation in which the United States has played a leading role, the United Nations General Assembly approved an Outer Space Liability Convention. . . .

"Last year also brought a new definitive charter for the operation of INTELSAT. . . ."

President said NASA-Soviet Academy of Sciences space cooperation had been among series of U.S.–Soviet agrements, "striking both in their diversity and in their promise of mutual advantage."

But in "changed world" conditioning U.S. foreign policy, President noted "the end of an indisputable U.S. superiority in strategic strength," and its replacement by a strategic balance in which the U.S. and Soviet nuclear forces were comparable. U.S.S.R. had continued to improve its capability in "virtually every category of strategic offensive and defensive weapons. . . . Soviet strategic forces, even at current levels, have the potential of threatening our land-based ICBMs if the Soviets choose to make certain qualitative improvements. They have the necessary technological base." People's Republic of China was "continuing to develop a strategic offensive capability. The possibility of accidental attacks remains." President said it was wise for U.S. to begin Safeguard ABM deployment, but "we may soon complete a SALT agreement with the USSR which will limit ABM deployment."

President commended 1970 Hague and Montreal International Civil Aviation Organization conventions to deter aircraft hijacking and sabotage: "These two conventions will increase the likelihood that hijackers, saboteurs, and persons committing other attacks against civil aircraft will be punished. . . . Universal ratification would ensure that air pirates could find no place to hide.

"We intend to press for wide adherence to these agreements and for continued international cooperation, including exchanges of information on security measures. We will also continue to urge international agreement to suspend air service to countries which refuse to cooperate in the release of hijacked aircraft and in the punishment of hijackers." (*PD*, 2/14/72, 235–411)

- Dr. George M. Low, NASA Deputy Administrator, warned of challenges to U.S. leadership in technology if U.S. did not proceed in space shuttle development. In speech before Advertising Club of Baltimore he said: "The problem is very real. Last year, the United States for the first time had an annual trade deficit, a negative balance of trade, of over $2 billion. But this deficit would have been three times that amount—$6 billion—had it not been for a favorable balance of almost $4 billion achieved in the aerospace field." Decision to proceed with space shuttle "provides a needed tool for the space program, and it will keep the United States strong in advanced technology." (Text)

- President Nixon's decision that proposed joint Federal Aviation Administration and European Space Research Organization aeronautical satellite program should be "substantially redirected" was conveyed to Secretary of State William P. Rogers and Secretary of Transportation John A. Volpe in memo from Presidential Assistant, Dr. Henry A. Kissinger. Memo said joint program was "not in accord with Administration policy." Every effort would be made to reassure ESRO governments that U.S. "fully supports international cooperation in space even though it cannot accept the specific proposals contained in the current draft FAA/ESRO Memorandum of Understanding." (*Av Wk*, 2/21/72, 17; Aug, W *Star*, 2/11/72)

- Edgar M. Cortright, Langley Research Center Director, advocated development of supersonic transport in speech before National Security Industrial Assn. in Washington, D.C.: "The day will come when virtually all intercontinental and some transcontinental air traffic will be supersonic—or faster." U.S. "must develop a supersonic transport or abdicate its enviable position as master builder of the world's commercial aircraft." If U.S. reentered SST competition, next SST "must be better than the one we canceled. It must be quieter, have a greater payload fraction, and have longer range—transpacific if possible." Because of lost time, competition "will not be the Concorde, but a follow-up super-Concorde of unknown characteristics."

  U.S. should "proceed with a sense of urgency" in developing necessary technology to produce a superior aircraft. "I believe that there is a remarkable unanimity in the aircraft and air transport industries that a good SST would be a winner for the United States—creating jobs and National income." (Text)

- Manned Spacecraft Center announced appointment of Richard S. Johnston, Deputy Director for Biomedical Engineering, as Acting Director of Medical Research and Operations. Johnston would assume responsibilities formerly held by Dr. Charles A. Berry, who became NASA Director of Life Sciences Sept. 1, 1971. (MSC Release 72-36)

- Pupils from seven suburban schools in Cleveland, Ohio, area visited Lewis Research Center to discuss participation in joint LeRC and Cleveland Div. of Air Pollution Control project to measure trace elements in air. NASA would supply equipment to schools for sampling outside air. Samples would be analyzed at LeRC and compared with Cleveland air. (Cleveland *PD*, 2/10/72)

February 10: Apparent failure of electronic component had caused loss of major portions of data from two of three scientific experiments on board particles and fields *Subsatellite* placed in lunar orbit by *Apollo 15* Aug. 4, 1971, NASA announced. Affected experiments—magnetometer and particles experiment—appeared to be functioning normally, but not all data could be received from them. Data controlled by 2 of 12 gates which released data in proper sequence were lost Feb. 3 and engineers had been unable to restore the data paths that handled orientation for analyzing magnetometer results, synchronizing pulses that allowed ground-based computers to decipher telemetry data, and portions of data collected by experiments. Manned Spacecraft Center personnel had been manually analyzing thousands of bits of data to determine extent of problem and to work out alternate procedures for obtaining the missing data. Analysis showed that failure was probably due to breakdown of some electronic part. (MSC Release 72-37)

- Flight Research Center received F-111 aircraft from Air Force for joint NASA-USAF supercritical flight program. Aircraft, 13th F-111 built, would be flown by FRC project pilot Einar Enevoldson to check out systems before NASA F-111 went off flight status until mid-June for instrumentation. It would then be flown with conventional wing to collect

baseline flight data for comparison with future supercritical wing data. (FRC *X–Press*, 2/18/72, 2; FRC PAO)
- Four-stage Javelin sounding rocket, launched by NASA from Wallops Station, carried 59-kg (130-lb) National Oceanic and Atmospheric Administration and Goddard Space Flight Center payload to 852-km (529.4-mi) altitude to study upper atmosphere. Onboard instruments measured ion composition and concentration and solar-wind neutral hydrogen atoms during 16½-min flight. Data would be correlated with ground observations. Rocket and instruments functioned satisfactorily. (WS Release 72–1)
- Air Force Systems Command announced award of $142 000 to North American Rockwell Corp. and $53 000 to Sperry Univac Systems Div. for study of man-machine relationships in remotely piloted vehicles (RPV) designed for air-to-ground missions. Study would define advantages and disadvantages of man's presence in RPV concept. (AFSC Release 252.72)
- National Science Foundation submitted to President annual report *Federal Support to Universities, Colleges, and Selected Nonprofit Institutions, Fiscal Year 1970*. Between FY 1969 and FY 1970, Federal funds to universities and colleges declined by $227 million, or nearly 7% to $3.227 million—lowest level since 1966 and first decline in actual dollars in direct Federal support since 1963. NSF attributed much of decline to "recent shift in Government policy away from direct Federal grants for facilities construction to subsidized interest charges on loans from non-Government sources." Of 13 agencies included in study, 6 had allocated 97% of all Federal obligations to universities and colleges in 1970. Health, Education, and Welfare continued as principal source of Federal agency funds—64%. NSF and Dept. of Defense together accounted for about 20% of Federal total; fluctuations in funding levels of NSF, DOD, Dept. of Agriculture, NASA, and Atomic Energy Commission were small. (Text)

February 11: Dr. James C. Fletcher, NASA Administrator, refuted economics of space shuttle opponents in speech before Commonwealth Club of San Francisco: "First, they inflate development costs; second, they say that the number of launches per year must be greatly increased before there are savings, which is not true; third, they incorrectly include the cost of payloads as part of the cost of the Shuttle. In this way, they reach the totally false conclusion that the Shuttle program will cost $30 or $40 billion over the next two decades.

"Their figures are wrong, and their logic is wrong. It is against common sense to add the operational costs of the Shuttle to the development costs . . . . The cost of using the Shuttle to carry out a space mission should be added to the cost of the mission. Actually we will be saving money, not spending money, every time we use the Shuttle for a space mission."

Shuttle development cost would be "about $6 billion to be spent over the next six or seven years" not $30 or $40 billion "as some Shuttle critics say." (Text)
- Mariner Mars '71 primary mission—to explore Mars from Martian orbit for period sufficient to observe planet's surface and view selected areas during dynamic changes, photograph planet, and obtain infrared and ultraviolet data on atmosphere and surface characteristics—was ad-

judged successful by NASA. Mariner 8 launch May 8, 1971, was not successful because of malfunction of Atlas-Centaur booster. *Mariner 9* was successfully launched May 30, 1971, and reached Mars Nov. 13, 1971. All *Mariner 9* instruments operated successfully and were continuing to transmit data. When Mariner 8 was lost, mission plan for *Mariner 9* had been altered to meet primary mission objectives in best way. Plan had been altered second time when dust storm on Mars delayed systematic mapping of planet. (NASA proj off)

- NASA announced new assignments in Office of Public Affairs. Alfred P. Alibrando, Director of Public Information, was named Deputy Assistant Administrator of Public Affairs. He would be succeeded by Richard T. Mittauer, Public Affairs Officer for Space Science and Applications. Robert J. Shafer, Assistant to Assistant Administrator for Industry Affairs and Technology Utilization, was named Deputy Assistant Administrator for Public Affairs (Television). (NASA Release 72–28)

- Dept. of Transportation announced award to Rohr Industries, Inc., of $5-million contract to construct 60-passenger prototype of tracked air-cushion vehicle (TACV). Vehicle was part of $115-million urban transportation research and development program sent to Congress by President Nixon. (DOT Release 12–72)

February 12: Naval Research Laboratory announced its analyses had indicated that lunar soil "had been oxidized to some extent in the nearly nonexistent lunar atmosphere." All lunar soil returned by Apollo astronauts had displayed a characteristic resonance when examined by electron-spin-resonance spectrometry. Source of resonance had been thought to be iron metal. By oxidizing powder of simulated lunar glass, NRL scientists had produced for first time in the laboratory a resonance resembling characteristic resonance of lunar soils. Findings showed "conclusively that the 'characteristic' resonance arises from chemical compounds involving titanium dioxide and ferric oxide produced by oxidation of fine soils in the lunar environment." (NRL Release 9–72–2)

- Dr. John Teegan, Chief of Launch Site Medical Operations at Kennedy Space Center, announced his resignation to press at Cape Kennedy. Dr. Teegan, who would return to private practice, said he had resigned because his patients—astronauts—were too healthy. Astronauts were unusually healthy because they were in excellent physical condition when they were chosen and made it a point to stay that way. "I was educated to care for people who were ill and I want to get my feet wet again in clinical medicine. This chapter in space history was a very valuable experience that few people have had the opportunity to experience." (Reuters, B *Sun*, 2/13/72; KSC PAO)

February 14–25: U.S.S.R. launched *Luna 20* unmanned lunar probe from Baykonur at 8:28 am local time (10:28 pm EST Feb. 13). Tass said objective of mission was "further exploration of the moon and near lunar space."

On Feb. 18 *Luna 20* entered near-circular lunar orbit with 100-km (62-mi) altitude, 1-hr 58-min period, and 65° inclination. Engine firing on Feb. 19 placed spacecraft in elliptical orbit with 100-km (62-mi) apolune and 21-km (13-mi) perilune. Main retroengine burn for 267 sec on Feb. 21 thrust spacecraft toward moon and *Luna 20* free-fell to 760-m (2490-ft) altitude. Spacecraft was then guided by braking rocket to softlanding northeast of moon's Sea of Fertility at 12:19 am Baykonur time Feb. 22 (2:19 pm EST Feb. 21). Landing coordinates were 3°32′ north latitude and 56°33′ east longitude. Spacecraft began transmitting radio signals and conducting experiments.

After drilling lunar rock, collecting lunar samples, photographing lunar surface, and performing other undisclosed experiments, *Luna 20* lifted off moon at 3:58 am Baykonur time Feb. 23 (5:58 pm EST Feb. 22), after 27 hrs 39 min on lunar surface. Tass said samples had been obtained with earth-operated, percussion-rotary drill designed to handle hard and loose rock samples simultaneously.

Luna 20 reentered atmosphere Feb. 25 and parachuted to landing at 12:12 am Feb. 25 Baykonur time (2:12 pm EST Feb. 24) 40 km (24 mi) northwest of Djezkazgan, Kazakhstan, after 11-day 16-hr mission. Capsule containing lunar samples was recovered despite "extremely unfavorable weather conditions—strong wind, blizzard, and low clouds."

Luna 20 was second unmanned spacecraft to land on moon and return to earth with lunar samples. First had been U.S.S.R.'s *Luna 16* (Sept. 12–24, 1970). *Luna 17* had landed on moon Nov. 17, 1970, and had released *Lunokhod 1* lunar rover. *Luna 18* (launched Sept. 2, 1971) had crashlanded on moon Sept. 11, 1971, and *Luna 19* (launched Sept. 28, 1971) was still in lunar orbit on photography and mapping mission. (Tass, FBIS–Sov, 2/14/72, L1; 2/22–24/72, L1; 2/28/72, L1; *SBD*, 2/15/72, 226)

February 14: *Intelsat-IV F-4* comsat (launched Jan. 22) began full-time commercial service over Pacific with acceptance of 850 circuits between 15 earth stations. (ComSatCorp Release 72–10)

- Aerospace contractor representatives were scheduled to testify in person before House Committee on Science and Astronautics Subcommittee on Manned Space Flight, *Aviation Week & Space Technology* reported. "Unique move aimed at speeding the legislative process this election year" supplanted system in which committee members visited major contractors' plants for briefings and reviews of facilities. (*Av Wk*, 2/14/72, 13)

- Use of laser holography as nondestructive test technique for inspecting case-bonded solid-propellant rocket engines would be evaluated for NASA in program under negotiation with Lockheed Space & Missiles Co., *Aviation Week & Space Technology* reported. Technique consisted of viewing stressed test article through hologram produced when article was in unstressed condition. (*Av Wk*, 2/14/72, 11)

- Atomic Energy Commission was "encouraging" NASA to examine cost of permanent removal from environment of wastes from atomic plants by "shooting these high-level radioactive wastes into the sun—taking them right out of the world," Dr. James R. Schlesinger, AEC Chairman, said in interview published by *U.S. News & World Report*. Method would "depend on development of the space shuttle . . . a decade away." (*US News*, 2/14/72, 46–51)

- NASA launched Nike-Tomahawk sounding rocket from Andoeya, Norway, carrying Norwegian auroral energy experiment. Rocket and instrumentation performed satisfactorily. (SR list)

- Medium short-takeoff-and-landing-transport development program, included in FY 1973 Dept. of Defense budget, was among programs developed by Nixon Administration's Domestic Council to meet threat of erosion of U.S. technology by foreign competition and lack of domestic market, *Aviation Week & Space Technology* said. Program was aimed "at direct competition with the Boeing-Italian agreement in which Boeing is selling its technology in this area for a STOL transport to be built in Italy" by an Italian consortium. (Hotz, *Av Wk*, 2/14/72, 9)

February 15: Secretary of Defense Melvin R. Laird, testifying before Senate Committee on Armed Services, said Communist China "could begin deployment of intercontinental ballistic missiles with a range of 3,000 nautical miles [5600 kilometers] or more, capable of striking all or most of the U.S.S.R. by 1975." Chinese satellites *Chicom 1* and *2* (launched April 24, 1970, and March 3, 1971) and 12 nuclear tests since 1964 indicated "fairly high degree of sophistication in both missile and warhead development." (C *Sun Times*, 2/16/72, 18)

- Nike-Apache sounding rocket, launched by NASA from Churchill Research Range, carried Univ. of Pittsburgh experiment to 141-km (87.6-mi) altitude. Primary objective was to confirm initial observation of nitric oxide in auroral zone and collect additional information on processes that produced nitric oxide in an auroral arc. Secondary objective was to measure density of major atmospheric constituents in D and E regions. Rocket and instruments functioned satisfactorily. (NASA Rpt SRL)

- NASA issued technical memorandum "Description, Dissection and Subsampling of Apollo 14 Core Sample 14230," describing techniques used to preserve lunar sample intact. Techniques—developed originally by Dr. Roald Fryxell of Washington State Univ. to preserve sediments surrounding bones of "Marmes Man," oldest well-documented human remains in Western Hemisphere—had produced first permanent record of layering in dust that covered lunar surface. Scientists first removed 55 tiny subsamples from core length, working through rubber gloves in sealed nitrogen cabinets to protect samples from earth atmosphere. They then impregnated with resin three successive thin strips of lunar core with all layers intact and grains undisturbed. Stabilized deposits were then mounted on plexiglass as first permanent record of deposits beneath moon's surface. (NASA Release 72-30; NASA PAO)

February 16: NASA announced selection of moon's Taurus-Littrow region as site for Apollo 17 manned lunar landing in December. Region, named for Taurus Mountains and Littrow Crater and located just beyond southeast edge of Mare Serenitatis (one of largest mascons), had been selected to help fill in major gaps in development model of moon based on data from previous Apollo missions. Lunar Module landing point 20° north and 30° east of moon's center as viewed from earth would be one prime sampling objective. It contained very dark nonmare material believed to have eroded off mountains and volcanic-looking cinder cones which suggested material was explosively produced volcanic ash. One sampling site would be rock slide containing debris that had fallen into valley from high up on 2000-m (7000-ft) mountain.

Astronauts would travel over surface on foot and on lunar roving vehicle (LRV) Rover. They would deploy advanced Apollo lunar surface experiments package (ALSEP), containing heat flow experiment similar to that on *Apollo 14* and four new experiments, and two new surface traverse experiments not powered by ALSEP central station. Experiments were traverse gravimeter to measure vibrations in and provide data on subsurface, seismic profiling and surface electrical properties investigations to measure physical properties of lunar interior down to depth of one kilometer (six tenths mile) and indicate subsurface electrical and mechanical properties, tidal gravimeter to study response of moon to earth's tidal pull and its response to gravity waves, mass spectrometer to measure constituents of lunar atmosphere, and lunar ejecta and meteorite experiment to determine frequency and energy of small meteorites impacting moon.

Orbital science payload would include three new experiments: laser lunar sounder would identify electrical properties and layering of lunar crust overflown by spacecraft, infrared scanning radiometer would provide high-resolution thermal map of portions of moon for first time, and far-ultraviolet spectrometer would measure compositional and density variation of lunar atmosphere. (NASA Release 72-33)

- U.S.S.R. launched *Cosmos 474* into orbit with 365-km (226.8-mi) apogee, 179-km (111.2-mi) perigee, 90-min period, and 65° inclination. Satellite reentered Feb. 29. (GSFC *SSR*, 2/29/72)

- President Nixon signed H.R. 11487, Chapel of the Astronauts bill, which authorized sale of land near Kennedy Space Center Visitor Bureau to private corporation for construction of chapel. Bill became P.L. 92-227. (*PD*, 2/21/72, 448)

- Manned Spacecraft Center announced issuance of requests for proposals to 13 aerospace firms to study, design, and test polymer seal materials for storing space shuttle propellants. Firm selected would receive $100 000, firm-fixed-price contract. (MSC Release 72-43)

- Former scientist-astronaut Brian T. O'Leary, associate professor of astronomy at San Francisco State College, questioned compatibility of space shuttle with national goals and priorities in *New York Times* article: ". . . balanced presentation to decision-makers of technical information bearing on the costs and benefits of the shuttle cannot be accomplished without some degree of effort outside the perimeter of NASA and its contractors. Such effort is now almost entirely missing and is badly needed before a national decision can be made." NASA needed to answer questions: "What is the shuttle and what will it do? What precisely are NASA's goals for shuttle use? Will the shuttle be primarily a workhorse for the Department of Defense?" Dr. O'Leary asked, "Wouldn't this be an auspicious time for the Administration to set a goal more in tune with human values, which would make good use of . . . aerospace workers?" (*NYT*, 2/16/72, 37)

- World's record for accelerating atomic particles was claimed by officials at world's largest atom smasher—200-bev facility at Batavio, Ill. U.S. atom smasher had accelerated beam of protons to 100 bev, topping 76-bev record established by Soviet accelerator at Serpukhov. (AP, *NYT*, 2/20/72, 2)

- Adm. Thomas H. Moorer (USN), Chairman of Joint Chiefs of Staff, warned of possible U.S.S.R. nuclear superiority in testimony before closed session of Senate Committee on Armed Services considering FY 1973 military authorization bill: "The mere appearance of Soviet strategic superiority could have a debilitating effect on our foreign policy and our negotiating posture. It could erode the confidence of our friends and allies in the deterrent power of our strategic forces, upon which the entire structure of free world defense essentially rests." (Corddry, B *Sun*, 2/17/72, A4)

February 17: First fully automatic landings by full-scale manned helicopter at predetermined spot had been made by team of Langley Research Center engineers and pilots at Wallops Station, NASA announced. Accomplishment was milestone in research program aimed at permitting helicopters and future vertical takeoff and landing aircraft to fly routine missions under poor visibility. Landing approaches were fully automatic from 3 to 5 km (2 to 3 mi) from intended landing spot. Automatic system was engaged in level flight at 96.5 km per hr (60 mph) at 244-m (800-ft) altitude. When helicopter intercepted landing guidance path, system automatically locked on to start landing approach. At predetermined range helicopter began automatic deceleration to zero ground speed, came to hover 15 m (50 ft) above landing spot, and descended vertically to touchdown. (NASA Release 72-34)

• Dr. John E. Naugle, NASA Associate Administrator for Space Science, testified on planetary exploration program at opening session of FY 1973 NASA authorizatization hearings before House Committee on Science and Astronautics' Subcommittee on Space Science and Applications. Dr. Naugle presented plan under study for "much more modest outer planets exploration program" instead of originally proposed two-mission, four-spacecraft Grand Tour of all outer planets, canceled because of budgetary restrictions. Jupiter exploration would continue. "We are studying a mission for the initial exploration of Saturn. We would take advantage of the rare alinement of the planets to launch two Mariner-class spacecraft in 1977 on a trajectory which would carry them first past Jupiter and on to Saturn to investigate that strange planet. . . . At present, this program appears to be an excellent outer planets mission within the current constraints. However, we are studying other options with our science advisory groups to determine the set of outer planets missions which provide the most significant scientific information." Viking project to softland spacecraft on Mars in 1976 remained "most significant element of the ongoing planetary exploration program."

Mariner 9's orbiting of Mars had been major highlight in 1971 planetary program. To date, "Mariner 9 . . . has been in orbit 96 days, is still healthy and operating nominally. . . . it has transmitted back to Earth over 5700 pictures of Mars, a number of closeups of its two moons, and measurements of surface features and atmospheric properties by ultraviolet and infrared spectrometers." NASA had mapped approximately 66% of Martian surface at resolution of 1 km (0.6 mi) or better.

"To date we have visited Venus and Mars, and will soon reach Jupiter and Mercury. Within the next year we shall have studied Mars in detail from orbit and within 5 years shall have begun an intensive automated exploration of its surface. By the end of the decade we expect to know whether life exists, has existed, or could exist on Mars."

Three spacecraft to be launched in 1972 would open "brand new areas of research or application." Pioneer-F would be launched in March to begin study of Jupiter. Next, Earth Resources Technology Satellite would open new field of earth study. In fall, astronomy Explorer satellite would begin to study gamma rays from stars. Major

February 17: *The first fully automatic landings by a manned helicopter at a predetermined spot were announced by* NASA. *A Langley Research Center team had made the landings at Wallops Station the second week in February in a research program to permit helicopters and future* VTOL *aircraft to fly under conditions of poor visibility.*

highlight in 1971 astronomy program had been observations made by x-ray astronomy satellite *Uhuru* (*Explorer 42*), which was making map of universe in x-ray wavelengths. One hundred sixteen x-ray sources had been studied.

"Space" science, Dr. Naugle said, was no different from research man had been doing on earth . . . since he became man: "Man does love to wonder—to explore and understand his environment. The ability to wonder—the desire to extend one's horizon—is a measure of the mental health and vigor of an individual or a society. The ability of a nation to educate its youth and provide them with the necessary tools to explore our environment, whether it be the innermost secrets of the nucleon or the outermost regions of the universe, is one measure of that Nation's vigor and its capability to better understand, and in turn guide and control, its destiny." (Transcript)

- House Committee on Science and Astronautics' Subcommittee on Manned Space Flight began hearings on H.R. 12824, FY 1973 NASA authorization bill.

  Dale D. Myers, Associate Administrator for Manned Space Flight, described *Apollo 15* mission (July 26–Aug. 7, 1971) as "most rewarding mission so far." Mission "demonstrated the operational maturity of the [Apollo] program . . . . Because of its greatly increased capability for astronaut mobility, surface exploration, sample return, and orbital science, *Apollo 15* added as much new data as that gathered by the three previous missions combined." Mission had marked first use of Lunar Rover and of orbital science experiments to collect lunar-wide information. "The scientific returns from lunar orbit and surface have significantly increased our accumulated knowledge of the moon and by extrapolation of the data from this vacuum-packed laboratory, knowledge of the solar system, including our earth."

Upon completion of Apollo program with Apollo 17 in December, "we will have accumulated a vast treasure of lunar data. Many exciting scientific results have already been published; however, an adequate assessment of the total body of data will require scientific effort for years to come.

"As Apollo draws to a close, we are concentrating our activities in earth orbit—the primary field of manned space flight during this decade." Skylab was "first step in this direction" and "one of the most significant benefit-oriented programs of the space age." Skylab was four times larger than Soviet Salyut spacecraft and weighed 77 180 kg (170 000 lbs) with total 344 cu m (12 150 cu ft) pressurized volume. Skylab typified variety of activities that could be accomplished in space. "Several years ago we began to look at this total space arena in a different way. . . . While we were driving toward our goal of a manned lunar landing . . . a distinction was made . . . between manned and unmanned flight programs." Distinction "really does not exist" in NASA's program for 1970s. "The transition is already underway in Skylab."

Survey of total space activity had also led to shuttle concept. "We started with a fully reusable system . . . using winged flyback booster and orbiter stages with all fuel carried in internal tanks." Studies confirmed feasibility. Estimated development costs were close to $10 billion, but cost per flight, at $4.1 million, "made it extremely attractive." Problem was lack of funds. NASA "learned that there were cost advantages in using an expendable liquid hydrogen tank for the orbiter." This became baseline with booster configuration unchanged except for modification. While process reduced development costs, expendability of tanks increased flight costs. NASA found "additional savings could be made if we placed both the hydrogen and the oxygen tanks outside the orbiter." In efforts to economize on booster development, NASA first considered "flyback booster incorporating F–1 engines and phasing orbiter systems" and later "unmanned ballistic boosters." Leading candidates were recoverable and reusable pressure-fed liquid-propellant booster and booster with solid-fueled rocket motors. Contractors were completing studies, and decision would be made shortly.

Mathematica report of study for NASA in May 1971 had shown shuttle with flyback manned booster was cost-effective when averaging 39 flights a year. Myers submitted new edition of report, incorporating refined data and changes in shuttle configuration, which showed shuttle cost effective with traffic as low as 30 flights per year.

"I believe that practically all of this country's future earth-orbital space activities will use the Shuttle's capability in exploiting the potential benefits of space and providing the taxpayer as large a return of his space investment as possible." (Transcript)

- Roy P. Jackson, NASA Associate Administrator for Aeronautics and Space Technology, during hearings by House Committee on Science and Astronautics' Subcommittee on Aeronautics and Space Technology on FY 1973 NASA authorization, testified that recently developed NASA technology could have increased supersonic transport's investment return rate. Calculations had been made showing that "incorporation of now emerging technology for static stability augmentation and active flutter control could have allowed a weight savings, and thus a payload increase, of better than 16,000 pounds, or an additional return on investment of about 7.5%."

    NASA research and technology in aerodynamics, configurations, materials, structures, propulsion, and avionics applicable to advanced long-haul transport aircraft design showed "real promise." Application of automated active control concepts "could result in considerable weight savings by eliminating the need for inherent aerodynamic stability and thereby reducing the size of control surfaces. It could also result in reduced structural fatigue for longer life, lower structural loads for lighter structures, and ride quality control and active damping of aircraft flutter for a more comfortable ride." (Transcript)
- Dr. George M. Low, NASA Deputy Administrator, said "real increase in productivity in space" would come with shuttle, in speech before National Space Club in Washington, D.C. ". . . we have built more than 80 launch stands in this country, 50 of which are still either active or on standby . . . [and] we are using 17 different combinations of boosters and upper stages . . . today. The shuttle will allow us to consolidate most of these. . . ." Shuttle, "major investment in a new tool for space exploration," had been designed for productivity "in terms of launch services, . . . access it will provide to space, . . . [and] the way it will let us operate in space." (Text)
- President Nixon spoke at White House ceremony before his departure for state visit to People's Republic of China: ". . . if there is a postscript that I hope might be written with regard to this trip, it would be the words on the plaque which was left on the moon by our first astronauts when they landed there. 'We came in peace for all mankind.'" (*PD*, 2/21/72, 443–4)
- European Launcher Development Organization deferred firing of Europa II (F.12) launcher to first half of 1973, to allow more time for investigation of June 17, 1970, Europa I (F.11) failure and remedial action. (*SF*, 4/72, 161)

February 18: Notices to terminate work on NERVA (nuclear engine for rocket vehicle application) program by June 30 were issued by AEC–NASA Space Nuclear Systems Office to major NERVA contractor Aerojet Nuclear Systems Co. and major subcontractor Westinghouse Astronuclear Laboratory. Active development of flight-model 333 600-newton (75 000-lb)-thrust nuclear rocket engine for future space missions had been suspended in 1971 because of funding limitations and space program stretchouts. Work on long-lead-time components had continued. NERVA cancellation would affect about 300 Aerojet employees in California and 330 Westinghouse employees in Pennsylvania. Employees at Space Nuclear Rocket Development Station in Nevada would continue reactor testing and nonnuclear component testing. (NASA Release 72–36)

February 18

- *Heos 2* Highly Eccentric Orbit Satellite, launched Jan. 31, was adjudged successful by NASA. Spacecraft had entered planned orbit and all experiments were functioning satisfactorily. (NASA proj off)
- U.S.S.R.'s *Mars 2* (launched May 19, 1971) had completed 111 revolutions of Mars and *Mars 3* (launched May 28, 1971) had completed 6 revolutions, Tass announced. Radio contact with spacecraft remained stable and onboard instruments were continuing to function satisfactorily. (FBIS–Sov, 2/22/72, L2)
- Portable, high-intensity light to illuminate Skylab workshop interior for motion picture photography during long space missions was being tested at Marshall Space Flight Center. Developed by IOTA Engineering, 8-kg (17-lb) light produced concentrated illumination of proper color balance for use with fast daylight film scheduled for Skylab. (MSFC Release 72–15)
- NASA announced award of 150-day, $199 000 contract under Minority Business Enterprise Program to RO&AS Joint Venture of Houston for construction of trainer hardware support facilities building. RO&AS was partnership formed by Roy Owens Interests, Inc., a Black-owned construction firm, and Advance Systems Construction, Inc., a Mexican-American-owned company. Manned Spacecraft Center's Contractor Equality Opportunity Programs Office had played key role in bringing the two firms together for joint venture, which marked first time in NASA history that two minority businesses had pooled resources for a joint contract effort. (MSC Release 72–45)
- Senate confirmed nomination of *Apollo 10* Astronaut Thomas P. Stafford as U.S. Air Force brigadier general. (*CR*, 2/18/72, S2125)
- Administration had summoned 400 leading U.S. scientists and engineers to series of White House meetings on nation's domestic and economic problems, *Washington Post* reported. Presidential Science Adviser, Dr. Edward E. David, Jr., had exchanged ideas with physicists, engineers, chemists, and aerospace officials on urgent situations in meetings intended "to rally the country's science-engineering community behind other goals than weapon-making." (Cohn, *W Post*, 2/18/72, A1)
- Group of 49 industrialists and educators from England, Sweden, Denmark, Germany, France, Italy, Holland, and Belgium visited Kennedy Space Center during week-long orientation arranged by London *Financial Times*. Group also visited NASA Hq., Goddard Space Flight Center, Marshall Space Flight Center, and Manned Spacecraft Center. (KSC Release 40–72)
- Fact that all spacecraft orbiting earth or moon traveled "oval shaped path called Keplerian elipse" testified to "durability" of Johannes Kepler's laws of planetary motion, *New York Times* said. In article on 1971 celebration of 400th anniversary of mathematician-astronomer's birth, Dec. 27, 1571, *Times* noted, "Whether it is the space program or a new fashion in historical research, the life and works of the sickly son of a mercenary soldier who rose to be Imperial Mathematician of the Holy Roman Empire are undergoing a kind of rehabilitation at the hands of scholars." (Reinhold, *NYT*, 2/18/72)

February 19: *Intelsat-IV F–3* comsat, launched by NASA for Communications Satellite Corp. on behalf of International Telecommunications Satellite Consortium Dec. 19, 1971, began full-time commercial service over Atlantic with acceptance of 700 circuits between 8 earth stations.

Satellite had been held in orbital position between Atlantic and Pacific pending successful launch and checkout of *Intelsat-IV F-4* (launched Jan. 22). (ComSatCorp Release 72-12; ComSatCorp PIO)
- NASA launched Aerobee 170 sounding rocket from White Sands Missile Range carrying Univ. of Wisconsin x-ray astronomy experiment. Rocket and instrumentation performed satisfactorily. (SR list)
- NASA had spent $800 000 on North American Rockwell Corp. Space Div. study and design of system to permit rendezvous and docking of U.S. and Soviet spacecraft in orbit, Associated Press reported. Study envisioned 14-day mission with Apollo command and service module linking special docking module with Salyut orbiting laboratory and Soyuz command spacecraft. NR had received $300 000 for method study and $500 000 for hardware design. No contract had been issued for hardware. (*CSM*, 2/19/72)

February 20: Tenth anniversary of *Friendship 7* mission, first U.S. manned orbital space flight. Piloted by Astronaut John H. Glenn, Jr., *Friendship 7* had circled earth three times during 4-hr 55-min mission. Since Glenn's flight—which was third U.S. manned space flight and sixth flight in Project Mercury—30 U.S. astronauts had flown in space total of 7700 hrs, 51 min—nearly 46 wks.

Anniversary was observed at Kennedy Space Center Feb. 22, with 3000 guests and dignitaries invited to attend ceremony at Launch Complex 14, where commemorative plaque was unveiled. Glenn said space flight was "not a pleasure cruise for an individual. It's research at the highest level and it is difficult to tell where it will go from here." (KSC Release 37-72; B *Sun*, 2/23/72)

- Proposals for Skylab experiments and demonstrations had been submitted by 3409 U.S. secondary school students to date, NASA announced. Response to Skylab student project had been among largest encountered by National Science Teachers' Assn. in 20 yrs of sponsoring student science projects. (NASA Release 72-35)
- Air Force HC-130H piloted by L/C Ed Allison established world record for nonstop, unrefueled flight by turboprop aircraft of 14 052.94 km (8732.09 mi) in flight from Ching Chuan Kang Air Base on Taiwan to Scott Air Force Base, Ill. (*Airman*, 7/72, 2-8)
- Dr. Maria G. Mayer—professor of physics at Univ. of California at San Diego, originator of "shell model" of atom nucleus, and winner of 1963 Nobel Prize in physics—died of heart failure at age 65. (AP, B *Sun*, 2/22/72, A8)

February 21: Classified advertising for aerospace engineers had shown "consistent week-to-week increase" during January and February over same 1971 period in Southern California, *Aviation Week & Space Technology* reported. Trend indicated upturn in "Nation's largest aerospace manufacturing complex." (*Av Wk*, 2/21/72, 11)
- *Apollo 15* Astronaut Alfred M. Worden—who had orbited moon alone in command module while Astronauts David R. Scott and James B. Irwin explored lunar surface during July 26-Aug. 7, 1971, mission—read number of his space poems before meeting of Poetry Society of Texas. (*W News*, 2/18/72; Worden)
- *Aviation Week & Space Technology* editorial described "tremendous new thrust of avionics into electronic countermeasures" (devices and systems for countering enemy electronic defenses) as "most dynamic

element of modern warfare." ECM had "expanded its sphere from ground operations to the sea, air and even the space environment" of satellites and ICBMs. There was "half-billion-dollar market for hardware with over another billion dollars expended on operational use of ECM equipment." There was "ample evidence . . . that the Soviet Union has also embarked on a major ECM effort." There were still elements of "legitimate military secrecy" in ECM design and operation "but most of its functions are now well known to both sides." Electronic operations had expanded to "full range of a finite definable spectrum that both sides can scan fully and detect what is being done." (Hotz, *Av Wk*, 2/21/72, 9)

- *Twenty-First Annual Report of the Activities of the Joint Committee on Defense Production, Congress of the United States* was transmitted to Congress. Report said U.S. might soon be entirely dependent upon U.S.S.R. and Japan for titanium used in jet aircraft construction. Last titanium plant in U.S.—Reactive Metals, Inc., in Niles, Ohio—had closed Dec. 13. It was third plant to go out of business in 1971 because of lack of titanium market, oversupply of titanium sponge, increased foreign competition, and critical financial losses. U.S. Bureau of Mines later reported that two U.S. titanium plants reopened during February 1972. (Text; Bureau of Mines PIO)

February 22: Apollo 16 press conference was held at NASA Hq. Dr. Rocco A. Petrone, Apollo Program Manager, said Apollo 16 lunar landing site would consist of volcanic highland material. "The highlands which we are landing on . . . are composed of two types of units. One type is material formed early in lunar history and subsequently fragmented and redistributed by the numerous meteorite impacts which have scarred the lunar highlands. These units are predominantly ejecta deposits now. And on Apollo 15 with the genesis rock we hopefully have sampled some of these early deposits. The second major unit . . . is materials which have modified these early deposits by flooding them or building up on top of them predominantly by volcanic processes. The primary objective of Apollo 16 is the exploration and sampling of this lighter type of highland terrain with hopefully a fragment or two of the old original stuff brought in by impact from . . . the old original crust which is probably still exposed down in the lower southern hemisphere."

Dr. M. E. Langseth of Lamont-Doherty Geological Observatory explained changes in drill design, with joints to permit core samples to be brought out more readily.

Dr. H. J. Moore of U.S. Geological Survey described orbital science experiments, including photographic system to determine moon's geometric shape. Apollo 16 CM would also photograph floor of Alphonsus Crater—where *Ranger 9* impacted March 24, 1965—to provide positive identification of *Ranger 9*'s impact.

Apollo 16 lunar module pilot Charles M. Duke, Jr., said astronauts would begin lunar surface extravehicular activity (EVA) as soon after landing on moon as they could complete postlanding checkout. They would conduct 7-hr EVA, and reenter LM about 17 hrs after previous sleep period. Crew would conduct three EVAs On first, primary objective was to deploy Apollo lunar surface experiments package; on second, to sample as high on Descartes and Stone Mountains as possible; and, on third, to reach North Ray Crater. (Transcript)

- Dr. Rocco A. Petrone, Apollo Program Director, appraised scientific contributions of Apollo program in testimony during House Committee on Science and Astronautics' Subcommittee on Manned Space Flight hearings on NASA FY 1973 authorization bill.

". . . Apollo 11 crew crossed over four hundred thousand kilometers (250,000 miles) of space to touch another planetary body. Four teams of astronauts, from Apollos 11, 12, 14 and 15 . . . have now made the epic journey and walked the lunar surface. They have returned some 176 kg (388 lbs.) of lunar rocks and soils for analysis in earth-based laboratories and have established scientific stations on the moon that are continuously transmitting scientific and engineering data back to earth." Apollo data, with information expected from continued sample analysis and lunar data analysis in the post-Apollo period, "surely will be of practical value to man, even helping him to cope with current environmental problems on earth."

NASA sample analysis program had "required unprecedented advancement in instrumentation and techniques to perform multiple element and isotope analyses with extreme precision on minute amounts of sample." Improved high-resolution instruments made "detailed investigations of surface features and internal structure of materials."

Other related important breakthroughs had been computer hardware and software for processing and reducing large amounts of analytical data in a very short time. "These advances, in large part stimulated by NASA-funded meteorite studies, were absolutely essential, because of the scientifically unique, but limited, lunar materials. . . . these high precision research methods can now or will soon be routinely utilized in industrial processes and laboratories." They could also "be immediately applied to a variety of environmental and biomedical problems by providing the capability to check and monitor levels of pollutants and toxic elements in our atmosphere, water, and foods."

Skylab Program Director William C. Schneider outlined program's progress: "During the coming fiscal year testing and checkout will be completed and operation of Skylab will have started. Within 2 years, the first Skylab . . . will have become part of history, having contributed new knowledge in many fields."

Skylab offered "an earth observation capability never before available" to U.S. manned spacecraft. During eight-month mission, Skylab would fly over entire U.S. except Alaska, over much of Europe, all of Africa, Australia, China, and almost all of South America—covering 75% of earth's surface and passing over each point every five days. By end of 1971, 288 investigations requiring Skylab data had been submitted, 249 U.S. and 39 foreign. Of these, 164 had been identified for further study. Skylab was "first manned space flight program designed specifically to carry activities and equipment explicitly aimed at improving man's life on earth. It will contribute significantly to the increase of knowledge of pure science and is also an experimental space station; a forerunner of permanent space stations of the future." Earth-oriented sensors would test technology for synoptic surveys of many environmental and ecological systems. Solar and astronomical

February 22

observations and other science experiments would expand knowledge of solar system, universe, and near-earth space. Biomedical experiments would inform how man's well-being and ability to function were affected by living in space.

Harry H. Gorman, Deputy Associate Administrator for Management in Office of Manned Space Flight, testified on $305-million FY 1973 budget request for development, test, and mission operations in manned space flight research and development program: "This is the first year in which this activity is presented as a separate item. . . ." With near-completion of Apollo program, "we believe it important to separately identify and control this basic capability which has been built up. . . . This work is essential to the support of the Nation's space programs for the 1970's and beyond." Requested funding would provide for contractor support of in-house capabilities for Apollo and Skylab programs; for definition, design, development, and subsystem testing activities in shuttle program; and for pre-definition and definition of future programs, including High Energy Astronomy Observatory, Stratoscope 1 and 2, and Earth Observation Satellites. (Transcript)

- National Academy of Sciences and National Research Council Space Science Board "unanimously and warmly" endorsed proposed NASA program to explore Jupiter and Saturn in late 1970s as "strong and flexible program which is a reasonable next step in planetary exploration." Endorsement was expressed in letter from Chairman, Dr. Charles H. Townes, to Dr. James C. Fletcher, NASA Administrator. Program had been presented by NASA to Space Science Board meeting Feb. 8–9 as possible alternative to canceled Grand Tour missions. (Text)

- NASA announced selection of 13 scientists, including two from Europe, to participate in definition phase of proposed missions to Venus with Pioneer-class spacecraft beginning in late 1976. Scientists, selected from 109 scientists who had submitted proposals in response to NASA invitation in July 1971, would work closely with mission engineering team to define typical scientific payload and design features for initial missions and to make recommendations for subsequent missions. NASA planned to initiate funds for spacecraft development in FY 1974. (NASA Release 72–37)

- Annual Dept. of Defense report to Congress, *National Security Strategy of Realistic Deterrence*, was presented by Secretary of Defense Melvin R. Laird to House Committee on Appropriations' Subcommittee on Dept. of Defense Appropriations during hearings on FY 1973 DOD budget and FY 1973–1977 program. U.S.S.R. had reached position "where—unless we take appropriate action—there could be new surprises and new 'sputniks.' But they are less likely to be in areas such as the peaceful exploration of space; rather they are more likely to be a part of a major new Soviet military capability." (Text)

- National Science Foundation published *Federal Scientific, Technical, and Health Personnel in 1970* (NSF 71–47). Federal scientists, engineers, and related nonprofessional personnel numbered 271 000 in October 1970, 1% below 274 100 employed in October 1969. Figure for 1970 included 79 300 scientists and 83 000 engineers—about 10% of total scientists and engineers in national economy. Decline of professional and nonprofessional personnel in Federal Government in

1970 followed several years of relative stability. Decrease in Federal scientists and engineers between 1969 and 1970 centered in Dept. of Defense, largest employer of such personnel. In 1970 DOD employed 74 400 scientists and engineers, 3% below 76 000 in 1969. NASA decline was from 13 900 in 1969 to 13 400 in 1970. Engineers accounted for almost all of 1% decrease in Federal scientists and engineers from 1969 to 1970. Figures for engineers were 84 100 in 1969 and 83 000 in 1970. DOD engineers declined 3%—from 53 800 to 52 300. Largest decline among other agencies was at NASA and Dept. of Interior. DOD scientists declined from 22 200 in 1969 to 21 200 in 1970. NASA scientists declined from 5100 to 4800 (NSF *Highlights*, 2/22/72, 1)

- Dept. of Defense was trying to develop nuclear warhead that could evade Soviet missile defenses, *Wall Street Journal* reported. DOD spokesman had said USAF had been working on "evasion technology" for several years but program—budgeted for $42.1 million over past two years—was in early R&D stages. (*WSJ*, 2/22/72, 7)

February 22–24, 29: Dr. John E. Naugle, NASA Associate Administrator for Space Science, continued testimony on FY 1973 authorization before House Committee on Science and Astronautics' Subcommittee on Space Science and Applications. He announced Feb. 24 NASA's decision to substitute two flybys of Jupiter and Saturn for Grand Tour of all five outer planets.

New plan—necessitated by budgetary restrictions and congressional recommendation, and endorsed by National Academy of Sciences—National Research Council's Space Science Board—would launch two spacecraft in 1977 on trajectory that would carry them by Jupiter and then Saturn during rare alignment of planets. Spacecraft for mission would be based on Mariners developed to explore inner planets Mars, Venus, and Mercury. They would have large antennas for long-distance communication, be powered by radioisotope thermoelectric generators, carry more than 68 kg (153 lbs) of instrumentation including TV cameras, and be launched by Titan-Centaur-Burner III boosters and accelerated by Jupiter's gravity and orbital velocity for flight to Saturn. Possible Saturn encounters included flying close to planet's satellite Titan—larger than planet Mercury and only satellite known to have atmosphere. Spacecraft would encounter Jupiter in about $1\frac{1}{2}$ yrs and Saturn about $3\frac{1}{2}$ yrs after launch. Mission plan flexibility would permit selection of different flyby trajectory at Saturn for second spacecraft based on data from first. Plan would maintain option to use Pioneer-class spacecraft for Jupiter and Saturn orbiters in late 1970s. Project responsibility had been assigned to Jet Propulsion Laboratory.

In other testimony, Dr. Naugle said NASA had developed heat sterilization techniques for Viking lander and new thermoradiation techniques expected to have future space and industrial applications. Program initiated measures to ensure that all planetary missions had "acceptably low probability of impacting or contaminating the target planets." (Transcript; NASA Release 72–42)

February 23: Two volunteers began pilot investigation at U.S. Public Health Service Hospital in San Francisco to retrace *Apollo 15* mission profile and determine cause of astronauts' slower-than-usual return to physiologic norm. Program, conducted for Manned Spacecraft Center, included set periods of vigorous physical exercise, bed rest, and posttest analysis. Bed rest portion was considered analogue of weightlessness. Potassium intake would be closely controlled because NASA doctors postulated *Apollo 15* crew had not consumed enough potassium for their workload and adjustment to weightlessness. Program consisted of 9-day equilibration, 7-day control phase, 12-day bed rest, and 14-day recovery period. (MSC Release 72-39; MSC PAO)

- Lewis Research Center issued requests for proposals for first phase of three-phase effort to demonstrate that lower aircraft engine emissions could be reached without sacrificing combustion efficiency or combustor's ability to reignite in flight. Firm selected would receive 15-mo contract to conduct screening tests of various combustor designs and modifications. Contractor would consider LeRC-developed "swirl can" design, which provided many fuel-injection points and premixed some fuel and air and could be operated to temperatures twice as hot as standard combustors with significant reductions in pollutants. Second phase of project would test best designs from phase one. Third phase would test best design as part of a complete engine. (NASA Release 72-38)
- Marsall Space Flight Center issued requests for proposals for development of visible laser communications experiment for ATS–G Applications Technology Satellite mission in 1974. Contractor selected would begin designing experiment in July and would build and test experiment and provide ground support through experiment's two-year lifetime. (MSFC Release 72-17)
- Senate approved S.R. 261, making minority changes in committee assignments. Sen. James L. Buckley (R-N.Y.) was removed from Committee on Aeronautical and Space Sciences and Sen. Karl E. Mundt (R-S.D.) was added. (*CR*, 2/23/72, S2379-80)

February 23-24: NASA officials continued testimony before House Committee on Science and Astronautics' Subcommittee on Aeronautics and Space Technology during FY 1973 authorization hearings.

Associate Administrator for Aeronautics and Space Technology Roy P. Jackson reported "excellent results" in first acoustic tests, completed in November 1971, of experimental quiet engine for conventional takeoff and landing (CTOL) aircraft. "In an unsuppressed baseline configuration . . . the noise levels were the equivalent of at least 3 EPNdB [epndb, effective perceived noise in decibels] lower than the original targets. In a first cut simulation of an acoustically suppressed nacelle which was not an optimum design, the noise levels were from 7 to 9 EPNdB lower yet, with aerodynamic performance losses no greater than expected." It appeared possible to design an advanced version of CTOL engine "which will incorporate all the advances in technology and will have a more nearly optimum trade-off between noise reduction and installed weight penalty."

In FY 1973, $2.5 million was requested for program to reduce noise of short takeoff and landing (STOL) aircraft to 95 epndb at 500 ft (150 m). Tests at Flight Research Center of quieted T-34 high-bypass-ratio engine operating with simulated wing-flap system would develop design data for both quiet, experimental STOL (QUESTOL) aircraft program and quiet, clean, STOL experimental engine program. Vertical takeoff and landing aircraft (VTOL) noise reduction program, with budget request of $2.5 million, focused on lift fans with additional work on helicopter rotor noise. Acoustic and aerodynamic data would continue to be developed in tests at Langley, Ames, and Lewis Research Centers.

George W. Cherry, Deputy Associate Administrator for Programs in Office of Aeronautics and Space Technology (OAST), described NASA role in Joint NASA, Federal Aviation Administration, and Dept. of Defense microwave landing system (MLS) development plan to provide landing system for civil and military use to replace current very-high-frequency (VHF) landing system. NASA would provide personnel in FAA MLS Project Office, chairmanship of the MLS Interagency Advisory Group, flight-test validation of the MLS for STOL operations, and extension of our work in FY 1973 to include CTOL and VTOL aircraft. "We are convinced the national microwave landing system program will make significant contributions toward noise abatement, airport congestion relief, and, especially, safety enhancement in the airport area."

Gerald G. Kayten, Director of Transport Experimental Program Office in OAST, said on Feb. 23 that NASA programs would "provide a foundation upon which industry can base the design and development of the new STOL systems in the late 1970's, and upon which the Government can establish criteria for certification and regulation of such systems." Programs also would enable Government to make 1976 decision on vertical short takeoff and landing (V/STOL) incorporation recommended in Dept. of Transportation Northeast Corridor Report. "NASA's main focus is on propulsive-lift technology for effective and economical STOL capability, engine technology for development of quiet and clean STOL propulsion systems, and airborne STOL avionics and system technology." Technology represented "totally new capability based on major advances in each of several technical disciplines."

Albert J. Evans, Director of Military Aircraft Programs Office in OAST, on Feb. 24 described support of military aircraft development programs. NASA had assigned research engineers to F-15 and B-1 aircraft and Air Force prototype systems program offices. "This has put NASA 'on the scene' as technical problems become recognized and defined, and brings NASA's technical capabilities . . . to bear on the problems at an early date." F-15 would make first flight early in FY 1973 "with about 12,000 NASA wind tunnel hours behind it, and much technical consultation by NASA staff members with the Air Force and the industry that is developing the F-15." F-15 operational engine would be installed at Lewis Research Center to analyze air flow through engine components, combustion efficiency, and performance characteristics. NASA would analyze F-15 engine and airframe performance and flight-test aircraft at LeRC, where similar tests would be run on Air Force AX aircraft in cooperation with USAF. In B-1 program, four FRC engineers would be assigned to USAF Systems Program Office and contractor to assist in establishing flight-test requirements and instrumentation to verify analytical design techniques and ground-test data obtained at small scale in wind tunnels and information from simulator studies. (Transcript)

February 24: Dr. Charles A. Berry, NASA Director for Life Sciences, discussed spinoff of NASA biomedical research in testimony before House

Committee on Science and Astronautics' Subcommittee on Manned Space Flight during FY 1973 authorization hearings. Ames Research Center (ARC) was evaluating changes in lung tissue of coal miners in collaboration with Dept. of Health, Education, and Welfare to pinpoint possible damage to astronauts' lungs from particulate contaminants in atmosphere. Study results suggested "cell destruction can be induced by release of intracellular digestive enzymes into the cell cytoplasm following penetration of the cells by particles, in the case of coal miners, silicates." Joint NASA and U.S. Bureau of Mines program would develop carbon monoxide detecting system. ARC and U.S. Dept. of Agriculture would develop hydroponic system for spacecraft to accelerate plant growth under space flight conditions.

"The medical information needed to assure man's safe journey into and return from space has led to the development of many items of information and equipment of great value to terrestrial medicine." NASA life sciences program "exemplifies the success which can be achieved by combining the special talents of Government, industry, and university scientists."

Dr. Berry also reported there had been "no findings in the manned space flights to date by the United States or U.S.S.R. that would preclude the commitment of man to longer duration missions. Therefore . . . we feel confident in committing man to a 28-day mission." Apollo 16 and 17 and Skylab experiments would study physiological response of body before commitment to 56-day Skylab mission.

Philip E. Culbertson, Director of Advanced Missions in OMSF, testified on space shuttle orbital systems and advanced missions definition and planning: In areas of payloads, research and applications modules (RAM), advanced missions, and advanced development, "principal effort is directed toward developing . . . understanding and planning the use of the shuttle." In RAM program, "we will be continuing the study of concepts of a family of payload carrier modules suitable for transport in the Shuttle Orbiter cargo bay." Modules would range from "austere shuttle 'Sortie Can' with relatively simple laboratory equipment . . . through more sophisticated, dedicated laboratory and observation facilities, including automated free-flyers serviced by the Shuttle."

Sortie mission using orbiter as platform for short-duration experiments was among most promising shuttle operating modes. Mission would use advantages of shuttle in payload design, development, and operation, and, "for the first time, permit the investigator to accompany his experiment into space." Sortie missions would be modeled after ARC Airborne Research Program in which Convair 990 aircraft was used as airborne laboratory. "The goal in developing procedures for managing the Space Shuttle Sortie Missions will be to retain the relatively simple, flexible and highly responsive aspects of the '990' program." (Transcript)

- Deployment tests of Skylab Workshop meteoroid shield were under way at Astronautics Laboratory, Marshall Space Flight Center announced. Shield, thin sheet of aluminum wrapped around outer wall, would protect Skylab crewmen from micrometeoroids and ensure comfortable temperature in space. (MSFC Release 72–19)

February 25: U.S.S.R. launched *Cosmos 475* from Plesetsk into orbit with 1003-km (623.2-mi) apogee, 967-km (600.9-mi) perigee, 104.8-min period, and 74.1° inclination. (GSFC *SSR*, 2/29/72; *SBD*, 2/29/72, 295)

- Reporting on Jan. 10–13 Third Lunar Science Conference results, *Science* said "far more comprehensive picture of the moon but far fewer claims to understand how the moon evolved were evident." Conflict in geochemical evidence suggested "an initially cold moon in which partial melting of its outer layers took place and magnetic evidence that suggests an initially hot moon with a molten core during the early part of its history." Other puzzles were indication that radioactive materials on moon's surface were concentrated in one area and evidence of unexpectedly high heat flux from moon's interior at *Apollo 15* site. Seismic evidence that moon had layered crust 65 km (40 mi) thick had been unexpected. More details on lunar chronology had been determined, "such as the date of formation of the Imbrium basin about 4.0 billion years ago and of the Copernicus Crater only 900 million years ago." It also appeared that, "because of the initially high rate of meteoroid impacts on the moon, very little if any of the original crustal rocks have survived." Lunar-sample chemistry was "proving far from easy—the possibility that volatile elements have migrated on the lunar surface . . . is complicating the analysis of trace constituents." Despite complications, "scientific understanding of the moon is much further advanced than . . . even a year ago. The systematic coverage of the moon's surface provided by the orbital experiments on Apollo 15, in combination with the lunar samples, laboratory work, and the experiments on the lunar surface have provided new data to replace what sometimes proved to be unfounded speculations." (Hammond, *Science*, 2/25/72, 868–70)

- Dr. Eberhard F. M. Rees, Director of Marshall Space Flight Center, received Hermann Oberth Award of Huntsville, Ala., Chapter of American Institute of Aeronautics and Astronautics for outstanding scientific achievement in aeronautics and astronautics. He was cited at awards banquet in Huntsville, for his "many valuable contributions to lunar science, planetary formation, and better understanding of the origin of the solar system through the Apollo program; high resolution astronomy through the Stratoscope and LST (large space telescope) programs; the science of our sun through the ATM . . . program; stellar astronomy, earth resources, and life sciences through the Skylab program; the future of high energy astronomy through HEAO . . . and the promising possibilities for scientific payloads through the Shuttle program." (MSFC Release 72–14)

- Dr. John S. Foster, Jr., Director of Defense Research and Engineering in Dept. of Defense, testified on early warning satellite system before Senate Committee on Appropriations' Subcommittee on Defense during hearings on FY 1973 DOD appropriations: "We now have a satellite that is capable of immediately reporting ICBM launches from the Sino-Soviet area. Additional satellites will be deployed to give us effective warning. This year, our early warning system will provide coverage of the entire threat area by at least two unique types of sensor." (Testimony)

February 25

- Senate Committee on Armed Services rejected Administration request for immediate $35-million appropriation in supplementary funds to accelerate development of undersea long-range missile system (ULMS). Committee decided to consider ULMS program as part of FY 1973 Dept. of Defense budget and to hold further hearings on project. (Middleton, *NYT*, 3/4/72, 11)
- NASA announced selection of 10 scientists and their proposals for Skylab ground-based astronomy program to acquire corollary solar data from ground simultaneously with Skylab observations from earth orbit.

  In related program, Apollo Telescope Mount principal investigators had offered guest investigator opportunities to all interested astronomers worldwide. Principal investigator and guest investigator would arrange for access to selected ATM data of mutual interest, and guest investigator would publish scientific paper on analysis of ATM data he received. Guest investigators would be approved by NASA but would have no automatic claim on NASA funding. Some astronomers interested in program had proposed that rocket and balloon flights be conducted in addition to ground-based observations. Acquisition of simultaneous solar data from many of these observations, with Skylab data, was expected to greatly enhance development of models of solar atmosphere. (NASA Release 72-39)
- Lockheed Missiles & Space Co. scientists had reported development of "uniquely simple" power cell to produce more electricity per pound and up to 100 times more energy than lead-acid storage battery, Associated Press said. Device could operate without releasing harmful pollutants and was fueled with water and alkali metals like sodium or lithium. Scientists said they could control usual violent reaction of lithium and sodium to contact with water. (W *Star*, 2/25/72, C10)
- NASA launched Nike-Tomahawk sounding rocket from Poker Flat Rocket Range, Fairbanks, Alaska, carrying Rice Univ. auroral particles experiment. Rocket and instrumentation performed satisfactorily. (SR list)

February 26: France's *Eole* satellite (launched by NASA Aug. 16, 1971) would be used by Australian marine biologists to track ocean movements of crayfish larvae, Reuters reported. Biologists would attempt with tracking buoy to solve mystery of crayfish cycle in which crayfish drifted during its plankton stage. Minute crayfish larvae in free-floating stage had been found up to 1100 km (700 mi) from their hatching grounds outside coastal reefs, but there was little information on currents that carried them there. (*CSM*, 2/26/72; French Embassy)
- *Noaa 1* Improved Tiros Operational Satellite (launched by NASA for National Oceanic and Atmospheric Administration Dec. 11, 1970) was adjudged successful by NASA. Mission objectives of placing spacecraft in desired sun-synchronous orbit and collecting daytime and nighttime cloud-cover data had been achieved and spacecraft had been officially deactivated Aug. 19, 1971, when attitude control could no longer be exercised. (NASA proj off)
- ITOS-B Improved Tiros Operational Satellite (launched by NASA for National Oceanic and Atmospheric Administration Oct. 21, 1971) was adjudged unsuccessful by NASA. Satellite had failed to reach orbit when pitch-and-yaw attitude control jets began pulsating to counteract tumbling force from leak in 2nd-stage oxidizer system. Jets kept spacecraft in proper attitude until control gas was expended and vehicle

tumbled out of control and impacted above Arctic Circle. ITOS-B was to have been placed in sun-synchronous orbit to observe daytime and nighttime cloud cover. (NASA proj off)

February 27: Launch of NASA's Pioneer-F Jupiter probe was postponed because of momentary power shutdown at Eastern Test Range launch pad. (*NYT*, 3/1/72, 4)

- Lunar samples returned to earth by U.S.S.R.'s unmanned *Luna 20* (launched Feb. 14) were removed from core tube and examined by scientists at Soviet Academy of Sciences. Tass said lunar material consisted of small round anorthosite stones in ash-colored dust significantly lighter than slate-colored dust recovered by *Luna 16* (Sept. 12–24, 1970). Core material in drill tube was placed in huge cylindrical chamber. After chamber had been sealed, pumps produced vacuum and then introduced an unidentified sterilizing gas which was later replaced with helium gas. After 14-hr procedure, samples were removed from tube and placed on steel tray for examination. (Shabad, *NYT*, 2/28/72, 17)

- More than 70 industrial firms, universities, laboratories, and government agencies were using NASTRAN (NASA's structural analysis computer program) to solve their structural engineering problems, NASA announced. NASTRAN was general-purpose program originally designed to analyze behavior of elastic structures in space program. One major use had been in design of space shuttle. More than 600 engineers had been made familiar with NASTRAN since it was made available to public in November 1970 and program was being used in more than 185 different applications, with at least 55 more planned. Firms using NASTRAN predicted many improvements in product safety, reliability, and quality as direct benefit to their customers. (NASA Release 72–40)

- *Apollo 11* Astronaut Edwin E. Aldrin, Jr., blamed pressures of public life following participation in first lunar landing July 20, 1969, for mental condition that had forced him to seek psychiatric help. In interview published by *Los Angeles Times,* Aldrin—who had announced retirement as Commandant of Aerospace Research Pilot School at Edwards Air Force Base, Calif., effective March 1—said he was writing book about "personal experiences." (*LA Times*, 2/27/72, A1)

- U.S.S.R. had tested new SS-NX-8 underwater ballistic missile with 5600-km (3000-nm) range, according to sources quoted by *New York Times*. Adm. Thomas H. Moorer, Chairman of Joint Chiefs of Staff, had said weapon "might soon be ready for operational deployment" if tests were successful. (*NYT*, 2/27/72, 10)

February 28: Launch of NASA's Pioneer-F Jupiter probe from Eastern Test Range, originally scheduled for Feb. 27, was postponed for second time because of severe high-altitude wind conditions. Launch was rescheduled for March 1. (UPI, *W Post*, 2/29/72, A3)

- *On the Moon with Apollo 16* by Dr. Gene M. Simmons, former Manned Spacecraft Center Chief Scientist, was available for sale from Government Printing Office, NASA announced. Book described landing site and scientific aspects of mission scheduled for launch to moon's Descartes region April 16. (NASA Special Release)

- Declining costs of food for space flights was cited by Manned Spacecraft Center Chief of Food and Nutrition, Dr. Malcolm C. Smith, in interview

February 28

published by *Los Angeles Times*. Daily rations for each pre-Apollo astronaut in space cost about $300; first Apollo rations, $190; *Apollo 15*, $142; and Apollo 16, $75. Costs for packaging, inspection, and preparation accounted for major share of expense. Food itself was donated by manufacturers and producers. About 100 companies participated. (Voltz, *LA Times*, 2/28/72)

- Results of survey of Boeing 747 in third year of operation were reported by *Aviation Week & Space Technology:* Passenger appeal had fluctuated but was on upturn. Aircraft was expected to have long life with "unquestioned airworthiness and performance." Introduction of huge aircraft during declining traffic growth had caused order cancellations, groundings, and delays in delivery acceptance. Size and flexibility had enhanced profit potential. Criticism of mechanical reliability had focused on JT9D engine but complaints had dropped with introduction of improved versions of engine and airframe; ground handling problems were less severe than had been predicted. (*Av Wk*, 2/28/72, 35–9, 42–3)

- White House Office of Telecommunications Policy would review all major Government communications satellite programs to avoid duplication and to influence preparation of FY 1974 budget, *Aviation Week & Space Technology* reported. (*Av Wk*, 2/28/72, 11)

- Army selected McDonnell Douglas Corp. to receive five-year $382-million contract to develop and demonstrate site defense of Minuteman, missile-defense system to supplement Safeguard antiballistic missile system. McDonnell Douglas received initial $10 million for first five months. (AP, *W Post*, 2/29/72, A2)

- One consequence of President Nixon's Feb. 17–28 state visit to People's Republic of China could be indefinite ban on overflights of mainland China by reconnaissance drones and high-altitude U–2 aircraft engaged in intelligence-collecting missions, *Aviation Week & Space Technology* article said. "If the anticipated decision is made, it presumably would end any consideration of flying Boeing's upcoming long-endurance Compass Cope drone on proposed 24-hr missions over Communist China." (*Av Wk*, 2/28/72, 11)

February 28–March 24: City of Memphis and Shelby County, Tenn., conducted "Space and Science Awareness Month" to inform citizens about national space program. Project—sponsored by greater Memphis area school systems and supported by NASA Hq. and Marshall Space Flight Center—was opened by Astronaut Russell L. Schweickart and NASA Assistant Administrator for Public Affairs John P. Donnelly. Program exhibited space-related displays—including moon rock, lunar roving vehicle mockup, and models and displays on several NASA programs—at Pink Palace Museum. Also included were displays of educational literature on space technology in 250 school libraries; lectures at 65 schools and at various club meetings by NASA officials; 14 seminars and workshops for teachers; showing of 25 films at elementary schools, on Memphis educational TV station, and to adult groups; and presentation of series of programs on ETV station. Memphis awareness month was largest space information program ever held by U.S. city. (MSFC Release 72–16)

February 29: Symposium "A Progress Report on Flight Investigations of Supercritical Wing Technology" was held at Flight Research Center for engineers and scientists from Government, industry, and airlines. Meeting included interim report of NASA supercritical wing flight-test program and report on evolution of F–8 supercritical wing configuration. Dr. Richard T. Whitcomb, head of 8–Foot Tunnels Branch at Langley Research Center and developer of supercritical wing concept, commented on correlation of wind-tunnel and flight-test results.

Flight tests of NASA supercritical wing on TF–8A aircraft had demonstrated that new airfoil shape increased flight efficiency at speeds near mach 1. Since its first flight March 9, 1971, TF–8A had carried supercritical wing on 27 flights and had reached mach 1.2 (1274 km per hr; 792 mph) and 15 500-km (51 000-ft) altitude. NASA test pilot Thomas C. McMurtry said flight-test program had indicated that piloting procedures and tasks at near-sonic cruise speeds should be as routine as present jet transport operations and that introduction of supercritical wing was not expected to cause any serious problems. Future plans for supercritical wing included addition of side fairings for increased area ruling, determination by simulation of effects of wing roughness from manufacturing imperfections, and development of plans for follow-on flight programs to further the readiness of new technology for industry applications. (FRC Release 2–72; FRC *X–Press*, 2/18/72, 2)

- Large storage battery at Manned Spacecraft Center exploded, killing one man and seriously injuring another. Battery was near underwater tank where astronauts trained in simulated weightlessness. Training Supervisor Kirby C. Dupree, who was killed, and James E. Scott—both employed by Brown & Root-Northrop—apparently had been performing routine maintenance on the 240-cu-m (64 000-gal) tank. No astronauts were in the building. (*W Post*, 3/1/72, A4; Reuters, *NYT*, 3/1/72, 24)

- Single-stage Nike rocket, carrying 47.2-kg (104-lb) payload, was launched by NASA from Wallops Station to 11 600-m (38 000-ft) altitude to obtain precise records of vibrations and accelerations caused by small variations in burning process of solid-fueled rocket motor. Flight data would be compared with similar measurements made in ground vibration tests and captive firings and was expected to provide improved knowledge of effects of rocket-motor burning on payload dynamics for more efficient payload design. Rocket and instruments functioned satisfactorily and payload was recovered as planned. (WS Release 72–2; WS PAO)

- Gerald M. Truszynski, NASA Associate Administrator for Tracking and Data Acquisition, described telemetry on-line processing system (TELOPS)—NASA's planned "new approach to data processing"—in testimony during House Committee on Science and Astronautics' Subcommittee on Aeronautics and Space Technology hearings on NASA FY 1973 authorization. TELOPS, planned to replace NASA data processing system that took "considerable number of weeks between the acquisition of data . . . and the actual delivery of that data to the experimenter," would increase processing speed and efficiency so that "both the greater real-time requirements and the greater volume can be accommodated." With TELOPS "scientific and tracking data will be sent via the communications lines in near real-time where it will be prepared for entry into a large mass storage device" with "capacity of storing online all telemetry data acquired over a 6- to 12-month period. Existing computer would separate data and merge orbit and attitude. "The output product of this processing will be a combination of real-time data sent via communications terminals to the experimenters as necessary, and the generation of data tapes which can be forwarded to the experimenters and used for permanent record." Advantages would be elimination of many tape-handling procedures, processing of more data on demand basis, and processing of data in near real-time to increase interaction of experimenters with flight experiments.

Daniel J. Harnett, NASA Assistant Administrator for Industry Affairs and Technology Utilization, reported that program included "cooperative efforts with over 150 public sector agencies, groups, and institutions, which have derived over 300 public sector problems currently under consideration; over 130 technology applications projects are underway, with projects at nine NASA Field Centers and activity supported by nine major contractors." Interagency cooperative efforts included work with Dept. of Health, Education, and Welfare; Dept. of Housing and Urban Development; Dept. of Interior; Dept. of Justice; Dept. of Transportation; Environmental Protection Agency; National Oceanic and Atmospheric Administration; U.S. Postal Service; Veterans Administration; more than 77 medical schools and health care institutions; some 25 criminalistics laboratories; and 6 state highway departments. Efforts included projects in air pollution control, biomedicine, law enforcement, mine safety, postal services, urban construction, water pollution control, and transportation. (Transcript)

- House by vote of 356 to 32 passed H.R. 11021, to control emission of noise detrimental to human environment. House rejected amendment—after consultation with NASA Administrator, Secretary of Transportation, and Federal Aviation Administration—that would have authorized Environmental Protection Agency to provide for control and abatement of aircraft noise and sonic boom. (CR, 2/29/72, H1508-39)
- About 150 engineering students from universities in Alabama, Georgia, Tennessee, and Mississippi attended special Fluid and Thermal Engineering Seminar at Marshall Space Flight Center, held to acquaint them with hardware and system developments in fluid and thermal engineering related to their studies. Group included 22 foreign students representing 6 foreign countries, who were students at Univ. of Tennessee Space Institute. (MSFC Release 72-21)
- Smithsonian Institution opened exhibit "Ballooning: 1782–1972" at Arts and Industries Building in Washington, D.C. Exhibit, depicting ballooning and brave aeronauts, was described by *New York Times* as prototype show for National Air and Space Museum, "appropriately headed by Michael Collins, the 'anchor astronaut' for man's first landing on the moon in 1969." (Robertson, *NYT*, 2/29/72, 14)

During February: NASA selected three life scientists to receive first grants under NASA life scientist program. Dr. Roger P. Maickel, professor of pharmacology at Indiana Univ., would study effects of stressful or otherwise abnormal environment, such as space travel, on action of selected therapeutic drugs. Dr. Walther Stoeckenius of San Francisco Medical Center Cardiovascular Research Institute would conduct exobiology research to determine physiologic role of bacterial membrane dependent on high salt concentration. Dr. William A. Bonner, professor of chemistry at Stanford Univ., would conduct exobiology research to develop means for detecting measurable, optically active molecules in remote regions of space. Program had been established in July 1971 to increase cooperation among university and NASA life scientists in contributing to advancement of life sciences disciplines related to NASA mission. Scientists and their graduate students would spend one third of their time at a NASA center. (NASA Release 72–41)

- First European Conference on Space and Youth was sponsored in Paris by National Assn. of Aerospace Clubs of France and International Coordinating Committee (CIC) for out of school scientific activity. Delegates from aerospace groups in 23 European nations were invited to conference to plan first program of intra-European cooperation for youth in aerospace field. Meeting established conference as annual event and approved resolutions on future cooperation in technical meetings, exchange of materials, establishment of secretariat, and origination of joint youth rocket projects. (*Government Executive*, 4/72, 28–9)
- Office of Management and Budget issued charter establishing Interagency Coordination Committee for Earth Resources Survey Program (ICC:ERSP) to coordinate and integrate Federal plans, policies, and programs of remote sensing of earth's resources. (*Fed Rpt on ERSP*, 8/30/72, 5)
- Chairman M. G. K. Menon of Indian Space Research Organization (ISRO) laid foundation stone for propellant and fuel complex to produce polymeric binders at Thumba, India, and discussed extended India-U.S.S.R. space cooperation with delegation from Soviet Academy of Sciences visiting Thumba. (*SF*, 7/72, 261)
- George Washington Univ. released *Applications of Aerospace Technology in the Public Sector* (GW BSCP 72–02R). Semiannual review of Biomedical and Public Sector Technology Application Team Program, published for NASA, summarized projects of NASA technology utilization program active from June 1, 1971, through November 30, 1971. Biomedical applications of NASA technology included treatment and research in cancer, cardiovascular disease, kidney function disorders, and rehabilitation medicine. Public sector applications included housing and urban construction, fire safety, law enforcement and criminalistics, transportation, marine science and engineering, air and water pollution, and mine safety. (Text)
- *Spaceflight* magazine reviewed *Vanguard—A History*, by Constance M. Green and Milton Lomask. Book presented "fine coverage of the project [first U.S. satellite program] which should be of interest at all levels. The subject is still controversial and though the present treatment will not meet with full approval everywhere, they have shown that without Vanguard a manned landing on the Moon in 1969 would have been highly unlikely." (*SF*, 2/72, 77)

March 1972

March 1: Air Force launched unidentified reconnaissance satellite from Air Force Eastern Test Range by Titan IIIC booster at 4:39 am EST. Satellite entered orbit with 35 964-km (22 347-mi) apogee, 35 418.5-km (22 008-mi) perigee, 1429.9-min period, and 0.2° inclination. (Pres Rpt 73; *SBD*, 3/2/72, 15)

- U.S.S.R. launched *Cosmos 476* from Plesetsk into orbit with 634-km (394-mi) apogee, 615-km (382.1-mi) perigee, 97.1-min period, and 81.2° inclination. (GSFC *SSR*, 3/31/72; *SBD*, 3/2/72, 16)

- NASA announced initiation of cooperative program with Air Force to establish technology base needed by USAF and industry for development of military and civil short takeoff and landing (STOL) aircraft. Memorandum of Understanding signed by Dr. James C. Fletcher, NASA Administrator, and Dr. Robert C. Seamans, Jr., Secretary of the Air Force, called for close coordination of USAF advanced, medium STOL transport (AMST) program and NASA quiet, experimental STOL (QUESTOL) program to ensure that "the STOL technology developed will . . . serve both areas of interest without undue compromise to either." Program would be managed by new Air Force/NASA STOL Coordinating Council, co-chaired by Assistant Secretary of the Air Force for Research and Development Grant L. Hansen and NASA Associate Administrator for Aeronautics and Space Technology Roy P. Jackson. (NASA Release 72-44)

- *Nimbus 4* meteorological satellite (launched by NASA April 8, 1970) was continuing to gather data, working with ground IRLS (interrogation, recording, and location system). Satellite earlier had tracked weather balloons floating around world, floating ocean buoys, wild animals, and British Aviatrix Sheila Scott, who set several aviation records while circling globe in 1971. In October 1971 IRLS platform had been placed in buoy off Arctic ice island T3 to measure atmospheric pressure, surface and water temperatures, and, in tandem with *Nimbus 4*, platform position. In April six more buoys would be installed on ice floes in Arctic Ocean north of Alaska to obtain basic weather information during summer. Data had never before been obtained because ice islands melted into brackish ponds and streams covered by fog and rain and were impossible to reach.

    Monitoring platform in Ohio was transmitting water-quality data from Great Miami River near Cincinnati. IRLS platforms would also be attached to fixed stations in Lake Erie to monitor water oxygen content, electrical conductivity, alkaline and acid content, and temperatures and to test efficiency of collecting water data quickly and repetitively from many different areas simultaneously. Drifting platform in Lake Ontario would collect water data and information on currents. About 15 sensors on Mt. Kilauea, Hawaii, volcano would monitor temperatures to determine relationship between temperature rise and volcano eruption. (NASA Special Release)

March 1

- Atomic Energy Commission's new particle accelerator near Batavia, Ill., reached design energy of 200 bev on schedule and within $250 million cost estimate. (AEC Release P-52)
- Lewis Research Center engineers were running first pure jet engine designed and assembled at LeRC to power single- and twin-engine light aircraft, *Cleveland Press* reported. Prototype of engine—expected to usher in generation of safe, easy-to-fly aircraft that could cruise at 644 km per hr (400 mph) and costing less to produce than current piston-engined aircraft—was about 31 cm (12 in) in diameter and 122 cm (48 in) long. (*Cl Press*, 3/1/72)
- NASA Deputy Associate Administrator for Space Science Vincent L. Johnson testified before House Committee on Science and Astronautics' Subcommittee on Space Science and Applications during hearings on NASA FY 1973 authorization. Johnson reported that in 1971 Thor-Delta launch vehicle had had five launches with four successes. One failure, on ITOS-B mission Oct. 21, had been caused by leak in 2nd-stage oxidizer system. Review board recommendations had been carried out and Jan. 31 launch of *Heos 2* had been successful. Thorough design review and certification was in process for launch by new-configuration Thor-Delta of ITOS-C in April. All critical components of old and new configurations had been completely and intensively reviewed by independent group and most previous qualifications had been verified. Estimated cost of responding to review board recommendations was $2 million for FY 1972, $3 million for FY 1973, and $5 million per year thereafter. (Transcript)
- Harry H. Gorman, NASA Deputy Associate Administrator for Management, Manned Space Flight, testified on in-house portion of manned space flight R&D budget, during FY 1973 authorization hearings before House Committee on Science and Astronautics' Subcommittee on Manned Space Flight: In FY 1973 Manned Spacecraft Center, Marshall Space Flight Center, and Kennedy Space Center capabilities specializing in project management would be "in the midst of a changeover as current programs are phasing down and the Shuttle program and related activities are getting underway. Our laboratory and technical facilities will be dedicated to developing the technology for on-going programs and to payload planning."

In FY 1973, manned space flight's civil service strength at the three Centers would be 11 350, "a continuing reduction from previous years. This level is the minimum necessary for carrying out our responsibilities for Apollo, Skylab, and Space Shuttle, while still providing support to new technology development and planning for future missions." On top of planned reduction of 619 positions in FY 1972 budget, Centers were further cut by 636 positions under Federal employment control in FY 1972 and 1973, or 10% loss in those two years. Added reduction of 361 positions in FY 1972 and 275 more positions in FY 1973 would require further reduction-in-force procedures and continue to contribute to aging of overall personnel complement.

Of civil service total, 5% were in program planning and definition, 15% in program management, 30% in engineering and development, 5% in science and medicine, 20% conducting flight and launch operations, with rest in general management and administration. (Transcript)

- Adelbert O. Tischler, Director of Shuttle Technologies Office in NASA Office of Aeronautics and Space Technology (OAST), testified on progress of shuttle transportation system before House Committee on Science and Astronautics' Subcommittee on Aeronautics and Space Technology during NASA FY 1973 authorization hearings: "Methods for rapidly assessing the aero-heating and flight handling characteristics of the shuttle are being developed. A base for applying new design methods and for use of new materials in the sensitive problem of structural mass has been laid. Thermal protection systems, some comprising materials which were virtually unknown at the start of the program, will be brought to . . . technological readiness. The questions of long-lived propulsive and power equipment, suitable for repeated re-use, will be answered." Computerized index and search procedure for access to all published shuttle technology was nearing completion. "While technology issues remain, the continuing progress of the technology program conducted by . . . NASA centers and industrial contractors is building a technological base that is far deeper and stronger than that which has heretofore underlain new development programs. This base . . . will assure successful Shuttle development."

David S. Gabriel, Manager of AEC-NASA Space Nuclear Systems Office, described NASA-funded electrophysics research program conducted by Ames, Langley, and Lewis Research Centers and Jet Propulsion Laboratory. Program was concerned "with understanding and, ultimately, using electrons, nuclei, atoms and molecules. Such work is of central importance to future sciences, for example, to laser beam generation, transmission and conversion; superconductivity; or plasma power devices." In FY 1972, research had been completed on continuous-flow plasma accelerator to simulate spacecraft reentry conditions accurately, new contributions to understanding plasma turbulence had been made, knowledge of uranium plasma properties had been improved, and carbon monoxide lasers had been demonstrated to be highly efficient. New superconductor made up of thin films of vanadium and silicon had been discovered. For FY 1973 $3.4 million had been requested for programs in plasma dynamics, lasers, superconductivity, and nuclear physics.

Roy P. Jackson, Associate Administrator for Aeronautics and Space Technology, testified that "clearly the space shuttle orbiter, a rather sophisticated piece of machinery, is not only a launch vehicle and a spacecraft but also a high speed airplane, with all the attendant considerations of aircraft-type control and landing. It offers NASA an opportunity to apply aerodynamic, structural, and materials technology that has been developing over the past 10 or 15 years for high performance aircraft. Reciprocally the shuttle work offers much to benefit aircraft in the next 10 to 15 years." NASA had also capitalized on early Air Force technology oriented toward manned high-speed flight in the atmosphere as well as reentry into the atmosphere, such as Dyna-Soar and X-15 projects.

Francis J. Sullivan—Director of Guidance, Control, and Information Systems in OAST—testified on nuclear electric power system research for spacecraft: Nuclear electric power systems offered "attractive alternatives" to batteries for long-life orbiting systems, "in terms of minimizing structural constraints and the economic considerations of supplying

March 1

several kilowatts of power. Current research is directed primarily toward two types of nuclear systems, isotope systems for low power levels of about 1 kilowatt, and the intermediate reactor power systems for levels in the range of several kilowatts up to 10-20 kilowatts." Isotope systems were radioisotope thermoelectric generators (RTGs). Intermediate reactor power system consisted of zirconium hydride reactor to generate heat and thermoelectric power-conversion components to convert this heat into electricity. Program was directed toward reactor test in 1976-1977 period to demonstrate long-life capability. Stable performance had been achieved on thermoelectric converter test modules. Effort in FY 1973 would concentrate on verifying this performance. Conversion system would be developed for testing with the reactor to verify the integrated system performance. (Transcript)

- Appointment of Kenneth B. Gilbreath, Manager of White Sands Test Facility, as Deputy Director of Center Operations at Manned Spacecraft Center was announced by MSC. He would be succeeded at White Sands by Chief Jesse C. Jones of MSC's Engineering and Development Directorate's Laboratory Operations. Appointments would be effective April 1. (MSC Release 72-50)
- Atomic Energy Commission published *Global Inventory and Distribution of Pu-238 from SNAP-9A*, report on study of plutonium accidentally injected into stratosphere when U.S. Navy Transit satellite failed to reach orbit after launch from Vandenberg AFB April 21, 1964. Accident had caused threefold increase in global fallout of plutonium isotope over that usually injected into atmosphere by atmospheric nuclear weapon tests. SNAP-9A nuclear generator, designed to convert heat from Pu-238 into electrical energy, had burned up during reentry and had ablated into small particles at 46-km (28.6-mi) altitude. Concentrations of Pu-238 in stratosphere had been measured through end of 1970 and soil samples had been collected at more than 60 sites to estimate global distribution of fallout from the accident. Of the 17 kilocuries of Pu-238 originally in the generator, 13.4 plus or minus 2.2 kilocuries had been deposited, with 3.1 plus or minus 0.8 in northern hemisphere and 10.3 plus or minus 2.1 in southern hemisphere. Although plutonium was one of most enduring and deadly radioactive poisons known, amounts inhaled when Pu-238 settled to earth were so small they were not considered hazardous. (Text)
- *Los Angeles Times* editorial said, despite insistence of military authorities that regulations restricted supersonic flights to isolated areas, sonic booms were menacing Death Valley National Monument in California. It was "continuing sport for military pilots" to "make low level supersonic assaults on the peace and tranquility of that remarkable recreation and sightseeing area. Rangers counted 35 booms in January, three times the number [of] the preceding January." (*LA Times*, 3/1/72, 8)

March 1-2: Fourteenth Israel Annual Conference on Aviation and Astronautics was held in Tel Aviv and Haifa. Fifth Theodore von Kármán Memorial Lecture, "Perspectives in Aeroelasticity," was delivered by Langley Research Center scientist I. Edward Garrick. (Off Naval Research London, *European Scientific Notes*, ESN-26-4, 4/28/72, 91)

March 1-3: NASA-wide conference on equal employment opportunity was held at Kennedy Space Center at request of Dr. James C. Fletcher, NASA Administrator. During conference chaired by Dr. George M. Low, NASA

Deputy Administrator, Dr. Fletcher said: "Our successes in space exploration have been outstanding . . . . However, in spite of the efforts of many and a number of significant accomplishments, our achievements in equal employment in NASA, here on earth, do not match what we have done beyond our planet."

Dr. Fletcher said he would insist "that as fast as humanly possible, by transfer, training, promotion, and whatever other means are available, and appropriate, our offices move toward the goal of the balanced staffing that will achieve true equal employment opportunities."

NASA was planning to get members of minority groups into space. Space shuttle would be important factor in accomplishing this goal. NASA also would consider establishing annual awards for outstanding performance in Office of Equal Opportunity and would insist that "spirit as well as the letter" of contractors' equal employment obligations be fulfilled. (Text)

March 2–24: NASA's *Pioneer 10* (Pioneer-F) Jupiter probe was launched from Eastern Test Range at 8:49 pm EST by three-stage Atlas-Centaur–TE–M–364–4 booster, used for first time. Spacecraft reached highest launch velocity ever attained, 51 500 km per hr (32 000 mph) relative to earth. Primary objective was to obtain precursory scientific information beyond Mars orbit with emphasis on investigation of interplanetary medium, asteroid belt, and Jupiter and its environment. Secondary objective was to advance technology for long flights to outer planets.

The 258-kg (569-lb), spin-stabilized spacecraft carried 11 scientific experiments to provide new knowledge about Jupiter, solar system, and Milky Way galaxy. It would return first closeup images of Jupiter and make first measurements of Jupiter's twilight side, never seen from earth. Two additional experiments—celestial mechanics and S-band occultation—would use communication signal and earth-based equipment. *Pioneer 10* would be first spacecraft to observe Asteroid Belt between orbits of Mars and Jupiter, first man-made object to escape solar system, first spacecraft to use orbital velocity and gravity of Jupiter for escape, and first NASA spacecraft powered entirely by nuclear energy—four radioisotope thermoelectric generators developed by Atomic Energy Commission. *Pioneer 10* would make 20 kinds of measurements of Jupiter's atmosphere, radiation belts, heat balance, magnetic field, moons, and other phenomena. It would also characterize solar atmosphere, interstellar gas, cosmic rays, asteroids, and meteoroids.

Spacecraft also carried pictorial plaque designed to show scientifically educated inhabitants of another star system—who might intercept it millions of years later—the time spacecraft was launched, from where, and by whom. Plaque design was etched into gold-anodized aluminum plate 15 by 23 cm (6 by 9 in) and 1.27 mm (0.05 in) thick attached to spacecraft's antenna support struts. Radiating lines on left of plaque represented positions of 14 pulsars, with mathematical binary code "1-" symbols representing pulsars' frequencies relative to hydrogen atom, which was used as "universal clock." Hydrogen atom was also used as "universal yardstick" for sizing human figures and outline of spacecraft on right of plaque. Man's hand was raised in good-will gesture. Across bottom of plaque were planets ranging outward from sun, with spacecraft's trajectory arching away from earth, passing Mars, and swinging past Jupiter.

March 2-24: NASA's Pioneer 10 was launched with the highest velocity yet attained and began its journey toward the Asteroid Belt, the planet Jupiter, and eventual escape from the solar system. By March 24 all instruments had been turned on and were returning data. The probe carried a plaque to show any inhabitants of another star system—who might intercept it millions of years in the future—where Pioneer 10 had been launched, when, and by what kind of beings. In the photo above, technicians adjusted the spacecraft at TRW, Redondo Beach, Calif., facilities.

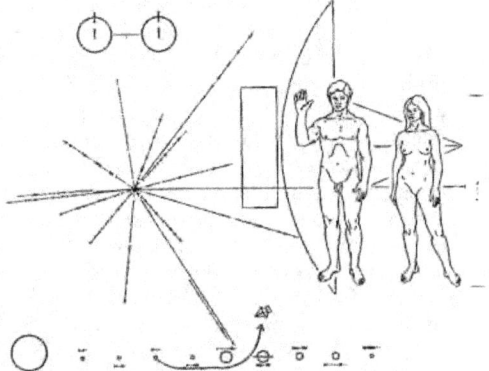

Initial orientation of *Pioneer 10* was achieved six hours after launch so that communications high-gain antenna was directed toward earth. By March 3, four scientific instruments—magnetometer, charged particles experiment, geiger tube telescope, and meteoroid detector—had been turned on. First midcourse maneuver, March 7, increased velocity by 14 m per sec (45.9 fps), adjusted trajectory, and shortened flight time to Jupiter by nine hours. At 3:00 pm EST March 8, *Pioneer 10* was 4 570 000 km (2 840 000 mi) from earth traveling at 32 800 km per hr (20 400 mph). Round-trip radio communication time was 31 sec.

By March 10 trapped radiation detector and ultraviolet photometer had been turned on and meteoroid detector had recorded several hits. Second midcourse maneuver, with two brief thruster firings March 23 and 24, delayed arrival time by 2½ hrs and moved arrival point at Jupiter about 6400 km (4000 mi) closer to planet. Maneuver was also attempt to fly *Pioneer 10* behind Jupiter's moon Io. Whether flight path behind Io had been achieved would not be known until spacecraft neared Jupiter. To go behind Io, spacecraft had to arrive within period of less then eight minutes. By March 24 remaining instruments—plasma analyzer, cosmic ray telescope, imaging photopolarimeter, infrared radiometer, and asteroid/meteoroid detector—had been turned on and all experiments were recording good data. Meteoroid detector had recorded 10 impacts, and asteroid/meteoroid telescope had seen one meteoroid. Spacecraft was expected to enter Asteroid Belt in early July and to reach Jupiter Dec. 3, 1973.

Pioneer 10 was first spacecraft to attempt Jupiter probe. *Pioneer 5* was launched March 11, 1960, to study heliocentric space environment inward toward sun. *Pioneer 6* (launched Dec. 16, 1965) and *Pioneer 7* (launched Aug. 17, 1966) were studying heliocentric space environment. *Pioneer 8* (launched Dec. 13, 1967) and *Pioneer 9* (launched Nov. 8, 1968) were studying interplanetary phenomena. Last four were still operating satisfactorily. (NASA proj off; NASA Releases 72–25, 72–32, 72–50, 72–68)

March 2: *Isis 2* International Satellite for Ionospheric Studies (launched by NASA March 31, 1971, in joint U.S. and Canadian program) was adjudged successful by NASA. Spacecraft had exceeded mission objectives. All 12 scientific instruments were operational and were still acquiring detailed information on latitudinal and diurnal variations of ionosphere. Spacecraft was obtaining radio-sounder and correlative direct measurements to allow continuation and extension of ionospheric studies during period of declining solar activity, and long lifetime was anticipated.

Isis 2 was fourth and final mission in U.S.-Canadian program to study ionosphere that began with launch of *Alouette 1* Sept. 28, 1962. All four spacecraft were operational and were providing useful data. ISIS program had acquired wealth of knowledge on worldwide morphology of topside ionosphere, virtually unknown region before Alouette-ISIS program, and in discoveries in plasma physics and radio-wave propagation. (NASA proj off)

- Bellcomm, Inc., scientist Dr. Farouk El-Baz had suggested that water vapor detected on moon March 7, 1971, could have come from water and urine dumped into lunar orbit from *Apollo 14* spacecraft, Reuters reported. He had shown particles of waste water could have combined, orbited moon for several weeks at gradually decreasing altitudes, and finally descended to lunar surface, where they were detected as water vapor. Bellcomm's findings were rejected by Dr. H. Kent Hills of Rice Univ., who said waste discharge would have spread out in orbit and would not have produced as high a reading on two suprathermal ion detectors as he and codiscoverer Dr. John W. Freeman had found. Discovery of water vapor had been announced Oct. 15, 1971 [see also March 3]. (*NYT*, 3/3/72, 22; Bellcomm PIO; NASA proj off)

- Roy P. Jackson, NASA Associate Administrator for Aeronautics and Space Technology, discussed role of aircraft technology in alleviating airport terminal congestion in speech before American Institute of Aeronautics and Astronautics' National Capital Section in Washington, D.C.: Total number of civil air operations was expected to double approximately every decade throughout the rest of this century. "The busiest airports will have to handle twice to three times their present operations. Since most jetports are too noisy today, the capacity increase must be accomplished with an actual integrated decrease in noise. Steeper operational procedures, retrofitted engines, and quieter engines in new terminal-configured aircraft must all be considered as companions to dual runways, improved automation, and success in the Discrete Addressable Beacon and Microwave Landing Systems Programs.

"To design the next generation of civil aircraft as terminal-configured vehicles, the designer will need technology advancements that will permit steep and curved vertical profiles for noise abatement, curved ground tracks for noise abatement and airspace efficiency, and an ability to satisfy a specified time-of-arrival at the runway threshold to maximize runway capacity." NASA had "a new start in FY '73 to lead the way in technology, . . . the Terminal Configured Vehicles and Avionics Operating Experiments Program, for which funding of $4.2 million will be programmed." (Text)

- NASA applications program was reviewed by Associate Administrator for Applications Charles W. Mathews in testimony before House Committee on Science and Astronautics' Subcommittee on Space Science and Applications during FY 1973 NASA authorization hearings. Objectives of program were "to establish useful applications of space and space know-how through the development of user relationships, through the development of requisite technology, and through the conduct of appropriate ground, airborne and space flight investigations."

  Space was "a magnificent vantage point to survey the surface of the earth and to expedite and enhance the flow of information around the globe." Core of program was new applications of earth observations and communications, but "other areas associated with the unique environment of space, particularly weightlessness, show promises which must be explored, developed and exploited." Space manufacturing processes offered good example. "Space applications, however, should not be envisioned just as activities that lead to satellites or laboratories in space, for . . . applications relate to contributions to men and women who . . . reside right here on the ground. Know-how provided by experience with the development of space systems can be applied to . . . areas involving housing, transportation, health care and many others."

  NASA participation in Global Atmospheric Research Program (GARP) would increase significantly in FY 1973 as transition occurred from planning to carrying out Tropical Experiment and Data Systems Test, both planned for CY 1974. Planning would be accelerated for Global Experiment during 1976–1977. GARP was being conducted by World Meteorological Organization and International Council of Scientific Unions to increase understanding of circulation of atmosphere and provide mathematical and physical basis for long-range weather prediction, determination of feasibility of large-scale climatic modification, and assessment of consequences of man's pollution of atmosphere. (Transcript)

- Officials of NASA Office of Aeronautics and Space Technology (OAST) testified during House Committee on Science and Astronautics' Subcommittee on Aeronautics and Space Technology's final hearings on NASA FY 1973 authorization.

William H. Woodward, Director of Space Propulsion and Power, cited importance of isotope power systems to NASA planetary program: "The ability of nuclear electric power systems to operate for long periods . . . without consideration to distance from the sun is the primary feature which makes nuclear systems mandatory for outer planet missions. This self-contained energy input characteristic also provides a relative insensitivity to the operating environments encountered and this capability may be important on many missions, e.g., lander missions on planetary surfaces or missions which encounter high natural radiation fields." Power requirements of most missions could be satisfied with low-level isotope power systems using radioisotope thermoelectric generators (RTGs). "Currently, RTG's are scheduled for flight on the Pioneer Jupiter Flyby and the Mars Viking Lander missions and in addition, are planned for use in future outer planet programs."

Associate Administrator for OAST Roy P. Jackson testified that solar energy could prove important energy resource. National Science Foundation's RANN (Research Applicable to National Needs) program had budgeted $4 million for solar energy investigations in FY 1973. Atomic Energy Commission, Dept. of Interior, and NASA were conducting in-house studies. Solar Energy Subcommittee had been established by Federal Council on Science and Technology Committee on Energy Research and Development Goals to assess benefits to recommend R&D funds and programs. Subcommittee was first time solar energy had been considered at this level of national planning. NASA was evaluating concept for generating electric power from solar cells in space and transmitting power to earth via microwave beam, which would deliver more concentrated solar energy than naturally received on earth's surface. Collecting in space also could be continuous, rather than having to stop at night or in cloudy weather.

Edwin C. Kilgore, Deputy Associate Administrator for Management, described OAST manpower problems and methods to alleviate them. Though OAST manpower applied to aeronautics had increased 90% since 1966, manpower for all other efforts had decreased by 56%. Growth in aeronautics could not be provided continuously by transferring personnel from space projects. "In the short term we will continue to meet the increased need for aeronautics manpower by retraining and reassignment of personnel at the Research Centers. For the long term we have initiated a program to encourage the training of students interested in aeronautical sciences. . . . we are continuing the cooperative NASA/university graduate research and study program begun in FY 1971 where faculty and students are working with the Research Centers."

R. D. Ginter, Director of Technology Applications, testified on NASA and National Science Foundation cooperation in applications: NSF's RANN program contained "two broad areas of mutual interest: Advanced Technology Applications and Environmental Systems Resources. One of our senior people has been assigned to focus this effort, coordinate task selection, and provide program management. During the next year we intend to expand our problem definition activity and prepare specific proposals for NSF consideration." (Transcript)

- Report of Soviet progress in space presented at annual meeting of Soviet Academy of Sciences appeared in *Sotsialisticheskaya Industriya*.

Academician M. D. Millionschikov said 1971 had been year of achievement in space. Creation of "the world's first manned orbital scientific station, Salyut, was a major event. . . . Exploration of the moon and the planets with the help of automatic space vehicles has continued. Recently, a magnificent space experiment was completed—the Luna-20 automatic station's delivery to earth of soil samples from the almost inaccessible' lunar mountain region. Two automatic 'Mars' stations were created. . . . Separating from the Mars-2 station, a capsule landed a pennant . . . and the Mars-3 descent vehicle made the first soft landing on the surface of Mars. In addition to the scientific exploration of space, broad plans were elaborated for utilizing satellites for extra-atmosphere astronomy. Work began on the utilization of satellites for exploring natural resources in the interests of the national economy including geological survey work, agriculture, forestry, water economy, fishing, the hydro-meteorological service, and geodesy." (FBIS–Sov, 3/13/72, L2)

- Cutbacks in NASA budget would necessitate dismissal of some 40 civil service employees at Kennedy Space Center during March, KSC Director, Dr. Kurt H. Debus, announced. Cutback was part of NASA-wide work force reduction. Similar reduction might be required in 1973. (*Today*, 3/2/72)
- World's first nuclear-powered artificial heart was exhibited for first time by National Institutes of Health in Bethesda, Md. Dr. Theodore Cooper, Director of National Heart and Lung Institute in NIH, said heart might be perfected for human use toward end of decade. Electrically powered version and atom-powered heart-helper had already been implanted in calves. System was expected to operate for 10 yrs. (Asher, *W Post*, 3/3/72, A1)

March 3: U.S. had obtained clearance from Coordinating Committee on Trade in Strategic Materials (COCOM), whose members represented North Atlantic Treaty Organization countries and Japan, for sale of satellite ground station to Communist China, *New York Times* reported. Japan and other allies had interpreted sale as political act and "made it clear that they intend to use the same criterion. Japan . . . was reported to have insisted that the political aspect of the deal be entered into the record," and had accused U.S. of "selling equipment of advanced technology because it was politically advantageous but restraining others from doing the same thing." Station had been installed at Shanghai Airport to cover President Nixon's visit to Communist China [see March 13]. (Giniger, *NYT*, 3/4/72, 2)

- Robert Stern of Bellcomm, Inc., said in Washington, D.C., that his studies "tended to agree" with those of Rice Univ. scientists who had reported on Oct. 15, 1971, detection of possible water vapor being vented from lunar surface. Reuters had reported release March 2 of Bellcomm computer analysis suggesting vapors could have come from urine dumped into lunar orbit by *Apollo 14* astronauts. (*NYT*, 3/4/72)
- Physicist Dr. Ralph E. Lapp discussed space shuttle economics in letter published in *Science*. Cost and purpose of shuttle's projected 900-metric-ton (1000-short-ton) payload over 10 yrs needed careful evaluation "before the nation plunges ahead on a space project whose cost may well eclipse that of the Apollo program." Assuming NASA could

slash payload costs to $2000 per 0.5 kg (1 lb), "then 20 million pounds [9 million kg] in orbit represent a national investment of $40 billion." Shuttle development cost of $5.5 billion plus deployment and operation of shuttle system would cost from $11 to $16 billion. NASA estimate of less than $100 for lifting 0.5 kg (1 lb) of payload from earth to orbit was "easily disputed." Estimate of $11 billion for shuttle cost divided by 9 million kg amounted to $550 per 0.5 kg and not much less than the $700 per 0.5 kg that existing NASA launch vehicles had cost. Only "if orbital payloads can be reduced so they cost only several times more per pound than gold, is it meaningful to seek cheaper space transportation. Even then the nation ought to be asking what kind of space program is the space shuttle designed to support." (*Science*, 3/3/72, 944)

- Manned Spacecraft Center would have reduction in force of 75 to 100 Civil Service positions by June 30, MSC announced. RIF was part of NASA-wide personnel reduction and would reduce MSC complement from 3944 to 3817 Civil Service positions. (MSC Release 72–52)
- U.S.S.R. had dismantled about 100 of its older SS–4 and SS–5 medium- and intermediate-range missiles that could strike China and Europe, Dept. of Defense announced. Newer SS–11 ICBMs that could reach U.S. were replacing older missiles and were being stored in underground silos. DOD spokesman Jerry W. Friedheim said dismantling and storing appeared to be "a force modernization on their part" that had been going on for about five years at the rate of about 20 missiles per year. (*W Post*, 3/4/72)

March 4: U.S.S.R. launched *Cosmos 477* from Plesetsk. Satellite entered orbit with 309-km (192-mi) apogee, 214-km (133-mi) perigee, 89.5-min period, and 72.8° inclination and reentered March 16. (GSFC SSR, 3/31/72; SBD, 3/7/72, 36)

- U.S.S.R.'s meteorological satellite system was described by Tass as "a turning point in the history of Soviet meteorology." Three operating Meteor spacecraft transmitted data on area from pole to pole every 1½ hrs in band up to 1500 km (900 mi) wide—covering over two thirds of globe daily—to network of ground stations covering about one fourth of earth. Meteor system had helped accumulate valuable data for further perfection of onboard equipment and ground reception, registration, and processing of information. "The operation of sputniks outside the boundaries of the atmosphere enabled a new approach to the study of the air ocean of the planet [and] led to interesting discoveries in the sphere. The sputniks . . . made it possible for the first time 'to cast a glance' on strong currents in the upper layers of the atmosphere." (FBIS–Sov, 3/7/72, L2)
- *New York Times* editorial commented on plaque launched on *Pioneer 10* Jupiter probe March 2: "Despite the uncanny mastery of celestial laws that permits man to shoot his artifacts at the stars, we find ourselves still depressingly inept at ordering our own systems here on earth. Even as we try to find a way to insure that sapient man will not consume his planet in nuclear fires, a rising chorus warns us that man may very well exhaust his earth either by overbreeding or by inordinate demands on its resources, or both. "So the marker launched into space is at the same time a gauntlet thrown down to earth: that the

gold-plated plaque convey in its time the message that man is still here—not that he has been there." (*NYT*, 3/4/72, 26)

March 5: Apollo Program Director Rocco A. Petrone, in interview published by *Washington Post*, described Apollo 16 experiment to determine whether moon had its own magnetic field. Apollo 16 astronauts would carry to moon a magnetic rock collected by previous Apollo landing to see if it changed when returned to moon. "We want to see if the magnetism that we have found in moon rocks was not somehow picked up on the voyage back to earth. The whole purpose of this exercise is to prove the existence of a magnetic field on the moon, which in effect proves that the moon was once hot like the earth. Most scientists believe there will be no change, that the magnetism was locked forever into these rocks when they solidified in the moon's magnetic field."

Astronauts would also set up magnetometer on moon and place a moon rock on head of instrument for instant reading of its magnetism. When rock was brought to earth, it would be read again for comparison with reading on moon. (O'Toole, *W Post*, 3/5/72, A3)

- Columnist Hank Burchard criticized message on plaque launched on *Pioneer 10* Jupiter probe March 2, in *Washington Post*: "The supplying of the map implies the invitation for a visit." Invitation was being made "on behalf of all men who may ever live, by a few men who happen to be living now and who happen to be Americans. Nobody asked the Russians or the Chinese or the Masai, or me, whether we wanted anybody from outer space dropping in for dinner." (*W Post*, 3/5/72)

- *New York Times* editorial praised accomplishments of Atomic Energy Commission's new nuclear accelerator at Batavia, Ill. [see March 1]: It was at accelerator "that some of the most illuminating experiments of the next decade are likely to be carried out. For this powerful instrument may have the resources—the penetrating power of its incredible proton beam—to answer some of the ultimate questions about the structure of nuclear particles. If so, man's understanding of the material universe and his powers over it may well be revolutionized in the years immediately ahead." (*NYT*, 3/5/72, 4:12)

March 6: Dr. James C. Fletcher, NASA Administrator, said in letter to NASA staff that *Pioneer 10* (launched March 2) if successful would be "technological and scientific achievement of the first magnitude." During launch it had set spacecraft speed record of some 51 500 km per hr (32 000 mph). "One small detail that impressed me was . . . Pioneer 10 passed the Moon only 11 hours after launch. It makes the Moon seem very close at hand. So we have a new measure for the nearness of what we used to call 'outer space.' It brings home, even as we begin the great adventure of exploring the outer planets, the primary importance of moving forward with our preparations for the routine exploitation of inner space." (*NASA Activities*, 3/15/72, 44)

- NASA Associate Administrator for Organization and Management Richard C. McCurdy, on behalf of NASA, received award from Small Business Administration (SBA) for NASA's efforts in assisting minority business development. Award was presented by Marshall Parker, Associate

Administrator for Procurement and Development at SBA. (NASA Hq WB, 3/20/72; Release 72-46)
- *Chicago Daily News* editorial on *Pioneer 10* said: "After several moon landings and all the fuss over spending on space ventures, the public is probably a bit jaded by now with outbound rockets. Yet there is something special, and more than a little bit eerie, about this one." Only "scientists whose minds can absorb the concept of infinity can really appreciate the possibilities wrapped up in the voyage of Pioneer 10. But the rest of us can share a sense of awe as the spacecraft plunges into eternity, and think a bit about the specks called Man, lost in the vastness of the Universe." (*C Daily News*, 3/6/72)
- Former NASA Administrator James E. Webb was appointed by Speaker of House to succeed Joseph M. Barr on Commission on Government Procurement. Barr had resigned. (*CR*, 3/6/72, H1752)
- U.S.S.R. was deploying multiple warheads on its intercontinental ballistic missiles, Secretary of Defense Melvin R. Laird said in speech before Veterans of Foreign Wars in Washington, D.C. (Text)

March 7: Dr. James C. Fletcher, NASA Administrator, compared NASA FY 1973 budget program with FY 1972 plans in testimony before opening session of House Committee on Appropriations' Subcommittee on Department of Housing and Urban Development-Space-Science-Veterans hearings on FY 1973 budget: "We have reconfigured our plans so that a substantially higher level of appropriations will not be required in future years to complete the programs in our FY 1973 budget" and to continue "useful and significant . . . new projects in space science, exploration, and practical applications, in advanced technology, and in aeronautics." FY 1972 program would have required NASA appropriations in future years to approach $4 billion per year. Run-out costs to complete program as planned would have risen to $3.7 billion in FY 1973 and $3.95 billion in FY 1974. Revised program estimated run-out costs at $3.37 billion in FY 1974, $3.3 billion in FY 1975, $3.2 billion in FY 1976, and $3.1 billion in FY 1977. NASA could "by properly phasing-in the start of needed future new programs . . . hold NASA appropriations in future years to approximately the current total appropriations level in current dollars."

Previous program would have committed U.S. to "higher NASA budget in future years, or to the waste from termination of programs in midstream," but revised plan would "give the Nation a good, viable, and balanced program in aeronautics and space at a cost it can afford." Dr. Fletcher believed "realistic long-term plan in which the Nation's commitment is limited to budgets of approximately the current size is the proper posture for NASA from the standpoint of responsible management" and "should go far in alleviating the concerns that have been expressed that in embarking on the new space programs of the 1970's we are committing the Nation to a program that it cannot afford.

"To achieve this posture, we have had to make some basic changes in our planning and accept yet another stretch-out of the period over which our continuing and long-term objectives in space exploration and space science will be achieved." Principal change has been in space shuttle program. Shuttle now planned would cost $5.5 billion, about half what configuration envisaged year before would have cost. Second significant change was cutback in planned program for exploration of outer planets, reducing plans to explore all five outer planets on Grand Tour missions in late 1970s to new focus on Jupiter and perhaps Saturn, with less expensive spacecraft. Third significant change was termination of nuclear engine for rocket vehicle application (NERVA) program, now including in budget request $8.5 million for defining with Atomic Energy Commission smaller nuclear rocket engine and making trade-off studies of preferred propulsion system for missions to distant planets some time in 1980s.

Final major change was increased emphasis on aeronautics, in response to strong urgings of committees of Congress. (Transcript)

- Tenth anniversary of *Oso 1* Orbiting Solar Observatory, first U.S. satellite devoted entirely to studying sun. Satellite had carried 13 experiments to collect data on solar radiation in ultraviolet, x-ray, and gamma ray regions. After one year of operation it had provided more data on behavior and composition of sun than any single ground-based observatory and all previous rocket, balloon, and satellite flights combined and had measured 75 solar flares and subflares. *Oso 1* had stopped operating Aug. 6, 1963, exceeding 6-mo design lifetime by almost 11 mos. (GSFC PAO; *A&A 1963*)

- MSFC announced it had extended interim contract with North American Rockwell Corp. Rocketdyne Div. for space shuttle main engine design. One-month, $1-million extension was awarded pending completion of review by General Accounting Office of original $500-million contract awarded July 13, 1971. Review had been requested Aug. 3, 1971, by competing contractor United Aircraft Corp. Pratt & Whitney Div. (MSFC Release 72-23; *A&A 1971*)

- NASA launched two Nike-Tomahawk sounding rockets from Poker Flat Rocket Range, Fairbanks, Alaska, carrying Goddard Space Flight Center fields and neutral winds experiments. Rockets and instrumentation performed satisfactorily. (SR list)

- Photographic equipment carried by U.S.S.R.'s *Mars 2* and *Mars 3* spacecraft in Mars orbit were described by Tass: "Automatic devices take pictures on the film which is automatically processed after photographing and then, on command from earth, the pictures are sent by television channels to the flight control centre." Each spacecraft carried two "photo-television devices with wide-angle and long-focus objectives. . . . all units and elements are placed together in one rigid body with the objective fixed outside and in the wide-angle camera—together with the objective and a device for changing colour filters." Pictures were taken synchronously and transmitted with "an optico-mechanical television device." (FBIS–Sov, 3/8/72, L1)

- Progress made in 1971 strategic nuclear arms talks with U.S.S.R., and on many subjects from scientific cooperation to trade, indicated that "1972 will be a year of substantial developments in bilateral cooperation," Secretary of State William P. Rogers said in Washington, D.C., news conference. (Marder, *W Post*, 3/8/72, A2)

March 8: *Mariner 9* probe (launched by NASA May 30, 1971) completed 232 orbits of Mars. Primary mission objective of photographing Mars from south pole to northern hood had been completed. Jet Propulsion Laboratory was preparing to assemble map of Mars, using more than 100 overlapping pictures taken by *Mariner 9* and data from onboard

instruments for corresponding surface and atmospheric characteristics. (NASA Release 72-48)
- NASA Associate Administrator for Applications Charles W. Mathews testified on earth resources program during FY 1973 authorization hearings before House Committee on Science and Astronautics' Subcommittee on Space Science and Applications: Program was probably most significant present element of earth observations program, using advanced satellite and ground-based techniques. FY 1973 funding of $48.4 million would go for procurement (including spacecraft development) and flight operations, data-handling system development, and aircraft program already underway. Queried on pace of program, Mathews said NASA had "good basic program" that would merit expansion. Results from ERTS-A satellite—to be flown with high degree of participation by users from Government, industry, and universities—should show program merited extension and how it should be done. Mathews also cited need for regulation and control of information as it was gathered more nearly completely and accurately. Gathering information was "less of a problem and less of a challenge than the proper use of this information." (Transcript)
- Manned Spacecraft Center announced award of $299 250 contract to Lockheed Missiles & Space Co. to study and develop program for weld-bonding materials for space shuttle. (MSC Release 72-55)
- Attorney for widow of Astronaut Virgil I. Grissom, killed in Jan. 27, 1967, Apollo spacecraft fire, said Mrs. Betty Grissom had agreed to $350 000 out-of-court settlement of her $20-million suit against North American Rockwell Corp. and three subsidiaries for negligence in construction of spacecraft. (AP, *W Post*, 3/9/72, A17)
- NASA launched two sounding rockets. Nike-Tomahawk was launched from Poker Flat Rocket Range, Fairbanks, Alaska, carrying Goddard Space Flight Center fields and neutral winds experiment. Rocket and instrumentation performed satisfactorily.

  Nike-Tomahawk was launched from Wallops Station, carrying Univ. of New Hampshire energetic particle experiment. Rocket performed satisfactorily. Scientific requirements were not satisfied. (SR list)
- President Lyndon B. Johnson had made "impulsive decision" in "intensity of emotion" when he changed name of Cape Canaveral to Cape Kennedy, Circuit Judge James R. Knott of West Palm Beach, Fla., said in testimony before Senate Committee on Interior and Insular Affairs. Committee began hearings on S.R. 193, to redesignate Cape Kennedy as Cape Canaveral. Judge Knott testified that President Johnson had acted on request by widow of President John F. Kennedy following Kennedy assassination and said original name should be restored. (*W Post*, 3/9/72, A25; *CR*, 3/8/72, D224)
- Completion of satellite communications station at Emeq Haela, Israel, was announced by Jerusalem Domestic Service. Station would become operational in about three months. Israeli Post Office Director General Simha Sorokov said opening of station "will speed up the possibility that every resident in the state will be able to dial directly from his house to anyone abroad without . . . an operator. When the system goes into operation, Israeli residents will be able to watch a man walking on the moon at the exact moment he is doing it." (FBIS-Israel, 3/9/72, H3)

March 8

- Canadian-built Black Brant VC rocket was launched from Natal, Brazil, carrying West German and Brazilian payload. Rocket and instruments functioned satisfactorily. (Reuters, *W Post*, 3/10/72, F3)
- Germ-control and dust-purging technique developed by NASA and aerospace industry was being used by St. Luke's Hospital in Denver, Colo., to lower infection risk in surgical procedures. Concept, based on technique to sterilize spacecraft assembly and self-contained life support systems, used portable equipment to remove dust and germs continually from surgical area. Equipment included helmets like those worn by astronauts and specially treated surgical garments that bacteria could not penetrate, collapsible plexiglass and aluminum enclosure, and air-circulating units that forced air through filters before air flowed from rear of enclosure to front. Enclosure fitted inside conventional surgical room. (NASA Release 72-47)
- Federal Aviation Administration announced award of five-month $414 226 contract to Xonics, Inc., for additional wake-turbulence studies leading to design of ground-based acoustic wake-turbulence detection system. Xonics would analyze results of previous studies and recommend design of system capable of providing wake-vortex surveillance along airport instrument-landing-sytem approach path. (FAA Release 72-37)
- President Nixon issued Executive Order 11652 establishing new classification system for Government documents related to national security.

March 8: *Germ-control and dust-purging techniques developed for spacecraft sterilization were being used in St. Luke's Hospital in Denver, Colo. Below, a surgery team performed a hip-joint replacement in a new clean-room facility that could be folded for storage when not in use. Air was forced in a gentle breeze from the rear of the room to the open front. Team members "upwind" of the patient wore astronaut helmets and garments impermeable by bacteria, while air around the downwind team members was carried away from the patient. Application of the technique was developed for NASA by the Martin Marietta Corporation's Denver Division.*

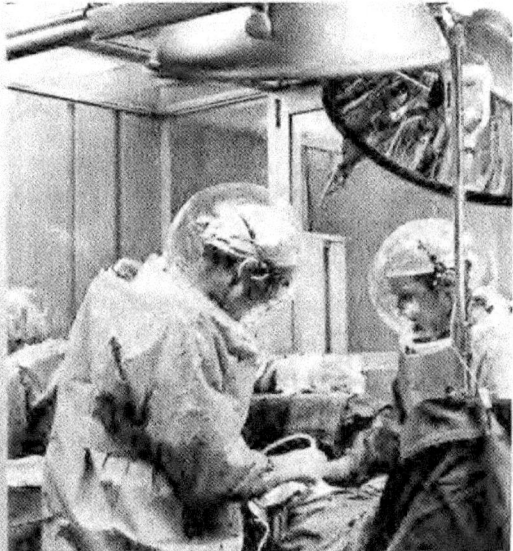

System ordered materials classified Top Secret, Secret, or Confidential only if their unauthorized disclosure "could reasonably be expected" to cause exceptionally grave damage, serious damage, or damage to national security. NASA, Atomic Energy Commission, Dept. of Defense, and United States Arms Control and Disarmament Agency were among Government departments whose heads and their senior deputies were empowered by order to classify material or information as Top Secret. (*PD*, 3/13/72, 542–50)

- American Society for Testing and Materials—non-profit organization that developed standards for more than 4700 materials, products, systems, and services—urged adoption by U.S. of metric system during Washington, D.C., workshop. In interview Society's President Erle I. Shobert II said: "Basically, the reason for changing . . . to the metric system is to enhance our ability to communicate with others, but that ability . . . is reflected in hard dollars and cents." Failure of U.S. to convert had been large factor in U.S. trade deficit, he said. (Kadis, W *Star*, 3/9/72, C8)

March 9: President Nixon ordered immediate implementation of new Federal Aviation Administration regulations to prevent carrying of weapons or explosives aboard civil aircraft and unauthorized access to aircraft, tighten baggage check-in procedure, and improve cargo and baggage loading security. Dept. of Transportation would expedite rule-making for new regulations governing airport operators to complement measures required of airlines. Action had been prompted by discovery of bombs aboard aircraft at New York, Las Vegas, and Seattle during 48 hrs. (*PD*, 3/13/72, 553)

- Use of remote TV to monitor weather conditions in mountain pass used by general-aviation pilots operating under visual flight rules (VFR) would be evaluated by Federal Aviation Administration, FAA Director John H. Shaffer announced. Program would be conducted for FAA by National Weather Service of National Oceanic and Atmospheric Administration under $108 000 interagency agreement. (FAA Release 72–38)

- Financial problems affecting development of new civil aircraft in U.S. were described by William M. Magruder, adviser to President Nixon on technology, in speech before National Security Industrial Assn. in Washington, D.C.: "Nature of the development base applicability of military programs to civil transports has been altered drastically since the propeller transports and early jets. Those civil transports enjoyed literally 'off-the-shelf' engines, tires, brakes, instruments, hydraulic, electrical and air conditioning systems and nearly exact aerodynamics and controls. The cost of a civil transport development project under those conditions was about one time the manufacturers' net worth, or about $300 million. During this period, it was not unusual for the military to be developing as many as one or two dozen related bombers and transports of equal size and technology. Today, a single civil air transport development program, such as the B–747, DC–10 or L–1011, will cost about five times the net worth of the manufacturers, or nearly $1.5 billion. At the same time, the Defense Department has only the C–5A, F–14, F–15 and B–1A in development, with little direct . . . fallout into the civil programs.

March 9

"Simultaneously with the skyrocketing of civil transport development costs, foreign nations have noticed that it takes a family of transports to dominate the marketplace. Where the U.S. offered twenty or thirty different models and captured 85% of the civil transport market, we now are isolated to the trijets and the B-747, while Europe offers Concordes, A-300 airbuses, Mercures for short range; VFW-614s and Falcon 20Ts for commuter traffic; and have started their STOL projects for urban traffic. To do this, they have used multinational treasuries for subsidized financing." (Text)

- Arrival of seven Soviet scientists at National Accelerator Laboratory near Batavia, Ill., was announced by Atomic Energy Commission. Team would join U.S. scientists in probing fundamental nature of matter. (AEC Release P-66)
- Washington *Evening Star* editorial commented on plaque launched on *Pioneer 10* Jupiter probe March 2: "Like a note in a bottle cast on an outgoing tide, or a relic in a cornerstone, or a time capsule, this message to whom it may concern affirms mankind's insistence on its precious and distinctive identity. It is behind the concept of pyramids, eternal vigil flames, and ancient cathedrals. More than an instinct to survive, it includes a demand that its existence be known and acknowledged." *Pioneer 10* launch "serves to transmit an ancient challenge: 'Who's there?' Any authenticated reply would be the biggest news in quite a spell." (W *Star*, 3/10/72)

March 10: Apollo 16 Flight Readiness Review was conducted at Kennedy Space Center to gather status reports on preparation for lunar landing mission scheduled for launch April 16. (KSC Release 55-72)

- NASA TF-8A aircraft, equipped with supercritical wing and piloted by Thomas C. McMurtry, resumed flight program at Flight Research Center after three-week halt for improvement of fuselage area distribution and installation of pressure survey rakes. Purpose of flight—28th in series—was to check out new instrumentation and angle-of-attack and airspeed calibrations. Aircraft reached 10 700-m (35 000-ft) altitude, instrumentation was activated, and calibrations were made at mach 0.90, 0.95, 0.97, 0.99, and 1.2. (NASA proj off)
- U.S.S.R.'s *Luna 19* lunar probe completed 1810 orbits of moon and was operating satisfactorily. Spacecraft had conducted 516 radio communications sessions since launch Sept. 28, 1971, and had collected data on moon's gravitational field. (Tass, FBIS-Sov, 3/13/72, L1)
- Lewis Research Center's 60-megawatt reactor, used to study radiation effects on materials and components for future space nuclear systems, was also being used to solve local environmental problems. Reactor was analyzing trace elements in air, coal and other fuel samples, and corn damaged by birds. LeRC scientists at Plum Brook Station, Ohio, were making analyses for LeRC Environmental Research Office to identify and catalog trace elements in air. Other government agencies and universities had requested irradiation of samples for environmental studies. (NASA Release 72-49)
- NASA announced issuance by Ames Research Center of request for proposals for studies on four proposed Pioneer-class missions to Venus in late 1970s. Proposals would cover design studies for entry probes and orbiting spacecraft. Two contractors would be selected to perform parallel, $500 000 design studies. (NASA Release 72-53)

- Marshall Space Flight Center announced award of $26 923 contract to Huntsville (Ala.) Hospital to equip room with NASA-developed devices to test applications in aiding quadriplegics. When room was completed, immobile patient would be able to open and close doors and windows, control room temperature, change channels and control volume on TV set, dial telephone, adjust bed position, signal nurse at remote station, and turn pages in a book. Devices provided by MSFC included sight switch operated by looking into sensor on eyeglass frame, panel switch operated by touching panel with head, foot switch, breath switch operated by blowing against small paddle, and pneumatic switch operated by rolling head against pillow to depress airbag behind ear. Contract expenses would be shared, with MSFC contributing $15 450 and hospital and Univ. of Alabama at Huntsville contributing balance. (MSFC Release 72-25)
- Progress in nuclear explosion seismology was reported in *Science:* "Some experts now believe that explosions in hard rock with yields as small as 2 kilotons could be identified on a global scale with no more than a dozen high-quality seismograph stations." In 1958 "there seemed to be little prospect . . . of identifying shots with yields smaller than 50 kilotons" at distances greater than 2500 kilometers (1553 miles) from the center of the blast. (Wick, *Science*, 3/10/72, 1095-7)
- Membership in National Academy of Sciences ranked "second only to the Nobel Prize as recognition of achievement in American science," Univ. of Oklahoma political scientist Dr. Don E. Kash and team said in *Science* article. Striking aspect of composite picture of NAS membership was "extent to which members are educated in, elected from, and employed by a relatively small number of universities." Academicians from top 10 universities made up 54.8% of NAS membership, averaged 48 yrs at election time, and were nearly 3 yrs younger on election to NAS than those with degrees from other universities. (Kash et al., *Science*, 3/10/72, 1076-83)
- *Catholic Review* article by Frank Moriss criticized lack of religious reference on *Pioneer 10* plaque: Plaque summarized "our place in the universe, the ingenuity of our escape from our own planet's gravity, the physical nature of man and woman." Unrevealed was "that people sending that probe for the most part acknowledge a Divinity. . . . It includes everything but God, which is to say it includes only that which is irrelevant in the long run." (*Catholic Review*, 3/10/72)

March 11-24: NASA launched European Space Research Organization's *TD-1A* (Thor-Delta 1A) astronomical observatory satellite from Western Test Range at 5:55 pm PST (8:55 pm EST) by long-tank, thrust-augmented Thor-Delta booster. *TD-1A* entered sun-synchronous orbit with 539-km (334.9-mi) apogee, 531-km (329.9-mi) perigee, 95.3-min period, and 97.6° inclination. NASA objective was to place spacecraft in earth orbit that would permit successful achievement of scientific objectives and to provide tracking and telemetry support. By March 24 all experiments had been turned on and mission was adjudged successful by NASA.

Box-shaped, 472-kg (1038-lb) spacecraft carried seven experiments from six European universities and research organizations to study high-energy emissions from stellar and galactic sources and sun that were not visible to ground-based observatories. *TD–1A* was first European-built spacecraft with three-axis stabilization system, which would permit directional viewing by experiments and permit solar panels to point at sun for electrical power to operate spacecraft. *TD–1A* was sixth scientific spacecraft launched by NASA for ESRO in program that began with orbiting of *Iris 1* May 16, 1968. Most recent launch in program had been *Heos 2* launched Jan. 31. ESRO would reimburse NASA for Thor-Delta booster, launch, and launch-support services. Spacecraft would be controlled from European Space Operations Center (ESOC) and tracked by European Space Tracking (ESTRAK) stations with NASA support. (NASA proj off; NASA Release 72–60; ESRO PIO)

- *Apollo 11* Astronaut Neil A. Armstrong, first man to walk on moon, was made honorary Freeman in Langholm, Scotland, ancestral home of Armstrong family. Armstrong, now professor of aerospace engineering at Univ. of Cincinnati, had been invited to Langholm by town council after July 20, 1969, moonwalk. (AP, *W Star*, 3/13/72, A3)

March 12: Philippines successfully launched Bong Bong II, its first liquid-propellant rocket. Rocket was successfully retrieved from South China Sea. (*SBD*, 3/22/72, 121)

- U.S.S.R. hoped to land robot life-detection laboratory on Mars within few years, according to *Leninskoye Znamya* (Lenin's Banner) newspaper article by Soviet space experts. Automatic Microbiological Laboratory would scoop up sample of Martian soil and examine it for microorganisms such as bacteria, yeasts, and fungi that might be able to survive in thin Martian atmosphere. Laboratory would also draw Martian air through filter to trap and analyze microorganisms and would detect emission of heat and carbon dioxide. (Wilford, *NYT*, 3/27/72, 7; UPI, *W News*, 3/13/72, 5)

- *Bridgeport* (Conn.) *Post* commented on March 1 launch of Air Force reconnaissance satellite. Satellite "is intended to monitor rocket tests of the Soviet Union and Communist China. And it is to remind us all not to get so carried away in the new approaches to the two Communist giants that we will let down our guard against them. . . . informed guesses are that the new spy satellite's major job is to monitor submarine launching missiles. In its high position, the sensors would be able to observe one third of the globe. Another outpost would complete surveillance of the whole globe." Despite President Nixon's visit to China and prospective visit to Soviet Union "spy satellites are needed precautions." (*Bridgeport Post*, 3/12/72, 23)

March 13: Two press briefings were held at NASA Hq., Apollo 16 Mission Director's press briefing and press conference on medical aspects of *Apollo 15* and 16.

In first, press was told hypergolic loading of propellants had begun and countdown demonstration test with fueled spacecraft would begin March 29. MSC was continuing simulations, and prime recovery ship would leave San Diego March 23 and Pearl Harbor April 8 to take position. Launch preparations were proceeding on schedule toward liftoff at 12:45 pm EST April 16.

Apollo Mission Director Chester M. Lee said one purpose of microbial ecology evaluation device (MEED; experiment to carry fungi and bacteria) on Apollo 16 was to prepare for future Skylab missions. In Skylab "we are going to have some EVAs and some of these bacteria, fungi, are on the suit before they go out. So the idea is to find out is there anything when we bring those suits back in and stow them for long periods of time like we are going to have on Skylab, do we have anything that we should be considering or concerned about as far as mutants and so forth."

Lunar roving vehicle (LRV) Rover had been improved with new seatbelts that could be hung up out of way when not in use and with extra circuit breaker to permit deployment of TV after third extravehicular activity period. LRV would not cover as much distance on Apollo 16 as on *Apollo 15*. "We are planning 25.2 kilometers [15.7 miles] . . . because that gets us to our primary objectives," not including 10% wander factor for missing craters. (Transcript)

In second briefing, Dr. Charles A. Berry, NASA Director of Life Sciences, explained medical problems and procedures on *Apollo 15* and changes for Apollo 16. Major concerns on *Apollo 15* were cardiac arrhythmia—irregular heart rhythm—and fluid-electrolyte imbalance, with severe loss of potassium. Cardiac arrhythmia occurred occasionally in normal persons and frequently in college students, apparently as a result of fatigue from all-night studying, adrenalin increase from apprehension about grades, drug effect from drinking coffee to stay awake, and heavy workload. Fatigue, adrenalin increase, drug effect, and heavy workload also acted on astronauts. Fluid-electrolyte imbalance was dependent upon intake in food and drink and output in breath, sweat, and body wastes. In space, balance was affected by weightlessness, acceleration, 100%-oxygen atmosphere, abnormal work-rest cycle, and sustained work loads. Countermeasures that could be taken were better control of intake, supplementing electrolytes with intravenous fluids, coordinating exercise, providing lower body negative pressure, and optimizing work-rest cycles.

Medical objectives for Apollo 16 were essentially same as for *Apollo 15:* to ensure crew safety medically, to improve probability of mission success by providing necessary medical information, and to detect and understand changes incident to space flight and learn more about effects of space flight on man. To improve medical wellbeing of Apollo 16 astronauts, 72-hr preflight controlled-diet period would determine amount of potassium needed by each crew member. Since potassium could not be preloaded, small amounts of extra potassium would be given in daily diet in flight to ensure adequacy. In addition to regular preflight cardiovascular examination, x-rays of astronauts' hearts would be taken before and after mission to determine whether heart size decreased during flight. Skylab equipment would be used in flight for electroencephalogram, electrocardiogram, and electro-oculogram. Other inflight medical changes were daily sample of improved electrocardiogram data, option to alter bioharness wearing schedule, expansion of crew status report, high-potassium diets with potassium-enriched beverages and snack supplements, measurement of food and fluid intake and urine and fecal output, return to earth of urine and fecal samples for analysis, and provision of anti-arrhythmic drugs lidocaine, procainamide, and atropine.

Dr. Berry also announced that Astronaut Donald K. Slayton, who had been removed from *Aurora 7* mission (May 24, 1962) and NASA flight status 10 yrs ago because of possible heart malfunction, had been returned to NASA and Federal Aviation Administration flight status. Evidence of atrial fibrillation found in 1959 and treated until 1969 had disappeared and tests completed recently had shown "perfectly normal cardiac function and cardiac anatomy." First space flight for which Slayton could be eligible would be proposed joint mission with U.S.S.R. (Transcript)

- U.S. Dept. of Commerce issued permanent export license to RCA Global Communications, Inc., authorizing sale of earth station installed at Shanghai Airport in People's Republic of China to cover President Nixon's February visit. Transaction was first direct export sale from U.S. to PRC since trade relations were liberalized. (RCA Release 3/15/72)
- Sen. Hubert H. Humphrey (D-Minn.), campaigning for Presidential nomination, told street crowd in Titusville, Fla., that U.S. should be first in space and science. "We can't afford to have the Soviet Union first in space." Later, rival Democratic candidate Sen. Henry M. Jackson (D-Wash.) told Orlando, Fla., crowd that Humphrey had never been recorded on space shuttle issue "until he entered the Florida primary." (C Trib, 3/14/72)

March 13–14: American Astronautical Society (AAS) sponsored 1972 Goddard Symposium, "Transfer of Space Technology to Community and Industrial Activities," in Washington, D.C. Dr. Wernher von Braun, NASA Deputy Associate Administrator for Planning, in keynote address stressed importance of high technology development to U.S. standard of living. He compared favorable balance of trade in high-technology aerospace products with unfavorable balance in low-technology products. Annual U.S. exports in aerospace products were valued at $3.6 billion over imports; computers, $1.1 billion; and machinery, $1 billion. Motor vehicle imports were $3.3 billion over exports, clothing and textiles, $1.9 billion, and iron and steel, $1.9 billion. (NASA Release 72–56)

National Space Club awards were presented at Goddard Memorial Dinner at Space Club on evening of March 14.

Dr. Robert H. Goddard Memorial Trophy, NSC's highest award, was presented to Sen. Clinton P. Anderson (D-N. Mex.), Chairman of Senate Committee on Aeronautical and Space Sciences, by former NASA Administrator James E. Webb. Citation read: "His deep knowledge of the space program and his effective leadership in the Senate and the Nation have been invaluable in fostering Congressional understanding of and support for this program. The benefits to mankind from the U.S. space program flow in large measure from the dedicated efforts of this great American."

Astronautics Engineer Award was presented to Jet Propulsion Laboratory engineer Dan Schneiderman. Citation read: "For his management of the Mariner Venus '67 and Mariner Mars '71 missions. His perception, decisiveness, and inspiration have made it possible for man to place a measuring device in orbit around another planet for the first time in the history of the world." Hugh L. Dryden Memorial Fellowship was presented to C. Thomas Snyder of Ames Research Center "so that he may continue to seek and maintain pre-eminence in outer space for the United States."

Robert H. Goddard Historical Essay Award for 1971 competition was given Albert B. Christman of Naval Weapons Center for "Robert H. Goddard and the Military." National Space Club Press Award was presented to Howard Benedict of Associated Press for "giving the public authoritative yet highly readable chronicles of the unfolding Space Age."

Nelson P. Jackson Aerospace Award was presented to Boeing Co. by *Apollo 15* Astronaut David R. Scott. Citation read: "In recognition of great mobility provided by the ingenious Lunar Roving Vehicle. . . . Exploration of the moon by the Apollo 15 crew, its cameras, and its equipment, covering 28 kilometers [17.3 miles] over the rough terrain, was made possible by this masterful 500-pound [227-kilogram] stowable vehicle." (Program)

March 14: President Nixon transmitted to Congress *Aeronautics and Space Report of the President: 1971 Activities.* In transmission message President said: "Aided by the improvements we have made in mobility, our explorers on the moon last summer produced new, exciting and useful evidence on the structure and origin of the moon. Several phenomena which they uncovered are now under study. Our unmanned nearby observation of Mars is similarly valuable and significant for the advancement of science. During 1971, we gave added emphasis to aeronautics activities which contribute substantially to improved travel conditions, safety and security, and we gained increasing recognition that space and aeronautical research serves in many ways to keep us in the forefront of man's technological achievements. There can be little doubt that the investments we are now making in exploration of the unknown are but a prelude to the accomplishments of mankind in future generations."

Report said U.S. was "proceeding into this decade with clearly defined goals and objectives for its civilian space program." U.S. continued to invest "substantial resources" in space activities because of "national thrust to explore the unknown, the desire to acquire new knowledge, and the realization that space activity has practical and widely beneficial applications. Space is increasingly coming to be seen as an arena of human activity, rather than national activity, and therefore a particularly congenial focus for cooperative undertakings that bring the people of the world together." (*CR*, 3/14/72, S3885; Text)

- Dr. James C. Fletcher, NASA Administrator, and other officials testified before Senate Committee on Aeronautical and Space Sciences during FY 1972 NASA authorization hearings. Dr. Fletcher said: "I regard the fiscal year 1973 budget . . . as marking the end of a difficult transition period for NASA. For the past 2 or 3 years, NASA, the executive branch, and the Congress have been grappling with questions on the course the Nation's space program should take in the 1970's and beyond. The main goals in science, exploration, and applications, and the advanced system and technology required to achieve these goals, have been identified . . . and generally approved. . . . But there has remained a basic uncertainty . . . the question of cost. Can the Nation afford the space program of the 1970's? Will the approval of the Space Shuttle and the other programs . . . *commit* the Nation to a large increase in future spending levels?

"This year we can give clear answers to these questions. We can assure the committee, the Congress, and the public that the program we are proposing does not commit the country to higher budget levels in future years, at least measured in current dollars . . . . The revised plan is, I believe, more realistic and will give the Nation a good, viable, and balanced program in aeronautics and space at a cost it can afford."

Dr. George M. Low, Deputy Administrator, reported substantial progress in two basic cooperative activities with U.S.S.R.—joint work on compatible docking systems and exchanges in scientific areas, including lunar sample exchanges. In response to question, Dr. Low said he thought it likely U.S.S.R. would continue lunar program after U.S. completed Apollo program. "They may even try to land men on the moon in the latter part of this decade. That will be a catching-up process."

To further question, Dr. Fletcher said that in late 1970s U.S.S.R. would be "doing things in the manned space program that are beyond what we would be able to do." From end of Skylab in 1973 until 1978, "it is likely that they will be carrying on some important activity and we will have no opportunity to respond until 1978" [with space shuttle].

Richard C. McCurdy, Associate Administrator for Organization and Management, submitted statement on NASA's work force: "When several reductions are made serially, as they have been in NASA over the past several years, the effects compound, and the healing processes, such as the resumption of hiring of younger people, the filling of vacancies via merit promotion . . . do not have an opportunity to function." NASA was "feeling these disadvantageous effects severely." Most worrying was "dwindling of the supply of new blood." It was obvious "trend cannot continue without the health and vigor of the organization being undermined." NASA needed "to embark upon a period of stability" in work force to "return to normal personnel development practices, heal the damage of the past several years, and be prepared to meet the challenges of the 70's." With the "prospect of a period of relative stability in the general level of spending" and "clearer view of the NASA forward mission," such a policy was indicated.

Statement submitted by Daniel J. Harnett, Assistant Administrator for Industry Affairs and Technology Utilization, described NASA Regional Dissemination Centers' assistance to users "ranging from small businesses to the largest companies in our Nation's economy—across a broad spectrum of business sectors including manufacturing services, transportation and communications. The Centers also serve university faculty and students; varied national, state, and local organizations; and governmental entities at all levels." NASA RDCs had served more than 2000 organizations in 1971, of which over 48% of industrial users were small businesses.

M/G Robert H. Curtin (USAF, Ret.), NASA Director of Facilities, submitted statement on facilities. In $77.3-million FY 1973 request for construction of facilities, $27.9 million was for modifications and additions to facilities for space shuttle, $11.2 million for facilities supporting scientific investigations in space, $12.9 million for facilities supporting aeronautical research and technology, and $25.3 million for smaller supporting activities and planning and design. (Transcript)

- Dr. Thomas O. Paine, Vice President and Group Executive of General Electric Co. and former NASA Administrator, delivered 1972 Tizard Memorial Lecture at Westminster School in London. Discussing man's future in space, Dr. Paine evaluated role of applications satellites: Their "great economic and social value" was apparent "although their full potential is not yet appreciated. Men on every continent regularly use these services today; shares can be bought in the enterprises that provide them. We must not, however, overlook the equal value of the new information about the earth's origin, history, environment, and destiny that our more basic lunar and planetary scientific programs are providing. Already studies of radiative energy transport in the atmosphere of Mars have contributed to terrestrial meteorology; lunar geological surveys have provided new clues to the early formation and structure of earth. The line between science and application in this rapidly advancing area is continually shifting and new scientific understanding of such questions as the origin of the Ice Ages might be of incalculable human importance." (Text)

- NASA released findings of U.S. authorities who met at Langley Research Center Aug. 16–20, 1971, to discuss detection of air and water pollution by remote sensing, especially from aircraft or satellites. Group had reported that remote sensing produced essential information unobtainable by other means. Measurements could be obtained by aircraft, high-flying balloons, and satellites and from ground, to provide picture of global or regional pollution problems. Group had recommended that new techniques and instrumentation be used as fully as possible with special emphasis on remote sensing potential of space technology. (NASA Release 72-58)

- California Lt. Gov. Ed Reinecke in press release announced progress in application of NASA-developed technology to management of California's environment. Reinecke, chief aide to Gov. Ronald Reagan in developing use of new technology for solution of state problems, said Gov. Reagan had pronounced NASA technology vital to resource management. Series of symposia held in cooperative effort by State of California, Ames Research Center, and Univ. of California at Berkeley was examining environmental problems, including smog. (ARC Release 72-16)

- Development by Ames Research Center scientists of simple aperture method that made microscopic details of atomic dimensions clearly visible under transmission electron microscopy was announced by NASA. Method permitted perception of distances of one angstrom instead of resolutions of three angstroms normally obtained with standard electron microscopes. Cancer researchers had obtained best results using NASA technique. (NASA Release 72-59)

- U.S. patent No. 3649921 was issued to Langley Research Center technician David F. Thomas for backpack harness that future astronauts could fasten with one hand, leaving other hand free for stabilization purposes in zero-g environment. Harness was pressure plate device with reinforced straps that encircled user's body automatically when his back was pressed against plate. Commercial applications might include use as safety belt in automobiles and aircraft and use by handicapped persons. (Biggins, Newport News, Va, *Times-Herald,* 4/6/72; Pat Off PIO)

March 15: Cosmos 478 was launched by U.S.S.R. from Plesetsk. Orbital parameters: 282-km (175.2-mi) apogee, 176-km (109.4-mi) perigee, 89-min period, and 65° inclination. Satellite reentered March 28. (GSFC *SSR*, 3/31/72; *SBD*, 3/17/72, 98)

- Decision that space shuttle booster stage would be powered by recoverable, reusable, solid-fueled rocket motors in parallel burn configuration, rather than by pressure-fed liquid-fueled rocket motor, was announced by Dr. James C. Fletcher, NASA Administrator, to Senate Committee on Aeronautical and Space Sciences during FY 1973 authorization hearings. Announcement was also released to press during morning and Dr. Fletcher held press conference in afternoon. Dr. Fletcher said both solid- and liquid-fueled boosters had been found technically feasible but "choice was made in favor of the solid parallel burn because of the lower development cost and the lower technical risks. Cost estimate for shuttle development was now $5.15 billion instead of earlier $5.5-billion estimate. Request for proposals would be issued about March 17. NASA expected to select shuttle launch and landing site shortly. At press conference he said: "I think we have made the right decision at the right time. And I think it is the right price." Dr. Fletcher said it was likely NASA would want to launch and land at same coastal site and possibly from sites on both coasts. Inland sites were unlikely because of requirement for booster recoverability.

    Dale D. Myers, Associate Administrator for Manned Space Flight, said at conference there would be eight-week response time to request for proposals to become prime contractor for shuttle development. NASA would evaluate proposals in another eight weeks and would be ready for choice of contractor in early July. Requirements had been developed jointly with Air Force and "we have a shuttle that meets the user's requirements, both in NASA and in the Air Force."

    Dr. George M. Low, NASA Deputy Administrator, estimated breakdown of 580 shuttle flights projected over 12-yr period for newsmen. Out of 100% of missions, 26% would be manned or man-tended, 74% unmanned. Of that 74%, 27% would be for applications (including NASA and other agencies or commercial applications), 25% for science, and 22% for Dept. of Defense missions. (Hearings transcript; Press conference transcript)

- NASA Associate Administrator Dale D. Myers—following detailed presentation on space shuttle decision announced by Administrator James C. Fletcher [see above]—moved to testimony on Apollo program, during Senate Committee on Aeronautical and Space Sciences hearings on NASA FY 1973 authorization. In new Apollo 16 lunar surface experiments, far-ultraviolet camera/spectograph would photograph celestial objects or areas emitting or absorbing energy from atomic hydrogen. Data would provide information on "composition, density, distribution, and motion of interplanetary, interstellar, and intergalactic gas clouds which will . . . enhance our understanding of the structure and evolution of the galaxy." Data, recorded on film and returned from lunar surface by astronauts, would be "first attempt to evaluate the Moon as a platform from which to make future celestial observations." Second new experiment, cosmic ray detector, would acquire data on origin and source mechanism of high-velocity cosmic rays and solar particles beyond effects of earth's atmosphere and magnetic field. (Transcript)

March 15: NASA *Associate Administrator for Manned Space Flight Dale D. Myers, Administrator James C. Fletcher, and Deputy Administrator George M. Low (left to right) explained space shuttle plans at a press conference following the announcement that the shuttle booster stage would be powered by recoverable, reusable, solid-fueled rockets. In the artist's drawing, the rockets are dropping away for recovery.*

NASA TF–8A aircraft, equipped with supercritical wing and piloted by Thomas C. McMurtry, completed 29th flight from Flight Research Center. Purpose of flight was to check out new instrumentation and angle-of-attack and airspeed calibrations. Fixed rakes and rotating survey probes were checked out and angle-of-attack calibrations were obtained at 13 700-m (45 000-ft) altitude from mach 0.9 to 1.1. (NASA proj off)

- Federal Communications Commission made public recommendations by its staff for domestic communications satellite system. Recommendations were for "policy of limited open entry permitting all qualified applicants to own and operate domestic satellite systems." Applicants for related uses of satellite systems would be expected to join forces. FCC spokesman said proposals for satellite use might ultimately be grouped into as many as four independent systems, each requiring its own satellite. Cost of complete system was estimated at $60 million to $100 million. Satellites would cost between $11 million and $26 million. Cost-sharing plan among applicants was recommended. Hearings on FCC staff report were scheduled to start May 1. (Schmeck, *NYT*, 3/16/72, 1)

- Communications Satellite Corp. and Cunard Line, Ltd., jointly announced beginning of two-month test to demonstrate high-quality, reliable communications between *Queen Elizabeth 2* at sea and ComSatCorp Laboratories in Clarksburg, Md., via *Intelsat-IV F–2* over Atlantic. Test was first transmission of voice and data communications via satellite with commercial passenger liner at sea. Ship carried 2.4-m (8-ft) parabolic antenna, antenna control and stabilization system, and digital communications terminal for communications in C-band frequencies (400–600 mhz). When satellite communications became operationally available for use with ships at sea, service would be provided in L-band frequencies (1500–1600 mhz). (ComSatCorp Release 72–16; ComSatCorp PIO)

- National Science Foundation announced selection of 121-hectare (300-acre) desert site west of Socorro, N. Mex., as site for $76-million very-large-array (VLA radiotelescope—most sensitive and accurate such instrument in world. VLA—which would consist of 27 dish-shaped radiotelescopes, each 25 m (82 ft) in diameter and movable along three 20.9-km (13-mi) Y-shaped tracks—would pick up naturally produced radio signals from objects within and outside Milky Way galaxy. NSF had included $3 million in FY 1973 budget request for initial phases of project. Subject to successful negotiations for land use and availability of funds, work on facility would begin during 1972. (NSF Release 72–127)

- House Committee on Government Operations in 10th report, *Aircraft Collision Avoidance Systems*, criticized Federal Aviation Administration's failure to develop "vigorous and meaningful program for preventing mid-air collisions." Committee recommended FAA research and development program to provide "unified, coordinated approach encompassing all elements of air traffic control, including ground-based air traffic control, airborne collision avoidance, the structure of the Nation's airspace." Committee gave FAA Administrator until June 30 to report development by FAA, NASA, and aviation community of program to provide U.S. with "acceptable level of collision avoidance capability at the earliest possible date." (Text)

March 15–21: NASA launched 29 sounding rockets from Wallops Station in cooperative international program to test weather data systems for France, Japan, and U.S. Project was sponsored by United Nations World Meteorological Organization's Commission for Instruments and Methods of Observation. Objective was to compare data for better understanding of upper-air temperature and wind measurements and to encourage international cooperation in exchanging data obtained from meteorological rockets. Rockets—which carried radiosonde payloads ejected at apogee and descending on parachutes—included 12 U.S. Loki-Datasondes carrying NASA payloads, 10 Japanese MT–135s carrying Japan Meteorological Agency payloads, and 7 U.S. Super Arcas rockets carrying French National Meteorological Agency payloads. Twenty-four flights were successful, three were failures, and two were partially successful. Second series of launches was planned for late 1973 with U.S., U.S.S.R., United Kingdom, Brazil, France, and India. (WS Release 72–3; WS PAO)

March 16: President Nixon praised ability of Americans to "harness the discoveries of science in the service of man" in message to Congress on science and technology: "They have found a way of preventing polio, placed men on the moon, and sent television pictures across the oceans." But, "when other countries are rapidly moving upward on the scientific and technological ladder, challenging us both in intellectual and in economic terms," U.S. position in aircraft, steel, automobile, and shipbuilding industries "is not as strong as it once was. A better performance is essential to both the health of our domestic economy and our leadership position abroad." U.S. should "combine the genius of invention with the skills of entrepreneurship, management, marketing and finance;" ensure favorable environment for technological innovation; realize "that mere development of a new idea does not necessarily mean that it can or should be put into immediate use;" encourage young people to become "dedicated scientists and engineers;" continue priority assigned to basic research and to "exploratory experiments which provide the new ideas on which our edifice of technological accomplishment rests;" and "appreciate that the progress we seek requires a new partnership in science and technology—one which brings together the Federal Government, private enterprise, State and local governments, and our universities and research centers in a coordinated, cooperative effort to serve the national interest."

President had increased FY 1973 Dept. of Defense budget to "ensure our strategic deterrent capability, continue the modernization of our Armed Forces, and strengthen the overall technological base that underlies future military systems."

To "apply our scientific resources in meeting civilian needs," President had developed "overall strategic approach in the allocation of Federal scientific and technological resources." He urged Congress to support new strategy that included reorientation of space program "to focus on domestic needs—such as communications, weather forecasting and natural resource exploration. One important way . . . is by designing and developing a reusable space shuttle, a step which would allow us to seize new opportunities in space with higher reliability at lower costs."

President also advocated "providing new sources of energy without pollution;" developing "fast, safe, pollution-free transportation;" reducing losses from natural disasters; improving programs to curb drug abuse; increasing biomedical research; providing better health care; and drawing more directly on "capabilities of our high technology agencies"—NASA, Atomic Energy Commission, and National Bureau of Standards—in applying R&D to domestic problems.

To improve "climate for innovation" President had proposed National Science Foundation support of "assessments and studies focused . . . on barriers to technological innovation and on the consequences of adopting alternative Federal policies which would reduce or eliminate these barriers." Additionally he would "submit legislation to encourage the development of the small, high technology firms" and "provide additional means for the Small Business Investment Companies (SBICs) to improve the availability of venture capital to such firms."

Among "cooperative international efforts" in science and technology President cited exchange of lunar samples with U.S.S.R. and exploration of "prospects for closer cooperation in satellite meteorology, in remote sensing of the environment, and in space medicine. Beyond this, joint working groups have verified the technical feasibility of a docking mission between a SALYUT Station and an Apollo spacecraft." (PD, 3/20/72, 581–90)

- NASA officials testified before Senate Committee on Aeronautical and Space Sciences during FY 1973 NASA authorization hearings.

  Roy P. Jackson, NASA Associate Administrator for Aeronautics and Space Technology, presented program request with increased emphasis on technology for "solving urgent domestic civil air transportation problems of noise and congestion," continuing concern for "rapidly growing foreign competition" to U.S. leadership in world aircraft markets, greater attention to "countering the erosion of this Nation's preeminence in military aeronautics," phase-over in space shuttle technology from exploratory activities to development support, continuing focus on space technology for advanced earth-orbit systems and outer planet exploration, change in emphasis from development of large nuclear rocket to continuing the technology base for nuclear propulsion, with focus on small high-energy system. Budget also included continuing reduction of research staff and emphasis on upgrading existing facilities, with no new facilities requested.

  In aeronautics, advanced system studies in FY 1971 and FY 1972 had examined application to new subsonic and sonic aircraft designs of projected advances in aerodynamics, structures, propulsion, controls, and avionics; identified technology advances offering greatest economic and competitive position benefits; and defined actions to "bring the advanced technologies to a state of readiness for industry utilization by the late 1970s." Follow-on studies would be conducted in FY 1973 on "additional noise alleviation possibilities and on terminal compatibility." Application of supercritical aerodynamics and configuration treatment, engine noise and emission suppression techniques, composite structures, and use of active control systems "were shown to result in a quieter, safer, more economical subsonic/sonic transport having greater productivity than the current wide-body jet transports." In FY 1973 "aerodynamics effort will concentrate on refinements of supercritical wing and related configuration treatment for transport concepts capable of increased flight efficiency."

David S. Gabriel, Manager of AEC–NASA Space Nuclear Systems Office, testified on office's on-going work after termination of nuclear engine for rocket vehicle application (NERVA) program June 30, 1972: "To reach the outer planets in reasonably short trip times during the 1980's requires more advanced propulsion systems than are currently available. Among the candidate propulsion systems for these missions is a small, approximately 15,000-pound-thrust [66 720-newton-thrust] nuclear rocket engine, more suited to these unmanned outer planet missions than the large NERVA engine. Other candidates include high-performance chemical rockets and solar-and-nuclear-electric propulsion.

"For the balance of this fiscal year and during the next, we will compare . . . candidate propulsion systems for a variety of future potential missions of the 1980's and 1990's including the outer planet exploration missions." (Transcript)

- Dr. John E. Naugle, NASA Associate Administrator for Space Sciences, reviewed space science program before Senate Committee on Aeronautical and Space Sciences during FY 1973 authorization hearings: Major accomplishment in astronomy in 1971 had been mapping of universe at x-ray wavelengths with first x-ray astronomy satellite, *Explorer 42* (*Uhuru*, launched Dec. 12, 1970). Two dozen scientific articles had been published on observations by satellite; 116 objects emitting x-rays had been studied and positions accurately determined. Analysis of rocket data on one object indicated diameter was less than 20 km (13 mi), yet it emitted much larger amount of x-radiation than did sun. "The discovery of such X-ray sources is of fundamental importance because matter in ordinary states . . . cannot account for such intense radiation."

In 1971 NASA had launched eight missions in physics and astronomy program. Most important was Orbiting Solar Observatory *Oso 7* (launched Sept. 29, 1971), "which carried into orbit a set of solar telescopes more powerful than any launched before. One instrument—a coronagraph—revealed visible evidence of three luminous jets of intensely hot solar plasma being hurled out into space at a speed of more than two million miles per hour [some 3 million km per hr]. These luminous jets were 20 times the size of the earth, and represented the energy of 100 million atomic bombs. A peak temperature of 30 million degrees must have been attained in this event. Again . . . matter must have been produced on a scale man cannot produce on earth."

International cooperation continued major role in space science program; five of eight 1971 missions had had significant international participation.

During year experiments and prime contractor, TRW Inc., had been selected for "newest and most significant program in astronomy"—High Energy Astronomy Observatory, designed to "unlock some of the most profound and puzzling secrets of modern physics." Pulsars and quasars had given evidence of new concept of energy generation that produced power at levels unexplainable by modern nuclear physics. (Transcript)

- Tenth anniversary of U.S.S.R.'s first Cosmos satellite, *Cosmos 1*, launched March 16, 1962.

  Cosmos satellites' wide variety of functions were described in *Trud* by Soviet engineer T. Borisov. Primary objectives of Cosmos program —which had included almost 500 spacecraft—were "profound and comprehensive scientific investigations of circumterrestrial space" and earth, improvement of space vehicles, and application of experiments "connected with the extensive use of space for national economic and cultural purposes. Very appreciable results have been achieved in . . . the historically short span of 10 years." Cosmos satellites had studied circumterrestrial plasma and its interconnection with earth's magnetic field, previously unknown dynamics of processes in upper atmosphere and space, and relationship between sun and earth. "Cosmos-261 and Cosmos-348 explored the plasma ocean. Cosmos-166, which spent a long time patrolling the sun, made a careful survey of the solar disc. Cosmos-92 and Cosmos-149 could be described as modern geophysical laboratories. The astronomers' automatic laboratory Cosmos-215 had eight telescopes on board to observe hot stars in visible and ultraviolet rays. It also had an X-ray telescope and two photometers to record solar radiation dispersed by the earth's upper atmosphere."

  Series had also contributed to "the further improvement of space technology" and led to creation of first orbital station. "It was only after testing all the complex mechanical, electrical, and radio engineering systems and installations, and after two Cosmos satellite automatic dockings that the way was opened to Soyuz . . . and Salyut. The Cosmos satellites have tested many systems, units and assemblies of lunar and interplanetary automatic apparatuses. . . . the thermoregulating system assemblies, the solar batteries, means of communication, and many scientific instruments for the Luna, Venera, Zond, and Mars apparatuses were created on the basis of or having regard to experience gained in the operation of similar installations in Cosmos satellites." (FBIS–Sov, 3/30/72, L1)

- European Launcher Development Organization (ELDO) Council announced decision to begin development of Europa III launch vehicle Oct. 1, subject to approval by European Space Conference meeting in Brussels in October. (*SF*, 7/72, 241)

- Sprint antiballistic missile launched by Dept. of Defense from Kwajalein Atoll and guided by missile site radar on Meck Island successfully intercepted Minuteman warhead over Pacific. Test was 24th in series to check out Safeguard antiballistic missile components. (UPI, *NYT*, 3/19/72, 15)

March 17: Air Force launched unidentified reconnaissance satellite on Titan IIIB-Agena booster from Vandenberg Air Force Base at 9:00 am PST. Orbital parameters: 396-km (246-mi) apogee, 141-km (88-mi) perigee, 89.8-min period, and 110.9° inclination. Satellite reentered April 11. (Pres Rpt 73; *Sov Aero*, 3/27/72, 91; *Av Wk*, 3/27/72, 17)

- NASA issued requests for proposals to industry for development of orbiter vehicle and systems integration for space shuttle orbiter's external propellant tank, solid-fueled rocket booster, orbiter main engine, and orbiter air-breathing engine [see March 15]. (NASA Space Shuttle Fact Sheet, 5/72)

- Apollo 16 astronauts John W. Young, Charles M. Duke, Jr., and Thomas K. Mattingly II announced code names for Apollo 16 lunar module and command module at Manned Spacecraft Center press conference. LM would be called "Orion" for constellation—also known as the Hunter—which would be visible to crew throughout mission. CM would be called "Casper" after cartoon character Casper the Friendly Ghost. In response to press comment that crew members were all from the South and that exploration site terrain had been designated with Southern names, Duke said only other Southern thing planned would be "a new first ... to take along some grits for breakfast." (Transcript; *NYT*, 3/18/72, 21)
- NASA TF–8A aircraft, equipped with supercritical wing and piloted by Thomas C. McMurtry, completed 30th flight from FRC. Purpose of flight was to check out new instrumentation and angle-of-attack and airspeed calibrations. Fixed rakes and rotating survey probes were checked out and angle-of-attack calibrations were obtained at 13 700-m (45 000-ft) altitude from mach 0.9 to 1.1. (NASA proj off)
- *Vanguard 1*, second U.S. satellite, began 15th year in orbit. When launched March 17, 1958, satellite's lifetime was estimated at 5–10 yrs; current estimate was 200 to 2000 yrs. *Vanguard 1* was one of 5865 man-made space objects cataloged by North American Air Defense Command Space Defense Center. Of these, 2659—including 504 payloads and 2155 pieces of debris—were still orbiting earth. (Miles, *LA Times*, 3/13/72; GSFC PAO)
- Cameras on board NASA's *Mariner 9* Mars probe (launched May 30, 1971) were turned off to check out malfunction in onboard computer. (NASA Release 72-67)
- NASA launched Aerobee 150 sounding rocket from Churchill Research Range, Canada, carrying Johns Hopkins Univ. auroral studies experiment. Rocket and instrumentation performed satisfactorily. (SR list)
- NASA Associate Administrator for Organization and Management Richard C. McCurdy testified on decrease in research and program management (R&PM) budget in House Committee on Science and Astronautics hearings on FY 1973 NASA authorization: Requested $700.8 million was decrease of $26.5 million from FY 1972. Ceiling of 26 850 was proposed for Civil Service component of NASA in-house work force, 1500 lower than original 1972 ceiling—bringing total reduction since 1968 peak to 8300 positions, or 25%.

    NASA had planned to achieve 1973 ceiling by normal attrition, but rate of attrition was running 22% lower than required and Center directors had been instructed to use reduction-in-force procedures as necessary to reach ceilings.

    Deputy Associate Administrator for Organization and Management Bernard Moritz testified that under FY 1973 authorization bill all NASA facility projects would be presented for approval and funding under full disclosure concept.

    M/G Robert H. Curtin (USAF, Ret.), Director of Facilities in Office of Administration, entered prepared statement detailing $77.3-million FY 1973 request for construction of facilities and explained $8 million request for facility planning and design within that total. First $3.5 million was for "regular recurrent" requirements and $4.5 million for "special" requirements. Of the $4.5, $3.75 million was for space shuttle facilities, including preliminary engineering reports for larger 1974 and 1975 programs. Second portion was for large aeronautical facility —$750 000 for design of full-scale subsonic wind tunnel that would ultimately cost $150 million to $200 million if approved, with total design cost of $7 million to $10 million. Requested FY 1973 funds would complete design on drive section of tunnel.

Deputy Associate Administrator for Aeronautics and Space Technology George W. Cherry described FY 1973 disbursements for aeronautics: Important part dealt with reduction of jet aircraft noise, with $28.5 million requested for research and technology to continue this work. Roughly one half was for research and technology attacking noise generation and noise propagation; $12.4 million was for programs for near-term relief from jet aircraft noise. Of the $12.4 million, $9 million was requested for engine-modification development program and $3.4 million for development of operational procedures for noise abatement.

Cherry presented audiovisual demonstration of effectiveness of engine modifications and operational procedures. (Transcript)

- *Philadelphia Inquirer* quoted Lockheed spokesman as saying last C–5 Galaxy military transport aircraft to be built under Government contract was on assembly line and was expected to be completed in mid-1973. Lockheed was selling 81 C–5s to Air Force for $4.9 billion. (*P Inq*, 3/17/72, 89)
- Assignment by President Nixon of Gen. Horace M. Wade, Chief of Staff, Supreme Hq. Allied Powers Europe, as Vice Chief of Staff, United States Air Force, was announced by Secretary of Defense Melvin R. Laird. (DOD Release 195–72)
- President Nixon signed S. 3244, bill increasing funds for international aeronautical exposition from $3 million to $5 million. Bill became P.L. 92–251. (*CR*, 3/21/72, D293)

March 18: NASA was being embarrassed by management, planning, and production problems incurred by Fairchild Industries (formerly Fairchild Hiller Corp.) in implementation of disputed Applications Technology Satellite F (ATS–F) contract, *Business Week* reported. Contract had been awarded originally to General Electric Co. in 1970. Protest by Fairchild Hiller that its proposal was superior to GE's had led to full-scale review of award by General Accounting Office. NASA had reawarded contract to Fairchild Hiller. Now, 18 mos after re-award, NASA had noted that cost of ATS–F had increased from $50 million to $65 million "and is still climbing." Satellite's spring 1973 launch date was said to be slipping badly and NASA had moved new manager, John M. Thole, into program "to see if he can get things back into line." (*Bus Wk*, 3/18/72)

March 19: Shuttle Program Office, managed by Roy E. Godfrey, was formed at Marshall Space Flight Center. Office would be staffed by Shuttle Task Team until July 1, when team would be abolished and some members assigned permanently to office. (*Marshall Star*, 5/17/72, 1; MSFC PAO)

- *New York Times* science reporter John Nobel Wilford toured Zvezdny Gorodok (Star City), near Moscow, as first Western correspondent to be invited to center of U.S.S.R. cosmonaut activities. Guide for visit arranged by Soviet Novosti News Agency was Soyuz Cosmonaut Vladimir A. Shatalov, chief of cosmonaut training. Shatalov said U.S.S.R.

did not announce space mission dates "because we don't want to bind the people who are preparing the flight." Cosmonaut flight training areas were closed to most visitors "because we just don't want journalists to interfere with the working atmosphere of the cosmonauts." Area contained medical center with centrifuge that cosmonauts called "Devil's merry-go-round."

From size and number of apartment buildings, Wilford estimated Star City population at 1500 to 2000. City was "smaller than Houston space center because it does not also serve as the mission control complex and apparently has no primary responsibility for the management of spacecraft design and development." Like NASA Manned Spacecraft Center at Houston, Star City contained facilities for premission training of cosmonauts and postmission physical examinations and recreation. Like U.S. astronauts, Soviet cosmonauts were trained in aircraft "that make deep dives to simulate weightlessness." Parachuting tested ability to handle high-stress situations. First simulator for Salyut orbiting laboratory was being installed in new building to be completed at year's end. Wilford said that "this could mean that no major advance beyond the present version of Salyut should be expected until well after the simulator is ready."

Of about 50 cosmonauts in Soviet space program, one third were civil engineers living in Moscow, Shatalov said; remainder were Soviet air force pilots who lived in Star City. No women were preparing for space missions but U.S.S.R. had "many women who specialize in meteorology and medicine, which are professions necessary on the orbital stations."

From descriptions by Shatalov and Cosmonaut-engineer Aleksey S. Yeliseyev, Wilford had constructed outline of preparations for Soviet space mission: Cosmonauts were assigned to mission one year in advance of launch date by commission of medical and engineering experts and Shatalov. There were two backup crews. After learning spacecraft mechanics and electronics and practicing in spacecraft simulator at Star City, crew went to Baykonur Cosmodrome in Kazakhstan three weeks before launch. There they completed training in actual spacecraft, checked instruments, stocked food, adjusted couches, and recommended minor changes in housekeeping. "Unlike Cape Kennedy," Baykonur had no simulators for last-minute practice. For two days before launch, "after the state commission approves the spaceship's readiness," crew relaxed. On launch morning cosmonauts were examined physically and biomedical sensors were attached to their bodies before they donned spacesuits. Traditionally, cosmonauts then gathered in quarters for few moments of silence broken by shouts of "off we go!" They signed their names on door before leaving for launch pad. Between missions, cosmonauts were expected to devote time each month to "social work"—visits to schools, collective farms, and factories.

Shatalov had suggested that new Soviet manned missions were possible within months and that U.S.S.R. was studying mission longevity in preparation for earth-orbiting laboratories. Of next Salyut mission, Shatalov said: "We are going to make it better . . . . We will prolong the visits of men, and the number of expeditions will grow. This is the difference now between our program and yours." Yeliseyev had said his countrymen would "probably" be on the moon by 1975.

At tour's completion, Shatalov had offered toast: "To cooperation between our peoples, to working together in space, which is our future and the future of the world." (Wilford, *NYT*, 3/22/72, 30; 3/26/72, 7; *Huntsville Times,* 4/2/72, 17)

March 20: Engineers at Kennedy Space Center pumped 805 cu m (212 000 gals) of kerosene into 1st stage of Apollo 16's Saturn V booster in preparation for scheduled April 16 launch toward moon. Countdown demonstration test would begin March 22, with crew completing simulated liftoff at 12:54 pm EST March 29. (UPI, *NY News,* 3/21/72)

- National Science Foundation issued *Manpower and Financial Resources Allocated to Academic Science and Engineering Activities, 1965–71* (NSF 72–302). Universities and colleges employed 273 800 full- and part-time scientists and engineers in January 1971, 6% per year increase from January 1969. Annual increase rate had been 8% 1965–1969. Current and capital expenditures for science and engineering totaled $7.9 billion in 1970, 6% increase per year over 1968. Annual increase rate had been 15% 1964–1968. In constant dollar terms based on Gross National Product, 1968–1970 increase averaged only 1% per year, while 1964–1968 rate of growth averaged 12% per year. Annual growth rate of 6% in expenditures for scientific and engineering activities in current dollars during 1968–1970 was considerably lower than the 16% "nonscientific" activities in universities and colleges 1968–1970. (NSF *Highlights,* 3/20/72, 1)

March 20–23: New uses of electrical energy developed from NASA research were exhibited at Institute of Electrical and Electronics Engineers convention in New York. Exhibit included patient-assist devices for human rehabilitation, "life signs" monitoring systems for hospitals and nursing homes, powered prosthetic hand replacement, portable light indicator for blind persons, sight-switch-operated wheelchair, system for recording school attendance, private alarm systems, gas analysis system for checking room or container atmospheres, system of easily concealed circuitry for controlling lights and appliances, bacteria detection equipment, and long-range laser surveying system. (NASA Release 72–57)

March 20–24: American Astronomical Society Planetary Sciences Div., meeting in Hawaii, heard report of oxygen discovery on Mars and discussed 1971 Mariner Mars mission. Univ. of Texas astronomer Dr. Ed Barker described finding, by examining Martian light through 272-cm (107-in) telescope at MacDonald Observatory, of average of one tenth of one percent oxygen in Martian atmosphere. Since earth atmosphere contained about 10 000 times that amount, Dr. Barker concluded that "there is not enough oxygen in the Martian atmosphere to sustain human life, but I can't rule out a bacterial type of life." Mariner Mars mission was expected to return data through November 1972. (*H Chron,* 4/2/72; JPL Release 611)

March 21: President Nixon sent letter to Dr. Rocco A. Petrone, Apollo Program Director: "As we approach the final countdown for Apollo 16, I want you and all the men and women of Apollo to know how much this nation values your splendid efforts. The moon flight program has captured the imagination of our times as has no other human endeavor. You and your team have, in fact, written the first chapter in the history of man's exploration of space, and all future achievements must credit all of you for having blazed the path. Countless people throughout the world will soon be sharing with you the excitement of Apollo 16's voyage, and I know I speak for all of them in conveying to you my warmest best wishes for a safe and successful flight. Good luck!"

Dr. Petrone later sent copy of President's letter to Apollo team. (Petrone letter)

- Soviet scientists V. Moroz and L. Ksanfomaliti described in *Izvestia* findings on Mars atmosphere from *Mars 2* and *3* spacecraft. Difference in contrast on two wavelengths between continents and seas during dust storm and change in contrast in same field had permitted evaluation of dust particle size and cloud thickness. Average radius of particles over Iapigya region Dec. 15, 1971, was estimated to be less than one micrometer and density of particles to be 100 million per sq cm (645 million per sq in). "If clouds of such density uniformly envelop the entire planet, its atmosphere must have contained several thousand million tons of dust." Estimates suggested that particles have mostly silicate composition, "which agrees with the infra-red spectra obtained by Mariner 9. Particles of such a tiny size . . . fall very slowly—for about a month even in the absence of supporting vertical currents in the atmosphere. Hence the conclusion that no prolonged storm in the true sense of the word, i.e. constant winds in the period of investigation, occurred." Unlike Venus, where cloud layer and atmosphere were more pervious and created "hothouse effect," surface of Mars "gets cool rather than heated, giving rise to what may be described as an 'anti-hothouse effect.' When the storm ended, the temperature rose." Water vapor content of atmosphere was low during and after the dust storm (Tass, FBIS-Sov, 3/22/72, L1)
- Jet Propulsion Laboratory had issued report calling for analysis of potential hazards of 64-m (210-ft) dish antenna at Goldstone Tracking Station near Barstow, Calif., *Los Angeles Times* reported. Beam could pose electromagnetic threat to electronics components in aircraft and to passengers on board who had cardiac pacemakers. Preliminary estimates indicated high potential hazard within tubular, 400 000-w invisible beam out to 16.9-km (10.5-mi) altitude, where its power diminished to safe levels. (Miles, *LA Times*, 3/21/72)
- Air Force announced award of $1 353 205 firm-fixed-price contract to Space Data Corp. for production of meteorological probes and launch support equipment. (DOD Release 204-72)

March 22: U.S.S.R. launched *Cosmos 479* from Plesetsk into orbit with 541-km (336.2-mi) apogee, 514-km (319.4-mi) perigee, 95.1-min period, and 74° inclination. (GSFC SSR, 3/31/72; SBD, 3/24/72, 138)

- Apollo 16 countdown demonstration test (CDDT) without propellants began at Kennedy Space Center in preparation for launch toward moon April 16. (KSC Release 55-72)
- NASA TF-8A aircraft, equipped with supercritical wing and piloted by Thomas C. McMurtry, completed 31st flight from Flight Research Center. Primary objective was to evaluate aircraft's longitudinal stability with augmentation off at altitudes from 11 300 to 13 700 m (37 000 to 45 000 ft) and mach 0.8 to 1.0. Survey equipment was removed. Angles of attack up to 17° and normal accelerations as high as 3.7 g were covered. (NASA proj off)

March 22

- *Apollo 15* Astronauts David R. Scott, James B. Irwin, and Alfred M. Worden and Manned Spacecraft Center Director, Dr. Robert R. Gilruth, were named by National Aeronautic Assn. to receive 1972 Robert J. Collier Trophy May 31. (AP, W *Star*, 3/22/72, A25)
- Charles W. Mathews, NASA Associate Administrator for Applications, testified on future applications of NASA research in domestic communications before Senate Committee on Aeronautical and Space Sciences during NASA FY 1973 authorization hearings: "Domestic communications systems . . . will make available many television-bandwidth channels, for both public and commercial use, to all the population . . . including remote and sparsely populated regions. Several . . . channels could be available for educational and instructional use. There are likely to be universities of the air which will permit many students . . . to achieve college degrees while spending only a small . . . time in academic residence." Improved systems would "permit an increase in the number of people who could receive actual pre-school training and training in early-childhood education and health-care practices."

  Health care could be improved and costs reduced through broad public access to health-care services and information, while centralized health-record-keeping could aid hospital personnel, administrators, and private physicians. Two-way television teleconferences between health professionals could become possible. In law enforcement, "fingerprints and other criminal records and data could be almost instantaneously available anywhere in the country, and television conferencing could make possible the remote taking of testimony as well as remote identification procedures." Postal service would improve "with the advent of electronic mail handling. Satellite links may be expected to carry a majority of the long distance letter mail, making inexpensive overnight delivery possible to all but the most remote locations. New satellite systems for aviation and maritime traffic management will reduce the accident rate and increase the efficiency of transoceanic and coastal transportation of goods and people.

  NASA Deep Space Network support of *Mariner 9* mission (launched May 30, 1970) was described by Gerald M. Truszynski, Associate Administrator for Tracking and Data Acquisition: "From lift-off on May 30 to insertion into its egg-shaped orbit around Mars on November 13 and continuing during orbital operations around Mars, Mariner 9 has been monitored and controlled by the deep space network. The navigational accuracy required to insert Mariner 9 into Martian orbit was unprecedented in prior flights into deep space. The aiming point . . . after a flight of about 380 million kilometers (240 million miles) was an area only 160 kilometers (100 miles) square. This represents an aiming area some 100,000 times smaller than was attainable for the Mariner 2 flight to Venus in 1962. Since the network issued the commands for retroengine firing and Mariner 9 was injected into orbit . . . the spacecraft has been completing two orbits of Mars each earth day and recording some 30 television pictures of the Martian surface on each orbit. After each orbit, when the spacecraft can be 'seen' by the Goldstone 64-meter [70-yard] antenna, Mariner 9 transmits back the tape recorded pictures upon command from the network." (Transcript)

- Langley Research Center awarded 31-mo, $4 128 860 contract to Univ. of Iowa for design, construction, integration, test, and launch support for Injun-F spacecraft and onboard experiments. Injun-F would be launched from Western Test Range by Scout booster in 1974 to study interaction of solar winds with geomagnetosphere at large radial distances over earth's north polar cap. Orbit would be highly elliptical and would traverse an area of space not previously surveyed. Spacecraft would weigh 27 kg (60 lbs) and would carry three principal experiments: magnetometer to map geomagnetic field over polar caps, low-energy proton electron differential energy analyzer to provide data on population density of charged particles, and extremely-low-frequency and very-low-frequency (ELF–VLF) experiment to determine nature of electrostatic and electromagnetic fields. Experiments might lead to understanding of way in which natural radiation belts surrounding earth maintained their supply of charged particles. Principal investigator would be Dr. James A. Van Allen, discoverer of Van Allen belts. (NASA Release 72–65)
- Director Bruce T. Lundin of Lewis Research Center announced major organizational changes to strengthen and streamline LeRC for its assumption of greater leadership in aeronautics, space communications, and electrical power field. Changes—which had yet to be formally approved by NASA—included creation of new position of Deputy Director, Management, filled by Henry C. Barnett, and creation of two new directorates: Space Communications, headed by J. Howard Childs, and Launch Vehicles, headed by Edmund R. Jonash. Dr. John C. Evvard was named Chief Scientist to provide "greater organizational visibility" in basic research. Laser power group and applications group were established, strengthening of computer management was begun, several divisions were restructured, and new offices were established. Lundin said changes would strengthen LeRC coordination and parallel research organization at NASA Hq. (*Lewis News*, 3/24/72, 1; Lundin Off)
- *Apollo 10* command module and mock-up of Apollo lunar roving vehicle were part of exhibition "Research and Development in the U.S.A." that opened in Moscow. Exhibit had been in Ibilisi, capital of Soviet Georgia, earlier, in scheduled tour of six Soviet cities. (Reuters, *NYT*, 3/23/72, 6)
- Soviet tests of three-part warhead for giant SS–9 intercontinental ballistic missiles had failed to demonstrate sufficient accuracy to seek out and destroy three separate U.S. Minuteman missiles in their steel and concrete silos, Dr. John S. Foster, Director of Defense Research and Engineering in Dept. of Defense, said in testimony before Senate Committee on Armed Services. U.S. intelligence had concluded that U.S.S.R.'s projected ability to destroy large part of U.S. Minuteman force by mid-1970s might not materialize until 1980s. (Beecher, *NYT*, 3/22/72, 1)

March 23: House Committee on Science and Astronautics unanimously approved H.R. 14070, $3.429-billion FY 1973 NASA authorization bill, $50 million above NASA budget request of $3.379 billion. Bill, reported to House April 11, allocated $2.651 billion for research and development (R&D), $77.3 million for construction of facilities, and $700.8 million for research and program management.

NASA budget request of $163.4 million for aeronautical research and technology was increased $48.5 million to $211.9 million. In House report, Committee scored previous cut in funding for aeronautics and recommended increase to support 1971 Civil Aviation Research and Development (CARD) Policy Study of NASA and Dept. of Transportation by expediting solution of "major problems in aviation today: noise pollution and safety." Bill allocated $41 million of the increase (for $50-million total) toward R&D to retrofit civil aviation fleet with noise reduction modifications and $7.5-million increase (for $24.7-million total) for aviation safety. Of safety funds, $3.8 million was for technology to modify civil aircraft to use Federal Aviation Administration's new microwave landing system by 1978, $700 000 for turbulence research, and $3 million for research on aircraft collision avoidance.

Bill increased NASA budget request for technology utilization from $4 million to $5.5 million. Of increase, $1.2 million was to be spent in applying NASA technology to problems in urban structures, fire safety, transportation systems, and energy conversion and $300 000 for applications engineering products and patent licensing.

Committee deleted $4 million from $59.6-million NASA request for High Energy Astronomy Observatory (HEAO) and restored $4 million deleted by Office of Management and Budget for Earth Observation Satellite (EOS), follow-on development to Earth Resources Technology Satellite (ERTS). Committee said it was expressing concern that NASA placed "higher priority upon certain expensive scientific projects . . . than it does upon space applications project which Congress considers most important." Committee considered level of funding for space applications "inadequate" and urged "a substantial increase in the budget for fiscal year 1974." (H Rpt 92–926; Com Off)

- Dr. John S. Foster, Jr., Director of Defense Research and Engineering in Dept. of Defense, testified before Senate Committee on Aeronautical and Space Sciences on DOD coordination with NASA on space shuttle development during hearings on NASA FY 1973 authorization: "I firmly believe that the U.S. must retain a strong national commitment to scientific and technological progress. The NASA space program is a vital element of this overall effort."

In space area, DOD was "becoming increasingly dependent on space systems, particularly in . . . communications. Other agencies and the public are becoming increasingly dependent on the use of space for communications, weather forecasting, and natural resource exploration. The Department of Defense supports the NASA space shuttle as an important part of the President's program, which will lead to a greater and more effective use of the space environment in the future."

Dr. Foster discussed Defense Satellite Communications System (DSCS) for military needs not satisfied economically and reliably by more traditional communications systems. First DSCS, currently operational, employed "simple, low-power satellites and ground terminals originally designed for experimental and test programs. To replace the existing satellites . . . we are deploying new DSCS satellites that have roughly 10 times the communications capability. . . . We are also developing improved terminals which will supplement the existing DSCS terminal complex." (Transcript)

- Collection of sketches and notes on rockets and spacecraft and original manuscript on astronomy by Dr. Wernher von Braun, NASA Deputy Associate Administrator for Planning, were placed on permanent exhibition at Alabama Space and Rocket Center in Huntsville. Display opening coincided with rocket pioneer's 60th birthday. (*Birmingham Post-Herald*, 3/24/72)
- *New York Times* editorial commented on March 19 visit of its science writer John N. Wilford to U.S.S.R.'s Star City, home of cosmonauts. Visit "implied a major change in Soviet policy toward space cooperation. No Western journalist had previously been permitted inside this key center of Soviet manned space activities." Wilford's report had suggested U.S.S.R. might be willing to explore moon jointly with U.S. "If so, the idea catches the United States at an embarrassing point in development of its space program. After the scheduled flights of Apollos 16 and 17 . . . this country plans to focus most of its future activity in space on earth-orbiting space stations and development of a space shuttle." (*NYT*, 3/23/72)

March 24: Air Force launched unidentified satellite from Vandenberg Air Force Base at 12:45 pm PST by Thor-Burner II booster. Orbital parameters: 884-km (549-mi) apogee, 802-km (498-mi) perigee, 101.7-min period, and 98.7° inclination. (Pres Rpt 73; *SBD*, 3/29/72, 156)

- NASA's *Mariner 9* Mars probe—orbiting Mars since Nov. 13, 1971—had resumed its scientific examination of Mars following engineering analysis of onboard computer, NASA announced. Cameras had been turned off March 17 while trouble-shooting procedures were conducted. Cause of problem had not been determined, but systems were working properly and problem had not recurred.

  Mariner 9 was 288 million km (179 million mi) from earth and had returned 7000 pictures of Mars and mapped 85% of Mars since launch May 30, 1971. Last pictures to be taken by *Mariner 9* until first week in June had been recorded March 22 and 23 and science instruments would be turned off March 30. From April 2 to June 4 *Mariner 9* would pass through Mars shadow once during each orbit for periods of up to 100 sec, during which spacecraft would operate on battery power. Limited science operations would resume after June 4, with spacecraft recording data and taking pictures about once a week until mission ended in November. Primary objectives of mission had been completed Feb. 11. (NASA Release 72-67)

- Future need for public administrators who understood science and technology was forecast by Dr. James C. Fletcher, NASA Administrator, in speech before American Society of Public Administrators in New York. Sociological and technological components of world problems raised problems to "new level of complexity."

  Management and public administration teams were examining "successful conduct of the Apollo program—the most complicated and difficult technical task ever undertaken by man." However, "in Apollo we were carrying out a highly complex, but basically technical program. Essentially we were dealing with machines, not people."

  Asked if U.S. should use Apollo program approach to solve environmental problems, *Apollo 8* Astronaut Frank Borman had told interviewer: "I don't think that I would care to live in a country that solved its social problems in the same way we solved the Apollo problems."

Dr. Fletcher agreed "our Science-Administrator needs for the future should not be equated with needs for better technical program managers, although this too will become increasingly important. In developing and formulating wise and sound policies and programs to tackle our emerging problems, it will be essential to have individuals within all governmental levels who are capable of understanding the potentials of science and technology . . . and who are alert to their ultimate purposes and consequences. . . . It is in this role that I conceive of the new Science-Administrator. (Text)

- U.S.S.R.'s *Cosmos 476* (launched March 1) was believed by U.S. military and space experts to be a new-generation electronic-intelligence satellite, Associated Press reported. *Cosmos 476* payload was estimated at between 4500 and 5400 kg (10 000 and 12 000 lbs), about 10 times heavier than earlier generation ferrets designed to eavesdrop on other nations' radio traffic and to monitor radar. Heavier weight suggested *Cosmos 476* incorporated much more equipment that was probably more technically sophisticated. (W *Star*, 3/24/72, A5)

- NASA announced issuance of *Monitoring Earth Resources from Aircraft and Spacecraft* (NASA SP-275). Illustrated technical report had been prepared by Univ. of California at Berkeley conservationist Dr. Robert N. Colwell and team to describe potential uses for photos of earth from satellites. Text stressed need for reliable current inventories of agricultural and vegetational resources. Illustrations compared views of sections of U.S. photographed from Gemini spacecraft, from aircraft, and on the ground. (NASA Release 72-70)

- National Science Foundation and National Center for Atmospheric Research planned to launch small rockets carrying silver iodide into thunderstorms suspected of being potential hail producers to see if seeding could cut hail formation, Associated Press reported. Rockets would be launched from aircraft flying through hailstorms to measure wind patterns and other atmospheric conditions. Researchers hoped seeding would disrupt hail growth so that moisture would fall as rain or small ice pellets. Experiment was part of $2.5-million-per-year Joint Hail Research Experiment conducted by NSF, NCAR, National Oceanic and Atmospheric Administration, Colorado Air National Guard, and seven universities. (W *Star*, 3/24/72)

- Short-haul transportation was "essential to a complete national transportation system," Aerospace Industries Assn. executive Jean Ross Howard said in speech before STOL [short takeoff and landing aircraft] Seminar sponsored by Aviation Committee of St. Louis, Mo., Chamber of Commerce and St. Louis Research Council. Although 70% of U.S. travel market was composed of distances up to 800 km (500 mi), air transportation system as yet was not oriented in this direction. Survey had shown that 30% of U.S. air passengers traveled less than 500 km (300 mi). Air congestion delays, caused by lack of airport capacity, cost airlines $180 million each year and figure would increase to $1 billion by 1981. NASA's $100-million program to develop short-haul, short takeoff and landing airliner was "welcome recognition of the urgent requirement for more efficient and convenient short-haul air service." (Text)

March 25: U.S.S.R. launched two Cosmos satellites from Plesetsk. *Cosmos 480* entered orbit with 1202-km (746.9-mi) apogee, 1174-km (729.5-mi) perigee, 109.1-min period, and 82.9° inclination. *Cosmos 481* entered orbit with 511-km (317.5-mi) apogee, 269-km (167.2-mi) perigee, 92.3-min period, and 71° inclination and reentered Sept. 2. (GSFC *SSR*, 3/31/72; 12/31/72; *SBD*, 3/28/72, 149)

- NASA launched first in series of five Nike-Javelin III sounding rockets carrying Naval Research Laboratory cesium cloud experiments from Wallops Station at 5:15 am EST. Payloads in series would be released at 97-km (60-mi) altitude, producing yellowish clouds visible for several hundred kilometers along East Coast. NRL scientists would make radio propagation, photographic, and geomagnetic observations to determine role of wind motions, magnetohydrodynamic effects, and chemical-thermal processes governing atmosphere. Second launch would be conducted March 27. (WS PAO; WS Release 72-4)

March 26: NASA–AEC Nuclear Rocket Development Station at Jackass Flats, Nev., was "first ghost town of the space age," *New York Times* reported. Since announcement (in Administration's FY 1973 budget request on Jan. 24) of termination of nuclear engine for rocket vehicle application (NERVA) project, 259-sq-km (100-sq-mi) area had been littered with "$200-million worth of deserted engine test stands, enormous assembly bays with birds nesting in the rafters, empty administration buildings, and aluminum and steel chards of pumps, turbines and motors." Only 200 employees made 290-km (180-mi) round trip between station and Las Vegas each day. "Eight years ago 56 buses provided commuting service for the station's 3,000 workers; today there are only three." (Lyons, *NYT*, 3/26/72, 57)

Parade magazine printed letter to Dr. James C. Fletcher, NASA Administrator, from Chairman Raymond L. Bisplinghoff of Aeronautics and Space Engineering Board, National Academy of Engineering: "The board members would like to reaffirm their continuing belief that a shuttle-like vehicle is the key to a forward moving and economical space program beyond 1980." (*Parade*, 3/26/72)

- National Book Committee included Norman Mailer's *Of a Fire on the Moon*, account of *Apollo 11* lunar landing mission, among 102 books it announced as nominees for National Book Awards. Final choice of 10 winners for $1000-prizes would be announced April 11. (Raymont, *NYT*, 3/27/72, 42)

March 27: U.S.S.R. launched 1180-kg (2600-lb) *Venus 8* unmanned probe into earth orbit from Baykonur at 9:15 am local time (11:15 pm EST March 26). At 10:42 am Baykonur time (12:42 am EST) 243-sec burn placed spacecraft on trajectory to Venus, where spacecraft would study physical characteristics of interplanetary space and continue research of Venus. Spacecraft would reach Venus in July after traveling 312 million km (194 million mi). Five hours after launch *Venus 8* was 65 000 km (40 000 mi) from earth and all systems were operating satisfactorily.

Venus 8 was eighth spacecraft in Venus series. *Venus 4* (launched June 12, 1967), *Venus 5* (launched Jan. 5, 1969), *Venus 6* (launched Jan. 10, 1969) had ejected capsules onto surface of Venus. *Venus 7* (launched Aug. 17, 1970) had ejected capsule which transmitted data for few second after reaching surface. Orbit decayed July 22.

(FBIS–Sov, 3/27/72, L1; *SBD*, 3/28/72, 147; *A&A 1970*; GSFC *SSR*, 3/31/72, 12/31/72)

- Apollo 16 astronauts John W. Young, Charles M. Duke, Jr., and Thomas K. Mattingly II began three-week preflight quarantine at Kennedy Space Center to minimize exposure to disease or illness that could delay launch toward moon, scheduled for April 16. (AP, B *Sun*, 3/28/72, A6)
- First launch of British Skylark sounding rocket to test techniques for high-altitude photography of earth resources was made from Woomera, Australia. Primary objectives were to test stabilization control system and remote sensing hardware.

    Advantages over conventional methods of high-altitude photography were that single frame was equivalent to 650 frames photographed from 15 000 m (50 000 ft), making assembly of photographic mosaics easier and more accurate, and that entire area could be photographed under identical conditions. Project was developed by United Kingdom's Dept. of Trade and Industry to assess mineral, agricultural, and other resources inexpensively. (*SF*, 8/72, 282–286)
- Nike-Javelin III sounding rocket, carrying Naval Research Laboratory cesium cloud experiment, was launched by NASA from Wallops Station at 5:08 am EST. Launch was second in series of five to collect atmospheric data; first launch had been conducted March 25. (WS PAO; WS Release 72–4)
- Sen. William Proxmire (D-Wis.) released statement saying he would ask Dept. of Justice to investigate alleged overpayment for building Lockheed C–5 military transport aircraft. GAO study made at his request had found that Air Force paid some $400 million in excess progress payments to Lockheed because Lockheed overstated value of work in progress. (Text)
- *St. Louis Globe-Democrat* editorial commented on high level of Soviet space activity: "It is fine to talk about devoting activities in space to peaceful purposes for the benefit of all mankind, as the United States has done. But the talk is hollow when the space scorecard shows the Soviets at an all-time high in military launches, and the United States at a low point, with the trend still further downward." There was urgent need for U.S. "to develop military space weapons capable of neutralizing Soviet supremacy." (*St Louis G–D*, 3/27/72)

March 27–April 3: Working Group of NASA and Soviet engineers met at Manned Spacecraft Center to discuss technical details of docking mechanisms. Talks continued U.S.–U.S.S.R. study of compatible rendezvous and docking systems for spacecraft. (NASA Release 72–69; MSC PAO)

March 28: Space leaders endorsed shuttle in supplement published by *Washington Daily News*:

Dr. James C. Fletcher, NASA Administrator, said shuttle would "produce substantial flight economies, ease strictures on spacecraft design, lower manufacturing cost, remedy malfunctions in orbit, and afford a built-in versatility that will alter the nature of space planning and operations in major ways." Editorial writer had described shuttle's impact on space program as transition from dramatics of a countdown to an airport routine, Dr. Fletcher said. "The dramatics will be fondly remembered by many of us, but nostalgia will not obscure the hard fact that the routine and economic access to space, of which the shuttle is the hinge pin, is the practical and commonsense means by which we shall reap the great harvest of benefits that space holds for mankind."

Supplement interview quoted Dr. Wernher von Braun, NASA Deputy Associate Administrator for Planning, as eagerly looking forward to shuttle flight. "By the time the first passenger shuttle flies I'll be 68. I think flying into space in a shuttle will be just like being a passenger in an airliner, only smoother." Dr. von Braun foresaw use of shuttle for travel to Mars "probably around 1990, or at least before the end of the 20th Century." Flight probably would begin with Mars spacecraft being hauled, piece by piece, in space shuttle cargo bays, to be assembled in orbit. "I think the only way to fly to Mars is with nuclear power, and because we would never want to ignite a nuclear rock on earth, we'll have to begin the journey from orbit." Dr. von Braun saw shuttle preparing way for manned exploration by ferrying small, unmanned spacecraft into orbit and launching them toward Mars in late 1970s and 1980s.

Article by Rep. James W. Symington (D-Mo.), member of House Committee on Science and Astronautics, said shuttle offered opportunity "not only to the United States to advance its own aerospace science and technology, but also to countries which could not finance a complete shuttle system. The advantages will be mutual and conducive to needed international scientific cooperation." Proper use of shuttle "can help us in our efforts toward the resources management necessary to make this planet fit to house future generations of man."

James J. Harford, Executive Secretary of American Institute of Aeronautics and Astronautics, said in article that guessing shuttle's significance was "like trying to predict what the DC-3 would lead to in air transportation six years before its first flight. The space shuttle could be that powerful and more." If shuttle performed the way NASA intended it to—"and NASA's Apollo track record has to impress even the cynical"—shuttle would "help to create entire new industries and . . . hundreds of thousands of new jobs." (*W News*, 3/28/72, 27, 29)

- *New York Times* published response by Univ. of Michigan astronomer James A. Loudon to Jan. 16 letter in which former scientist-astronaut Dr. Brian T. O'Leary questioned compatibility of space shuttle with national goals. O'Leary letter had been "strange document from a man who wrote a whole book to complain that scientists don't have access to space." Shuttle's most important aspect, "aside from its economy," was its ability to carry passengers. "For the first time, scientists will be able to perform experiments in space without spending years in irrelevant pilot training first." Shuttle's effect would be "to make space very much more available, with results for a dozen different branches of science that are now incalculable." (*NYT*, 3/28/72)
- U.S. and U.S.S.R. began seventh round of Strategic Arms Limitation Talks in Helsinki, Finland. (Hamilton, *NYT*, 3/29/72, 3)

March 29: Apollo 16 spacecraft, loaded with propellants, completed countdown demonstration test at Kennedy Space Center. (KSC Release 55–72)

- Convention on International Liability for Damage Caused by Space Objects was signed simultaneously in Washington, Moscow, and London.

In statement following signing, Secretary of State William P. Rogers said convention had been formulated to compensate, fairly and promptly, if space object of one party caused injury or damage to citizen of another party. "I believe that the convention which we have signed today will accomplish that purpose. I concur in Admiral Alan Shepard's statement in the U.N. General Assembly last year that it is a 'sound treaty based on realistic perceptions of mutual interest and mutual benefit.' Like its predecessor, this treaty is the result of intensive work in the United Nations Outer Space Committee and represents both a practical and an imaginative step in developing just international rules governing man's activities in space. Many countries have had an active part in its formulation during years of intensive negotiations. The conclusion of the agreement has again demonstrated what noteworthy results a cooperative approach to international negotiations can achieve." (Tass, FBIS–Sov, 3/29/72, A1; Rogers Text)

- Dept. of Justice charged 20 major U.S. aircraft companies and Manufacturers Aircraft Assn., Inc., with violating Sherman Antitrust Act by agreeing to suppress competition in research and development of aircraft and patentable components. Alleged agreement "hindering and delaying" development of new inventions dated back to 1928, Justice Dept. charged. Civil suit filed in U.S. District Court in Manhattan, N.Y., asked dissolution of manufacturers' association and injunction against aircraft companies entering into future agreements to restrict research or reduce competition in R&D by patent policies. Suit said U.S. manufactured nearly $8 billion worth of aircraft in 1967, more than half of which was military, and that four largest companies accounted for 60% of work, and eight largest for 88%. Manufacturers Aircraft Assn., Inc., and Boeing Co. later said charges were "completely unfounded in fact and law." (Shanahan, *NYT*, 3/30/72, 1)

- Nike-Javelin III sounding rocket, carrying Naval Research Laboratory cesium cloud experiment, was launched by NASA from Wallops Station at 5:07 am EST. Huge yellowish clouds formed were visible along East Coast. Lauch was third in series of five to collect atmospheric data; previous launches had been conducted March 25 and 27. (WS PAO; WS Release 72–4)

- President Nixon sent message to Congress urging Executive reorganization: "The product would be four entirely new, goal-oriented departments concerned with our communities, our earth, our economy, and our potential as individuals—plus a revitalized fifth department concerned with keeping America in food and fiber." New Departments of Community Development, Natural Resources, Economic Affairs, and Human Resources would replace Departments of Interior, Commerce, Labor, Health, Education, and Welfare, Housing and Urban Development, and Transportation. Dept. of Agriculture would be "streamlined." Several independent Federal agencies "would be drawn into the consolidation process as appropriate." (*PD*, 4/3/72, 708–14)

- Award of $361 494 contract to EGNG Inc. Bedford (Mass.) Div. for production and testing of prototype cold fog dissipation (CFD) system was announced by Air Force Systems Command. CFD system, designed by Air Force Air Weather Service for global use, would be deployed to U.S.-operated bases in United Kingdom, Germany, Alaska, and Netherlands. CFD system used liquid-propane spray vaporized from tanks at preselected sites upwind of runway complex to disperse fog affecting aircraft at airports. (AFSC Release 023.72)

- Dr. Donald Blessing Rice, Assistant Director of White House Office of Management and Budget, was named President of RAND Corp. (Belair, *NYT*, 3/30/72, 17)

March 30: U.S.S.R. launched *Meteor 11* meteorological satellite into orbit with 890-km (553.0-mi) apogee, 867-km (538.7-mi) perigee, 102.5-min period, and 81.2° inclination. (GSFC *SSR*, 3/31/72)

- Apollo 16 Astronauts John W. Young, Thomas K. Mattingly II, and Charles M. Duke, Jr., successfully completed countdown demonstration test at KSC on board unfueled spacecraft. Final countdown would begin April 10 in preparation for April 16 launch to moon. (KSC Release 55-72)

- NASA's TF-8A aircraft, piloted by Thomas C. McMurtry and equipped with supercritical wing, completed 33rd and 34th flights from Flight Research Center. Unaugmented stability and control data were obtained from mach 0.8 to 0.99 at dynamic pressure of 19 152 newtons per sq m (400 lbs per sq ft). Wing loads were measured from mach 0.5 to 0.99 at 9576 newtons per sq m (200 lbs per sq ft). Targets placed along 35% chord line and at wing tip were photographed to determine wing deflections. (NASA proj off)

- Soviet scientists at Krym Astrophysical Observatory had discovered unidentified astronomical objects, Tass announced. Using radiotelescope with 200-mm (8-in) mirror, scientists had located radio emitters emitting 3½-cm signals that "do not belong to the category of pulsating stars (pulsars), which emit short impulses only in a meter-long (three-feet-long) wavelength. For their properties the new astronomical objects are closer to the class of quasi-stellar radioemitters—quasars which are 1,500 million light-years from the earth." (FBIS-Sov, 4/3/72, L4)

- Arthur D. Little, Inc., had announced signing of agreement with Raytheon Co., Grumman Corp., and Textron, Inc., to study feasibility of large-scale satellite system to harness solar energy for earth use, *Wall Street Journal* reported. (*WSJ*, 3/30/72)

- Aerosonic Corp. President Herbert J. Frank, testifying before Subcommittee on Priorities and Economy of Joint Economic Committee of Congress, said Navy had paid $1700 each for altimeters for 300 aircraft without competitive bidding. Almost identical instruments had cost Air Force $565 each after competitive bidding. (Witkin, *NYT*, 3/30/72, 12)

March 31: U.S.S.R. launched *Cosmos 482* into orbit with 9798-km (6088.2-mi) apogee, 208-km (129.3-mi) perigee, 201.4-min period, and 52.1° inclination. (GSFC *SSR*, 3/31/72)

- U.S. Comptroller General Elmer B. Staats said in letter to NASA Administrator, Dr. James C. Fletcher, that General Accounting Office had determined July 13, 1971, NASA award of $500-million space shuttle engine contract to North American Rockwell Corp. Rocketdyne Div. had been fair. "We believe the procurement was conducted in a manner which was consistent with applicable law and regulations and was fair to all proposers." Award had been challenged by competing firm, United Aircraft Corp. Pratt & Whitney Div. (Text)

- Progress in NASA development of technology to alleviate aircraft noise pollution and ensure quieter aircraft operation was described by Walter A. Pennino, Public Affairs Officer in Office of Aeronautics and Space Technology (OAST), at NASA Hq. press briefing. Research had shown aircraft noise could be reduced if pilot flew aircraft "so that it climbs quickly" to be "as far from a ground observer as possible." When wing flaps were retracted as aircraft accelerated, aircraft required "less thrust to climb. When a safe altitude is reached, the pilot can reduce the engine thrust from that required for takeoff to the lower thrust required for climb." NASA and Federal Aviation Administration were encouraging airlines to adopt two-segment approach to reduce landing noise. Pilot would adjust controls to descent first at 6° glide slope. About one mile from the runway he would change path to slope. In final stages there would be no difference in the landing procedure. Aircraft would approach with engine power somewhat reduced and would be at higher altitude. Noise reduction would be three to five effective perceived noise in decibels (epndb) one nautical mile from the end of the runway and even greater reduction farther out from the runway. Ames and Langley Research Centers were studying requirements to make two-segment approach safe standard procedure. In Lewis Research Center Quiet Engine Program, experimental engines suitable for conventional takeoff and landing aircraft were being built. Test had shown reduction of noise by using quiet engine with and without acoustically treated nacelle. Goal chosen for short takeoff and landing (STOL) aircraft was 95 epndb at 152 m (500 ft). "We don't have an engine that can do that job right now, but NASA is working on it."

    Follow-on programs for new family of aircraft were described by James O. Kramer, Chief of Noise and Pollution Reduction Branch in OAST Aeronautical Propulsion Div. "Preliminary design study contracts will be let in the next month or two to study a low-noise, low-pollution engine . . . suitable for STOL. In addition, we have been carrying on design study for an advanced technology transport . . . appropriate for the 1980-timeframe as a replacement for 707 or for the DC-8. We are . . . planning to initiate in FY 73 preliminary design studies of a Quiet Engine . . . appropriate for such an advanced technology airplane." (Transcript)

- Analysis of "exceptional" satellite picture of Alaskan topography obtained by image dissector camera system (IDCS) aboard *Nimbus 4* (launched April 8, 1970) was described by U.S. Geological Survey scientist Ernest H. Lathram in *Science*. Major structural features had been identified despite relatively low degree of ground resolution of image, which had been exposed at altitude of 1100 km (680 mi). Regional lines not yet recognized in surface mapping had been perceived. Synoptic view provided by image showed orthogonal fractures not heretofore apparent in regional maps of Alaska that might reflect "conjugate set of fractures within the crust which has exerted significant control over the geological history of the state." Scientists had concluded increased resolution in other images from space platforms, like resolution of 60 to 200 m (200 to 650 ft) planned for Earth Resources

Technology Satellite program TV cameras, "will permit the discernment of finer detail and a greater accuracy in identifying and locating geologic features." (*Science*, 3/31/72, 1423–7)

- NASA launched Aerobee 150 sounding rocket from Churchill Research Range, Canada, carrying Univ. of Michigan auroral aeronomy experiment. Rocket and instrumentation performed satisfactorily. (SR list)
- Expansion of environmental and resource base beyond earth was urged by North American Rockwell Corp. scientist Dr. Krafft A. Ehricke in *New York Times* article. Great crisis of the time was fact "man's nature is attuned to an infinite, indestructible world" but earth had ceased to be indestructible. Global development, therefore, "must be based on an open-world concept and include both the development of extraterrestrial resources and the wiser management of our terrestrial resources." Extraterrestrial imperative was "indivisibility of Earth and space." (*NYT*, 3/31/72)
- Sophisticated TV systems were being used by some of world's largest telescopes to record planetary atmospheres, faint galaxies, and quasistellar objects, *Science* reported. Astronomers had been criticized for being less receptive to new technologies than other physical scientists. "At a time when automation of many stages of data-taking is standard procedure . . . automation of large telescopes is just becoming fashionable." (Metz, *Science*, 3/31/72, 1448–9)

During March: NASA released *Funds for Research, Development, R&D Plant and Scientific Technical Information, Fiscal Years 1971–1973: Annual Report to the National Science Foundation*. Report, for inclusion in government-wide survey published by NSF for Congress, White House, Government agencies, and scientific community, showed distribution of NASA funds based on FY 1973 budget estimates. It reflected NASA's total direct obligations incurred in FY 1971 and budget estimates for FY 1972 and FY 1973. In FY 1973 NASA would fund approximately 19% of Federal Government's total research and development effort and provide 23% of funds for research, alone. When Dept. of Defense was excluded from computation, NASA was expected to conduct 36% of total civilian R&D effort. Total research in FY 1973 was expected to show 4% increase over FY 1971.

For first time since FY 1962 research would represent over 50% of NASA funding for Research and Development. Increase in research resulted in decreased funding for development, with impending completion of Apollo and Skylab development; decrease was offset to a degree by funds applied to begin space shuttle development. Analysis of funding showed decline in engineering funds of about 7% of total, with increases in physical and life sciences. Reduction in engineering reflected anticipated completion of Apollo. High Energy Astronomy Observatory (HEAO) program effected increase in astronomy, and Viking program, increase in chemistry and biology. (Text)

- Scope of United Kingdom participation in space was described by A. V. Cleaver in *Spaceflight* article. British annual space budget was $70 million, "little more than half of the French or German," and "about an eighth of the American" on gross national product (GNP) basis. "While the effort on space is markedly increasing in France, Germany and Japan, is decreasing in the USA only from an initially very high level, and is beginning to be significant even in India and China, for the UK there is a decrease in real terms." U.K. had canceled Black Arrow launch vehicle program and ceased participation in main programs of European Launcher Development Organization (ELDO). Geostationary Test Satellite project to replace Black Arrow "has not yet been firmly initiated—though . . . official intention to do so has at least been indicated." U.K. space policy had "moved in recent years from a modest interest in all aspects of the subject to an almost complete concentration on satellites or spacecraft, to the exclusion of launch vehicles." Policy was based on assumptions that satellite was most technically interesting and commercially profitable aspect of space activities; that U.K. was "not too far behind to catch up in this field, whereas we are on launchers"; and that there was no need to duplicate U.S. rocket achievements because U.S. would sell Europe necessary launch vehicles. (*SF*, 3/72, 99–102)

- First successful transmission of fingerprint records via satellite, using *Ats 1* (launched Dec. 6, 1966) was reported by *Government Photography*. Fingerprints and photos filed by Videofilm system equipment in Sacramento, Calif., were transmitted as video signals to Videofile system in Tallahassee, Fla.; converted to video recordings; and retrieved by "instant replay" as TV pictures and printed copies in experiment to demonstrate possible state-to-Federal Government fingerprint transmission under permanent satellite system. Experiment had been conducted in December 1971 by Project Search (System for Electronic Analysis and Retrieval of Criminal Histories), funded by Law Enforcement Assistance Administration and administered by California Crime Technological Research Foundation. (*Govt Photo*, 3/72, 24–5)

- Development of small-scale water-reclamation units for home use from NASA-developed technology was forecast by Langley Research Center scientist Richard H. Weinstein in *Astronautics & Aeronautics*. NASA had begun in late 1950s to develop regenerative systems for purifying and recycling water for long-term space missions. "Research programs have been concentrated on . . . physio-chemical systems. . . . Using largely existing water-processing concepts, the development has gone toward making the concepts efficient on a small scale and adapting them, where necessary, to zero gravity." NASA systems were "far cry from the integrated, simplified systems needed for domestic water reclamation of the future" but provided "base from which to work." (*A&A*, 3/72, 45–51)

April 1972

April 1: Soviet news agency Tass reported Washington, D.C., interview with Dr. Philip Handler, President of U.S. National Academy of Sciences. Dr. Handler had praised scientific cooperation between U.S. and U.S.S.R: "Collaboration with their Soviet colleagues in so many fields—high energy physics, space research, and medical sciences, among others—promises markedly to benefit the work of U.S. scientists. . . . We are pleased that new opportunities are being arranged for joint enterprises . . . in space exploration, cardiovascular and cancer research, environmental monitoring, while we continue to explore the unique opportunities for mutual benefit in the broader enterprises in the international scientific unions." (FBIS–Sov, 4/1/72, L1)

- NASA launched Aerobee 170 sounding rocket from White Sands Missile Range, N. Mex., carrying American Science and Engineering, Inc., x-ray astronomy experiment. Rocket and instrumentation performed satisfactorily. (SR list)

April 2: Interview of Boris N. Petrov, Chairman of Soviet Academy of Sciences' Council for International Cooperation in Investigation and Utilization of Outer Space, on U.S.–U.S.S.R. joint space mission was published in *New York Times*. Interview, described as "first full public discussion of the negotiations from the Soviet point of view," had been obtained in Moscow by *Times* science writer John N. Wilford. Petrov had emphasized that "tangible progress" had been achieved and he was "optimistic" about prospects of joint space flight, technically feasible in "the middle of the nineteen-seventies." Government approval would "depend much" on April meetings in Moscow between NASA and Soviet Academy officials and May meeting in Moscow of Working Groups on rendezvous and docking. Petrov said compatible docking system would allow "docking of spaceships of our countries with the humane aim of rendering help to crews of spaceships and space stations in distress. These systems would also allow joint experiments during the flights in space." Petrov said negotiations had occurred in "atmosphere of mutual understanding and business-like discussion" that "gives us the hope that all the difficulties connected with . . . such a sophisticated scientific and technological undertaking could be overcome." (*NYT*, 4/2/72)

April 3: *Cosmos 483* was launched by U.S.S.R. from Plesetsk into orbit with 290-km (180.2-mi) apogee, 203-km (126.1-mi) perigee, 89.5-min period, and 72.8° inclination. Satellite reentered April 15. (GSFC SSR, 4/20/72; *Sov Aero*, 5/1/72, 9)

- NASA launched two Nike-Javelin III sounding rockets carrying Naval Research Laboratory cesium cloud experiments from Wallops Station at 5:04 am and 5:06 am EST. Huge pinkish-white clouds were formed and were visible along East Coast. Launches were last in series of five that began March 25 to study atmosphere in 97-km (60-mi) region. (WS Release 72–4; WS PAO)

April 3–20: Ames Research Center's Convair 990 flying laboratory participated in Arctic Ice Dynamic Joint Experiment (AIDJEX), international cooperative research program to obtain quantitative data on interactions of pack ice and ocean currents. AIDJEX team of scientists from U.S., Canada, and Japan made studies from test site in Beaufort Sea north of Point Barrow, Alaska. Convair 990 carried microwave radiometers, infrared imager, infrared radiometers, and laser geodolite above site in precise pattern to obtain remotely sensed data for correlation with measurements on ground. Data would help solve problems related to Arctic ice cover, its influence on global ocean circulation, and ship passage through ice-covered seas. (ARC *Astrogram*, 3/30/72, 1; ARC Release 72–22)

April 4: U.S.S.R. launched *Molniya I–20* comsat and France's *Sret 1* (Satellite de Récherches et d'Environment Technique; Satellite for Environmental and Technical Research) from Baykonur on one RD–107B booster.

Molniya I–20 entered orbit with 39 911-km (24 799.6-mi) apogee, 442-km (274.7-mi) perigee, 717.7-min period, and 65.5° inclination. Satellite would ensure "operation of a remote telephone-telegraph radio-communication system, as well as . . . the transmission of Soviet central television programs to stations of the Orbita network."

Sret 1 entered orbit with 39 250-km (24 388.8-mi) apogee, 458-km (284.6-mi) perigee, 704.6-min period, and 65.6° inclination. The small, 15-kg (33-lb), autonomous satellite would study characteristics of solar batteries for space operations and degeneration of solar cells from cosmic ray exposure in Van Allen Belts. It was first of three SRET satellites planned for launch in French-Soviet cooperative program. SRET 2 would be launched in late 1974 or early 1975 and SRET 3 one year later. (GSFC *SSR*, 4/30/72; *Sov Aero*, 4/10/72, 108; FBIS–Sov, 4/5/72, L1)

- Dr. James C. Fletcher, NASA Administrator, told press in Washington, D.C., he believed U.S. and U.S.S.R. would sign agreement during 1972 for joint mission in which Apollo spacecraft would dock with Soviet Salyut orbiting laboratory in 1975. At dinner party for seven newsmen Dr. Fletcher said: "We certainly don't believe that the first mission of American and Soviet astronauts together will be the last." Mission would "take a lot of planning and some hardware changes. . . . I also think the crews will need time to train together for the first flight." Crews would be bilingual and familiar with each other's spacecraft. Astronauts Thomas P. Stafford, John L. Swigert, Jr., and Donald K. Slayton were already studying Russian. First joint docking mission would cost U.S. as much as $300 million, most of it to keep manned flight facilities intact after conclusion of 1973 Skylab mission. Dr. Fletcher foresaw series of joint U.S. and U.S.S.R. missions, each longer and more ambitious. U.S.S.R. was emphasizing space station while U.S. emphasized shuttle. "I think it only logical that the two hook up together, especially since the shuttle will have rescue capability." U.S.S.R. was expected to continue Lunokhod lunar rover and lunar sample return programs and "probably marry the two some time in the next two years. I also think they will attempt a manned lunar landing, but not until late in the '70s, perhaps 1977." (O'Toole, *W Post*, 4/6/72, A2)

- Chances of joint U.S. and U.S.S.R. space mission, "probably in 1975," were "75-to-25," Dr. Christopher C. Kraft, Jr., Manned Spacecraft Center Director, said in press interview. NASA had spent about $700 000 on technical study for flight. Additional hardware would cost between $100 million and $200 million. (*W Post*, 4/5/72, A16)
- NASA announced award of 90-day, $9 800 000 letter contract to North American Rockwell Corp. Rocketdyne Div. to begin work on development and production of engine for space shuttle orbiter stage. Rocketdyne would begin work immediately, while negotiating $450 000 000, cost-plus-award-fee contract with NASA for engine. (NASA Release 72-73)
- Marshall Space Flight Center announced establishment of Sortie Can Task Team to plan project definition phase for sortie can—low-cost, space shuttle payload carrier for manned research and applications studies in earth orbit. Team would be managed by Fred E. Vreuls, with Hans R. Palaora as chief engineer. (MSFC Release 72-41)

Marshall Space Flight Center announced request for proposals for nine-month space tug operations and payload support study. Contractor selected would define flight operations to be conducted by tug after deployment from space shuttle and define interface between tug and automated payload or satellite that tug was delivering or returning. Proposals were due by April 10. (MSFC Release 72-39)
- National Academy of Sciences published *Human Factors in Long-Duration Spaceflight*. Report of study made for NASA by Space Science Board of NAS and National Research Council analyzed behavioral, psychological, and medical factors of manned space flights lasting up to two years. Object was to assess "whether man's participation is possible and feasible and to identify major obstacles and unknowns that must be resolved." Common problems were life support in space environment, confinement of crew in spacecraft and isolation from outside world except for communication with ground, maintenance of health without access to elaborate medical facilities, provision of crew training facilities, and crew selection with regard to ability to tolerate long-term isolation psychologically. Unknown factors included possible hazardous effects of prolonged weightlessness and of high-energy cosmic ray particles (high Zs) in deep space. Major recommendations were: research to determine course of physical deconditioning over two years of weightless and restricted activity; study of effects of radiation from high-Z particles; research program on effects of confinement to emphasize cognitive functioning; measurement and test of crew's physiological, psychological, and performance status in flight; incorporation of life-support and safety requirements, including medical and dental facilities, into spacecraft engineering and emphasis on habitability of spacecraft; adoption of systems analysis approach to mission to provide common frame of reference for interdisciplinary planning and implementation; and priority study of crew composition and criteria. (Text)
- Marshall Space Flight Center began crafts apprentice program in which 20 persons would work at MSFC and study at Calhoun Technical College in Decatur, Ala., during alternate periods. MSFC would pay for books and tuition during college period and would pay wages for work done at Center. Participants in program would be predominantly veterans, with many from minority groups. They would be trained to become

machinists, sheet metal workers, or electrical technicians. (*Marshall Star*, 3/29/72, 1; MSFC PAO)

- President Nixon accepted resignation of John H. Chafee as Secretary of the Navy, effective at undetermined date. President also submitted to Congress nomination of R/A Allen L. Powell to succeed R/A Don A. Jones as Director of National Ocean Survey in National Oceanic and Atmospheric Administration in Dept. of Commerce. R/A Jones was retiring effective April 30. Senate confirmed nomination of Powell May 3. (*PD*, 4/10/72, 722, 725, 737; *CR*, 5/3/72, D484)

April 4–6: Moscow meeting of NASA and Soviet Academy of Sciences delegations confirmed desirability of joint U.S.–U.S.S.R. rendezvous and docking test mission and established understanding on management and operation of mission to be conducted in 1975, subject to government approval. Dr. George M. Low, NASA Deputy Administrator, led U.S. delegation. Soviet delegation was headed by Vladimir A. Kotelnikov, Acting President of Soviet Academy. Parties agreed to use of U.S. Apollo spacecraft and Soviet Soyuz spacecraft with systems developed by both sides, to specific documents for development of joint documentation, and to principles for preparatory and operational phases of mission. (Summary of results; NASA Release 5/24/72)

April 5: Senate Committee on Aeronautical and Space Sciences continued hearings on NASA FY 1972 authorization. Roy P. Jackson, NASA Associate Administrator for Advanced Research and Technology, testified that LeRC quiet engine program was proceeding on schedule: "Tests of full-scale fans for the engine have revealed much new information on the interrelationships among noise, aerodynamic performance, and structural integrity. The complete experimental engine will be operated in about 6 months. Results . . . indicate that we may expect to better the original noise abatement goals. Program results are given to industry on a continuing basis. . . . We can expect some of our improvements to be adopted by industry in new engine designs before completion of our program."

NASA–AEC Space Nuclear Systems Office Manager Milton Klein testified: "Reactor power sources will be required to provide power in amounts greater than a few kilowatts for many of the missions of the future. For unmanned military satellites, radiation hardening—mandatory for certain military missions—and low recurring costs for multiple missions are important advantages of small reactor systems. For unmanned NASA earth orbit missions, reactors become cost effective above a few kilowatts, a power range needed for such applications missions as communications satellites. In manned missions . . . overriding advantage of reactor systems is their ability to accommodate unforeseen large increases in power demand." Advanced reactor systems were needed for electric propulsion systems capable of high energy missions to far planets.

Associate Administrator for Tracking and Data Acquisition Gerald M. Truszynski reported that satellite network included 10 electronic ground stations operated by NASA, supplemented by optical tracking network operated by Smithsonian Astrophysical Observatory. Electronic stations provided worldwide system that tracked each satellite, determined status of onboard systems, commanded satellite functions, and acquired data. Optical stations, which included laser capabilities at selected sites, provided specialized services in precision orbital tracking. (Transcript)

- *Ats 1* applications technology satellite (launched by NASA Dec. 6, 1966) had again helped save lives of two Alaskans, NASA reported. Satellite had been used in 1971 to summon medical aid for two stricken residents of remote sections of state. Recently, satellite had been used to alert Public Health physicians to plights of 11-yr-old girl with appendicitis and man severely injured in snowmobile accident. (NASA Release 72-74)

- Finding by California Institute of Technology astronomers of waves 1000 km (600 mi) from crest to crest undulating outward from center of sunspots at speeds to 40 000 km per hr (25 000 mph) was reported at meeting of American Astronomical Society in College Park, Md. Astronomers believed waves—named Stein waves after Alan Stein, Cal Tech student who discovered them by accident in motion pictures taken through telescope—were sound waves caused by thunder on sun. (AP, B *Sun*, 4/6/72, A3)

April 6: U.S.S.R. launched *Cosmos 484* from Plesetsk into orbit with 236-km (146.6-mi) apogee, 177-km (110-mi) perigee, 88.6-min period, and 81.5° inclination. Satellite reentered April 18. (GFSC *SSR*, 4/20/72; *Sov Aero*, 5/1/72, 9)

- Establishment of Urban Systems Project Office was announced by Manned Spacecraft Center. Office would apply space technology and techniques to urban problems and systems in effort by NASA, Dept. of Housing and Urban Development, and Atomic Energy Commission to develop new methods of servicing urban areas. Modular-sized integrated utility system (MIUS) would investigate and demonstrate feasibility of servicing limited-size communities with complete line of utilities generated by single processing plant. (MSC Release 72-75)

- U.S. planned to commemorate 200th anniversary of Nation's birth by attempting to land unmanned spacecraft on planet Mars July 4, 1976, according to NASA spokesman quoted in *Washington Post*. Attempt would be made by Viking spacecraft scheduled to enter Mars orbit about July 1, 1976. (*W Post*, 4/6/72, A2)

- NASA and National Science Teachers Assn. announced selection of 25 finalists in Skylab Student Project to propose flight experiments and demonstrations for performance aboard Skylab in 1973. Finalists' proposals had been selected from 3409 entries from students throughout U.S. and overseas. Selection of limited number of these proposals for actual inclusion on Skylab would be completed by June 1. (NASA Release 72-71)

- Sen. William Proxmire (D-Wis.) conveyed to Senate report by General Accounting Office on C-5A aircraft program that showed Air Force was continuing to accept delivery of defective C-5A aircraft. From February through September 1971 Air Force had accepted 15 C-5As with average of 251 deficiencies per aircraft. Most serious defects were in wings, landing gear, and radar. (*CR*, 4/6/72, S5510-7)

April 6-7: Outstanding Handicapped Federal Employee of the Year Award—given annually by Civil Service Commission—was presented in Washington, D.C., to Shirley Price, armless 1.2-m-tall (4-ft-tall) administrative aid at Manned Spacecraft Center. Miss Price had graduated from Texas Southern Univ. with honors and had attended graduate school at Univ. of Wisconsin. At MSC she controlled 10 000-event data library, used and interpreted UNIVAC 1108 computer runs, filed, answered telephones, and typed 45 to 50 words per minute. Her hobbies included drawing, embroidering, crocheting, and knitting, all of which she did with her toes. Mayor Walter E. Washington presented keys to city of Washington, D.C., to Miss Price on April 7. (W *Star*, 4/8/72; *NASA Activities*, 5/15/72, 95–6)

April 7: U.S.S.R. launched *Intercosmos 6* from Baykonur Cosmodrome into orbit with 334-km (207.5-mi) apogee, 197-km (122.4-mi) perigee, 90-min period, 51.8° inclination. Purpose of mission was to study "particles of primary radiation with an energy range of 10^{12} to 10^{13} electron volts; the chemical composition and energy spectrum of space beams in the high energy sphere; [and] meteoric particles in inner space." Instruments—including 1070-kg (2359-lb) photo-emulsionary block and ionization calorimeter—had been designed by U.S.S.R., Hungary, Mongolia, Poland, Romania, and Czechoslovakia. Satellite reentered April 11. (GSFC *SSR*, 4/30/72; *SBD*, 4/25/72, 301; Tass, FBIS–Sov, 4/7/72, L1)

- NASA's *Mariner 9* completed 146th day in Mars orbit. Since launch May 30, 1971, *Mariner 9* had responded to 37 764 commands and had taken 6876 pictures. Spacecraft was undergoing solar occultation (passing into shadow of Mars) and would resume acquisition of data in early June when occultation ceased. Spacecraft resources were being budgeted to keep *Mariner 9* active through superior conjunction of Mars and sun in September when spacecraft signals would be deflected by solar gravity. Analysis of deflection could provide accurate check on theory of relativity. (NASA proj off)

- Manned Spacecraft Center announced redesignation of Medical Research and Operations Directorate as Life Sciences Directorate. Reorganized directorate would consolidate all biomedical research into Biomedical Research Div.; all medical, operational, and environmental health activities into Health Services Div.; and all equipment development, testing, and integration into Bioengineering Systems Div. Acting Director Richard S. Johnston of MR&OD would be Director of Life Sciences, Dr. Lawrence F. Detlein Deputy Director, and Dr. Willard R. Hawkins Deputy Director of Medical Operations. (MSC Release 72–76)

- Manned Spacecraft Center announced it was negotiating parallel contracts for space-walk studies with LTV Aerospace Corp. and Hamilton Standard Div. of United Aircraft Corp. Contracts were to investigate emergency, contingency, and normal extravehicular and intravehicular activities by astronauts and protective equipment required. Firm-fixed-fee contracts would amount to $126 000 for LTV and $98 511 for Hamilton Standard. (MSC Release 73–78)

- NASA launched Nike-Apache sounding rocket from Thumba, India, carrying Indian langmuir probe experiment. Rocket and instrumentation performed satisfactorily. (SR list)

April 8: Atomic Energy Commission named its Los Alamos, N. Mex., Meson Physics Facility for Sen. Clinton P. Anderson (D-N. Mex.), member of Joint Committee on Atomic Energy, and celebrated "Clinton P. Anderson Day" with ceremonies in Los Alamos. History of space and nuclear rocket programs was recounted by AEC Commissioner James T. Ramey in welcoming speech: "Senator Anderson with some help from the Chairman of the AEC almost had the Joint Committee and AEC responsible for the emerging space program after Sputnik." Joint Committee "held one of the first series of hearings on the development of rocket technology for outer space. These hearings represented one of the times . . . that Senator Anderson . . . and Admiral [Lewis L.] Strauss of the AEC were in entire public agreement on a project. Senator Anderson was thinking of putting space under the Joint Committee and Admiral Strauss was interested in the space development program being under AEC. Their mutual interest was short-lived, however, since the then Senate Majority Leader Lyndon [B.] Johnson apparently had different ideas on this subject!" (AEC Release P-95; Text)

April 9: Americans were wondering why U.S. spent "$40 billion" to land on moon rather than on programs for social action, Jeffrey St. John commented in *Los Angeles Times* article. "The fact remains, however, that long before the landing of Apollo 11 public funds for social problems on earth had risen far beyond the annual space budgets. In 1972, the federal government has earmarked $100 billion for social action programs, as opposed to $3.2 billion for space. The ratio of expenditures for space and social action programs during the 15 years of the space effort has been roughly 11 to 1 in favor of social action." Lunar landing had been "triumph of 400,000 scientists, engineers, and technologists and the sophisticated expertise of 20,000 private companies, large and small." Apollo had been "clear-cut example of a successful government venture" but on earth Government had "made a mess of the social landscape." Compounding irrationality was refusal of liberals to understand why "space program succeeded, largely as a private endeavor coordinated by government, and why social problems grow worse under government's guidance."

To accomplish lunar mission, space program participants had needed two crucial elements: "intellectual freedom" to analyze a problem correctly and "strict observance of natural laws." In social problem-solving, intellectual freedom was abandoned for political expediency that played one pressure group against another and "there is a continual violation of certain immutable natural laws. Namely, the refusal of the social planners to believe the slums and poverty can be the products of an absence of individual initiative and enterprise." (*LA Times*, 4/9/72)

- *The Kremlin & The Cosmos* by Nicholas Daniloff was reviewed in *New Haven Register*. Book by UPI correspondent who worked in Moscow from 1961 through 1966 detailed Soviet space achievements and forecast U.S.-U.S.S.R. cooperation in space through which "new side to the many facets of Soviet-American understanding will develop." Author had drawn information from public and private sources including "defectors to the West." Chapter was devoted to "intriguing subject of the Soviet press' handling of Russian ventures into space." (Desruisseaux, *N Hav Reg*, 4/9/72)

- Congressional restraints on sale of military aircraft and equipment to developing nations were "not working," and had cost U.S. aerospace industry "more than $800 million and 70,000 to 90,000 jobs in the last two years," Copley News Service reported. Denied credit purchases of jets and other modern aeronautical equipment, nations once dependent on U.S. had turned to United Kingdom, France, Italy, Scandinavian countries, and Japan. U.S.S.R. was "trying to push its way into the market, particularly in Latin America." (*San Diego Union*, 4/9/72, 20)

April 10: Six-day countdown for Apollo 16 began at 8:30 am EST at Kennedy Space Center. Spacecraft was scheduled for launch at 12:45 pm EST April 16. (AP, *NYT*, 4/11/72, 15)

- Legal Subcommittee of United Nations Committee on the Peaceful Uses of Outer Space began four-week session in Geneva to study Soviet draft treaty for cooperation in exploration and use of moon. Draft treaty had been submitted to U.N. in June 1971. At opening session U.S. representative Herbert K. Reis proposed that nations planning manned or unmanned lunar missions be required to notify U.N. in advance and that Soviet draft treaty be amended to permit use of military personnel or facilities on moon for peaceful purposes. (*NYT*, 4/11/72, 10)

- Convention on the Prohibition of the Development, Production and Stockpiling of Bacteriological (Biological) and Toxin Weapons and on Their Destruction was signed simultaneously in Washington, London, and Moscow. In Washington, D.C., ceremony President Nixon said Convention meant "that all the scientists of the world, certainly a universal community, whatever their language, whatever their race, whatever their background, instead of working to develop biological weapons which one nation might use against another nation, now may devote their entire energy toward working against the enemy of all mankind—disease." (*PD*, 4/17/72, 744–5)

- Tenth anniversary of Navy Astronautics Group at Point Mugu, Calif. Group operated Navy Navigation Satellite System, by which Navy ships at sea pinpointed their positions. Group had five navigation satellites in orbit. (PMR *Missile*, 4/21/72, 1)

- National Science Foundation released *National Patterns of R&D Resources: Funds and Manpower in the United States 1953–1972* (NSF 72-300). Research and development expenditures were expected to reach $28.0 billion in 1972, up from $26.8 billion in 1971, increase of 4.3%; 1970 to 1971 increase had been 2.1%. R&D was expected to account for 2.5% of estimated 1972 U.S. gross national product (GNP), down from 2.6% in 1971. In 1964, R&D and GNP ratio reached 3.0%. Approximately 54% of national R&D effort was supported by Federal funds, primarily from Dept. of Defense and NASA. Industrial sector was second largest source. Companies were expected to provide two fifths of total U.S. R&D funds in 1972, up steadily from 31% in 1964.

    Estimated 519 000 scientists and engineers in U.S.—more than one third—were employed in R&D during 1971, 5% fewer than in 1970. Nearly 70% of these professionals worked for industrial firms. Basic research expenditures for 1972 were projected at $4.1 billion, with applied research at $6.4 billion. In 1971, $4.0 billion and $6.1 billion had been spent on these activities. Universities and colleges were expected to perform more than 55% of national basic research effort during 1972, up slightly from 1971. (Text)

- "Space in the Age of Aquarius," hour-long special on space program, was shown on nationwide Columbia Broadcasting System TV. Program, produced and emceed by actor Hugh O'Brian, included interviews with NASA scientists on benefits of space exploration. Comedian Jonathan Winters played part of numerous skeptics opposed to space program and new technology.

In Baltimore *Sun* Judy Bachrach later commented that "gist of program could be found in Mr. O'Brian's very last rhetorical question, 'Do we go or do we stop?' We go on, we go on." By end of program "you could have just about died of shame for ever having dared to think anything perverse about our space program. NASA couldn't have got better propaganda if it had produced the show itself." (UPI, *C Trib*, 4/10/72; B *Sun*, 4/14/72)

April 10–11: NASA Aerospace Safety Advisory Panel reviewed all aspects of Skylab program during meeting at Marshall Space Flight Center. Skylab program managers and key personnel briefed panel. Panel would later advise NASA Administrator on any hazards in program's operation or facilities. (MSFC Release 72–40; MSFC PAO)

April 11: U.S.S.R. launched *Cosmos 485* from Plesetsk into orbit with 462-km (287.1-mi) apogee, 267-km (165.9-mi) perigee, 91.8-min period, and 70.9° inclination. Satellite reentered Aug. 30. (GSFC SSR, 4/30/72; 8/31/72; *Sov Aero*, 5/17/72, 120)

- Apollo 16 astronauts John W. Young, Thomas K. Mattingly II, and Charles M. Duke, Jr., passed final physical examination at Kennedy Space Center. Dr. Charles A. Berry, NASA Director of Life Sciences, said results showed astronauts "in good shape and ready to fly." (AP, W *Star*, 4/11/72, A3)

- House Committee on Science and Astronautics favorably reported, without amendment, H.R. 14070, $3.429-billion FY 1973 NASA authorization bill [see March 23]. (H Rpt 92–976)

- Dr. James C. Fletcher, NASA Administrator, discussed relevancy of space program to communications field and scored antitechnologists as "misguided," in speech before National Assn. of Broadcasters engineers' luncheon in Chicago: "Our success in gathering information from space is forcing a revamping of surface communications as they relate to data transmission, and a substantial upgrading of the capacity of the system's components. The echoes of our revolutions will be heard in many sectors. . . . They will be as beneficial as they will be far-reaching."

Disputing anti-technologists, Dr. Fletcher said that "they argue that, if the space program is done away with, the problems that are rooted in technological advance will disappear as well . . . and . . . funds now allocated to space can be applied to poverty, aid to the inner city and to other national needs." This was nonsense. "In so arguing, the base for planned error has been laid. A fundamentally irrational position has been taken. Critics of this stripe are unable or unwilling to recognize that technology is a tool—and nothing more." As space program progressed, Dr. Fletcher said, "I am confident that . . . we will be able to confront the anti-technologist with a sequence of results from our space work . . . that have a direct impact on the solving of some of the social problems that concern him—and the rest of us.

"While this erosion of the position of the anti-technologist is in process, it is important that we take every precaution that his campaign does not impede the formation of the wise policies of federal support.

April 11

> Our immediate future depends on it, our long range future will be determined by it." (*NASA Activities*, 5/15/71, 92–4)

- U.S. National Academy of Sciences and Soviet Academy of Sciences signed agreement providing for exchange of 216 man-months of visits by scientists between U.S. and U.S.S.R., or 180 man-months in each direction, over two-year period retroactive to Jan. 1, 1972. Agreement renewed and expanded interacademy exchange program that began in 1959. (NAS–NRC–NAE *News Rpt*, 4/72, 1; NAS Off of Foreign Sec)

- Airborne laser system developed by NASA's Wallops Station and Langley Research Center was being used to measure presence of drifting plankton in offshore areas of Chesapeake Bay, Md., and Chincoteague Bay, Va. System, mounted on helicopter, used orange light flash from laser to induce fluorescence in plankton. Fluorescence then was measured and recorded through telescope on helicopter. Test results had confirmed feasibility of using laser system on low-flying aircraft or surface vessel for observation of chlorophyll-bearing marine microorganisms. (Wallops Release 72–6)

- Tracked air cushion research vehicle (TACRV) constructed for Dept. of Transportation by Grumman Aerospace Corp. was officially displayed for first time at Grumman facility in Bethpage, N.Y. Vehicle 15.5 m (51 ft) long could achieve speeds to 480 km per hr (300 mph). (DOT Release 34–73)

- NASA and Federal Aviation Administration announced they would jointly sponsor major aeronautics exhibit at world's first international transportation exhibition, TRANSPO '72, to be held at Dulles International Airport in Chantilly, Va., May 27 to June 4. Exhibit would illustrate how two agencies were working with aviation community to overcome problems of congestion, noise, pollution, and inadequate airport facilities. (FAA Release 72–57)

- Leesburg, Fla., high school students Dale Delpit and Mike Holloway emerged after 6 days and 16 hrs in capsule built by their science class to simulate spacecraft. They had walked and slept on waterbed to simulate weightlessness and had eaten pre-prepared foods. Both boys, aged 14, emerged with wobbly knees and with charcoal on faces to look like astronauts' beards. (*W Star*, 6/12/72, 20)

April 12: Dr. James C. Fletcher, NASA Administrator, and other NASA officials testified before Senate Committee on Appropriations' Subcommittee on HUD-Space-Science-Veterans Appropriations during hearings on NASA FY 1973 appropriation. Dr. Fletcher and Dr. George M. Low, Deputy Administrator, demonstrated model of space shuttle and explained configuration and costs. Dr. Fletcher said that in halving cost of original plans main sacrifice was that "we are not reusing all the parts. We use this expendable gas tank to supply the orbiter. . . . That, however, had an enormous effect on the size. We were able to bring the size of the orbiter down so that we were able to use solid fuel rockets." Dr. Low added, "We sacrificed nothing in terms of what kinds of satellites, what kinds of payloads we can put into orbit."

Estimates of larger costs by shuttle opponents had been derived by adding in cost of "all the satellites it will carry, the weather satellites, the communications satellites, and all of those things that would be carried anyway, whether there is a Shuttle or not," Dr. Low said.

"And this could be likened to charging to the cost of a freight train the cost also of all of the freight that it is going to carry."

In reply to question on cooperation with U.S.S.R., Dr. Low cited decision to build docking and rendezvous systems compatible between U.S. and U.S.S.R. manned spacecraft. Shuttle "will have a system on it so it can attach to another satellite and we will make that system so it is the same as the Russians will use on their satellites, so that if we need to rescue each other . . . or if we want to conduct joint experiments in space, we can do that."

Queried on comparison of U.S. and U.S.S.R. space programs, Dr. Fletcher said, "It is quite apparent that over the last 3 or 4 years, the Soviet space effort has grown in dollar volume and in number of launches per year, during the period that ours has declined. So that . . . they are spending quite a bit more in space than we are. It is estimated that perhaps 2 percent of their gross national product is spent for space, whereas a half of a percent of our GNP is spent on space. . . . Our number of launches per year has steadily declined from a peak of something [like] almost 80 in 1966, down to . . . the order of 30 or 35 in 1970 and 1971; whereas, the number of their launches has steadily climbed . . . and is now higher than even our peak. In 1971, they launched over 80 spacecraft, whereas our launches were down to less than half that." (Transcript)

- Physicist Dr. Ralph E. Lapp testified on space shuttle economics before Senate Committee on Aeronautical and Space Sciences during FY 1973 NASA authorization hearings. Following testimony by Dr. Oskar Morgenstern, Chairman of Board of Mathematica, Inc., and Dr. Klaus P. Heiss of Mathematica, on corporation's study of space shuttle economics [see Jan. 31], Dr. Lapp criticized calculations and said entire issue of launch vehicle choices for future should be restudied. He said shuttle payloads would average "10 times more expensive than their weight in solid gold." Cost would be $5100 per 0.5 kg (1 lb) for launches 1979–1990. Cost projection prepared for NASA had indicated $7-billion saving for shuttle launches in program that would cost $42 billion using current methods, Dr. Lapp said. He added that no such $42-billion program had been authorized for the 1980s. U.S. taxpayers could save $19 billion by not building space shuttle and by using existing rockets for space program. (Transcript)

- NASA Space Shuttle Technology Conference was held in San Antonio, Tex., in conjunction with 13th Structures, Structural Dynamics and Materials Conference of American Institute of Aeronautics and Astronautics, American Society of Mechanical Engineers, and Society of Automotive Engineers. Topics discussed included aeroelasticity and loads, structure and liquid interactions, vehicle dynamics test and analysis, thermal protection systems, and structural design. *Apollo 8* Astronaut James A. Lovell, Jr., was member of panel for high school students that discussed benefits to be derived in space shuttle use. (AIAA Release)

- U.S.–U.S.S.R. talks on space cooperation were progressing successfully and internationally manned space laboratories might be launched in "nearest future years," Soviet space expert Leonid I. Sedov said in article published in Soviet defense ministry newspaper *Krasnaya Zvezda (Red Star)*. (AP, *W Post*, 4/13/72, A11)

April 12

- U.S.S.R. celebrated Cosmonauts Day with annual memorial meeting on 11th anniversary of *Vostok 1*, first manned space flight, piloted by Cosmonaut Yuri A. Gagarin. Soviet Academy of Sciences president Mstislav V. Keldysh summarized Soviet space achievements and described Soviet space programs. (Tass, FBIS–Sov, 4/18/72, L1)
- India's Minister of State for Home Affairs K. C. Pant, told Indian Parliament, Lok Sabha, that India had developed large number of high-precision and sophisticated components and assemblies needed for control and guidance system of satellite launch vehicle and had flight-tested some components. Sriharikota launch site had become operational, but use of facilities at Thumba Equatorial Rocket Launching Station (TERLS) would continue. (FBIS–India, 4/14/72, P2; State Dept PIO)
- Kennedy Space Center announced award of grants to Florida Technical Univ (FTU) in Orlando and Florida Institute of Technology (FIT) in Melbourne for ecological studies of spaceport area, much of which was a wildlife refuge. Grants, each totaling $90 000, were for one year. FIT would study living processes in ecosystem which could be used as indicators of significant changes in environment. FTU would divide land areas and associated waters at KSC into geographical zones that would be intensively sampled for life forms for comparative academic studies. (KSC Release 80–72)
- National Science Foundation announced issuance of *Solar Energy in Developing Countries: Perspectives and Prospects*. Report of NAS Board on Science and Technology for International Development said solar energy had "capability to meet energy needs substantially beyond the applications now being made." But best course was to study general energy needs of developing nations and methods of satisfying these needs without placing excessive hope in promising but unproved applications of sun's energy. Report recommended establishment of regional centers in developing nations to evaluate supplies and uses of all kinds of energy and to begin research and development efforts on applications suitable to local conditions. (NAS Release)

April 13: Soviet and NASA scientists exchanged lunar soil samples in ceremony at Soviet Academy of Sciences in Moscow. NASA representatives Dr. Paul W. Gast, Chief of Manned Spacecraft Center Planetary and Earth Sciences Div., and Dr. Michael B. Duke, Lunar Sample Curator at MSC, presented Soviet scientists with 1 g (0.04 oz) of soil brought to earth by *Apollo 15* astronauts. Exchange brought total weight of *Apollo 15* lunar material given to Soviet scientists to 3 g. Soviet scientists gave NASA representatives 2 g (0.07 oz) of soil brought to earth by *Luna 20*, several photos of *Luna 20* landing site, and pictures of separate rock fragments. (NASA Release 72–77; FBIS–Sov, 4/13/72, L4; NASA OMSF)

- NASA announced selection of RCA Aerospace Systems Div. for negotiation on $175 000 fixed-price contract to produce long-range laser traversing system. Nicknamed "Smokey," system had been developed at Goddard Space Flight Center for U.S. Forest Service. Two back-packable, 18-kg (40-lb) units contained vertical-firing laser and telescope receiver that viewed laser pulse above all obstructions, permitting surveyors to lay out property lines over terrain obstacles faster and cheaper than by previous methods. (NASA Release 72–79)

- Delaware Valley Council presented its 1972 Aviation Award to NASA in ceremony in Philadelphia. Award, for outstanding achievements in aviation and aerospace, was accepted by Richard C. McCurdy, NASA Associate Administrator for Organization and Management. (*P Bull*, 4/14/72, 8)
- Transmitting to Congress fourth annual report of National Science Foundation [see Feb. 10] President Nixon said: "I have great hope that we can realize the full potential of American technology for serving our national purposes. The commitment of this Administration to continue progress toward that goal is clearly reflected in the array of programs which I have detailed in my latest Budget Message." (*SPD*, 4/17/72, 750)

April 14: U.S.S.R. launched two satellites.

Prognoz 1 (Forecast), launched from Baykonur Cosmodrome, entered orbit with 199 667-km (124 067.3-mi) apogee, 1005-km (624.5-mi) perigee, 5782.1-min period, and 65.0° inclination. Primary objective was "to study processes of solar activity, their influence upon inter-planetary medium and the earth's magnetosphere." *Prognoz 1* weighed 845 kg (1863 lbs) and carried instruments to study corpuscular, gamma ray, and x-ray solar radiation; solar plasma flows and their interaction with magnetosphere; and magnetic fields in near-earth space.

Cosmos 486, launched from Plesetsk, entered orbit with 253-km (157.2-mi) apogee, 218-km (135.5-mi) perigee, 89.1-min period, and 81.3° inclination. Satellite reentered April 27. (GSFC *SSR*, 4/30/72; Tass, FBIS–Sov, 4/17/72, L1; *Sov Aero*, 5/1/72, 9; *SBD*, 4/17/72, 120)

- Selection of Kennedy Space Center in Florida and Vandenberg Air Force Base in California as launch and landing sites for space shuttle was announced by Dr. James C. Fletcher, NASA Administrator, in Washington, D.C. KSC site would be used for research and development launches, to begin in 1978, and for all operational flights into easterly orbits. NASA would provide facilities for shuttle users at KSC by modification of existing facilities. Vandenberg AFB site would be phased in toward end of decade for shuttle flights requiring high-inclination orbits. Dept. of Defense would provide basic shuttle facilities. NASA Site Review Board studies had shown Kennedy-Vandenberg combination had cost, operational, and safety advantages over all other U.S. sites. Cost of establishing facilities at KSC was estimated at $150 million; at Vandenberg, $500 million.

    Decision to use KSC was "most gratifying to NASA's launch organization" and ensured "continued utilization of this base and the supporting Eastern Test Range," Dr. Kurt H. Debus, KSC Director, said in statement read at KSC press briefing following announcement of shuttle site selection.

    Dr. George M. Low, NASA Deputy Administrator, explained at KSC briefing why site construction at Vandenberg AFB would cost more than at KSC: "At Kennedy we can make full use of the Saturn V launch complex—Complex 39. The Shuttle will be brought back in there after flight, . . . refurbished there, . . . mated with its booster . . . in the VAB [Vehicle Assembly Building], taken on the trawler transporter to the existing launch pad and launched." KSC modifications would be minor but at Vandenberg "launch facilities . . . are not yet as well suited to the shuttle." Larger modifications would be required, "and this is where the cost is."

Selection between KSC and Vandenberg for a particular launch would depend on required launch azimuth, Dr. Low said. "If we launched something toward the east . . . then it makes sense to launch it from the east coast, whether it is for the Department of Defense or for NASA or for civilian users. When you launch into a polar orbit, or a high-inclination orbit, then it makes more sense to launch from the west coast so you don't overfly land."

In response to question, Dr. Low said NASA did not plan to mate Air Force-supplied booster with shuttle. "NASA will develop, or have developed under NASA contract by industry, the full shuttle including the orbiter, its large tanks, and the solid rocket motor." NASA did have "clear understanding" with DOD that it would use shuttle when it became available and if it met requirements. (NASA Release 72-81; Transcript)

- Final Senate Committee on Aeronautical and Space Sciences hearing on NASA FY 1973 authorization bill heard testimony on space shuttle by Dr. Brian O'Brien, physicist and Chairman of NASA Space Program Advisory Council and of National Academy of Sciences' Advisory Committee to Air Force Systems Command. Committee also heard shuttle testimony from Dr. Courtland D. Perkins, Chairman of Princeton Univ. of Dept. of Aerospace and Mechanical Sciences and Chairman of NASA Space Systems Committee.

  Dr. O'Brien said he believed shuttle would provide "space transportation system which can be available on short notice to meet unforeseen conditions which may result from world events, and . . . great flexibility in our space operations, whether civil or military." It "should make possible a drastic reduction in cost of launches and in cost of payloads destined for earth orbit; while providing the opportunity for refurbishing or repairing satellites in orbit, or returning them to earth. In short, it should prove an invaluable tool in keeping us at the forefront in our operations in space."

  Dr. Perkins said nothing NASA could do "could have a more profound effect on the total space program in the years ahead." Shuttle would provide launch capability for nearly all programs, manned and unmanned, permitting elimination of large stable of rocket boosters and large complex of launch sites. Program provided strong signal to young scientists and engineers that U.S. wasn't "about to throw away our carefully developed competence in space" and it was one of only few new national programs continuing "growth and stretch" in high technology. (Transcript)

- John N. Wilford contrasted U.S. and U.S.S.R. space programs in *New York Times* article: "As American astronauts prepare for another Apollo voyage to the moon . . . their counterparts in the Soviet Union are training for long-duration flights in earth-orbiting laboratories, leaving their own lunar exploration to automatic probes. As the American Pioneer 10 heads for Jupiter, a Soviet spacecraft, Venus 8 is traveling toward a planned softlanding on Venus. As American engineers design a reusable shuttle for ferrying men and satellites into space, the Russians apparently are still trying to develop a giant rocket to give them lifting power as great or greater than the Saturn V. As

American launchings run about 30 a year and the space budget is under public attack, the Russians are launching nearly three times as many spacecraft and, if anything, are expanding their space program." Contrasts had been highlighted in interviews with space officials during Wilford's month-long visit to U.S.S.R. during March. According to Dr. Boris N. Petrov, director of Soviet Academy of Sciences Intercosmos Branch, "the priorities of the Soviet space program will remain the systematic research of near-earth space with the help of automatic stations, manned spaceships and orbital stations. For the moon and the nearest planets, the priority will continue to be research with the help of automated means." (*NYT*, 4/14/72, 10)

- Appointment of Dr. Noel W. Hinners as Deputy Director and Chief Scientist, Apollo Lunar Exploration, in Office of Manned Space Flight was announced by Manned Spacecraft Center. Dr. Hinners had headed lunar exploration department of Bellcomm, Inc. (MSC Release 72–82)
- Use of energy in cosmic plasma to launch spacecraft inexpensively was discussed in *Science* by Dr. Hannes O. Alfven, Univ. of California at San Diego physicist. Energy for space propulsion was being used in "extremely inefficient" way because of low exhaust velocity of rocket gases. Total start weight of spacecraft "must be two or three orders of magnitude larger than the payload." By plasma ejection, "one can easily reach exhaust velocities of several tens of kilometers per second" so that "the ejected mass may be of the same order of magnitude as the payload and the efficiency very high." Energy for launch could be supplied from solar wind, tapping voltage of 1000 v if transfer resistance between ends of conductors and solar wind could be "made small." Or energy could be supplied by transfer from ground through plasma channel in atmosphere. "When Apollo 12 was launched, there were low clouds but no thunderstorm. When the spacecraft had disappeared in the clouds, lightning was observed . . . where the spacecraft had last been seen. This effect was interpreted as due to the exhaust gases which facilitated a discharge of electrostatic charges in the clouds down to the ground. One cannot avoid thinking that it would be preferable to transfer power up to the spacecraft in this way." If spacecraft at launch was "accelerated by two plasma guns at a large distance from each other, and electric arcs are produced between each . . . and the corresponding electrodes on the ground, it is possible that these arcs can be maintained during the launch through the atmosphere." (*Science*, 4/14/72, 167–8)
- Federal Aviation Administration announced publication of 1972 edition of *National Aviation System (NAS) Ten Year Plan and Policy Summary*. Summary detailed FAA's long-range program for development of balanced airport-airways system to meet projected demand for air transportation through 1982. Plan objectives included: completion of semiautomation of en route air-traffic control facilities; provision of basic automation equipment for radar-equipped terminal facilities and automatic metering and spacing of approaches in medium- and high-density terminal areas; increase in instrument-landing-system (ILS)-equipped runways and installation of improved ILS at high-density airports; introduction of microwave ILS; investigation of use of satellite technology for air-traffic control communications, navigation, and surveillance and implementation of satellite communications in oceanic

areas; expansion of designated area navigation routes to shorten en route and terminal flight paths; and plans for landing facilities and systems for vertical or short takeoff and landing (v/stol) aircraft. (FAA Release 72-63)

April 15: Apollo 16 preflight briefing was held at Kennedy Space Center. Apollo Spacecraft Program Manager James A. McDivitt said Apollo 16 countdown was "probably the smoothest" to date. McDivitt announced that Apollo 16 would be last flight with which he would be associated: "I really don't have any plans for the future because I've really been concentrating on Apollo 16, and while I made the decision that it would probably be a good time for me to leave the program, I have not had time to really sit down . . . [and] think through what my future would be." Primary reason for leaving program after Apollo 16, rather than after Apollo 17, was so that if change meant move to new location McDivitt's children could enter new schools at beginning of new school year. McDivitt was Air Force brigadier general and had been commander of *Apollo 9* mission (March 3–13, 1969) and command pilot of *Gemini 4* (June 3–7, 1965) during seven years as NASA astronaut. (Transcript)

- Washington *Evening Star* editorial commented on Apollo 16: While space program had been "worth the money," it was "just a well that this is the next to last manned mission; the most promise henceforth will be in earth-orbital projects. The Soviets appear to be investing much more heavily in those than the United States." (W *Star*, 4/15/72, A4)

April 15–16: Wallops Station held open house as part of joint program of Federal activities. Displays included moon rock and Apollo spacesuit. Television set was available for visitors to watch Apollo 16 launch on April 16. (Wallops Release 72-5)

April 16–27: NASA's *Apollo 16* (AS–511) carried three-man crew on fifth successful lunar landing mission. Lunar module (LM–11) *Orion* landed in moon's Descartes region and two astronauts conducted experiments, rode lunar roving vehicle (LRV), and explored lunar surface. After 71 hrs 14 min on lunar surface *Orion* rejoined orbiting command and service module (CSM–113) *Casper* and astronauts transferred for safe return to earth with lunar samples.

April 16–18: Spacecraft—carrying Astronauts John W. Young (commander), Thomas K. Mattingly II (command module pilot), and Charles M. Duke, Jr. (lunar module pilot)—was launched from Kennedy Space Center Launch Complex 39, Pad A, on time at 12:54 pm EST April 16 by Saturn V booster. Launch was watched by group of invited guests—including Vice President Spiro T. Agnew, Soviet poet Yevgeny Yevtushenko, King Hussein of Jordan, and President Jose Figueres Ferrer of Costa Rica—and estimated half million other viewers near KSC and 38 million TV viewers.

Spacecraft and S–IVB combination entered parking orbit with 175.9-km (109.3-mi) apogee and 166.7-km (103.6-mi) perigee. Insertion into trajectory toward moon was achieved at 2:34 GET in spite of minor anomalies—leak in instrument unit (IU) temperature control system's gaseous nitrogen bottle pressure and malfunction of S–IVB auxiliary propulsion system (APS) helium regulators, which caused continuous

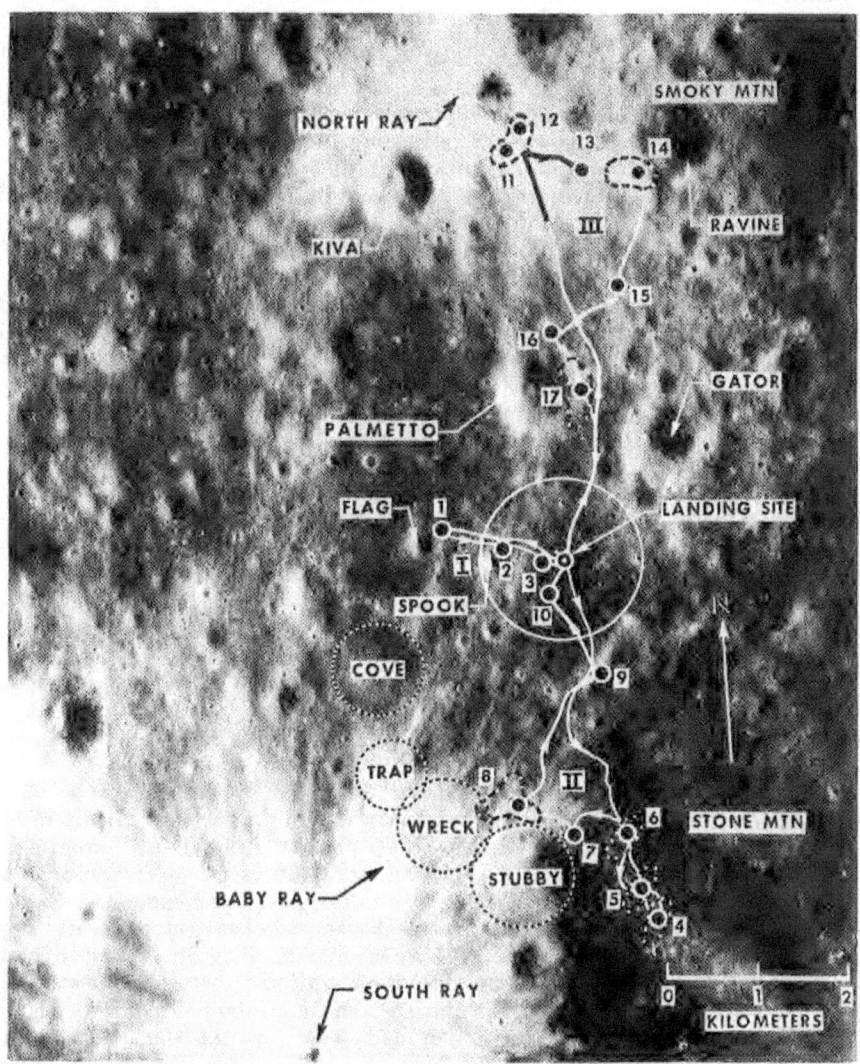

April 16-27: Apollo 16 *carried a three-man crew on the fifth successful mission to explore the surface of the moon. The traverse routes planned for Astronauts John W. Young and Charles M. Duke, Jr., were outlined on photograph of the Descartes landing site obtained by Apollo 14, with geographical names to be used by the astronauts.*

venting. CSM separation from LM/S–IVB/IU at 3:05 GET and docking with LM was shown on color TV for 18 min.

First S–IVB APS burn was near nominal, but because of helium depletion and potential trajectory disturbances, second burn was not made and S–IVB lunar impact operations were terminated. Tracking of stage ended at 27:09 GET when IU signal was lost. S–IVB impacted lunar surface at 75:08 GET (4:02 pm EST April 19) at point estimated to be 1.8° north latitude and 23.3° west longitude. Impact was detected by seismometers at *Apollo 12, 14,* and *15* sites.

Unexplained light-colored particles streaming from LM close-out panel during docking were reported by crew at 7:18 GET. Young and Duke entered LM, powered up, and found all systems normal. Grass-like particles, shown on CM TV, were later identified as shredded thermal paint, but degraded thermal protection was not expected to affect LM operations. First midcourse correction (MCC-1) was canceled because spacecraft trajectory was near nominal.

Electrophoresis demonstration began on schedule at 25:05 GET and appeared to be successful, from crew's description of reaction of charged particles in liquid. Ultraviolet (UV) photographs of earth from 107 300 km (66 700 mi) and 216 600 km (134 600 mi) were taken as planned. Duke reported "spectacular" view of half earth from 201 000-km (125 000-mi) altitude when he could see sun shining on only half of planet. During MCC-2, made at 30:39 GET, services propulsion system (SPS) tank pressure anomaly that had occurred earlier in mission was tracked and minor leak in transducer reference cavity was located. Procedures were sent to crew to account for transducer reading. Visual light flash experiment was started about two hours late, at 49:10 GET, because of problem in aligning spacecraft. Duke counted 70 white, instantaneous light flashes that left no after-glow. UV photography of earth at 327 800 km (203 700 mi) was completed as planned.

Young and Duke entered LM at 53:50 GET for two-hour checkout of LM systems. All system checks were nominal. Scientific instrument module (SIM) door jettisoned at 69:59 GET.

April 19–24: *Apollo 16* entered lunar orbit with 314-km (195.9-mi) apolune and 107.7-km (66.9-mi) perilune after lunar orbit insertion (LOI) burn at 74:28 GET (3:22 pm EST April 19). Young, who had orbited moon in *Apollo 10* (May 18–26, 1969), said moon was "just as fantastic as it ever was." He described black mounds like volcanic craters with white central peaks and said central peak of Tsiolkovsky Crater looked like "a white marshmallow in a sea of hot chocolate." Astronauts also described whitish fracture patterns that looked like chalk scribbling and surface markings like chicken tracks. Mattingly said Humboldt Crater had "every contrast in color on the moon." Material overflowing from Descartes crater looked like cinder fields and submerged craters looked like coral atolls of South Pacific. Young and Duke reentered LM, powered up LM systems, and noted pressure rise in reaction control system (RCS) helium pressure regulator. About 25 kg (54 lbs) of fuel and oxidizer were transferred to LM APS tank in attempt to prevent rupture of burst disc, but disc ruptured at 95:03 GET, decreasing helium source pressure.

CSM–LM undocking and separation maneuver was conducted on schedule at 96:14 GET. When Young and Duke came from behind moon in LM, Duke told Mission Control his helmet was full of orange juice. Duke's microphone apparently had become entangled with tube through which he drank juice stored in his spacesuit. "Every time I turn my head I get orange juice. It's delicious . . . but it's better in your mouth than floating around the cockpit. I've already had an orange shampoo."

April 16–27: *Astronaut John W. Young (above) walked away from the deployment site of the lunar surface experiments package during Apollo 16's first EVA April 21. Components in the background are central station, radioisotope thermoelectric generator, heat flow experiment, and lunar surface magnetometer. The lunar surface drill is at right center and its stems at left center. Below, Young—photographed by Charles M. Duke, Jr., while Thomas K. Mattingly II took lunar and astronomic photos from the command module in lunar orbit—replaced tools in the lunar roving vehicle beside Stone Mountain during EVA-2 April 22. Smokey Mountain is in the background.*

When Mattingly came from behind moon in CSM he reported he had not conducted circularization burn because of malfunction in SPS yaw gimbal servo loop. While flight controllers evaluated problem, LM-powered descent-initiate maneuver was delayed, and LM and CSM maneuvered into station-keeping mode prepared to dock or continue landing procedures. Analysis of problem identified malfunction in secondary circuit that would not preclude lunar landing. Second separation maneuver by CSM with RCS burn was performed and LM began descent to moon on 16th revolution.

LM *Orion* touched down in moon's Descartes region 230 m (755 ft) northwest of planned target point at 104:30 GET (9:23 pm EST April 20). "Wow, down. Old Orion is finally here, Houston. Fantastic," Duke exclaimed. "All we've got to do is jump out the hatch and we've got plenty of rocks." Duke and Young said site was rolling country surrounded by mountains and covered with big boulders. Sunlight was so bright it washed out moon's colors and made lunar surface appear white. Since LM had stayed in lunar orbit six hours longer than planned it was powered down to conserve electrical power and first extravehicular activity (EVA-1) was rescheduled to follow sleep period.

EVA-1 began at 119:05 GET (11:59 am EST April 21). Young climbed down ladder and stepped onto lunar surface. "Okay, Cayley Plains," he said, "Apollo 16 is gonna change your image." Duke followed and astronauts began deploying experiments. Because of problems with LM antenna astronauts' first steps on moon—for first time—were not beamed back to earth. Once Young and Duke set up portable antenna, TV signal and communication with ground controllers were received clearly on earth. TV pictures showed bleak, pale surface resembling *Apollo 15* site, with flat-topped Smoky Mountain to north and Stone Mountain to south. Astronauts said site was rockier, more hilly, and more cratered than previous landing sites. Soil was firm, with rocks scattered over 30% of the surface and small craters covering 70%. During deployment of Apollo lunar surface experiments package (ALSEP) Young tripped over heat flow experiment (HFE) cable, pulling it from connector after Duke had drilled first bore hole. Mission Control later developed way to repair $1.2-million experiment, but procedure was not used because it would have been too complex and too time consuming and included some risk to other ALSEP components. U.S. flag and remaining ALSEP components were deployed successfully and functioned nominally.

Crew deployed LRV and found ampere readings for battery No. 2 off-scale low and rear steering inoperable. About 40 min later, after loading, all meters and rear steering operated properly and LRV performed nominally throughout remainder of EVA.

Young and Duke explored Flag, Spook, Plum, and Buster Craters and collected 20 kg (44 lbs) of samples, including number of white rocks. One crystalline rock coated with bluish glass appeared to have same texture as anorthosite "Genesis rock" collected on *Apollo 15* mission; another was pure white. Some were flecked with green and black glass, some were black and gray, and one weighed about $6\frac{1}{2}$ kg (14 lbs). EVA-1 terminated at 126:16 GET after 7 hrs 11 min 11 sec.

After resting inside LM and reviewing plans for second EVA, astronauts left LM at 142:51 GET (11:33 am EST April 22). They explored

April 16-27: *Stars near the center of the Milky Way galaxy were photographed from the surface of the moon during Apollo 16. An ultraviolet camera/spectrograph, designed and built by the Naval Research Laboratory, made photos in far ultraviolet light with hydrogen light filtered out, recording only the very hot, blue stars (streaked in the photo because of the moon's rotation during the 30-minute exposure).*

Survey Ridge and Stone Mountain and said ridge was pockmarked with small, subdued craters, most about one meter (three feet) wide and some containing smaller craters within main crater. Area around South Ray Crater was strewn with sharp boulders and resembled crater in Nevada where astronauts had trained. Young and Duke turned on LRV TV so viewers could watch them walking and collecting samples high up on slope of Stone Mountain. They reported seeing more of previously discovered white soil under gray surface layers and said soil was very loosely consolidated.

While ascending ridge and traversing very rocky terrain, LRV rear wheels failed to respond at full throttle. LRV continued to move, but front wheels were digging in. Troubleshooting procedures identified mismatch of power mode switching. After change in switch configuration LRV worked properly. Astronauts took core samples and collected rocks that appeared to be volcanic, a greenish rock shaped like a snake's head, and pure crystalline rocks. EVA was televised in color with LRV camera and was extended 20 min. During crew's return to LM, 5-cm (2-in) portion of Young's portable life support system (PLSS) antenna broke off, causing 15- to 18-db drop in signal strength. EVA-2 ended at 150:14 GET after 7 hrs 23 min 26 sec.

Third EVA began 30 min early at 165:45 GET (10:27 am EST April 23) to allow additional time near North Ray Crater and two other stations. Astronauts chipped off samples from enormous basalt rock the size of a house. All planned activities were accomplished, but astronauts had trouble configuring cosmic ray detector for stowage and return to earth. LRV covered 9½ km (6 mi) and reached 17 km per hr (11 mph) going down 15° slope. TV coverage was excellent throughout EVA. Total EVA-3 time was 5 hrs 40 min 14 sec—bringing total EVA time to 20 hrs 14 min 54 sec and total distance traversed to 27.1 km (16.8 mi).

While LM was on moon, Mattingly, orbiting moon in *Casper*, completed lunar and astronomic photography and prepared for rendezvous. He radioed information to geologists at Mission Control to help them plan EVA-3 exploration of Stone Mountain and determine best way to approach North Ray Crater. He located radioactive hot spot on eastern edge of Ocean of Storms south of *Apollo 14* landing site and gamma spectrometer identified thorium, potassium, and uranium. Mattingly also discovered, on far side of moon, volcanic lava flows that resembled maria flows and spilled out over northeast rim of crater high in mountains near lunar equator. Two areas resembled volcanic flows around Flagstaff, Ariz., where crew had trained. Many large basins on moon's far side appeared to be covered with same kind of material that covered area around *Apollo 16* site.

Young and Duke depressurized LM, discarded excess equipment, repressurized LM, and lifted off lunar surface with 96.6 kg (213 lbs) of lunar samples at 175:44 GET (8:26 pm EST April 24). Liftoff was photographed in color by camera on LRV left on moon and was seen by millions of TV viewers. Spacecraft docked successfully and LM crew transferred samples, film, and equipment to CSM. For first time docking was not seen on TV, because LM steerable antenna was inoperable. LM jettison was delayed one day so that astronauts could rest. After LM jettison at 195:12 GET, LM lost altitude and began tumbling, apparently because circuit breaker in attitude-and-translation-controller assembly's primary guidance and navigation system had accidentally been left open. LM ascent stage remained in lunar orbit and was expected to impact lunar surface in about one year. Boom that carried instrument to measure atmosphere of moon was jettisoned because it would not retract and might have broken off. Orbit-shaping maneuvers were canceled to avoid firing of SPS because of degraded backup SPS thrust vector control.

Scientific *Subsatellite* was launched at 196:14 GET (4:56 EST April 24) into lunar orbit with 123.3-km (76.6-mi) apolune and 97.9-km (60.8-mi) perilune. Hexagonal, 40-kg (90-lb) satellite 77 cm long and 36 cm in diameter (30 x 14 in) carried three experiments. *Subsatellite* extended three 1½-m (5-ft) booms, two with magnetometer sensors to measure interplanetary and earth magnetic fields near moon and third for stabilization. Spacecraft also carried particle shadows and boundary layer experiment to collect data on plasmas and solar flares and S-band transponder to detect variations in lunar gravity caused by mascons. Expected lifetime was 6 to 9 mos—shorter than 12 mos planned because shaping burn to optimize orbit had not been performed—but *Subsatellite* crashed into moon May 29.

Transearth injection maneuver at 200:33 GET put CSM on trajectory for earth after 114 hrs 5 min (65 revolutions) in lunar orbit.

April 25–27: Good-quality TV pictures of receding moon from inside CM and of lunar surface from LRV on moon were transmitted between 227 and 229 GET. At 243:35 GET (3:43 pm EST April 25) Mattingly left CSM for inflight EVA to retrieve panoramic and mapping camera film cassettes from SIM on SM. Mattingly made two trips, during which he observed condition of instruments, reported that insulation paint near rocket nozzle was blistered, and deployed and exposed microbial ecological evaluation device (MEED) experiment for 10 min. He reentered CSM after 1 hr 24 min. Scheduled TV press conference began at 268:13 GET and ended at 268:31 GET. Astronauts briefly described moon's far side, including crater Guyot, which seemed to be so full of material that it overflowed and spilled down side of the crater.

CM separated from SM at 290:08 GET, 15 min before entry interface at 121 900-m (400 000-ft) altitude. Drogue and main parachutes deployed normally and CM splashed down and flipped upside-down in mid-Pacific about 5 km (3 mi) from recovery ship U.S.S. *Ticonderoga* at 2:44 pm EST April 27, after 11-day 2-hr (265-hr 51-min) mission.

After flipping spacecraft upright by inflating air bags, astronauts, wearing fresh flight suits, left CM and were carried by helicopter to biomed area on recovery ship for postflight examinations. After being declared in excellent condition, astronauts were flown on following day to Hickam Air Force Base, Hawaii, and to Ellington AFB, Tex. Young had lost 3½ kg (7½ lbs); Mattingly, 3 kg (6½ lbs); and Duke, 2½ kg (5½ lbs), but all were "euphoric, hungry and thirsty, and walking well." CM was retrieved and placed in dolly on board recovery ship. Lunar samples, data, and equipment were flown to Ellington AFB and CM was off-loaded at San Diego.

Mission achieved primary *Apollo 16* objectives: to make selenological survey and sampling of materials in preselected area of Descartes region; emplace and activate surface experiments; evaluate capability of Apollo equipment to provide extended lunar surface stay time, increased EVA operations, and surface mobility; and conduct inflight experiments and photographic tasks from lunar orbit. Launch vehicle and spacecraft systems performance were near nominal except for lack of rate feedback in thrust vector control and yaw oscillations that delayed lunar landing, caused revision of lunar surface activities, and necessitated end of mission one day earlier than planned. All anomalies were quickly analyzed and resolved or compensated for by workaround procedures developed to permit mission to continue. Flight crew performance was excellent. Minor errors made by crew were attributed to fatigue and did not preclude mission success. Mission was officially judged a success.

Accomplishments included largest payload (34 518 kg; 76 100 lbs) placed in lunar orbit, first cosmic ray detector deployed on lunar surface, first use of far-UV camera on lunar surface, longest lunar surface stay time (71 hrs 14 min), longest lunar surface EVA (20 hrs 15 min), and largest amount of lunar samples brought to earth (96.6 kg; 213 lbs). In addition, astronauts returned 3793.5 m (12 446 ft) of exposed film used on lunar surface and in lunar orbit.

Apollo 16 was 13th Apollo mission to date, 10th manned Apollo mission, and 5th successful manned lunar landing mission. *Apollo 15* had been conducted July 26–Aug. 7, 1971. Apollo program was directed by NASA Office of Manned Space Flight. Manned Spacecraft Center was responsible for Apollo spacecraft development, Marshall Space Flight Center for Saturn V launch vehicle, and Kennedy Space Center for launch operations. Tracking and data acquisition was managed by Goddard Space Flight Center under overall direction of Office of Tracking and Data Acquisition. (NASA proj off; NASA Release 72–64K; *NYT*, 4/16–30/72; *W Post*, 4/16–28/72; *W Star*, 4/16–28/72; *B Sun*, 4/18/72, A1; AP, *New Haven Register*, 4/21/72)

April 16: Soviet poet Yevgeny Yevtushenko—first Soviet dignitary to accept invitation to watch Apollo launch—described his reaction to launch to press at Kennedy Space Center. "It was beautiful, it was poetry. It was a great impression." Yevtushenko said he had been even more impressed on his midnight visit to the floodlit launch pad with astronaut David R. Scott, "It's really a beautiful show, this white tender body of a rocket, supported by the clumsy but sometimes tender hands of the red gantry tower. I absolutely had the feeling of one big brother embracing his sister before a long way, a long road. It was wonderful. Silence, not people. No press. Nothing. The sky, the ground, the rocket. It was so beautiful." (Witkin, *NYT*, 4/17/72, 24)

• Astronaut John W. Young had not had "command pilot syndrome"—fastest peak heart rate of crew at launching—when *Apollo 16* was launched, Dr. Charles A. Berry, NASA Director of Life Sciences, said in interview. "The commander has the greatest responsibility and . . . usually has the fastest rate." During *Apollo 16* launch lunar module pilot Charles M. Duke, Jr., had fastest peak rate, 130 beats per min. Command module pilot Thomas K. Mattingly II registered 115 beats per min, and Young—who was making fourth space flight and second trip to moon—registered 108 beats per min. Record had been set by Astronaut Charles Conrad, Jr., whose rate had reached 166 beats per min just before launch of *Gemini 11* Sept. 12, 1966. (Altman, *NYT*, 4/17/72, 24)

• Tass announced *Apollo 16* launch 22 min after U.S. news agencies reported blastoff and issued brief report when spacecraft entered orbit. UPI said coverage was quick by Soviet standards and was part of "gradual expanding coverage the Russians have been giving to U.S. space shots which were all but ignored ten years ago." (*W Star*, 4/17/72, A6)

• Jan. 25 fuel leak in *Apollo 16* spacecraft had added extra $200 000 to cost of mission, *Washington Post* reported. Most of money went to pay overtime to technicians who had worked two weekends bringing vehicle with CSM/LM from launch pad to Vehicle Assembly Building (VAB), where command module had been partly disassembled, and returning spacecraft to pad after repair work on launch vehicle and command module had been completed. (*W Post*, 4/16/72, A3)

- NASA had invited 63 ambassadors to *Apollo 16* launch but had canceled flight scheduled to carry them to Cape Kennedy because only 10 had accepted, *Washington Post* reported. (*W Post*, 4/16/72)
- U.S. was "running out of blank spaces on the map" and needed last frontier of space "to keep the human spirit up," Dr. Wernher von Braun, NASA Deputy Associate Administrator for Planning, said in interview published by *National Enquirer*. "Huge tasks" set for computer industry by Apollo program had boosted that industry from $1-billion-per-year earnings in 1969 to $8 billion in 1972. "The aerospace industry itself is a $27 billion business. These and associated industries which the space program stimulates . . . produce the base for the taxes to support all programs. Let the tax base erode and you will very soon see that we won't be able to afford the social programs." (Adler, *National Enquirer*, 4/16/72)
- *New York Times* editorial commented on space cooperation: "These past few years astronomy has probably advanced further than in any previous century since the invention of the telescope. Fantastic progress has been made through American and Soviet exploration of the moon, Mars and Venus." Need was "to build on this new base of knowledge with a program of further manned and unmanned space exploration conducted under a system of international cooperation which combines the resources, manpower and knowledge of many nations. No generation has ever faced a nobler challenge than space now poses before all humanity." (*NYT*, 4/16/72, 5:12)

April 17: Newspaper editorials commented on *Apollo 16* mission as astronauts headed toward moon from April 16 launch.

Philadelphia *Evening Bulletin*: "The Apollo missions are still high adventure. And if the firing sequence is now familiar, there was still suspense yesterday as the huge Saturn V rocket engines lifted another brave crew of American astronauts . . . . But as the thunder faded away . . . there was the same old question every recent space probe has inspired: What is to be the future of U.S. space research when the Apollo missions are completed?" (*P Bull*, 4/17/72)

New York Times: "The Apollo effort began in 1961 in an aura of cold war Soviet-American rivalry. But yesterday the Moscow television news gave the Apollo 16 launch equal time with events in Vietnam. Soviet-American space cooperation—perhaps a prelude to truly international team-work—could help greatly in easing world tensions. It would be an unexpected but welcome by-product of the Apollo program." (*NYT*, 4/17/72, 33)

- *St. Louis Post-Dispatch* editorial commented on Soviet announcement of plans for Soviet-American space mission: "If successful, the way would be opened for further joint exploratory missions, and for emergency rescues. Hopefully, the experiment would provide a stimulus for further combining . . . space efforts. This would provide a valuable interchange of scientific data and a saving in money. The goal should be co-operation in spacefaring by all the nations of the earth." (*St Louis P-D*, 4/17/72)

April 17–19: Senate Committee on Armed Services' Ad Hoc Subcommittee on Tactical Air Power held hearings on F-14 naval aircraft program. Grumman Corp. Chairman E. Clinton Towl in April 17 testimony said Grumman Aerospace Corp. was "financially unable" to build 48 more

F–14 jet fighter aircraft to complete Navy's order for 134 unless Government restructured contract. Grumman would require at least $2.2 million more per aircraft to compensate for expected losses, plus additional amount to provide profit. Navy was seeking $734 million in FY 1973 for 48 F–14s at current unit cost of $16.8 million. Adm. Elmo R. Zumwalt, Chief of Naval Operations, testified April 19 that Government should "stick to the contract." F–14 was "real winner, one the country really needs," but aircraft could be produced profitably under current contract terms. (CR, 4/17/72, 4/19/72; Madden, NYT, 4/18/72, 1; Vasquez, NYT, 4/20/72, 5)

April 18: Science briefing on magnetic enigmas of moon was held at Manned Spacecraft Center. Finding that rock samples of different kinds collected at all four Apollo landing sites were imprinted with magnetic field from 3 billion to 4 billion yrs ago was reported by Dr. David W. Strangway, Chief of MSC Geophysics Div. Laboratory tests had shown 400 to 500 gammas of local magnetic field would have been needed to leave magnetism observed in lunar rocks—$1/70$ to $1/100$ as strong as magnetism on earth's surface. Earth's magnetism was believed to be caused by earth's rapid spin and molten core but moon spun only once a month and had been assumed to be solid throughout. Analysis had determined that lunar rocks cooled from molten state between 3.6 billion and 4 billion yrs ago and it was "almost certain" that rocks had acquired their magnetism at that time.

Dr. Paul J. Coleman of Univ. of California at Los Angeles reported that findings obtained with *Subsatellite* launched from *Apollo 15* Aug. 4, 1971, had shown marked variations in lunar magnetism on moon's far side, with sharp, even peak near crater Van de Graaf close to center of moon's far side.

Dr. Palmer Dyal, Ames Research Center physicist, said *Apollo 16* efforts to find explanation for observed lunar magnetism would include survey of Palmetto Crater north of landing site to assess possibility that impacts of large meteorites could leave residue of local magnetism. Equipment included magnetometers left on moon by previous missions and lunar subsatellite to be ejected from *Apollo 16* command module. (Transcript)

- *Apollo 16* astronauts' electrocardiograms would be transmitted from space to heart specialists at Stanford Univ. and Univ. of Wisconsin if astronauts experienced irregular heartbeats and potassium loss during mission, Dr. Charles A. Berry, NASA Chief of Life Sciences, said in interview. Dr. Berry said he would stop extravehicular activity if *Apollo 16* astronauts developed symptoms experienced by *Apollo 15* crew. (Altman, NYT, 4/19/72, 1)

- *New York Times,* in editorial on *Apollo 16*'s technical problems, praised entire Apollo effort: "Whatever the outcome of the present Apollo expedition . . . enough has already been accomplished by these vehicles to assure them an imperishable place in the history of technology and science. Each Apollo rocket . . . has been the most complex and complicated machine ever made by man. Each has been composed of millions of parts, all or almost all of which have had to work perfectly for the voyage to succeed." Apollo was "most impressive example of what American engineering can do at its best, and an implicit

rebuke to the more routine areas of technology where far lower standards are too frequently the rule." (*NYT*, 4/18/72)
- Langley Research Center was making major effort to formulate advanced supersonic technology program "to help get the United States back into the competition," LaRC Director Edgar M. Cortright said in speech before National Space Club in Washington, D.C. Center was "one of the last remaining strongholds of hypersonic research" in U.S., with broad-based program that represented "national trends in aerospace research and development." LaRC goals were to help solve critical transportation problems, retain U.S. leadership in civil aviation, improve U.S. military aircraft, develop space shuttle technology, successfully complete Viking project, assist other agencies on civil sector problems in expanding program of space applications, and maintain strong foundation of research in basic disciplines of aerospace sciences. LaRC advanced supersonic technology program envisioned U.S. supersonic transport for 1980s that was one third faster than Concorde, carried three times more passengers and twice Concorde payload, had 50% longer range, and was quieter and 30% cheaper to operate. In hypersonic field, Cortright envisioned aircraft that could fly to Australia in 2½ hrs carrying 300 passengers. (Text)
- Kennedy Space Center announced award of nine-month $140 090 contract to Martin Marietta Corp. to study operational processing of space shuttle payload carriers at KSC space shuttle launch site. Study would develop objectives, technical information, and recommendations for flexible, cost-effective procedures for shuttle processing. (KSC Release 86–72)
- Washington *Evening Star* article cited results of "intolerable pressure" on astronauts' domestic life. Of 73 original astronauts, 2 had remained bachelors and 6 had received divorces. Astronauts were away from families on average three out of four weeks. As mission time approached they were away continuously. Danger in training and "almost constant cross-country flying" added to pressures; 8 married astronauts had died in accidents. (Recer, W *Star*, 4/18/72, C1)
- Belgian artist Paul Van Hoeydonck had sculpted small statue "Fallen Astronaut" that was left on moon by *Apollo 15* astronauts during July 26–Aug. 7, 1971, mission, *New York Times* reported. Name of artist had not previously been disclosed, but Smithsonian Institution recently had asked Van Hoeydonck for replica. (*NYT*, 4/18/72, 51; Smithsonian NASM)
- *Christian Science Monitor* editorial commented on National Academy of Sciences–National Research Council Space Science Board's report, *Human Factors in Long-Duration Spaceflight* [see April 4]: "What intrigues us most is that the board insists on the need for mental self-knowledge and control. To successfully undertake long space missions, men must make togetherness work in confined quarters." Board had noted significantly that "men must learn to recognize when their thinking is manipulated by the fantasies of a brain challenged by feelings of isolation and boredom." Requirements sounded like "basis for more harmonious living on Space Ship Earth. Perhaps this is space flight's larger challenge: to expand our range even a little way into the cosmos requires us to transcend human shortcomings through deeper self-knowledge." (*CSM*, 4/18/72)

April 19: Air Force launched unidentified reconnaissance satellite from Vandenberg Air Force Base by Thorad-Agena booster. Satellite entered orbit with 247-km (153-mi) apogee, 155-km (96-mi) perigee, 88.4-min period, and 81.4° inclination and reentered May 12. (Pres Rpt 73; *SBD*, 4/28/72, 325)

- *Chicago Tribune* editorial commented on *Apollo 16:* "Obviously so high a number as 16 accounts for our headline Monday: 'Blastoff Fails To Excite City.' Human nature is so constructed that novelty stimulates interest, but interest diminishes quickly with repetition." But "however commonplace the recently incredible has become . . . some part of every American is with our latest flying astronauts. We all identify with them as they attempt the successful accomplishment of their exacting undertaking." (*C Trib*, 4/19/72)

- NASA and Air Force had initiated program and developed contract with Lockheed-Georgia Co. to design and install composite-reinforced center wing boxes in two USAF C–130E aircraft, NASA announced. Objective of program was to evaluate concept of selective composite reinforcement of primary aircraft structures. Concept would increase strength and stiffness of metal structures by bonding composite materials to metal. (NASA Release 72–84)

- Filter cassette—filter-paper holder developed by Lewis Research Center and Cleveland (Ohio) Air Pollution Control Div. and used in high-volume air samplers—was being used to simplify and improve collection of air samples over Cleveland area. Cassette had been developed to avoid contamination of filter papers used to collect trace materials. They were being used in 16 Cleveland air samplers as parallel network to city monitoring stations and in 7 LeRC samplers in Cleveland suburbs. (NASA Release 72–82)

- Role of satellite in weather forecasting was described by Howard W. Pollock, Deputy Administrator of National Oceanic and Atmospheric Administration, in speech before Ninth Space Congress at Cocoa Beach, Fla. Invention of earth satellites "might well be called the greatest single technological advance in meteorology since the invention of the balloon." Geostationary and orbiting satellites had provided National Weather Service with "fantastic tools, immense amounts of information, and far greater capabilities than it has ever had before." With orbiting satellites "we now have the potential for vertical temperature profiles at a theoretically infinite number of points around the world." With satellite data "we foresee the day when we will . . . determine . . . the initial state of the world atmosphere in all essential details."

  NOAA was studying disaster warning satellite able to reach every home in every part of U.S. Every home could have a small inexpensive receiver. When natural disaster was forecast, satellite would be instructed to turn on receivers in area to be affected and broadcast appropriate warning with instructions about actions people should take. Fundamental technology for disaster warning satellite in coming decade was "largely available now" and planning was under way. (Text)

April 19–21: Youth Science Congress at Lewis Research Center was sponsored by NASA and National Science Teachers Assn. to stimulate interest in science at high school level. Panel of LeRC scientists and engineers and science teachers from Cleveland, Ohio, schools evaluated papers of 20 students selected from 69 who applied. (LeRC Release 72-26; LeRC PAO)

April 20: NASA TF-8A aircraft, piloted by Thomas C. McMurtry and equipped with supercritical wing, completed three performance evaluation flights from Flight Research Center. Aircraft flew at speeds ranging from mach 0.60 to 1.10 at 10 700-m (35 000-ft) altitude on one flight. On another it tested performance from mach 0.70 to 0.99 at 7300-m (24 000-ft) altitude and from mach 0.90 to 0.99 at 11 000 m (36 000 ft). On third flight, aircraft flew at mach 0.96 and 0.98 at both 7600 m (25 000 ft) and 10 700 m (35 000 ft) and mach 0.95 to 1.18 at 13 700 m (45 000 ft).

Aircraft, with experimental wing designed to decrease buffeting at high subsonic speeds, would continue test flights through May 19 and would have new side fairings installed beginning May 22. (NASA proj off)

- House approved H.R. 14070, $3,429-billion FY 1973 NASA authorization bill [see March 23] by vote of 277 to 60 after defeating amendment by Rep. Les Aspin (D-Wis.) to defer shuttle development until completion of study by National Academy of Sciences. House also defeated amendment by Rep. Charles D. Rangel (D-N.Y.) to delete $3 million for support of NASA tracking station in South Africa. (CR, 4/20/72, H3353-89)

- NASA's *Pioneer 10* Jupiter probe, launched March 2, was continuing toward encounter with Jupiter planned for 9:35 pm EST Dec. 3, 1973. Spacecraft was 36.6 million km (22.7 million mi) from earth and 691.4 million km (429.6 million mi) from Jupiter and was traveling at 10 km per sec (6 mps) relative to earth. Meteoroid detector had recorded 22 hits and asteroid/meteoroid detector had observed 14 objects. Data from other eight instruments were being furnished to investigators. Infrared radiometer would be turned on for checkout April 21 and then turned off until encounter with Jupiter. (NASA proj off)

- Goddard Space Flight Center was awarding $100 000 contract to RAND Corp. to plan and coordinate study for Broadcasting Satellite System. RAND would provide improved analytical tools for technical guidance and support of broadcasting satellite services development. (GSFC PAO; SBD, 4/20/72, 273)

- Dept. of Transportation award of $1.3-million grant to Atomic Energy Commission's Lawrence Livermore Laboratory to assess effects of proposed 1985-1990 high-altitude aircraft on environment was announced by Secretary of Transportation John A. Volpe. Project was part of DOT's Climatic Impact Assessment Program (CIAP). Laboratory would construct sophisticated computer models to predict effect of subsonic and supersonic aircraft on temperature, cloud formation, and stratospheric shielding of potentially hazardous ultraviolet radiation from sun. (DOT Release 38-72)

April 21: *Cosmos 487* was launched by U.S.S.R. from Plesetsk into orbit with 504-km (313.2-mi) apogee, 267-km (165.9-mi) perigee, 92.2-min period, and 70.9° inclination. Satellite reentered Sept. 24. (GSFC *SSR*, 4/30/72; 9/30/72; *SBD*, 4/25/72, 301)

- Plaque placed in auditorium of National Academy of Sciences bore inscription: "The National Academy of Sciences Auditorium is dedicated to all individuals who through the Academy have devoted their talents and knowledge to the service of the Nation and mankind. It perpetuates the memory of Hugh L. Dryden whose friends, by establishing the Dryden Memorial Fund, helped to make this auditorium possible." Dr. Dryden was NASA Deputy Administrator 1958–1965, former Director of National Advisory Committee for Aeronautics (NACA), and former Associate Director of National Bureau of Standards. (NAS Pres Off; NASA Hist Off)

- *Washington Daily News* editorial urged space shuttle funding by Congress: Shuttle was "gamble worth taking—not just because it will provide 50,000 new jobs in the depressed aerospace industry, but because it offers a window on the universe at a relatively modest cost." To retreat from space "after so many billions have been spent and so many men have risked their lives" would be "small credit to a nation that prides itself on perseverance, a pioneering spirit and common sense." (*W News,* 4/21/72)
- Continuation of space program was urged by Edwin McDowell in *Wall Street Journal* editorial: "For our periodic thrusts aimed at solving the age-old mysteries of the cosmos are not just an expensive plaything of the Silent Majority, but a quest whose revelations probably will benefit all mankind." Perhaps "temporary moratorium the administration has placed on space flights is advisable, even though each moon mission is far more productive and scientifically rewarding than each of its predecessors. But abandonment of America's space venture . . . would be disastrous." (*WSJ,* 4/21/72, 8)
- NASA launched Aerobee 150 sounding rocket from White Sands Missile Range, N. Mex., carrying National Center for Atmospheric Research aeronomy experiment. Rocket performed satisfactorily. Scientific objectives were not satisfied. (SR list)

April 22: Richard Witkin commented on clarity of *Apollo 16* TV pictures in *New York Times* article: "The space agency has a $46,008 contract with a three-month-old California company to process signals from the moon and improve the quality of Apollo television reception. The verdict so far from *Apollo 16* scientists, the primary beneficiaries, is that the money is being well spent. And there appears to be no quarrel from the viewing public." What company, Image Transforms, Inc., had been doing to signals from *Apollo 16* was being kept confidential—even from NASA—to protect seven existing and two pending patents. Signals were sent from *Apollo 16* to Houston and were immediately sent to Image Transform's North Hollywood, Calif., plant and back before being displayed on NASA screens and being sent to networks. Extra delay was only one fifth second. NASA had arranged image processing so geologists could give astronauts better input on which rocks to collect on moon. Better reception for home TV viewers was side benefit. Process had been developed by Canadian John Lowry, coowner of company. (*NYT,* 4/22/72, 16; *LA Times,* 4/23/72)

- Change in *Apollo 16* schedule necessitated by command module problem had at least tripled cost of beaming TV coverage of mission to Europe, *Washington Post* reported. European Broadcasting Union had planned to record coverage in New York and beam pictures to viewers via satellite. When CM problem developed, TV officials had decided situa-

tion warranted immediate "live" coverage. Later network decided to beam moonwalks live, as well. (*W Post*, 4/22/72, A10)
- Newspaper editorials commented on *Apollo 16* mission as astronauts conducted second extravehicular activity (EVA) on lunar surface.

  New York Times commented on April 20 decision to continue *Apollo 16* mission despite technical problems: "Thursday's dilemma and its solution were particularly useful in reminding all concerned that there are two kinds of errors that can be made in situations of this type." There was possibility of mission directors "deciding to continue the journey in the face of a danger that brings disaster. Against that is the alternate risk of being overly cautious and aborting a mission unnecessarily, at huge financial and scientific loss." It was "delicate dilemma" and "tribute to the Project Apollo leadership and organization that to date these voyages have foundered on neither rock." (*NYT*, 4/22/72, 30)

  Cleveland *Plain Dealer:* "Science has been well served. Mankind has made its giant steps. The pure, fresh enthusiasm of Astronaut Charles M. Duke, Jr., with his 'Yahoo!' and his 'Fantastic!' is infectious, but it cannot waken again the public's wonder over the first moonwalk." Everyone wished *Apollo 16* well but "they have seen this show before. What refinements have been added in this latest version are obscurely technical. The Apollo program is aging, and the American public . . . is quite willing to see it come to a close." (Cl *PD*, 4/22/72)
- Montana State Low Income Organization, activist group, had sent box of rocks to President Nixon to protest *Apollo 16* mission, UPI reported. Group had said purpose of mission appeared to be merely collection of rocks; Montana rocks were free and represented areas in state where social problems needed Government attention. (*NYT*, 4/23/72, 60)
- Among strongest arguments for funding space shuttle was its great military usefulness, D. J. R. Bruckner said in *Washington Post* column. Argument might persuade Congress, "but do they really need another weapon?" Shuttle's greatest use would be "as a tool supporting research, including the construction and maintenance . . . of a research station on the moon's surface." But that was insufficient reason for Congress. "If they cannot convince themselves that it will help destroy something, they won't put up the money for it." Apollo program had destroyed nothing. "It only added a cubit to the stature of every American in the eyes of mankind, . . . benefited the understanding, the imagination and the spirit of humanity . . . [and] opened up enormously fruitful areas of knowledge about how the world we live in was formed and how it works." (*W Post*, 4/22/72, A14)
- Belgium released annual Postage Day stamp, depicting *Apollo 15* Astronaut David R. Scott postmarking U.S. lunar rover stamp on moon Aug. 2, 1971. (AP, *P Bull*, 4/23/72)

April 23–29: Anglo-French Concorde supersonic airliner and Soviet Tu-144 supersonic transport were displayed at Hannover Air Show in West Germany. Reportedly they would not appear at U.S. TRANSPO '72 exposition scheduled to be held at Dulles Airport in Virginia in May. (*W Star*, 4/30/72, C12)

April 24: Second set of reports by NASA and Soviet Academy of Sciences Working Groups which met in Moscow Nov. 29 to Dec. 6, 1971, had been approved by NASA and Soviet Academy, NASA announced. Meetings, to define technical requirements for rendezvous and docking of U.S. and U.S.S.R. spacecraft, had not resulted in definite decision to commit either U.S. or U.S.S.R to joint manned space mission. Reports indicated Working Group 1 had completed general documentation on life support systems, coordinate systems, and constraints on spacecraft configuration and had agreed on objectives and preliminary documentation for possible test mission, exchanges on test windows, program elements, and required communications channels. Working Group 2 had listed guidance and control systems and onboard spacecraft equipment to be compatible and had nearly completed documentation on lights, docking targets and contact conditions, control systems, and radio tracking. Working Group 3 had agreed on basic values for compatible docking system, including diameter of tunnel, and to create scale model of docking system that would verify parameters and ensure compatibility in early development stage. (NASA Release 72-88)

- Possible discovery of historical point of origin of life from study of rocks believed to be 3.4 billion yrs old was announced by Ames Research Center scientist Dr. Keith A. Kvenvolden. Dr. Kvenvolden had conducted research with Dr. J. William Schopf and Dorothy Oehler of Univ. of California at Los Angeles on rocks Dr. Kvenvolden brought back in 1968 from Onverwacht Strata in Barberton Mountain Land near South Africa's border with Swaziland. Rocks showed fossil evidence of beginnings of photosynthesis—life process of plants. Ancient carbon found in rocks which could predate photosynthesis was similar to organic carbon in meteorites believed to have come from asteroid belt. Immediately above was layer of identical pre-Cambrian rock with evidence of carbon produced by early photosynthetic organisms. Lack of photosynthesis in lowest layers of rock fitted theory of chemical evolution of origin of life. (UPI, *W News*, 4/25/72, 10; *Time*, 6/19/72)

- NASA launched Aerobee 170 sounding rocket from White Sands Missile Range, N. Mex., carrying Lockheed Missiles & Space Co. solar physics experiment. Docket and instrumentation performed satisfactorily. (SR list)

- *New York Times* editorial commented on *Apollo 16* mission now in progress: "It will take many months and even years before the full harvest of Apollo 16's scientific contribution is in. But even now there is much material for scientists to analyze. It is information made available by the superb television pictures received here on earth, by the verbal descriptions of the astronauts and by the data radioed back to this planet from the various automatic sensors . . . on the lunar surface." Preliminary conclusion was "that in some important ways the moon is now even more mysterious, even harder to understand, than it was before Apollo 16 took off. The additional knowledge of lunar conditions available from last weekend's exploration indicates that the reality of the moon's origin and history is even more complex than previous theories had assumed." (*NYT*, 4/24/72, 34)

April 24: *Dr. Keith A. Kvenvolden, organic geochemist at Ames Research Center, displayed a 3.4-billion-year-old sedimentary rock thought to contain evidence of the beginning of photosynthesis on earth—the historical point of the origin of life.*

- Edward P. Andrews, Manager of Skylab Spacecraft and Skylab Launch Vehicles at NASA Hq., became Director of Systems Operation in Space Shuttle Program. (*NASA Activities*, 5/15/72, 105)
- West German government was preparing blueprint for aerospace development for remainder of decade, *Aviation Week & Space Technology* reported. Basic document for government financing and support would establish priorities in military and civil aerospace development and provide government and industry with "idea of how much room remains for maneuver in regard to implementing new projects." Theme would be "multi-national cooperation." (*Av Wk*, 4/24/72, 32-41)

April 24-25: American Physical Society held annual spring meeting in New York. Team of Columbia Univ. scientists reported observations made with 1971 rocket experiment to study x-rays from Crab Nebula had supported view that pulsars were hurling high-energy particles, or cosmic rays, into space constantly at close to speed of sound. Experiment had shown that rays were polarized. X-rays from space could not penetrate atmosphere. (Sullivan, *NYT*, 4/26/72, 11)

April 24-26: At annual meeting of National Academy of Sciences (NAS) in Washington, D.C., members voted to reorganize National Research Council (NRC) to deal more effectively with major problems of U.S. society. Members also adopted resolution that NAS projects and studies should be unclassified. If leadership approved contract for classified project, membership would receive unclassified résumé explaining proposed study. If 10 or more members from at least 2 scientific institutions considered project inappropriate, they could present case to NAS's governing body. Resolution on foreign policy asked President Nixon and Congress to develop and apply science and technology in industry, agriculture, and health for furtherance of human welfare and called for de-emphasis on reliance on military force, direct or indirect.

NAS James Craig Watson Medal of Science was presented to Dr. André Deprit, NRC postdoctoral resident research associate at NASA Goddard Space Flight Center. Citation was for "adaptation of modern computing machinery to algebraic rather than arithmetical operations." Award consisted of gold medal and $2000 honorarium. (Program; Schmeck, *NYT*, 4/27/72, 27; *W Star*, 4/23/72, D3)

April 25: Senate Committee on Aeronautical and Space Sciences approved H.R. 14070, $3.420-billion FY 1973 NASA authorization bill after inserting amendment that decreased by $9 million authorization of $3.429 billion passed by House April 20. As approved by Senate Committee, authorization was $41 million above original NASA budget request of $3.379 billion and $12.5 million above amended budget request of $3.420 billion. Bill, reported to Senate May 3, allocated $2.613 billion for research and development (R&D), $77.3 million for construction of facilities, and $729.5 million for research and program management.

House recommendation of $211.9 million for aeronautical research and technology was reduced to NASA request level of $163.4 million. In Senate report, Committee praised increase in aeronautical R&D motivated by 1971 Civil Aviation Research and Development (CARD) Policy Study of NASA and Dept. of Transportation, but said it doubted additional funding of $48.5 million approved by House could profitably be spent during FY 1973.

Committee disapproved cancellation of development of 333 600-newton (75 000-lb)-thrust rocket engine in Nuclear Engine for Rocket Vehicle Application (NERVA) program and recommended new authorization of $8.5 million as requested by NASA, plus allocation of $16.5 million in unused FY 1972 funds, for engine's development in FY 1973. Senate bill restored $4 million deleted by House from funds for High Energy Astronomy Observatory (HEAO) and allocated total of $156.6 million—NASA-requested level—for Physics and Astronomy program. Bill added $12.5 million to NASA request of $194.7 million for space applications and stipulated $5 million out of total of $207.2 million be used for small applications technology satellite project. But it disapproved $4 million House had added for Earth Observatory Satellite project. Committee recommended that NASA present space shuttle as separate program item on FY 1973 budget request, restructure manned space flight budget for "better visibility of total effort," and formulate vigorous program to study development of terrestrial uses of solar energy. (S Rpt 92-779; Com Off)

- Newspaper editorials continued comment on *Apollo 16* mission, as astronauts headed back to earth from lunar surface.

  Christian Science Monitor commented on efficiency with which emergencies had been handled: "We very much appreciate the fine job the Apollo 16 crew has done. Nevertheless, for us, the star of the mission has been the impersonal intelligence expressed throughout the Apollo program. This is what underlies Apollo 16's success and has added another important reference point to our growing knowledge of the moon." (*CSM*, 4/25/72)

Chicago Daily News: "This expedition, the fifth and next to last of the moon landings, revealed growing confidence and ability to improvise, and an increasing breadth of engineering genius that will carry over into many Earthbound projects. The advance in 'pure science'—mainly in knowledge as to how the universe is put together—is impossible to estimate. Somewhere down the line, whatever is learned on the moon trips should add to man's ability to cope with his environment. Once more, gratitude is due to the courageous astronauts . . . who carried out their assignment so superbly." (C *Daily News*, 4/25/72)

- West German pilot Hans-Werner Grosse set new world record in long-distance straight-line glider flying with 12-hr, 1450-km (900-mi) flight from Luebeck Airport, West Germany, to Biarritz on Spanish border. Flight, in ASW-12 all-plastic glider, was 273.5 km (170 mi) longer than previous record set by U.S. pilot Wallace Scott. Grosse cruised at altitudes to 2300 m (7500 ft) pushed by 56-km-per-hr (35-mph) winds. (Stueck, *NYT*, 5/28/72, 42)
- House passed by vote of 329 to 16 H.R. 14108, $673.8-million National Science Foundation FY 1973 authorization. Bill included additional $7 million in foreign currencies for expenses incurred outside U.S. (*CR*, 4/25/72, H3494–3508)
- Secretary of Transportation John A. Volpe announced award of two Federal Aviation Administration contracts for development of plans for quiet, short-haul air transportation system. Battelle Memorial Institute was awarded $243 600 for structured program planning based on wide range of viewpoints within public and private sectors. Urban Systems Research and Engineering, Inc., was awarded $202 400 to analyze strategies for development of vertical or short takeoff and landing (v/stol) aircraft system in terms of decision making under uncertain conditions. (FAA Release 72-71)
- Federal Aviation Administration released aviation forecasts for FY 1972 through FY 1983: U.S. passenger emplanements would increase from 170 million in FY 1971 to 174.9 million in FY 1972 and 189.7 million in FY 1973—gain of about 3% in FY 1972 and almost 9% in FY 1973. By end of 1983, emplanements were expected to reach 475.5 million—almost three times FY 1971 level. Revenue passenger-miles recorded by U.S. airlines were expected to grow at annual rate of about 10.5%, to reach 445 billion in FY 1983, more than three times total of 132.4 billion in FY 1971. Total hours flown by general-aviation aircraft was expected to increase almost 4% in FY 1972 and 5% in FY 1973, to 28.8 million. By 1983, total hours were expected to reach 46.2 million. General-aviation fleet was expected to grow from 131 407 aircraft in January 1971 to 212 000 in January 1983. (FAA Release 72-67)
- National Science Foundation reported rise in share of Federal research and development (R&D) expenditures on functions other than space and defense. During 1963–1973, 10 other budget functions shifted

upward in R&D program emphasis. Until 1966 space research and national defense spent 90% of Federal R&D total. Joint share of these two functions was expected to be 77% in FY 1973. Share of other 10 functions had risen 5% between 1970 and 1973 alone. (NSF *Highlights*, 4/25/72, 1)
- Importance of high technology to U.S. future was discussed by Dr. Karl G. Harr, Jr., President of Aerospace Industries Assn. of America, Inc., in speech before Aero Club of Washington, D.C.: "If we as a nation are to continue to base our national security on modern forces-in-being, relying on qualitatively superior weaponry, then we must adequately fund both research & development and procurement." In space exploration "our national willingness or unwillingness to proceed with this effort so rich in demonstrated benefits of all kinds, will foretell our spiritual strength to face the future." (Text)
- Environmental Action, group of 31 scientists and professionals, issued statement in Washington, D.C., urging Congress to deny Administration's request for $500 million to build demonstration model of nuclear breeder reactor to generate electricity. Group, which included Nobel Prize winners Dr. Harold C. Urey (physicist) and Dr. Linus C. Pauling (chemist), questioned safety of nuclear reactors and favored concentration of coal to generate power. (Cowan, *NYT*, 4/26/72, 7)

April 26: Air Force SR-71 reconnaissance jet aircraft set record for sustained speed at high altitudes in 10½-hr flight from Beale Air Force Base, Calif. Aircraft, piloted by L/C Thomas B. Estes (USAF), flew 24 100 km (15 000 mi) at speeds over mach 3 and altitudes above 24 400 m (80 000 ft). Flight plan included two round trips across northern and central U.S. and complete circle of western states, equivalent to nonstop flight from San Francisco to Paris and return. (Haughland, AP, *W Post*, 7/17/72, A3)
- Lockheed L-1011 TriStar wide-body jet transport flew first passengers on Eastern Airlines flight between Miami and New York, via Atlanta, Ga. (Witkin, *NYT*, 4/27/72, 85)
- *Apollo 16* astronauts were "klutzes on the moon," unable to repair damage done to their equipment through "clumsiness," controversial columnist Nicholas von Hoffman said criticizing Apollo program in *Washington Post*. "As more and more people come down with raging cases of dull skull from watching this repetitious ennui they become converts to the no longer very heretical proposition that this money might be better spent on schools or sewers. An alluring but flawed idea. The dough would be shifted into bomb procurement." (*W Post*, 4/26/72, B1)
- *San Jose* [Calif.] *Mercury* editorial, commenting that *Apollo 16* astronauts were safely on their way home "after a hugely successful exploration of the highlands of the moon," said: "Space exploration should continue to command a relatively high place in the nation's list of priorities; five trips to the moon have only served to whet the appetite for more and better knowledge of the earth's nearest neighbor." (*San Jose Mercury*, 4/26/72)

April 27: President Nixon telephoned *Apollo 16* astronauts aboard U.S.S. *Ticonderoga* following splashdown, to offer congratulations on successful completion of mission. He invited astronauts to be special guests at state dinner for Mexican President Luis Echeverria Alvarez in June and said he would nominate L/Cdr Thomas K. Mattingly II (USN) for rank of commander and L/C Charles M. Duke, Jr. (USAF), for rank of colonel. President then issued statement from Key Biscayne, Fla.: "The journey of Apollo 16 has ended, but the contributions of this mission to scientific progress have only begun. Rarely if ever has so much new information been made available to science in such a brief period. As the work of evaluation and analysis goes forward, the impact of Apollo 16 will be felt for many years to come.

"On behalf of all Americans, I am pleased to welcome Astronauts Young, Duke, and Mattingly back to earth and to salute them for a job well done. Their skill, their courage, and their enthusiasm have written another proud chapter in the stirring story of mankind's struggle to unlock the mysteries of the unknown." White House press officer later said President had asked possibility of promoting Capt. John W. Young to admiral. He was told Capt. Young had received two recent promotions and Navy would not permit third at this time. (PD, 5/1/72, 796; NYT, 4/28/72, 22)

- NASA's TF-8A aircraft, equipped with supercritical wing, completed two flights from Flight Research Center. Objective was to check repeatability with two different pilots—Thomas C. McMurtry and Gary Krier. Wake surveys and pressure distributions were run at constant dynamic pressure of 9600 newtons per sq m (200 psf) and mach 0.90, 0.95, 0.97, and 0.99. (NASA proj off)

- Low-altitude qualification test of Viking parachute—decelerator system to slow Viking spacecraft for softlanding on Mars 1976—was conducted by NASA at Dept. of Defense Joint Parachute Test Facility in El Centro, Calif. Parachute, 16-m (53-ft) in diameter and deployed by mortar, was dropped from B-57 aircraft. It opened on schedule, but dropped payload. Test was first of two low-altitude tests planned for 1972 and followed series of six development tests. High-altitude tests with balloon-launched parachute would be conducted during summer. Investigation was under way to determine cause of failure. (FRC Release 6-72; W News, 4/28/72, 45)

- Marshall Space Flight Center announced selection of Ball Brothers Research Corp. for negotiation of contract to develop heavy nuclei experiment for High Energy Astronomy Observatory (HEAO-A), scheduled for launch in 1975. Experiment would measure charge-to-mass ratio, abundance, and energy of cosmic ray particles. (MSFC Release 72-53)

- Secretary of Defense Melvin R. Laird announced appointment of L/G Samuel C. Phillips (USAF) as Director of National Security Agency, effective Aug. 1. Gen. Phillips, assigned to NASA in 1964, had been Apollo Program Director at NASA Hq. until Sept. 1, 1969, when he became Commander of Air Force Space and Missile Systems Organization. He would replace Adm. Noel A. M. Gayler (USN), who would become U.S. commander in chief in Pacific area. (W Star, 4/27/72, A2)

April 28: Soviet President Nikolay V. Podgorny sent message congratulating President Nixon on successful completion of *Apollo 16* mission and safe return of astronauts. He asked President to convey "best of wishes to the members of the spaceship crew, courageous cosmonauts

April 28

John Young, Charles Duke and Thomas Mattingly." (FBIS–Sov, 4/28/72, G2)

- Dr. Charles A. Berry, NASA Director of Life Sciences, told press at Manned Spacecraft Center briefing *Apollo 16* astronauts had returned from mission in excellent condition. "They could walk well. And there wasn't any obvious effect . . . on the ability to stand up well without having blood pressure drop. And . . . there was no motion sickness" even though astronauts had spent about 30 min in spacecraft in fairly rough sea after splashdown. All three astronauts had minor irritation around sensor attachments on the skin. All three had normal white blood cell counts rather than increased counts usually seen on previous missions. Their response to postflight tests showed they were in "better shape" than *Apollo 15* crew had been. Cardiac silhouette size appeared to have decreased; only astronaut to have cardiac arrythmic contractions was Charles M. Duke, Jr., who had three. Data on potassium levels was not yet available. Astronauts would be on rigid diet and undergo extensive tests for three days so doctors could compare postflight medical measurements with those taken during three-day preflight controlled diet period. (Transcript)

- U.S. press commented on safe April 27 return of Apollo 16 astronauts.

  New York Times: Mission had "reminded us once again that man's study of the moon is but in its infancy, and that the great bulk of the moon's territory is still completely unexplored." Yet, "only one more manned lunar mission is now scheduled." As Apollo program ended, "its enormous successes underline the opportunities on the moon and compel all men to lift their eyes to the heavens if they would understand whence and how our species, our world and our solar system came into being." (*NYT*, 4/28/72)

  Atlanta Journal Constitution: "These three strong young men are heros in the truest sense of the term. They have pitted their courage, their physical strength and their minds against a multiplicity of awesome forces which singularly or collectively could totally obliterate them. And they did it for the noblest of reasons—the pursuit of knowledge. If mankind has any claim to greatness, surely it is men like this that give it credence." Editorial also approved space funding: "It is appropriate to note, just when the next-to-last planned Apollo spacecraft is safely home from the moon, that a few days ago the U.S. House voted approval of a $3.4 billion space agency budget bill. Part of the money will go for the development of a space shuttle, the next logical step in the great adventure of exploring space. The technological advances that are part of the space program have already proven themselves of value here on earth. It's likely that this kind of extra technical dividend will continue to lead to break-through areas not even related to space." (*Atlanta JC*, 4/28/72)

  San Francisco Examiner: ". . . Apollo 16 may well turn out to be the most rewarding of all our moon expeditions. All Americans can take pride in a feat so remarkably compounded of scientific precision and human courage." (*SF Exam*, 4/28/72)

  Chicago Daily News: "With experience, the crews and the thousands of technical experts supporting them from the Earth have learned to correct the minor flaws and juggle the entire intricate timetable when necessary. It seems a pity that much of the experience so dearly gained is soon to be put on the shelf. . . . But the lessons of Apollo can be applied at least in part to the portions of the space program that will remain. . . . And if the pioneering aspects of the space program are about over, its benefits will continue—most spectacularly in the weather and communication satellites, but extending also into more mundane improvements in daily living." (*C Daily News*, 4/28/72)

- Proposed general reorganization of Flight Research Center was approved by Dr. James C. Fletcher, NASA Administrator. Under new organization FRC would be divided into four directorates and subdivided into divisions and branches. (FRC *X-Press*, 5/12/72, 2)
- Existence of 10th planet in solar system, three times as large as Saturn and twice as far as Neptune from sun, was suggested by scientists from Univ. of California's Lawrence Livermore Laboratory. Mathematician Joseph L. Brady told press team had used computer to process mathematical observations on deviations in orbit of Halley's Comet. Presence of 10th planet, far beyond Pluto, had been predicted from mathematical computations. (UPI, *W Post*, 4/29/72, A3; *Publications of the Astronautical Society of the Pacific*, 5/72)

April 29: Apollo 16 astronauts John W. Young, Charles M. Duke, Jr., and Thomas K. Mattingly II were greeted by crowd of 2500 persons at Hickam Air Force Base, Hawaii, and presented copper plaques bearing etching of Hawaii's state capitol. Young said mission's success "shows the real teamwork that's going to keep this country great. We really enjoyed the trip . . . but we sure are glad to be back." (AP, *W Star*, 4/30/72, A3)

- *Washington Post* editorial commented on *Apollo 16:* ". . . what has struck us about the moon flights is that so little has gone wrong and how much of what has gone wrong has been correctable. It is a sad commentary about the quality of life on earth when it takes days to fix a television set or a refrigerator while it takes only hours to analyze and overcome an electronic flaw in a craft that is circling the moon." (*W Post*, 4/29/72, A18)

April 30: Apollo 16 discoveries were described in *New York Times* article. Mission had been "longest, most ambitious and, in many respects, most productive" in Apollo series. Discoveries included fact that Descartes region of lunar highlands contained almost no specimens of unaltered volcanic rock despite region's age, "older than the seas and the more prominent craters." Region had been found to contain breccia rock that had been "fragmented, jumbled, and cemented together again." High magnetism observed at sites visited by *Apollo 16* astronauts might provide clue to moon's early history, since analysis of lava rocks on earth had indicated earth's magnetic field flipped over at intervals of thousands or millions of years, with shift requiring centuries. Reverse polarity discovered in rocks near *Apollo 16* landing site might mean moon had done magnetic flip at some time in its early history. Among experiments whose outcome was still uncertain was attempt to photograph and analyze ultraviolet light from hydrogen clouds thought by some scientists to exist and to be "missing mass" that would explain gravity assumed to be keeping clusters of galaxies from flying apart. (Sullivan, *NYT*, 4/30/72, 3)

- Marshall Space Flight Center announced it had designed and built compact shower assembly for use on Skylab earth-orbital missions

April 30

beginning in 1973. Astronauts would step inside ring on floor and raise fireproof beta cloth curtain on hoop and attach it to ceiling. Flexible hose with push-button shower nozzle could spray 2.8 liters (3 quarts) of water from personal hygiene tank during each bath. Used water would be vacuumed from shower enclosure into disposable bag and deposited in waste tank. (MSFC Release 72–38)

- Full-scale model of Boeing supersonic transport costing $10 million would be stored at Boeing facility near Seattle, Wash., another year for further "technical evaluation," Washington *Evening Star* reported. Federal Aviation Administration had said technical benefits to be gained by retaining mockup would outweigh proceeds from its sale. (W *Star*, 4/30/72, C12)
- *Apollo 16* astronauts John W. Young, Charles M. Duke, Jr., and Thomas K. Mattingly II underwent detailed medical examination and began technical debriefing at Manned Spacecraft Center. Dr. Willard Hawkins of MSC said astronauts looked good, were "in good physical shape," and had normal responses. (AP, B *Sun*, 5/1/72, A3)

During April: *Soviet Space Program, 1971* was released by Senate Committee on Aeronautical and Space Sciences as supplement to Senate Document 92–51 [see Jan. 6 and 22]. Report—prepared by Library of Congress under direction of Dr. Charles S. Sheldon II, Chief of Science Policy Research Div.—said 1971 had been peak year in Soviet space activity, with 83 successful launches against 31 for U.S. Soviet successes had included "remarkable performance of the roving lunar vehicle, Lunokhod 1, the versatile manned space station Salyut in which a manned world duration record was set, the arrival in Mars orbit of two very heavy payloads, Mars 2 and 3, and the placing in lunar orbit of the heavy automated laboratory Luna 19." On negative side had been inability to develop large G booster, death of Soyuz cosmonauts, crash of *Luna 18* on moon, failure of first Mars probe, and failure of *Mars 2* and *3* landers to carry out planned experiments. U.S.S.R. in 1971 had flown manned-precursor spacecraft "with greater maneuverability than the present ones, quite possibly testing the advanced propulsion needed for manned lunar flight." Soviet spokesmen had reiterated commitment to manned space stations, applications satellites, and unmanned planetary exploration throughout solar system. Manned lunar and planetary flights "continued to get mention as a later goal." U.S.S.R. was expected to launch another Salyut to be visited by Soyuz 12 and 13 for long stay. Thus far U.S.S.R. had given space program "continued verbal support" and had demonstrated "commitment of hardware which has been rising for the 15 years flights have been undertaken." (Committee Print)

- Space shuttle was "investment in the future," Rep. Olin E. Teague (D-Tex.), Chairman of House Committee on Science and Astronautics' Subcommittee on Manned Space Flight, said in *Aerospace* article. Controversy over manned versus unmanned space flight had been "negated" by advancing technology that provided "entirely new approach to space operations." Shuttle combined "advantages of man in the cycle with a degree of economy hitherto unobtainable." Fundamental reason for shuttle program was "to make available a means of routine access to space, to remove the constraints imposed by an earlier level of technology, to progress from space adolescence to full maturity." (*Aerospace*, 4/72, 3–9)

- Relation of U.S. space program to national interest was discussed in *Foreign Affairs* article by Dr. Robert Jastrow, Director of NASA's Goddard Institute for Space Studies, and Dr. Homer E. Newell, NASA Associate Administrator: "To some, space is, or should be, pure science; to others, it is prestige and the American image; to still others, space means national security." Some space program students had said "its impact will be felt in other respects than the cost-accounting of economic productivity versus technological investment. There has been much discussion of the influence of space exploration on the mind and spirit of man, as an extension of the revolution in thought that was initiated by Copernicus and continued by Newton and Darwin. A study of the history and pre-history of man suggests that the human drive is expressed through just such tentative movements out of the world of tried experience into a new world of untested promise.

  "Sputnik and the moon landing signaled the opening of a new frontier across which man can now travel into the endless reaches of outer space. The new frontier of space will not be closed quickly, for astronomical knowledge shows that billions of stars, some undoubtedly accompanied by earthlike planets, surround us in the galaxy. Space exploration has brought home to more people than ever before the reality of this vast complex of stars and planets. 'Out there' is really there, a place that one can get to. Never again can we make the mistake of identifying our speck of planetary matter as the universe." (*Foreign Affairs*, 4/72, 532–44)

- British press correspondent Arthur Smith presented British view of U.S. and U.S.S.R. space programs since *Apollo 11* (July 16–24, 1969) in *Bulletin of the Atomic Scientists*: "Two major events in space marked 1971—both of them signs that America's formidable lead in space exploration is being whittled away by the incredibly dynamic program" of U.S.S.R.: launch of *Soyuz 11*, which put U.S.S.R. ahead of U.S. in total launches for first time since 1958, and U.S.S.R.'s landing of *Mars 3* capsule on Mars—"relatively unproductive from the scientific point of view," but impressive. "These two events demonstrate that the Soviet space program is now making great strides after many years of relatively minor achievement in both manned and unmanned exploration and application. The characteristic pattern was a Soviet achievement announced to a surprised world, followed some months later by a more sophisticated and more useful U.S. version of the same mission, yielding far more scientific data, but little prestige."

  U.S.S.R. program was characterized by excessive secrecy because, according to Soviet officials, rockets used in space program were similar to those used in military program. Smith said genuine reason for secrecy "is the desire to guard against a failure in the full blaze of world publicity." U.S.S.R. "guards against ever having to admit failure by announcing launches only post facto. If there is a launch pad failure there is no announcement at all. If orbit is reached by the spacecraft, but it does not function as planned, no mention is made of this in the curt bulletin." If lunar or planetary probe failed to enter planned trajectory, mission objective "is not mentioned, and the spacecraft is merely allocated a number in the catch-all Cosmos series. . . . It is impossible to be absolutely certain how many of these escape stage failures there have been, but there may have been as many as 10 or 15 aimed at Venus and Mars alone."

Soviet program had "little . . . of the technical virtuosity of the American approach," but had more impetus and would "obviously continue on a growing scale. Apart from the progress in individual projects, the ever increasing number of Soviet launches demonstrates that the Russian equivalent of NASA is financially more secure. And the speed with which the Soviet space establishment can respond to a failure is also a pointer to the massive scale of the back-up precautions which are taken." Soviet program, by its nature, was ensured greater financial security. "By their very success in being able to reach an unparalleled level of technology in their spacecraft, the American engineers have been working against their own best interests." (*Bull Atom Sci*, 4/72, 18–24)

- NASA published *Apollo 15 Preliminary Science Report* (NASA SP–289) containing mission description, summary of scientific results, photographic summary, crew observations, findings from geologic investigation of landing site, findings from lunar sample examination, and results of specific experiments on July 26–Aug. 7, 1971, mission. (Text)
- *Finance* magazine devoted entire issue to "Corporate Continuity in Space: The Case for NASA's Future." Preface noted that in 1960s, NASA had "achieved its mandate to the moon and back with time to spare. For the Seventies and beyond, its new goals are a balanced mix of manned aeronautical work in near-earth orbit and unmanned research flights into deep space. While NASA administrator James C. Fletcher works with Congress to implement more stable funding concepts, the agency and its industrial contractors continue to provide security and scientific benefits for an earth-bound society while pushing forward the frontiers of aerospace technology." (*Finance*, 4/72)
- Lunar Nomenclature Committee of International Astronomical Union, meeting in Paris, decided to name lunar crater after internationally known astrophysicist Dr. Charles G. Abbot to honor Abbot's 100th birthday May 31. Abbot became 13th exception to rule that lunar sites must be named only for persons deceased. Other 12 exceptions were astronauts and cosmonauts. Soviet members of Committee had proposed Abbot's name in 1967 under mistaken impression he had died. (Casey, *W Post*, 5/11/72, C1)

May 1972

May 1: Ames Research Center C-8A Buffalo augmentor-wing research aircraft completed first flight from Boeing Co. Seattle, Wash., plant. Aircraft, piloted by Boeing test pilot Thomas E. Edmonds and equipped with wing planned for use on short takeoff and landing aircraft, reached 1980-m (6500-ft) altitude during 51-min flight. Primary objectives—to functional checkout and evaluate systems operation, aircraft's structural integrity, and flight characteristics—were achieved. Aircraft would be used in cooperative project between NASA and Canadian Dept. of Industry, Trade and Commerce. (ARC *Astrogram*, 5/11/72, 1; *AF Mag*, 8/72, 46)

• Marshall Space Flight Center announced completion of largest solar-cell-array system for electric power ever devised for spacecraft. Two arrays, with almost 236 sq m (2540 sq ft) of surface area, would use sunlight to power electrical systems of Orbital Workshop, Apollo Telescope Mount (ATM), and other major components of Skylab cluster scheduled for launch in 1973. Each array could provide 10 500 w of power—more than twice average level needed for three-bedroom house—at 328 K (130°F) during 58- to 69-min portion of each 94-min orbit. (MSFC Release 72-54)

• NASA released photos of *Apollo 16* astronauts on lunar surface during April 16-27 mission. One was of Astronaut John W. Young holding hammer as he searched for samples along edge of crater, with lunar roving vehicle (LRV) visible in background. Another showed Charles M. Duke, Jr., carrying sample scooper near huge crater. (*NYT*, 5/2/72, 1, 28)

• Dismantling of Saturn Launch Complexes 34 and 37, sites of Saturn I and IB launches at start of Apollo program, was begun at Kennedy Space Center by Southern Contractors Service. Firm had bid $15 051 for structures built at cost of $147 990 581. Dismantling would be completed within seven months. NASA spokesman had said total of $53 856 403 had been saved in equipment and structures from complexes, "one of the highest returns ever" for such a program. NASA Centers would receive salvaged equipment for active programs. Equipment valued at $5 891 866 would be given to Atomic Energy Commission; Depts. of Commerce, Interior, and Agriculture; Dept. of Health, Education, and Welfare; and universities with Government research grants.

Launch control rooms in Launch Complex 34 and 37 blockhouses would be retained in launch configuration for historical and public information purposes. Historical marker outside would inform visitors of past events. Seven Saturn I and IB vehicles had been launched from Complex 34 culminating in *Apollo 7*, first manned command and service module launch, Oct. 11, 1968. Complex also had been scene of tragic Jan. 27, 1967, fire that took lives of three astronauts. Eight Saturn I and IB vehicles had been launched from Pad 37B, terminating

May 1

with *Apollo 5*, first Apollo lunar module launch, Jan. 22, 1968. (*M Her*, 4/29/72; KSC Release 88–72; KSC Hist Off; KSC memo to NASA Hist Off)

- Appointment of Owen G. Morris as Manager, Apollo Spacecraft Program, was announced by Dr. Christopher C. Kraft, Jr., Manned Spacecraft Center Director. Morris, former manager for lunar module in program office, succeeded B/G James A. McDivitt (USAF), who became Special Assistant to the Center Director for Organizational Affairs. Appointments were effective immediately. (MSC Release 72–85)
- NASA announced publication of *The Experiments of Biosatellite II* (NASA SP–204), edited by Joseph F. Sanders. Book described first successfully completed Biosatellite mission, which carried plants, frog eggs, amoeba, and seeds on earth orbital mission Sept. 7–9, 1967, to determine effect of space flight on living organisms. (NASA *WR*, 5/1/72, 5; *A&A 1967*)

May 1–4: Fourth meeting of Joint U.S.–U.S.S.R. Editorial Board for preparation and publication of space biology and medicine review convened in Washington, D.C. Cochairmen were Dr. Melvin Calvin of Univ. of California at Berkeley and Prof. Oleg G. Gazenko, Director of Soviet Ministry of Health. Under Oct. 8, 1965, Memorandum of Understanding between NASA and U.S.S.R. Academy of Sciences, Board reviewed research from the two countries for publication. (NASA Release 72–89)

May 2: Apollo 16 rock and photography briefing was held at Manned Spacecraft Center. Dr. Paul W. Gast, Chief of MSC Planetary and Earth Sciences Div., said first few weeks of preliminary examination of *Apollo 16* samples would be occupied with opening bags and weighing, dusting, photographing, and examining rocks. Catalog, including photos, of all samples should be completed by end of May. Rocks examined to date appeared to be breccias. One rock—collected during third extravehicular activity and described on moon by Astronaut Charles M. Duke, Jr., as "an honest to goodness igneous rock"—was different in texture and appearance from other typical breccias, but had not yet been precisely classified. "Zap pits" on the surface indicated rock had been lying on one side. It was flattish with very fine, silverish crystals in finer grain matrix with no vesicles. Dr. Gast said rock could be rapidly crystallized igneous rock, very-high-grade metamorphic rock, or metamorphic rock that had been at temperatures of 1000°C (1300 K) long enough to be completely recrystallized.

Dr. William R. Muehlberger of Univ. of Texas and U.S. Geological Survey expressed surprise at "the tremendous number of clasts [rock fragments]" within rocks. Clasts represented "crustal material hauled in from somewhere and that's a tremendous plus beyond what we had originally conceived of as getting. . . . So whether it's volcanic in origin and therefore . . . this is crust from straight under somewhere or whether it's impact origin and therefore crustal material thrown in from some distance doesn't matter." By selecting Descartes as landing site, NASA had obtained extra sampling, "maybe even a better sampling," than in other highlands missions. "And we may even have sampled another quadrant of the Moon that we wouldn't have gotten in any other way. . . . we may have a good deal more than the 15 percent or so of the highlands represented in these rocks and we may

- NASA Associate Administrator for Applications Charles W. Mathews briefed House Committee on Science and Astronautics' Subcommittee on Space Science and Applications on Applications Technology Satellites F and G and Synchronous Meteorological Satellite. Projects had been reported to be experiencing technical and cost problems. NASA review had found in ATS project "considerable progress has been made in . . . completion of . . . thermal structural model. Other elements, such as the integrated communications transponder, were not keeping pace . . . significant imbalance had developed in the total project. It has been necessary then to reschedule and reestablish the proper phasing between project elements." Manpower and expenditure rates had been reduced, management changes had been made, and reporting arrangements had been strengthened. Reduced project content would provide for protoflight ATS–F and required integration and testing.

  "These actions are expected to be completed this summer. Our appraisal . . . is that the ATS–F launch will occur in the early spring of 1974 with ATS–G to follow in a year to eighteen months." Costs were expected to remain within FY 1973 estimate of $180 million to $215 million exclusive of $48.4-million vehicle costs.

  Review of SMS had confirmed need for correction in funding and scheduling because of introduction of advanced design features not adequately assessed in program definition. "Primary emphasis is now being placed on completing engineering model verification tests during . . . this year to insure an orderly initiation of the production phase in early FY 1973." NASA had established and was monitoring milestone schedules, had strengthened contractor's project management organization—including change in project manager and transfer of key NASA personnel to organization—and had improved flow of information between NASA and contractor to improve project visibility. Parts procurement was proceeding satisfactorily after "special remedial attention." SMS development schedule would be 33 mos, placing launch of SMS–A in early fall of 1973. Second spacecraft, SMS–B, was scheduled to be launched five months later. SMS–B total runout cost was estimated at maximum $50 million exclusive of $8.4-million vehicle cost, reflecting FY 1973 reprogramming requirement of $4.5 million. "I believe that this project is now proceeding on a sound basis and will prove to be a very significant addition to NOAA's [National Oceanic and Atmospheric Administration's] weather satellite system." (Testimony)

- Discovery that calcium chloride concentration in brackish waters of ponds in Victorialand region of Antarctica—most Mars-like area on earth—prevented growth of microorganisms was announced by NASA. Jet Propulsion Laboratory scientists Dr. Roy E. Cameron and Frank A. Morelli had concluded that any water found on Mars might be as salty and sterile. They had successfully grown microorganisms in Antarctic soil samples, but no form survived when kept in pond water. "In this environment," Dr. Cameron said, "the capacity of life to adapt and survive is pushed to its limit and beyond. The concentration of living things around water sources in the dry valleys and their thinning out and final disappearance on the cold, dry slopes may provide the model for the distribution of life we may . . . someday find on Mars." (NASA Release 72–90)

- Manned Spacecraft Center announced Deputy Director of Flight Operations Howard W. Tindall, Jr., had been named Director of Flight Operations effective April 30. He succeeded Sigurd A. Sjoberg, who had been acting Director of Flight Operations and Deputy MSC Director. (MSC Release 72-88)
- NASA announced award of two cost-plus-fixed-fee contracts for six-month studies of advanced propulsion systems for short takeoff and landing aircraft. General Motors Corp. Diesel-Allison Div. would receive $574 000 and General Electric Co. Aircraft Engine Group, $567 000 to evaluate and submit designs on different propulsion systems—including augmentor wing, internally blown flap, and externally blown flap. (NASA Release 72-92)
- Senate received President Nixon's nominations of *Apollo 16* Astronaut Charles M. Duke, Jr., for promotion from lieutenant colonel to full colonel in Air Force and of *Apollo 16* Astronaut Thomas K. Mattingly II for promotion from lieutenant commander in Navy to permanent grade of commander. Senate confirmed both nominations May 16. (CR, 5/2/72, S7135; 5/16/72, S7969)

May 3: *Apollo 16* Astronauts John W. Young, Thomas K. Mattingly II, and Charles M. Duke, Jr., held press conference at Manned Spacecraft Center and showed films and still photos taken during April 16-27 mission. Crew's physical condition was already back to normal. Mattingly's exercise-response tests had been normal one day after landing, and Young's and Duke's had been normal two days after landing. Tests to determine potassium levels were still under way.

Young said engine problem that had delayed lunar landing six hours had made *Apollo 16* "a cliff-hanger of a mission from where we were sitting in the cockpit; but the ground came through . . . with a couple of clutch hits and put us right back in the ball game. I know there was a lot of work being done on the ground and, speaking for the flight crew, we certainly appreciate it." Young also praised lunar roving vehicle which carried him and Duke over lunar surface. "The suspension system on the Rover is absolutely remarkable. We never had the feeling of spinning out or anything, but sometimes we went in some deep holes and sometimes the vehicle bounced up in the air and came down on a rock . . . [but] it didn't seem to affect the performance of the machine."

Duke said lunar surface was so clear that "features look almost like you're right next door to them," like a clear day in western U.S. desert. "Some mountains you think are right next to you and [they] end up 20 miles [32 km] away. You have the same feeling on the lunar surface. Looking out the window from the lunar module . . . South Ray Crater . . . looks like it's just right next to you, but it was 6 kilometers [3½ miles] away." (Transcript)

- Senate Committee on Aeronautical and Space Sciences reported to Senate H.R. 14070, $3.420-billion FY 1973 NASA authorization bill [see April 25]. (S Rpt 92-779)
- President Nixon sent message to Congress transmitting *World Weather Program, Plan for Fiscal Year 1973*, report of U.S. participation in program. He said: "Through new satellites, telecommunications, and computer technology, global information for early predictions and hazardous weather warnings is being acquired, processed, and then

distributed in increased volume and detail." Program was "essential to a total environmental monitoring system for our planet" and could "serve as a model . . . for other environmental systems." Other environmental data could be "collected and exchanged through a vehicle like the World Weather Program." (*PD*, 5/8/72, 819)
- Two-month program to demonstrate feasibility of transmitting TV and voice communications direct via satellite to six Alaskan communities began at Juneau, Alaska. ComSatCorp would install small transportable ground station at six locations—Juneau (May 3–5), Kodiak (May 16–18), Bethel (May 29–31), Nome (June 7–9), Barrow (June 20–22), and Ft. Yukon (July 3–7). TV programming by Alaskan Educational Broadcasting Commission would be transmitted from ComSatCorp's Bartlett ground station in Talkeetna, Alaska, via *Intelsat-IV F–4* over Pacific, to transportable station. (ComSatCorp info sheet, 5/1/72; ComSatCorp PIO)
- Lockheed Aircraft Corp. Chairman Daniel J. Haughton told annual meeting of shareholders in Van Nuys, Calif., that he expected higher net earnings on lower sales volume in 1972. He predicted that Lockheed would receive substantial orders for L–1011 TriStar transport in coming years. (Pearlstine, *WSJ*, 5/2/72, 17; Lockheed PIO)

May 3–11: Scientific and Technical Subcommittee of United Nations Committee on the Peaceful Uses of Outer Space met to plan program for parent Committee. Italian space expert Franco Fiorio asked 28 members present if scientists from smaller countries would be allowed to stand in line and buy tickets to board Soviet or U.S. manned space laboratory. Dr. Fernando Mendonca of Brazil said advances in space technology were providing data that could help speed economic development in poorer countries but U.N. itself had neither technical manpower nor finances to assist these countries in using data. During debate over remote sensing by satellite, Sweden tried to open discussion of what U.N. could do to protect economic interests of small nonspace power from possible commercial exploitation by countries collecting data from satellites. Only Argentina supported Swedish view; majority, including U.S. and U.S.S.R., said such discussions were premature. Arnold W. Frutkin, NASA Assistant Administrator for International Affairs, renewed U.S. pledge that data from satellites would be shared with other countries without discrimination. All members of Subcommittee were invited by NASA to inspect data processing center at Goddard Space Flight Center but only six countries accepted. (Teltsch, *NYT*, 5/14/72, 15)

May 4: Arcas ozone sounding rocket launched by NASA from Wallops Station carried Goddard Space Flight Center payload to 60-km (37.3-mi) altitude to measure performance of new parachute deployment technique and measure payload pendulation below parachute. Technique provided greater parachute opening shock and was expected to reduce parachute fouling experienced with older techniques. Parachute and payload performed nominally and experimental objectives were met. (NASA proj off)
- Appointment of Clifford E. Charlesworth, Deputy Manager of Manned Spacecraft Center Skylab Program Office, as Manager of newly created Earth Resources Program Office was announced by MSC. (MSC Release 72–94)

May 5: U.S.S.R. launched *Cosmos 488* from Plesetsk into orbit with 317-km (197-mi) apogee, 185-km (115-mi) perigee, 89.5-min period, and 65.2° inclination. Satellite reentered May 18. (GSFC *SSR*, 5/31/72; *SBD*, 5/9/72, 48)

- United Nations Legal Subcommittee on the Peaceful Uses of Outer Space ended four-week session in Geneva with agreement on some articles in Soviet draft treaty to ensure free access to moon. Subcommittee had agreed to uphold right to explore and use moon and right of lunar explorers to use facilities placed on moon by another country in distress situations. U.S.S.R. had refused to accept U.S. proposals that treaty apply to other celestial bodies and that every country give 60-day notice before launching space mission. U.S. and U.S.S.R. had expressed reservations on draft convention introduced by France and Canada that would compel publication of launch information. (*NYT*, 5/6/72, 3:9)

- Plans for Upper Atmosphere Observatory for U.S. and Canadian scientists were described in *Science* by Massachusetts Institute of Technology scientist J. V. Evans. Ground-based observatory would be close to U.S.-Canadian border, near magnetic 60° latitude. It would provide first opportunity to measure all parameters important in upper atmosphere—altitude to 1000 km (620 mi) over north-south expanse of 1000 km. Users could observe two distinct parts of upper atmosphere and transition between them, one part coupled to earth by magnetic field and second part loosely held and subject to particle bombardment from outside earth. Observatory's applications would include long-distance radio communications, long-range weather prediction, and international cooperation in research and education. Facility, costing $14 million, would complement Global Atmospheric Research Program (GARP). (*Science*, 5/19/72, 463–72)

- Radar-guided USAF Sprint missile launched from Kwajalein Atoll intercepted multiple-warhead missile launched from Vandenberg Air Force Base over Pacific in second successful test of antiballistic missile against multiple targets. (UPI, *NYT*, 5/9/72, 33)

- Award of $112 000 contract to McDonnell Douglas Corp. for visual system definition study leading to development of space shuttle mission simulator was announced by Manned Spacecraft Center. (MSC Release 72–92)

May 6: U.S.S.R. launched *Cosmos 489* from Plesetsk into orbit with 1004-km (623.9-mi.) apogee, 967-km (600.9-mi) perigee, 104.7-min period, and 74° inclination. (GSFC *SSR*, 5/31/72; *SBD*, 5/9/72, 48)

May 7: Tank cart defueling recovered *Apollo 16* command module for storage at San Diego Naval Air Station exploded because of overpressurization. Forty-six persons were hospitalized. All were suspected of having inhaled toxic fumes, but examination revealed no symptoms of inhalation. One suffered broken kneecap and two broken toes. Six were treated for lacerations and bruises. (UPI, *W Post*, 5/8/72, A2; AP, B *Sun*, 5/8/72, A1; MSC PAO)

- NASA had definite plans for international rendezvous and docking mission to link three Soviet cosmonauts with three U.S. astronauts in 1975, *Washington Post* reported. Tentative date was June 14, four days after U.S.S.R. orbited Salyut station and three days after cosmonauts docked Soyuz spacecraft at one end of Salyut 15 m (50 ft) long. Apollo spacecraft would dock with orbiting Salyut on 15th Apollo orbit of earth, day following Apollo launch. For 56 hrs astronauts and cosmonauts would work and live together under guidance of mission control centers in Houston and in Baykonur. There were "no longer any technical doubts that the feat of flying an American and a Russian spacecraft together can be accomplished." Dr. Christopher C. Kraft, Jr., Director of Manned Spacecraft Center, had said, "We haven't seen anything in a technical sense that says we should stop." (O'Toole, *W Post,* 5/7/72, B2)

- Soviet scientists were studying "collective effect" accelerators with high-current electrons to achieve very high energies needed to split atoms at lower cost and effort than by larger traditional accelerators. Traditional circular accelerators raised energy of electrons 50 mev per m. Collective-effect principle increased energy gain to 1000 mev per m. Very-high-current electron accelerators had been developed in U.S. and U.S.S.R. to generate beams of relativistic electrons with speeds close to velocity of light. (Shabad, *NYT,* 5/7/72, 64)

May 8: Dr. James C. Fletcher, NASA Administrator, discussed space prospects in interview published in *U.S. News & World Report:* Chances of joint U.S. and U.S.S.R. mission were "quite good." It hadn't been decided "exactly how it would work, but there might be two or three astronauts and the same number of cosmonauts." U.S.S.R. probably would be "very active" in manned space flight until U.S. space shuttle was flying in 1978. "So it may appear that we have lost our leadership during that period." U.S.S.R. would "try to be first in something—probably with manned and unmanned missions" during 1974 to 1978. "If and when they land men on the moon—even if they land on the same side . . . where we landed—that would attract a lot of public attention. . . . If they were to land on the far side of the moon, that would be even more spectacular." It would also be spectacular, Dr. Fletcher said, if U.S.S.R. announced it was "putting up a space station as the first step in a program to get to Mars."

Gap in U.S. space program had been caused by "series of things" including cancellation of two Apollos. "We probably should have worked it so that the shuttle would be operating earlier, but we spent a lot of time studying the shuttle—and that was good. . . ." Only possible fill-in was to "fly more Apollo components, and we are looking at that possibility. We have leftover Saturn rockets and leftover command-and-service modules, and we're looking to see if they can be used between 1974 and 1978." It would take another Sputnik to accelerate U.S. space program to peak force, "but I don't see that it's necessary. . . . I honestly think that we could do a reasonably good job on the budget that we've programmed for the '70s. I think the main problem with the five-year gap is going to be loss of national prestige during the period." Space program critics and decreasing space budget "may have a negative impact on people wanting to stay with the space program." If there were no gap-filling missions, "we may have to phase part of the astronaut group into some other kinds of work while we wait for the shuttle." When shuttle was operational it would "enable us to make space flight a routine event. Maybe we'll have an every Monday-morning flight out to space. This will change the nature of what we do in space completely. When we can get into

May 8

space cheaply, easily, quickly and routinely, this will open up space for new ventures at present not predictable." (*US News*, 5/8/72)

- Decision to plan large multipurpose optical telescope to be launched and serviced by space shuttle in 1980s was announced by NASA. Large Space Telescope (LST) to study energy processes in galactic nuclei, early stages of stellar and solar system formation, supernova remnants and white dwarfs, and other phenomena relevant to origin of universe would also provide long-term monitoring of atmospheric phenomena on Venus, Mars, Jupiter, and Saturn. Telescope, able to observe galaxies 100 times fainter than those seen by most powerful ground-based optic telescope, would weigh between 9000 and 11 000 kg (20 000 and 25 000 lbs) and have length of 12 to 16 m (40 to 52 ft) and diameter of 3.6 to 4 m (12 to 13 ft). It would include diffraction-limited mirror with 3-m (10-ft) diameter. LST project management had been assigned to Marshall Space Flight Center. (NASA Release 72–98)

- NASA announced formation of JT3D/JT8D Refan Program Office in Office of Aeronautics and Space Technology. New Office—to be managed and directed by James J. Kramer, Chief of Noise Pollution Reduction Branch in Aeronautical Propulsion Div.—would direct program to develop and demonstrate modifications to JT3D and JT8D engines produced by United Aircraft Corp.'s Pratt & Whitney Div. Larger diameter, quieter, single-stage fans, together with suitable acoustic nacelles, would seek maximum practicable quieting at takeoff and landing approach. Refan program was part of joint effort by NASA, Dept. of Transportation, and Federal Aviation Administration to define retrofit changes that would produce significant aircraft noise reduction. (NASA Release 72–96)

- Scientists at Soviet Academy of Sciences' Ioffe Physical-Technical Institute had reported detecting seven cosmic ray events involving particles identified as antiprotons, *New York Times* reported. Experiment—in which balloon-borne detecting devices had drifted at 32-km (20-mi) altitude for 20 hrs before being recovered—appeared to strengthen hypothesis that universe was composed symmetrically of ordinary matter and antimatter. According to *Izvestia* description, scientists had devised cosmic ray counter sensitive only to positively charged protons and negatively charged antiprotons. (Shabad, *NYT*, 5/9/72, 33)

- *Aviation Week & Space Technology* editorial adapted from paper by Dr. Homer E. Newell, NASA Associate Administrator, presented views on future trends in space: Decade of 1970s would see emphasis on "beneficial returns from our investments in the first decade of the space age. It will also be a decade in which we will consciously seek to avoid the large funding peaks that characterized the Apollo era." Number of achievements possible within limited budgeting would be limited. "In the science arena, and in particular in the exploration of the solar system, this will mean some effort on extending our exploration of earth-like planets. And that will be about it, unless some circumstances again change our emphasis on space science." In 1980s NASA would "be getting to those very special, very important missions to the smaller bodies of the solar system . . . which will give us probably our best look at the very earliest days of the solar system. Then, in addition, we should during the 1970s look to the best husbanding and application of our space sciences resources to see if we can't get in one or two early missions of this type." At any progress rate, "the nation, and indeed the world, is not going to run out of important and exciting and inspiring things to do in exploring and seeking to understand this wonderful universe." (*Av Wk*, 5/8/72, 7)

- Swiss federal aircraft factory in Emmen had developed hot-water rocket for assisted aircraft takeoffs, *Aviation Week & Space Technology* reported. Pohwaro (pulsated, overheated, water rocket) had been tested on Swiss air force Dassault-Breguet Mirage 3 and Pilatus Porter aircraft. (*Av Wk*, 5/8/72, 9)

May 8–10: U.S.–U.S.S.R. Joint Working Group on the Natural Environment met at NASA Hq. to carry out recommendations of Joint Working Group which met in Moscow Aug. 2–6, 1971. The 1971 meeting had recommended experiments in remote sensing of environment at analogous and complementary sites in U.S. and U.S.S.R., with each country carrying out research at its own sites, and joint efforts in remote sensing of ocean to relate satellite measurements to sea-surface measurements. Cochairmen of Group were Leonard Jaffe, NASA Deputy Associate Administrator in Office of Applications, and Yu. K. Khodarev, Deputy Director of Soviet Institute of Outer Space Research. (NASA Release 72–97)

May 8–12: First U.S.–U.S.S.R. meeting on lunar cartography was held in Washington, D.C., to implement recommendations of Joint Working Group on Exploration of Near-Earth Space, the Moon and the Planets. Group at Aug. 2–6, 1971, meeting in Moscow had recommended exchange of lunar maps and meetings of U.S. and Soviet experts to discuss lunar map preparation using common lunar coordinate system. (NASA Release 72–93; NASA PAO)

- NASA officials met with 25 national winners in Skylab student project competition at Marshall Space Flight Center to discuss design of students' space experiments and demonstrations. During visit students toured MSFC laboratories and Alabama Space and Rocket Center. (NASA Release 72–94; MSFC PAO)

May 9: Sen. Walter F. Mondale (D-Minn.) said in Washington, D.C., that arguments for space shuttle development had become "classic case of bureaucratic deception and waste." He would offer amendment to cut off all shuttle funds. (*NYT*, 5/11/72)

- U.S.S.R. announced appointment of Gen. Vladimir F. Tolubko, rocket specialist serving near Chinese border, as commander of Soviet strategic missile forces. (Shabad, *NYT*, 5/10/72)

May 10: NASA and European Space Research Organization (ESRO) were jointly soliciting proposals from scientists for planning experiments that could be carried on Explorer-class spacecraft in late 1970s. Scientists whose proposals were chosen would work in teams to aid in planning mission and help NASA and ESRO decide whether to conduct missions. Current plans called for one Explorer spacecraft to fly in circular, solar orbit with 2-million-km (1.2-million-mi) radius while dual mother-daughter spacecraft flew close together in elliptical earth orbit. Three spacecraft would be launched by Thor-Delta booster from Kennedy Space Center between 1976 and 1978. Mother-daughter spacecraft would make repeated passes through earth's magnetic field to measure fine-scale space and time variations while interplanetary spacecraft recorded simultaneous variations of incoming solar wind. (NASA Release 72-100)

- *Apollo 12* commander Charles Conrad, Jr., parachuted safely from NASA T-38 jet aircraft just before it crashed near Bergstrom Air Force Base, Tex. Cause of crash was being investigated. Conrad was on routine flight from Dobbins AFB, Ga., to Ellington AFB, Tex. (UPI, *W News*, 5/11/72, 20)
- Dr. William W. Duke, twin brother of *Apollo 16* Astronaut Charles M. Duke, Jr., replied in letter published by *Washington Post* to April 26 article by *Post* columnist Nicholas von Hoffman that criticized astronauts and *Apollo 16* mission. "My brother stated publicly that he would go to the moon anonymously. He sought no personal gain. No one will recall the names of these space explorers, but each time one mission is completed, the next is easier. I am sure the second, third, or one hundredth man to sail around the world found it easier than the one preceding. Who is to judge what is to be gained now or one hundred years hence in and from space?" (*W Post*, 5/10/72, A15)

May 10–17: U.S.-U.S.S.R. Working Group II met in Moscow to discuss compatible rendezvous and docking of spacecraft in space. (NASA Off Int Aff)

May 10–24: Committee on Space Research (COSPAR) held 15th meeting in Madrid. First firm indication of water ice on Mars from photos returned by *Mariner 9* (launched by NASA May 30, 1971) was announced by Jet Propulsion Laboratory scientist Dr. James A. Cutts. He and team had noted that residual south polar cap 320 km (200 mi) wide was unchanged throughout southern summer but was hidden by thin carbon dioxide frost that covered region 3200 km (2000 mi) in diameter during winter. In spring, it appeared that dry ice again evaporated and cap shrank to 320-km zone of water ice.

Working sessions on terrestrial planets and outer planets were held by NASA and Soviet scientists to discuss programs for exploration of Venus, Mars, Saturn, and Jupiter. Discussions fulfilled recommendation of August 1971 NASA–U.S.S.R. Working Group that planetary exploration be discussed jointly at international colloquia.

Space Science Board of National Academy of Sciences and National Research Council submitted *United States Space Science Program: Report to COSPAR*. Report, covering 1971, summarized analyses and observations from experiments on board spacecraft, sounding rockets, balloons, and aircraft. (Miles, *LA Times*, 5/11/72; Text; NASA Int Aff; NASA Release 72-106; *W Post*, 5/23/72, A10)

May 11: Senate passed H.R. 14070, $3.444-billion NASA FY 1973 authorization bill, after adopting amendment by Sen. Howard W. Cannon (D-Nev.) that increased allocation for aeronautical research and development by $24 million. Increase included $21 million for program to design and demonstrate certifiable modifications to retrofit existing civil aircraft for quieter performance and $3 million for aircraft collision avoidance. Senate rejected, by vote of 61 to 21, amendment by Sen. Walter F. Mondale (D-Minn.) to delete $228 million requested for space shuttle development. (*CR*, 5/11/72, S7698–7763)

- NASA-developed battery technology had been used by McCulloch Electronics Corp. to design high-energy-output batteries for sure, fast starts for power tools and sports equipment. Medical and photographic applications were also being developed. New batteries could be recharged 90 to 100 times faster than existing batteries—15 to 20 min instead of 14 to 16 hrs. (NASA Release 72-95)

- Sen. William Proxmire (D-Wis.) sent letter to NASA Hq. requesting NASA comments on General Accounting Office report that recommended Manned Spacecraft Center contracting officer "devote more attention" to performance of NASA contractor Service Technology Corp. Firm, subsidiary of Ling-Temco-Vought, Inc., had received average annual income of $12 million from extensions to $16-million contract signed with NASA in 1967. STC work included building and ground maintenance, engineering design and construction of hardware, technical writing and editing, and publications distribution service. GAO had found STC practices "doubtful." (*H Post*, 5/14/72)
- Dr. Aleksander P. Vinogradov, Vice President of Soviet Academy of Sciences, presented report on *Luna 20* samples to Academy. He had earlier, on May 4, published data on soil sample in *Pravda*. *Luna 20* had landed near moon's Apollonius Crater in mountainous area between Sea of Crises and Sea of Fertility and returned to earth with lunar samples. *Luna 20* soil was light grey, porous consertal (of texture in which irregular crystals interlocked) material containing fewer glassy, fused particles than *Luna 16* samples from Sea of Fertility. Average particle size was 70–80 millimicrometers with largest particle in excess of one millimeter. Albedo value was higher than for samples returned by *Luna 16, Apollo 11,* or *Apollo 12.*

  Basic mass of particles were anorthositic rock containing large amount of feldspar, metallic iron, and olivine. "The lunar samples' surprise proved to be the presence of pulverized metallic iron, which . . . does not oxidize, as distinct from terrestrial iron. This discovery may be of great practical significance. We have shown experimentally that under certain conditions it is possible to obtain such iron in the laboratory from basalts. If we succeeded in developing cheap technology for the industrial production of inoxidizable iron, this would give technology a magnificient construction material." Understanding nature of formation of anorthosites would be of even greater significance. "Perhaps it will even elucidate the most ancient history of our planet. A popular hypothesis once existed among geologists that the earth's crust originally consisted of anorthosite rocks. Now it has unexpectedly received strong 'corroboration' from the moon.

  "After study of the lunar rock the nocturnal luminary appears before us as a 'twofaced Janus.' On the moon's visible side vast territories are occupied by basaltic seas, while the entire far side is practically a solid continent, apparently formed from anorthositic rocks. Such a sharp difference is one of the moon's most exciting secrets." (*Izvestiya*, FBIS-Sov, 5/17/72, L1; *Sov Rpt*, 5/19/72, 1)
- Astronaut Stuart A. Roosa was named chairman of Manned Spacecraft Center board to investigate cause of May 10 T-38 aircraft accident in which Astronaut Charles Conrad, Jr., parachuted to safety at Bergstrom Air Force Base, Tex. (MSC Release 72-100)

May 12: Dr. Thornton Page of Naval Research Laboratory (NRL) showed photos and described results of *Apollo 16* far-ultraviolet camera and spectrograph experiments at Manned Spacecraft Center press conference. Photos had revealed presence of three atmospheric rings around earth, rather than two as previously believed. Newly discovered ring was probably composed of oxygen or emissions from molecular nitrogen. Photo also showed shape of geocorona. "The geocorona is well known, and . . . measurements had been made from below. But . . . this is the first time, except for the work on OGO-V results, where the geocorona has been viewed from outside. From these photographs, we'll be able . . . to give the amount of Lyman-Alpha light emitted by each cubic mile at each distance away from the earth on all sides . . . and get a three dimensional model of the geocorona." (Transcript)

- Four companies submitted technical proposals to Manned Spacecraft Center for NASA space shuttle contract: Grumman Aerospace Corp., Lockheed Missiles & Space Co., North American Rockwell Corp. Space Div., and McDonnell Douglas Astronautics Co. (NASA Release 72-103; *Marshall Star*, 5/17/72, 1)
- Extension of McDonnell Douglas Corp. contract for design study of space shuttle orbiter auxiliary propulsion systems to include study of earth-storable propellants was announced by Manned Spacecraft Center. Extension increased contract value to $576 000 and carried it through Oct. 15, 1972. (MSC Release 72-101)
- *Apollo 16* Astronauts John W. Young, Thomas K. Mattingly II, and Charles M. Duke, Jr., completed 13 days of postflight debriefing at Manned Spacecraft Center and prepared to spend first weekend in seven weeks with their families. (Reuters, *NYT*, 5/13/72, 33)
- NASA launched Aerobee 170 sounding rocket from White Sands Missile Range, N. Mex., carrying Naval Research Laboratory astronomy experiment. Rocket and instrumentation performed satisfactorily. (SR list)
- Impact of computers on society was discussed in *Science* by Massachusetts Institute of Technology computer scientist Joseph Weizenbaum. Computer had "very considerably less societal impact than the mass media would lead us to believe." Space travel could not have been undertaken without computers; computer and computer education industries had "grown to enormous proportions." But "much of the industry is self-serving . . . like an island economy in which the natives make a living by taking in each other's laundry." Part that was not self-serving was supported largely by Government agencies and other gigantic enterprises that knew "short-range utility of computer systems but have no idea of their ultimate social cost." Airline reservation systems and computerized hospitals "serve only a tiny, largely the most affluent, fraction of society. Such things cannot be said to have an impact on society generally." (*Science*, 5/12/72, 609–14)

May 12–19: U.S.–U.S.S.R. Joint Working Group on Space Biology and Medicine met for second time at Manned Spacecraft Center. Meeting discussed data from Soviet *Soyuz 11–Salyut I* mission (June 6–30, 1971), preliminary data from *Apollo 16* (April 16–27, 1972), pre- and postflight examination procedures, and methods of predicting inflight status of crew members. Previous meeting was held in Moscow Oct. 9–13, 1971. (NASA Release 72-101; *W Post*, 5/14/72, A28)

May 13: Meteorite crashed into moon near *Apollo 14* Fra Mauro landing site with force estimated to be equivalent to 8900 kilonewtons (1000 tons of TNT)—100 times larger than impact of Saturn V S-IVB stage, larger than any previous seismic event recorded on moon, and large enough to record reflections from lunar core if it existed. Dr. Gary V. Latham, principal investigator for Apollo passive seismic experiments, said meteorite must have been about 3 m (10 ft) in diameter and probably created new crater as large as a football field. Signal recorded by four Apollo seismometers on moon was so strong that *Apollo 14* seismometers were knocked completely off-scale for 16 min. Total signal lasted more than three hours and signals that might correspond to rain of debris thrown out by impact were recorded for about one minute. (MSC Release 72-105)

- Supernova, or exploding star, 97 billion billion km (60 billion billion mi) from earth was discovered by Hale Observatories astronomer Charles Kowal. Kowal, who confirmed discovery on May 15, said supernova, brightest discovered in 35 yrs, could become "one of the most studied objects in astronomy." Marshall Space Flight Center astronomers Dr. Thomas Wdowiak and Joseph Michlovic used 1.5-m (5.6-ft) telescope at Univ. of Arizona's Catalina Mountain Observatory to observe supernova in infrared, optical, and near-ultraviolet wavelengths immediately following first sighting and distributed findings. MSFC astronomer Dr. Thomas Parnell and Dr. G. J. Fishman of Brown Engineering Co. launched balloon over Lubbock, Tex., to make gamma ray observations of supernova for six hours. (AP, *NYT*, 5/26/72, 9; MSFC Release 72-83; *Marshall Star*, 6/28/72, 1)

- AP quoted scientists as saying new mirror being installed in Boyden Observatory at Mazelspoort, South Africa, would provide most powerful telescope in Southern Hemisphere. Scientists planned to use telescope to flash laser beam to moon. (*W Post*, 5/14/72, D5)

- Delegation from Federation of American Scientists (FAS), nongovernment organization, left for People's Republic of China to explore means of improving contacts between Chinese and American scientists. (*Science*, 5/19/72, 783)

May 14: Physicist Dr. Ralph E. Lapp said in *Washington Post* article that NASA followed "cart-before-the-horse" or "rocket-before-payload policy" in figuring shuttle costs. Question to be debated was not shuttle transportation "but the basic purpose and worth of payloads." NASA was "sorely perplexed as a federal agency whose basic mission was accomplished when men landed on the moon." Skylab mission in 1973, "when a temporary, makeshift station is placed in orbit," would end NASA manned space flight "unless the shuttle project goes ahead. In a sense, it's a shuttle to nowhere unless a permanent space station is also approved. This project, last costed at $10 billion, was once the basic rationale of the shuttle, but its price tag scared off congressmen. Now NASA downplays the space station, but the Jan. 31, 1972, Mathematica study includes 62 shuttle trips to space stations in one scenario." (*W Post*, 5/14/72, B3)

- Role of Vandenberg Air Force Base in space shuttle program was doubtful, Frank Macomber said in *San Diego Union*. California politicians were claiming "partial victory" in NASA's selection of West Coast site for second shuttle launch and landing site. But "clear-cut win went to Florida's Cape Kennedy." Dr. James C. Fletcher, NASA Administrator, had "made it clear it will be another eight years before space shuttle launch base work would begin at Vandenberg." In eight years "entire

U.S. space picture will have changed. There will be a whole new set of American leaders, and a Congress which today would approve a space shuttle program for the West Coast might be so changed by 1980 that it would be voted down." (Copley News Service, *SD Union*, 5/14/72, 23)
- U.S. was moving towards conversion to metric system despite opposition from some industries concerned about resultant confusion and cost, *New York Times* reported. American chemical and pharmaceutical industries were now largely on metric system, as were NASA and "important parts" of weapon and aircraft industries. General Electric Co., whose products for foreign consumption were already made to metric dimensions, had come out for conversion. In "carefully timed manner," each industry was following pace suited to its needs. (Sullivan, *NYT*, 5/14/72, 4:6)

May 15: NASA Flight Research Center was testing new curved, steep, instrument landing approach for short takeoff and landing (STOL) aircraft to reduce noise discomfort and airport congestion. Approach could begin at various locations close to airport and would provide better aircraft separation, more landings per given time, less en route time, and lower airline operating costs by shortening terminal-to-terminal times. Approaches began at altitudes between 1200 and 1500 m (4000 and 5000 ft) and were flown to approximately 15 m (50 ft) above ground. (FRC Release 7-72)
- National Oceanic and Atmospheric Administration (NOAA) had indefinitely postponed launch of ITOS-C meteorological satellite, originally scheduled for last February, because of NASA's delay in updating Thor-Delta launch vehicle with onboard computer for inertial guidance, *Aviation Week & Space Technology* reported. Spacecraft would be retrofitted with improved instrumentation and launched after ITOS-D, scheduled for launch in fall with improved Thor-Delta. (*Av Wk*, 5/15/72, 11)
- Two proposed flight plans, both with nighttime launch, for December Apollo 17 mission were being studied by Manned Spacecraft Center, *Houston Post* reported. Proposed launch times—selected for optimum sun angle during lunar landing and surface activities—were 8:39 pm EST Dec. 6, with lunar landing at 1:47 am EST Dec. 11, and 8:48 pm EST Dec 6, with lunar landing at 1:47 pm EST Dec. 11. Nighttime liftoff would be first in U.S. manned program. (Maloney, *H Post*, 5/15/72)
- Termination notices effective June 29 were sent to 198 employees at Marshall Space Flight Center; 115 were downgraded or reassigned in manpower reduction that would bring MSFC employment level to 5317 by June 30. Further reduction of 103 was planned for June 1973. (*Huntsville Times*, 5/11/72)
- "Rendock" was new word "trying to enter space lexicon," *Aviation Week & Space Technology* reported. Engineers studying proposed U.S.–U.S.S.R. joint rendezvous and docking mission had tired of repeating phrase in its entirety and had coined "what they hope will become the standard, short version." (*Av Wk*, 5/15/72, 13)
- Physicists were studying possibility that magnetic substorm was created by experiment conducted by Los Alamos Scientific Laboratory and Univ. of Alaska Geophysical Institute. Experiment attempted to trace

line in earth's magnetic field from pole to pole by injecting barium ions into upper atmosphere by rocket. Launch of Sandhawk Tomahawk sounding rocket from Poker Flat near Fairbanks, Alaska, March 6 had caused unusual one-hour auroral display, accompanied by sudden appearance of very-low-frequency (VLF) radio signals, 17 min after barium injection. (Av Wk, 5/15/72, 11; Los Alamos PIO)

- Sole reason for nonappearance of Soviet supersonic transport Tu-144 at Dept. of Transportation's TRANSPO '72 exhibition at Dulles International Airport May 27–June 4 was fact that pre-production prototype, only version cleared for flights outside U.S.S.R., had range of less than 6000 km (4000 mi), *Aviation Week & Space Technology* reported. (Av Wk, 5/15/72, 13)
- National Academy of Sciences was seeking Congressional approval of project to provide funds to Poland for construction of $1-million astronomical research center as gift to mark 500th anniversary of birth of Copernicus in 1973, *New York Times* reported. Additional funds for equipment would be sought in U.S. from Polish-American community. Facility would be in Warsaw. (Feron, NYT, 5/15/72, 81)

May 16: Press briefing on *Apollo 16* preliminary science results was held at Manned Spacecraft Center. Dr. David W. Strangway, Chief of MSC Geophysics Branch, said magnetic sample returned to moon on *Apollo 16* "did pick up fairly strong component of soft magnetization as the result of exposure to the spacecraft systems. On return to the earth, however, we were able to clean that out and to recover the original hard component of magnetization which is what . . . we were hoping to do." Experiment had worked out "quite successfully and suggests that the soft components that we're seeing in the rocks is indeed of spacecraft origin and is not an artifact of some kind that came from the lunar surface. . . . if it did come from the lunar surface we've contaminated it so much that we can't sort it out afterwards. But the hard component then, we believe, is real and it came back through the whole journey and . . . is essentially unchanged and unmodified."

Dr. Paul W. Gast, Chief of MSC Planetary and Earth Sciences Div., said qualitative conclusion from chemical analysis of few samples examined to date was "that the soil is certainly quite homogeneous from point to point." Homogeneity suggested that rock at Cayley Plains might not be different from rock in Descartes and that neither formation might have been formed from volcanic lava. Samples had very high concentrations of aluminum and calcium—26½% aluminum oxide and 15% calcium—"very different from what we see everywhere on the surface of the earth."

X-rays of core tubes showed dispersion throughout soil of metal particles—probably metallic iron—up to 0.5 cm (0.2 in) in diameter. Tentative conclusion was "that metallic iron, both in the soil and the rocks, is very much more abundant at this site than it's been at any other site on the moon and this may be a qualitative characteristic of the *Apollo 16* site and perhaps a qualitative characteristic of the highlands is that rocks there contain bits of metallic iron. It is tempting to speculate that this has something to do with the high magnetic field at this region." Although scientists couldn't tell just from looking at rocks whether any were volcanic, "the opinion is evolving that they are not." Rocks appeared, to naked eye, to be impact breccias. (Transcript; O'Toole, *W Post*, 5/17/72)

- Congress received *Apollo 16* Astronauts John W. Young, Charles M. Duke, Jr., and Thomas Mattingly II. In welcoming speech Sen. Strom Thurmond (R-S.C.) said, "The trail they took into space has further inspired all of us to reach for greater accomplishments." Astronauts urged continued support of space program.

  Astronauts earlier in morning reported preliminary results from April 16–27 lunar landing mission to House Committee on Science and Astronautics. Slides shown included first photo showing intensity of ultraviolet light around earth and also aurora belts, including magnetosphere on equator never seen before [see May 12]. They also showed photo of geocorona protecting earth from UV rays from sun. Mattingly said of lunar exploration. "I think as we go further . . . into our exploration of the moon, we'll find it more and more complex than we suspected it would ever be. In our fondest dreams, I believe we never suspected the hidden treasures that are buried underneath the lunar soil." (*CR*, 5/16/72, S7839–40; *W Post*, 5/17/72, A1; Hearing transcript)

- NASA's TF–8A aircraft, equipped with supercritical wing and piloted by Thomas C. McMurtry, completed one of two flights scheduled from Flight Research Center. Pressure distributions were measured from mach 0.90 to 0.99 at altitudes from 12 800 to 13 960 m (42 000 to 45 800 ft) with flaps closed and at 5° and 10°. Second flight, an excess-energy maneuver, was canceled because DC generator light came on at end of first flight. Problem was later corrected and second flight was rescheduled. (NASA proj off)

- Arcas sounding rocket was launched by NASA from Antigua, West Indies, to 52.3-km (32.5-mi) altitude. Primary objective was to obtain ozone measurements in upper atmosphere in conjunction with overpass of *Nimbus 4* satellite and with Nike-Cajun ozone launch (May 17). Secondary objective was to validate further performance of new parachute deployment system tested May 4. Payload was ejected near apogee and descended successfully by parachute; good ozone data were acquired. (NASA Rpt SRL)

- Marshall Space Flight Center announced agreement with U.S. Army Corps of Engineers Huntsville Div. to provide facility design and construction support as needed by MSFC for space shuttle program. (MSFC Release 72–61)

- Canadian government purchase of video equipment valued at $500 000 to receive direct TV transmissions from NASA's Earth Resources Technology Satellites (ERTS) was announced by RCA. First data-acquisition facilities installed outside U.S., equipment would be at central receiver site in Prince Albert, Saskatchewan, and Canadian Centre for Remote Sensing in Ottawa. Facilities would go into full-scale operation when first experimental ERTS was launched from Western Test Range. (RCA Release)

- U.S. patent No. 3 662 744 was awarded to General Electric Co. scientists Robert W. Richardson and David B. Wright for esthesiometer invented under NASA contract. Instrument, for possible use on Skylab, could also help physicians test patients' skin perception. User touched skin with flexible nylon wire projecting from instrument to measure degree of sensitivity. (*NYT*, 5/20/72; Pat Off PIO)

- U.S. and U.S.S.R. were nearing "historic agreement to begin joint ventures in space," *Wall Street Journal* reported. "For the first time, the Soviets are dropping their secrecy curtain enough to provide detailed engineering data on their spacecraft—a sign that Brezhnev, Kosygin & Co. really want to go ahead." U.S. officials—including Dr. James C. Fletcher, NASA Administrator, and Dr. Edward E. David, Jr., Presidential Science Adviser—"are eagerly committing themselves to cooperation; American spacemen, feeling pinched for funds, now count on accord to keep them flying in the future." Political decision on joint space mission "is in the hands of the Kremlin and the White House." Some officials believed "announcement of this Soviet-American engagement could come before June—perhaps in Moscow if President Nixon goes ahead with his scheduled trip there this month." (Spivak, *WSJ*, 5/16/72, 1)

- Former scientist-astronaut Dr. Brian T. O'Leary continued dialogue with Univ. of Michigan astronomer James A. Loudon [see Jan. 16 and March 28] in letter published by *New York Times*. "I basically agree with Mr. Loudon's defense of the human value of communications and other applications satellites, but we do not need enormous shuttles to transport such minuscule payloads. In addition, the shuttle juggernaut is forcing NASA to cancel, postpone or trim much of its space science and applications activity. This is regrettable. We need to take a very careful look at the space shuttle before Congress appropriates billions of dollars and we are irrevocably committed to an apparent boondoggle." (*NYT*, 5/16/72, 40)

- Collection of 28 paintings by artist-architect-astronomer Chesley Bonestell and 3 original drawings by Dr. Wernher von Braun, NASA Deputy Associate Administrator for Planning, were placed on exhibition at American Museum-Hayden Planetarium in New York for indefinite period. Bonestell paintings of early 1950s depicted future space explorations. Von Braun paintings suggested suit for astronaut extravehicular activity on moon. (Am Mus Release)

- Redesignation of undersea long-range missile system (ULMS) as "Trident" was announced by Secretary of Defense Melvin R. Laird during Washington, D.C., press conference. (UPI, W *Star*, 5/17/72, A6)

May 16–17: Conference on Aircraft Engine Noise Reduction was held at Lewis Research Center to present latest results of NASA Quiet Engine Program and efforts to reduce noise of future short takeoff and landing (STOL) aircraft to airline industry representatives. Roy P. Jackson, NASA Associate Administrator for Aeronautics and Space Technology, said future of civil aviation depended on "making the airplane acceptable to the community . . . in noise level and . . . engine emissions." NASA's goal was to provide technology to make aircraft "unobtrusive in its environment. Noise levels beyond the airport boundaries should one day be indistinguishable in the ambient level." Primary work in propulsion-source noise reduction was "technology to modify existing engines and . . . to design new propulsion systems . . . that operate at significantly lower noise levels." NASA also was working on steep approaches in landing to reduce ground noise and studying community response to noise.

Flight tests in 1971 with modified American Airlines Boeing 720 aircraft from Stockton (Calif.) Metropolitan Airport in clear weather had shown "cockpit modifications made procedure of reducing ground noise by effecting steeper approaches in terminal area safe, repeatable, and reasonable under clear-weather conditions."

Carl C. Ciepluch, Deputy Manager of LeRC Quiet Engine Project Office, discussed first tests, in May, of experimental engine A with acoustic nacelle and 97 900-newton (22 000-lb) thrust. LeRC had tested engine A at takeoff and landing speeds when noise was most noticeable. Tests had shown that if four such engines were installed on DC-8 or Boeing 707 aircraft takeoff noise would be 90 effective perceived noise in decibels (epndb) instead of 116 on takeoff and 118 on approach with current DC-8 and Boeing 707 engines. Although not suitable for wide-bodied jets, quiet engine and nacelle was 8 to 13 epndb quieter than engine used on Boeing 747s and DC-10s. Testing of engine with nacelles would continue through fall to measure more precisely effectiveness of noise suppression techniques and measure engine's internal noise sources other than the fan. Late in year, engine would be tested in a new altitude chamber at Lewis. Second quiet engine was undergoing aerodynamic tests by manufacturer General Electric Co. at Peebles, Ohio, before being delivered to LeRC at end of year for acoustic tests. NASA would use evaluation of both engines in designing quiet experimental engine Mark II for "optimum trade-off period noise reduction and installed weight penalty for aircraft application."

Newell D. Sanders, Chief of LeRC V/STOL and Noise Div., chaired conference and discussed noise levels, methods for noise reduction, and quiet engine principles. (LeRC Release 72-29; Text; *Lewis News*, 5/5/72, 1; 5/19/72, 1; NASA SP-311)

May 17: U.S.S.R. launched *Cosmos 490* from Plesetsk into orbit with 339-km (210.6-mi) apogee, 202-km (125.5-mi) perigee, 89.9-min period, and 65.4° inclination. Satellite reentered May 29. (GSFC *SSR*, 5/31/72; *Sov Aero*, 5/22/72, 32)

- NASA's TF-8A aircraft, equipped with supercritical wing and piloted by Thomas C. McMurtry, completed flight from Flight Research Center. Performance data were obtained at 10 700-m (35 000-ft) altitude and mach 0.95–0.99. Wing pressure distributions were recorded at altitudes from 8800 to 9100 m (29 000 to 30 000 ft) at mach 0.95–0.99. (NASA proj off)

- NASA launched four Nike-Cajun sounding rockets from Point Barrow, Alaska. First rocket carried Goddard Space Flight Center grenade payload to 110.2-km (68.5-mi) altitude to provide temperature, pressure, density, and wind data in support of closely coordinated ozone and atomic oxygen flights. All 31 grenades were ejected and detonated as planned and sound arrivals were recorded on ground.

  Second rocket carried GSFC payload to 84.6-km (52.6-mi) altitude to measure ozone distribution in mesosphere and stratosphere in Arctic after spring wind reversal and in conjunction with overpass of *Nimbus 4* satellite. Payload deployed properly and on time and descended by parachute. Excellent ozone data were acquired.

  Third rocket carried experiment developed by Univ. of Michigan and National Oceanic and Atmospheric Administration to 127-km (78.9-mi) altitude. Objective was to measure atomic oxygen distribution in mesosphere in proximity to ozone and grenade flights. Rocket and instruments functioned satisfactorily and good atomic oxygen data were acquired.

Fourth rocket carried 31-grenade GSFC payload to collect atmospheric data in support of ozone and atomic oxygen flights. All grenades were ejected and detonated as planned and sound returns were recorded on ground. (NASA proj off)

- *Apollo 16* Astronauts John W. Young, Thomas K. Mattingly II, and Charles M. Duke, Jr., were honored during Astronaut Day celebration in Chicago. They rode in tickertape parade, attended ceremonies at City Hall escorted by Chicago Mayor Richard J. Daley, and visited Michael Reese Hospital. At hospital they met Dr. Samuel Natelson who had invented instruments and techniques by which astronauts' body fluids were monitored by Manned Spacecraft Center Mission Control. (*C Daily News*, 5/17/72)
- U.S. and Soviet scientists were designing interferometer—system comprising two radiotelescopes—with one radiotelescope that could be emplaced in outer space, Soviet astrophysicist Josif Shklovskiy reported in *Izvestiya*. Instrument would be used to examine quasars and formation regions of stars and planetary systems. (FBIS–Sov, 5/18/72, L3)
- President Nixon announced appointment of John S. D. Eisenhower as Chairman of new Classification Review Committee to implement security classification system established March 8. President also released text of National Security Council directive governing classification, downgrading, declassification, and safeguarding of national security information. Government departments were requested to establish computerized data index system for classified material. President said: "This application of computer technology across the board should lead to a much more manageable classification system and greatly enhance the flow of information to the public." Executive Order 11652 of March 8 included NASA among agencies originating classified material. (*PD*, 5/22/72, 870–1)

May 17–19: American Helicopter Society held 28th Annual National V/STOL Forum in Washington, D.C. Society's Dr. Alexander Klemin Award was presented to Bell Helicopter Chairman Edwin J. Ducayet for "notable achievement in the advancement of rotary wing aeronautics." (*NASA Activities*, 5/15/72, 104; *Av Wk*, 5/22/72, 9)

May 18: House Committee on Appropriations favorably reported H.R. 15093, FY 1973 Dept. of Housing and Urban Development-space-science-veterans appropriations bill that included $3.349-billion NASA appropriation. Appropriation—$58 million less than budget estimate of $3.407 billion, but $39 million above FY 1972 appropriation of $3.310 billion—allocated $2.55 billion for research and development (R&D), down $50.9 million from budget request; $69.7 million for construction of facilities, $7.5 million below budget estimate; and $729.45 million for research and program management as requested.

Committee did not distribute R&D funds but recommended balanced program to continue "useful and significant . . . projects in space science, exploration, practical applications of advanced technology, and aeronautics"; approved authorization bill's levels for aviation safety and noise reduction program; and deferred request of $5.54 million for construction of facilities for manufacturing and final assembly of space shuttle at "various locations" until "such sites are decided upon and plans are developed." (H Rpt 92-1071)

- Canada and European Space Research Organization (ESRO) signed space cooperation agreement under which ESRO would fly equipment on board Canada's Communications Technology Satellite, scheduled for launch by NASA in 1975. Equipment to be supplied by ESRO included traveling-wave tube, parametric amplifier, and solar cells. (*SBD*, 5/30/72, 150; Canadian Embassy PIO)

- NASA Equal Opportunity Council to review, advise, and recommend on implementation of NASA Equal Employment Opportunity Program was established by Dr. George M. Low, NASA Deputy Administrator. Membership of full-time NASA employees from key staff and program offices would serve two-year terms. Chairman, Bernard Moritz, would serve one-year term. (*NASA Activities*, 6/15/72, 119)

- Dr. George M. Low, NASA Deputy Administrator, discussed cost of doing business in space in speech before Aerospace Industries Assn.'s 26th Williamsburg Conference at Williamsburg, Va. NASA was developing "steady flow of products in the space business." With shuttle "we will have a real opportunity to do something about the high cost of doing business in space." Formula was to identify areas with greatest potential payoff for cost improvement, determine standard requirements for systems with highest potential payoff for future spacecraft, develop these standard items with low cost and high reliability, and ensure that items were used. "Our work to improve the costs of the things we will do in space is an essential adjunct to the Shuttle—not only for our own projects, but to help pave the way for many more commercial uses of space." (Text)

- International treaty banning emplacement of nuclear and other weapons of mass destruction on seabed came into force with signing of articles of ratification in Washington, London, and Moscow. (*W Post*, 5/17/72, A20; Tass, FBIS-Sov, 5/19/72, A7)

- *Apollo 14* Astronaut Edgar D. Mitchell described tests of extrasensory perception (ESP) he made during Jan. 31–Feb. 9, 1971, mission at 15th annual conference of Spiritual Frontiers Fellowship in Chicago. Mitchell had attempted to send messages from spacecraft to four "receivers" on earth and had reported later that some "exchanges" had been made. (*C Daily News*, 5/13/72; *A&A 1971*; MSC PAO)

- General Electric Co. and Esso Research & Engineering Co. would support $3-million research project by Univ. of Rochester to explore possible use of laser beams to ignite hydrogen fusion fire for electric power production, *Wall Street Journal* reported. (*WSJ*, 5/18/72)

May 19: U.S.S.R. launched *Molniya II–2* comsat from Plesetsk into orbit with 39 295-km (24 416.8-mi) apogee, 433-km (269.1-mi) perigee, 705.1-min period, and 65.3° inclination. Objective was to maintain "a system of long-distance telephone and telegraph radio-communications" in U.S.S.R., relay programs of Central Television Service to stations of Orbita network, and further international cooperation. Satellite operated in centimeter wave band. (GSFC *SSR*, 5/31/72; *SBD*, 5/23/72, 123; Tass, FBIS-Sov, 3/22/72, L1)

- President Nixon signed H.R. 14070, $3.444-billion FY 1973 NASA authorization bill. Bill, approved by Senate May 11, became Public Law 92–304. (*PD*, 5/29/72, 900; NASA Off Admin)
- President Nixon briefed press on forthcoming state visit to Austria, U.S.S.R., Iran, and Poland and outlined three areas on agenda "in which there is a possibility . . . of agreement" with U.S.S.R., "provided we can break some bottlenecks which still exist." Areas were arms limitation, trade, and cooperation in space. President recalled his TV speech to Soviet people during 1959 visit to U.S.S.R.: "I said at that time . . . let us go to the moon together." While he would not suggest that joint missions to moon or to Mars would be discussed during forthcoming visit, he said, "I do know that considerable progress has been made over the past several months with regard to cooperation in . . . exploration of space. We are going to try to see a culmination of that progress in this area." President said "NASA people" had been among those working "toward the time when this meeting would take place." (*PD*, 5/22/72, 876–9)
- NASA announced appointment of Robin K. Ransone as Director of Aeronautical Operating Systems Office in Office of Aeronautics and Space Technology. Ransone—American Airlines Development Engineer, V/STOL Technology—would replace Deputy Associate Administrator (Programs) George W. Cherry, who had been Acting Director of office. (NASA Release 72–107)
- Skylab statistics were released by NASA. Spacecraft, to be launched by two-stage Saturn V rocket in spring 1973, would contain 370 cu m (13 000 cu ft) of working and living space. More than 13 000 individual items weighing total of 5000 kg (11 000 lbs) for long-duration space mission would be stowed, including 910 kg (2000 lbs) of food, more than 2700 kg (6000 lbs) of water, 60 changes of astronaut jackets and shirts and trousers, 210 pairs of shorts, 30 constant-wear garments, 15 pairs of boots and gloves, 55 bars of soap, 96 kg (210 lbs) of towels, 1800 urine and fecal bags, 156 rolls of teleprint paper, 104 film magazines, medical kit, 108 pens and pencils, and vacuum cleaner. (NASA Special Release)
- NASA launched two Aerobee 170 sounding rockets from White Sands Missile Range, N. Mex. One carried Massachusetts Institute of Technology x-ray astronomy experiment. Second carried Goddard Space Flight Center cosmic ray experiment. Rockets and instrumentation performed satisfactorily. (SR list)
- President Nixon had approved assignment of M/G Kenneth W. Schultz (USAF) as commander of Space and Missile Systems Organization (SAMSO) effective Aug. 1, Dept. of Defense announced. Gen. Schultz, Deputy Chief of Staff for Air Force Systems Command, was nominated for promotion to lieutenant general. (Reuters, *NYT*, 5/20/72, 8)

May 19–20: Working session of U.S. and U.S.S.R. scientists met in Madrid to discuss scientific results, objectives, and strategy for planetary exploration. Meeting fulfilled August 1971 recommendations by joint U.S. and U.S.S.R. working group for discussions to propose complementary activity by one party during planetary investigations conducted by the other and for exchange of information from planetary experiments. Dr. John E. Naugle, NASA Associate Administrator for Space Science, headed U.S. group; G. I. Petrov, Director of U.S.S.R.

Science Institute of Space Research, headed Soviet group. (NASA Release 72-106)

May 20: Unidentified satellite launched by Air Force from Vandenberg Air Force Base at 8:30 am PDT by Titan IIIB-Agena booster failed to achieve orbit. (*Av Wk*, 6/5/72, 14)

- President Nixon and party left Andrews Air Force Base, Md., for Salzburg, Austria, first stop on scheduled 13-day visit to Austria, U.S.S.R., Iran, and Poland. Visit would include summit meetings with Soviet officials in Moscow to discuss possible cooperation in joint manned space flight mission, strategic arms limitation, environmental and health problems, reducing risk of incidents at sea, international trade, and other ventures in "peaceful coexistence." (*PD*, 6/5/72, 912; Marder, *W Post*, 5/21/72, A1)

- Administration sources quoted in *New York Times* said U.S. and U.S.S.R. would pledge not to interfere with each other's reconnaissance satellites as part of agreement to limit strategic weapons that President Nixon expected to initial on forthcoming Moscow visit. U.S. had insisted on "open skies" arrangement because, with U.S.S.R. ban on inspection, spy satellites were primary means of checking on compliance with terms of arms control agreements. (Beecher, *NYT*, 5/20/72, 1)

- Secretary of Transportation John A. Volpe announced award of $12.1-million contract to Boeing Co. to secure benefits from valuable work done in supersonic transport (SST) development program. Contract called for completion of seven advanced technological projects to enhance flight safety and efficiency and make aircraft more compatible with environment. Volpe said, by completing projects already begun, "funds already spent on them will not be wasted and the resulting technology will be available for advancement of aircraft design." Congress had voted against further funding of SST program during 1971. (FAA Release 72-88)

May 21: Proposal for domestic communications satellite system was reviewed by John J. O'Connor in *New York Times*. Federal Communications Commission was considering several possible structures for system: completely open entry for applicants, with restrictions limited to antitrust considerations; approval of one applicant or consortium of applicants, subject to restrictions; or plan between these two. Spokesman for one independent group, Network Project, had testified in May congressional hearings protesting that ownership and management by private industry would make granting access for either educational institution or individual "the prerogative of a specialized interest which has historically subordinated public service to private gain." Network Project viewed "public access to, and control of, satellite facilities as a fundamental right," particularly since "American people have already invested more than $20-billion in the space program through their taxes." Question should be opened to debate, including public.

O'Connor asked: "Do we need a domestic satellite system? Do we want a 'wired nation'? Who precisely is articulating these needs? And why?" Satellite research "dominated by the Defense Department" had "involved lasers, computers, sensors and a wide variety of surveillance technology. Cameras on helicopters have already been used to record unaware citizens. Will these techniques become a standard element in a domestic satellite system? And, if so, will the public have access similar to that of a local government or police department?" Questions accumulated. "The answers are, or should be, of public interest." (O'Connor, *NYT*, 5/21/72, 21)

- White House announced plans for new series of awards, Presidential Prizes for Innovation. Prizes, to be distributed Sept. 15, would consist of medal and $50 000. Money would come from National Science Foundation budget for Research Applied to National Needs (RANN). Recipients would be chosen by Prizes Staff of White House Office of Science and Technology, headed by National Bureau of Standards physicist Carl M. Muelhause. Prizes would honor technological advances dealing with "significant problem" in domestic area. (*Sci & Gov Rpt*, 6/15/72, 2)

May 22: President Nixon and party flew from Salzburg, Austria, to Moscow for first official visit of U.S. President to U.S.S.R. He met for two hours with Communist Party leader Leonid I. Brezhnev and later attended Kremlin dinner given by Presidium of U.S.S.R.

During toast, Soviet President Nikolay V. Podgorny said: ". . . great importance is attached in the Soviet Union to Soviet-American talks which should cover a wide range of questions. We approach these talks from realistic positions and will make every effort in accordance with the principles of our policy to achieve positive results and try to justify the hopes placed in our countries and beyond them."

President Nixon responded: "Because we are both prepared to proceed on the basis of equality and mutual respect, we meet at a moment when we can make peaceful cooperation a reality." Summarizing objectives of summit talks, President Nixon said: "First, we want to complete work on matters that years of patient negotiations have brought to the decision point: Bilateral matters will serve as our point of departure: Our two nations can work together in the exploration of space, the conquest of disease, the improvement of our environment." (*PD*, 6/5/72, 915–7)

- Award of $400 000 cost-plus-fixed-fee contract to Bell Aerospace Corp. for space shuttle orbiter reaction control system (RCS) was announced by Manned Spacecraft Center. Bell would supply Air Force Minuteman III post-boost propulsion system technology to RCS requirements. (MSC Release 72-109)

- Ames Research Center scientist Dr. Robert T. Jones was testing theory that antisymmetrical wing pivoting on center point might convert aircraft to supersonic transport without high fuel consumption and noise of conventional SST, NASA announced. Aircraft would have conventional straight wing at right angles to fuselage during takeoff on medium-length runways and would require about one fourth the takeoff energy needed by comparable delta-wing jet transports with similar payloads. As aircraft reached speed and altitude where swept wings were efficient, entire wing would rotate about 45° so that wing on one side would point in direction of flight and wing on other side would trail. Studies had indicated that cruising in this configuration at supersonic speeds to mach 1.2 would not produce sonic boom and could be accomplished with fuel economy similar to that of current jet aircraft at subsonic speeds. Conceptual aircraft could cruise up to mach 1.5 over ocean where sonic effects were not detrimental. ARC

May 22: *Dr. Robert T. Jones of Ames Research Center stood by the wind-tunnel model of a conceptual aircraft being developed in computer and wind-tunnel studies. The fan jet transport's wing would rotate 45° or more above a center pivot as the speed increased, in attempt to achieve supersonic speed without high fuel comsumption and noise. By taking off with the wing perpendicular to the fuselage, the aircraft would require one fourth the takeoff energy of comparable delta-wing transports.*

would continue theoretical computer and wind-tunnel studies using models "flown" at speeds to mach 1.4. (NASA Release 72-104)

- Navy was seeking approval from Dept. of Defense to develop advanced Transit navigation satellite to meet USN operational needs during last half of 1970s and early 1980s, *Aviation Week & Space Technology* reported. Spacecraft would be made less vulnerable to nuclear weapon radiation and to Soviet "killer-satellites." Navy also hoped to maintain more operational satellites in space to provide more frequent navigation fixes. (*Av Wk*, 5/22/72, 20)
- *Apollo 7* Astronaut Donn F. Eisele was sworn in as a Peace Corps director. He would supervise 315 volunteers and 14 paid staff members in Bangkok, Thailand, beginning assignment in mid-June while on terminal leave from Air Force. His retirement from Air Force, after 20 yrs service, would be effective July 1. (*W News*, 5/23/72; *NYT*, 5/23/72, 27)

May 22–25: Aviation/Space Writers Assn. held annual convention in New York. Remarks of Dr. James C. Fletcher, NASA Administrator, on significance of first decade in space were read May 24 by Dr. Homer E. Newell, NASA Associate Administrator: "Our first manned lunar landing was . . . realization of 'the impossible dream.'" With *Apollo 11*, "man on Earth was forced to reassess his potential for achievement—including the dream of universal peace, of brotherhood." Joint U.S.–U.S.S.R. rendezvous and docking agreement under discussion in Moscow "may be another small but significant step in this direction." Dr. Fletcher recalled December 1969 *Life* magazine article had said that "decade ended with an adventure so fantastic as almost to overshadow and redeem all the turmoil that had scarred it." Redeeming adventure had been *Apollo 11* moon landing, which Dr. Fletcher called "the one great event of the decade that gave the American people renewed faith in their destiny."

William M. Magruder, special consultant on technology to President Nixon, said U.S. had lost any chance of competing for first airline supersonic transport orders. He predicted U.S. would produce "second generation" SST that would begin carrying passengers in a decade.

L/G Otto J. Glasser (USAF), Deputy Chief of Staff, Research and Development, said in speech that U.S. aerospace industry had become "overbuilt, overmanned, and overmanaged. Not only are there too many companies, but collectively they have more production capacity than we have any conceivable future need for. The passing of relatively ineffective . . . DC-3s and P-51s has gradually rendered obsolete and archaic our production oriented industry." Gen. Glasser recommended that U.S. firms follow European approach of "very small, tightly integrated design teams manned by top-of-their-graduating-class engineers." (Texts; *NYT*, 5/23/72)

May 23: Scientists at Manned Spacecraft Center detonated by radio command three of four explosive mortar charges left on moon by *Apollo 16* astronauts. Fourth grenade was not detonated because data indicated attitude sensor might have been knocked out of position by previous firings. Analysis of data on pitch sensor was under way to determine if sensor had failed or if mortar package had shifted its position. First grenade contained 270 g (0.6 lb) of explosive with predicted range of 910 m (3000 ft). Second had 45 g (0.1 lb) and 150-m (500-ft) predicted range. Third had 140-g (0.3-lb) charge and 300-m (1000-ft) predicted range. Signals from three explosions, recorded by geophones left on lunar surface, indicated that moon's surface layer was at least 50 m (164 ft) thick. (MSC Release 72-112; O'Toole, *W Post*, 5/24/72, A3; UPI, *C Trib*, 5/24/72)

- NASA's TF-8A aircraft, equipped with supercritical wing and piloted by Thomas C. McMurtry, completed flight from Flight Research Center. Flight was made with wing vortex generators off at constant g of 200 from mach 0.50 to 0.99 and altitudes of 4800 to 14 000 m (16 000 to 46 000 ft). Windup turns and a few stick pulses with augmentation system off were performed. Pilot reported no significant differences in handling qualities and buffeting between flight with vortex generators on and off but said pitch seemed softer with generator off. (NASA proj off)

- House by vote of 367 to 10 passed H.R. 15093, FY 1973 Dept. of Housing and Urban Development-space-science-veterans appropriations bill that included $3.349-billion NASA appropriation [see May 18]. House rejected amendment by Rep. Bella S. Abzug (D-N.Y.) to delete $200 million for space shuttle. (*CR*, 5/23/72, H4896-924)

- U.S. and U.S.S.R. signed two agreements during Moscow summit meetings. Agreement on Cooperation in the Field of Environmental Protection, signed by President Nixon and Soviet President Nikolay V. Podgorny, called for cooperation to prevent pollution, to study pollution and its effect on environment, and to develop basis for controlling "impact of human activities on nature." Agreement established Joint

May 23

Committee on Cooperation in the Field of Environmental Protection to meet each year alternately in Washington and Moscow.

Agreement on Cooperation in the Field of Medical Science and Public Health was signed by Secretary of State William P. Rogers and Soviet Minister of Health Boris V. Petrovsky. Parties agreed on joint efforts to combat most widespread and serious diseases, solve problems of effect of environment on man's health, and resolve other important health problems. Agreement would be implemented by coordinated research, exchange of specialists and delegations, organization of colloquia, scientific conferences and lectures, exchange of information, and familiarization with technical aids and equipment. (*PD*, 6/5/72, 97–20)

- NASA announced start of development of space experiment to test equivalence principle of Einstein's general theory of relativity. Principle asserted that pull of gravity and oppositely directed mechanical acceleration could not be distinguished in small region of space. Thus, when two identical clocks were in areas with different gravitational pull, clock rates would appear to be different, although rates themselves would not change. Gravitational Redshift Space Probe Equipment would be launched from Wallops Station on Scout-D rocket in late 1974. "Gravitational redshift" was term for decrease or shift toward red spectrum as a light wave escaped from strong gravitational field of sun or other massive body toward weaker gravitational field.

  Clock in experiment would follow 3½-hr elliptical flight trajectory over Atlantic and would always be in weaker gravitational field than identical clock at Bermuda ground station. Frequency of clock in probe, as observed by telemetry, should appear greater than frequency of clock on ground and rate should appear to increase as strength of gravitational field decreased up to 18 000-km (11 000-mi) altitude and decrease as it returned to stronger field at lower altitudes. Difference between experimental and ground-based clock rates would be compared with shifts predicted by Einstein's theory. Hydrogen maser clocks for experiment would be developed by Smithsonian Astrophysical Observatory under direction of Marshall Space Flight Center. (NASA Release 72–105)

- Astronauts Edgar D. Mitchell and James B. Irwin announced plans to resign from NASA and military. Mitchell, Navy captain, had been lunar module (LM) pilot on *Apollo 14* (Jan. 31–Feb. 9, 1971). Irwin, Air Force colonel, had been LM pilot on *Apollo 15* (July 26–Aug. 7, 1971) and was backup pilot for Apollo 17, scheduled for launch in December. Because of Irwin's plans to retire, new backup crew for Apollo 17 had been named, effective July 1: John W. Young, Stuart A. Roosa, and Charles M. Duke, Jr. Original Apollo 17 backup crew was David R. Scott, Alfred M. Worden, and Irwin. Scott, Worden, and Thomas K. Mattingly II had been assigned to Space Shuttle Program in Manned Spacecraft Center Astronaut Office. Mattingly, who recently had become a father, had requested assignment which would allow him more time with his family during next six months. (NASA Release 72–110)

- Nuclear system for recycling all human wastes into potable water and converting spacecraft cabin wastes to easily stored ash during extended missions was ready for laboratory testing, NASA and Atomic Energy

Commission announced. Water-recovery and waste-incinerator system fueled by plutonium 238 had been developed by NASA, AEC, and Dept. of Defense. Flight-model system built by General Electric Co. under AEC contract would be tested for 108 days, beginning in June, at GE Space Div. Manned tests in space simulator would be held later. System collected and recycled wash water, cabin condensate, human waste, cabin trash, and foods, along with containers, clothing items and other nonmetallic scrap. At least 98% of all liquid waste processed by system was recoverable as sterile water. Solid waste was incinerated into ash with volume 100 times less than original product. (NASA Release 72-108; AEC Release P-153)

- Arcas sounding rocket launched by NASA from Antigua, West Indies, carried Goddard Space Flight Center payload to 55.8-km (34.6-mi) altitude to measure ozone in upper atmosphere in conjunction with overpass of *Nimbus 4* satellite and to validate new parachute deployment system further [see May 4]. Payload was ejected successfully, parachute inflated properly, and good ozone data were acquired. (NASA Rpt SRL)

May 24: President Nixon and U.S.S.R. Premier Aleksey N. Kosygin in Moscow signed Agreement Concerning Cooperation in the Exploration and Use of Outer Space for Peaceful Purposes. Agreement read: "The United States of America and the Union of Soviet Socialist Republics; Considering the role which the U.S.A. and the U.S.S.R. play in the exploration and use of outer space for peaceful purposes; Striving for a further expansion of cooperation between the U.S.A. and the U.S.S.R. in the exploration and use of outer space for peaceful purposes; Noting the positive cooperation which the parties have already experienced in this area; Desiring to make the results of scientific research gained from the exploration and use of outer space for peaceful purposes available for the benefit of the peoples of the two countries and of all peoples of the world; Taking into consideration the provisions of the Treaty on Principles Governing the Activities of States in the Exploration and Use of Outer Space, including the Moon and Other Celestial Bodies, as well as the Agreement on the Rescue of Astronauts, the Return of Astronauts, and the Return of Objects Launched into Outer Space; In accordance with the Agreement between the United States of America and the Union of Socialist Soviet Republics on Exchanges and Cooperation in Scientific, Technical, Educational, Cultural, and Other Fields, signed April 11, 1972, and in order to develop further principles of mutually beneficial cooperation between the two countries: Have agreed as follows:

"Article 1. The parties will develop cooperation in the fields of space meteorology; study of the natural environment; exploration of near earth space, the moon and the planets; and space biology and medicine; and, in particular, will cooperate to take all appropriate measures to encourage and achieve the fulfillment of the Summary of Results of Discussion on Space Cooperation Between the U.S. National Aeronautics and Space Administration and the Academy of Sciences of the U.S.S.R. dated January 21, 1971.

"Article 2. The parties will carry out such cooperation by means of mutual exchanges of scientific information and delegations, through meetings of scientists and specialists of both countries, and also in

May 24

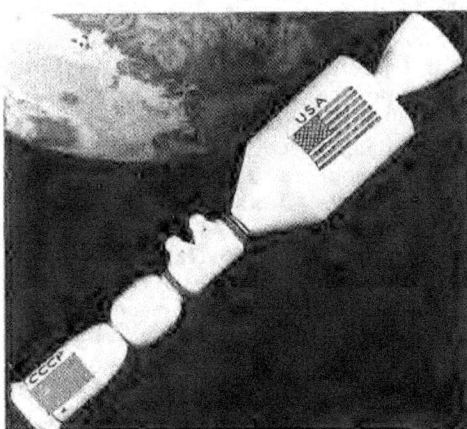

May 24: *President Richard Nixon and Premier Alexsey N. Kosygin in Moscow signed the U.S.-U.S.S.R. Agreement Concerning Cooperation in the Exploration and Use of Outer Space for Peaceful Purposes. The agreement included a project to rendezvous and dock an American and a Soviet spacecraft in earth orbit and exchange astronaut and cosmonaut visits in each other's spacecraft during 1975. In the photo of the docked model, the first two segments at lower left are the Soviet Soyuz spacecraft, the cylindrical structure in the center is the docking module, and the Apollo command and service module is at the upper right.*

such other ways as may be mutually agreed. Joint working groups may be created for the development and implementation of appropriate programs of cooperation.

"Article 3. The Parties have agreed to carry out projects for developing compatible rendezvous and docking systems of United States and Soviet manned spacecraft and stations in order to enhance the safety of manned flights in space and to provide the opportunity for conducting joint scientific experiments in the future. It is planned that the first experimental flight to test these systems be conducted during 1975, envisaging the docking of the United States Apollo-type spacecraft and a Soviet Soyuz-type spacecraft with visits of Astronauts in each other's spacecraft. The implementation of these projects will be carried out on the basis of principles and procedures which will be developed in accordance with the Summary of Results of the Meeting Between Representatives of the U.S. National Aeronautics and Space Administration and the U.S.S.R. Academy of Sciences on Question of Developing Compatible Systems for Rendezvous and Docking of Manned Spacecraft and Space Stations of the U.S.A. and the U.S.S.R. dated April 6, 1972.

"Article 4. The Parties will encourage international efforts to resolve problems of international law in the exploration and use of outer space for peaceful purposes with the aim of strengthening the legal order in space and further developing international space law and will cooperate in this field.

"Article 5. The Parties may by mutual agreement determine other areas of cooperation in the exploration and use of outer space for peaceful purposes.

"Article 6. This Agreement shall enter into force upon signature and shall remain in force for five years. It may be modified or extended by mutual agreement of the Parties."

Plans for joint U.S.–U.S.S.R. rendezvous and docking mission were outlined in Fact Sheet issued by White House Press Secretary in Moscow: "Under today's agreement both sides commit to the development of a complete project schedule, and to meeting that schedule. Arrangements will be made for the necessary contact and understanding between specialists. . . . Training exercises will be conducted in [each] country for the other country's flight crew and ground operations personnel." Working groups would meet in July to plan engineering aspects of mission. Persons directly participating in flight operations of test mission would be included in working groups two years before mission. During mission each country's control center would be expected to control its own craft, but on decisions affecting joint elements of program—including countdown coordination—consultations would occur with other country. "There will be preplanned contingency courses of action. Television down-link will be transmitted to the other country's control center, and voice communications between vehicle and ground will be available to the other country's control center on a pre-planned basis." Flight crews would be trained in other country's language at least well enough to understand it and act in response during normal and contingency courses of action. Apollo command and service module (CSM) and Saturn IB booster would be used. Additional cost of mission to U.S. would be about $250 million. Project Directors would be Glynn S. Lunney, Assistant to Apollo Program Manager for Operations, Experiments, and Government Furnished Equipment at Manned Spacecraft Center, and Soviet Academician K. D. Bushuyev. (Texts)

- Signing of space cooperation agreement in Moscow was announced by Vice President Spiro T. Agnew at Washington, D.C., press briefing attended by Dr. James C. Fletcher, NASA Administrator; Dr. Edward E. David, Jr., Presidential Science Adviser; Glynn S. Lunney, Assistant to Apollo Program Manager for Operations, Experiments, and Government Furnished Equipment at Manned Spacecraft Center; and *Apollo 9* Astronaut James A. McDivitt, Special Assistant for Organizational Affairs to MSC Director. Dr. Fletcher said: "We . . . are very pleased that the President has been able to meet with the officials of the Soviet Union and to provide what we think is by far the most meaningful cooperation in space achieved ever by these two nations."

Dr. Fletcher described rendezvous and docking of Apollo space craft and Soviet Soyuz spacecraft as "perhaps the most dramatic" commitment under cooperation agreement. Spacecraft were at present incompatible "because they have different atmospheres. Theirs is a normal atmosphere, ours is a low pressure oxygen . . . atmosphere and there are certain communications and electronics incompatibilities." Spacecraft would be made compatible by docking module, new device constructed by NASA to link spacecraft. Apollo spacecraft would be launched on Saturn IB rocket. "Then we will make a maneuver at 100 miles [160 kilometers] or so altitude, . . . go in and pick out the docking adaptor from the Saturn and then continue on in orbit. Meanwhile the Soviets will launch their Soyuz . . . [and] rendezvous or at least move their altitude to something like 150 miles [240 kilometers]. We won't be at the same altitude, but hopefully we will be far enough north so too much maneuvering will not be required." Rendezvous would be performed primarily with U.S.-developed guidance and instrumentation. Then spacecraft would dock with "new kind of . . . mechanism we call androgynous, because it doesn't consist of a prop and a drogue. . . . It is an inverse docking mechanism so they can dock with us or we can dock with them either way." System "will apply not only to this . . . mission, but all succeeding spacecraft . . . so that we do have the capability, over a long period, of rendezvousing and docking with each other's spacecraft."

Dr. Fletcher said original plan to attempt rendezvous and docking with Soviet Salyut space station had been canceled when U.S.S.R. "found that was a very complicated mission because it involved not just two launches, their Soyuz and our APOLLO, but a third of their Salyut." Universal docking system using docking adaptor between Apollo and Soyuz spacecraft had been selected because "there has been concern for many years that we did not have a rescue capability . . . in space [so] that we each would have the option of rescuing the other. Future spacecraft beyond 1975, which is the planned date for this mission, will all have the capability."

Total cost of project was estimated at $250 million. Fringe benefits from U.S.-U.S.S.R. joint program were: "It does have the impact of requiring 4,400 people to be employed . . . primarily from the aerospace industry, partly to prepare the command service model and the Saturn 1B launch device, but also to construct the new docking adaptor," and "it will keep the APOLLO team together . . . through 1975 in preparation for the first launch of the shuttle . . . in 1978." NASA was optimistic that cooperative space effort "may lead to greatly increased cooperation on still other programs." Joint mission would be "most visible" U.S.-U.S.S.R. cooperative effort in history, "since it will involve cosmonauts and astronauts working together on a very complex mission in space while the whole world is watching, presumably on television via satellite relay." (Transcript)

- Second press briefing on U.S.-U.S.S.R. space cooperation agreement was held at NASA Hq., following White House briefing. Dr. James C. Fletcher, NASA Administrator, said: "This is not the culmination but a very major milestone in the long period of negotiations with the Soviet Union on space cooperation. And in my judgment this is at least one of the most important missions that NASA has and that is to promote international cooperation." Mission was, as well, "an important part of the new program of NASA for the 70's." Rendezvous and docking mission could be accomplished in context of constant NASA budget. "This constant budget . . . and the programs . . . now planned, the Skylab, the shuttle and now the international rendezvous and docking mission show I think the kind of program that we envision for the rest of this decade." Program had congressional support "at least partially because of the strong international flavor that NASA has in manned space." Dr. Fletcher said crew for joint mission had not yet been selected. Assistant Administrator for Manned Space Flight Dale Myers and Glynn S. Lunney, Assistant to Apollo Program Manager, explained details of plans in response to press questions. (Transcript)

- Agreement on Cooperation in the Fields of Science and Technology was signed in Moscow by U.S. Secretary of State William P. Rogers and U.S.S.R. State Committee for Science and Technology Chairman Vladimir A. Kirillin. Main objective was to "provide broad opportunities for both Parties to combine the efforts of their scientists and specialists in working on major problems, whose solution will promote the progress of science and technology for the benefit of both countries and of mankind." Cooperation might include exchange of scientists and specialists; exchange of scientific and technical information; joint development and implementation of programs and projects in basic and applied sciences; joint research, development, and testing, and exchange of research results and experience between scientific research institutions and organizations; organization of joint courses, conferences, and symposia; and help in establishing contacts and arrangements between U.S. firms and Soviet enterprises. Agreement would establish U.S.-U.S.S.R. Joint Commission on Scientific and Technical Cooperation to meet at least annually, in Washington and Moscow alternately.

    Dr. Edward E. David, Jr., Presidential Science Adviser, said at Washington, D.C., press conference following Moscow signing that Agreement would "enable Soviet and American specialists to solve some major common problems." Cooperation would produce "more vigorous activities here . . . on research, management and systems science, wide use of natural resources, weather modification, superconductivity, high energy physics and basic science." U.S. and U.S.S.R. had already derived major benefits from cooperation in fusion technology and "in the magneto hydrodynamic area, the breeder reactor area and in the solar energy area, we are doing things . . . that are of interest to the Soviet Union." (PD, 6/5/72, 921-2; Transcript)

- Vladimir A. Kirillin, Chairman of U.S.S.R. State Committee for Science and Technology, and Soviet space expert Academician Boris Petrov, held Moscow press conference following signing of U.S.-U.S.S.R. space and science and technology agreements. Petrov said: "It is difficult to overestimate the importance of these works, this research, which has pursued humane goals of ensuring the safety of space flights. The concluded agreements give ground for the belief that these directions in Soviet-American cooperation in outer space will develop successfully in the future." Kirillin said of science and technology agreement: "I think that if we get down to the realization, to the implementation of this agreement, it will definitely yield positive results." (Transcript)

May 25: NASA's *Mariner 9* spacecraft, in its seventh month in Mars orbit, was in period of both sun occultation (April 2–June 4) and earth occultation (May 6–June 24) for short periods of each orbit. Sun occultations—when spacecraft passed into shadow of Mars and was powered by onboard battery—were declining from peak total time of 97 min. Only data from celestial mechanics and earth occultation experiments and engineering telemetry were being received. Battery was recharging satisfactorily twice daily and spacecraft was in good condition.

Earth occultation—when Mars was between spacecraft and earth—varied from 43 to 90 min orbit. Resumption of limited data acquisition from scientific instruments would begin in June to conserve attitude control gas. (NASA Special Release)

- NASA's *Pioneer 10* spacecraft, en route to Jupiter, crossed orbit of Mars and entered space never before visited by spacecraft. *Pioneer 10* had safely crossed region of "Great Galactic Ghoul" 209 million km (130 million mi) from sun where two Mars-bound spacecraft had been lost and third had stopped transmitting for seven hours, apparently because of impact of high-velocity meteoroids. Since launch March 2 *Pioneer 10* had traveled 249 million km (155 million mi) on its 998-million-km (620-million-mi) flight path to Jupiter and was 75 million km (47 million mi) straight-line distance from earth. It would enter Asteroid Belt in early July, reach Jupiter in December 1973, and cross orbit of Pluto in about 15 yrs. (NASA Release 72–111)
- Air Force launched unidentified reconnaissance satellite from Vandenberg Air Force Base at 11:41 am PDT by Thorad-Agena booster. Satellite entered orbit with 259-km (161-mi) apogee, 154-km (96-mi) perigee, 89.2-min period, and 96.4° inclination. Satellite reentered June 4. (Pres Rpt 73; *Av Wk*, 6/5/72, 14)
- U.S.S.R. launched *Cosmos 491* from Baykonur into orbit with 369-km (229.3-mi) apogee, 173-km (107.5-mi) perigee, 89.9-min period, and 64.9° inclination. Satellite reentered June 8. (GSFC *SSR*, 5/31/72; 6/30/72; *SBD*, 5/30/72, 153)
- NASA test pilot Gary E. Krier flew modified F–8 jet fighter equipped with digital-fly-by-wire control system from Flight Research Center in first flight test of aircraft that was completely dependent on electronic control system. Aircraft had no mechanical backup control system. Purpose of flight, limited to maximum speed of 680 km per hr (425 mph) and maximum altitude of 6000 m (20 000 ft), was to check out aircraft's primary and secondary flight control system. Fly-by-wire system with digital computer and inertial measuring unit had been developed originally for Apollo lunar module and was designated for future use in space shuttle to ensure smoother flight. (NASA Release 72–112)
- *Apollo 16* astronauts were honored with luncheon and hour-long ceremony at Kennedy Space Center. Astronauts greeted KSC employees who had worked to make mission successful. (*O Sen*, 5/18/72; KSC PAO)
- British Overseas Airways Corp. placed first firm order for Anglo-French Concorde supersonic transport. British national airline would take delivery of first of five Concordes early in 1975. Total cost would be $299 million. (*W Post*, 5/26/72, A21)
- Sen. George S. McGovern (D-S. Dak.) addressed aerospace workers in presidential campaign speech at Redondo Beach, Calif. He felt "American people are getting sour" on science. "They're weary of watching

billions of dollars going into what they see as exotic projects, when their basic needs and concerns are still not being met. They're tired of hearing about 'spin-offs.' And they're wondering why we can't go after some of these problems directly instead of just hoping for beneficial effects from other projects to do the job." (McGovern Release)
- Symposium on use of electronic computing machines in production management was held in Moscow by Soviet specialists and U.S. engineers from Digital Equipment Corp. Digital Equipment Corp. Vice President Theodore Johnson later told Tass correspondent that agreements signed between U.S. and U.S.S.R. in science and technology May 24 would widen firm's contacts in U.S.S.R. (FBIS–Sov, 5/26/72, L1)
- First-year, full-time graduate science enrollment in doctorate-granting institutions had decreased 5% between 1970 and 1971 after decrease of 2% in 1969, National Science Foundation reported. First-year, full-time enrollment at "top 20" graduate institutions had decreased at greatest rate—8%. Reductions had been experienced in all areas of science and number of full-time graduate students supported primarily by fellowships and traineeships had declined nearly 10% from 1970 to 1971. Proportion receiving primary support from Federal Government had declined from 37% in 1969 to 32% in 1971. (NSF *Highlights*, 5/25/72, 1)
- Newspaper editorials commented on U.S.–U.S.S.R. space cooperation agreement.

    New York Times: "In effect it announces to the world that the leaders of the two nations expect their relations to be sufficiently amicable that each is prepared to open its space installations to the other and let Soviet cosmonauts be trained in this country and their American opposite numbers in the Soviet Union." Corollary was "growing likelihood that the chief dividends from space programs will be political gains here on earth. Born in the mad competition for status characteristic of the cold war, manned and unmanned space research has taught both sides how puny are man's resources in facing the mystery and challenge of the universe. As that lesson has sunk in, both sides have come to understand the advantages of cooperation as against useless and wasteful rivalry." (*NYT*, 5/25/72, 42)

    St. Louis Post-Dispatch: "Cooperation in solving medical and environmental problems and the exchange of personnel and information in space research should stand as predecessors for other joint efforts at solving problems faced by both nations. And it should go without saying that the more both countries pursue common objectives together, the more they ought to trust each other and the less likely they should be to resort to rash belligerency." (*St Louis P–D*, 5/25/72)

May 26: U.S.–U.S.S.R. Treaty on the Limitation of Anti-Ballistic Missile Systems and Interim Agreement With Respect to the Limitation of Strategic Offensive Arms were signed in Moscow by President Nixon and Communist Party Leader Leonid I. Brezhnev. ABM treaty—requiring Senate ratification—limited each country to two ABM deployment areas with radius of 150 km (93 mi) each and with no more than 100 ABMs in each. One area would be centered on national capital and one would protect field of intercontinental ballistic missiles (ICBMs). Each country was permitted six modern ABM radar complexes within capital area. Interim agreement froze production of ICBMs for five years or until full treaty was negotiated, if earlier. Possession of ICBMs was limited to those already deployed or under construction—1054 for U.S. and 1618 for U.S.S.R. Agreement also froze construction of submarine-launched ballistic missiles for nuclear submarines beyond those already begun. New construction of ICBMs, submarine missiles, or modern ballistic missile submarines would require dismantling equal number of older ones.

Agreements left both nations with near-parity in nuclear weapon strength. U.S. would lead in strategic bombers with 460, with 140 for U.S.S.R.; both countries would have 2500 delivery vehicles; and Soviet megatonnage would be triple that of U.S. but U.S. would have some advantage in warheads. Each country would use "national technical means of verification," rather than on-site means, to monitor compliance with treaty terms. Baltimore *Sun* said later that means intended were "obviously spy satellites that use photographic and other means to check on developments in each country."

Statement released by White House Press Secretary in Moscow following signing of agreements said President Nixon was "deeply gratified that now, for first time since the advent of nuclear weapons a generation ago, the two most powerful nations . . . have taken the lead in a direction away from wasteful, dangerous, and self-perpetuating competition in armaments and toward a mutual restraint on weapons systems." President intended "to press ahead . . . in building on that foundation laid down by this initial SALT [Strategic Arms Limitation Talks] accord, and to seek both further progress in the limitation of offensive weapons, and more comprehensive agreements across the whole range of issues confronting our two countries." (*PD*, 6/5/72, 925–37; Kumpa, B *Sun*, 5/27/72, A1)

- NASA anounced retirement of Dr. Wernher von Braun, Deputy Associate Administrator for Planning, effective July 1 when he would become Corporate Vice President for Engineering and Development at Fairchild Industries. Dr. von Braun had joined NASA in 1960 when Army Ballistic Missile Agency development team which he headed was transferred to NASA and formed nucleus of Marshall Space Flight Center.

  Born in Germany, Dr. von Braun became director of German army portion of Peenemuende Rocket Center. Near end of World War II he led group of scientists to West, surrendering to allies. In 1945 he went to U.S. under contract to U.S. Army to direct high-altitude firings of V–2 long-range ballistic missile at White Sands Missile Range. As Director of Development Operations Div. of ABMA, Dr. von Braun led development of Pershing Army missile and Jupiter intermediate-range ballistic missile, whose nosecone reentry test vehicle Jupiter-C was used to launch first U.S. satellite, *Explorer 1* (Jan. 31, 1958). Dr. von Braun was MSFC Director from July 1960 to February 1970, when he assumed his current position. He had received numerous awards and honors, including Dept. of Defense Distinguished Civilian Service Award, Dr. Robert H. Goddard Memorial Trophy, British Interplanetary Society Gold Medal, Herman Oberth Award, NASA Medal for Outstanding Leadership, Galabert International Astronautical Prize, NASA Distinguished Service Medal, American Society of Mechanical

Engineers Man of the Year Award, and Associated Press Man of the Year in Science Award.

Dr. von Braun said he would leave NASA "with a deep feeling of gratitude for the wonderful and unique opportunities the agency has given me during the last 12 years . . . [and] with the knowledge that NASA has enough well thought out plans to keep it moving ahead for many years to come."

Dr. James C. Fletcher, NASA Administrator, said Dr. von Braun's retirement was "a source of regret to all of us. . . . For more than a quarter of a century, he has served the United States as a leader in space rocket development. His efforts first put the United States in space" with *Explorer 1* and as director of MSFC "he directed the development of the world's most powerful rocket, the Saturn V. All of us at NASA will miss the daily stimulation of his presence, but we are confident that we will continue to have the benefit of his inspiration and counsel in the continuing exploration and use of space." (NASA Release 72–113; NASA biographical data)

- Soviet Premier Aleksey N. Kosygin said in toast to President Nixon at Moscow dinner party given by President: "A number of Soviet-American agreements, bound to serve peaceful aims, have been signed. . . . We have agreed . . . on pooling the efforts of our countries in environmental protection, in peaceful exploration and mastering of outer space, in cooperation in the fields of science and technology, medicine and public health." Strategic arms agreement was "great victory" in "easing international tension . . . , victory for all peace-loving peoples, because security and peace is their common goal." (*PD*, 6/5/72, 924–5)
- House Committee on Science and Astronautics' Subcommittee on International Cooperation in Science and Space announced it would hold hearings in July on U.S.–U.S.S.R. agreements for cooperation in space, medicine, science, and environment. (Reuters, *W Post*, 5/27/72, A23)
- Tracks attributed to spontaneous fission of plutonium 224 and uranium 238 had been detected in large whitlockite crystal in lunar breccia No. 14321 from Fra Mauro formation brought to earth by *Apollo 14* astronauts, Univ. of California at Berkeley physicists reported in *Science*. It was first time evidence for decay products of plutonium 244 had been detected in objects other than meteorites and indicated age of rock could be 3.95 billion yrs. No positive evidence of terrestrial or lunar rocks with antiquity comparable to that of meteorites had yet been found. (Hutcheon, Price, *Science*, 5/26/72, 909–11)
- Award of $168 000 cost-plus-fixed-fee contract to General Dynamics Corp. for design study of low-density ablative materials was announced by Manned Spacecraft Center. Materials could be used to protect space shuttle orbiter from heat. (MSC Release 72–117)
- Dept. of Transportation said it was contributing $250 000 toward 16-mo joint study by NASA, Dept. of Transportation, and National Oceanic and Atmospheric Administration of effect on climate of water vapor pumped into stratosphere by high-altitude jet aircraft. NASA was contributing $100 000 and NOAA $50 000. (DOT Release 48–72)
- Dept. of Transportation would begin priority program to establish quiet, short-haul air transportation system to relieve airport congestion and noise, Secretary of Transportation John A. Volpe said in speech before

May 26

Air Line Pilots Assn. in Washington, D.C. Federal Aviation Administration would designate certain existing U.S. airports as short-haul airports and develop strict standards on noise and pollution compatible with the community. (DOT Release 49-72)

- Aerospace Industries Assn. of America, Inc. (AIA), announced formation of International Coordinating Council of Aerospace Industries Associations (ICCAIA) to pursue "development and advancement of international civil aeronautical and astronautical arts and industries." ICCAIA would also petition for observer status for its members in proceedings of International Civil Aviation Organization (ICAO). ICCAIA members—representing manufacturers of 12 countries—were AIA, Air Industries Assn. of Canada, Society of Japanese Aircraft Constructors, and Association Internationale des Constructeurs de Materiél Aérospatial (AICMA). AIA President Karl G. Harr, Jr., was elected Chairman and AICMA President R. J. L. Diepen was elected Vice Chairman. (AIA Release 72-11)

- Harrison H. Huntoon of Manned Spacecraft Center Management Analysis Office had received first NASA assignment under Intergovernmental Personnel Act of 1970, MSC announced. He would serve one to two years as chief of planning and evaluation for Galveston, Tex., County Health District under law to facilitate mobility of trained personnel between national, state, and local governmental agencies. (MSC Release 72-116)

- Formation of General Aviation Accident Prevention Industry Advisory Committee to work with Federal Aviation Administration in developing techniques for preventing general-aviation accidents was announced by Secretary of Transportation John A. Volpe. (FAA Release 72-96)

- Newspapers commented on U.S.–U.S.S.R. space cooperation agreement.

    Wall Street Journal: "Whereas past space efforts have been an extension and symbol of Soviet-American competition, the proposed new effort will attempt to extend and symbolize cooperation." Question was "whether this change is contrived and doomed to fail as an expression of national needs and attitudes or whether it signifies the dawning of a new era in the progress of man. A case may be made either way. . . ." (*WSJ*, 5/26/72)

    Christian Science Monitor saw agreement as "commitment to peace, to friendly relations in the years ahead on the part of the world's two superpowers." Agreement meant extension of Apollo program to bridge gap between last Skylab mission in 1973 and 1978 space shuttle flights. "It will mean an estimated 4,400 more jobs in the aerospace industry. It will mean that Soviet space crews will come to the United States for training and vice versa. And once the rendezvous is made televiewers in both countries will be able to watch their astronauts moving from one spacecraft to another and conducting joint scientific experiments." (*CSM*, 5/25/72)

    Miami Herald: "Both sides may learn a bit from each other . . . . What most people will learn is likely to be more meaningful because emotions are the mainspring of human attitudes and actions. National prestige has been a foremost consideration in space for more than 15 years. Preparation for the joint flight promises to remold the effort into a symbol of international cooperation." (*M Her*, 5/26/72)

Washington Post: "The promise is that this first docking and the steps that lead to it may be the key to a new coordinated joint effort to explore and exploit those parts of the universe within man's reach." If mission was successful, "it ought to be possible for [U.S. and U.S.S.R.] to cooperate on most other aspects of space operations. The benefits to each country from such an arrangement would be great. By curtailing both competition and duplication in space activities, scientists on both sides would be able to do in good time the things that ought to be done without straining the financial resources of their governments nearly as much in the process." (*W Post,* 5/26/72)

May 27: Donald K. Slayton, Director of Flight Crew Operations at Manned Spacecraft Center, said in telephone interview with *New York Times* that NASA astronaut corps would be cut because it had "three times as many people as are needed." If all astronauts had stayed in program, many "would just be on dead-end streets. There have been retirements right along, and there will be further reductions. . . . It's just a fact of life that we have only one Apollo, three Skylab and one 'ren-dock' flights ahead of us, and our manpower is over our known requirements by a factor of three."

Of 27 astronauts who had flown on Apollo missions, 12 had already left flying status. As many as 10 astronauts reportedly already had been asked to find other employment as soon as possible, *Washington Post* said astronaut corps had confirmed. Two had recently resigned—Edgar A. Mitchell and James B. Irwin—and many more of the 43 astronauts remaining were planning resignations. Philip K. Chapman would join staff of Massachusetts Institute of Technology and Anthony W. England would join Science and Applications Directorate at MSC. Donald L. Holmquest and John S. Bull, both on temporary leave to universities for past year, were also expected to resign. According to one unidentified astronaut, "Slayton really feels that he doesn't need any more than 15 men" for shuttle and joint Soviet missions. In 1967 and 1968, astronaut total had reached peak of 63 men. (Lyons, *NYT,* 5/28/72, 1; O'Toole, *W Post,* 5/28/72, A14)

- Dr. Wernher von Braun's resignation from NASA May 26 was "reminder of how many others who forged the American space effort in the last 15 years have left the scene," *New York Times* commented. Mercury Astronaut John H. Glenn, Jr., was "Ohio businessman with political ambitions"; *Apollo 8* Astronaut Frank Borman was Eastern Airlines Vice President; *Apollo 11* Astronaut Neil A. Armstrong was Univ. of Cincinnatti professor; former NASA Administrator James E. Webb, "who built the Apollo team," was lawyer-consultant in Washington, D.C.; Dr. Thomas O. Paine, "administrator in charge at the time of the moon landing," was General Electric Co. Vice President. Dr. Robert R. Gilruth, "who gathered the manned space flight team while Dr. von Braun was building the rockets," had "recently stepped down as director of the Manned Spacecraft Center in Houston." Others of "space's first generation have scattered to industry or are close to retirement." Of original 118 German-born engineers who came to U.S. with Dr. von Braun, only 35 remained at Marshall Space Flight Center. Dr. Eberhard F. M. Rees, "another of the Peenemuende V-2 rocket Germans," was MSFC director. Six of original team—including Dr. Kurt H. Debus, director of Kennedy Space Center—were still

May 27

with NASA at other Centers. Of others of team, 22 were in industry, 12 were deceased, 16 had returned to Europe, and 26 had retired. (Wilford, *NYT*, 5/27/72)

- *Apollo 15* Astronaut James B. Irwin received honorary Doctor of Science degree from Samford Univ. in Birmingham, Ala. (*Birmingham News*, 5/14/72)

May 27–June 4: U.S. International Transportation Exposition TRANSPO '72 was sponsored by Dept. of Transportation at Dulles International Airport in Chantilly, Va. More than 400 exhibitions from 10 countries included military and civil aircraft and aerial demonstrations, experimental safety vehicles developed by automobile industry, *Apollo 12* command module with *Apollo 12* moon rocks, and four working models of independently powered 6- and 12-passenger "people movers" operated by pushbutton for urban mass transportation. NASA theme, "Aviation Serves the Community and the Nation," was illustrated with exhibits on aviation passenger growth, airport planning, aircraft noise reduction, pollution control, aircraft safety, long-haul aircraft for 1980s, new vehicle technology, general aviation, nagivation and communications, and equation of ecology with technology.

Exhibition was marred by three accidental deaths during aerobatic and precision flying demonstrations. (*Langley Researcher*, 5/12/72, 1; Grubisch, *W Post*, 5/4/72, H1; *Transpo 72 News*)

May 28: President Nixon discussed Moscow summit meetings and resultant agreements in radio and TV address to Soviet people from Kremlin in Moscow. He said: "We have agreed on joint ventures in space. We have agreed on ways of working together to protect the environment, to advance health, to cooperate in science and technology." Most important agreement was "historic first step in the limitation of nuclear strategic arms." By settling arms limitation, "people of both of our nations, and of all nations, can be winners. If we continue in the spirit of serious purpose that has marked our discussions this week, these agreements can start us on a new road of cooperation for the benefit of . . . all peoples." (*PD*, 6/5/72, 939–41)

- Press commented on U.S.–U.S.S.R. space cooperation agreement.

U.S. would "be doing virtually all of the cooperating" in joint space mission called for under new U.S.–U.S.S.R. space agreement, William Hines said in Washington *Sunday Star*. "While negotiations were underway . . . it was assumed that the 1975 joint manned space flight envisioned . . . would be a 50–50 proposition. The Americans would build the hardware . . . and the Russians would provide a small space station in which experiments could be carried out." Instead, U.S.S.R. had "pulled back their item—the Salyut station— and changed the nature of the mission." This had relieved U.S.S.R. of considerable expense and "left Uncle Sam holding a $250 million bag, this sum being the estimated cost of the hardware our side will have to build anyway." (*W Star*, 5/28/72, A6)

St. Louis Post-Dispatch: "An agreement between the U.S. and the U.S.S.R. . . . ought to be just the first step toward full international co-operation in space. Countries which lack space programs of their own should be encouraged to participate in space ventures as limited partners." Other countries had scientific expertise to contribute. "Beyond this they ought to receive the full benefits of whatever information is gained through space research. Historically, science never has been the exclusive domain of a single nation or even a small group of nations. It has served all mankind and this is the way it should be in space." (*St. Louis P–D*, 5/28/72)

- Achievements of U.S.–U.S.S.R. summit meetings in Moscow to date were summarized in *New York Times* article by Max Frankel. U.S. and Soviet officials had "duly signed the agreements that had been specially packaged for the occasion. Some, on cooperative ventures in health research, environmental control, scientific cooperation and a link-up of astronauts in space, had been merely rewritten or upgraded from earlier agreements, to add to the cumulative sense of achievement. One of the accords, to avoid naval harassment and incidents at sea, had been negotiated under the deadline of the summit." Another, on trade, had become "too complicated to resolve" and was "earmarked for energetic followup this summer." Agreement to limit construction of defensive and offensive nuclear weapons "was finally finished in a whirlwind round of all-night sessions." Main premises behind summit meeting were "that the two superpowers are now truly strategic equals, that they both have enough invulnerable weapons to deter nuclear attack, that they need to stop adding more weapons before they can reduce the number of weapons, that they both can use the money they might eventually have for better purposes, that they must avert even indirect conflicts that might draw them into war, and that they must collaborate more broadly in other fields to avoid the intrusions of inevitable conflicts of interest and conflicts among smaller nations." (*NYT*, 5/28/72, 4:1)

- President Nixon signed H.R. 14582, Second Supplemental Appropriations Act, 1972. Act, which became Public Law 92–306, contained $12.087-million NASA appropriation to cover increased pay costs. (PL 92–306; *PD*, 6/5/72, 938)

May 29: President Nixon and Secretary General Leonid I. Brezhnev of Soviet Communist Party's Central Committee signed "Basic Principles of Relations" outlining agreements made between U.S. and U.S.S.R. during May 22–29 summit meetings in Moscow. They then released joint communique at conclusion of President's visit to U.S.S.R. Communique described agreements on cooperation in science and technology and space: "It was recognized that the cooperation now underway in areas such as atomic energy research, space research, health and other fields benefits both nations and has contributed positively to their over-all relations. It was agreed that increased scientific and technical cooperation on the basis of mutual benefit and shared effort for common goals is in the interest of both nations and would contribute to a further improvement in their bilateral relations." In peaceful exploration of space, "both Sides emphasized the importance of further bilateral cooperation. . . . In order to increase the safety of man's flights into outer space and the future prospects of joint scientific experiments, the two Sides agreed to make suitable arrangements to permit the docking of American and Soviet spacecraft and stations. The first joint docking experiment of the two countries' piloted spacecraft, with visits by astronauts and cosmonauts to each other's spacecraft is contemplated for 1975." NASA and Soviet Academy of Sciences would plan and implement flight "according to principles and procedures developed through mutual consultations." (*PD*, 6/5/72, 945–51)

- *Apollo 16 Subsatellite* for study of particles and fields apparently crashed onto lunar surface at about 4:00 pm EST after 425 revolutions of moon. Radio contact with satellite, launched from *Apollo 16* orbiting moon April 24, was not reestablished when it should have reappeared on moon's near side on 426th revolution. Last tracking data, from 416th revolution, showed *Apollo 16 Subsatellite* in orbit with 4.8-km (3-mi) perilune on moon's far side near point 10.2° north latitude and 111.9° east longitude. Subsatellite had been designed for 12-mo lifetime, but had been placed in orbit closer to moon than desired because of problems with *Apollo 16* command module engine. (NASA Release 72–115)

- Development Test Satellite (DTS), full-scale ground-test vehicle characteristic of high-capacity next-generation communications satellites, had been built by 15-member international consortium led by Lockheed Missiles & Space Co., *Aviation Week & Space Technology* reported. Privately funded vehicle embodied concepts and technologies that Lockheed considered essential for profitable or cost-effective operations 1976–1985. Features included three-axis spacecraft stabilization; electric and hydrazine propulsion; sun-tracking, flexible, solar substrate solar arrays; multiple narrow antenna beams; onboard transponder/antenna switching; and advanced thermal control. (*Av Wk*, 5/29/72, 41–7)

May 30: Dr. James C. Fletcher, NASA Administrator, told meeting of White House Council that U.S.–U.S.S.R. agreement on joint space mission would lead to some 4400 new jobs in California's aerospace industry by 1974 and retention of some 1500 jobs at NASA Centers in Florida and Alabama. (Gwertzman, *NYT*, 5/31/72, 9)

- NASA announced issuance to scientists of requests for proposals for International Ultraviolet Explorer (IUE) satellite that would be built and operated jointly by NASA, United Kingdom Science Research Council, and European Space Research Organization (ESRO). Plans called for launch of 303-kg (669-lb) spacecraft in 1976 to make high- and low-resolution ultraviolet observations of stars, planets, and other celestial objects. Satellite would transmit to ground stations in U.S. and Europe and make observations 24 hrs daily. Soviet observers would be able to direct and monitor own programs. (NASA Release 72–114)

- Boris Petrov, Chairman of Soviet Council for International Cooperation in Investigation and Utilization of Outer Space (Intercosmos), commented on U.S.–U.S.S.R. space cooperation agreement in Tass interview: Agreement was difficult to overestimate. It would open new opportunities for developing further activities for Soviet and U.S. scientists in space. Outer space was becoming arena for broad international cooperation and demanded joint efforts of many countries. U.S.S.R. had been cooperating with other countries for many years in Intercosmos program. Broad program was being implemented under Soviet-French agreement. Space relationships were being developed between Soviet and Indian scientists. Soviet Academy of Sciences and European Space Research Organization (ESRO) had agreement on exchange of scientific information. (FBIS–Sov, 5/31/72, L2–3)

- U.S.S.R.'s Special Astrophysical Observatory—world's largest astronomical observatory—had entered final stage of construction with installation of huge telescopic mirror 600 cm (236 in) in diameter, *Pravda* reported. The 730-metric-ton (800-U.S.-ton) telescope had been installed in azimuthal position in dome 40 m (130 ft) high. Second largest reflecting telescope was U.S. reflecting telescope with 500-cm (200-in) diameter at Mt. Palomar, Calif. (Shabad, *NYT*, 5/31/72, 7)
- Federal Grand Jury in Dallas, Texas, indicted General Dynamics Corp. and four officials for conspiring to defraud Government by charging Air Force for defective parts for F-111 fighter-bomber aircraft. Indictment said defendants had concealed from Air Force unauthorized production procedures by subcontractor Selb Manufacturing Co. Dept. of Justice said defective parts—carry-through plates for center section of wings and longerons or main fuselage braces—had not been turned over to Air Force. (AP, B *Sun*, 5/31/72, A9)

May 31: NASA announced completion of major preflight verification test of Skylab Workshop at McDonnell Douglas Corp. Huntington Beach, Calif., plant. Two teams of six astronauts each had performed checkout activities in two six-hour shifts daily for three days, activating Workshop to demonstrate that it could support all activities planned for missions. Test was one of last two major tests for Workshop, 14.6 m (48 ft) long, 6.7 m (22 ft) in diameter, and scheduled for launch in early 1973. Flight demonstration would be conducted before spacecraft was shipped to Kennedy Space Center during summer. (NASA Release 72-117)

- Senate Committee on Appropriations favorably reported H.R. 15093, FY 1973 Dept. of Housing and Urban Development-space-science-veterans appropriations bill that included $3.432-billion NASA appropriation. Senate appropriation was $24 million above budget request of $3.408 billion and increase of $82.4 million over House appropriation of $3.349 billion voted May 23. Senate version raised appropriation for research and development (R&D) $74.9 million above House version, to $2.624 billion. Construction of facilities funds were increased by $7.5 million, to $77.3 million. Research and program management funds remained unchanged. Of R&D appropriation increase, $24 million was for aeronautical research in noise abatement and aviation safety and $50.9 million to restore funds deleted in House bill from other NASA research programs. Construction of facilities increase restored funds deleted in House bill—$5.5 million requested by NASA for modification of space shuttle manufacturing and final assembly facilities and $2 million for planning and design. Shuttle funds would support orbiter assembly and external hydrogen and oxygen tank manufacture, baselined "for estimating purposes" at Michoud Assembly Facility, one of possible sites for either or both of these functions. (S Rpt 92-820)
- House Committee on Science and Astronautics' Subcommittee on Manned Space Flight met for briefing by NASA on joint U.S.-U.S.S.R. rendezvous and docking mission in 1975. Officials—Dale D. Myers, Associate Administrator for Manned Space Flight; Arnold Frutkin, Assistant Administrator for International Affairs; and Glynn S. Lunney, Assistant to Apollo Program Manager—presented chronology of major events leading to mission, preliminary mission description, description of

May 31

hardware elements, and data on management of engineering phase and of real-time operation. (Transcript)

- Two NASA aircraft equipped with camera systems similar to those on board Earth Resources Technology Satellite A (ERTS–A; scheduled for June or July launch) were flying from Ames Research Center in preparation for ERTS flights. ERTS flights would be first U.S. attempt to monitor earth resources from space. Purpose of NASA aircraft flights at 20 000-m (65 000-ft) altitude was to acquire photographic data in color, infrared, and other spectral bands at same time of day as ERTS–A would photograph from 885-km (550-mi) altitude. Data from 50 test sites in continental U.S. would be processed and distributed by ARC to 47 ERTS investigators who required seasonal terrain-change data in agriculture, hydrology, and forestry for analysis of early growing season. Seasonal change data from aircraft would be correlated with ERTS–A data after launch to monitor healthy crops and forests and disease infestation. (NASA Release 72–116)
- Robert J. Collier Trophy for 1971 was presented by National Aeronautic Assn. to *Apollo 15* Astronauts David R. Scott, James B. Irwin, and Alfred M. Worden and former Manned Spacecraft Center Director, Dr. Robert R. Gilruth, during Washington, D.C., ceremony. (Abrams, *W Post*, 6/1/72, D1)
- Two sounding rockets were launched from Andoeya, Norway, carrying Univ. of Maryland and Norwegian experiments to measure, *in situ*, electron and proton spectra in relativistic electron precipitation events. Launches were part of cooperative effort between NASA and Royal Norwegian Council for Scientific and Industrial Research (NTNF). NTNF was responsible for payload construction, integration, and launch. Nike-Apache reached 165-km (102.5-mi) altitude and good data were received from all experiments. Nike-Cajun reached 110.5-km (68.7-mi) altitude, but payload doors failed to eject, causing loss of Univ. of Maryland data. Good data were received from NTNF ion mass spectrometer. (NASA Rpts SRL)
- NASA launched Black Brant VC sounding rocket from White Sands Missile Range, N. Mex., carrying Univ. of Colorado solar ultraviolet spectrometer. Rocket performed satisfactorily. Scientific objectives were not satisfied. (SR list)
- Langley Research Center awarded $9 400 000 contract to LTV Aerospace Corp. Vought Missiles & Space Co. for 15 Scout launch vehicles, to be delivered one per month beginning in November 1973. Rocket motors for boosters would be furnished under separate $3 819 000 contract with LTV Aerospace Hampton Technical Center. (LaRC Release 72–7)
- U.S. sources quoted in *Washington Post* said U.S.S.R. had begun building modified class of missile-firing submarines to carry fewer but longer range missiles than current Soviet undersea fleet. New submarines would each carry 12 new SSN–8 missiles with estimated range about 5600 km (3500 mi), or more than twice that of missile currently installed on "Y" class Soviet submarines. (Getler, *W Post*, 5/31/72, A12)

During May: Air Force Materials Laboratory (AFML) was preparing experiment to determine stability of thermal-control coatings or paints and thin films of polymeric materials in space environment. Samples of materials and films would be carried in trays on discs 25 mm (1 in) in diameter on NASA Skylab mission in 1973. After exposure to space, samples would be returned to earth for evaluation of degradation. (AFSC *Newsreview,* 5/72, 14)

- Smithsonian Institution displayed collection of 60 works by astronomical artist Chesley Bonestell in Aerospace Art Hall of Arts and Industries Building. Artist had been commissioned by *Collier's* magazine in 1952 to illustrate series of articles by rocket expert Dr. Wernher von Braun, later NASA Deputy Associate Administrator, and by rocket historian Willy Ley. Articles had resulted in Viking Press books: *Conquest of Space, Conquest of the Moon,* and *Exploration of Mars.* Smithsonian exhibition was coincidental with American Museum showing of Bonestell's works in New York [see May 16]. (Smithsonian Release; Smithsonian PIO)

June 1972

June 1: President Nixon returned from state visits to Austria, U.S.S.R., Iran, and Poland and reported to Congress on trip. Commenting on agreements signed with U.S.S.R. during May 22–29 Moscow summit meetings, he said: "Recognizing that the quest for useful knowledge transcends differences between ideologies and social systems, we have agreed to expand United States–Soviet cooperation in many areas of science and technology.

"We have joined in plans for an exciting new adventure . . . in the cooperative exploration of space, which will begin—subject to Congressional approval of funding—with a joint orbital mission of an Apollo vehicle and a Soviet spacecraft in 1975." President asked for "fullest scrutiny" of antiballistic missile treaty signed May 26, "because we can undertake agreements as important as these only on a basis of full partnership between the executive and legislative branches of our Government." (*PD*, 6/5/72, 975–81)

- Astronomy Survey Committee of National Academy of Sciences and National Research Council issued *Astronomy and Astrophysics for the 1970's*, Vol. 1. Committee assessed state of astronomy and astrophysics and recommended priorities for Federal funding in 10-yr, $1.2-billion program to study "entirely new class of objects . . . undreamed of ten years ago." New technologies applied during decade had revealed new types of worlds. Previously well organized universe had "exploded into a bewildering universe" of objects with "exotic new names and marvelous new natures." Opening of radio sky had suggested presence of undiscovered physical laws and requirements for new observations and explanations.

Committee defined four programs of highest priority: (1) very large radio array to attain resolution equivalent to that of single radiotelescope 42 km (26 mi) in diameter, plus increased support of smaller radio programs at universities or research laboratories; (2) optical program to increase efficiency of existing telescopes with electronic auxiliaries, while creating new large telescopes to research limits of known universe; (3) increased support and development of infrared astronomy, including construction of large, ground-based infrared telescope, high-altitude balloon surveys, and design studies for very large atmospheric telescope; and (4) program for x-ray and gamma ray astronomy from High Energy Astronomy Observatories (HEAOs) supported by ground-based optical and infrared telescopes.

Committee recommended for secondary priority: construction of very large millimeter-wavelength antenna to identify and study complex molecules and to study quasars in their most explosive stages; doubling of support for astrophysical observations from aircraft, balloons, and rockets, at wavelengths from far infrared to gamma rays; continuation of Orbiting Solar Observatories through OSO–L, –M, and –N, with updating of ground-based solar facilities; increased support for theoretical investigations; expansion of optical space astronomy, including high-resolution imagery and ultraviolet spectroscopy, leading to launch of large space telescope at beginning of next decade; and construction of large, steerable radiotelescope to obtain observations with high angular resolution and to record emission from more distant objects than now possible. (NAS–NRC–NAE *News Rpt,* 6/7/72, 1, 6–7)

- Manned Spacecraft Center announced renegotiation and extension of thermal protection contract with LTV Aerospace Corp. for advanced development of reinforced pyrolyzed plastics to protect space shuttle orbiter. About $493 500 would be added to previous funding for total study cost of $767 500. (MSC Release 72–121)

- NASA board investigating Jan. 20 crash of NASA T–38 aircraft at Matagorda Island off Texas coast had found accident was due to "inability of the pilot to attain visual flight conditions at sufficient altitude to effect a pullout," NASA announced. Test pilots Stuart M. Present and Marck C. Heath had been killed in crash. (MSC Release 72–120)

- Sen. William Proxmire (D-Wis.) said in letter to Dr. James C. Fletcher, NASA Administrator, that NASA should cut costs to take advantage of U.S.–U.S.S.R. space treaty signed May 24. Sen. Proxmire asked for estimates of savings possible through space cooperation and suggested that proposed joint space mission "would make it possible for us to postpone development of certain aspects of our manned space program." (UPI, *NYT,* 6/2/72)

- William P. Lear, Sr., pioneering designer of autopilots and executive jet aircraft, told press conference at TRANSPO '72 at Dulles Airport, Va., that he hoped to sell aircraft with supercritical wing that would increase speed capability from mach 0.80 to 0.92 without increasing fuel consumption. NASA engineers at Langley Research Center had developed original supercritical wing. (AP, *LA Times,* 6/27/72)

- Nixon Administration was reported in press to be considering public display of "super-secret satellite spy system" to counter congressional criticism of U.S.–U.S.S.R. strategic arms limitation agreement. (Sloyan, *Baltimore News-American,* 6/1/72, 3)

- Soviet officials in Washington, D.C., said they would be prepared to begin next round of Strategic Arms Limitation Talks (SALT) in September to maintain momentum generated by May 22–26 summit talks. (Berger, *NYT,* 6/2/72, A1)

- *Science & Government Report* said space shuttle was like "goldplated limousine to deliver small bundles; once built, its existence becomes the justification for delivering lots of bundles." Analogy explained "shuttle passion of the financially depressed civilian space establishment and its parasitic affiliate, the bashful but booming military space establishment, which is planning a California shuttle port to complement NASA's Cape Kennedy selection." That, and space spending "shrewdly dispersed around the country," explained congressional willingness to approve $227 million downpayment on "what NASA— frontman for getting the money—estimated will be a $5.5 billion to $6.5 billion system, excluding expenses for launch costs, refurbishment of reusable boosters and orbiters, and payloads." Publication had impression that "opponents, though riddling the constantly shifting arguments of NASA and its hired consultants, might just as well have spared the effort." Assistant to shuttle opponent Sen. Walter F. Mondale (D-Minn.) had said: "It's like gun control. Plenty of support but no votes." (*Sci & Gov Rpt*, 6/1/72, 1)

- Naval Missile Center at Point Mugu, Calif., held dedication ceremonies for versatile avionics shop test (VAST), general-purpose, computer-controlled, automatic test system for checking avionics equipment in fleet-operated aircraft. System combined functions of individual specialized avionics support equipment into one system compatible with all future avionics packages. (PMR *Missile*, 6/2/72, 1)

- Boeing Co. Chairman William M. Allen announced his retirement, effective in September. He would remain consultant to company. (*NYT*, 6/2/72, 48)

June 2: General Accounting Office released *Cost-Benefit Analysis Used in Support of the Space Shuttle Program* (B-173677). Review of analysis prepared for NASA by Mathematica, Inc., and used by NASA to justify shuttle development economics, had been requested by shuttle-foe Sen. Walter F. Mondale (D-Minn.). GAO said it had "worked with estimates received from Mathematica for two representative configurations of the space shuttle," solid-fuel and liquid-fuel reusable-booster shuttles; "made computations using NASA's cost model developed by Mathematica to show the effect of increasing or decreasing selected critical areas within their plausible boundaries"; and "found that the two configurations were economically justified in terms of the 10-percent investment criterion proposed by Mathematica as the basis for evaluating the Space Shuttle Program." Investment criterion used was "space transportation system having the lowest total space program cost, considering the time value of money." Use of criterion required selection of discount rate to consider time value of money. Mathematica had proposed 10% discount rate. (Text)

- Lunar Samples Analysis Planning Team, using information imparted at Third Lunar Science Conference [see Jan. 10–13], reported in *Science* on advances made and problems still outstanding in lunar science. Most important gain had been that "we now know there is a lunar crust." Primal igneous activity in outer layers of moon had generated feldspathic crust 40 km (25 mi) thick. *Apollo 15* core samples had shown that material at all levels had resided within few centimeters (one inch or so) of lunar surface at some time, but gradient in some cosmic ray product allowed tentative conclusion that soil in lower half of core had been stratified for 0.5 billion yrs. *Apollo 15* mission and Third Lunar Science Conference had answered many fundamental questions, but each answer raised new questions. "What are the nature and chronology of rock types in the lunar highlands? Answers to date have been indirect or broadbrush. The most interesting epoch of lunar history, the first half-billion years, is recorded in highland rocks if it is preserved anywhere. Why is the occurrence of lunar norite or KREEP so sharply restricted to the northwest quadrant of the moon's nearside? Is an exceptionally mighty impact . . . required to raise noritic rock to the surface? This would be inconsistent with the supposedly shallow depth of origin of the Fra Mauro Formation cratering debris."

Last two Apollo missions were targeted to regions that offered prospect of answering these questions. Most "profound question of all" was origin of earth-moon system. Apollo science had eliminated once-popular hypothesis that moon had been captured 1 to 2 billion yrs ago, "but beyond this the question remains unanswered. If the nature of compositional heterogeneities in the moon at the time of its accretion can be inferred correctly from chemical and petrologic studies . . . this information will go far toward answering the question." Goal of understanding moon would be furthered by international cooperation. Third Lunar Science Conference had been "minor landmark" because of participation of Soviet scientists and exchange of samples between U.S. and U.S.S.R. (*Science*, 6/2/72, 975–81)

- Protocol on U.S.–U.S.S.R. scientific and technological cooperation in meteorology was signed in Moscow by Director Ernest Ambler of U.S. National Bureau of Standards and Chairman Boris Isayev of Soviet Standards Committee. (Tass, FBIS–Sov, 6/5/72, L1)
- NASA U–2 reconnaissance aircraft on loan to Virginia Institute of Marine Science would begin 45-day study of Chesapeake Bay area within two weeks to seek photographic evidence of path of Hurricane Agnes, VIMS Director, Dr. William J. Hargis, announced. Aircraft cameras, with high-resolution lenses, would photograph bay and adjacent continental shelf from 18 000-m (60 000-ft) altitude to trace flow of fresh water dumped over Chesapeake basin by storm, which Dr. Hargis described as "a once-in-200-years phenomenon." (Nunes, *W Post*, 7/6/72, B2)
- Air Force Systems Command announced plans for summer launch of particle identifier developed by Air Force Cambridge Laboratories to measure energetic particles across polar regions and South Atlantic Anomaly over Brazil. Measurements would permit more accurate prediction of ionization buildup that interfered with communications systems. Experiment would be carried on satellite launched from Vandenberg Air Force Base into polar orbit with 740-km (460-mi) altitude. It would transmit data twice daily for at least one year. (AFSC Release 059.72)
- NASA and Dept. of Interior dedicated Merritt Island National Wildlife Refuge at Cape Kennedy. Under agreement signed by NASA Associate Deputy Administrator Willis H. Shapley and Assistant Secretary of the Interior Nathaniel P. Reed, all land not used for rocket launch and launch support at Cape Kennedy would be used for wildlife management. Citrus grove and fish camp leases formerly managed by NASA for Army Corps of Engineers would be administered by Interior Dept.'s Bureau of Sport Fisheries and Wildlife. Agreement added to existing refuge 16 200-hectare (40 000-acre) property at northern tip of Kennedy Space Center that included Indian burial ground, 1830 Army fort, and Civil War sugar mill. (*Today*, 6/3/72; M Her, 6/4/72)
- Sen. John G. Tower (R-Tex.) introduced S.J.R. 238 requesting President Nixon to designate July 20, 1972, as "National Moon Walk Day." Resolution was referred to Senate Committee on the Judiciary. (*CR*, 6/2/72, S8712)

June 3: Lewis Research Center announced award of four fixed-price contracts for four-month studies of methods to measure pollution due to aircraft emissions in upper atmosphere. American Airlines would receive $11 697, Trans World Airlines $14 177, and United Air Lines $19 699 to study costs of installing and servicing pollution-measuring instruments on commercial aircraft. Boeing Co. would receive $31 350 to determine aircraft modifications needed to install instruments, data recorders, and support systems. Studies were in support of Global Air Sampling Program (GASP), managed by LeRC. (LeRC Release 72-39)

- *Washington Post* editorial commented on retirement of Dr. Wernher von Braun, NASA Deputy Associate Administrator, announced May 26: "During the closing months of World War II, when his handiwork was employed on behalf of Nazi Germany, he was our enemy." Yet, "25 years later, Mr. von Braun had become something of an American hero; the Saturn 5, . . . rocket used to propel the missions, was his ultimate contribution to the space program." Most constructive conclusion to be drawn from record "has to do with the transferability of knowledge across international boundaries and between warlike and peaceful purposes." Advantages to mankind were great in breaking barriers to flow of knowledge "and focusing that knowledge on its peaceful user. That, we trust, is what the agreement on space cooperation . . . is all about. If so, it comes at a particularly appropriate time when the era of space competition—the drive to do things bigger and faster—is ending and the hard work of reaping the benefits of space operations is beginning." (*W Post*, 6/3/72, A14)
- First anniversary of dedication of Visitor Center at Langley Research Center. Since June 8, 1971, opening, Center had attracted more than 175 000 visitors from all 50 states and more than 20 foreign countries. (*Langley Researcher*, 6/23/72, 1)

June 4: *New York Times* editorial commented on United Nations Conference on the Human Environment, opening in Stockholm June 5: Conference would not be "the kind of body that can pass laws to reduce the further polluting of air and water, neither is it to be a forum for free-flowing discussion. The first would imply a yielding of national sovereignty not remotely in sight; the second, a gathering of individuals responsible only to themselves, rather than an assembly of official delegations. But their coming together has rightly raised the world's expectations, and there are significant gains that the conference can and should deliver." Conference, without plenary powers, would attempt to "pave the way for a worldwide treaty . . . to arrest the polluting of the oceans and to fix criteria for the tolerance of man and his world to certain pollutants, leaving governments to apply those criteria to their own countries." Conference would disappoint if it failed to "set up international machinery, flexible and capable of growth, to establish and expand the interest of the entire international community in the environmental problems of any part of it." (*NYT*, 6/4/72, 4:14)

- Number of visitors taking Kennedy Space Center tours had risen from 47 220 in May 1971 to 72 890 in May 1972, *Today* reported. Total of 519 000 visitors had taken tour in first five months of 1972; 398 000 had taken tour in same 1971 period. (*Today*, 6/4/72)

June 5: President Nixon signed Executive Order 11671, "to establish general standards for formation, use, conduct, management and accessibility to the public" of committees appointed to advise or assist Federal Government. Order directed Federal agencies to open meetings of their advisory boards to public. (*PD*, 6/12/72, 999–1001)

- *New York Times* editorial criticized Secretary of Defense Melvin R. Laird for insisting that U.S. press on with large defense budget increases that were requested before conclusion of U.S.-U.S.S.R. arms limitation pacts, including funds for accelerated buildup of strategic arms not covered in treaties. "If this were needed to insure Congressional approval of the Moscow agreements—the overriding need at the moment—the cost might not be too high. But the real aim seems to be to create 'bargaining chips' for the next round of negotiations to limit strategic arms. And the first round of . . . SALT talks suggests that the Soviet Union will insist on matching or exceeding the additional American buildup before further agreement can be reached. Both sides already have more than ten times the number of warheads they need to deter or destroy each other. Mutual restraint now could speed up a SALT II agreement and, more important, assure much lower limits on offensive strategic forces than would be possible if both sides now press ahead to add to overkill as bargaining chips for the next negotiating round." (*NYT*, 6/5/72, 30)
- Report by Italian Research Organization SORIS was quoted as saying U.S. and U.S.S.R. were rocketing out of Europe's reach in technical and managerial sense as result of the success of their space technologies. Report, prepared for Commission of the European Community, recommended that high priority be given to strengthening and unifying European aerospace industry. (CSM, 6/5/72)

June 5–16: United Nations Conference on the Human Environment, meeting in Stockholm, was attended by 1200 delegates from 114 nations, including U.S. U.S.S.R. and its Eastern European allies boycotted conference because of exclusion of East Germany. Conference produced 200-point program of international action, with proposed funding of $100 million to be used in next five years to stimulate environmental preservation efforts. Program included establishment of "Earth-watch" global atmospheric monitoring system, to be coordinated by World Meteorological Organization, with 10 baseline stations in remote areas to detect long-term trends and 100 stations to measure air pollution. Conference approved establishment within U.N. of Governing Council for Environmental Programmes (GCEP), as urged by President Nixon Feb. 8, and "Declaration on the Human Environment" that embodied code of 26 principles. Principles asserted man's "fundamental right to freedom, equality and adequate conditions of life," called for prompt agreement on "elimination and complete destruction of nuclear weapons and other means of mass destruction," and urged free flow of scientific and technical information to solve environmental problems of developing nations.

World Bank President Robert S. McNamara said during June 9 session: "Ecological considerations have made us all more aware of the interdependencies of our world. We have come to see our planet as a 'spaceship earth.' But what we must not forget is that one-quarter of the passengers . . . have luxurious first-class accommodations and the remaining three-quarters are traveling in steerage. That does not make for a happy ship—in space or anywhere else."

Major conflict of Conference was over resolution to halt nuclear weapon testing. Resolution was approved by majority but opposed by Communist China and France, with 14 nations—including U.S. and United Kingdom—abstaining. Conference recommendations, subject to ratification by U.N. General Assembly in September, were not legally binding. (*NYT*, 6/5–17/72; Hawkes, *Science*, 6/23/72, 1308–10)

June 6: Dr. James C. Fletcher reviewed space program accomplishments during his first year as NASA Administrator in speech before Salt Lake City, Utah, Rotary Club. He had been asked at his first NASA Hq. press conference if he "did not feel like the captain of the ill-fated ship Titanic." Ten "good reasons" why reporters no longer asked that question were: (1) NASA had defined space program for 1970s that was "as challenging, as exciting, and as rewarding as our space achievements of the last decade, but less costly." (2) New program had been approved by President and Congress "with a strong bi-partisan majority." (3) Program could be carried out over next six or eight years at present $3.4-billion NASA budget level "which calls for steady space progress at a cost the country can afford." (4) Balanced program meant increased capability for "useful work in all the major areas of space activity." (5) "We have chosen the re-usable Space Shuttle as our major investment in new space technology . . . and . . . have succeeded in redesigning the Shuttle to cut estimated development costs in half without significant losses in performance." (6) "We have settled the argument over whether to stress manned or unmanned space missions in the Seventies. The shuttle will be used for both." (7) Space shuttle in military service "will make an important contribution to our national security." (8) Program was "relevant to the needs of modern America." Information culled from Earth Resources Technology Satellites would "help protect the environment of spaceship Earth." (9) Space shuttle program to develop future uses of space provided "necessary base for increased international cooperation," adding significance to U.S.–U.S.S.R. space cooperation agreement. (10) NASA again had "national commitment to make space progress a clear expression of our national character and a symbol to the world of what America stands for and works for." (Text)

- Crew compartment fit and function test of Skylab airlock module (AM) and multiple docking adaptor (MDA) was completed at McDonnell Douglas Astronautics Co. in St. Louis, Mo. Astronauts had activated AM and MDA to demonstrate that modules would support all activities planned for Skylab mission in 1973. They worked with experiments, installed and stowed hardware to verify mechanical and electrical functions, and verified on-orbit operations. Skylab crew members participating in test included Charles Conrad, Jr., Paul J. Weitz, Dr. Joseph P. Kerwin, Russell L. Schweickart, Dr. Story Musgrave, Bruce McCandless II, and Dr. William B. Lenoir. (MSFC Release 72–72; MSFC, *Skylab Chron*, 83)

- Dr. Gary V. Latham of Columbia Univ. Lamont-Doherty Geological Observatory, principal investigator for Apollo seismic experiments, said seismic signals from May 13 crash of meteorite onto moon had confirmed that moon had crust 61 km (38 mi) thick over a mantle and, possibly, a core. Lunar crust measured near *Apollo 14* site in Fra Mauro region was about twice as thick as crust on earth. Lunar highlands appeared to have solid rock beneath thick regolith. Change in reflected seismic waves below crust to 8.2 km per sec (5.1 mps) showed moon had mantle similar to earth's. Measurement of same layer during *Apollo 15* mission had been 9 km per sec (5.6 mps), but 8.2 km per sec was more nearly valid and more consistent with known rock types at pressures expected at that depth. Rocks rich in ferromagnesian minerals such as olivine and pyroxene were dominant in earth's mantle and also produced seismic velocities of about 8.2 km per sec. Signals also appeared to show change of state in lunar material at depth of 1000 km (600 mi), suggesting possibility of core. (MSC Release 72–126)

- Rep. George P. Miller (D-Calif.), Chairman of House Committee on Science and Astronautics since 1961, was defeated as candidate for reelection to Congress during Democratic primary in California. (Dem Natl Com; *Science*, 6/16/72, 1219)
- Manned Spacecraft Center announced contract activity: TRW Systems Group had received two study contracts totaling $134 160, one for safety criteria study for space shuttle sortie payload and one for payload systems compatibility criteria study, both to be concluded by May 18, 1973. Grumman Aerospace Corp. had received $248 500 contract to study and develop test samples of closed pore insulation for possible thermal use on space shuttle orbiter. (MSC Releases 72–124, 72–125)
- RCA announced it had received $360 000 NASA contract to develop system to reduce, by 10 times, the size and weight of signal cabling in spacecraft onboard data-management systems. RCA would demonstrate use of common data bus to transmit signals to large number of onboard systems. Commands and data would be transferred from central computer through a twisted shielded-pair cable instead of bundle of cabling. (RCA Release)
- First fighter squadrons of Navy F–14 jet fighters would be formed "sometime this fall," Grumman Corp. Vice President Michael Pelehach said at meeting of Long Island Assn. of Commerce and Industry in Woodbury, N.Y. Eleven test F–14 prototypes had logged 400 flights and nearly 1000 hrs. (*NYT*, 6/7/72, 11)
- Patent No. 3 667 358 for cold camera for photographing stars was awarded to William D. Williams, Jr., 19-yr-old Univ. of Michigan student, and Scott Usher, packaging engineer for Lehn & Fink Products Co. Camera, mounted on telescope and loaded with film supersensitized after chilling with dry ice, permitted deep sky photography in very dim light with brief exposure time. Camera was simpler and less expensive than other cold-emulsion cameras and was expected to appeal to schools and clubs. (Jones, *NYT*, 6/10/72, 37; Pat Off PIO)

June 7: Congressional resolutions were introduced to honor Dr. Wernher von Braun, Deputy Assistant Administrator for Planning, on his retirement from NASA. Sen. Hugh Scott (R-Pa.) submitted S.R. 84 expressing "gratitude and appreciation" to Dr. von Braun for "Outstanding contributions to and achievements in the space program" and "recognition of the benefits which his contributions and achievements have conferred and will continue to confer upon mankind." Rep. J. Irving Whalley (R-Pa.) introduced H.R. 628, "expressing the Nation's gratitude and appreciation." (*CR*, 6/7/72, S8885; H5391)

- Former astronaut James A. McDivitt—Special Assistant to Manned Spacecraft Center Director for Organizational Affairs since May—announced his resignation from NASA and Air Force, effective Sept. 1, when he would become Senior Vice President of Consumer Power Co. in Jackson, Mich. He would be on terminal leave from July 1 until Sept. 1. McDivitt, Air Force brigadier general, had commanded *Gemini 4* (June 3–7, 1965) and *Apollo 9* (March 3–13, 1969) and had left Astronaut Office in June 1969 to become Manager for Lunar Landing Operations in Apollo Spacecraft Program Office. (NASA Release 72-131)
- Benjamin O. Davis, Jr., Assistant Secretary of Transportation for Safety and Consumer Affairs, met with aviation representatives in Washington, D.C., to discuss joint Government and industry action to combat aircraft hijacking and extortion which were threatening civil aviation. At press conference following meeting he said: "The success rate for the hijacker has been dropping sharply over the past 3-½ years—from 85 percent in 1969 to 37 percent thus far this year." Government would "continue to press for full compliance with aviation security relations . . . , for increased resistance to hijack demands, and for full cooperation with the FBI [Federal Bureau of Investigation]." (Text)

June 8: Secretary of Defense Melvin R. Laird, in secret testimony before Senate Committee on Armed Services, had disclosed that U.S.S.R. was flight-testing multiple independently targetable reentry vehicle (MIRV). Dept. of Defense spokesman Jerry W. Friedheim later confirmed that Laird had made statement. To date, MIRVs had remained a U.S. monopoly. (Gwertzman, *NYT*, 6/9/72, 1)
- President Nixon accepted resignation, effective July 15, of Kenneth M. Smith as Deputy Administrator of Federal Aviation Administration. (*PD*, 5/12/72, 1008)
- Institute of High Energy Physics at Serpukhov, near Moscow, inaugurated experimental complex which could extract proton cluster with maximum energy from 70-bev proton accelerator in millionth fractions of a second. (Tass, FBIS–Sov, 6/9/72, L1)

June 8–9: NASA's *Mariner 9* spacecraft—orbiting Mars since Nov. 13, 1971—began photographing Mars after two-month rest period. Clear photos of Mars south and north poles—not clearly visible with earth-based telescopes and obscured from spacecraft by clouds in winter and spring—were taken June 8 and stored before transmission June 9 to Jet Propulsion Laboratory's Goldstone Tracking Station. Pictures were taken during 418th revolution in orbit with 16 900-km (10 500-mi) apogee and 1650-km (1000-mi) perigee. *Mariner 9* would take about 500 pictures during next nine weeks. It had taken 6876 photos covering 85% of planet between Nov. 13, 1971, and April 2, 1972, before it entered occultation period. (NASA Release 72-123)

June 9: U.S.S.R. launched *Cosmos 492* from Baykonur into orbit with 316-km (196.4-mi) apogee, 202-km (125.5-mi) perigee, 89.7-min period, and 65° inclination. Satellite reentered June 22. (GSFC *SSR*, 6/30/72; *SBD*, 6/26/72, 241)
- Award of estimated $1.8 million cost-plus-incentive-fee contract to Itek Corp. Optical Systems Division for three multispectral camera systems was announced by Manned Spacecraft Center. Cameras would be used on MSC's earth resources aircraft in conjunction with Skylab missions. Airborne multispectral photographic system would obtain photos from

altitude of 378 km (235 mi), each showing more than 20 000 sq km (8000 sq mi) of earth's surface. Photos would be used in assessing urban and metropolitan growth and land use patterns and in inventorying crop, range land, and forest resources. Contract called for first delivery by Jan. 29, 1973; second by Mar. 26, and third by May 21. (MSC Release 72-130)

- NASA launched two sounding rockets from White Sands Missile Range, N. Mex., Black Brant VC carried Univ. of California galactic astronomy experiment. Nike-Apache carried California Institute of Technology galactic astronomy experiment. Rockets and instrumentation performed satisfactorily. (SR list)
- NASA announced that NASA Assistant General Counsel Arthur D. Holzman had been detailed to Federal City College in Washington, D.C., to develop seminar courses in government. Courses would provide minority group students with background in government management and enhance their competitive position in government service. Assignment of Holzman was made under Intergovernmental Personnel Act of 1970, which provided for temporary exchanges of personnel between Federal agencies and state and local governments and institutions of higher learning. (NASA Release 72-125)
- Atomic Energy Commission's Los Alamos, N. Mex., Meson Physics Facility linear proton accelerator produced its first beam at full design energy of 8000 mev. (AEC Release P-171)
- Science agreements signed during U.S.-U.S.S.R. summit meetings May 22-26 were praised in *Science* article: "Taken together, they represent not only a significant expansion of programs already in existence, but forays into areas hitherto unexplored." Agreements were designed to "insulate cooperative scientific endeavors from the stresses of international politics by putting authority into the hands of appropriate government agencies rather than the foreign ministries." In space agreement, "NASA is operating on the assumption that the joint docking is only the first in a series of increasingly ambitious mutual experiments—serving both the causes of economy and international understanding—which could conceivably lead to something as grandiose as a joint landing on Mars 20 years hence." (Holden, *Science*, 6/9/72, 1106-8)
- *Washington Post* editorial criticized May 27-June 4 TRANSPO '72: "On three of the nine days of America's first, federally financed and operated international transportation exhibition, horrified crowds watched death in the sky." Air show—in which aerobatics stunt man, racing pilot, and Air Force pilot had been killed—"helped swell Transpo's attendance to an impressive 1.5 million." But "we hope that neither the attendance nor the as yet unspecified sales figures will mislead anyone to believe that Transpo 72 was the kind of hit that calls for an instant repeat performance." That kind of exhibition "should not need to be linked to an aerial circus. Nor would there be any reason to place it way out at Dulles airport—unless . . . some form of rapid transit transportation is first built to get us there." (*W Post*, 6/9/72)

June 9-16: Hugh O'Brian Foundation sponsored annual space seminar at Kennedy Space Center for youths from 50 states and several foreign countries. Dr. Kurt H. Debus, KSC Director, discussed Skylab mission with honor students selected in competition among 10 000 U.S. high schools, under program administered by NASA, National Assn. of Student Councils, and National Secondary Principals Assn. Dr. Debus said Skylab program would be first space venture "to prove and make it very, very clear in people's minds that space is here to stay and that they need space to achieve a tool to help mankind to survive." Sixty-five youths watched June 13 launch of *Intelsat-IV F–5*. (*O Sen*, 6/11/72, 6/15/72; *M Her*, 6/18/72)

June 10: Arthur B. Freeman, Director of Administration at Ames Research Center and ARC's first employee, died at age 62. Freeman had begun his Government career with National Advisory Committee for Aeronautics (NACA) at Langley Memorial Aeronautical Laboratory (predecessor of Langley Research Center) in 1935 and had joined ARC when it was founded in 1941 as Moffett Field Laboratory. (*Marshall Star*, 6/14/72, 1)

June 11: India would launch satellite on Soviet booster by end of 1974, Delhi General Overseas Service reported. Construction of facilities to develop and test rockets and satellites was expected to create more jobs, help locate natural resources, and strengthen industrial infrastructure. (FBIS–India, 6/12/72, O2)

- C. L. Sulzberger commented on shortcomings of United Nations Conference on the Human Environment (June 5–16) in *New York Times* article: "There can never be major ecological reform until world government limits population and polices the earth's surface. Nor is there any prospect of such world government in time to act." Day approached "when technical accomplishment makes it not only possible to export mass quantities of men and women in fleets of immense spacecraft but also to seed the distant planets and still more distant stars so that they become habitable for human beings." This was "logical, ultimate goal of all ecological crises. Within sensible limitations of cleanliness, the aim must be not to reduce by Gandhian means of zero growth but to produce—to mass produce—new Noah's arks and thrust them into the infinity of the universe." (*NYT*, 6/11/72, 4:15)

June 12: Dr. James C. Fletcher, NASA Administrator, spoke on cooperation between NASA and small research and development businesses before Conference on Survival and Growth of Small R&D Firms, in Washington, D.C.: "New technology is the area of greatest potential growth in U.S. industry and the key to a favorable international balance of trade. Small business may be the key element which preserves American free enterprise." NASA's "broad and vigorous" R&D effort generated "literally hundreds of new processes, new products and new materials." Transfer of technology to areas unrelated to space meant "profits for private enterprise, jobs for labor, and benefits for the consumer." Transfers benefited small entrepreneur because "he is flexible—and can take advantage of the financing specifically made available for this kind of enterprise." NASA maintained contract category especially for small business—contracts between $2500 and $500 000 for construction work. "In 1971, this resulted in 73% of NASA's expenditures for construction." (Text)

- NASA had asked private industry to help assess space shuttle booster-recovery system, Marshall Space Flight Center announced. Large parachutes would lower 90-metric-ton (100-U.S.-ton) spent booster

stages to ocean. Six proposals for study of parachute deceleration system had been received. (MSFC Release 72-70)
- Marshall Space Flight Center announced award of $1 323 565 contract to Air Products and Chemicals, Inc., to provide liquid hydrogen to East Coast space program users from June 1, 1972, through March 31, 1973. Liquid hydrogen would be distributed to Government agencies and contractors in Saturn V booster and space shuttle activities. Under interagency agreement, NASA contracted for all liquid hydrogen used by NASA and Dept. of Defense in eastern U.S. (MSFC Release 72-69)
- TRANSPO '72, held May 27–June 4, had been "surprising success," *Aviation & Space Technology* editorial said. It had been "tremendous public report" on advanced technology, particularly aerospace, to the American public; surprising business stimulant; and "triumph of construction expertise, logistics and short-term management by the team brought in at the last possible minute . . . to salvage what was shaping into a major disaster." Exhibition—plagued in beginning by construction holdups due to inclement weather, lack of cooperation from exhibitors, and "fuzzy management"—had been "major success against formidable odds." But "to mold its successors into the type of operation that will effectively achieve the multiple goals established for it will require the organization of a competent, permanent management right now that can begin to work on Transpo 74 next week." (Hotz, *Av Wk*, 6/12/72, 7)

June 13–14: Intelsat-IV F-5 comsat was launched by NASA for Communications Satellite Corp. on behalf of International Telecommunications Satellite Consortium (INTELSAT). Satellite, launched from Eastern Test Range at 5:39 pm EDT by Atlas-Centaur booster, entered elliptical transfer orbit. Primary NASA objective was to place satellite into transfer orbit accurate enough for spacecraft onboard propulsion systems to place it in planned synchronous orbit. ComSatCorp objectives were to fire apogee motor, position satellite in its planned geostationary orbit, and operate and manage communications system for INTELSAT.

Apogee-kick motor was fired at 9:00 pm EDT June 14 and *Intelsat-IV F-5* entered circular orbit with 36 824.2-km (22 882.1-mi) altitude at 124° east longitude. It would drift westward 4.4° per day to reach station at 61.4° east longitude over Indian Ocean by end of July. Satellite's first major transmission would be coverage of Olympic Games in Munich in August.

Intelsat-IV F-5 was fourth comsat in Intelsat IV series. Satellite was 238 cm (93.7 in) in diameter and 528 cm (208 in) high and weighed 1387 kg (3058 lbs) at launch. It had 12 transponders, providing 12 TV channels and 3000–9000 telephone circuits, and was capable of multiple-access and simultaneous transmissions. Expected lifetime was seven years. *Intelsat-IV F-4* had been launched Jan. 22 and was operating satisfactorily over Pacific. (NASA proj off; NASA Release 72-119)

June 13: President Nixon transmitted to Congress U.S.–U.S.S.R. Treaty on Limitation of Anti-Ballistic Missile Systems and also Interim Agreement on Strategic Offensive Arms, both signed in Moscow May 26. In filmed TV broadcast from White House President said: ". . .

these agreements are in the security interest of the United States." They would enable U.S. "to maintain defenses second to none." In transmission message he said agreements "open the opportunity for a new and more constructive U.S.-Soviet relationship, characterized by negotiated settlements of differences, rather than by the hostility and confrontation of decades past." (*PD*, 6/19/72, 1026)

- NASA launched Aerobee 170 sounding rocket from White Sands Missile Range, N.Mex., carrying Univ. of Colorado solar physics experiment. Rocket and instrumentation performed satisfactorily. (SR list)
- Claire Sterling commented in *Washington Post* article on progress of United Nations Conference on the Human Environment [see June 5-16]: At halfway point "nobody is signing anything. We may all be in favor of what the Chinese speak of here as 'a beautiful environment for mankind,' but neither China nor any other state is giving up a shred of sovereignty, or binding itself on a single proposition by treaty or covenant." Conference was not discussing some of worst environmental problems—"worldwide population explosion, rapid depletion of the planet's non-renewable resources, and an approaching global energy crisis"—except at counterconferences, "where nondelegates are talking their heads off." Everyone was talking about money, "but nobody is . . . coming up with enough of it to matter." (*W Post*, 6/13/72, A20)

June 13-15: National Symposium on Technology Transfer was sponsored in Washington, D.C., by American Chemical Society, American Institute of Chemical Engineers, American Society for Metals, American Society for Nondestructive Testing, American Society for Quality Control, and American Society for Testing and Materials. NASA officials led workshop on design techniques and reliability procedures on June 15. Cochairmen Howard M. Weiss, Deputy Director of Reliability and Quality Assurance Office, and Joseph H. Levine, Chief of Reliability Div. at Manned Spacecraft Center, discussed computer aids to structural design, principles of reliable design and systems safety, and role of testing in design process. (NASA Release 72-91)

June 14: Press conference on scientific results of NASA's *Mariner 9* Mars probe (launched May 30, 1971) was held at Jet Propulsion Laboratory. Dr. Robert H. Steinbacher, project scientist, said Mars "used to be likened to earth in science fiction, and was compared to the moon after the first pictures from *Mariner 4* in 1965. Now we are seeing that Mars has a character all its own. It is not earth-like or moon-like, it is Mars-like."

Harold Masursky of U.S. Geological Survey said Mars could be divided into four major geological provinces: Nix Olympica-Tharis volcanic province with volcanoes up to 8 km (5 mi) high and 500 km (300 mi) in diameter; Ophir-Eos equatorial plateau region with faults and rifts 5 km (3 mi) deep; cratered and smooth terrains with large circular basins like lunar impact basins; and south polar cratered terrain covered with glacial sediment layers up to 100 m (330 ft) thick. Convincing evidence that water had played active role in evolution of Mars was shown in *Mariner 9* photos of deep, winding channels that might have once been beds of fast-flowing streams. Masursky suggested two explanations for apparent heavy rains and floods on Mars in past: volcanic heat might have melted water ice stored below surface as permafrost or increased dosages of sunlight on polar caps every 50 000 yrs, because of axis wobbling, might have melted polar caps completely.

Dr. Geoffrey A. Biggs of JPL described north polar hood of variable clouds that appeared to have general west-to-east flow pattern, with some systems resembling small cyclones. Photos also showed "white cloud phenomenon"—brilliant white spots, believed to be water ice clouds, associated with volcanoes. "The clouds appear mostly in the afternoon, when warmer air is moving up the slopes and cooling as it goes." Pictures of north polar area showed wavelike clouds "formed by a simple, harmonic movement of air, of winds up to 115 miles per hour [185 kilometers per hour] . . . very much like the cold fronts shown in earth atmosphere pictures." (Summary statement and abstracts; Wilford, *NYT*, 6/15/72, 1; Swaim, Pasadena *Star-News*, 6/15/72)

- NASA's *Pioneer 10* (launched March 2) was continuing on trajectory toward planned Dec. 3, 1974, encounter with Jupiter. After 104 days in space *Pioneer 10* was 113 800 000 km (70 700 000 mi) from earth and 554 000 000 km (344 300 000 mi) from Jupiter, traveling at 23.4 km per sec (14.5 mps) relative to earth. It was 256 800 000 km (1.7 astronomical units) from sun, moving through region of space never before traversed by man-made object. All science instruments had been turned on and checked out and all other instruments were operating satisfactorily. Meteoroid detector had recorded 42 hits to date. When *Pioneer 10* entered asteroid belt between July 15 and 26, frequency of meteoroid encounters was expected to increase. Radioisotope thermoelectric generators were operating at power loss of about one watt per month—rate that would leave sufficient power remaining at Jupiter encounter. (NASA proj off)
- Senate by vote of 70 to 2 passed H.R. 15093, FY 1973 Dept. of Housing and Urban Development-space-science-veterans appropriations bill that contained $3.432-billion NASA appropriation, as approved by Senate Committee on Appropriations May 31. (*CR*, 6/14/72, 89371–97)
- European Space Conference delegation visited U.S. to complete resolutions for July 11 ESC conference in Brussels, which would discuss government action on European participation in U.S. post-Apollo space program. Brussels conference later was postponed. During Washington, D.C., meeting NASA and Dept. of State officials told ESC delegation that space tug was no longer being considered as item for European development. Possibilities for European participation had been reduced to four structural areas of space shuttle. U.S. urged ESC to undertake development of shuttle sortie laboratory. (NASA OMSF; *Av Wk*, 7/17/72, 19)
- Ecological applications of first Earth Resources Technology Satellite (ERTS), scheduled for launch July 21, were discussed by NASA Associate Administrator for Applications Charles W. Mathews in speech before National Space Club in Washington, D.C.: "This mission is an experimental one, but through its operation we expect to develop understandable and highly useful applications." ERTS could "obtain images from any area on the globe every eighteen days and . . . provide repetitive coverage of every segment of the United States and of many selected areas in other parts of the world." Images obtained by ERTS

would be "not just color photographs, but electronically derived, multispectral images amenable to many kinds of interpretive techniques developed recently by various agencies, universities and industrial organizations using data from Gemini, Apollo and flights of NASA earth resources airplanes." NASA had received more than 700 proposals from investigators for applications of ERTS data. From these, 300 were being selected; 50 investigations would be carried out in 35 other countries—about one third of the countries represented at United Nations Stockholm conference [see June 5-16]. Investigations reflected "an understanding of real needs and are aimed at meeting these needs." (Text)

- Dr. Wilmot N. Hess, Director of National Atmospheric and Oceanic Administration's Environmental Research Laboratories, outlined space modification techniques at international meeting of Society of Engineering Science in Tel Aviv, Israel. Techniques would modify near-earth space environment to make it safe for manned space flight.

  "In the last few years experimenters have artificially modified the space environment. We can now produce artificial aurorae. We can change the population of the Van Allen radiation belt. We can artificially modify the ionosphere from the ground and our other ideas about artificial experiments for the future stretch as far as trying to copy the sweeping action being carried on naturally by Jupiter's moons." Other possible man-made changes in near-earth space environment included releasing beam of energetic particles from satellite to produce electromagnetic waves in space, injecting thermal plasma at high altitudes to change radiation population in space, and transmitting electromagnetic waves from earth to disturb particles trapped in Van Allen belt. Experiments would provide data on natural processes that caused geomagnetic storms on earth, permit improved predictions of radar clutter and quality of polar radio communications in future naturally occurring storms, permit modification or attenuation of naturally occurring storms, and enable scientists to make near-earth space environment safe for prolonged manned space flight and study. (NOAA Release 72-83)

- Marshall Space Flight Center had awarded $11 165 375 contract extension to Boeing Co. for design changes on final five Saturn V 1st stages (S-IC-11 through S-IC-15), *Marshall Star* reported. Work would be completed by Boeing Aerospace Group at Michoud Assembly Facility by Feb. 28, 1973. (*Marshall Star*, 6/14/72, 4)

- Dept. of Transportation and Dept. of State said in joint statement on growth of air piracy, terrorism, and sabotage in air and on ground: "We believe the United States Government can and must take the lead in solving these problems in our own country and internationally. We are confident that with cooperation from other governments, the airlines and their crews, the airport operators, and other involved organizations and citizens, we will win the fight." U.S. was cabling resolution to Council of International Civil Aviation Organization (ICAO) directing immediate resumption of work on Convention to provide for joint actions against any state refusing to extradite or prosecute hijackers. Convention had been proposed by U.S. and Canada in April. (Text)

- Soviet oceanologists had explored underwater volcanoes in Mediterranean using mobile Crab submarine apparatus similar to *Lunokhod 1* lunar rover, Tass announced. Crab had TV system and system for transmitting control commands and was linked to control ship by cable for communications. Lowered 2000 m (6600 ft) below water from ship *Academician Sergei Vavilov,* Crab had examined underwater mountain north of Lipari Islands in Tyrrhenian Sea and collected samples of soil, algae, and marine animals. (FBIS–Sov, 6/15/72, L1)
- Soviet news agency Tass reported completion of Polar Morning—joint U.S.S.R. and France experiment at E. T. Krenkel Hydrometeorological Observatory in Arctic. During three-month expedition, series of meteorological rockets had carried Soviet and French payloads to altitudes between 90 and 180 km (56 and 112 mi) during period of transition from winter to summer and from polar night to polar day. Data were obtained on atmospheric temperature, wind speed and direction, cloud formations, corpuscular radiation, and solar activity. (FBIS–Sov, 6/16/72, L1)

June 15: Dr. James C. Fletcher, NASA Administrator, and *Apollo 16* Astronauts Charles M. Duke, Jr., John W. Young, and Thomas K. Mattingly II met with President Nixon at White House to discuss April 16–27 *Apollo 16* mission and joint U.S.–U.S.S.R. rendezvous and docking mission scheduled for 1975.

Following discussion Dr. Fletcher and astronauts held press conference. Dr. Fletcher said President was "very pleased" with progress made toward joint mission. "The President particularly wanted to emphasize that . . . this is only the beginning of what he conceives as a long association with the Soviet Union in many other areas of space cooperation. . . . we are busy working on other kinds of things that we can do together . . . perhaps through the turn of the century, doing things together for economic reasons and because he feels strongly that the international aspect of this is important in improving relationships. If we can do this together, we can do other things together." President also emphasized worldwide space cooperation.

As areas for U.S.–U.S.S.R. cooperation in future, Dr. Fletcher thought "the ones we can be sure of" were "the cooperative program on Mars [and] a new cooperative program on Venus, where we explore the upper atmosphere and the entire atmosphere and they explore the surface. We are going to be doing that together." (Other NASA officials explained later these would be coordinated efforts rather than joint missions.)

July meeting of principals at Manned Spacecraft Center would "iron out the management" of joint docking mission. Astronaut Young said that "real problem" was communication; "once we solve that problem we can prove that we can work together in a lot of other areas besides." Dr. Fletcher said plans for joint astronaut and cosmonaut training had not been concluded, "but it is going to be absolutely essential that the cosmonauts come over here and train on our equipment and vice versa. That much has been agreed to." Training program probably would start "within about a year." Dr. Fletcher summed up progress made on joint mission since May 24 signing of cooperation agreement: "We have set up the agenda for the meeting which will primarily involve the management aspects. . . . We also have, at home, worked out some of the details of the specifications . . . and now are waiting approval." Perhaps most "spectacular" progress had been "great enthusiasm from Congress for the mission, particularly when we told them this was not going to cost any more money." (Transcript; *WSJ*, 6/16/72)

- Dr. George M. Low, NASA Deputy Administrator, testified before House Committee on Science and Astronautics' Subcommittee on International Cooperation in Science and Space on U.S.–U.S.S.R. agreement on cooperation in space [May 24]. He described events that led to agreement on joint rendezvous and docking mission: Dr. Thomas O. Paine, then NASA Administrator, had suggested to Soviet Academician Anatoly A. Blagonravov during April 1970 meeting in New York "possibility of cooperation in the area of astronaut safety, including compatible docking fixtures for space stations and shuttles." National Academy of Sciences President, Dr. Philip Handler, had conveyed NASA's interest to Soviet Academy President Mstislav V. Keldysh. "In July, Dr. Paine suggested directly to President Keldysh that this possibility be considered." Dr. Low said if docking mission was successful, "both countries will have increased their chances of rescuing astronauts in distress without commensurate increase in the costs of a standby rescue capability. It is here that the joint test mission is particularly important. It will give us the opportunity to identify and resolve under the best possible conditions the problems we can expect to flow from differences in language, equipment, and operational procedures."

  Joint working groups were making significant progress. Space meteorology group was organizing experiments to advance temperature sounding from satellites and microwave measurement of precipitation zones, ice conditions, and sea surface roughness and temperature. Another joint working group had provided for coordination of meridional sounding rocket networks in Eastern and Western Hemispheres and had begun to exchange operational and scientific meteorological data. Working group on natural environment was defining coordinated experiments in remote sensing of the environment related to vegetation, geology, and the oceans. Oceanographic effort would continue satellite and ship measurements over same ocean area. In group on scientific investigation of near-earth space, moon, and planets, "experts from both sides have been working on a common system of lunar coordinates and a program for compiling a complete map of the moon on the scale of 1:5,000,000." Group on space biology and medicine was "considering certain common standards and procedures to increase the comparability of information."

  If U.S. and U.S.S.R. could work together to achieve "intimate and complex common goal" like joint space mission, "we will have built mutual confidence and trust. The symbol of US and Soviet spacemen meeting in orbit before the eyes of the entire world cannot help but ameliorate attitudes, viewpoints and expectations throughout the world." (Transcript)
- *Apollo 16* astronauts attended state dinner at White House honoring Mexican President Luis Echeverría Alvarez. (*PD*, 6/19/72, 1060)
- Dr. Henry A. Kissinger, Assistant to the President for National Security Affairs, rejected premise of critics that U.S.–U.S.S.R. arms limitation agreements would perpetuate strategic disadvantage to U.S. During

congressional briefing at White House he said: "Our present strategic military situation is sound. Much of the criticism has focused on the imbalance in number of missiles." To assess overall balance "it is necessary to consider those forces not in the agreement; our bomber force which is substantially larger and more effective than the Soviet bomber force, and our forward base system." U.S. had "major advantage in nuclear weapons technology and in warhead accuracy." With its multiple independently targetable reentry vehicles (MIRVs), U.S. had two-to-one lead in warheads. Lead would be maintained during agreement period, "even if the Soviets develop and deploy MIRV's of their own." U.S.S.R. had more missile launchers than U.S., "but when other relevant systems such as bombers are counted there are roughly the same number of launchers on each side. We have a big advantage on warheads. The Soviets have an advantage on megatonnage." What was disadvantageous to U.S. was "trend of new weapons deployment" by U.S.S.R. and "projected imbalance 5 years hence based on that trend." Relevant question was "what the freeze prevents; where would we be by 1977 without a freeze?" Considering current Soviet momentum in intercontinental ballistic missiles and submarine-launched ballistic missiles, "ceiling set in the Interim Agreement can only be interpreted as a sound arrangement that makes a major contribution to our national security." (Text)

- Convention on International Liability for Damage Caused by Space Objects was received by Senate, which removed injunctions of secrecy from treaty and referred it to Senate Committee on Foreign Relations. (*CR*, 6/15/72, D681)

- White House announced exchange of diplomatic notes on bilateral agreements between U.S. and Mexico in scientific and technological fields during state visit of Mexican President Luis Echeverría Alvarez to Washington, D.C. Remote Sensing Agreement extended until July 1, 1974, a 1968 agreement covering cooperative research in remote sensing for earth surveys. Agreement also modified 1968 agreement by providing that NASA acquire and process Earth Resources Technology Satellite data over Mexico and train qualified Mexican technicians in sensing techniques. Agreement for Scientific and Technological Cooperation called for commission for formulation, orientation, and review of programs to strengthen economic and social development, intensify relations between scientists and technicians, and facilitate and increase exchange of persons, ideas, skills, experience, and information between U.S. and Mexico. (*PD*, 6/19/72, 1054)

- Two-man crew of Air Force F–111 fighter-bomber aircraft bailed out safely after explosion in aircraft during flight from Mountain Home Air Force Base, Idaho. Aircraft crashed in Nevada. Accident was being investigated. (*W Post*, 6/21/72, A5)

June 16: Federal Communications Commission adopted by 4-to-3 vote policy of "multiple entry" of firms into domestic satellite communications field. Commission rejected proposal by its Broadcast Bureau for "limited entry" that would have required applicants with similar technology to combine efforts. FCC said that multiple-entry policy would require applicants to demonstrate that they were financially and technically qualified to provide domestic satellite service in public interest. American Telephone & Telegraph Co. and ComSatCorp would be subject to certain additional conditions. (B *Sun,* 6/17/72)

- *Science* commented on June 6 defeat of Rep. George P. Miller (D-Calif.) as candidate for reelection to Congress: Since 1961, when Miller became Chairman of House Committee on Science and Astronautics, "Apollo program has reached fruition and starts have been made on Skylab and the shuttle. Miller was NASA's foremost advocate in the House and took pride in announcing every latest achievement of the space program." Miller's committee had been "influential in having NASA expand its investment in . . . scientific satellites." *Science* cited Miller's age, 81, as "decisive factor" in his loss of election. (*Science,* 6/16/72, 1219)
- NASA awarded 18-mo, $1 236 500, cost-plus-fixed-fee contract to General Electric Co. Aircraft Engine Group to test inlet choking—new method to quiet short takeoff and landing aircraft. GE would build and test two choked inlets with large two-stage fans. Inlet air flow, accelerated to nearly mach 1 at inlet throat, would be nearly choked so that noise generated by engine fan and compressor could not propagate forward through choked region and out inlet. Studies with scale models of choked inlets were being made under separate contract with Boeing Co. Commercial Airplane Group. (NASA Release 72-126)
- *Wall Street Journal* article criticized "strange aspects to NASA's push for the shuttle." Economic estimates of cost were "troublesome." Some critics had charged that development costs might be "two to four times the $6 billion estimate of NASA. And the operating cost may not be as cheap as NASA says." "Strangest" aspect of NASA argument was "agency's almost total silence on the one space project for which the shuttle makes real sense—the permanent orbiting space station." It had been assumed space station would be "keystone of any long-range space effort involving men" from beginning of space program. Project had been "shoved aside" in favor of manned lunar landing but "the moment that race was won . . . NASA pulled the space station plans back out." It was easier to sell public $6-billion research and development program for "useful" space shuttle than $40-billion program for space station. But NASA's "selling spiel" was "somewhat misleading." Once $6-billion development program was finished shuttle would go into routine operation. "Plans call for 30 to 40 flights a year for 10 years. At $10.5 million a flight this comes to $30 billion to $40 billion—about the same in time and money as proposed for the space station." Paper later acknowledged its arithmetical miscalculation and corrected total to "$3 billion to $4 billion," but said corrected figure also was "misleading." NASA estimate had projected $35 billion for shuttle program through 1990, including $12 billion for shuttle and $23 billion for payloads. (*WSJ,* 6/16/72, 10; 6/20/72)
- President Nixon signed S. 3607, $2.6-billion FY 1973 Atomic Energy Commission appropriations bill. Bill became Public Law 92-314. (*PD,* 6/26/72, 1089)

June 17: Prospects of U.S. astronauts and Soviet cosmonauts "chummily tooling around" in orbit had "hardly raised a ripple when it was announced at the Moscow Meetings," author Robert Sherrod commented in *New York Times* article. But attempts of President John F. Kennedy to "cooperate with the Soviets in space nine years ago" had failed.

Failure showed "differing methods of Kennedy the idealist and Nixon the pragmatist." President Nixon had "smoothed the way" with seven meetings between U.S. and Soviet engineers, beginning with April 24, 1970, meeting between then NASA Administrator, Dr. Thomas O. Paine, and Soviet Academician Anatoly A. Blagonravov. President Kennedy's attitude toward space program had been "ambivalent" until Sept. 20, 1963, when he proposed "joint expedition to the moon" in speech before United Nations. Proposal had come as surprise and had "staggered" anti-Soviet space supporters in Congress, who proceeded to pare NASA budget. House had voted 125 to 110 against using NASA funds for joint lunar landing with Communist country. Kennedy had "forgotten that politics is the art of the possible, forgotten that a leader should never surprise his own close support," Sherrod said. "In the sunnier climate of 1972, Richard Nixon duly observed the axiom and the rule." (*NYT*, 6/17/72)

June 18: There was "general satisfaction" among ordinary Russians with "the improved atmosphere fostered by President Nixon's visit to Moscow" in May, *New York Times* reported. Since visit, Soviet press had hailed arms-limitation agreements as triumphs for Soviet foreign policy. But privately editors had reported some criticism of government for having conducted meetings despite U.S. escalation in Vietnam war. (Smith, *NYT*, 6/18/72, 1)

Ending of Moscow summit talks (May 22–26) just as United Nations Conference on the Human Environment (June 5–16) opened was "coincidental," Stanley Karnow commented in *Houston Chronicle* article. "But historians may one day regard the timing of the two meetings as symptomatic of a significant change that is now taking place in the world." Moscow summit "essentially spelled the end of an era in which the powers viewed military rivalries as the main threat to their security." Stockholm conference "appears to mark the start of an era in which nations will view ecological disequilibrium as an increasingly dire menace to their security." (*H Chron*, 6/18/72)

• U.S. Postal Service officials had been "somewhat upset" by reports that *Apollo 15* Astronauts David R. Scott, James B. Irwin, and Alfred M. Worden had taken 400 unauthorized covers—stamped envelopes with postal markings—to moon, Washington *Sunday Star* reported. German philatelic magazines had reported that Herman E. Sieger, German stamp dealer, had sold 100 *Apollo 15* covers at 4850 German marks each for total $150 350 at official exchange rate. Under NASA regulations, astronauts had been permitted to carry on mission small mementos of flight for friends and relatives but not for sale. Official "moon mail" covers were distributed to purchasers by U.S. Postal Service. Specially struck medals had been carried on several Apollo missions and later sold to public through Government channels. NASA officials had identified unidentified "friend of crew now in Germany" as link between *Apollo 15* astronauts and German dealer. Friend had told dealer that 20 covers had gone to astronauts' friends and relatives; remaining covers had been turned over to NASA management. Donald K. Slayton, Director of Flight Crew Operations at Manned Spacecraft Center, said that no member of crew had profited directly or indirectly from covers sold. (Faries, W *Star*, 6/18/72, 1)

- Thomas O'Toole commented on June 1 report of Astronomy Survey Committee of National Academy of Sciences and National Research Council in *New York Times* article: "Through all this discovery runs a . . . central theme that points up the countless findings of the last decade," discovery by astronomers "that the universe is more intricate, more dynamic and more energetic than even the most fanciful scientists dared to dream." Findings in cosmology were "stunning surprise." For first time "science is able to measure things within the universe that had lain hidden from men's minds, measurements that have brought cosmology out of the realm of the unreal." Findings facilitated by creation of radiotelescope, rockets, and satellites had included discovery of pulsars, black holes—"theoretical stars that have gone from the pulsar into an even denser state of collapse"—and "eclipsing stars" that cohabited with second stars that eclipsed them at regular intervals. (*NYT*, 6/19/72)
- Air Force F–111 fighter bomber aircraft plunged in Choctawhatchee Bay after takeoff from Eglin Air Force Base, Fla., killing both crew members. Cause of accident was being investigated. (AP, *W Post*, 6/21/72, A5)
- John Stack, engineer who had helped develop first high-speed wind tunnel in 1930s, died at age 65 after fall from horse. Stack had joined Langley Memorial Aeronautical Laboratory (predecessor of Langley Research Center) in 1928 and had been head of high-velocity airflow research at Laboratory in 1939, NASA Director of Aeronautical Research from 1961 to 1962, and Vice President for Engineering at Fairchild Industries until his retirement in 1971. He had received numerous awards for his work in supersonic technology, including Robert J. Collier trophy in 1947 for work with X–1 rocket aircraft. (AP, *NYT*, 6/20/72, 38; FRC *X–Press*, 7/21/72, 2)

June 18–July 30: Exhibit of kinetic art of Dr. Frank J. Malina—artist, astronautical engineer, and past Director of Jet Propulsion Laboratory (1944–1946)—was held at Univ. of Texas Institute of Texan Cultures in San Antonio. Catalog noted Dr. Malina was "pioneering on the new frontier of man's adjustment to the explosive impact of science, technology and the exploration of space on the world." Artist believed "man must understand this world in order to survive in it, and that this understanding must be both intellectual and emotional." Paintings included impressions of space phenomena. (Catalog)

June 19: *Explorer 45* Small Scientific Satellite (launched Nov. 15, 1971) was adjudged successful by NASA. Spacecraft had carried eight experiments into highly elliptical orbit to study dynamic processes in inner magnetosphere from two to five earth radii. After six months in orbit all experiments and spacecraft systems were operating satisfactorily, excellent data had been received, and mission had exceeded its objectives. Scientific results of experiments had been presented at American Geophysical Union April 17. (NASA proj off)
- Air Force would develop data-relay satellite to enable early warning satellites stationed over Indian Ocean to flash warning of missile attack directly to continental U.S., *Aviation Week & Space Technology* reported. At present, signals were received by ground stations in Guam or Australia and were then relayed by comsat. New data-relay satellites also would make possible transmission of search-and-find

reconnaissance-satellite photos directly to U.S. for speedy analysis and transmission of commands back to satellite to take close-look pictures of targets of opportunity discovered in earlier photos. Air Force had reportedly requested $23 million in FY 1973—up from $17.8 million in FY 1972 and $13.3 million in FY 1971—to continue system's development. "Closely guarded and heavily censored references to the . . . system and its intended functions" had appeared in recently released congressional testimony in which Air Force official was questioned about $21.7 million requested for continued research and development of early warning satellites and $128.9 million estimated to be required for them through FY 1978. (Klass, *Av Wk*, 6/19/72, 12)

- Willis H. Shapley, NASA Associate Deputy Administrator, described functions of NASA advisory committees in testimony before House Committee on Government Operations' Subcommittee on Foreign Operations and Government. Subcommittee was conducting hearings on public access to information under Freedom of Information Act, which required public accessibility to advisory committees [see June 5]. NASA Historical Advisory Committee, established in 1964, advised and assisted NASA management in implementing NASA historical program. Committee met annually to review program and its relation to academic and public activities. Research and Technology Advisory Council and Related Committees (RTAC), established in 1967, provided consultation and advice on advanced research and technology goals in aeronautics and space and reviewed work in progress. NASA Space Program Advisory Council and Related Committees (SPAC), established in 1971, provided comprehensive space program advisory structure for NASA management on goals and objectives.

  NASA Tracking and Data Acquisition Panel, established in 1972, reviewed and advised on potential application of new technologies in carrying out tracking and data acquisition functions. Panel would be disbanded following its final report in near future. Aerospace Advisory Panel, established by NASA Authorization Act of 1968, reviewed safety studies and operation plans and reported on them to NASA Administrator. Membership of NASA advisory committees was drawn largely from universities, nonprofit organizations, industry, and other Government agencies. Scientific and technical subjects addressed by committees were open to public access through meetings and public had direct access to NASA officials and committee members. (Testimony)

- Langley Research Center awarded 32½-mo, $1 717 200 cost-plus-fixed-fee contract to United Aircraft Corp. United Technology Center (UTC) Div. to demonstrate technology of high-performance hybrid rocket system. UTC would build and test-fire seven rockets about 90 cm (36 in) in diameter and 2.6 m (101.5 in) long with 48 930-newton (11 000-lb) thrust. (NASA Release 72-127)

- *Orlando Sentinel* editorial praised NASA Earth Resources Technology Satellites (ERTS) and Earth Resources Observation System (EROS) programs: "Without more comprehensive knowledge we are limited in the decisions we must make to exploit or conserve the resources upon which our continued existence depends." ERTS and EROS "working together is probably the most far reaching earth-science experiment ever undertaken by man." (*O Sen*, 6/19/72)

June 20: NASA held press conference on ERTS-A Earth Resources Technology Satellite (to be launched July 23) at Goddard Space Flight Center. Dr. Arch B. Park, Chief of NASA Earth Resources Survey Program, described experiments that "typify . . . a direct relationship between the results of the experiment and the interest . . . of the general public." Most important experiments were in land use and understanding geology, soils, vegetation, and engineering aspects of man's mass works. Second was environmental awareness, including experiments in biological control of pests, water quality, mine safety, forest conservation, and resources management.

Primary objective of ERTS was described by project manager Wilfred E. Scull as "the acquisition of multi-spectral images" over U.S. and rest of world and use of data for various disciplines. Spacecraft, scheduled for launch July 21, would weigh 939 kg (2070 lbs) and would be launched into sun-synchronous, circular earth orbit at 920-km (570-mi) altitude, repeating coverage under same conditions of illumination over entire globe every 18 days. Onboard equipment included three return-beam-vidicon (RBV) cameras that would take pictures every 25 sec, multispectral scanner subsystem (MSS) that would continually scan ground directly beneath satellite, and two wide-band video tape recorders (WBVTR) to record and store images for later playback. Data-collection system consisted of remote platforms for measuring soil, water, and air quality and other environmental data and transmitting data to ERTS as it passed overhead. Data would be telemetered to three main stations (Fairbanks, Alaska; Goldstone, Calif.; and Greenbelt, Md.) and processed at GSFC. All images received by NASA would be sent to Dept. of Agriculture, Dept. of Interior, and National Oceanic and Atmospheric Administration. (Transcript)

- Two NASA research aircraft, each equipped with five cameras, were photographing State of Arizona—area of more than 295 000 sq km (114 000 sq mi)—from 20 000-m (65 000-ft) altitude. NASA, Dept. of Interior, and State of Arizona project to develop land use inventory system for natural resource management included acquisition of film for development of orthophotoquads—large photographic maps of entire state—for use by Arizona agencies. Arizona Land Use Experiment would make comprehensive analysis based on remotely sensed data; document social and economic benefits of experiment; prepare manual on application of remote sensing to solving resource management problems; and provide plan for updating information by including data from spacecraft, high- and low-flying aircraft, and ground surveys. Arizona had appropriated $400 000 to project. Film processed by Ames Research Center was sent to U.S. Geological Survey laboratories, where orthophotoquads were to be prepared. (NASA Release 72-129)
- U.S.S.R.'s *Intercosmos 6* (launched April 7) had detected particle in space with highest energy yet recorded near earth, U.S.S.R. announced. Laboratory analysis of spacecraft's photoemulsion plates, recovered after four-day earth orbital mission, showed track of cosmic ray particle with impulse of 1 million bev—at least 1000 times more than maximum anticipated by Soviet scientists. (FBIS–Sov, 6/21/72, L2; Shabad, *NYT*, 6/21/72, 10)
- Faulty design of water-tight battery enclosure used in underwater astronaut training at Manned Spacecraft Center had been primary

contributing factor to February 29 explosion which killed one man and injured another, MSC announced. In report to Dr. James C. Fletcher, NASA Administrator, accident investigation board said explosive mixture of hydrogen and oxygen in battery was detonated when switch was closed. It recommended use of power supplies other than batteries. (MSC Release 72-133)
- Signing of $125 000 firm-fixed-price contract with Florida Operations of Honeywell, Inc., St. Petersburg, for study of inertial measurement unit program for space shuttle orbiter was announced by Manned Spacecraft Center. Honeywell would develop alternative approaches to allow NASA to assess feasibility of adapting existing units for space shuttle. (MSC Release 72-132)
- Secretary of Defense Melvin R. Laird, testifying before Senate Committee on Armed Services hearings on military implications of strategic arms limitation agreements, said he would recommend disapproval of agreements unless Congress increased funding for new weapon projects permitted under pacts. (CR, 6/20/72, D703; Getler, W Post, 6/21/72, A1)
- President Nixon said in statement on June 5-16 United Nations Conference on the Human Environment: "I believe that the deepest significance of the conference lies in the fact that for the first time . . . the nations of the world sat down together to seek better understanding of each other's environmental problems and to explore opportunities for positive action, individually and collectively." (PD, 6/26/72, 1078)
- Air Force grounded entire fleet of U.S.-based F-111 fighter-bomber aircraft "as a precautionary measure" until cause of June 15 and 18 F-111 crashes had been determined. (AP, W Post, 6/21/72, A5)
- President Nixon nominated Adm. Thomas H. Moorer (USN) for reappointment as Chairman of Joint Chiefs of Staff for two-year term ending July 2, 1974. (PD, 6/19/72, 1087)
- Atomic Energy Commission designated nearly 81 000 hectares (200 000 acres) of land in South Carolina as first environmental research park in U.S. Land surrounded AEC facilities near Aiken. (AEC Release P-181)

June 20-23: Three Federal scientists attended meeting of Project CICAR (Cooperative Investigation of the Caribbean and Adjacent Regions) in Havana, Cuba, on first official U.S. mission to Cuba since 1959 break in U.S.-Cuban diplomatic relations. State Dept. later said trio, from National Oceanic and Atmospheric Administration, could attend only because meeting was held under aegis of United Nations. (Cohn, W Post, 7/11/72, A4)

June 21: U.S.S.R. launched *Cosmos 493* from Baykonur into orbit with 266-km (165.3-mi) apogee, 201-km (124.9-mi) perigee, 89.1-min period, and 64.9° inclination. Satellite reentered July 3. (GSFC SSR, 6/30/72; 7/31/72; SBD, 6/26/72, 241)
- White House announced that Dr. Edward E. David, Jr., Presidential Science Adviser, would lead scientific delegation to Moscow July 12 for formal discussions on implementation of Moscow summit agreement on cooperation in science and technology [see May 24]. Dr. David met at White House later with 21 scientific advisers who would accompany him to U.S.S.R. He told press following meeting that, instead of previous exchange of small numbers of scientists and engineers, there now would be "joint programs and common operations"

in U.S. and U.S.S.R. Specific areas had not yet been selected, but Dr. David saw among possibilities joint research on electrical power, solar power, and exploitation of geothermal energy (heat inside earth) and magnetohydrodynamics (MHD). MHD generators burned fuel and directed resulting hot gas through magnetic field to produce electricity. Other areas of joint research might be crop genetics, atmospheric sciences, earthquake prediction, fisheries management and production, metallurgy, forestry, urban planning, computer sciences, and weather manipulation. (PD, 6/26/72, 1087; Cohn, *W Post*, 6/22/72, A2)

- NASA launched Aerobee 150 sounding rocket from White Sands Missile Range, N. Mex., carrying Manned Spacecraft Center atmospheric composition and photography experiment. Rocket and instrumentation performed satisfactorily. (SR list)

June 22: Press conference on preliminary results of *Apollo 16* experiments was held at Manned Spacecraft Center. Dr. Farouk El-Baz of photo-geology team described metric camera photo of old, three-ring basin about 960 km (600 mi) in diameter on far side of moon. Eastern part of ring coincided with "what the Soviets had called some 10 years ago right after the Luna 3 photographs . . . the Soviet Mountains." NASA had concluded from Lunar Orbiter and *Apollo 8* photos that formation was not mountains but "the coincidence of very bright rays from two relatively young craters," and U.S.S.R. had conceded to elimination of Soviet Mountains from lunar dictionary. *Apollo 16* photos showed, however, that "what we have here is really the Soviet Mountains rediscovered, so to speak, that is part of a middle ring of a very old basin." Dr. El-Baz said that, pending approval of International Astronomical Union, large basin would be named Arabia in honor of Arab contributions to astronomy, eastern range of middle ring of basin would be called Soviet Mountains, and long western range of middle ring discovered by *Apollo 16* would be called American Mountains.

Dr. Gary V. Latham, principal investigator for passive seismic experiment, said May 13 meteorite impact on moon "may well turn out to be the most important single seismic event the Apollo Network will record in its life unless the good fairy sends us another fragment of rock. It provided very important confirmation of the structure that we have proposed based on the S–IVB impact." Seismometers had also recorded other meteorite impacts and dozens of moonquakes. "We've been able to establish at least the presence of 20 different moonquake source locations. They . . . have the familiar pattern which can be correlated with lunar tides and hence are triggered by lunar tides." Most puzzling question was what energy source within moon caused moonquakes. "We don't have the answer to that. We do have the puzzling fact that the two source zones we've located are nearly 900 kilometers [560 miles] deep . . . the depth at which our very tentative evidence places a reflector. . . . there seems something special about that depth. All of the Moon above that depth within our network area, at least, must be . . . solid. Very high frequency seismic waves can propagate through it. Such high frequencies are eliminated from terrestrial earthquake signals by a partially molten low velocity zone." If part of moon were molten, "it's either in very, very small isolated pockets or at greater depth than 1000 kilometers [620 miles]."

June 22

Dr. Paul W. Gast, Chief of MSC Planetary and Earth Sciences Div., said about 90% of *Apollo 16* lunar samples consisted of aluminum- and calcium-rich material. "The aluminum and calcium concentration . . . approach the composition of pure plagioclase . . . a name component of . . . anorthosite. They are very typical of what one finds in the highlands anywhere on the moon where we have any chemical data. Now, the negative side of that conclusion is that these rocks are not as had been thought prior to the mission, a kind of volcanic rock that is rich in silica . . . a viscous volcanic rock. It's quite definite that the so-called Cayley formation . . . is not a volcanic formation and I think we can generalize with some confidence that . . . it probably is not a volcanic formation in the other places on the moon where it has been identified or mapped in, either. There are a few rocks . . . which are distinctly different" from the rest in lower aluminum concentration, 15 to 18 percent, and higher magnesium-iron concentration with composition that could be that of volcanic rocks. They were also high in potassium and phosphorus. From Apollo data scientists had identified four major types of rocks: grey breccias that contained black and white clasts (rock fragments); black feldspar rock that appeared under microscope to be igneous because it was tough and dense; very white cataclasites of crushed plagioclase-rich material; and crystalline rocks. One rock in *Apollo 16* samples, "rusty rock," was covered by red, rust-like coating. Identification of rust, so far, was only visual; x-ray and mineralogical tests had not yet been made. "It may turn out to be . . . a more exotic type of iron compound than simple rust. It's quite definite that this . . . is something that's rusting; it's something that took place on the Moon . . . not . . . in the spacecraft on the way back or from handling the rock. The coatings are dispersed through the inside of the rock and too thick and too diverse to have been formed in the spacecraft." (Transcript)

- U.S. had withdrawn invitation to 10 European nations to build space tug, *Today* reported. Paper said European Space Research Organization (ESRO) and European Launcher Development Organization (ELDO) representatives had been told by U.S. that U.S. preferred to build tug (reusable propulsion system to be put into low earth orbit by space shuttle, to deliver and retrieve payloads) [see June 14]. Paper quoted NASA sources as saying European firms lacked funds and advanced technology to construct tug that could be launched from shuttle to send a satellite into synchronous 35 000-km (22 000-mi)-altitude orbit. (*Today*, 6/22/72)

- Dr. James C. Fletcher, NASA Administrator, and *Apollo 16* astronauts presented 50 awards at Marshall Space Flight Center to Government and industry personnel who assisted successful *Apollo 16* mission (April 16–27). (MSFC Release 72–75)

- U.S.S.R. and India began third phase of cooperative meteorological rocket program with launch of 76th M–100 rocket from Thumba Equatorial Rocket Launching Station (TERLS). Rockets launched during third phase would measure atmospheric temperature, density, and pressure. (Delhi Domestic Service, FBIS–India, 6/22/72, O1)

- Kennedy Space Center announced award of $19.9-million contract extension to ITT Federal Electric Corp. to provide communications and instrumentation services at KSC. Contract would expire June 30, 1973. (KSC Release 145–72; DJ, W Star, 6/26/72, A13)

- Was it "reasonable to expect that Europe's disorganized space effort can make a positive contribution to a major American manned space programme?" English aerospace expert Kenneth W. Gatland asked in *New Scientist* article. Poor showing of European Launcher Development Organization (ELDO) to date had "done nothing to inspire confidence" since its 1962 inception. Despite 11 firings with last 4 intended to achieve orbit, "we have yet to see an ELDO satellite circling the Earth." In same period U.S. had "developed Apollo and landed 10 men on the Moon!"

  European specialists had been working with U.S. aerospace companies on shuttle components and it would be possible, "depending on the financial arrangements," for Europe to build complete unit like sortie can or space tug. "Whatever Europe produced, in effect, would be supplied free of cost to NASA. Actually, there should be a bonus as labour costs here are about half those in America, which would ... offset the extra costs involved in collaboration." In return "Europe would gain access to the fully operational system, opening the way for our satellites to be launched cheaply and allowing European scientists and engineers to conduct their own experiments in space." Money might better be spent by Europeans on commercial satellites but "we might then have to pay whatever price is demanded for launching and become dependent on America for all advanced commercial services in space." Biggest problem was "to assess precisely what importance space will have in the last quarter of the 20th Century. We need—and have never had—an in-depth assessment of the scientific and commercial potential." Possibly post-Apollo period held "beginnings of a new growth industry which ... will rival commercial air transport. If so, America's offer to internationalise her manned space programme provides an excellent opportunity to get in on the ground floor." (*New Scientist*, 6/22/72)

June 23: U.S.S.R. launched two Cosmos satellites from Plesetsk. *Cosmos 494* entered orbit with 804-km (499.6-mi) apogee, 789-km (490.3-mi) perigee, 100.7-min period, and 74.0° inclination. *Cosmos 495* entered orbit with 320-km (198.8-mi) apogee, 171-km (106.3-mi) perigee, 89.4-min period, and 65.4° inclination and reentered July 6. (GSFC SSR, 6/30/72; 7/31/72; SBD, 6/26/72, 241)

- Apollo Telescope Mount was shipped from Marshall Space Flight Center to Manned Spacecraft Center, where it would undergo alignment tests. The 11 800-kg (24 000-lb) spacecraft, housing telescopes and other instruments for studying sun, would be launched on Skylab Orbital Workshop in April 1973. (MSFC Release 72–77)

- U. Alexis Johnson, Under Secretary of State for Political Affairs, and Dr. George M. Low, NASA Deputy Administrator, testified on space agreements with U.S.S.R. signed in May, before Senate Committee on Aeronautical and Space Sciences' special hearing.

  Johnson said agreement signed May 24 endorsed at highest level understandings reached over past 18 mos between NASA and Soviet Academy of Sciences for cooperation in space science and applications such as meteorology, study of earth's environment, further exploration of moon and planets, and space biology and medicine. It permitted development of compatible rendezvous and docking systems which "should enhance safety and value of space flight" and opened possibility of further areas of cooperation. It demonstrated "in full view of the world that the two great space powers have both the will and the capability to work together on important and difficult tasks. We have high hopes . . . that this agreement will be a milestone in our relationship with the Soviet Union. . . . It serves our broad national purposes, as well as our specific foreign policy objectives."

Dr. Low outlined earlier agreements and gave details of joint rendezvous and docking mission agreed on for 1975. During April visit to Moscow, agreement had been reached that astronauts must understand Russian to extent normally used in both regular and contingency operations. Soviet cosmonauts also would need to understand other language and be able to respond. Dictionary would be developed of commonly used phrases. Some astronauts, all interested in flying this mission, had begun studying Russian on their own. Astronauts selected would take formal language training. Others in NASA would also learn Russian. (Transcript)

- F-5E international fighter aircraft was unveiled in ceremony at Northrop Corp. Aircraft Div. in Hawthorne, Calif. Aircraft, not part of Air Force inventory, would be available to U.S. allies through Military Assistance Service funding program. Aircraft was powered by J-85-21 engines and had greater maneuverability, speed, payload, and range than previous F-5s, with new fire control system that included search-and-track radar. F-5E was scheduled for delivery in 1973. (AFSC *Newsreview*, 8/72, 1)

June 24: Kennedy Space Center announced award of $82 000 fixed-fee contract to Howard, Needles, Tammen and Bergendorff for preliminary survey of Launch Complex 39 as proposed site for Space Shuttle Landing Facility. (KSC Release 150-72)

- Soviet nuclear scientist Andrey D. Sakharov had asked Soviet Communist Party Leader Leonid I. Brezhnev and other Soviet leaders to engage in dialogue on program to liberalize "every facet of Soviet life," *Christian Science Monitor* reported. He had urged that U.S.S.R. announce it would never be first to use weapons of mass destruction and would permit on-site inspection of armament facilities. He also had urged creation of international council to consider questions of peace, disarmament, economic aid, human rights, and environment. (Saikowski, *CSM*, 6/24/72)

June 25: Soviet engineer V. Borisov commented in *Sotsialisticheskaya Industriya* on U.S. and U.S.S.R. cooperation: "The agreement on cooperation in researching and utilizing space for peaceful purposes concluded between the governments of the USSR and United States opens up broad prospects. . . . Compatible docking systems on all spacecraft will not only open up the path to joint space flights. What is possible and no less important is the fact that such systems will enable the spacecraft of one country to come to the aid of crews of another country's spacecraft should they be in distress. World science will be enriched not only by the experiment on the docking . . . but also by cooperation in many other fields. This includes joint work in the sphere of space biology. Perhaps we shall succeed in solving many

problems more rapidly—such as . . . the influence of weightlessness on the human organism." (FBIS–Sov Int Aff, 6/30/72, G3)
- President Nixon was riding "crest of popularity" following his Moscow summit meeting, with 60% of U.S. public approving his performance in office, pollster George H. Gallup reported in *Philadelphia Inquirer*. (*P Inq*, 6/25/72, 14A)
- Soviet newspaper *Sotsialisticheskaya Industriya* had published article on "curious stories" from Cape Kennedy, Philadelphia *Sunday Bulletin* reported. Article had said angry alligator had tripped alarm system protecting rocket at launch site few seconds before launch; Polaris rocket had exploded about 0.8 km (0.5 mi) from launch site causing bush fire that frightened poisonous snakes into buildings; and woodpecker near electronic equipment had caused launch delay. (*P Bull*, 6/25/72)
- Univ. of Washington architectural student R. Danning Roberts had won $3000 first prize in competition for interior design of orbiting space station sponsored by NASA and California Council of American Institute of Architects, *San Diego Union* reported. (*SD Union*, 6/25/72)
- Unidentified flying objects (UFOs) were still being reported despite Air Force study that had found in 1968 that further study "could not longer be justified," *New York Times* said. Spokesman for Aerial Phenomenon Research Organization—private investigation group—had said, "We still get about 1,000 reports a year . . . about 60 per cent of them lights and others sightings of disks." (Waldron, *NYT*, 6/25/72, 40)

June 26: U.S.S.R. launched *Cosmos 496* from Baykonur into orbit with 253-km (157.2-mi) apogee, 176-km (109.4-mi) perigee, 88.8-min period, and 51.6° inclination. Satellite reentered July 2. (GSFC *SSR*, 6/30/72; 7/31/72; *Sov Aero*, 7/3/72, 6)
- President Nixon discussed future meaning of Moscow summit meetings in exclusive article published by *U.S. News & World Report*: "Our agreements for joint American-Soviet efforts to combat pollution, advance medical science and public health, and work together in science and technology can be expected to broaden with the passage of time, creating a steadily growing vested interest in peace between our two countries. The dramatic space cooperation agreement, including an Apollo-Soyuz rendezvous and docking mission in 1975, points in this same direction. So does the comprehensive trade agreement which should be worked out in a matter of months." (*US News*, 6/26/72)
- Marshall Space Flight Center announced issuance of requests for proposals for design and construction of mobile ground-station receiver for visible laser communications experiment, scheduled to fly on ATS-G Applications Technology Satellite in mid-1975. Proposals were due July 17. (MSFC Release 72–78)
- Air Force unveiled F–15 air superiority fighter aircraft in ceremonies at McDonnell Douglas facility in St. Louis, Mo. F–15 was fixed-wing, single-place aircraft in 18 000-kg (40 000-lb) class, with maximum mach 2 speed. (AFSC Release 080.72)

June 26–29: Dale D. Myers, NASA Associate Administrator for Manned Space Flight, and NASA Sortie Lab Team held discussions in Noordwijk, Netherlands, with officials of ESTEC, European Space Research

Organization's Research and Technology Center. ESTEC had been asked by ESRO to determine resources required by Europe if European Space Conference decided to develop space shuttle sortie lab [see June 14]. NASA told ESTEC officials Europe should view sortie lab as nearly uniquely European as possible, with dependence on U.S. held to minimum. If U.S. agreed to let Europe build system to U.S. design, U.S. would insist on completely new qualification program to qualify source and system. NASA found ESRO and ESTEC representatives "eager" to undertake sortie lab program. (NASA OMSF, Memo for Record, 9/7/72)

June 27: Eole transponder—radio receiver/transmitter developed for French-built weather satellite *Eole* (launched by NASA for France Aug. 16, 1971)—had sailed on American yawl *Foolscap* during Newport Bermuda race, Goddard Space Flight Center announced. Objective was to test its potential for locating lifeboats by satellite. Transponder had sent position data to *Eole* as satellite passed 800 km (500 mi) over *Foolscap*. Information, transmitted to tracking station operated by French Center for Space Studies (CNES), was processed and teletyped to GSFC. *Foolscap*'s up-to-date location was then plotted and sent on to U.S. Coast Guard in Miami, Florida, and race committee in Bermuda. Test, sponsored by GSFC engineers, had proved highly successful and indicated that small inexpensive transponders might become standard equipment on ship and aircraft lifeboats. (GSFC Release G–10–72; GSFC proj off)

- NASA launched Aerobee 170 sounding rocket from White Sands Missile Range, N. Mex., carrying Univ. of Colorado solar physics experiment. Rocket and instrumentation performed satisfactorily. (SR list)

June 28: *Intelsat-IV F–5* comsat, launched by NASA for Communications Satellite Corp. June 13, was adjudged successful by NASA. Satellite would reach permanent station over Indian Ocean at 61.4° east longitude in late July. (NASA proj off)

- Institute of Navigation presented Thomas L. Thurlow Award to John P. Mayer, Chief of Manned Space Center's Mission Planning and Analysis Div., in recognition of his outstanding and continuing efforts in all facets of navigation from NASA Mercury missions through Apollo manned lunar landings. (*Av Wk*, 7/10/72, 9; Inst of Nav PIO)

June 29: U.S.S.R. launched *Prognoz 2* from Baykonur into highly elliptical orbit to continue research of "processes of solar activity and their influence on interplanetary space and the earth's magnetosphere." Orbital parameters: apogee, 200 000 km (124 000 mi); perigee, 550 km (340 mi); period, 97 hrs; and inclination, 65°. *Prognoz 2* carried Soviet equipment to study corpuscular radiation, gamma ray and x-ray solar emissions, and solar plasma fluxes and their interaction with earth's magnetosphere. Spacecraft also carried French equipment to study solar wind, outer regions of magnetosphere, and gamma rays from sun and to search for solar neutrons. *Prognoz 1* had been launched April 14. (*Spacewarn*, 7/11/72; Tass, FBIS–Sov, 6/29/72, L1; *Sov Aero*, 7/3/72, 6; GSFC SSR, 6/30/72)

- Federal District in Washington, D.C., issued temporary order restraining NASA from reducing staff at Marshall Space Flight Center on petition of American Federation of Government Employees. MSFC reduction in force had been scheduled for June 29. Judge Joseph C. Waddy scheduled hearing for July 10 to determine if restraining order would become permanent injunction. (MSFC Release 72–80)

- Cuban Premier Fidel Castro and Soviet Communist Party Leader Leonid I. Brezhnev visited Yuri Gagarin Cosmonaut Training Center near Moscow during state visit of Castro to U.S.S.R. Castro placed flowers on monument to Gagarin, first man in space (April 12, 1961). Visitors were shown automatic linking of model spacecraft, Salyut space station, and spacecraft that had returned from space. (Tass, FBIS–Sov, 6/30/72, G1)

June 30: U.S.S.R. launched three satellites. *Intercosmos 7,* launched from Kapustin Yar, entered orbit with 550-km (341.8-mi) apogee, 260-km (161.6-mi) perigee, 92.7-min period, and 48.4° inclination. Objective of mission was "to continue joint studies of the sun's ultraviolet and X-ray radiation and the influence of these radiations on the structure of the earth's upper atmosphere." Spacecraft carried equipment designed and built by specialists from East Germany, U.S.S.R., and Czechoslovakia. It reentered Oct. 5.

Cosmos 497, launched from Plesetsk, entered orbit with 787-km (489.0-mi) apogee, 271-km (168.4-mi) perigee, 95.2-min period, and 71° inclination.

Meteor 12 weather satellite was launched from Plesetsk to obtain "meteorological information necessary for swift forecasting." Orbital parameters: 904-km (561.7-mi) apogee, 888-km (551.8-mi) perigee, 81.2° inclination, and 102.8-min period. (GSFC *SSR,* 6/30/72; 7/31/72; 10/31/72; Tass, FBIS–Sov, 6/30/72, L1; 7/12/72, L1; *Sov Aero,* 7/3/72, 7; 7/10/72, 15; *SF,* 12/72, 46)

- NASA announced award of $83 645 871 cost-plus-award-fee contract to TRW Systems Inc. for development of two automated High Energy Astronomy Observatory (HEAO) satellites to study high-energy radiation from space. First HEAO would be launched in late 1975; second would be launched 18 mos later. (NASA Release 72–133)

- Clocks were turned back one second at midnight Greenwich Mean Time at Royal Greenwich Observatory, England; Naval Observatory in Washington, D.C.; National Bureau of Standards in Boulder, Colo.; and NASA tracking stations around the world. Radio time signals worldwide inserted extra second. Action was taken to align clocks and signals with atomic time scales adopted internationally in 1967 as official time measurement. Under atomic scale, 9 billion internal vibrations of cesium atom equaled atomic second. Because earth was rotating at relatively slow rate, occasional addition of one second to GMT was necessary. Adjustment was particularly important to seamen who navigated by earth time and position. (AP, *W Post,* 6/21/72, 2; *W News,* 6/28/72, 7)

- Marshall Space Flight Center announced plans to implement Center reorganization, pending NASA Hq. concurrence. Reorganization, necessitated by changing MSFC roles and missions, included elimination of Program Management (PM) with major program offices reporting to Office of MSFC Director; creation of Space Science Projects Office to furnish management and technical direction to High Energy Astronomy Observatory and Large Space Telescope projects; reassignment of Michoud Assembly Facility to Saturn Program Office; redesignation of Administration and Technical Services Directorate as Administra-

tion and Program Support; and consolidation of Contracts Office with Purchasing Office to become Procurement Office. (MSFC Release 72-82)
- Soviet scientists writing in magazine *Zemlya I Vsellennaya* had suggested that canals on Mars were optical illusions created by dark features of Mars surface relief, Tass reported. Scientists—using data from U.S. *Mariner 9* and U.S.S.R. *Mars 2* and *3* spacecraft (all three launched in May 1971 and still in orbit around Mars)—had said large-scale pictures of Mars showed no signs of canals, but showed planet was pockmarked with craters. They had concluded Mars surface relief was about 300 million yrs old and that craters were destroyed twice as fast as those of moon because Mars atmosphere encouraged erosion. (FBIS-Sov, 7/5/72, L4)
- Four officials retired from Lewis Research Center: Irving I. Pinkel, Director of Aerospace Safety Research and Data Institute, who had begun working at Langley Memorial Aircraft Laboratory in 1940 and transferred to LeRC (then named Aircraft Engine Research Laboratory) in 1942; Irving Johnsen, Chief of Chemical Propulsion Div., who had transferred to LeRC in 1943 after three years at Langley; Newell D. Sanders, Chief of V/STOL and Noise Div., who had served at Langley, LeRC and NASA Hq. since 1938; and Leslie F. Hinz, Chief of Finance Div., who had joined LeRC in 1942. (*Lewis News*, 6/16/72, 1)
- NASA announced that joint U.S.-U.S.S.R. rendezvous and docking mission would be designated Apollo Soyuz Test Project (ASTP). (PAO memo)
- European Sounding Rocket Range (ESRANGE) in Kiruna, Sweden, was handed over officially by European Space Research Organization to Swedish authorities. Ownership had been transferred to Sweden following December decision by ESRO Council to abandon sounding rocket activities. (*SR*, 9/72)
- Navy F-14 fighter aircraft crashed into Chesapeake Bay during test flight from Patuxent River Naval Air Station, Md. Cause was unknown. Pilot, listed as missing, was identified by Navy as William H. Miller, who had piloted first F-14 to crash, Dec. 30, 1970. (AP, *NYT*, 7/2/72, 10)

During June: Discovery of new energy source from data returned by Office of Naval Research's ONR-001 ionospheric research experiment launched on Air Force *Sesp 1971-2* satellite Oct. 17, 1971, was announced by ONR. Experiment's low-energy energetic-ion spectrometer had observed intense fluxes of heavy ions precipitating into atmosphere at low-altitude edge of auroral zone during Dec. 17, 1971, magnetic storm. Fluxes were observed over entire energy rate of spectrometer on 24 satellite passes in two days. New source could aid understanding of Van Allen Belts and ionospheric processes affecting radio and radar propagation. (*Naval Research Reviews*, 6-7/72, 32-33)
- *Astronautics & Aeronautics* editorial by Courtland D. Perkins, Chairman of Princeton Univ. Dept. of Aerospace and Mechanical Sciences, urged support of space shuttle program because it was "natural development of the NASA mission to advance space technology," which "takes advantage of the high competence in technical and operational management of large space-oriented systems now fully developed within NASA." Program provided launch capability for nearly all manned and unmanned space programs, would eliminate "patchwork, disparate stable

of boosters and facilities," and was "one of only a few national programs that will continue growth in the various fields of high technology." (*A&A*, 6/72, 20–1)

- Air Force and Dept. of Defense viewed space not as a mission but "as a place that offers more effective ways of accomplishing certain defense missions," Grant L. Hansen, Assistant Secretary of Defense for R&D, said in published interview. There was no "space program in the sense of NASA" within DOD. During first phase of space exploration and use for national defense, Air Force has been assigned role as DOD's executive agency for military space efforts, from operation of Joint DOD and Atomic Energy Commission Vela space surveillance system to launches of payloads for other military services. New instructions from Dr. John S. Foster, DOD Director of Defense Research and Engineering, in 1971 had encouraged all military services to "look for ways of using space systems to perform their jobs better and cheaper." Air Force would continue launches but "will no longer operate space systems of other services, nor will there be the requirement to develop their systems for them. The cooperation among the services on space matters is excellent, and, I believe, this change, which recognizes that space as a military medium has reached maturity, benefits everybody." (*AF Mag*, 6/72, 35–40)

- National Science Foundation released *Federal Funds for Research, Development, and Other Scientific Activities, Fiscal Years 1970, 1971, 1972* (NSF 71–35). Federal obligations for research and development (R&D) in 1970 totaled $15.3 billion excluding plant; slight rise to $15.4 billion was expected for 1971. Figure was decrease of 5% in constant dollars. Basic research obligations were expected to rise from $2.1 billion in 1970 to $2.2 billion for 1971 and $2.4 billion for 1972. Federal support for basic research declined in 1970 but was expected to reach new record highs for 1971 and 1972. Bulk of Federal basic research, 81%, was funded by NASA; Dept. of Health, Education, and Welfare; National Science Foundation; Atomic Energy Commission; and Dept. of Defense. NASA had led in basic research support since 1961, with 29% of total in 1971.

    Applied research obligations totaled $3.5 billion in 1970 and were expected to rise to $3.8 billion for 1971, increase of 8%. Total was expected to reach $4.2 billion in 1972—all-time high in current dollars. DOD, HEW, and NASA continued to dominate Federal applied research effort and would account for 78% in 1971. Obligations for development totaled $9.7 billion in 1970 and would decline 3%, to $9.4 billion, in 1971. In 1972, Federal development total was scheduled to rise to $10.0 billion, or 1966 level. DOD, NASA, and AEC were most prominent in Federal development effort in 1960–1972, accounting for 92% in 1971. DOD funded 63%; NASA share had declined since 1965 to expected 19% for 1971.

    Although total Federal R&D obligations were expected to increase 9% from 1970 to 1972, NASA and AEC decrease continued. NASA estimate for 1972, however, was smallest reduction since downward trend began in 1967. Report noted increases in key programs—Skylab, space shuttle, Viking Mars explorations, and unmanned missions to outer planets—had almost offset decreases in Apollo program. (Text)

During June

- Tenth NASA Summer Seminar on History and Space began, with three history professors and four graduate students participating. Each participant would prepare a historical report. (NASA Hist Off)
- NASA released *Physics of the Solar System* (NASA SP-300), edited by Dr. S. Ichtiaque Rasool of Goddard Institute for Space Studies. Book was based on lectures given at Fourth Summer Institute for Astronomy and Astrophysics at Stony Brook, N.Y., June 17–July 15, 1970. It covered broad range of topics in solar physics, planetary structure and atmospheres, and origin and evolution of solar system. (Text)

July 1972

July 1: Proposal to produce rare bloodclot-dissolving enzyme urokinase aboard Apollo spacecraft during 1975 joint U.S.–U.S.S.R. rendezvous and docking mission would be made to NASA by university-industry scientific team, *Washington Post* reported. Scientists from Abbott Laboratories, General Dynamics Corp. Convair Div., and Univ. of California at Los Angeles had suggested use of device developed by UCLA biophysicist Alexander Kolin that separated the one kidney cell out of every 19 that made urokinase. Device had been used successfully on earth, but team believed urokinase-producing cells could be isolated more effectively in space, where gravity did not hinder separation. Cells would be separated aboard spacecraft, ampuled, frozen, and brought back to earth, where enzyme would be produced by pharmaceutical plant. If successful, method could reduce high cost of urokinase to $50 per dose. National Heart and Lung Institute had treated 236 victims of pulmonary embolism with enzyme at more than $1000 per patient. (O'Toole, *W Post*, 7/1/72, A3)

- Manned Spacecraft Center awarded two parallel, firm-fixed-price contracts for study of aircraft that could simulate subsonic flight characteristics of space shuttle orbiter. Grumman Aerospace Corp. would receive $150 000 and Lockheed Aircraft Corp. Georgia Div. would receive $149 000 to provide sufficient data to assure NASA that proposed shuttle training aircraft was technically feasible and acceptable and that recommended design could meet specified requirements. (MSC Release 72–147)

- Swiss Transport Museum inaugurated its Air and Space Wing in Lucerne with ceremonies attended by Dr. James C. Fletcher, NASA Administrator; *Apollo 11* Astronaut Neil A. Armstrong; and Mercury Astronaut John H. Glenn. (Swiss Transport Museum)

- Achievement Award of Small Business Assn. was presented to Kennedy Space Center for its aid to President Nixon's Minority Business Development Program for FY 1971. (*NASA Activities*, 8/1/72, 171)

July 2: U.S. had been using weather control for military purposes, *Washington Post* reported. Previously ignored passage in Pentagon papers Dept. of Defense report on Vietnam war released by press had indicated Indochina had been "test battleground, the site of purposeful rain-making along the Ho Chi Minh trails." Report by DOD's Advanced Research Projects Agency (ARPA) called Nile Blue had been cited as prime evidence of DOD rainmaking. *Post* said report was computer study of how man-made changes might affect world's year-to-year climate. ARPA Director Stephen J. Lukasik had told Senate Committee on Appropriations that Nile Blue was established in FY 1970 to achieve U.S. capability to evaluate consequences of possible actions to manipulate climate, detect changes in global circulation, and determine means to counter deleterious climatic changes. (Cohn, *W Post*, 7/2/72, C2)

July 3: Dr. James C. Fletcher, NASA Administrator, believed U.S.S.R. was working toward year-long manned space mission in earth-orbit, *Aviation Week & Space Technology* article reported. Opinion was based largely on "intuition," Dr. Fletcher had said, but long-duration flight would be necessary before either U.S. or U.S.S.R. could attempt manned planetary mission. "I don't know what the optimum flight time would be before we could be sure man could survive extremely long missions." (Strickland, *Av Wk*, 7/3/72, 21)

- Nike-Apache sounding rocket launched by NASA from Arenosillo, Spain, carried Spanish and Goddard Space Flight Center experiments to 122-km (75.7-mi) altitude to study ion composition in D- and E-regions of ionosphere. Rocket and instruments performed satisfactorily and good data were obtained. Launch was fifth of six launches to study ionosphere under July 30, 1970, Memorandum of Agreement between NASA and Comisión Nacional de Investigación del Espacio (CONIE). (NASA Rpt SRL)
- Manned Spacecraft Center announced appointment of Lynwood C. Dunseith as Deputy Director of Flight Operations. Dunseith had joined NASA in 1959 as member of Space Task Group at Langley Research Center and had been Assistant Director for Computation and Flight Support at MSC since 1970. (MSC Release 72–137)
- Firm guarantees of smokeless engines and noise levels equal to Boeing 707 and McDonnell Douglas DC-8 jet transports were being written into sales contracts for Anglo-French Concorde supersonic transport, *Aviation Week & Space Technology* reported. Sixteen production versions of aircraft were under construction, with deliveries scheduled for thirty aircraft to customer airlines by end of 1975. (*Av Wk*, 7/3/72, 31)
- Atomic Energy Commission announced it had issued license to Veterans Administration Hospital in Buffalo, N.Y., authorizing implant of nuclear-powered cardiac pacemakers for clinical evaluation in patients. (AEC Release P–203)
- Federal Electric Corp. had received 1-yr, $19.9-million extension of contract with NASA, *Aviation Week & Space Technology* reported. Contract was for instrumentation and communications support services to Kennedy Space Center for Apollo and Skylab launches. Extension brought total value of contract to $113.9 million. (*Av Wk*, 7/3/72, 20)

July 4: NASA released mid-year review of items of collateral space technology which had been adapted to improve quality of life on earth. Items included compact, automatic gas analyzer that measured air in lungs to monitor human pulmonary and cardiovascular activity; temperature-activated remote thermistor alarm that sounded at nurse's station when infant's breathing stopped in hospital room; electroencephalograph helmet to diagnose hearing defects in children; powered prosthetic hand; portable light indicator for blind persons; ear oximeter to measure oxygen content of blood by noting red and infrared light absorption; analyzer to record intensity of sleep; and computer motion pictures that projected three-dimensional image of heart areas for study by physicians. Items used in industry included compound to prevent fogging of lenses; low-voltage switching circuits; fire-retardant or nonflammable foams, paints, fabrics, and glass fiber laminates; rapid-scan infrared tire tester; land-surveying system using laser and telescope that viewed laser pulse above obstructions; and management system devel-

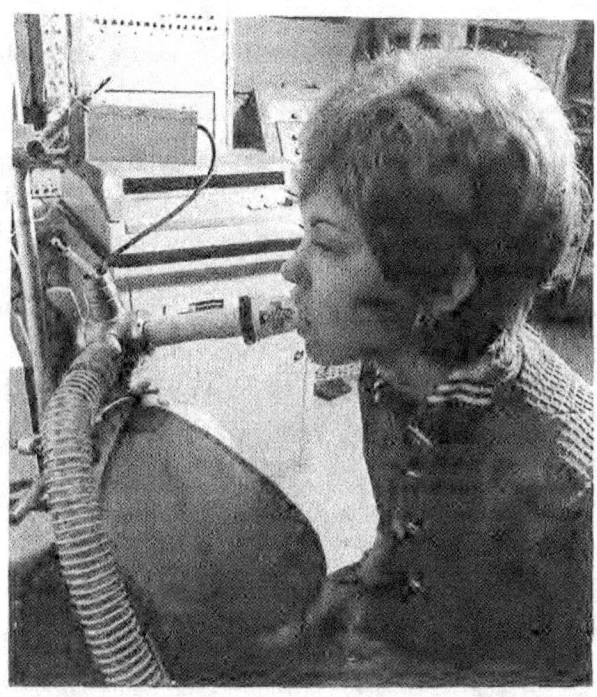

July 4: NASA's *mid-year review of collateral space technology benefits on the earth described a compact, fully automatic, mobile medical gas analyzer (above). The analyzer measured composition of air breathed in and exhaled from the lungs to monitor pulmonary and cardiovascular activity in patients. Industrial items included an ultrathin, two-volt switching circuit for buildings. Low-volt circuits were displayed at the Second Annual Urban Technology Conference in San Francisco July 24–26.*

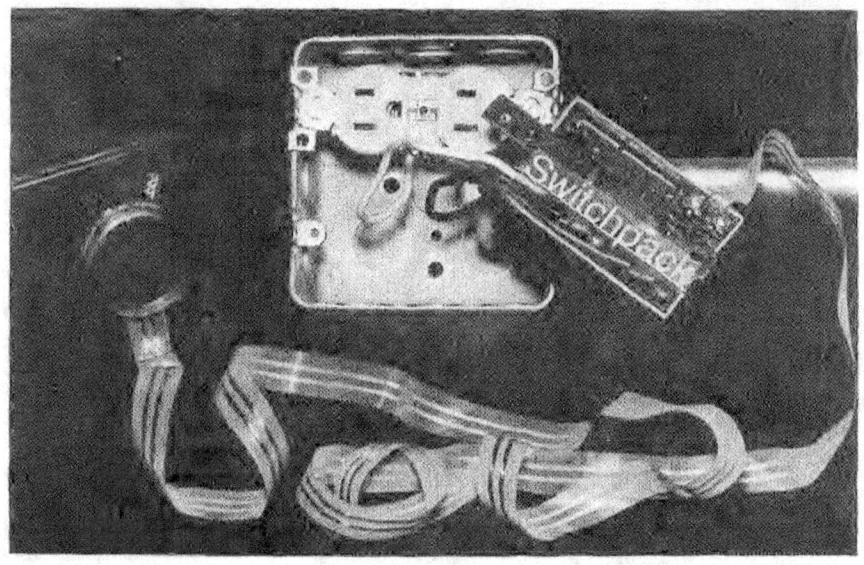

July 4

oped by NASA to supply daily updated information in Apollo program. (NASA Release 72-132)

- U.S. Patent No. 3 675 026 was issued to International Business Machines Corp. scientist Jerry M. Woodall for solar cell that turned about 18% of sun's light into power with output one and one half times that of commercially available solar cells. *New York Times* later said NASA had expressed great interest in cell, which required less shielding against radiation and could operate at higher temperatures than silicon cells. (Pat Off PIO; Jones, *NYT*, 7/8/72, 31)

- *New York Times* editorial commented on geophysical warfare—"act or acts of environmental engineering designed to change the flow of air and water in order to damage one side in a conflict and benefit the other": Revelation that U.S. had used rainmaking techniques for military purposes in Indochina had placed problem on world's agenda for international discussion. "Even those who may feel that dropping rain on an enemy is better than dropping bombs must realize that rainmaking is only the first step. Once accepted as a normal military technique, geophysical warfare may some day be capable of drowning vast continental coastal areas, turning fruitful areas into deserts and even perhaps ultimately of radically rearranging the entire world climate." It was "imperative that environmental engineering be placed under some kind of international control." (*NYT*, 7/4/11)

July 5: U.S.S.R. launched *Cosmos 498* from Plesetsk into orbit with 469-km (291.4-mi) apogee, 267-km (165.9-mi) perigee, 91.8-min period, and 70.9° inclination. Satellite reentered Nov. 25. (GSFC *SSR*, 7/31/72; 11/31/72; *SBD*, 7/7/72, 30)

- Director Glynn S. Lunney of Apollo-Soyuz Test Project (ASTP) briefed press at Manned Spacecraft Center on plans for joint U.S.–U.S.S.R. mission in 1975 to be finalized at meetings to begin July 6 at MSC. Delegation of 25 Soviet engineers would discuss with NASA "precisely how we are going to do the mission, when we are going to do things and who is going to be doing things." Groups would exchange documents proposing specific details of mission, including launch time, docking schedule, launch sequence, and communications. Spacecraft would be docked about 48 hrs and crews would make several exchange visits. Docking module would be used to transfer crews and as prebreathing chamber to protect crews from bends while changing from lower-pressured Apollo to higher-pressured Soyuz. Lunney said he expected U.S. crew or pool of crew candidates to be selected "within a year or so." (AP, B *Sun*, 7/6/72)

- France and U.S. had agreed on three-year project FAMOUS (Franco-American Midocean Undersea Study) to explore rift valley 2.4 km (1.5 mi) beneath Atlantic for volcanic activity that was slowly pushing America away from Europe and Africa, *New York Times* reported. Paper compared project, outgrowth of 1970 Franco-American Agreement on Scientific Cooperation, to Apollo program. Training and employment of international crews had anticipated U.S.-U.S.S.R. space cooperation. Similarities to Apollo were "use of special vehicles [deep-submergence craft] to carry men across otherwise inaccessible terrain, the development of special geologic tools and the training of participants in accessible areas thought to bear some resemblance to the forbidding target sites." Crews would begin training in August in Gulf of Maine. (Sullivan, *NYT*, 7/5/72, 12)

- Award of two research contracts to reduce aircraft noise were announced by Secretary of Transportation John A. Volpe. General Electric Co. had received $789 000 and Lockheed-Georgia Co., $77 042. Contractors would analyze and experiment with core engine noise control for propulsion systems of future aircraft. (FAA Release 72-131)
- Dept. of Commerce announced it had granted $150-million export license to Boeing Co. to cover sale of 10 Boeing 707 jet cargo aircraft to People's Republic of China. (DOT, "Export Licenses Approved," 6/30/72; Witkin, *NYT*, 7/6/72, 1)
- Dr. Reimar Lust had succeeded Dr. Adolf F. J. Butenandt as president of Max Planck Society for Advancement of the Sciences in Bremen, West Germany, *New York Times* reported. Society, founded as Kaiser Wilhelm Society in 1911, operated 52 institutes and research centers throughout West Germany. They embraced all physical and biological sciences, as well as law, teaching, theory, psychology, and guidance of technical societies. Society's annual budget was nearly $160 million, of which 85% came from West German government and remainder from industry and other private sources. (*NYT*, 7/5/72, 25)

July 6: U.S.S.R. launched *Cosmos 499* from Baykonur into orbit with 284-km (176.5-mi) apogee, 205-km (127.4-mi) perigee, 89.4-min period, and 51.9° inclination. Satellite reentered July 17. (GSFC *SSR*, 7/31/72; *SBD*, 7/7/72, 30)

- NASA announced award of six-month, $197 000 contract to Arthur D. Little, Inc.—acting for four-firm industry team—for feasibility study of using large satellites in synchronous orbit to beam down electrical energy for power needs on earth. Station would convert solar energy to electric power in space and then transmit power via microwave beam to earth. NASA was investigating concept for comparison with other methods of producing large amounts of power from sun for use on earth. Contract would be managed by Lewis Research Center. (NASA Release 72-135)
- Nike-Cajun sounding rocket launched by NASA from Arenosillo, Spain, carried Spanish and Goddard Space Flight Center experiments to 85-km (52.8-mi) altitude to study ion composition of D- and E-regions of ionosphere. Rocket and instruments performed satisfactorily except for short tone-ranging data dropout which was not expected to affect scientific results. Good data were obtained. Launch was made under April 21, 1972, modification of July 30, 1970, Memorandum of Agreement between NASA and Comisión Nacional de Investigación del Espacio (CONIE) for cooperative sounding rocket project to study ionosphere. (NASA Rpt SRL)
- Apparent U.S. decision not to seek European participation in shuttle development was discussed by Nicholas Valery in British publication *New Scientist:* "By all accounts it seems that just as Britain and her continental partners were deciding whether to accept America's invitation to join the post-Apollo space programme, Washington has slammed the door smartly in their face. Having used the idea of international participation—ostensibly to ease the burden on the US taxpayer—as a means of winning congressional support for its re-usable space shuttle, NASA is now quietly letting it be known that it has no

intention of letting Europe get its hands on anything technologically important." (*New Scientist* [London], 7/6/72)

- U.S.S.R., "in apparent attempt to display restraint," had "suspended space maneuvers with its fearsome SS-9 Scarp rocket," *Washington Post* reported. The United Nations "international space log shows that the Soviet Union has desisted from the SS-9 maneuvers in space so far this year, in contrast to earlier years." Since U.S. had not flight-tested any space systems comparable to U.S.S.R.'s fractional orbital bombardment system (FOBS), "the Russians might have reasoned it would have been impolite of them to stage maneuvers this year when the SALT [Strategic Arms Limitation Talks] agreement appeared in reach." (Wilson, *W Post*, 7/6/72, A2)

July 6–18: Series of meetings on Apollo-Soyuz Test Project (ASTP), joint U.S. and U.S.S.R. rendezvous and docking mission scheduled for 1975, were held at Manned Spacecraft Center. Meetings' objective was preliminary definition of all elements of test mission, including mission plan and detailed design requirements for compatible systems. Group had met at MSC in June 1971 to discuss feasibility of compatible systems, U.S. delegation had visited Moscow twice, and individual working groups had exchanged visits during past two years.

Communique on results of July 6–18 meetings was signed by MSC Director, Dr. Christopher C. Kraft, Jr., July 17. Agreements had been reached on launching Soyuz before Apollo, acceptable launch windows, necessary communications equipment, and development of technical requirements for compatible docking system. Areas for further study included possible emergency extravehicular activity, possible adjustment of cabin pressures, use of TV as docking aid, attitude control after docking, specific launch date, and development of public information plan that considered obligations and practices of both countries. [See also press conferences July 13 and 17.]

Senior members of Soviet delegation were Chairman Boris N. Petrov of Soviet Intercosmos Council and Konstantin D. Bushuyev, ASTP Director for U.S.S.R. NASA delegation was headed by Dr. Kraft and Glynn S. Lunney, ASTP Director. (NASA Release 72-134; *Marshall Star*, 7/12/72, 1; Text)

July 7: Air Force launched two unidentified satellites from Vandenberg Air Force Base by single Titan IIID booster. First satellite entered orbit with 257-km (160-mi) apogee, 173-km (108-mi) perigee, 88.7-min period, and 96.8° inclination and reentered Sept. 13. Second entered orbit with 501-km (311-mi) apogee, 499-km (310-mi) perigee, 94.5-min period, and 96.1° inclination. (Pres Rpt 73; *SBD*, 7/13/72, 57; 7/18/72, 78)

- Navy's 90.7-metric-ton (100-short-ton) surface effect testcraft *SES–100A* made underway test run on Puget Sound, Wash. Advanced design craft was built by Aerojet-General Corp. under direction of Navy's Surface Effect Ships Project Office. Craft was powered by four gas turbines and traveled over water on air cushion propelled by high-capacity waterjets. *SES–100A* would be used in designing larger surface effect ships capable of high speeds in ocean operations. (*SES–100A* first test run cover; Navy SES Proj Off)

- Experiment to plot evolution of sun and earth over aeons was described in *Science* by Cornell Univ. astronomers Dr. Carl E. Sagan and

Dr. George Mullen. Findings had indicated that sun's energy emission had increased 40% over last 4.5 aeons and would continue to increase until earth temperature reached boiling point, in about 4.5 more aeons. Infrared characteristics of atmospheres of other planets would determine whether life on them, if any, would survive. Mars atmosphere should be approaching required conditions for emergence of life at about same time earth temperature accelerated upward. (*Science*, 7/7/72, 52–5)

- Presidential Prizes for Innovation announced by White House May 21 were similar in objectives to "Olympics of Science" proposed in 1849 by U.S. Commissioner of Patents Thomas Ewbank, patent attorney Harry Goldsmith said in *Science* article. Ewbank had asked Congress to appropriate $100 000 as permanent "Inventors' Premium Fund," to be distributed once every four years for outstanding contributions by scientists and inventors. Proposal had not been adopted. (*Science*, 7/7/72, 35)

July 9: Article in Soviet space magazine *Aviatsiya i Kosmonavtika* (Aviation and Cosmonautics) had indicated U.S. was not only nation investigating space shuttle development, *New York Times* reported. Two Soviet space scientists using pseudonyms V. Vasilyev and L. Leonidov had called shuttle "logical continuation of the space exploration effort." *Times* said it was "first technical discussion" of the space shuttle concept to appear in Soviet press since President Nixon approved U.S. shuttle development in January. (Shabad, *NYT*, 7/10/72)

- Space stamps were most popular topical issues among collectors in America Topical Assn. directory for 1972–73, *Book World* reported. Space issues, collected by 10% of ATA members, had also led in popularity in 1970–71 directory. (*Book World, W Post*, 7/9/82, 12)

- Llewellen J. Evans, President of Grumman Corp., died at age 49. Evans had joined Grumman in 1951 after four years in Navy Office of General Counsel and had played important role in development of Apollo lunar module. (*W Post*, 7/17/72, C4; NASA Hist Off)

July 10: NASA's *Mariner 9* Mars probe (launched May 30, 1971) completed photo-mapping of entire Mars surface with acquisition of final elements needed in playback of 30 pictures, bringing total photo count to more than 7000. Spacecraft—in Mars orbit since Nov. 13, 1971—would make 500th revolution around Mars July 20. During next three months *Mariner 9* would provide precision tracking data to test relativity theory that electromagnetic radiation—spacecraft's radio signal—passing close to sun would be slowed by sun's gravitational field. Additional pictures would be taken once in August, once in September, and twice in October and stored for playback in late October. Lifetime was dependent upon supply of attitude-control gas. (NASA Release 72-143)

- U.S.S.R. launched *Cosmos 500* from Plesetsk into orbit with 545-km (338.7-mi) apogee, 508-km (315.7-mi) perigee, 95.1-min period, and 74° inclination. (GSFC *SSR*, 7/31/72; *SBD*, 7/12/72, 49)

- NASA's *Pioneer 10* Jupiter probe (launched March 2) was continuing on trajectory toward Dec. 3, 1973, encounter with Jupiter. Spacecraft performance—except for stellar reference assembly—had been satisfactory and science instruments in operation were providing good data.

Some polarization measurements had been lost because of imaging photopolarimeter anomalies. Meteoroid detector had recorded 53 hits, including four in 15-hr time span. *Pioneer 10* would enter as yet unexplored Asteroid Belt July 15. (NASA proj off)

- Photos taken by far-ultraviolet camera-spectrograph trained on earth from moon during April 16–27 *Apollo 16* mission had supported theory that photosynthesis by green plants was less important as oxygen source than commonly believed, Naval Research Laboratory reported. Experiment indicated "that solar effects on the earth's water that evaporates to the high atmosphere may provide our primary supply of oxygen." NRL principal investigator for spectrograph experiment, Dr. George R. Carruthers, said new photos provided much greater detail about geocorona than had been inferred from earlier measurements. (NRL Release 30–72–7; AP, *NYT*, 7/12/72, 24)

- Total eclipse of sun over northern Canada was studied by Marshall Space Flight Center physicists Dr. Robert J. Naumann and K. Stuart Clifton and five Johns Hopkins Univ. physicists. Observation period of 90 sec marked last opportunity in North America to observe total solar eclipse until 1979. Team recorded shadow bands on either side of totality in first attempt to use low-light-level TV to obtain actual images of bands. Bands were extremely difficult to photograph because contrast between bands and surrounding light was faint. MSFC participation had been requested because of Center's experience with low-light-level TV observations of transient phenomena during meteoroid research and study of barium clouds and aurora. Team made observations from schoolyard near Cape Chat, Quebec.

    In second experiment 24 scientists from Los Alamos (N. Mex.) Scientific Laboratory rendezvoused in Air Force cargo aircraft with eclipse totality at altitude of 12 000 m (39 000 ft) northwest of Churchill, Manitoba. Scientists observed solar corona using telescopes, cameras, and radiation-measuring devices. Eclipse extended from north of Japan, across Siberia and Alaska, eastward through Canada, and ended in Atlantic Ocean south of Azores. It was partial throughout U.S. (MSFC Release 72–84; Rensberger, *NYT*, 7/10/72, 35; O'Toole, *W Post*, 7/10/72, A3; AP, B *Sun*, 7/11/72, A3; MSFC PAO)

- NASA announced issuance of invitation to scientists in U.S. and other countries to propose additional uses of data from *Oao 2* Orbiting Astronomical Observatory (launched Dec. 7, 1968) and OAO–C (*Oao 3* when launched), scheduled for launch in late summer 1972. Data would be available from *Oao 2* Univ. of Wisconsin experiment (array of four photometers and three spectrophotometers to study stars, galaxies, and nebulae in ultraviolet portion of spectrum) and OAO–C Princeton experiment package (81-cm [32-in] reflecting telescope to study effects of interstellar space on UV light from stars). (NASA Release 72–138)

- Manned Spacecraft Center announced three awards for space shuttle work. Textron, Inc., Bell Aerospace Div. received $261 500 firm-fixed-price contract and North American Rockwell Corp. Space Div. received $258 000 for constructing and testing thrust chambers for space shuttle engines for in-space maneuvering. Bell Aerospace also received 15-mo, $540 000, cost-plus-fixed-fee contract to study and design hypergolic

bipropellant engine for space shuttle orbiter reaction control system. (MSC Releases 72-140, 72-142)
- Tenth anniversary of first transoceanic TV via satellite—*Telstar 1* comsat built by American Telephone & Telegraph Co. and launched by NASA from Cape Canaveral (now Cape Kennedy), Fla. Millions of TV viewers had watched taped picture of American flag transmitted from Andover, Me., via *Telstar 1* to Andover and Holmdel, N.J. Signals were also picked up by stations at Plemeur-Bodou, Brittany, and Goonhilly Downs, Cornwall. Within one week after launch *Telstar 1* had transmitted first TV pictures from Europe and first pictures in color and, within one month, first international exchange of live TV. During four months of operation before it was disabled by radiation, satellite handled more than 400 transmissions, including 50 TV demonstrations, telephone calls and data in both directions, and facsimile material. It was revived in January 1963 for communications tests.

    Telstar 1 was followed by *Telstar 2* (launched May 7, 1963), 2 Relays, 3 Syncoms, and 15 Intelsats and commercial traffic grew to more than 2500 hrs of TV per year and more than 4000 full-time, leased, two-way circuits. (NASA Special Release, 7/4/72)
- NASA launched two Aerobee 170 sounding rockets from White Sands Missile Range. First rocket carried American Science and Engineering, Inc., solar physics experiment. Rocket performed satisfactorily. Scientific objectives were not met. Second rocket carried Naval Research Laboratory solar physics experiment. Rocket and instrumentation performed satisfactorily. (SR list)
- U.S. District Court in Washington, D.C., dissolved temporary restraining order it had issued June 29 and refused to grant preliminary injunction against NASA reduction in force of 503 employees at Marshall Space Flight Center. Suit brought by American Federation of Government Employees charged NASA with violations of Civil Service Commission rules applicable to reductions in force. (Bell, *Birmingham* [Ala.] *Post-Herald*, 7/11/72)
- Soviet scientists had obtained spectrograms of stars Vega and Agena—unobservable from earth—using Orion astronomical system on *Salyut 1–Soyuz 11* orbital station, *Izvestiya* reported. Temperature of Vega was reported to be 10 000°C and of Agena, 24 000°C. Terrestrial atmosphere nearly totally absorbed their radiation. Orion system, controlled both automatically and manually, had been designed to consider "the strong vibration and overloads, the low temperatures and vacuum which the equipment would encounter in space. For this reason both mirrors of the telescope . . . were made from a strong crystallic glass—sital. The telescope itself was made in the form of a titanium alloy tube. The cosmonauts' task was to recognize the required celestial target and aim the sighting tube." (Tass, FBIS–Sov, 7/13/72, L1)
- Washington *Evening Star* editorial commented on use of U-2 aircraft to map storm damage in Chesapeake Bay area [see June 2]. Soviet Embassy had purchased residence in area and was improving property. "We just hope every precaution will be taken with the flood-mapping spy plane, so that the new era of sweetness and light between the United States and the Soviet Union will not come crashing down in

flames just because somebody at the CIA wants an aerial closeup of the Russians' new basketball court." (W *Star*, 7/10/72)

July 10–14: International Atomic Energy Agency and U.S. Atomic Energy Commission held international scientific meeting in Seattle, Wash., to exchange research information on radioactivity in the world's oceans and estuaries. Oceanographers and marine scientists from 15 countries presented technical papers. (AEC Release P–187; AEC PIO)

July 11: Viking parachute system to lower instruments onto Mars surface in 1976 was damaged during first in series of three high-altitude tests over White Sands Missile Range. Parachute was carried to 5900-m (19 500-ft) altitude by large balloon and then boosted upward to supersonic speed by rocket. Damage had occurred when parachute deployed at 43 300-m (142 000-ft) altitude, where dynamic pressure reached 689.5 newtons per sq m (14.4 psf), instead of in thinner air at 45 300 m (148 500 ft) with 598.5 newtons per sq m (12.5 psf) as planned. Objective was to demonstrate capability of parachute system at highest speed and most severe loading conditions expected at Mars. NASA said parachute would have functioned satisfactorily in actual Mars landing. Since launch was only partially successful, it would be repeated in August. [See July 26, Aug. 13 and 19, Dec. 14.] (NASA Release 72–118; NASA proj off; *Today*, 7/14/72)

- NASA announced it would reprimand *Apollo 15* crew for carrying 400 unauthorized postal covers on July 26–Aug. 7, 1971, mission. Astronauts' actions would be considered in future assignments. Astronauts David R. Scott, Alfred M. Worden, and James B. Irwin had acknowledged giving 100 covers to acquaintance, currently in Germany. Covers later were sold to stamp collectors for $1500 each. NASA inquiry had shown astronauts had at one time agreed to accept "trust fund" for their children in return for the 100 covers but had later realized impropriety of transaction and had voluntarily declined to accept fund or alternative offer of stamps. Dr. George M. Low, NASA Deputy Administrator, said in release to press: "Astronauts are under extreme stress in the months preceding a flight to the Moon and their poor judgment in carrying the unauthorized covers must be considered in this light. Nonetheless, NASA cannot condone these actions." (NASA Release 72–140)

- Manned Spacecraft Center awarded North American Rockwell Corp. Space Div. $200 000, fixed-price, research and development contract to study feasibility of manned and unmanned multipurpose spacecraft to provide support for spacecraft in geosynchronous orbit. Spacecraft in geosynchronous orbit were stationed at 37 800-km (23 500-mi) altitude, where they conducted communications, earth observations, and navigation experiments. (MSC Release 72–141)

- Vladimir N. Pavlov—director of Soviet exhibit scheduled for Seattle, Wash., Fair Aug. 11 to 20—said exhibit would include display of special-purpose plastic used in Soviet *Lunokhod 1* lunar rover and scale model of vehicle. (Shabad, *NYT*, 7/12/72)

July 12: U.S.S.R. launched *Cosmos 501* from Kapustin Yar into orbit with 2105-km (1308-mi) apogee, 215-km (133.6-mi) perigee, 108.5-min period, and 48.4° inclination. (GSFC *SSR*, 7/31/72; *SBD*, 7/14/72, 68)

- During five-day visit to Poland Dr. Edward E. David, Jr., Presidential Science Adviser, informed Polish government in Warsaw that U.S. would name OAO-C Orbiting Astronomical Observatory for Polish astronomer Nicolaus Copernicus. Launch of OAO-C, scheduled for August, would inaugurate U.S. contribution to worldwide Copernican celebration feting 500th anniversary of Feb. 14, 1473, birth of Copernicus. OAO-C would carry 81-cm (32-in) reflecting ultraviolet telescope developed by Princeton Univ. and battery of smaller x-ray telescope developed by Univ. College, Univ. of London, to study UV and x-ray emissions of celestial bodies. OAO-C would weigh 2200 kg (4900 lbs) and would be launched into earth orbit with 740-km (460-mi) altitude by Atlas-Centaur booster. (NASA Release 72-141)

July 13: U.S.S.R. launched *Cosmos 502* from Plesetsk into orbit with 248-km (154.1-mi) apogee, 208-km (129.3-mi) perigee, 89.1-min period, and 65.4° inclination. Satellite reentered July 25. (GSFC *SSR*, 7/31/72; *SBD*, 7/20/72, 98)

- Press conference on overall aspects of May 24 U.S.-U.S.S.R. space cooperation agreement was held at Manned Spacecraft Center during meetings of NASA and Soviet working groups [see July 6-18]. Participants included Dr. George M. Low, NASA Deputy Administrator; Academician Boris N. Petrov, Chairman of Soviet Intercosmos Council; Dr. Christopher C. Kraft, Jr., MSC Director; and Arnold W. Frutkin, NASA Assistant Administrator for International Affairs.

  Dr. Low said "great deal" had been accomplished in all areas covered by cooperation agreement. Working groups were meeting almost constantly. Nations had exchanged lunar samples from all flights through *Apollo 15* and *Luna 20* and were about to exchange samples on *Apollo 16*. Area of greatest interest was rendezvous and docking mission scheduled for 1975. Talks on joint mission were "going extremely well."

  Petrov said Soviet people had been informed of plans for joint space flight "at the same time as the American people"—when technical discussions were in progress. "Our press, just as your press, carried information on the discussions." Asked if foreign press would be permitted to witness launch of Soviet Soyuz during joint mission, Petrov said both Apollo spacecraft and Soyuz launch would be conducted "under the same conditions as all previous launches." Question on control of mission was deferred to conference on joint rendezvous and docking mission scheduled for July 17, but Dr. Low said "general principle" would be that, "while the Soviet cosmonauts are visiting in our Apollo command module, we will be in command; when we're visiting in their Soyuz, they will be in command of that ship . . . and we will be under their command." Petrov said he hoped, and was certain, "that this first test flight will not be the last one" and that there would be possibility of space rescue flights if needed, "regardless of what country's spacecraft are involved." If test flight was successful, Dr. Low said, "our plans are to equip the Space Shuttle with this kind of a docking system. We don't think that we will need a follow-on test but instead we will be flying operationally with compatible systems beyond that point."

  Asked to comment on Soviet Venus probe *Venus 8*, launched March 27, Petrov said spacecraft should reach planet at month's end. "That vehicle will release another one which should come down onto the surface of Venus." Soviet scientists had done everything possible to guarantee mission's success, "but what is not excluded is the possibility of a landing in such a position . . . or place where transmission of information will not last for very long." Probe was expected to provide data to increase understanding of planet's surface and structure, "which is not only a question of increasing our knowledge but which can . . . clarify the origins and structure of . . . Earth." Data already transmitted for *Venus 8* had been "so new and so unexpected and so interesting." (Transcript)

- NASA announced it was making available to U.S. industry patent rights for commercial production of two scientific devices developed at Goddard Space Flight Center. FLASH (fast luciferace automated assay of specimens for hospitals)—biochemical machine process for quantitative analysis of bacteria in urine samples—had been developed by Emett W. Chapelle and Dr. Grace L. Picciolo from technology to detect extraterrestrial life. FLASH process caused urine sample to glow if bacteria were present; qualitative readings were made automatically by detecting and recording emitted light.

  Smokey long-range laser traversing system—small back-packable surveying system—had been developed from optical system technology from spacecraft tracking and communications experiments [see April 13]. (NASA Release 72–139)

- *Apollo 16* Astronaut Charles M. Duke, Jr., received Distinguished Service Medal from Dr. Robert C. Seamans, Jr., Secretary of the Air Force, and Command Pilot Astronaut Badge from Gen. John D. Ryan, Air Force Chief of Staff, in Washington, D.C., ceremony. DSM was for "exceptionally meritorious achievement in a duty of great responsibility." Badge identified Duke, *Apollo 16* lunar module pilot, as one of 37 Americans who had piloted powered vehicles more than 80 km (50 mi) above earth's surface. Duke gave Dr. Seamans and Gen. Ryan lunar sample and silver medallion in recognition of Air Force role in space program and as NASA salute to Air Force's Silver Anniversary. (President Truman signed Armed Forces Unification Act creating Dept. of Air Force July 26, 1947.) (DOD Release 509–72)

- National Science Foundation released *Federal Funds for Academic Science, Fiscal Year 1970* (NSF 72–301). In FY 1970 total Federal support to universities and colleges declined 7% from FY 1969 funding level. Major portion of decrease was in academic science activities, which experienced 8% drop in 1970, contrasting with 3% in support for nonscience activities. Research and development and manpower development support accounted for bulk of total academic science funding. Obligations for these activities totaled $1.396 billion and $429 million. (Text)

- Philadelphia *Evening Bulletin* editorial commented on *Apollo 15* astronauts' carrying unauthorized postal covers on mission: It was to crew's credit that they withdrew from plan to profit from venture. "But recognition of their original misjudgment . . . does not completely erase the taint of commercialization in an otherwise heroic venture." (P *Bull*, 7/13/72)

- Eugene W. Wasielewski, Associate Director of Goddard Space Flight Center, died at age 59 after long illness. Wasielewski had been principal official for institutional management of GSFC and for operation of worldwide tracking and data-acquisition network. He had joined GSFC in 1960 after career in private industry. (*W Post*, 7/19/72, C4)

July 14: Astronauts Philip K. Chapman and Anthony W. England announced their resignations from NASA. Dr. Chapman, *Apollo 14* mission scientist, had accepted position as Principal Research Scientist for Avco Research Laboratories in Everett, Mass., and would also work as senior research associate in Measurement Systems Laboratory at Massachusetts Institute of Technology. His resignation was effective immediately.

Dr. England, *Apollo 13* and *Apollo 16* mission scientist, would accept position with U.S. Geological Survey Regional Geophysics Group in Denver, Colo., Aug. 14. He would assist in developing techniques using radar from aircraft and spacecraft to learn about surface and subsurface structure of earth and other planets. (NASA Release 72–142)

- NASA launched Aerobee 170 sounding rocket from White Sands Missile Range, carrying Univ. of Colorado astronomy experiment. Rocket and instrumentation performed satisfactorily. (SR list)
- U.S. delegation headed by Dr. David Ballentine of AEC's Div. of Applied Technology left for Moscow to tour industrial process radiation facilities in U.S.S.R. under exchange agreement in peaceful uses of atomic energy signed in 1959. (AEC Release P–218)
- France, in Bastille Day parade, exhibited intermediate-range ballistic missile of its second-generation nuclear strike force. Missile was followed by capsule designed to contain missile's 150-kiloton nuclear warhead. Three-stage missile had range of 3000 km (1900 mi). (Reuters, *B Sun*, 7/15/72, A4)
- Newspaper editorials commented on *Apollo 15* astronauts' carrying unauthorized postal covers on mission.

  Baltimore Sun said their actions in "novel attempt to capitalize" on mission could not be excused. "From the start of the nation's space program, astronauts have become instant celebrities and in some cases instant millionaires through endorsements, publications and investments. We're not opposed to a little free enterprise, but the space program, as envisioned by the late President Kennedy, was a noble exercise in exploration not exploitation." If space program was to be exploited at all, "profits should go back into the national treasury." (*B Sun*, 7/14/72, A12)

  Chicago Tribune: "It is disappointing to realize that man has already tainted the moon with moral pollution, but we are impressed less by the gravity of the crime than by the debasement of the stamp collection business which it reflects. If stamp collectors were more discriminating, and stamp dealers less greedy, the moon might not have been contaminated so soon." (*C Trib*, 7/14/72)

July 15: NASA's *Pioneer 10* Jupiter probe (launched March 2) entered Asteroid Belt, beginning first reconnaissance of huge region of dust and rocks that circled sun between orbits of Mars and Jupiter. Doughnut-shaped belt was about 3 billion km (1.8 billion mi) around, 280 million km (175 million mi) wide, and 80 million km (50 million mi) thick. Asteroids in belt ranged in size from dust particles to rock chunks as large as Alaska. Passage of *Pioneer 10* through belt would permit first measurements of asteroids too small to be seen with earth-based telescopes and of amounts and kinds of asteroidal material. During seven-month passage through belt, spacecraft was expected to pass within 8.9 million km (5.5 million mi) of asteroid Palomar-Layden Aug. 2 and asteroid Nike Dec. 2. *Pioneer 10* carried three experiments for measurements in Asteroid Belt—asteroid-meteoroid detector, gas-cell-array meteoroid detector, and imaging photopolarimeter. (NASA Release 72-136; ARC *Astrogram*, 7/20/72, 1)

- Sprint antiballistic missile successfully selected and intercepted dummy warhead from among several warheads over Pacific Ocean. Maneuver was 29th test of Safeguard system. Objective was to ensure that Safeguard radar could lock onto one target while disregarding others and guide missile to target intercept in space. (DOD press memo, 7/16/72)

- Sen. William Proxmire (D-Wis.) said Dept. of Defense had suppressed secret report showing that cost of Navy F-14 jet fighter aircraft would rise to $20.8 million each—$4 million above published estimates. In statement to press, Sen. Proxmire said he was "shocked and dismayed" at DOD's refusal to make report public. (Corddry, B *Sun*, 7/16/72, 17)

July 17: Apollo-Soyuz Test Project (ASTP) press conference was held at Manned Spacecraft Center. Participants were Dr. Christopher C. Kraft, Jr., MSC Director; Glynn S. Lunney, ASTP Director for U.S.; Academician Boris N. Petrov, Chairman of Intercosmos Council of Soviet Academy of Sciences; and Konstantin D. Bushuyev, ASTP Director for U.S.S.R. Lunney said that during series of meetings that had begun July 6, U.S. and U.S.S.R. had discussed and agreed on "the technical content of three very important project level documents": project technical proposal that described mission and hardware elements, organizational plan that described how countries would work together in controlling project before and during flight, and schedule of activities. Officials had also concluded "technically detailed documents describing the agreements that we've reached on various systems," such as docking aids, docking targets, control systems, and docking mechanism. Further technical discussions would be held in U.S.S.R. in fall.

Preliminary plans called for launch of Apollo spacecraft carrying three astronauts 7½ hrs after launch of Soyuz with two cosmonauts. Second Soyuz would be ready as backup if Apollo could not be launched during one of three launch opportunities. On Apollo's 14th revolution spacecraft would rendezvous and dock, using U.S.-built docking mechanism on Apollo. After docking, two astronauts would pass through airlock into Soyuz and spend several hours with cosmonauts before returning to Apollo for night. Next day one astronaut and one cosmonaut would exchange places for whole day. After returning to own spacecraft, crews would undock vehicles and complete separate missions. Apollo would probably remain in orbit nine more days and Soyuz about one more day. Docking would be televised.

Bushuyev praised "friendly atmosphere" and "wonderful conditions" of meetings and said friendliness was "the greatest achievement and the most important assurance for the success in our future work. The difficult tasks that we are called upon to solve . . . would be impossible unless these conditions existed. We are on the first stage of a very long and difficult journey. I believe that the most important thing that we succeeded in achieving so far is the following: We succeeded in identifying and solving those basic and important problems without the solution of which further work . . . would be impossible. These various problems which we have now solved will permit us to start the planning, design, and actual construction and manufacture of the various assemblies . . . necessary for the realization of our project." (Transcript; NYTNS, W Star, 7/18/72, A4)

- Discovery by instruments aboard *Uhuru* (*Explorer 42* Small Astronomy Satellite launched Dec. 12, 1970) of regular pulse rhythm in x-ray pulsar was reported by *Washington Post*. Pulsar discovered in constellation Hercules by *Uhuru* in November 1971 pulsed for 9 days; then stopped pulsing for 27 days, in repeated pattern. Cornell Univ. astronomer Dr. Frank D. Drake had said discovery was "every bit as bewildering" as discovery of pulsar itself. "There is no known reason why a star should disappear and then reappear at such regular and predictable times." Possibilities advanced by astronomers included one that pulsar was orbiting another pulsar whose radio waves were unobservable by *Uhuru* and that second pulsar overwhelmed first for 27 out of 36 days. Second possibility was that Hercules pulsar was outermost of two stars circling third star in egg-shaped orbits and that eccentric orbits and eclipsing behavior of second orbiting star kept Hercules pulsar hidden from earth three fourths of time. Third and most likely possibility was that pulsar wobbled off orbiting path around its fellow star so that it cast beam into space, away from earth, for 27 out of 36 days. (O'Toole, *W Post*, 7/17/72, A1)

- Manned Spacecraft Center announced contract awards for space shuttle work. Three companies had been awarded six-month extensions to contracts for continued development of new surface materials for shuttle orbiter stage. McDonnell Douglas Corp. received $350 000, General Electric Co. Aerospace Group $346 000, and Lockheed Missiles & Space Co. $345 000. Companies would deliver to MSC sample tiles of rigidized reusable surface insulation for special testing.

    CCI Aerospace Corp. Marquardt Co. had been awarded $181 846, cost-plus-fixed-fee, research and development contract to study helium regulator systems that could be used by shuttle orbiter.

    Martin Marietta Corp. had received $130 000, firm-fixed-price, R&D contract to design, build, and test couch to be used by passengers on shuttle. (MSC Releases 72-145, 72-146, 72-148)

- Completion of 18-mo Skylab vibration and acoustics test program at Manned Spacecraft Center by MSFC–MSC–industry test team was announced. Modal (resonance frequencies) survey operations were conducted using automatic modal tuning analysis system developed to permit control by computer of all modal excitation and data processing on payload assembly. Test marked first time such an extensive test operation had been conducted with computer-controlled system. Skylab hardware cluster would be dismantled and shipped from MSC to MSFC on NASA barge *Poseidon* around Aug. 4. (MSFC Releases 72-87, 72-90)

- American Telephone & Telegraph Co. and Communications Satellite Corp. asked Federal Communications Commission to reconsider its June 16 ruling permitting "multiple entry" of firms into communications satellite field but placing major operating restrictions on AT&T and

July 17

ComSatCorp. AT&T protested FCC's barring proposed agreement under which ComSatCorp would lease satellites from AT&T and protested limitations on its initial use of satellites. ComSatCorp asked FCC to eliminate what it contended were unjustified restrictive conditions that jeopardized "viable, competitive domestic satellite service." (Shifrin, *W Post*, 7/18/72, D8)

- Issuance of requests for proposals for new astronaut space suit was announced by Manned Spacecraft Center. Companies were asked to develop prototype space suit that would weigh less than 18 kg (40 lbs), be fully mobile with high degree of hand dexterity while pressurized at 55 200 newtons per sq m (8 psi), give trouble-free service in earth orbit for up to one year with 50, six-hour extravehicular activity (EVA) periods, and be repairable in space. (MSC Release 72-152)
- West German government would demand that European Launcher Development Organization (ELDO) be scrapped at international space conference scheduled for September in Brussels, *Wall Street Journal* reported. Spokesman for West German Science Minister Klaus von Dohnanyi had said West Germany would suggest that U.S. rockets be purchased for launching European spacecraft. (*WSJ*, 7/17/72)
- *Detroit News* editorial commented on *Apollo 15* astronauts' carrying unauthorized postal covers on mission: "The trio did not follow through with the script but for the men to even contemplate the scheme was dead wrong. The nation has expended considerable money on them, trained them for national duty and paid them well. No one doubts their courage and skill but the spirit of the whole enterprise is tainted now the truth is out." Paper said NASA should cancel astronaut privilege of carrying personal items to moon on condition they not be used later for commercial purposes. (*D News*, 7/17/72)

July 18: First Skylab command and service module (CSM-116), which would transport astronauts to and from Skylab orbital workshop in 1973, was shipped by aircraft from North American Rockwell Corp.'s Downey, Calif., plant to Cape Kennedy, Fla., for start of prelaunch operations. Other Skylab segments—including Orbital Workshop, multiple docking adapter, airlock module, and Apollo Telescope Mount—were undergoing final checkout at contractor plants and at Manned Spacecraft Center. (MSC Release 72-144)

- National Oceanic and Atmospheric Administration of Dept. of Commerce released *The World Weather Program for Fiscal Year 1973*. Major U.S. contributions to World Weather Watch in FY 1973 would include improvement of existing operational satellite system, development of next-generation system, establishment of expanded atmospheric-monitoring capability, increase in computer-processing capacity, and assistance to weather services of developing nations. NOAA-sponsored polar-orbiting TIROS satellites would be launched by NASA to provide temperature soundings and day and night cloud-cover images. NASA would launch Synchronous Meteorological Satellite (SMS) for 24-hr viewing of severe storms and cloud cover by spacecraft that was prototype of NOAA's Geostationary Operational Environmental Satellite (GOES) system. First Arctic baseline station for atmospheric measurement would be established at Barrow, Alaska, by NOAA; National Science Foundation would support expansion of Antarctic monitoring program; and Atomic Energy Commission would begin measurement of tropospheric radionuclides on Pacific island. U.S. would provide ground-receiving equipment for acquiring weather satellite photos to developing nations and establish regional telecommunications networks linking several South American cities with Washington, D.C., and national networks within six other nations. (NOAA Release 72-94)

- Dr. Robert C. Seamans, Jr., Secretary of the Air Force, discussed future air superiority aircraft in speech before Aero Club of Washington in Washington, D.C. Air Force had not developed new fighter aircraft that could maintain air superiority in battle area in 20 yrs, while U.S.S.R. had developed "whole family of sophisticated fighters." F-15 program should give U.S. "what we need for the air superiority mission." For allies, U.S. had developed F-5E international fighter, "a simple design that cost about half that of our present long-range, multipurpose aircraft." It would "play a particularly vital role in improving the self-defense capabilities of our Asian allies. . . . In modernizing our forces, we have learned an important lesson in the last few years. You cannot build an aircraft with every possible new technological device and expect to buy enough of them to get the job done. . . . The unit cost of high performance aircraft has increased tenfold every 18 years." Avionics was factor in driving up cost of new aircraft systems. F-111D included Mark II avionics system costing over $4 million. To lower unit cost of systems, Air Force was tailoring new aircraft for "somewhat narrower purposes." Through specialization "we hope to obtain aircraft that can both survive and perform more effectively in their individual mission and . . . we will be able to afford many more of these aircraft than we could obtain in the case of one super-capable, multipurpose plane." (Text)

- Mackay Trophy for 1971 was awarded to L/C Thomas B. Estes (USAF) and L/C Dewain C. Vick (USAF) by Air Force Chief of Staff, Gen. John D. Ryan, in Washington, D.C., ceremony. Award was for April 27, 1971, 24 000-km (15 000-mi) nonstop flight at speeds over mach 3 in Lockheed SR-71 strategic reconnaissance aircraft. (NAA News, 8/72, 1)

- General Accounting Office reported increase of $28.7 billion in original cost estimates of 77 major U.S. weapon systems. Report was based on latest quarterly figures supplied by Dept. of Defense. GAO said it had analyzed 78 major weapon systems but had not included proposed Trident long-range missile-firing submarine because project had not yet been given final approval. Overrun was 31%, down from 1971 overrun of 40%. (AP, W Star & News, 7/18/72, A6)

- U.S. Patent No. 3 678 180 was issued to Donald S. Bond of Radio Corp. of America (RCA) Electronics Div. for system to deliver third-class mail to multiple addresses on mailing lists via radio facsimiles relayed by satellite. (NYT, 7/22/72, 37; Pat Off PIO)

- U.S. Patent No. 3 677 502 was issued to Soviet aircraft designer Aleksey A. Tupolev and seven associates for Tu-144 Soviet supersonic aircraft. Tupolev had carried on work of his father Andrey N. Tupolev on Tu-144. Earlier applications had been filed for Soviet patents and other American patents were pending on specific features of aircraft. Patent indicated that aircraft design was intended to prevent engine exhaust from contacting tail portion of fuselage. (Jones, NYT, 7/22/72, 33; Pat Off PIO)

July 18

- NASA announced appointment of J. Lloyd Jones, Jr., as Director of Aerodynamics and Vehicle Systems Div., Office of Aeronautics and Space Technology, effective Aug. 6. Jones had been Research Assistant to Director of Ames Research Center since 1970. (NASA Release 72-147)
- Flight Research Center announced selection of RCA Service Co. for final negotiation of contract to support NASA Aerodynamic Test Range—radar/communication sites 550 km (340 mi) apart that provided high-speed flight corridor 1850 km (1150 mi) long from mid-Washington state to Edwards, Calif. Negotiations were for four-year, $359 000 cost-plus-fixed-fee contract with four one-year options. (FRC Release 13-72)
- Office of Senate Republican Leader Hugh Scott (R-Pa.) said Sen. Scott would nominate President Nixon for Nobel Peace Prize to be awarded in winter 1972. Scott aide said Senator was preparing documentation citing President's initiatives in People's Republic of China visit and Soviet summit meetings. (UPI, *NYT*, 7/19/72, 15)
- Lombardo Mint's *Apollo 16* medal was among "most exciting" commemorative issues on mission offered, Baltimore *Sun* reported. Obverse featured lunar roving vehicle (LRV) patrolling Descartes region of moon, with view of earth on moon's horizon and portraits of *Apollo 16* crew on lower portion. Legend read: "April 16, 1972—The Descartes Region Mission—Thomas K. Mattingly, John W. Young, Charles M. Duke, Jr." Reverse pictured official *Apollo 16* shoulder patch superimposed over lunar surface. (Gould, B *Sun*, 7/18/72)

July 19: U.S.S.R. launched *Cosmos 503* from Plesetsk into orbit with 308-km (191.4-mi) apogee, 170-km (105.6-mi) perigee, 89.2-min period, and 65.4° inclination. Satellite reentered Aug. 1. (GSFC *SSR*, 7/31/72; 8/31/72; *SBD*, 7/21/72, 104)

- NASA announced Marshall Space Flight Center award of contracts and fund transfers totaling more than $49 million for scientific experiments for first High Energy Astronomy Observatory (HEAO), scheduled for launch in 1975. Receiving awards were Columbia Univ., $8 929 059; Massachusetts Institute of Technology, two awards of $7 821 720 and $5 357 245; Univ. of California at San Diego, $5 150 000; California Institute of Technology, Washington Univ. of St. Louis, and Univ. of Minnesota, $6 198 683 for joint experiment; Naval Research Laboratory, $7 919 000; and Goddard Space Flight Center, $8 250 000. (NASA Release 72-145)
- Univ. of Tokyo Institute of Space and Aeronautical Science announced plans to launch radio explorer satellite from Kagoshima Prefecture Aug. 16. The 75-kg (165-lb) satellite, 68 cm (2 ft) long, would be boosted by Mu-4S launch vehicle into elliptical orbit, where it would study cosmic radiation. (Kyodo, FBIS-Japan, 7/20/72, C7; *SBD*, 7/21/72, 101)

July 20: U.S.S.R. launched eight Cosmos satellites from Plesetsk with single booster.

Cosmos 504 entered orbit with 1497-km (930.2-mi) apogee, 1323-km (822.1-mi) perigee, 113.9-min period, and 74° inclination.

Cosmos 505 entered orbit with 1498-km (930.8-mi) apogee, 1354-km (841.4-mi) perigee, 114.3-min period, and 74° inclination.

Cosmos 506 entered orbit with 1498-km (930.8-mi) apogee, 1384-km (860-mi) perigee, 114.6-min period, and 74° inclination.

Cosmos 507 entered orbit with 1497-km (930.2-mi) apogee, 1414-km (878.6-mi) perigee, 114.9-min period, and 74° inclination.
Cosmos 508 entered orbit with 1497-km (930.2-mi) apogee, 1445-km (897.9-mi) perigee, 115.3-min period, and 74° inclination.
Cosmos 509 entered orbit with 1500-km (932.1-mi) apogee, 1475-km (916.5-mi) perigee, 115.6-min period, and 74° inclination.
Cosmos 510 entered orbit with 1512-km (939.5-mi) apogee, 1496-km (929.6-mi) perigee, 116-min period, and 74° inclination.
Cosmos 511 entered orbit with 1547-km (961.3-mi) apogee, 1496-km (929.6-mi) perigee, 116.4-min period, and 74° inclination. (GSFC *SSR*, 7/31/72; *Sov Aero*, 7/24/72, 30)

- Third anniversary of first manned landing on moon. NASA's *Apollo 11* mission had proved that man could carry out scientific and technological work on another planet and had fulfilled U.S. goal established in 1961 of developing technology to transport man to moon and return him safely to earth within 1960s. Since Astronauts Neil A. Armstrong and Edwin E. Aldrin, Jr., spent 2 hrs 31 min on moon's Sea of Tranquility July 20, 1969, NASA had launched five more Apollo missions. Eight men had explored moon at four different sites for 55 hrs 34 min. Four geophysical stations placed on moon were transmitting scientific data, 272 kg (600 lbs) of lunar samples had been brought to earth, 16 different scientific experiments had been conducted on lunar surface 48 times, and 17 different experiments had been conducted in lunar orbit 41 times. Apollo 17, last mission in program, was scheduled for launch Dec. 6.

  Manned Spacecraft Center expanded regular public open house activities with special Mission Control Center program re-creating radio transmissions between *Apollo 11* crew and Mission Control Center during lunar landing, duplicating flight controller displays, and showing motion pictures taken during mission.

  Neil Armstrong Air and Space Museum in Wapakoneta, Ohio, home of *Apollo 11* astronaut who was first man on moon, held informal opening to mark anniversary. President Nixon's daughter, Tricia N. Cox, presented moon rocks to museum, which would be formally dedicated in early autumn when all exhibits had been completed. (NASA Release 72-144; MSC Release 72-151; Wapakoneta *Daily News*, 6/16/72; *H Post*, 7/21/72)

- NASA and National Science Teachers Assn. (NSTA) announced approval of experiments proposed by 19 high school students from 16 states for Skylab space station. Experimenters were from 25 finalists selected from 3409 proposals submitted to NASA by U.S. secondary school students as part of Skylab Student Project. Six other finalists could not be accommodated because of Skylab performance requirements and schedule constraints. Medallions would be awarded to finalists, sponsors, and schools during Skylab Educational Conference at Kennedy Space Center at Skylab launch time. (NASA Release 72-146)

- Senate Committee on Foreign Relations unanimously approved agreements on strategic arms control concluded by U.S. and U.S.S.R. in Moscow May 26. (*CR*, 7/20/72, D796; Finney, *NYT*, 7/21/72, 2)

- *Apollo 16* Astronaut Thomas K. Mattingly II received Distinguished Service Medal and astronaut wings from Secretary of the Navy

John W. Warner in Washington, D.C., ceremony. Mattingly was Navy commander. (DOD Release 526-72)

- Manned Spacecraft Center announced three contract awards. Grumman Aerospace Corp. had been awarded 15-mo, $200 000 fixed-price, research and development contract to continue to design, build, and test space shuttle orbiter heating systems. Singer Co. had been awarded $135 530 firm-fixed-price contract to study and define simulator requirements for space shuttle mission simulator. McDonnell Douglas Corp. Astronautics Co. had received $181 000 contract for design study of orbital maneuvering system (OMS) to maneuver space shuttle orbiter in space. (MSC Releases 72-157, 72-158, 72-159)

- Syndicated columnist William Hines criticized NASA earth resources survey program in Chicago *Sun-Times*. Program's rationale "leaps from an insupportable assumption (namely, that everything can be viewed better from space than close up) to an unjustifiable conclusion (that even if this is true the multi-purpose, multi-disciplinary satellite is the best way to do it)." (C *Sun-Times*, 7/20/72)

July 21: Saturn V 1st stage (S-IC-513), scheduled to boost Skylab into earth orbit in 1973, left Michoud Assembly Facility on board NASA barge *Orion*. It would arrive at Kennedy Space Center July 26. Stage had been turned over to NASA by Boeing Co. at Michoud ceremony. (MSFC Release 72-91)

- Jet Propulsion Laboratory announced formation of new flight project office for unmanned mission to Jupiter and Saturn in 1977. Mariner Jupiter/Saturn 1977 project would be managed by Harris M. Schurmeier, with Dr. Edward J. Smith as acting project scientist, Raymond L. Heacock as spacecraft system manager, and Dr. Ralph F. Mills as mission analysis and engineering manager. (JPL Release 621)

- Explosion and flash fire at Newhall, Calif., plant of Space Ordnance Systems, Inc.—manufacturers of Apollo spacecraft components—injured six workers. Facility made explosive bolts used to separate Apollo lunar module ascent stage from descent stage on liftoff from moon. Explosive substance causing fire was identified only as "commercial product." (*W Post*, 7/22/72, A4; NASA proj off)

- Senate passed and cleared for House S.R. 193 to return Cape Kennedy to original name of Cape Canaveral. (*CR*, 7/21/72, S11438-40)

- Award of cost-sharing contracts totaling nearly $9 million to Boeing Co. and McDonnell Douglas Aircraft Co. to determine feasibility of retrofitting jet aircraft with noise-reduction devices was announced by Federal Aviation Administration. Cost of Boeing contract was $3 365 000, of which Government would provide $2 771 549. Cost of McDonnell Douglas contract was $5 600 000, with Government share $2 771 549. (FAA Release 72-141)

- Belgian artist Paul Van Hoeydonck, who had sculpted small statue "Fallen Astronaut" that was left on moon by *Apollo 15* astronauts [see April 18], was selling 950 copies of statue at $750 each through New York art dealer Waddell Gallery, Washington *Evening Star and Daily News* reported. NASA spokesman had said that astronauts had carried statue to moon with agency's approval and felt they had "gentlemen's agreement" with Van Hoeydonck that precluded any commercialization of venture. Spokesman said he knew of no regula-

tions violated by astronauts in moon sculpture incident. (Delaney, W *Star & News*, 7/21/72, 1)
- European Launcher Development Organization (ELDO) signed contract with Hawker Siddeley Dynamics Ltd. for two additional Blue Streak first stages for Europa II launch vehicles. (*SF*, 9/16/72, 361)
- National Science Foundation published *Changes in Graduate Programs in Science and Engineering, 1970–1972 and 1972–1974* (NSF 72–311). Ratio in 1970–1972 of net additions in doctoral programs in science and engineering to existing doctoral departments was 1 to 26. Plans for 1972–1974 indicated growth rate would be cut to 1 to 66. Top 20 universities showed no net change in science and engineering doctoral programs in 1970–1972; developing universities showed net increase equivalent to about 1 per 2 universities. Planning for 1972–1974 indicated no change for top 20. During 1970–1972 greatest increase in doctoral programs was in computer science and psychology. In 1972–1974 plans computer science was field with greatest expected relative increase. (NSF *Highlights*, 7/21/72)

July 22: U.S.S.R.'s *Venus 8* spacecraft, launched March 27, reached atmosphere of Venus after 117-day flight covering 482 million km (300 million mi). Tass said descent module separated from spacecraft and descended to surface of Venus by parachute, softlanding at 2:29 pm Baykonur time (5:29 am EDT). "In the process of aerodynamic deceleration its speed dropped from 11.6 kilometers per second [7.2 miles per second] to 250 metres per second [820 feet per second]. Studies of the atmosphere and surface layer of the planet were conducted during the parachute descent for the fifty minutes after landing. Carried out for the first time were experiments to determine brightness, pressure and temperature in the atmosphere and on the surface of the planet on its day side. Data has been obtained on the nature of the rocks of the planet's surface layer."

Venus 8 was second Soviet spacecraft to complete softlanding mission to Venus successfully. First was *Venus 7*, which had landed capsule on Venus Dec. 15, 1970, that had transmitted data from surface for 23 min. (FBIS–Sov, 8/24/72, L1; *SBD*, 7/25/72, 118; GSFC *SSR*, 7/31/72)

- Report under preparation by General Accounting Office would urge Dept. of Defense to declassify more of technology developed by DOD research, *Business Week* reported. Report also would recommend that Office of Management and Budget establish technology transfer as Government-wide policy. Magazine said report would be "rousing vote of confidence" in and documentation for NASA technology transfer program. (*Bus Wk*, 7/22/72)
- NASA spokesman said agency was tightening flight rules for Apollo 17 mission scheduled for Dec. 6, following commercialism of souvenir articles carried on previous missions. Personal preference kits in which astronauts carried articles to distribute to family and friends might be eliminated. (AP, W *Star & News*, 7/22/72)
- Associated Press reported *Apollo 15* crew had been favored to receive Air Force Gen. Thomas D. White Space Trophy but that trophy would be withheld because of unauthorized postal cover incident [see July 11]. Trophy was awarded annually for most outstanding contribution to U.S. aerospace progress. AP said *Apollo 15* Astronauts David R. Scott,

July 23–25: NASA's Erts 1 *Earth Resources Technology Satellite was launched into near-polar earth orbit and began taking multispectral images for studies in agriculture, forestry, mineral, and land resources; land use; water and marine resources; mapping and charting; and environment. The flight configuration above showed the solar panels deployed after testing at the* GE *Valley Forge plant before launch. On the opposite page, the first* Erts 1 *photos were reviewed by Dr. John F. Clark, Goddard Space Flight Center Director; Dr. James C. Fletcher,* NASA *Administrator; and Charles W. Mathews,* NASA *Associate Administrator for Applications, in Dr. Fletcher's office.*

Alfred M. Worden, and James B. Irwin also had been invited to annual meeting of Fédération Aéronautique Internationale in Paris in October to receive other awards with Soviet cosmonauts and that Dr. James C. Fletcher, NASA Administrator, had advised FAI that efforts would be made to have astronauts accept invitation. (AP, W *Star & News*, 7/23/72, A23)

- Federal Aviation Administration announced award of $119 249 contract to Parsons, Brinckeroff, Quade and Douglas, Inc., for architectural and engineering plans for 600-by-90-m (2000-by-300-ft) elevated STOLport for short takeoff and landing aircraft. Exterior design of structure

supporting STOLport would be planned to provide approaching pilot with realistic cues resembling those included in actual landing facility on top of a multilevel structure. (FAA Release 72-142)

- Grove Webster, NASA Director of Personnel, died at age 48 after long illness. Webster had joined NASA in 1959 after working for Air Force and Civil Service Commission. (NASA Special Ann)

July 23-25: Erts 1 (ERTS-A) Earth Resources Technology Satellite was launched by NASA from Western Test Range at 11:06 am PDT by 2-stage Thor-Delta booster with nine strap-on rockets. Satellite entered circular, near-polar, sun-synchronous orbit with 906.9-km (563.6-mi) apogee, 899.7-km (559-mi) perigee, 103.2-min period, and 99.1° inclination. Primary objective was to acquire for three months synoptic, multi-spectral repetitive images from which useful data would be obtained for investigations in agriculture and forestry resources, mineral and land resources, land use, water resources, marine resources, mapping and charting, and environment. As secondary objectives *Erts 1* would acquire complete, largely cloud-free coverage of U.S. with multispectral scanner (MSS) or return-beam-vidicon camera (RBVC) system, or both; acquire coverage over major earth land masses with MSS or RBVC; and demonstrate relay of data from remote ground-based platforms.

The 941-kg (2075-lb), butterfly-shaped satellite carried RBV camera system, MSS subsystem, data-collection system (DCS), and two wide-band video tape recorders (WBVTR). Launch was first step in merging space and remote sensing technology to manage earth resources more efficiently. Satellite would be able to cover earth with 500 pictures—1000 times fewer than 500 000 needed to cover earth from high-altitude aircraft. Each picture would cover 34 000 sq km (13 000 sq mi). Satellite would view 185-km (115-mi) strip of earth with global coverage every 18 days, crossing equator daily at about 9:30 am local time.

Solar array lock-on to sun and three-axis stabilization were achieved as planned. By July 25 instruments had been turned on and were operating satisfactorily. Initial imagery from RBVC system and MSS was excellent and showed detail.

Data were being transmitted to stations at Fairbanks, Alaska; Goldstone, Calif.; and Greenbelt, Md.; and would be sent to data-processing facility at Goddard Space Flight Center. Ground station at Prince Albert, Saskatchewan, constructed and operated by Canadian government, would also receive data. Data would be distributed in form of high-quality film images or digitized data on computer-readable magnetic tape.

Unlike previous unmanned programs where principal investigators had own instrumentation on board spacecraft and were responsible for own data analysis, all *Erts 1* investigators had access to all data from onboard instruments. The 300 investigators from 43 states, District of Columbia, 31 foreign countries, and two international organizations would be assisted by 25 specialists assigned by NASA to analyze results. To carry out research required to evaluate satellite techniques, NASA was working with user agencies—including Departments of Agriculture, Commerce, and Interior, Environmental Protection Agency, and U.S. Corps of Engineers—and state and local organizations. Extensive cooperative earth resources projects had been established with Canada, Brazil, and Mexico, using *Erts 1* and NASA earth observations aircraft.

ERTS program was managed by GSFC under direction of NASA Office of Applications. (NASA proj off; NASA Release 72-137)

July 23: NASA Technology Utilization Office representatives met in San Francisco with Public Technology Inc. officials and representatives of cities to review two-year joint effort to apply new technology to selected city problems. Discussed were improved protective clothing and equipment, improved detection and locating equipment for underground piping, equipment to determine presence of illegal drugs in human blood, and improved pavement patching materials that used waste matter. Meeting also discussed civil cooperation to form mass market for goods and services produced by application of new technology. Meeting prefaced July 24-26 second annual Urban Technology Conference at which NASA, other Federal agencies, and business and industry exhibited contributions to new technology. (NASA Release 72-154)

- *Apollo 15* Astronaut James B. Irwin said in telephone interview with Baptist Press that *Apollo 15* crew "thought they were doing the best for our families" in carrying unauthorized postal covers to moon. Their decision not to accept share of proceeds from sale of covers had come about eight months before incident was reported in press [see June 18]. "We acted in haste and under the terrific pressures of the pre- and post-flight schedule—but that does not excuse it." NASA had had "no choice but to reprimand us." (AP, B *Sun*, 7/24/72, A5)

July 24: Dr. George R. Carruthers of Naval Research Laboratory received NASA Exceptional Scientific Achievement Medal at special NASA Hq. ceremony. Dr. Carruthers—outstanding black astrophysicist who had developed rocket-borne instrument that detected molecular hydrogen in interstellar space for first time in 1970—was cited for development of far-ultraviolet camera-spectrograph for *Apollo 16* mission. Instrument had taken nearly 200 pictures of earth, Milky Way, and other galaxies in far-UV wavelengths not observable from earth and was providing new information on structure of universe. (NASA Release 72-149)

- People's Republic of China signed preliminary agreement to buy two Anglo-French Concorde supersonic transport aircraft. Agreement was announced later at Paris press conference by Aérospatiale President Henri Ziegler, who said aircraft were for delivery in mid-1977. (Samuelson, *W Post*, 7/25/72, D8)

- Navy and Dept. of Defense officials were quoted in *Wall Street Journal* as saying preliminary investigation by Navy examining board into June 30 crash of F-14 Navy fighter aircraft had indicated accident was not caused by "anything wrong" with aircraft. (Levine, *WSJ*, 7/27/72, 6)

- Poll of 123 persons in 6 Moroccan villages had shown 63% thought U.S. lunar landings were hoax or had doubts about them, *Washington Post* reported. Of those polled, 88% said they had heard of manned lunar missions since 1969. Of 20 persons between ages of 14 and 25, 55% believed reports of moon landings; only 25% of 20 persons between ages of 61 and 90 thought reports were true. Of villagers who believed men had gone to moon, 20% thought landings had been made by Russians. (Aubin, *W Post*, 7/24/72, A14)

July 24–26: Second annual Urban Technology Conference and Technical Display was held in San Francisco, NASA exhibited new building trades materials, including flat electrical conductor cables and low-voltage switching circuits that had been developed for space use, fire resistant materials, and earth-resources surveillance display including cameras and sensors. Also shown were samples of pavement patching materials, medical aids, psychomotor response tester for use by traffic departments and educational institutions, and devices for checking blood circulation, electrocardiograph readings, and circulation in humans. Guest speakers included William M. Magruder, Special Counsel to President, and former astronaut Walter M. Schirra, Chairman of Environmental Control Co. Jeffrey T. Hamilton of NASA Office of Technology Utilization presented paper describing mechanisms evolved by NASA to transfer aerospace technology to nonaerospace use. Mechanisms ranged from specialized dissemination systems of published technologies to attempts to solve specific problems in public sector areas like biomedicine, environmental protection, fire safety, building construction, transportation, and law enforcement. (Program; NASA Release 72-154)

July 25: M2-F3 lifting body, piloted by William H. Dana, completed 14th flight from Flight Research Center after air-launch from B-52 aircraft. Objective of powered flight was to check out command augmentation system and engine accumulator modification. M2-F3 reached 17 300-m (57 000-ft) altitude and mach 0.86. Flight was first for M2-F3 since its removal from flight status in January for installation of command augmentation fly-by-wire system. (NASA proj off)

- Manned Spacecraft Center announced award of 10-mo, $98 000 contract extension to Martin Marietta Corp. Denver Div. for continued design of crew compartment configured for space shuttle, which would be

tested in neutral buoyancy tank. Extension brought total value of contract to $173 670. (MSC Release 72-161)
- Federal Aviation Administration had approved airport planning project funding of $9 061 104 for 181 projects in 39 states under its Planning Grant Program for FY 1971, Secretary of Transportation John A. Volpe announced. FY 1971 funding had been $3 636 035 for 42 projects in 27 states. (FAA Release 72-147)
- *Los Angeles Times* editorial commented on *Erts 1* Earth Resources Technology Satellite (launched July 23): "ERTS gives every sign of being an achievement of enormous potential. It is a demonstration of the value of the U.S. space program in applying technology both to extending the frontiers of knowledge, and in contributing to the improvement of life on earth." (*LA Times*, 7/25/72)

July 26: NASA selected North American Rockwell Corp. (NR) Space Div. for negotiation of six-year, $2.6-billion cost-plus-fixed-and-award-fee contract as prime contractor to begin developing space shuttle orbiter. Increment covering first two years was $540 million. NR—selected from four firms submitting proposals—would be responsible for design, development, and production of orbiter vehicle and for integration of all shuttle elements. Shuttle main engine was being developed by NR Rocketdyne Div. under earlier contract. External tank and solid-fueled rocket boosters would be procured by NASA after system engineering for orbiter had progressed sufficiently.

Shuttle, first reusable space vehicle, would have delta-winged, airplane-like orbiter able to land on conventional runways. It would have cargo compartment 18 m (60 ft) long and 4.5 m (15 ft) in diameter and would be able to place 29 500 kg (65 000 lbs) into 185-km (115-mi) due-east orbit. Reusable solid-propellant booster rockets and orbiter high-pressure, liquid-oxygen and liquid-hydrogen main engines would boost orbiter into space and booster rockets would detach at 40-km (25-mi) altitude and splash down in ocean for recovery and reuse. Orbiter, under own power, would continue into low earth orbit. Shuttle would be able to place satellites in orbit, return satellites from orbit, permit in-orbit repair and servicing of satellites, deliver propulsive stages and satellites to low earth orbit, and make short-duration science and applications missions with self-contained experiments in low earth orbit. Shuttle development schedule called for horizontal test flights beginning in 1976 and manned orbital test flights beginning in 1978, with complete system operational by 1980. Employment generated by orbiter development and shuttle integration was expected to reach 15 000 by 1975–1976 and then would gradually decrease. (NASA Release 72-153)

- Second in series of very-high-altitude flight tests of Viking parachute system was conducted over White Sands Missile Range. Helium-filled balloon lifted simulated Viking entry vehicle from Roswell, N. Mex., to 11 200-m (36 880-ft) altitude where payload was dropped and rocket motors ignited, boosting payload to 41 000-m (135 000-ft) altitude. Parachute deployed as planned and payload landed at WSMR about 56 km (35 mi) northwest of Holloman Air Force Base. Objective—to check parachute system at transonic conditions and at lowest parachute loading conditions expected over Mars—was met. Successful completion of three tests—one supersonic, one transonic, and one subsonic—

would qualify Viking parachute system for use on Mars mission scheduled for launch in 1975. [See July 11, Aug. 13 and 19, Dec. 14.] (NASA Release 72–118; NASA proj off)
- Skylab medical experiments altitude test (SMEAT) to collect medical data and evaluate medical equipment began at Manned Spacecraft Center. Astronauts Robert L. Crippen (commander), Karol J. Bobko (pilot), and Dr. William E. Thornton (science pilot) entered altitude chamber 6 m (20 ft) in diameter with atmosphere 70% oxygen and 30% nitrogen at 34 500 newtons per sq m (5 psi), where they would live and work for up to 56 days. Skylab, scheduled for launch in April 1973, would support three-man crews in earth orbit for total of 130 days. (NASA Release 72–131; UPI, *NYT*, 7/27/72, 18)
- President Nixon and Israeli Prime Minister Golda Meir exchanged remarks in telephone call inaugurating communications satellite TV service for Israel via ground station at Emeq Ha'ela. White House later made available text of letter written by President to Prime Minister Meir July 20: "As your nation inaugurates its station connecting Israel with the Intelsat system, you have the best wishes of the American people. Clear and effective communication is basic to building the generation of peace we so earnestly seek, and to ensure the kind of sound relationships among nations which will enable that peace to endure." (*PD*, 7/31/72, 1175–6)
- Intense solar storm in solar region 331 was observed by sensors on NASA's *Oso 7* Orbiting Solar Observatory. Warnings were issued July 28. [See Aug. 2, 7, 13; and During October.] (NASA Release 72–164)
- Experiments at U.S.S.R.'s Radio Physics Institute had suggested that upper Mars mantle was hard and porous like earth dendrites, Tass reported. Report on study had said small, isolated dust regions were source of Mars dust storms and there was no uninterrupted dust mantle. Dust over Martian surface was no more than 1 mm (0.04 in) thick. White polar caps were reported to be hard-frozen carbon dioxide. (FBIS–Sov, 7/31/72, L1)
- Manned Spacecraft Center announced appointment of Astronaut David R. Scott as Technical Assistant to Manager of Apollo Spacecraft Program, effective immediately. Scott, recently reprimanded for carrying unauthorized postal covers to moon on *Apollo 15* mission, succeeded Ronald W. Kubicki, who had been named Manager for Command, Service and Lunar Modules. (MSC Release 72–162)
- *New York Times* editorial praising *Venus 8* achievement [see July 22] said planetary exploration would be "ideal area" for U.S.–U.S.S.R. cooperation, "since it poses none of the political or human problems involved in cooperation in manned space ventures such as the docking maneuver now scheduled for 1975. Both Soviet and American planetary probes would be better if the two nations pooled information and experience on the design and construction of these vehicles. Such pooling would seem to be a natural step in the evolution of Moscow-Washington space cooperation. The sooner it is realized the more effective and less expensive both nations' space exploratory efforts are likely to be." (*NYT*, 7/26/72, 34)

July 27: House and Senate conferees submitted conference report on H.R. 15093, FY 1973 Dept. of Housing and Urban Development-space-science-veterans appropriations bill that contained $3.408-billion NASA appropriation. Conference recommended $2.601-billion appropriation for NASA research and development, instead of $2.550 billion proposed by House and $2.624 billion proposed by Senate, and included $24 million for aeronautical research in noise abatement and aviation safety as proposed by Senate. Recommended appropriation for construction of facilities was $77.3 million as proposed by Senate, instead of $69.8 million proposed by House—and included $5.5 million for modification of manufacturing and final assembly facilities for space shuttle as proposed by Senate and $8 million for facility planning and design as proposed by Senate, instead of $6 million proposed by House. Appropriation for research and program management remained unchanged at $729.45 million. (H Rpt 92–1261; NASA Off Administration)

- NASA launched Aerobee 170 sounding rocket from White Sands Missile Range, carrying Harvard College Observatory solar physics experiment. Rocket and instrumentation performed satisfactorily. (SR list)
- Grumman Aerospace Corp. President Joseph G. Gavin, Jr., said in interview that 1500 employees might lose jobs because of award of NASA space shuttle contract to competitor, North American Rockwell Corp. [see July 26]. First layoffs at Bethpage, N.Y., plant might come in three weeks, with 200 employees affected. Labor officials later noted that action could affect as many as 8000 workers, including fringe concerns. (Andelman, *NYT*, 7/28/72)
- Jet Propulsion Laboratory announced appointment of Dan Schneiderman, former manager of Mariner Mars 1971 Project, as manager of Civil Systems Program Office. He would manage JPL effort to develop applications of laboratory capabilities to problems in medical engineering, public safety, urban land use, and transportation. (JPL Release 622)
- Charles A. Lindbergh discussed supersonic transport aircraft in *New York Times* article: "For me, aviation has value only to the extent that it contributes to the quality of the human life it serves. Research in the fields of supersonic flight is, obviously, of great importance and should continue, but my personal conclusion is that the regular operation of SST's in their present state of development will be disadvantageous both to aviation and to the peoples of the world. I believe we should prohibit their scheduled operation on or above United States territory as long as their effect on our over-all environment remains unsatisfactory." (*NYT*, 7/27/72, 31)
- House passed H.R. 1026 requesting President to proclaim Feb. 19, 1973, "Nicolaus Copernicus Day," marking quinquecentennial of birth of Polish astronomer Copernicus. (*CR*, 7/27/72, D833)

July 28: U.S.S.R. launched *Cosmos 512* from Plesetsk into orbit with 273-km (169.6-mi) apogee, 202-km (125.5-mi) perigee, 89.2-min period, and 65.3° inclination. Satellite reentered Aug. 9. (GSFC *SSR*, 7/31/72, 8/31/72; *SBD*, 8/1/72, 163)

- NASA's TF–8A aircraft, equipped with supercritical wing and piloted by Thomas C. McMurtry, completed first flight with new fuselage fairings. Aircraft reached mach 0.95, 0.97, and 0.99 at altitudes between 13 470 and 13 960 m (44 200 and 45 800 ft) and constant dynamic pressure of 9600 newtons per sq m (200 psf). Pressure distribution data were obtained and 2 g wind-up turns to measure stability and control characteristics were made. Acceleration and deceleration runs were made at 10 700-m (35 000-ft) altitude from mach 0.86 to 1.10 with maximum dynamic pressure of 20 110 newtons per sq m (420 psf). Aircraft was flown with generators off in configuration planned for future flights. (NASA proj off)

- NASA launched four-stage Trailblazer II reentry vehicle from Wallops Station to test chemical-injection method of eliminating radio blackout during reentry. Vehicle, launched for Air Force Cambridge Research Laboratories, carried 34-kg (75-lb) payload to 322-km (200-mi) altitude over Atlantic. Payload injected chemical into plasma surrounding vehicle during reentry and measured effectiveness of injection in eliminating blackout. Objective of AFCRL Trailblazer II program was to study techniques for improving transmission and reception of radio signals from aerospace vehicles during reentry. (WS Release 72-8)

- Preliminary data from U.S.S.R.'s *Venus 8* spacecraft, which landed on planet Venus July 22, was published in Moscow. Landing site temperature was 738 K (465°C) and pressure was 93 atmospheres. Measurements of substances in clouds during descent suggested Venusian surface was undergoing "stormy volcanic action" and demonstrating same evolutionary characteristics as earth had many millions of years ago, which could explain high surface temperatures. *Venus 8*'s ability to transmit from surface 27 min longer than *Venus 7* had was attributed to freezing of descent module by ground command just before descent. Ground controllers had transmitted command to freeze module "from inside," making it "very much like an ice chest" so that outside temperature was felt gradually irrespective of thermal pressure. (FBIS-Sov, 7/28/72, L5; SBD, 7/26/72, 125; 7/31/72, 151)

- NASA announced issuance to scientists around the world of requests for proposals for new investigations based on data already acquired by Orbiting Geophysical Observatory (OGO), Orbiting Solar Observatory (OSO), and Explorer satellites. NASA would provide limited support to new investigations, possible because of satellites' extended lifetimes and increased volume of available data. Proposals were due by Oct. 31. Funding for selected U.S. proposals would be available April 1, 1973. Foreign investigators would make own arrangements without cost to NASA. (NASA Release 72-152)

- Protocol on U.S.-U.S.S.R. Joint Commission on Scientific and Technical Cooperation, established under May 26 agreement, was signed by Dr. Edward E. David, Jr., Presidential Science Adviser, at White House ceremony attended by President Nixon. Protocol was signed simultaneously in Moscow by Vladimir A. Kirillin, Deputy Chairman of U.S.S.R. Council of Ministers. (PD, 7/31/72, 1187; Tass, FBIS-Sov, 7/31/72, G1)

- International Aerospace Hall of Fame announced 1972 selection of members: Gen. Henry H. Arnold, wartime commander of U.S. Army Air Forces; John K. Northrop, president and cofounder of Northrop Aircraft, Inc.; Sir Geoffrey de Havilland, British aircraft designer and founder of de Havilland Aircraft Co.; and Otto Lilienthal, German glider aircraft pioneer. Investiture ceremony would be held Oct. 14 in San Diego. (NASA Hist Off)

- Grumman Aerospace Corp. President Joseph G. Gavin, Jr., said he hoped to obtain $1 billion worth of subcontracts in space shuttle program. *New York Times* said later that this was more than three times as much as North American Rockwell, winner of $2.6-billion prime contract,

had tentatively allocated to all of New York State. (Witkin, *NYT*, 7/29/72)
- *Los Angeles Times* editorial commented on July 26 award of space shuttle contract to North American Rockwell Corp.: Argument over shuttle economics continued but "decision has already been made, as a practical matter, in favor of the shuttle. That is good news for the California economy in general, and aerospace workers in particular. But, more than that, it signals the country's decision to continue the exploration of space." (*LA Times*, 7/28/72)
- *Atlanta Journal Constitution* editorial lamented assignment of *Apollo 15* Commander David R. Scott to desk job because of carrying postal covers to moon: It was "bitter blow to one who flew so high" but was "warranted." Scott was "a brave, tough man who is honest enough to know that he did wrong, and moral enough to know that he deserves the punishment." (*Atlanta JC*, 7/28/72)

July 29: Dr. James C. Fletcher, NASA Administrator, discussed equal employment opportunities in space program before National Technical Assn. in Baltimore: "Space is working for all people, and nothing is going to change this happy circumstance." But "there is another side of the coin that we should examine: that is, the people who work for space." Matching broad participation of women and of representatives of minority groups "is not there." Percentage of women and minorities on NASA payroll reached "presentable figure" but, "if we take the professional staffs alone, the percentage drops sharply." Some NASA offices were almost all white and male; others had overabundance of minorities. "I am less concerned with how this came about than I am with applying the corrective. My sincere concern . . . is not only that space shall continue to work for all people: but also that all people shall participate in the space effort." (Text)
- NASA launched Aerobee 170 sounding rocket from White Sands Missile Range, carrying Harvard College Observatory solar physics experiment. Rocket and instrumentation performed satisfactorily. (SR list)
- New Orleans *Times Picayune* editorial commented on U.S.–U.S.S.R. space cooperation: "The U.S.–Soviet studies . . . have proceeded smoothly and professionally for almost two years, apparently signaling a detachment of space affairs from political rivalry. This is a significant and welcome advance, long sought by the U.S., for the promise of space could hardly be fulfilled if it only increased the range of surface-style aggressiveness." (*T Picayune*, 7/29/72)

July 30: *Intelsat-IV F-5*, launched by NASA June 13 for Communications Satellite Corp. on behalf of International Telecommunications Satellite Consortium (INTELSAT), entered commercial service over Indian Ocean. (ComSatCorp Operations Off; *SBD*, 8/2/72, 167)
- Intense solar storm observed by NASA's *Oso 7* Orbiting Solar Observatory July 26 emerged from behind sun on east limb and was immediately studied by scientists at ground-based observatories. Storm had created waves in earth's magnetic field and had affected power systems around world. *Oso 7* would be able to follow storm region for several days after storm disappeared behind sun and would provide early warning of possible new eruptions. (NASA Release 72-164)

July 31: Volunteers at Lewis Research Center were being trained to recognize odor intensity and character of jet engine exhaust gases. Project—

Jet Combustor Exhaust Study—was part of overall program at LeRC to determine emissions of pollutants from jet aircraft and was being conducted by LeRC and Arthur D. Little Co. Findings were expected to help in reducing odors around airports caused by jet engine exhaust. Other studies were checking possible correlation between combustion efficiency and odor intensity. Project director Helmut F. Butz said best instrument "for attacking the odor problem . . . is still the human nose, but the nose has to be trained." Volunteers, on cue, opened sniffing port in special odor chamber, took short whiff, and recorded what they smelled. Classes of chemical compounds responsible for odor would be determined by analytical tests. (LeRC Release 72–78)

- Sen. Clinton P. Anderson (D-N. Mex.) told press in Washington, D.C., that Senate Committee on Aeronautical and Space Sciences would include probe of *Apollo 14* crew's decision to carry 200 privately minted silver medals on Jan. 31–Feb. 7, 1971, mission in scheduled Aug. 3 hearings on commercial exploitation of *Apollo 15*. (AP, *Pasadena Star-News*, 8/1/72)
- Group of 60 Civil Air Patrol cadets from 32 states and Puerto Rico arrived at Marshall Space Flight Center for week-long space flight orientation course. Cadets would attend lectures on NASA programs, tour MSFC laboratories, and visit Alabama Space and Rocket Center. (MSFC Release 72–92)
- Manned Spacecraft Center announced award of 18-mo, $238 000, cost-plus-fixed-fee contract to United Aircraft Corp. Hamilton Standard Div. to design and build prototype waste collection system—usable by both males and females—for space shuttle orbiter. (MSC Release 72–168)

During July: U.S.S.R. launched Salyut spacecraft which failed to achieve orbit, *Aviation Week & Space Technology* reported. Magazine said spacecraft penetrated atmosphere, where it was detected by over-the-horizon radars. One of two 2nd-stage engines stopped firing early; second stopped at end of programmed firing time with fuel remaining. (*Av Week*, 4/9/73, 21)

- Use of Federal laboratories for research applied to national problems was advocated by electronics executive J. Ross Macdonald in *Technology Review* article: "We should recognize that the frontier, the challenge of our times, has moved from the outer space of the external world to the vastly more complex inner space of men's minds and their interaction. Our present federal scientific establishment is ill prepared to meet this challenge, and here lies the argument for a system of national laboratories better organized to help meet pressing national and worldwide needs, especially the greatest need of all: people who understand their own natures well enough to control themselves and their technological tools, moderate self-interest, fight irrationality, and set human goals above all else." (*Tech Rev*, 7–8/72, 10–1)
- L/G John W. O'Neil, Vice Commander of Air Force Systems Command since 1969, would retire from Air Force Sept. 1, AFSC *Newsreview* reported. President Nixon had nominated as successor to Gen. O'Neil, M/G Edmund F. O'Connor, who had also been nominated for promotion to grade of lieutenant general. (AFSC *Newsreview*, 7/72, 1)

August 1972

August 1: Launch of first Earth Resources Technology Satellite, *Erts 1*, and selection of prime contractor—North American Rockwell Corp.—for space shuttle system demonstrated "how we have begun to turn the space program . . . around, how we are returning to the home seas of space after 12 years of strenuous and highly successful effort to explore the Moon," Dr. James C. Fletcher, NASA Administrator, said in letter to NASA staff. NASA would continue new direction with 1973 Skylab launch, joint U.S.–U.S.S.R. mission in 1975, and first orbital flight of space shuttle in 1978. Additionally NASA would explore solar system with improved unmanned spacecraft, build and orbit improved unmanned observatories, and accelerate efforts to develop "basic new technology for the most promising space missions of the Eighties and Nineties. We will . . . maintain a balanced national space effort with our main area of concentration shifted from the Moon to Earth orbit." (*NASA Activities,* 8/25/72, 160–1)

- NASA and Soviet Academy of Sciences approval of recommendations of Joint Working Group on Space Biology and Medicine was announced by NASA. Group—headed by Dr. Charles A. Berry, NASA Director of Life Sciences, and Soviet Academician Dr. O. G. Gazenko—had met at Manned Spacecraft Center May 12–18 to continue exchange of experiences in manned space flight begun at first meeting in Moscow in October 1971.

    NASA had presented pre- and postflight medical requirements and flight crew health stabilization program for *Apollo 16* (launched April 16), mission's preliminary physiological results, and glossary of space medicine for mutual understanding of technical and scientific exchanges. U.S.S.R. had detailed medical findings from June 6–30, 1971, *Soyuz 11–Salyut 1* mission, in which three cosmonauts died; pre- and postflight clinical-physiological examination procedures for cosmonauts; and theoretical aspects of predicting physiological responses of crew during flight. Soyuz-Salyut data had indicated no deterioration of crew's physiological status or performance efficiency before accident. Inflight data were similar to data from previous Soyuz missions and to Apollo data and were consistent with findings by both countries that general adaptive process occurred during weightlessness. U.S.S.R. had confirmed cosmonauts' deaths were due to hypoxia and gaseous embolism (dysbarism), caused by rapid decompression of landing capsule some half hour before return to earth. Review and evaluation of 24-day Soviet mission had indicated no need to modify medical aspects of Skylab plans.

    Joint Working Group had agreed to develop common medical examination procedures to compare U.S. and Soviet pre- and postflight data on crew body functions and increase information on physiological responses to space flight. Group would exchange correspondence on orthostatic tolerance (cardiovascular response), vestibular measurements, exercise-working capacity, and biochemical examination of body fluids and would discuss these areas in depth at Moscow meeting early in 1973, (NASA Release 72-157; Text)

- Marshall Space Flight Center announced completion of design certification reviews on Saturn boosters to be used on Skylab missions. DCRs on Skylab payload—including Orbital Workshop, airlock module, multiple docking adapter, and Apollo Telescope Mount—would be completed within three months. (MSFC Release 72-96)
- Photos from Earth Resources Technology Satellite *Erts 1* (launched July 23) would be available to public through imagery dissemination centers at Depts. of Interior, Commerce, and Agriculture, NASA announced. Public and private sector organizations participating in NASA Technology Utilization program could obtain photos, analyses, and background material from six NASA-sponsored regional dissemination centers. (NASA Release 72-158)
- Grumman Corp. officials said layoff of 300 employees, from top engineers to maintenance and security personnel, would begin Aug. 4 as direct result of NASA's award of space shuttle contract to North American Rockwell Corp. (Andelman, *NYT*, 8/2/72)
- *Apollo 15* astronauts had been "disciplined out of the astronaut corps" for ignoring warning by Chief of Flight Crew Operations Donald K. Slayton not to carry items to moon for commercial gain, *Washington Post* said. Slayton's warnings had been in personal briefings to crew and memorandum to all astronauts. Warnings had been inspired by discovery that Franklin Mint had profited from melting down coins taken to moon by *Apollo 14* astronauts during Jan. 31–Feb. 9, 1971, mission, mixing their molten silver with other metals, and producing 130 000 coins for sale. (O'Toole, *W Post*, 8/1/72, A8)
- Dept. of Transportation announced it had awarded $280 000 grant to Dr. Harold S. Johnston, Univ. of California at Berkeley chemist, to test his theory, propounded in 1971, that exhaust from supersonic transport aircraft could break up earth's ozone shield and let in "blinding and even lethal" radiation. Dr. Johnston would make chemical tests two years under conditions simulating those of stratosphere. Grant was part of DOT Climatic Impact Assessment program to assemble scientific data on environmental effects of high-altitude aircraft. (Russell, *W Post*, 8/5/72, A3; DOT PIO)
- President Nixon accepted resignation of James H. Wakelin, Jr., as Assistant Secretary of Commerce for Science and Technology, effective Aug. 1. (*PD*, 8/7/72, 1204)

August 2: *Cosmos 513* was launched from Baykonur by U.S.S.R. Satellite entered orbit with 322-km (200.1-mi) apogee, 201-km (124.9-mi) perigee, 89.7-min period, and 65° inclination and reentered Aug. 15. (GSFC *SSR*, 8/31/72; *SBD*, 8/4/72, 182)

- Most intense solar storm in two years, first observed by NASA's *Oso 7* Orbiting Solar Observatory July 26, produced three major explosions. Storm was being observed by *Oso 7* and *5*; *Explorer 41, 43*, and *45*; *Pioneer 6* and *10*; and ground-based observations. [See Aug. 7.] (NASA Release 72-164)
- NASA U-2 reconnaissance aircraft from Ames Research Center, equipped with high-resolution cameras using black-and-white and infrared film, photographed forest fire raging in Big Sur country of California's Mon-

terey County. It was first use of U-2 for forest fire reconnaissance. Photos made from altitudes of 20 000, 14 000, and 8000 m (65 000, 45 000, and 25 000 ft) were used by forestry officials to make plans for fighting fire. Photos and dense fog cover had helped firefighters contain 50% of fire by Aug. 3 and eventually fire was extinguished. (ARC Release 72-55; *LA Times*, 8/4/72, 18; JPL Hist Off)

- Skylab medical experiments altitude test (SMEAT), begun July 26, was progressing satisfactorily. Richard S. Johnston, Director of Life Sciences at Manned Spacecraft Center, said astronauts Robert L. Crippen, Karol J. Bobko, and Dr. William E. Thornton were in good spirits and good physical condition after seven days inside MSC chamber 6 m (20 ft) long. Minor equipment and procedural problems had been solved quickly without disrupting test timeline. Astronauts were scheduled to remain in chamber for up to 56 days. (MSC Release 72-170)

- Academician Boris N. Petrov summarized U.S.-U.S.S.R. meetings on Apollo-Soyuz Test Project (ASTP) [see July 6-18] and praised efforts for joint mission in *Pravda:* "To prepare for the planned flight both sides will have a lot of strenuous work to do and they will have a lot of difficulties to overcome. The implementation of this joint plan will be a major step forward in the development of international cooperation in the research and utilization of outer space for peaceful purposes. There is no doubt that cooperation in this field of technical progress will be an essential contribution to the conquest of space for peaceful purposes, in the interests of science, technology, and all peoples." (FBIS-Sov, 8/4/72, L1)

- Maritime Administration of Dept. of Commerce announced signing of $7.8-million cost-sharing research contract with General Electric Co. Space Div. to develop and test integrated vessel-control system using shipboard computer linked through orbiting satellites to shore-based computer. Maritime Administration would contribute $4.6 million and GE would pay balance. Project was part of program to adapt aerospace technology to marine operations and to equip fleet of 30-50 ships with satellite communications and navigation systems by late 1970s. (Maritime Release 72-32)

- Manned Spacecraft Center announced award of $365 000 contract supplement to LTV Aerospace Corp. for work on space shuttle orbiter thermal-protection system using reinforced pyrolyzed plastic. Supplement brought total funding for work to be done by April 20, 1973, to $859 000. (MSC Release 72-171)

- McDonnell Douglas Corp. announced in St. Louis that it was laying off 11 000 employees in next 17 mos, largely because of its failure to win prime contract for space shuttle. President and Chief Executive Officer Sanford N. McDonnell said 260 employees would be laid off immediately and that by year's end figure would reach 6000. Most of 11 000 layoffs—perhaps 9800—would be from California divisions of company that employed total of 92 000 persons. (Andelman, *NYT*, 8/3/72, 43)

- President Nixon submitted to Senate nomination of members to National Science Board: Dr. Wesley G. Campbell, Director of Hoover Institute on War, Revolution and Peace; Dr. T. Marshall Hahn, Jr., President of Virginia Polytechnic Institute and State Univ.; Dr. Anna J. Harrison, Mt. Holyoke College chemist; Dr. Hubert Heffner, Stanford Univ.

physicist; Dr. William H. Meckling, Univ. of Rochester economist; Dr. William A. Nierenberg, Director of Scripps Institute of Oceanography; Dr. Russell D. O'Neal, Executive Vice President for Aerospace with Bendix Corp.; and Dr. Joseph M. Reynolds, Louisiana State Univ. physicist (reappointment). Terms would expire May 10, 1978. (PD, 8/7/72, 1198, 1205)

August 3: NASA's *Erts 1* Earth Resources Technology Satellite (launched July 23) experienced unexpected power transient during 149th orbit. Ground controllers, suspecting problem in No. 2 tape recorder, turned off all experiment-associated power systems and placed spacecraft in stowed mode until problem was analyzed. Sequential turn-on of payload instruments—bypassing tape recorder—would begin Aug. 4 and transmission of photographic imagery would resume Aug. 5. (GSFC Release 15-72)

- Air Force F-15 fighter aircraft completed first supersonic flight, reaching mach 1.5 during 45-min test at Edwards Air Force Base, Calif. Flight test program was scheduled to end March 1, 1973, when decision on production of first 30 of 729 aircraft would be made. (SBD, 8/7/72, 188)

- Astronauts Charles Conrad, Jr., Paul J. Weitz, and Dr. Joseph P. Kerwin —prime crew for first Skylab mission—checked out Skylab airlock module (AM) and multiple docking adapter (MDA) in vacuum chamber at McDonnell Douglas Astronautics Co.'s St. Louis, Mo., facility. Crew, wearing pressurized spacesuits, entered chamber and took positions in AM and MDA. One hour after pumpdown began, chamber pressure was less than 138 newtons per sq m (0.02 psi)—equal to that at 45 700-m (150 000-ft) altitude. Pressure inside AM and MDA was held at 34 500 newtons per sq m (5 psi), high enough for crew to remove spacesuits and work in constant-wear garments. Crew connected biomedical equipment and sensors for repressurization but did not put on suits. (MSC Release 72-176)

- NASA and Soviet Academy of Sciences had approved recommendations of two joint working groups on compatible rendezvous and docking systems for manned spacecraft, NASA announced. Groups were Working Group on Guidance and Control Systems, which met in Moscow May 11-17, and Working Group on Compatibility of Docking Systems and Tunnels, which met at Manned Spacecraft Center March 27-April 3. Groups had considered lights, docking targets, communications systems, and other requirements of spacecraft control systems and also terminology, interface elements and assemblies, and docking system scale model. (NASA Release 72-160)

- Senate Committee on Aeronautical and Space Sciences met in closed session to hear testimony from Dr. James C. Fletcher, NASA Administrator; Dr. George M. Low, Deputy Administrator; and *Apollo 15* astronauts David R. Scott, Alfred M. Worden, and James B. Irwin on *Apollo 15* commercialism. Discussed were selling of stamp covers carried on mission and of replicas of statue "Fallen Astronaut" placed on moon during mission. Following meeting Committee Chairman, Sen. Clinton P. Anderson (D-N. Mex.), issued statement: "The testimony of NASA management and of the astronauts was forthright and complete. There is no doubt that certain provisions of NASA's Standards of Conduct for NASA Employees were violated by the astronauts for which they have

- been punished. The question of whether or not they violated any law is being examined. No conclusions were reached by the Committee at this meeting and the Committee will decide at a later time what further action it will take." (CR, 8/3/72, D869; Text)
- Senate agreed by vote of 88 to 2 to ratify U.S.-U.S.S.R. treaty on limitation of antiballistic missile systems. (CR, S12598–616, S12619–30, S12675–83)
- NASA announced selection of 106 principal investigators, including 83 scientists from U.S. and 23 from other nations, for Skylab earth observation experiments. Investigators would use data from earth resources experiment package (EREP), consisting of five sensors developed to observe earth simultaneously in visible, infrared, and microwave spectral regions. (NASA Release 72–150)
- NASA selected Lockheed Missiles & Space Co., Inc., for negotiation of $1-million, cost-plus-fixed-fee contract to design health-care delivery system using integrated medical and behavioral laboratory-measurement-system (IMBLMS) concept. IMBLMS would provide remote medical care by medical instrumentation, computer-based data-management techniques, and advanced communications techniques. NASA would evaluate IMBLMS concept for possible use in advanced long-duration manned space missions. After 12-mo design phase NASA could exercise option to provide additional funds for 12-mo fabrication and checkout phase and 24-mo operational-testing and system-evaluation phase. (NASA Release 72–159)
- Fairchild Industries, Inc., and Western Telegraph Co. announced formation of new company, as yet unnamed, to enter domestic satellite field. Spokesman for Fairchild and Western (which had no connection with Western Union Telegraph Co.) said new company would use Fairchild's satellite systems and development and manufacturing capabilities. (Smith, *NYT*, 8/9/72)

August 4: Nike-Apache sounding rocket, launched by NASA from Churchill Research Range, carried Goddard Space Flight Center payload to 159-km (98.8-mi) altitude. Objectives were to measure intensity and energy spectra of low-energy protons, helium nuclei, and heavier nuclei during polar cap absorption event; examine relative abundances of charge species; and study changes in intensity with time during event. Rocket and instruments functioned satisfactorily but telemetry signal was sporadic after 150 sec. Payload landed on ground and was recovered. Launch was first of two; second would be Aug. 5. (NASA Rpt SRL)
- Manned Spacecraft Center announced award of $92 800 firm-fixed-price contract to General Dynamics Corp. Convair Aerospace Div. to study effect jet thrusters would have on aerodynamics of space shuttle orbiter during reentry. Study would help in determining best configuration for orbiter's reaction control system. (MSC Release 72–173)
- American Association for the Advancement of Science was beginning broad new program under National Science Foundation grant to enhance public understanding of scientific enterprise, *Science* editorial said. Program would include seminars for government officials and managers of mass media, new approach to science on television, critical journal to examine science and mass media, experimental project to interest AAAS members in more collegial activities, and modest re-

search program into sources and cures of public misunderstanding of science. (*Science*, 8/4/72, 391)

August 5: *Pravda* published interview with unidentified designer of U.S.S.R.'s *Venus 8* Venus probe (launched March 27). "The vehicle was designed for prolonged work on the planet's surface. Everything is aimed at delaying the apparatus' 'thermal death.' New devices and additional scientific instruments were fitted to the descent apparatus. Therefore, the task was to reduce some of the weight and lighten the apparatus chiefly at the expense of the body. A new system for releasing the cover from the parachute section and an improved parachute system are used on the . . . descent apparatus. The parachutes were tested on a special stand . . . like a wind tunnel with a gas current heated to 500 degrees [centigrade; 773 kelvins] being blown through it. Carbon dioxide—the chief component of the Venusian atmosphere—was passed through the tube. With the help of sensors and a movie camera the testers thus had the opportunity of observing the behavior of the parachute system under near natural conditions." Spherical capsule had been repeatedly tested in high-pressure chamber that simulated physical conditions of Venus landing. (FBIS–Sov, 8/8/72, L2)

- Nike-Apache sounding rocket, carrying Goddard Space Flight Center payload, was launched by NASA from Churchill Research Range. Objectives were to measure intensity and energy spectra of low-energy protons, helium nuclei, and heavier nuclei during polar cap absorption event; examine relative abundances of charge species; and study changes in intensity with time during event. Rocket and instrumentation performance was not satisfactory. Radar beam malfunctioned at liftoff, telemetry signal was lost after 128 sec, 2nd stage apparently failed to ignite, and payload was not recovered. Launch was second of two; first had been Aug. 4. (NASA Rpt SRL)

August 6: *Erts 1* return-beam-vidicon (RBV) camera system was shut down after unexpected power surge. RBV system was second experiment to be shut down because of power surge; first had been No. 2 tape recorder Aug. 3. Although both systems could be operated, shutdowns were necessary to determine whether operation would damage experiments and to identify exact cause of problem. *Erts 1* (launched July 23) was providing excellent images from multispectral scanner (MSS) system and No. 1 tape recorder was being used as required. (GSFC Release 16–72)

- Reorganization of Marshall Space Flight Center, announced June 30, was implemented. (MSFC Release 72–98)

August 7: Intense solar storm detected by NASA's *Oso 7* July 26 produced fourth major explosion. First three explosions had been Aug. 2. During one hour, storm produced enough energy to meet U.S. demand for electrical power for 100 yrs at present rate of consumption. Storm had caused blackout of short-wave radio transmissions in polar regions, disruptions in long-distance U.S. telephone communications, and southward extension of aurora borealis. [See also Aug. 2 and 13 and During October.] (NASA Release 72–179; Reuters, B *Sun*, 8/7/72, A6)

- NASA held separate Headquarters debriefings for three losing aerospace firms in $2.6-billion space shuttle contract competition. North American Rockwell had won contract July 26. Grumman Aerospace Corp. President Joseph G. Gavin, Jr., told press after Grumman debriefing that his firm was unsuccessful largely because its shuttle design was more "conservative" than that of NR. Gavin said NASA officials had considered "mission suitability" and not management problems as weakness in Grumman bid. Grumman design had more complex swing arms to hold shuttle stationary before launch and lacked "background with large cryogenic tanking" used in shuttle's fuel supply. Grumman design had also been more expensive than NR version. Spokesmen for other losers, Lockheed Aircraft Corp. and McDonnell Douglas Corp., declined to comment after debriefings. [See also Oct. 4.] (Andelman, *NYT*, 8/8/72)

- President Nixon transmitted to Congress *Environmental Quality: The Third Annual Report of the Council on Environmental Quality—August 1972*. He said: "I am pleased that the data presented . . . indicate that the quality of the air in many of our cities is improving. Across the nation, emissions from automobiles—a significant portion of total emissions—are declining. We can expect these welcome trends to accelerate as the new standards and compliance schedules called for by the Clean Air Act of 1970 become fully effective." New legislation was "badly needed" in number of areas but international efforts toward "cleaner and healthier environment" were "hopeful sign that the productive pursuits of peace are coming gradually to command increasing attention in the discourse and competition among nations." President cited as examples of international efforts U.S.–U.S.S.R. Cooperative Agreement on Environmental Protection signed May 23, June 5–16 United Nations Conference on the Human Environment, and proposal for World Heritage Trust which he made in 1971. (*PD*, 8/14/72, 1216–9)

- *Pravda* published description by correspondent Pavel Barashev of his flight aboard first serial Tu-144 Soviet supersonic transport aircraft. Aircraft had reached speed of nearly 2500 km per hr (1500 mph). "When the engine started I expected a roar but even when all four of them were working it was unusually quiet in the cabin even for a jet plane." (Tass, FBIS–Sov, 8/7/72, L1)

August 8: Skylab medical experiments altitude test (SMEAT) crew press conference was held at Manned Spacecraft Center. Astronauts Robert L. Crippen, Dr. William E. Thornton, and Karol J. Bobko answered press questions from inside chamber they had entered July 26. Chamber had atmosphere of 70% oxygen and 30% nitrogen and pressure of 34 500 newtons per sq m (5 psi). Astronauts said they missed wives, families, and variety of outside life most. They had not experienced any stress from close confinement and had no trouble sleeping, but had recommended minor changes in food, waste collection system, and some medical equipment.

Dr. Thornton said he was surprised at crew's physiological condition: "There has been literally no change worth mentioning during this period, either physiological or psychological." One or two items "might have even shown a little improvement. There was a sniffle or two when we came in, and this atmosphere seems to have . . . cleared it up."

Dr. Thornton said $3-million cost of test was justified and was "a very, very small part of what could be considered insurance and success of the Skylab program." Money being spent on Skylab medical research could build a number of medical institutions with "a great deal of short-term benefits, properly utilized. But . . . what we're seeing in Skylab, from a medical viewpoint, is the first opportunity to medically study man in space. If the medical scientists that are assigned to Skylab obtain the data that they need and do their job properly, this is literally a building block for space medicine . . . . Regardless of what happens in the short run . . . the human race is going to be in space more and more as time goes on." (Transcript)

- U.S.S.R. requested that international convention on use of artificial earth satellites for direct TV broadcasting be placed on agenda of 27th session of United Nations General Assembly in September. Provisions of draft convention delivered to U.N. Secretary General Kurt M. Waldheim were: (1) Broadcasts were to be "in the interests of peace, progress, development of mutual understanding and stronger friendly relations" and to improve educational and cultural levels and effect wider international exchanges. (2) All states had equal right to receive and disseminate direct TV. (3) Broadcasts to foreign states would be made only with "definitely expressed consent" of receiving states. (4) Transmission to other states without their consent of programs that "prejudice the cause of safeguarding international peace and security, which represent interference into internal affairs of states, which encroach upon basic human rights, which contain propaganda of violence and horrors, which undermine the foundations of local civilization and culture and which misinform the population are considered to be unlawful and involving international liability." (5) States could act to counteract unlawful direct TV broadcasts beamed at them on their own territory, from outer space, and from other places outside national jurisdiction of any state. (6) A state was responsible for all national activities on its direct TV. (Tass, FBIS–Sov, 8/10/72, A1)

- Western Union Telegraph Co. outlined plans for possible start of domestic satellite communications system before mid-1974 at New York press briefing. Company said it had signed contract to purchase three satellites from Hughes Aircraft Co. for $20 706 500, with incentive payments to be made after satellites were orbited. Hughes would supply launch support and options for additional satellites and services. Western Union was first company to order satellites under waiver granted by Federal Communications Commission July 26 that permitted ordering spacecraft in advance of issuance of construction permits. (Smith, *NYT*, 8/9/72)

- Reception in House Rayburn Building honored retiring Chairman George P. Miller of House Committee on Science and Astronautics. Rep. Miller (D-Calif.)—defeated June 6 in California primary for nomination for reelection to House—had presided over House Committee in overseeing and funding U.S. space program since 1961. House Speaker, Rep. Carl Albert (D-Okla.), said that "never had so much been accomplished in the national interest in so short a time" as by Committee. Dr. James C. Fletcher, NASA Administrator, and Secretary of the Air Force, Dr. Robert C. Seamans, Jr., praised work of Committee in evolution of entire U.S. space effort. Rep. Miller responded that U.S. space achievements were "a beginning and not an end" of the national enterprise required to serve Nation and society in the future. National Space Club President Donald R. Rodgers presided over ceremony attended by 300 guests, including Congressmen, NASA and industry representatives, and Senate and House Space Committee staffs. (NASA Hist Off)

- High-powered telescopes at Univ. of California at San Diego picked up satellite Toros, "earth's other moon," at distance of 20 million km (12.4 million mi)—closest approach of Toros to earth in centuries. (UPI, W Star-News, 8/9/72, A15)
- House Committee on Science and Astronautics' Subcommittee on NASA Oversight held hearing on policies and procedures governing disposal of real property under NASA Administrator's control. Purpose was to determine whether legislation should be enacted to require formal reports to Congress on real estate disposal. General Services Administration surveys had determined that certain lands at Wallops Station, Langley Research Center, and Lewis Research Center were excess to NASA needs. Congress had not previously been apprised of plans for property's disposal by GSA. Subcommittee heard testimony of NASA Deputy Associate Administrator for Organization and Management Bernard Moritz. GSA was surveying LeRC's Plum Brook Station, Wallops Station, LaRC, and Marshall Space Flight Center. Final decisions had not been made. (Transcript)
- National Academy of Sciences released findings of two-year study of science and its relation to society, made by Physics Survey Committee chaired by Yale Univ. physicist Dr. David A. Bromley. Report, tentatively titled "Physics in Perspective," was in press at release time. It rated 69 program areas in physics in order of priority and emphasized cost to U.S. science and society and to physics if programs were not adopted. Most "striking" aspect of survey had been "renewed discovery of the over-all power and vitality of U.S. physics," but "this strength is in danger." Committee found danger most immediate and obvious in subfields like elementary particle and nuclear physics where major facilities approved in mid-1960s were just becoming operational. Report named 15 programs that had growth potentials warranting high priority for incremental support: macroscopic quantum phenomena including superfluidity and superconductivity, quantum optics, scattering in solids and liquids, heavy-ion interactions, higher-energy nuclear physics, national accelerator laboratory, Stanford linear accelerator, controlled fusion, turbulence, nonlinear optics, lasers and masers, atomic and molecular beams, biophysical acoustics, very large radio array, and x-ray and gamma ray astronomy.

Committee concurred with recommendation of Astronomy Survey Committee [see June 1] that High Energy Astronomy Observatory (HEAO) should be prominent in national effort in astrophysics and astronomy. "Because of the opaqueness of the earth's atmosphere to both X and gamma radiation, these windows on the universe have only recently been opened through rocket and satellite astronomy. Any reasonable extrapolation from the preliminary soundings that have been possible so far suggests a larger return in fundamental insight into the structure and history of the universe. In the total national physics program, the estimated cost of this facility—$400 million . . . is extremely high. To be considered in proper perspective it must be viewed in the context of the total expenditures of the U.S. space program. In anticipated scientific return—both short- and long-range—

it merits high priority in that program." (NASA–NRC–NAE *News Rpt*, 8–9/72, 1–7)

- British Airways Board had decided to order six Lockheed L–1011 TriStar jet airliners and to place options on another six, *Wall Street Journal* reported. British Aerospace Minister Michael Heseltine had told House of Commons that British government would aid Rolls-Royce (1971) Ltd. to develop more powerful version of aircraft's RB–211 jet engine. Lockheed Aircraft Corp. Chairman Daniel J. Haughton had said later that order was "single most significant overseas order in Lockheed's commercial aircraft programs during the past 25 years." Lockheed estimated value of order at $150 million including spares. (*WSJ*, 8/8/71, 7)
- President Nixon submitted to Senate nomination of Dr. James R. Schlesinger, Chairman of Atomic Energy Commission, to be U.S. representative at 16th session of General Conference of the International Atomic Energy Agency. Among alternates nominated was Dr. T. Keith Glennan, first NASA Administrator. (*PD*, 8/14/72, 1225)

August 8–9: Skylab program managers reviewed design of mission's experimental hardware with 11 high school student winners of NASA and National Science Teachers Assn. competition to propose experiments for 1973 Skylab mission [see July 20]. Students attending critical design reviews at Marshall Space Flight Center were those whose experiments required new scientific equipment. (NASA Release 72–161)

August 9: Army announced cancellation of controversial Lockheed AH–56A Cheyenne helicopter program and plans, subject to congressional approval, to develop new attack helicopter. Production contract for Cheyenne had been canceled in May 1969, after Lockheed failed to solve rotor problem. Since then firm had claimed progress in combating problem, *Wall Street Journal* later reported. (*WSJ*, 8/10/72, 3; Army PIO)

- RCA announced it had received NASA contracts totaling nearly $1.5 million for four Skylab color tape-recording systems. Recorders—similar to those on *Erts 1* (launched July 23)—would help provide home viewers with color TV coverage of astronauts conducting experiments. Contracts had been awarded to RCA Government Communications Systems by Marshall Space Flight Center. (RCA Release)

August 10: International Institute for Strategic Studies in London released *Reconnaissance, Surveillance and Arms Control.* Study found that verification by observation satellites and electronic reconnaissance systems of adherence to terms of Strategic Arms Limitation agreements had greatly reduced possibility of one nation's cheating on another. "Therefore the risk of the Soviet Union achieving a strategic advantage as a result of a . . . treaty is a very low one." (UPI, *NY News*, 8/11/72, 6)

- NASA launched two Aerobee 170 sounding rockets from White Sands Missile Range, N. Mex. First carried Massachusetts Institute of Technology astronomy and soft x-ray experiment. Rocket performed satisfactorily, but scientific objectives were not met. Second rocket carried Univ. of Wisconsin astronomy experiment. Rocket and instrumentation performed satisfactorily. (SR list)
- President Nixon sent *Convention on the Prohibition of the Development, Production, and Stockpiling of Bacteriological (Biological) and Toxin*

Weapons, and on Their Destruction [see April 10] to Senate for advice and consent to ratification. He said: "I believe this Convention will enhance the security of the United States and the world community. It will help ensure that scientific achievements in . . . biology will be devoted not to destruction but to the service of mankind. It represents a significant advance in . . . arms control and disarmament. (*PD*, 8/14/72, 1220–1)

- President Nixon signed H.R. 14108, $696.9-million National Science Foundation authorization act for 1973. Act became Public Law 92–372. (*PD*, 8/14/72, 1225)

August 11: NASA's M2–F3 lifting body, piloted by William H. Dana, completed 15th flight from Flight Research Center. Objective—to investigate stability and control at mach 0.95—was achieved. (NASA proj off)

- Sen. William Proxmire (D-Wis.), critic of aerospace spending, was named Chairman of Senate Committee on Appropriations' Subcommittee on Housing and Urban Development, Space, Science, Veterans. Appointment followed death of Subcommittee Chairman, Sen. Allen J. Ellender (D-La.). (*LA Times*, 8/17/72; *W Post*, 8/18/72, A3)

- First laboratory demonstration of controlled fusion—peaceful use of hydrogen-bomb reaction to make electricity—could occur within three years, *Washington Post* reported. Intensely concentrated laser light beam would be used to trigger fusing of atoms. Fusion would produce energy sufficient for conversion into plentiful electric power with potentially unlimited fuel, a form of hydrogen from sea. Physicist Edward Teller had said there was "outside chance" that laser-fusion would carry men to Mars and back by end of century. U.S.S.R. was believed to be two to three years ahead of rest of world in both laser-fusion and other fusion research. (Cohn, *W Post*, 8/11/72, A3)

- President Nixon designated Chairman Russell E. Train of Council on Environmental Quality to lead U.S. participation in U.S.–U.S.S.R. Joint Committee in the Field of Environmental Protection. Train would head U.S. delegation to Moscow later in year. (*PD*, 8/14/72, 1224)

- National Science Foundation published *Federal R&D Funding Continues To Rise* (NSF 72–314): Upward trend since 1970 in Federal research and development support was clearly established. Federal R&D obligations, exclusive of plant, were expected to rise from $15.5 billion in FY 1971 to $16.8 billion in FY 1972 and to all-time high of $17.8 billion in FY 1973. Figures showed increase of 8% in FY 1972 and 6% in FY 1973. Upward trend reversed downward trend from FY 1967 through 1970. (NSF *Highlights*, 8/11/72)

- R/A William M. Harnish (USN) assumed command of Pacific Missile Range in ceremony at Pt. Mugu, Calif. (PMR *Missile*, 8/11/72, 1)

August 12–13: *Chicago Daily News* editorial said there appeared to be "conspiracy" to keep details of *Apollo 15* commercialism from public. "Not only were public and press barred from the five-hour committee hearing [see Aug. 3]; a curtain of silence was imposed after the session." Paper reminded "both NASA and the committee that they work for the public. Let's quit playing cozy and bring the facts into the open so the public can make its own judgment." (*C Daily News*, 8/12–13/72)

August 13–28: *Explorer 46* Meteoroid Technology Satellite (MTS) was launched by NASA from Wallops Station by four-stage Scout booster at 11:10 am EDT. Spacecraft and booster, which included new Algol III 1st stage, weighed 21 650 kg (23.9 tons)—heaviest vehicle ever launched from Wallops. Orbital parameters: 814.5-km (506.1-mi) apogee, 495.8-km (308.1-mi) perigee, 98-min period, and 37.7° inclination.

Primary objective was to evaluate effectiveness of bumper-protected multisheet spacecraft structure for protection against meteoroid penetration in first inflight test. Data would be compared with data from single-sheet structures launched previously. Secondary objectives were to obtain data on meteoroid impact velocity and measure impact flux of small mass meteoroids.

Satellite had been designed, fabricated, and tested by Langley Research Center. It carried three experiments: bumper penetration experiment with 12 panels, each 48 cm wide and 3 m long (19 in x 10½ ft), containing 8 pressurized cells; impact flux experiment containing 65 thin-film capacitance penetration sensors arranged in groups of 3 and 5 on various parts of spacecraft; and meteoroid-velocity experiment consisting of 2 plates of sensors and timer to measure time meteoroid traveled distance between plates.

By Aug. 22 spacecraft experiments and telemetry systems were operating and good data were being obtained, but half of panels in bumper experiment apparently had not deployed. Spacecraft was spinning at 3 rpm as planned, but because of panel deployment anomaly and resulting spin axis 90° to that planned, spacecraft battery temperature had risen from 299 K (79°F) past 308 K (95°F) design temperature to 337 K (147°F) Aug. 22. Battery was still providing voltage and all experiments were functioning satisfactorily. During spacecraft's second pass Aug. 22, telemeter A, powered by battery, came on as commanded but telemeter B, backup system operated by solar-cell power, went off unexpectedly. Officials decided to discontinue interrogation of telemeter A so that data from telemeter B would not be jeopardized. As a result, no data were being received from secondary experiments and no housekeeping data were received. About 2000 impacts by submicrometer particles had been detected during telemeter A operation, providing sufficient data for impact experiment. No data had been obtained from meteoroid-velocity experiment. By Aug. 28, 3 of 48 pressurized detector cells successfully deployed had been punctured and primary experiment was continuing to provide good data. (NASA proj off)

August 13: Third in series of very-high-altitude flight tests of parachute decelerator system for NASA's Viking Mars lander, scheduled for 1975 launch, was held over White Sands Missile Range by Martin Marietta Corp. under Air Force Cambridge Research Laboratories direction. In repeat of July 11 test, simulated Viking entry vehicle was lifted by helium-filled balloon and then boosted to supersonic speed by rocket to check out parachute system at highest speed and most severe loading conditions expected on Mars landing. [See July 11 and 26, Aug. 19, and Dec. 14.] (AFSC Release 105.72; Martin Marietta Corp PIO; NASA Release 72–118)

- Recent solar flares had reminded scientists of "undependable" behavior of sun, Walter Sullivan said in *New York Times* article. Solar cycle was "only an approximation. Its 11-year periodicity is no more than average and some of the biggest eruptions, described as 'Class 4,' occur when the cycle is not at its peak." Largest of recent series [see Aug. 7] had been graded Class 3 "but they were probably the most intense ever recorded close to the period when the sun is 'quietest.'" Sun's undependable behavior could affect safety of astronauts exploring Mars. "They are likely to be sent on such a voyage, lasting more than a year, at a time of sunspot minimum. But if they were outside their craft when a major flare occurred, they might be exposed to a severe dose of radiation." (NYT, 8/13/72, 9)

- Data from 10-yr U.S. Supersonic Transport Development Program, terminated March 1971, had been cataloged and would be available for dissemination by National Technical Information Service of Dept. of Commerce, Secretary of Transportation John A. Volpe announced. (DOT Release 72-159)
- Smithsonian Institution had recommended California Museum of Science and Industry in Los Angeles as National Aerospace Museum of the West, *San Diego Union* reported. If Congress approved, funds from Federal agencies would be used for public exhibit of aircraft, rockets, spacecraft, and satellites. (Copley News Service, *SD Union*, 8/13/72)

August 14: President Nixon signed H.R. 15093, FY 1973 Dept. of Housing and Urban Development-space-science-veterans appropriations bill that included $3.408-billion NASA appropriation. Bill became Public Law 92-383. (*PD*, 8/21/72, 1246; PL 92-383)

- Soviet scientists who had examined lunar samples brought to earth by *Luna 16* and *Apollo 11* had reported findings on lunar iron, Tass announced. Lunar iron was almost completely void of impurities and had not been affected by corrosion, in spite of its long presence on earth. Scientists maintained "that lunar vacuum and high temperature during the supposed eruptions of volcanoes themselves served as reducers of pure iron out of compounds." (FBIS-Sov, 8/15/72, L2)
- Space researchers should look for insect life on Jupiter and for seeds and pollen on Saturn, Russian-born cosmologist Dr. Immanuel Velikovsky said during lecture sponsored by Ames Research Center's Biotechnology and Planetary Biology Divs. at ARC. He hypothesized that Venus and Mars had been in different orbits and had passed close enough to earth in centuries before Christ to cause cataclysms described in Old Testament. Dr. Velikovsky also suggested that several magnetic shells within solar system could return echoes and urged that spacecraft be sent to investigate them. (ARC *Astrogram*, 8/3/72, 1; Mead, San Jose, Calif, *Mercury*, 8/15/72)
- International Biophysics Conference ended in Moscow. During conference, attended by scientists from more than 40 countries, Moscow Univ. biophysicist Andrey Rubin reported spectrometry detection of simplest organic compounds in stellar dust. Discovery indicated possible existence of carbon-based organic life on other planets. (Tass, FBIS-Sov, 8/15/72, L1)

August 15: Special Achievement Awards to Naval Research Laboratory scientists Richard L. Statler and Dr. Bruce J. Faraday for technique to prevent solar-cell overheating in satellite power systems were announced by NRL. Technique improved satellite efficiency and could extend satellite life by fabricating solar panels of unique materials that allowed cooler operation. Technique—already incorporated on NRL satellites built by NRL Space Systems Div.—increased available power, decreased satellite weight, and promised cost savings for building and operating solar-cell power-supply subsystems. (NRL Release 34-72-7)

- NASA awarded three letter contracts for design of modifications to quiet engines and reduce exhaust emissions of U.S. commercial jet aircraft. United Aircraft Corp. Pratt & Whitney Div. received $1.2 million to modify JT3D and JT8D engines and McDonnell Douglas Corp. and Boeing Co. each received $800 000 to propose methods for acoustically treating nacelles to absorb fan noise. Work would be noncompetitive. (NASA Release 72-166)
- Univ. of California at Los Angeles reported that test conducted with 3-mev electron microscope in Toulouse, France, had contradicted theory that at electron energies high enough to magnify specimens down to molecular level specimen would be so severely damaged that results would be useless. Test had shown reverse to be true. As energy of microscope's electron beam increased, extent of damage had decreased. (Sullivan, *NYT*, 8/16/72, 14)
- President Nixon informed Congress of plans for six-month International Exposition on the Environment to be held in Spokane, Wash., in 1974. (*PD*, 8/21/72, 1237)

August 15-16: Symposium on composite structural materials for space shuttle application, at Marshall Space Flight Center, reviewed progress and facilitated exchange of space shuttle composites information originated by NASA Centers and contractors. (MSFC Release 72-107)

August 16: U.S.S.R. launched *Cosmos 514* from Plesetsk into orbit with 975-km (605.8-mi) apogee, 957-km (594.7-mi) perigee, 104.3-min period, and 82.9° inclination. (GSFC *SSR*, 8/31/72; *Sov Aero*, 8/21/72, 64)

- NASA announced signing of definitive contract with North American Rockwell Corp. Rocketdyne Div. for development and production of space shuttle main engine. NASA had announced selection of Rocketdyne for negotiation of contract July 13, 1971, but award had been delayed pending outcome of protest by United Aircraft Corp. Pratt & Whitney Div. Initial work had been under way for several months under temporary contracts during final negotiations. First increment of cost-plus-award-fee contract for primary development through Aug. 31, 1975, was estimated at $205 766 000. Second increment of $236 709 000 would be for engine production and remaining development from Sept. 1, 1975, to June 30, 1979. (NASA Release 72-167)
- NASA conducted successful trial run of airborne visible-laser optical-communications (AVLOC) flight tests to determine effects of atmosphere on vertical transmission of laser beams. Two-month series of tests using WB-57 aircraft at 18 000 m (60 000 ft), above 95% of earth's atmosphere, would be operated by Manned Spacecraft Center over ground station at Redstone Arsenal near Marshall Space Flight Center. Tests were part of overall NASA program to develop optical communications systems for use during 1970s. Similar visible-light experiment was planned for Applications Technology Satellite ATS-G, to be launched by NASA in 1975. In trial test, experiments, checking out pulsed-laser radar acquisition-and-ranging system, located aircraft

flown from Ellington Air Force Base, Tex., and tracked it to 18 000 km (58 000 ft) but did not attempt laser communications. (MSFC Release 72–109)

- NASA announced resignation of Daniel J. Harnett, Assistant Administrator for Industry Affairs and Technology Utilization, effective Sept. 1. Harnett would join Aeronca, Inc., in major executive capacity. He had joined NASA Oct. 1, 1969. (NASA Release 72–165; *A&A 1969*)
- Bathroom commode system for space shuttle passengers was described by Manned Spacecraft Center engineer A. F. Behrend, Jr., and J. E. Swider, Jr., of United Aircraft Corp. Hamilton Standard Div. during annual Environmental Control and Life Support System Conference in San Francisco. In space, high-velocity airstreams would compensate for earth's gravity and would assist water-flush mechanism. Waste would be vacuum dried and chemically treated. System, being built for further testing and evaluation by Hamilton Standard under $238 000 NASA contract, was of type being studied by commercial airlines to reduce maintenance and operating costs. (NASA Release 72–163)
- Wallops Station announced award of $1 074 000, one-year contract renewal to Computer Sciences Corp. for engineering support services at Wallops. (WS Release 72–9)

August 16–17: Dr. George M. Low, NASA Deputy Administrator, spoke on NASA's attack on costs during symposium "Cost—A Principal System Design Parameter" in Washington, D.C. Symposium was sponsored by National Security Industrial Assn. and Armed Forces Management Assn. NASA was facing major cost problems. If NASA did not "do something about the high cost of doing business in space, and do it soon, our nation's space program is in deep trouble. We are on the verge of exciting new discoveries in space science, but we cannot follow through as rapidly as we should because we can't afford it. We see before us many important space applications, but we cannot move out as rapidly as we should because we can't afford it. Most important of all, we may lose our hard-won worldwide leadership in space, if we don't find a way to do more for our money!" NASA could do little about budgetary restraints, "but there is a great deal we can do about costs. Doing something about the high cost of doing business in space is today's biggest challenge."

In past, NASA had been forced to develop expensive equipment for one-time use. "Today we have 52 operating civilian spacecraft. By the end of next year we will launch 35 more, for a total of 87" representing "43 different spacecraft types—on the average each spacecraft [type] is flown only twice!" But payloads had become more complex, while launch costs decreased. "We are no longer as limited in weight and volume as we were ten years ago; with the shuttle, weight and volume constraints will be non-existent for many missions. This means that we should now optimize our payloads for low cost and high reliability, and not for minimum weight and maximum performance. . . . I am convinced that if we do this, we can drastically reduce the cost of doing business in space. And this . . . is as great a technological challenge as everything else that we have done in space." (Text)

Air Force initiatives in controlling system cost were summarized by Gen. George S. Brown, Commander of Air Force Systems Command: prototyping to avoid high developmental costs, joint operational technical reviews to identify tradeoffs that would reduce acquisition costs, contracting techniques to reduce acquisition costs, full support of program managers to reduce acquisition costs, improvement in life-cycle costing technique to identify actions that would minimize total cost of ownership, and positive actions to achieve reliability and maintainability goals for increasing system effectiveness and decreasing cost of ownership. (Text)

August 17: Plans for regional earth resources information system were announced at Marshall Space Flight Center by team of MSFC Summer Faculty Fellows headed by Dr. Reginald I. Vachon of Auburn Univ. and physicist Herman G. Hamby of MSFC Environmental Applications Office. System—Earth Resources Information Storage, Transformation, Analysis and Retrieval (ERISTAR)—would gather and correlate data from instrumented packages into data cross-sections useful in industrial development, conservation of natural resources, and pollution reduction. Information would be made available to Alabama Earth Resources Information Committee. Developed by team during 11-week stay at MSFC, system would assist in disseminating information from *Erts 1* Earth Resources Technology Satellite, launched July 23. (MSFC Release 72-105)

- People's Republic of China signed $5.7-million contract with RCA Global Communications, Inc., to install satellite earth station outside Peking and enlarge existing earth stations near Shanghai. (Samuelson, *W Post*, 8/18/72, D8)

- NASA announced award of $445 669 contract to Univ. of Minnesota to build mass spectrometer measurement instruments for Dual Air Density (DAD) Explorer satellites, scheduled for launch in 1974. (NASA Release 72-168)

- *Pravda* editorial praised U.S.S.R.'s *Venus 8* Venus probe: "A striking new page has been written in the annals of Soviet and world cosmonautics. After a 4-month flight the Venus-8 automatic interplanetary station reached Venus and its descent apparatus made history's first floating descent through the atmosphere on the light side of the morning star, making a soft landing on the planet's surface. This remarkable scientific experiment is another vivid demonstration of the great potential of unmanned spacecraft in studying space and the planets of the solar system. The Venus-8 station has brilliantly continued the impressive sequence of . . . unmanned spacecraft, which have provided world natural science with so much valuable information about the universe." (FBIS–Sov, 8/21/72, L1)

August 18: U.S.S.R. launched *Cosmos 515* from Plesetsk into orbit with 286-km (177.7-mi) apogee, 179-km (111.2-mi) perigee, 89.1-min period, and 73° inclination. Satellite reentered Aug. 31. (GSFC *SSR*, 8/31/72; *Sov Aero*, 9/11/72, 70)

- NASA awarded $500 000 parallel study contracts to Hughes Aircraft Co. Space and Communications Group, TRW Systems Group, and AVCO Corp. Systems Div. to design system for series of proposed missions to study Venus with Pioneer-class spacecraft. Missions, to begin in January 1977, would include entry probes and orbiters. Contractors for final design, development, and manufacture would be selected after completion of studies in June 1973. (NASA Release 72-172)

- NASA announced award of $5-million, cost-plus-award-fee contract to ITT Gilfillan to develop visible laser communications experiment for ATS-G Applications Technology Satellite, scheduled for launch in 1975. Contract was for five years, including ground station operational support throughout experiment's two-year lifetime. (NASA Release 72-170)
- Senate passed S.R. 2483, bill to formulate plans and programs to convert U.S. to metric system of weights and measures within 10 yrs. (CR, 8/18/72, S1396-403)
- *Los Angeles Times* editorial commented on scientific forecasts that earth's surface would become too hot for habitation in 4 billion yrs: "While we work on problems of more immediate concern, let's not scrimp on the space shuttle. Down the line a few billion years, they may think well of us for that." (*LA Times*, 8/18/72)

August 19: Japan launched *Denpa* (*Radio Wave*) Radio Explorer Satellite (REXS) from Kagoshima Space Center by Mu-4S booster at 11:40 am local time (10:40 pm EDT Aug. 18). Because of unusually strong winds, satellite entered highly elliptic orbit with 6302-km (3915.9-mi) apogee, 239-km (148.5-mi) perigee, 157.2-min period, and 31° inclination. *Denpa* carried ionospheric plasma probes, electromagnetic and plasma-wave receivers, cyclotron instability experiment, electron flux analyzer, and fluxgate magnetometer.

Instruments functioned satisfactorily immediately after launch, but satellite later experienced power failure that prevented transmission of data.

The 75-kg (165-lb) satellite, 68 cm (22.3 ft) long, was fourth satellite to be launched by Japan and second to carry scientific payload. *Shinsei* (launched Sept. 28, 1971) had carried instruments to study cosmic and electric waves. *Tansei* (launched Feb. 16, 1971) and *Ohsumi* (launched Feb. 11, 1970) test satellites had returned performance data. (GSFC *SSR*, 8/31/72; FBIS-Japan, 7/20/72, C7; *Spacewarn*, 8/8/72, 8; NASA Int Aff; Japanese Embassy PIO)

- Last in series of very-high-altitude flight tests of parachute decelerator system for NASA's Viking Mars lander, scheduled for 1975 launch, was held over White Sands Missile Range by Martin Marietta Corp. under Air Force Research Laboratories direction. Decelerator was dropped from balloon at altitude of 27 000 m (88 000 ft) to check out parachute system at subsonic conditions. In previous tests rocket motors had boosted speed after balloon had carried system to desired altitude. [See July 11 and 26, Aug. 13, and Dec. 14.] (AFSC Release 105.72; Martin Marietta Corp PIO; NASA Release 72-118)

August 21-29: Oao 3 (OAO-C) (*Copernicus*) Orbiting Astronomical Observatory was launched by NASA from Eastern Test Range at 6:28 am EDT by Atlas-Centaur booster. Orbital parameters: 745-km (462.9-mi) apogee, 740-km (459.8-mi) perigee, 99.7-min period, and 35° inclination.

Primary objective was to obtain high-resolution spectra of stars in ultraviolet range between 1000 and 3000 A to investigate composition, density, and physical state of matter in interstellar space and stellar sources. Secondary objective was to evaluate onboard computer, light

August 21-29

August 21-29: Oao 3 (Copernicus) *Orbiting Astronomical Observatory was launched into orbit to observe stars, planets, nebulae, galaxies, and interstellar matter in the ultraviolet and x-ray regions of their radiation, from above the earth's atmosphere. By Aug. 29, checkout had been completed and Oao 3 was obtaining excellent data. In the photo, the spacecraft's Atlas-Centaur launch vehicle stages arrived at Cape Kennedy Skid Strip in the cargo hold of an Air Force C-5A aircraft June 15. Use of the world's largest transport aircraft permitted shipment of both stages in one aircraft for the first time. The Atlas 5004 and Centaur 19D stages were later mated on the pad at Launch Complex 36, to form the Atlas-Centaur 22 vehicle.*

baffle, inertial reference unit, electronically scanned star tracker, and heat pipes. Spacecraft carried two experiments—Princeton experiment package (PEP), 80-cm (32-in) Cassegrainian telescope and photoelectric spectrometer; and Univ. College London (UCL) x-ray experiment containing three telescopes and collimated proportional counter.

All spacecraft operations were carried out as planned, with spacecraft checkout completed Aug. 24, experiment checkout completed Aug. 26, and stabilization and control system checkout completed Aug. 28. By 2:00 pm EDT Aug. 29 *Oao 3* had completed 120 orbits with systems operating almost perfectly. PEP was viewing bright star Zeta in constellation Ophiuchus and obtaining excellent data. UCL experiment was undergoing calibration and alignment and would begin observing x-ray sources in September.

Fourth OAO launched by NASA, *Oao 3* at 2200 kg (4900 lbs), 2 m wide and 3 m high (7 x 10 ft), was heaviest and most complex automated spacecraft developed by U.S. It had new gyro inertial reference unit with four star trackers, electrically scanned star tracker, precision digital solar-aspect sensor, and onboard computer that could handle 16 000, 18-bit words and store 1024 ground commands. OAO-B had crashed Nov. 30, 1970, when shroud failed to separate after launch.

Oao 2 (launched Dec. 7, 1968) was still returning valuable data. *Oao 1* (launched April 8, 1966) had failed because of high-voltage arcing in star trackers and malfunction in power supply system. OAO program was managed by Goddard Space Flight Center under direction of NASA Office of Space Science. (NASA proj off; NASA Release 72–156)

August 21: *Cosmos 516* was launched by U.S.S.R. from Baykonur. Orbital parameters: 264-km (164-mi) apogee, 250-km (155.3-mi) perigee, 89.6-min period, and 64.9° inclination. (GSFC *SSR*, 8/31/72; *Sov Aero*, 9/11/72, 70)

- Dr. Rocco A. Petrone, Apollo Program Director, had been assigned additional responsibilities as Program Director of Apollo Soyuz Test Project (ASTP), NASA announced. Dr. Petrone would have overall responsibility for direction and management of U.S. portion of joint earth-orbital mission with U.S.S.R. (NASA Release 72–174)

- First pre-production Anglo-French Concorde 01 supersonic transport had completed first stage of performance and handling characteristics tests, including 80 flights, *Aviation Week & Space Technology* reported. Aircraft would be grounded for about 20 wks for installation of production Rolls-Royce/Snecma Olympus 593 Mk 602 engines and production variable-geometry intakes to replace fixed intakes. (*Av Wk*, 8/2/72, 22)

- President Nixon issued Proclamation 4146 proclaiming Oct. 9 in each year as Leif Erikson Day in honor of Norse explorer who crossed Atlantic to North America in year 1000. President said: "Now, more than nine hundred years later, we must summon those same qualities [of adventure and courage] to aid us in meeting the challenges of this world and exploring the unknown of outer space." (*PD*, 8/28/72, 1260)

- *Chicago Tribune* editorial said there was still "a lot of studying to be done" on U.S. conversion to metric system [see Aug. 18]. Question of financing was "sticky" one. "Many of the countries which have gone or are going metric have provided some sort of government help to finance the changeover." U.S. estimated conversion cost at between $10 billion and $40 billion. "Both labor and industry—especially small business—have been leery of undertaking this expense without government help. But the government is in no position at the moment to give much help; and if it were to convert its purchases to the metric system quickly, the result would be favorable to foreign manufacturers already on the metric system and detrimental to our own." (*C Trib*, 8/21/72)

August 22: NASA announced preliminary timeline for Apollo 17 manned lunar landing mission. Spacecraft—carrying Astronauts Eugene A. Cernan, Ronald E. Evans, and Dr. Harrison H. Schmitt—would be launched from Kennedy Space Center at 9:53 pm EST Dec. 6. Lunar module would land on moon's Taurus-Littrow region at 2:55 pm EST Dec. 11 and two crewmen would explore lunar surface for three extravehicular activity periods before LM lifted off moon at 5:56 pm EST Dec. 14. Spacecraft would splash down in Pacific at 2:24 pm EST Dec. 19, after 304-hr 31-min mission. (NASA Release 72–169)

- Revision of NASA's FY 1973 operating plan to adhere to $3.2-billion budget limitation would necessitate work reductions in manned space flight, NASA announced. Work reductions at Manned Spacecraft Center, Marshall Space Flight Center, and Kennedy Space Center would not

August 22

affect Apollo 17 or Skylab launch schedules or safety and quality assurance requirements. Support contractor manpower at MSFC would be reduced by 600 to 700 positions. Cost ceilings also would necessitate additional reductions in prime contractor personnel. (NASA Release 72-176; MSFC Releases 72-111, 72-113)

- Saturn IB stage for Skylab 2 arrived at Kennedy Space Center on NASA barge *Orion*. Saturn IB would launch command and service module to join orbiting Skylab Workshop day after Workshop launch as Skylab 1 in 1973. Stage would be processed in Vehicle Assembly Building and would be erected on Mobile Launcher Aug. 31 at KSC. Arrival of stage marked first time in nearly four years that a Saturn IB had been in processing at KSC and first time vehicle would undergo flight preparation in VAB. Last mission for a Saturn IB was *Apollo 7* earth-orbital mission Oct. 11, 1968. (KSC Release 246-72)

- NASA announced it would continue Life Scientist Program for 1972 with award of up to five three-year institutional grants for research relevant to NASA needs. Principal investigators and their designated graduate students, selected from proposals due Nov. 20, would devote one third of their time to research at Manned Spacecraft Center, Ames Research Center, or Langley Research Center. Study areas would include physiological and psychological processes and medical aspects of manned space flight, changes in man's health and safety, design of advanced life support systems, man-machine integration, human augmentation devices, bioinstrumentation, habitability considerations, and use of flight environment to understand living systems better. Other areas announced by Dr. Charles A. Berry, NASA Director for Life Sciences, were possible existence of life elsewhere in universe, scientific explanation of life's origin, and detection and characterization from space of ecological phenomena on earth or other planets. Step-funded grants would not exceed $50 000 annually and would support maximum of 50% of scientist's salary. (NASA Release 72-175)

- NASA and Soviet Academy of Sciences had approved Summary of Results of second meeting of Joint Working Group on the Natural Environment, NASA announced. Group had exchanged information in Washington, D.C., May 8-12 and defined further investigations in structural geology, surface geology and geomorphology, archeology, hydrology, agriculture, soil moisture, and oceanography. Coordinated satellite observations in these areas were contemplated. Next meeting of Group, established under NASA and Academy of Sciences agreement of January 1971, would be held in Moscow at year's end. (NASA Release 72-171)

- President Nixon signed H.R. 15097, $8.393-billion Dept. of Transportation and related agencies appropriations act that contained $1.671-billion Federal Aviation Administration appropriation. Act became Public Law 92-398. (*PD*, 8/28/72, 1288; FAA PIO)

- Lockheed Aircraft Corp. and Japanese government were discussing production in Japan of twin-engine version of Lockheed TriStar transport aircraft, *Wall Street Journal* reported. McDonnell Douglas had proposed production in Japan of smaller-version DC-10. (*WSJ*, 8/22/72, 8)

August 23: Republican National Platform for science and technology was published by Washington *Evening Star and Daily News:* "We will place special emphasis on these areas in which break-throughs are needed: Abundant, clean energy sources; safe, fast and pollution-free transportation; improved emergency health care; reduction of loss of life, health, and property in natural disasters; rehabilitation of alcoholics and addicts to dangerous drugs. We will press ahead with the space shuttle program to replace today's expendable launch vehicles and provide low-cost access to space for a wide variety of missions, including those related to earth resources." (W *Star & News*, 8/23/72, A10)

- U.S.–U.S.S.R. treaty on limitation of antiballistic missile systems was ratified at joint meeting of commissions for foreign affairs of U.S.S.R. Supreme Soviet. (Tass, FBIS–Sov, 8/23/72, H1)
- Soviet Politburo member Mikhail A. Suslov, in speech before foreign affairs commissions of Supreme Soviet, endorsed formal ratification of U.S.–U.S.S.R. antiballistic missile treaty, but warned against U.S. attempts "to distort the spirit and letter of the treaty and interim agreement." He said U.S.S.R. would "take into consideration in its policy all changes that may appear in the position of the American side." (*NYT*, 8/24/72, 6)
- National Academy of Sciences program of postdoctoral staff fellowships under $250 000 grant from Alfred P. Sloan Foundation was announced by Dr. Philip Handler, NAS President. Some eight scientists would be selected annually to study aspects of science and public policy while participating in staff work of their own choosing within NAS, National Academy of Engineering, Institute of Medicine, and National Research Council. (NAS Release)

August 24: NASA's M2–F3 lifting body, piloted by William H. Dana, completed 16th flight from Flight Research Center. Objectives—to expand flight envelope and to obtain stability and control data at mach 0.95—were achieved. M2–F3 reached mach 1.3 and 20 400-m (67 000-ft) altitude. (NASA proj off)

- Tass reported completion of U.S.S.R.'s *Mars 2* and *Mars 3* missions. Spacecraft, launched May 19 and May 28, 1971, had reached Mars in November and December 1971 and had transmitted data on planet and atmosphere. *Mars 3* instrumented capsule had landed on Mars and transmitted data from surface for 20 sec. Preliminary findings were: temperature ranged from 163 K (− 166°F) to 286 K (55°F); soil had low conductivity, with Martian seas warmer than continents; surface-level atmospheric pressure was 50–60 newtons per sq m (5.5–6 millibars), 200 times less than earth's; carbon dioxide atmosphere broke up into carbonic oxide molecule and oxygen atom at 100-km (62-mi) altitude; atmosphere was 10 times less dense than earth's and was closer to surface; and magnetic field charges exceeded interplanetary background level 8 times. (FBIS–Sov, 8/25/72, L1)
- NASA announced publication of *NASA Patent Abstracts Bibliography* (SP–7039). New semiannual publication contained abstracts for 1892 NASA-owned inventions available for licensing. Bibliography—first of its kind undertaken by Government agency—implemented Administration's liberalized Federal patent policy announced in August 1971. (NASA Release 72–177)
- Sir James M. Lighthill, Research Professor of Royal Society of London, was presented in Moscow with two capsules containing lunar rock samples brought from moon by U.S.S.R.'s *Luna 16* (launched Sept. 12,

1970). In making presentation Vice President Aleksander P. Vinogradov of Soviet Academy of Sciences expressed hope that gift would "facilitate a further development of contacts between Soviet and British scientists." (*SF*, 12/72, 459)

- Safeguard antiballistic missile system was successfully tested by Dept. of Defense in Pacific in repeat of previously unsuccessful test of missile site radar for launching Sprint missile against intercontinental ballistic missile. (*W Post*, 8/26/72, 7)
- Philadelphia *Evening Bulletin* editorial commented on launch of *Copernicus* (*Oao 3*) Orbiting Astronomical Observatory [see Aug. 21–29]. "It's fitting that the satellite should be named for Copernicus . . . father of modern astronomy. The new satellite should add appreciably to man's knowledge of the universe, but sophisticated as it may be, it all began with that incomparable genius almost five centuries ago." (P *Bull*, 8/24/72)

August 25: Observations by Pioneer spacecraft during major explosions on sun [see Aug. 2 and 7] had differed from scientists' expectations and had provided new data on solar atmosphere, NASA announced. Measurements made by *Pioneer 9* and *10*, 212 million km (132 million mi) apart on direct line to sun, had shown that ionized gases traveled more slowly and increased in temperature as they moved away from sun. *Pioneer 9*, orbiting sun inside earth's orbit, had observed record solar wind speeds of 3.6 million km per hr (2.2 million mph) and had counted particles at levels 4000 times higher than usual. By time solar winds reached *Pioneer 10*, en route to Jupiter and 328 million km (204 million mi) from sun, winds had slowed to half speed measured at *Pioneer 9* and temperatures had risen to 2 000 000 K (3 600 000°F), far above usual 100 000 K (180 000°F). Although solar region 331 had rotated out of sight of earth, data were still being collected by *Pioneer 6, 7,* and *8*. (NASA Release 72–179)

- Dept. of Justice Criminal Div. was studying report submitted to it by NASA and transcript of Aug. 3 closed hearing by Senate Committee on Aeronautical and Space Sciences, *Houston Chronicle* reported. Study was to determine whether *Apollo 15* astronauts had violated any Federal law in carrying unauthorized stamp covers to moon. Senate Committee spokesman had told *Chronicle* Committee had no plans for open hearing on incident. (*H Chron*, 8/25/72)
- *The UFO Experience: A Scientific Inquiry* by Northwestern Univ. astrophysicist Dr. J. Allen Hynek was reviewed in *Science* by Dr. Bruce C. Murray of California Institute of Technology Div. of Geological and Planetary Sciences: Book was "more than just an attempt to justify scientific interest in UFO [unidentified flying object] phenomena." Dr. Hynek had criticized 1968 report of Univ. of Colorado physicist Dr. Edward U. Condon and group that said further study of UFOs could not be justified "in the expectation that science will be advanced thereby." Air Force, which sponsored Condon study, had terminated UFO Project Blue Book investigation following Condon report recommendation. Hynek book was "Hynek's version of what the Condon report should have been." Book discussed 80 UFO sightings, drawn mainly from Project Blue Book files, many unexplained by Condon. Book advocated "sufficient scientific respectability for the UFO subject to permit modest federal research funds to be awarded to it and new

data to be gathered without fear of ridicule." Dr. Hynek had "won reprieve for UFO's with his . . . provocative unexplained reports and his articulate challenge to his colleagues to tolerate the study of something they cannot understand." (*Science*, 8/25/72, 688-9)

August 26: Work reductions in manned space flight area announced by NASA Aug. 22 would eliminate about 1400 contract support personnel positions at Manned Spaceflight Center, *Los Angeles Times* reported. Kennedy Space Center was expected to dismiss about 500 workers after December launch of Apollo 17, and another 2000 technicians and engineers at end of Skylab program. Reductions necessitated by budget restrictions would force closing of KSC flight crew training after Apollo 17, reduction in force at White Sands Missile Range, and eventual closing of Space Environmental Simulation Laboratory, astronauts underwater training facilities, and altitude test chamber at MSC. "Even distribution of moon rocks . . . for display purposes" would be curtailed. (Chriss, *LA Times*, 8/26/72)

August 28: Apollo 17 spacecraft atop Saturn V booster, scheduled for launch toward moon Dec. 6, was rolled to Launch Complex 39, Pad A, at Kennedy Space Center. Roll-out was watched by about 5000 persons, including Apollo 17 astronauts. (Benedict, *M Her*, 8/29/72)

- Manned Spacecraft Center announced award of $226 256 contract to Martin Marietta Corp. to study system for handling space shuttle cargo in space. MSC engineers envisioned manipulators to handle payloads in shuttle orbiter as being 9-12 m (30-40 ft) long, electrically powered, and free-moving like human limbs. (MSC Release 72-189; *Av Wk*, 9/18/72, 9)

- People's Republic of China signed preliminary purchasing contract with British Aircraft Corp. in Peking for third Anglo-French Concorde supersonic airliner. PRC had signed similar agreement for two Concordes July 24. (BAC Release 14C/72)

August 29: Low-cost navigation aid Transim had been developed by Johns Hopkins Univ. Applied Physics Laboratory for Navy Space Projects Office, Baltimore *Sun* reported. Instrument, for shipboard use, was no larger than "ham" radio and gathered data from five Navy Transit satellites in earth orbit. (B *Sun*, 8/29/72)

- Consolidation of all engineering and development functions of Federal Aviation Administration into new service, Systems Research and Development Service (SRDS), was announced by FAA. Spencer S. Hunn, head of FAA's National Airspace System Program Office, would head new service, which would incorporate NASPO. (FAA Release 72-170)

August 30: U.S.S.R. launched *Cosmos 517* from Baykonur into orbit with 285-km (177.1-mi) apogee, 203-km (126.1-mi) perigee, 89.3-min period, and 64.9° inclination. Satellite reentered Sept. 11. (GSFC *SSR*, 8/31/72; 9/11/72; *Sov Aero*, 9/11/72, 70)

- Interagency Coordination Committee for Earth Resources Survey Program (ICC:ERSP) released *Annual Federal Report on Earth Resources Survey Programs*. Committee, established in February, was chaired by Dr. Homer E. Newell, NASA Associate Administrator. Report described activities and future plans of eight Federal agencies participating in U.S. Federal program in remote sensing of earth resources and environment: NASA; Depts. of Agriculture, Commerce, Defense, Interior, and State; Environmental Protection Agency; and Corps of Engineers.

NASA was responsible for development of space and aircraft technology for peaceful applications. Work included development of procedures, instruments, subsystems, spacecraft, and interpretive techniques for increasing basic knowledge of earth's atmosphere, land areas, and oceans and man's effect on these resources.

Corps of Engineers would evaluate contribution of remote sensing to water resources, marine science development, and emergency operations.

Dept. of Agriculture would identify applications where remote sensing could benefit agriculture and related natural resources and gather remotely sensed data to identify major crops and forest species, insect damage, crop disease, soil salinity, and moisture differences. Agency would also map surface water, record snowpack and soil and water temperatures, and note changes in land use.

National Oceanic and Atmospheric Administration in Dept. of Commerce was evaluating feasibility and cost-effectiveness of new remote sensing techniques. Dept. of Commerce also would operate Earth Resources Survey Data Center at Suitland, Md., for secondary users, including public. Dept. of Interior's Earth Resources Observation System (EROS) would be enhanced by space-based and aircraft remote sensing in achieving its management, conservation, and environmental objectives.

Environmental Protection Agency would use remote sensing for identification, quantification, and monitoring of pollutants. Dept. of Defense would derive global information on past, current, and future states of earth's environments.

Agency for International Development in Dept. of State was exploring remote sensing in international development, clarifying issues related to use for development purposes, disseminating information on NASA's ERS program, assisting developing countries to participate in ERS programs, and supporting remote sensing training activities. (Text)

- Marshall Space Flight Center engineers were conducting research on laser-doppler clear-air turbulence-detection system. System, ground-tested and installed in NASA's Convair 990 at Ames Research Center for flight tests, could detect small-scale atmospheric motions. Laser system, which might lead to first commercially available instrument for air turbulence measurements, was designed and built by Raytheon Co. for NASA Office of Aeronautics and Space Technology and U.S. Air Force's Cambridge Research Laboratories. (NASA Release 72-182)
- Marshall Space Flight Center announced appointment of Dr. Charles R. O'Dell as project scientist for Large Space Telescope (LST). Dr. O'Dell was Chairman of Univ. of Chicago Dept. of Astronomy and Astrophysics and Director of university's Yerkes Observatory. (MSFC Release 72-118)
- Col. Robert R. Wessels (USA) had been assigned to Marshall Space Flight Center as Director of Shuttle Construction Office, *Marshall Star* reported. Col. Wessels previously had been Deputy District Engineer for NASA Support, responsible for monitoring and coordinating district's design, engineering, and construction program at MSC and Mississippi Test Facility. (*Marshall Star*, 8/30/72, 1)
- President Nixon, arriving in Hawaii for meeting with Japanese Prime Minister Kakuei Tanaka, recalled previous visit "at the time of the

Apollo 13 flight, when we welcomed back brave men who hadn't succeeded but who came back, and it was one of those epics in American bravery which all of us wanted to pay tribute to." (*PD*, 9/4/82, 1314)

- Army Aviation Systems Command awarded research and development, cost-plus-incentive-fee contracts to Boeing Co. Vertol Div. and United Aircraft Corp. Sikorsky Aircraft Div. for engineering development phase of prototype helicopters for Utility Tactical Transport Aircraft System (UTTAS). Boeing Vertol would receive $91 million and Sikorsky Aircraft $61 million. (DOT note to correspondents)

August 31: Reassignment of *Apollo 15* Astronaut Alfred M. Worden from Astronaut Office at Manned Spacecraft Center to Airborne Science Office Space Science Div. at Ames Research Center was announced by Manned Spacecraft Center. Worden would use astronaut experience in developing and evaluating systems and procedures for ARC airborne science and space-shuttle-vehicle simulation studies and in earth resources surveys and high-altitude astronomy using NASA aircraft. (MSC Release 72–197)

- *Apollo 9* Astronaut James A. McDivitt retired from Air Force and received Air Force Distinguished Service Medal with Oak Leaf Cluster during ceremony at Bolling Air Force Base, D.C. McDivitt, who held rank of brigadier general, had announced retirement from NASA effective Sept. 1 [see June 7]. (Boldt, *W Post*, 9/1/72, C1)

- Retirement ceremony at Andrews Air Force Base, Md., honored L/G John W. O'Neill, Vice Commander of Air Force Systems Command. Gen. O'Neill, who was succeeded by L/G Edmund F. O'Connor, received Oak Leaf Cluster to Air Force Distinguished Service Medal from AFSC Commander, Gen. George S. Brown. (AFSC *Newsreview*, 10/72, 4)

During August: House Committee on Science and Astronautics' Subcommittee on International Cooperation in Science and Space published *U.S.–U.S.S.R. Cooperative Agreements.* Report of June hearings to explore scope of scientific and technical agreements signed during May 22–30 summit meeting in Moscow included texts of agreements and comments of experts. Dr. Philip Handler, National Academy of Sciences President, had said of space cooperation agreement: "The symbolism of the proposed link-up in space may prove to be the most significant aspect of the venture." Agreement was "symbolic of a mutual awareness that the planet earth is a small and fragile spaceship, that all men are its fellow passengers, and that the fate of the people of all nations is essentially a common fate." Subcommittee had concluded: "While a gradual amelioration of fundamental political conflicts and disagreements is the most that can be reasonably hoped for, the genuine cooperative spirit that has made these agreements possible opens the door to more effective communication and deeper understanding between the world's two predominant powers that could have the most profound effect on the future of mankind." (Committee Print)

- Flying model of winged shuttle orbiter by ERNO Raumfahrt-technik GmbH of West Germany would be air-launched during year to study subsonic dynamic stability, *Spaceflight* reported. Program, begun in mid-1960s by Federal German Ministry for Education and Science, had led to development of small pilotless flight model called Bumerang, which made first unguided flight off Heligoland Aug. 12, 1971. Research to date by ERNO specialists, participating with space shuttle team led by McDonnell Douglas Corp. of U.S. had been on winged (nonlifting body) orbiter models at German Research and Test Establishment for Aviation and Spaceflight. (*SF*, 8/72, 301)

- *Spaceflight* reported development of high-pressure topping-cycle, liquid-oxygen and liquid-hydrogen rocket engine with thrust of 196 to 245 kilonewtons (44 000 to 55 000 lbs) by Messerschmitt-Boelkow-Blohm in Ottobrunn, West Germany. Engine was planned for second stage of European Launcher Development Organization's Europa III launch vehicle. (*SF*, 8/72, 301)
- Aerospace Employment Project to transfer idled talents from aerospace industry to other endeavors in public sector began phase-out. Project had started in 1971 under $1.3-million contract from Dept. of Labor, Dept. of Housing and Urban Development, National League of Cities, and U.S. Conference of Mayors. Project had found reemployment for 297 out of 371 candidates—194, or 65%, in city, county, or state government work. (*NYT*, 7/16/72, 15)

September 1972

September 1: Air Force launched unidentified reconnaissance satellite from Vandenberg Air Force Base on Titan IIIB-Agena booster into orbit with 381-km (237-mi) apogee, 142-km (88-mi) perigee, 89.7-min period, and 110.4° inclination. Satellite reentered Sept. 30. (Pres Rpt 73; *SBD*, 9/6/72, 3)

- NASA's *Pioneer 10* Jupiter probe (launched March 2) was 320 million km (200 million mi) from earth and had traveled 450 million km (280 million mi) on 1-billion-km (620-million-mi) flight path to Jupiter. Spacecraft was traveling 80 000 km (50 000 mi) per hr and round-trip communication time at speed of light was 43 min. All spacecraft systems were functioning satisfactorily and 10 of 11 experiments were operating as planned. The 11th—infrared radiometer—would be turned on during flyby of Jupiter Dec. 3, 1973. Spacecraft was transmitting data at 2048 bits per sec to Deep Space Network's 64-m (210-ft) dish antenna, and 512–1024 bits per sec to 26-m (85-ft) dish antennas. Gas reserves for course changes were about twice the amount needed. *Pioneer 10* had been traveling in Asteroid Belt since July 15 and had encountered no problems. (NASA Release 72–184)

- NASA announced that *Pioneer 7* flight directors had located and revived *Pioneer 7* after one month of radio silence. Spacecraft was on far side of sun, more than 312 million km (194 million mi) away—so far away that round trip time for radio communication to spacecraft and back to earth was 35 min at speed of light. Controllers had reacquired *Pioneer 7* "in the blind" without knowing spacecraft's exact position or radio receiver's exact frequency and had set long-distance record for finding and reviving spacecraft that had turned itself off. Scientists had theorized that at *Pioneer 7*'s farthest distance from sun electrical power from spacecraft's solar cells had fallen below level required to operate spacecraft and scientific instruments. By leaving instruments off temporarily and thus reducing power requirements, turn-on of radio transmitter had been possible. (NASA Special Release)

- Dr. James C. Fletcher, NASA Administrator, in letter to staff congratulated NASA, contractor teams, and participating scientists for successful performance of *Oao 3* Orbiting Astronomical Observatory launched Aug. 21, 1972. To date, performance indicated "we have given the world a wonderful new window on the universe and a magnificent new tool of science worthy of the illustrious name it now bears [*Copernicus*]." Copernicus, father of modern astronomy, had changed world's thinking about nature of solar system and earth's place in it. "I believe we are going to see advances of comparable magnitude in our understanding of the universe, thanks to the new technology brought into being in our OAO program." (*NASA Activities*, 9/15/72, 180–1)

- Approval by NASA and Soviet Academy of Sciences of report of May 8–12, 1972, joint meeting of experts on lunar cartography in Washington, D.C., was announced by NASA. Experts had exchanged documents on

September 1

lunar coordinating systems and map making and agreed on joint development of principles for compiling lunar maps, on joint program to compile complete lunar map on scale of 1 to 5 000 000, and on development of common basic system of selendetic coordinates. Experts would exchange drafts and proposals for review, then meet to make recommendations for submission to cochairmen of Joint Working Group on the Exploration of Near-Earth Space, the Moon and the Planets. (NASA Release 72-186)

- Results of *Apollo 16* far-ultraviolet camera and spectrograph experiment on lunar surface were described in *Science* by Naval Research Laboratory scientists Dr. George R. Carruthers and Dr. Thornton L. Page. A major objective of experiment had been to obtain spectra and imagery of terrestrial atmosphere and geocorona in 500 to 1600 A range. Among data obtained were images and spectra in wavelength below 1600 A. These had provided spatial distributions and relative intensities of emissions due to atomic hydrogen, atomic oxygen, molecular nitrogen, and other species—some observed by spectrograph for first time. (*Science*, 9/1/72, 788–91)

- President Nixon and Japanese Prime Minister Kakuei Tanaka issued joint statement following their meetings in Hawaii. Statement described subjects covered in talks. They had "discussed cooperation in space exploration including Japan's goal of launching geo-stationary communications and other applications satellites. The President welcomed Japan's active interest in and study of the launching of a meteorological satellite in support of the global atmospheric research program." (*PD*, 9/11/72, 1333–4)

- NASA launched Aerobee 170 sounding rocket from White Sands Missile Range, carrying Johns Hopkins Univ. aeronomy experiment. Rocket and instrumentation performed satisfactorily. (SR list)

- *Wall Street Journal* article asked if U.S. supersonic transport revival was imminent: "Administration planners insist the U.S. will eventually build a supersonic transport. NASA researchers work up advanced 'second generation' designs, differing from the downed Boeing version. Backers claim progress against the sonic boom problem, plus the promise of developing a more profitable plane." (*WSJ*, 9/1/72, 1)

September 2: Navy *Triad OI–IX* Transit satellite was launched by NASA for Navy from Vandenberg Air Force Base by four-stage Scout booster. Orbital parameters: 838-km (520.7-mi) apogee, 743-km (461.7-mi) perigee, 100.6-min period, and 90.1° inclination. Objective of mission was to simplify navigation procedures by correcting long-term drift in satellite oscillator, reducing data-gathering time for navigation fix, demonstrating capability of experimental disturbance compensation system (DISCOS), obtaining performance data on radioisotope thermoelectric generator (RTG), evaluating environmental survey panel and thermal coating experiment, and providing operational satellite for Navy Navigation Satellite System.

DISCOS forced satellite to adhere to highly predictable orbit uninfluenced by forces of external atmospheric radiation. System, developed by Johns Hopkins Univ. Applied Physics Laboratory and Stanford Univ., compensated for external forces by providing equal and opposite thrust force. Success of DISCOS might lead to navigation satellites with orbits so precise and unchanging that their paths could be published in reference tables that would facilitate calculation of navigational fixes.

Triad OI–1X weighed 93.9 kg (207 lbs), was 1.7 m long and 0.8 m wide (5.5 x 2.5 ft), and was powered by Atomic Energy Commission RTG that would provide 30 w of power for five years. With three components—spacecraft, DISCOS, and RTG—extended separately in orbit, overall length was 3.7 m (24 ft). Satellite was transmitting data on nominal frequencies of 400 and 150 mhz. After six-month experimental phase satellite would be available to all Transit system users. NASA would be reimbursed by Dept. of Defense for cost of launch vehicle and services. (GSFC *SSR*, 9/30/72; DOD Release 632–72; NASA Scout Prog Off; Pres Rpt 73)

- Discovery that radio waves in constellation Cygnus had increased in energy more than 200 times, to become one of six strongest sources of radio waves in sky, was made by Canadian astronomer Dr. Philip C. Gregory. Source, Cygnus X–3, had been discovered originally by *Uhuru* (*Explorer 42* launched by Italy for NASA Dec. 12, 1970). Dr. Gregory had used National Research Council radiotelescope at Algonquin Park, Canada, in first observation of more than two- to fourfold leap in radio energy. Astronomers at seven radiotelescopes in U.S. and Canada began to study unprecedented event. Dr. Robert Hjellming of U.S. National Radio Astronomy Observatory at Green Bank, W. Va., had said possible explanations were "some kind of energy event around a black hole" where ancient neutron star had collapsed or that energy emanated from object that "suddenly wants to make itself smaller because gravitation is pulling it inward." (Cohn, *W Post*, 9/4/72)

September 4–10: Farnborough (England) Air Show emphasized Anglo-European aerospace cooperation. For first time aircraft, engines, electronics, and equipment of European manufacturers were displayed at formerly predominantly British show. Some 300 companies from U.K., Europe, British Commonwealth, and U.S. were represented. Twenty official government missions attended, with 5000 official guests from 120 countries, including 17 Chinese—interested in Anglo-French Concorde supersonic transport—and large group of Soviet observers. On Sept. 4, demonstration flight by Lockheed TriStar jet transport with British Rolls-Royce engines preceded demonstration flight by Concorde. Member of U.K. Environment Ministry said TriStar was "pointing the way to a silent future in the sky." He wished Concorde were "as quiet as the TriStar." Spokesman for British Aircraft Corp. said quieter and cleaner engines would be fitted to operational model of Concorde. (Cerutti, *C Trib*, 9/6/72, 28; Pinder, AP, *P Inq*, 9/7/72, 27)

September 5: Details of Mariner 10 mission to Venus and Mercury, first dual-planet flight, to be launched from Eastern Test Range in October or November 1973, were released by NASA. Jet Propulsion Laboratory would manage Mariner Venus mission project for NASA Office of Space Science. Project would include first use of one planet's gravitational field (that of Venus) to propel spacecraft to another and first Mercury exploration. Spacecraft's 500-kg (1100-lb) weight would include 78 kg (170 lbs) of scientific equipment—two TV cameras to take at least 8000 photos of planets and six experiments to return planetary and interplanetary data, with emphasis on Mercury. Objectives of experiments would be to provide celestial mechanics information during flight and

September 5

- physical characteristics of planets and their atmosphere; study structure of solar wind and its interaction with Venus and Mercury; search for magnetic field at both planets and measure interplanetary magnetic fields between orbits of earth, Venus, and Mercury; search for Mercury atmosphere and obtain data on Venusian atmosphere; measure temperature emissions from both planets and their atmospheres and correlate data with visible features; and measure charged particles over wide energy range to study solar charged-particle bombardment of Mercury and its possible atmosphere. Spacecraft would fly by Venus in February 1974 and by Mercury in March 1974, with closest approach to Venus 5300 km (3300 mi) and to Mercury only 1020 km (635 mi). (NASA Special Release, 9/5/72)
- French scientist Michel Siffre emerged from cave beneath Edwards Plateau near Del Rio, Tex., after six-month experiment to determine whether human time cycle was product of planetary system or of man's physiology. Results of experiment sponsored by NASA, French Speleological Institute, and French Defense Ministry would be used to serve astronauts on long-duration missions, jet pilots crossing time zones, and nuclear submarine crews on long cruises. Siffre told news conference that he had achieved 48-hr-day cycle twice, though average daily cycle underground had been 25 hrs. (Anderson, *W Post*, 9/5/72, A3)
- North American Rockwell Corp., winner of $2.6-billion NASA space shuttle prime contract, might retain up to 70% of contract money for itself and four principal subcontractors, *New York Times* reported. Subcontractors were Honeywell, Inc., International Business Machines Corp., General Electric Co., and American Airlines. Acceptance by NASA of 70% figure could be "major blow" to Grumman Aerospace Corp., *Times* said. Only hope for Grumman, loser in shuttle contract competition, and other aerospace companies to obtain substantial subcontracting on shuttle project would be to enter competition with NR subcontractors for whom NR might already have preference. (Andelman, *NYT*, 9/5/72, 3)
- Data from geomagnetic sensors near multimegaton nuclear explosion Cannikin detonated on Amchitka Island, Alaska, by Atomic Energy Commission in November 1971 were reported by National Oceanic and Atmospheric Administration. NOAA scientists Dr. W. P. Hasbrouck and J. H. Allen had found evidence that changes in underground stresses could produce small but measurable changes in earth's magnetic field. Finding should provide key to whether geomagnetic signals could be used as earthquake predictors. (DOC Release NOAA 72-115)
- Appointment of Dr. Alan M. Lovelace, Director of Air Force Materials Laboratory, as Director of Science and Technology for Air Force Systems Command was announced by Gen. George S. Brown, AFSC Commander. (AFSC Release 104.72)

September 5-7: Erts 1 Earth Resources Technology Satellite, launched by NASA July 23, gathered data over Florida including impact of Disney World to aid urban and regional planning. Images would be used to observe changes in transportation, motels, and other tourist-related activities. *Erts 1* data on area's water resources would enable Brevard County to inventory water supply, quality, and fluctuation. (Benedict, AP, *Today*, 9/6/72)

September 6: Marshall Space Flight Center announced plans for series of 20 water-entry simulation tests with solid-fueled rocket casing assembly. Tests would provide valuable data for assessment of parachute water recovery of space shuttle booster and aid in preliminary solid-fueled rocket motor design. Rocket assembly—representing 77% scale model of shuttle booster—was 30 m (100 ft) long, weighed 39 000 kg (43 tons), and was from previously fired motor. (MSFC Release 72–120)

- Plans for jointly owned communications satellite system were announced by Communications Satellite Corp. and MCI–Lockheed Satellite Corp., formed by Microwave Communications, Inc., and Lockheed Aircraft Corp. ComSatCorp had conditioned agreement on Federal Communications Commission permission to its entry into long-term contract with American Telephone & Telegraph Corp. ComSatCorp would then cancel its application for separate, multipurpose satellite system and joint MCI–Lockheed Corp. Each of the three companies would own part of system, with each member having less than 50% ownership. (W *Star & News*, 9/7/72, C8)

- Formation of Hughes Aircraft Corp. and British Aircraft Corp. team to study development of new-generation communications satellite had been confirmed by Hughes, *Wall Street Journal* reported. Hughes spokesman had said team was being formed because current global satellite system was expected to reach capacity by 1975. ComSatCorp, acting for International Telecommunications Satellite Consortium, had contracted with Hughes to study means of meeting increasing demand. BAC, primary contractor to Hughes, had participated in Intelsat IV program. (*WSJ*, 9/6/72, 3; *SF*, 12/72, 458)

- Artificial (gas-flame) fireplace logs were being produced by Martin Stamping and Stove Co. from materials developed for NASA, Marshall Space Flight Center announced. Materials—manufactured by Universal Atlas Div. of U.S. Steel Corp., Foote Mineral Co., and Vulcan Materials Co.—were fibrous insulation for Skylab and lightweight, refractory concrete used in Saturn rocket test stands to deflect hot exhaust. (MSFC Release 72–119)

- Grumman Corp. and American Aviation Corp. had agreed in principle to merge American Aviation with Grumman's commercial aircraft programs, *Wall Street Journal* reported. American manufactured sport, utility, and training aircraft. Grumman manufactured, sold, and serviced Gulfstream II corporate jet, and Ag-Cat agricultural aircraft. (*WSJ*, 9/6/72)

September 7: Skylab Orbital Workshop was delivered to Marshall Space Flight Center Director, Dr. Eberhard F. M. Rees, by McDonnell Douglas Astronautics Co. President Walter F. Burke in Huntington Beach, Calif., ceremony attended by Dr. James C. Fletcher, NASA Administrator. Workshop would be moved to U.S. Naval Facility at Seal Beach Sept. 8 for transfer to U.S.S. *Point Barrow* and 14-day voyage to Cape Kennedy. Workshop, weighing 25 000 kg (55 000 lb), was largest section of Skylab experimental space station scheduled for 1973 launch. (NASA Off Admin; Miles, *LA Times*, 9/8/72)

- NASA released three photos of Mars taken by *Mariner 9* Aug. 7. Pictures showed that Mars north polar icecap was shrinking and showed sedimentary systems during late spring. Fractured terrains, partially flooded by volcanic extrusions, were visible in center of planet. (LA *Her-Exam*, 9/7/72)

- Special awards ceremony marking outstanding success of NASA's *Mariner 9* Mars mission was held at Jet Propulsion Laboratory. Dr. James C. Fletcher, NASA Administrator—assisted by Dr. William H. Pickering, JPL Director, and Dr. Harold Brown, President of California Institute of Technology—presented 32 NASA medals. Dan Schneiderman, Manager of Mariner Project, was awarded NASA Distinguished Service Medal for "creative leadership in conception, organizational planning, and management of a highly efficient Mariner 9 team which . . . achieved all objectives of the most successfully productive mission to Mars." (JPL Release 624)
- Democratic presidential candidate Sen. George S. McGovern (D-S. Dak.) said during campaign visit to Manned Spacecraft Center that "future American space effort should focus on unmanned exploration." In statement released to press before his MSC tour, Sen. McGovern said he favored going ahead with earth resources experiment package (EREP). He accused Nixon Administration of eliminating some 500 000 aerospace jobs and 1.8 million positions in defense-related industries "without even thinking about new jobs to replace the ones destroyed." (Kneeland, *NYT*, 9/8/72, 20)
- Report released by International Institute for Strategic Studies in London said full-range intercontinental-ballistic-missile test by People's Republic of China was probable "before long," since PRC had built instrumentation ship for monitoring ICBM test. Report, *Military Balance 1972–73*, said 1972 was turning point in global strategic balance because year had brought codification of nuclear parity between superpowers and shift in Middle East with withdrawal of Soviet military advisers from Egypt. Strategic arms limitation agreement had curbed numerically the strategic arms systems of both superpowers, but the qualitative arms race continued. (UPI, W *Star & News*, 9/8/72, A6)

September 7–8: Seventh Aerospace Mechanisms Symposium at Manned Spacecraft Center was sponsored by MSC, California Institute of Technology, and Lockheed Missiles & Space Co., Inc. Symposium documented and discussed operational mechanism problems and experiences from Apollo program and considered unique mechanism areas for future programs like Apollo Soyuz Test Project and space shuttle. Use of docking-system computer program to analyze dynamic environment produced by two impacting spacecraft and attitude control systems was discussed by MSC engineer John A. Schliesing. Performance studies had indicated capture latching was most sensitive to vehicle angular-alignment errors and least sensitive to lateral-miss error. Load-sensitivity studies had shown that peak loads acting on Apollo spacecraft were lower than Apollo design-limit loads. Computer simulation of docking dynamics had been developed to support design and development of compatible docking system for international rendezvous and docking mission (IRDM). (NASA Preprint of Papers: MSC–07219; MSC PAO)

September 8: NASA and U.S.S.R. Academy of Sciences had approved recommendations of joint working session of scientists on planetary exploration held in Madrid May 19–20, NASA announced. Session had recommended that Mars study try to determine physical and chemical properties of planet's surface, seek evidence of biological activity, determine precise composition of Mars atmosphere, and ascertain role of water in Mars' evolution. Venus study should determine how and why planet became hotter than earth by determining planet's atmospheric composition, cloud structure, depth of sunlight penetration in Venusian atmosphere, and characteristics of Venusian surface. Outer planets study should emphasize exploration of Jupiter and Saturn and environments. Session agreed meeting should be held before year's end to define promising Mars landing sites and small group of U.S. and Soviet scientists should meet to consider probe of Venusian atmosphere. Madrid session had been held in accordance with recommendation of Joint U.S.–U.S.S.R. Working Group on Exploration of Near-Earth Space, the Moon and the Planets, made Aug. 6, 1971. Cochairmen of Madrid session were Dr. Georgy I. Petrov of Soviet Academy and Dr. S. Ichtiaque Rasool, Deputy Director of Planetary Programs in NASA Office of Space Science. (NASA Release 72–187)

- Publication of *A Current Index of Technical Briefs* (C1–1) by NASA and Small Business Administration was announced. Brochure contained abstracts from NASA's technical briefs filed during 1971 and 1972. NASA had published nearly 5000 technical briefs since 1963. First edition of new publication had been mailed to more than 50 000 small manufacturing and research and development companies. Brochure eventually would be published monthly. (NASA Release 72–188; NASA PAO)

- President Nixon transmitted to Congress 26th annual report on U.S. participation in work of United Nations. He noted in transmission message that 26th U.N. General Assembly had endorsed two treaties sponsored by U.S., Convention on the International Liability for Damage Caused by Space Objects and Convention on the Prohibition of Development, Production and Stockpiling of Bacteriological (Biological) and Toxin Weapons and on Their Destruction. International Civil Aviation Organization conference had adopted Convention for the Suppression of Unlawful Acts Against the Safety of Civil Aviation. (*PD*, 9/11/72, 1351–2)

September 8–9: Twenty-fifth anniversary of NASA Flight Research Center at Edwards, Calif., celebration included open house and Sept. 8 ceremony honoring original 14 employees who had participated with Air Force in successful 1947–1951 X–1 rocket aircraft program to exceed speed of sound in level flight. Dr. James C. Fletcher, NASA Administrator, presented plaques to seven original employees who attended. Since creation in 1947 as Muroc Flight Test Unit of National Advisory Committee for Aeronautics, predecessor of NASA, FRC had participated in advanced aircraft research programs that had extended boundaries of manned aeronautical flight from subsonic speed to 7200 km (4500 mi) per hr achieved by X–15 aircraft between 1959 and 1968. Other FRC programs included rocket-powered D–558 II Skyrocket, first aircraft to fly twice the speed of sound; X–5, forerunner of F–111, F–14, and B–1 aircraft; lunar landing research vehicle, prototype of Apollo lunar landing trainer; and XB–70 aircraft that had takeoff weight of more than 226 800 kg (500 000 lbs) and mach-3 cruising speed. FRC programs were continuing effort to improve safety, utility, and flying qualities of general-aviation air-

September 8-9: Flight Research Center at Edwards, Calif., celebrated its twenty-fifth anniversary by holding open house. Displays included the Hyper-3 (right foreground), flown in 1969 to evaluate remotely piloted research vehicle techniques; PA-30 general-aviation aircraft (center); F-111 supersonic aircraft (right rear), to be fitted with NASA's supercritical wing for testing; wind-tunnel test model of a B-52 aircraft with the Air Force F-15 advanced air-superiority fighter (left foreground); and the Parasev (suspended at upper left), used beginning in 1962 to test paraglider concepts for possible recovery of spacecraft and rocket boosters.

craft; M2, HL-10, and X-24A lifting-body flights to validate concepts for future manned spacecraft and aircraft; YF-12 program to acquire inflight data for future supersonic military and commercial aircraft; supercritical wing program to evaluate new NASA-developed wing's ability to permit modified F-8 aircraft to cruise economically at higher speeds; and digital fly-by-wire aircraft program to evaluate use of electronic flight control system to control aircraft completely. Fly-by-wire system was being considered for space shuttle use. (NASA Release 72-178; FRC Release 15-72; FRC *X-Press*, 9/15/72)

September 9: Tass reported results of *Venus 8* probe that had landed capsule on Venus July 22 after four-month flight. Wind had moved descending capsule horizontally in same direction as planet's rotation on its axis. Wind speed had declined from 48 m (160 ft) per sec at 45-km (28-mi) altitude to 2 m (6 ft) per sec at 11 km (7 mi). Key question during landing was whether sunlight reached Venusian surface or was absorbed completely by atmosphere and clouds. Preliminary data on changes in illumination at different altitudes showed that Venusian atmosphere considerably weakened sunlight but some sunlight did reach surface. Atmosphere consisted of 97% carbon dioxide, less than 2% nitrogen, and less than 0.1% oxygen and contained 0.01–0.1% ammonia at altitudes between 33 and 46 km (20 and 30 mi). Surface in landing area resembled earth granite rock in radioactive composition with 4.0% potassium, 0.0002% uranium, and 0.00065% thorium. (FBIS-Sov, 9/13/72, L7)

September 10: Navy's Manned Spacecraft Recovery Forces in Pacific and Atlantic would again cachet and cancel philatelic mail for Apollo 17 mission scheduled for Dec. 6 launch, Washington *Sunday Star and Daily News* reported. Covers would be processed in Norfolk, Va., and Honolulu; Pacific covers would be postmarked on board prime recovery ship on splashdown date. Atlantic covers would be canceled at Norfolk. (W *Star & News*, 9/10/72, G19)

- Central Switzerland Cultural Foundation awarded annual cultural prize to Alfred Waldis, Director of Swiss Transport Museum in Lucerne. (Swiss Trans Museum Announcement, 11/72)

September 11: NASA released prints from infrared photos of New York area taken by *Erts 1* (launched July 23). Black-and-white photo of New York City taken from 915-km (569-mi) altitude covering area 185 km (115 mi) square was covered by clouds, but was valuable as a general test of *Erts 1* instrumentation. Images of infrared radiation from which prints had been made clearly showed distribution of vegetation in New Jersey.

Dr. Paul D. Lowman of Goddard Space Flight Center said in interview that pictures of other areas had been of higher quality. "Our first discovery from ERTS is that all our maps, topographic as well as geologic, are out of date." In California's Monterey Bay area, scientists had discovered "a whole series of unsuspected faults" in earth's crust. While major earthquake faults ran northwest and southeast, newly discovered faults ran at right angles to mountains, northeast and southeast. About 15 faults, each extending about 16 km (10 mi), had been discovered in Monterey region and about 12 faults had been discovered near Lake Tahoe, Calif. Satellite picture of Dallas-Fort Worth area showed many new roads, reservoirs, suburbs, and airports that were not on area's most recent map, made three years before. (NASA Photos 72-HC-711, 72-4-241; Wilford, *NYT*, 9/12/72, 26)

- Mission officials for Viking Mars mission were concerned about tenuous values of dielectric constant (measure of electrical charge a substance could withstand at a given electrical field strength) estimated from radar observations of Mars surface, *Aviation Week & Space Technology* reported. Massachusetts Institute of Technology investigators had found strong signal in 2-5 dielectric-constant range and weaker signal above that. Values suggested Mars had either hard bedrock surface impenetrable by spacecraft sampling equipment or deep dust surface that could not support landing vehicle. (*Av Wk*, 9/11/72, 11)

- *Apollo 11* Astronaut Edwin E. Aldrin, Jr., was inducted into Aerospace Primus Club of Air Force Systems Command by Gen. George S. Brown, AFSC Commander, in ceremony at Andrews Air Force Base, Md. Aldrin, retired Air Force colonel, became 28th member of club restricted to military and civilian members of Air Force who had accomplished significant aerospace firsts. He was recognized for his historic flight to moon as pilot of *Apollo 11* lunar module *Eagle* July 16-24, 1969. (AFSC *Newsreview*, 10/72/16)

- Gen. Thomas D. White Space Trophy for 1971 was presented to L/G Samuel C. Phillips (USAF) for his achievements as Commander of Air Force Space and Missile Systems Organization (SAMSO). Trophy was awarded annually by National Geographic Society to Air Force member who made most outstanding contribution to U.S. progress in aerospace. Gen. Phillips had been Apollo Program Director in NASA and was currently Director of National Security Agency. (MSC Release 72-197)

- Reassignment of *Apollo 15* Astronaut Alfred M. Worden from Astronaut Office at Manned Spacecraft Center to Airborne Science Office, Space Science Div., at Ames Research Center became effective. (MSC Release 72-197)
- Director Robert R. Wilson of National Accelerator Laboratory told 16th Rochester International Conference on High Energy Physics in Batavia, Ill., that $250-million accelerator at Batavia had reached 300-bev proton energy level during August. Figure was four times the energy level of world's second most powerful atom smasher at Serpukhov in U.S.S.R. (Sullivan, *NYT*, 9/12/72, 23)
- Switzerland had dropped plans to purchase U.S. or French fighter-bomber aircraft, Swiss Defense Minister Rudolf Gnaegi said at Berne press conference. Reuters said later that announcement had ended weeks of uncertainty over whether Swiss would buy U.S. A-7G Corsairs or French Dassault Milans. (Reuters, *NYT*, 9/12/72, 4)
- *Huntsville* (Ala.) *Times* editorial criticized attitude of Democratic presidential candidate Sen. George S. McGovern (D-S. Dak.) toward U.S. space program: "The McGovern space-budget figure of $2 billion may not seem a paltry sum to many taxpayers and voters. But it becomes a piddling figure . . . when compared with massive governmental outlays for farm price-supports and other questionable multibillion-dollar expenditures. And even in endorsing this minimal space effort, Senator McGovern stresses the negative aspect of not wanting to see additional thousands of aerospace employees thrown out of work, rather than emphasizing the positive contributions of space to the expansion of man's scientific knowledge, to the strengthening of the technological know-how of American industry, and to the enhancement of the nation's image throughout the world." (*Huntsville Times*, 9/11/72)

September 11-12: American Institute of Aeronautics and Astronautics and American Astronautical Society sponsored Astrodynamics Conference in Palo Alto, Calif. Langley Research Center scientist David R. Brooks said in paper that LaRC was studying precursor flyby mission to examine two asteroids and periodic comet Forbes in 1977 to gain support for planetary program in 1980s. Since mission would require only exceptionally low velocities it could be accomplished at minimum cost and risk with modified Pioneer-F and Pioneer-G spacecraft. (Text)

September 11-13: American Society of Mechanical Engineers held annual Aerospace Div. Conference at Anaheim, Calif. Application of aerospace technology to societal problems and future of space exploration were discussed by officials of NASA; National Science Foundation; Depts. of Commerce, Transportation, and Interior; state and local governments; industry; and education. Aviation's role in earth resources surveys was described in paper by Deputy Director Clarence A. Syvertson and Airborne Science Office Chief Donald R. Mulholland of Ames Research Center: "At the present time, 275 companies are engaged in aerial surveys, but less than 50 of these are equipped for nonphotographic remote sensing. New sensor technology growing from and stimulated by the space program has made it possible to obtain vastly improved data." Federal Aviation Administration data showed that in 1971 "some 230 aircraft were used for aerial surveys and another 31 for patrolling." ARC used general-aviation, high-altitude research, and four-engine jet transportation craft. Remote sensing from aircraft provided data to improve environment, conserve natural resources, and contribute to economic growth. "The sensors required . . . are evolving rapidly, in part from well-established aerial reconnaissance technology, and in part from technology developed for the space program."

ASME Spirit of St. Louis Award was presented to *Apollo 11* Astronaut Neil A. Armstrong, first man on moon, for "meritorious service in the advancement of aeronautics and astronautics." (ASME Release 7/18/72; *NASA Activities*, 10/15/72, 207–10)

September 11–October 1: Series of six stratospheric balloons were launched to altitudes from 38 000 to 41 000 m (125 000 to 135 000 ft) from Cape Girardeau Municipal Airport, Mo., in project directed by Navy's Office of Naval Research and Raven Industries, Inc., in cooperation with four universities and NASA. Balloons, with volumes from 310 000 to 937 000 cu m (11 000 000 to 33 100 000 cu ft), achieved all flight objectives.

Sept. 11 flight carried instruments from Univ. of New Hampshire to study charge and isotopical composition of cosmic radiation particles and two ionization chambers from Atomic Energy Commission and Rice Univ. Sept. 14 flight carried Washington Univ., St. Louis, experiments, including one to measure isotopic abundances of silicon, sulfur, calcium, and iron nuclei.

On Sept. 14 and 24, Enrico Fermi Institute, Chicago Univ., counter experiments were launched to measure nuclear composition of very-high-energy cosmic rays. Univ. of California, at Riverside experiment was launched Sept. 19 to measure albedo neutrons from earth and solar neutrons from solar flares and quiet sun.

Sixth flight, Oct. 1, carried Goddard Space Flight Center experiment to measure flux of very-high-energy cosmic rays. (Navy, ONR Field Representative [Navy Balloon Proj])

September 12: NASA's M2-F3 lifting body, piloted by William H. Dana, reached 14 300-m (47 000-ft) altitude and mach 0.8 during 17th flight from Flight Research Center. Objectives—to expand flight envelope and obtain stability and control data at mach 0.95—were not achieved because two of four chambers failed to ignite. (NASA proj off)

- NASA launched Javelin sounding rocket from Churchill Research Range, Canada, carrying Goddard Space Flight Center fields and particles experiment. Rocket performed satisfactorily but scientific objectives were not met. (SR list)

September 13: Office of Management and Budget issued Apportionment-Reapportionment Schedule that placed in reserve $44.9 million of $50.5-million FY 1973 NASA appropriation for aeronautics. OMB impounded $27.5 million for quiet, experimental, short takeoff and landing aircraft (QUESTOL) program, $2 million from STOL experimental engine program, and $15.4 from $21-million engine retrofit program. (OMB PIO; *CR*, 10/14/74, H10111)

- *New York Times* editorial commented on data returned by Soviet *Venus 8* probe, which had landed capsule on Venus July 22: "Since the first Sputnik went into orbit a decade and a half ago, the world has seen so many major accomplishments in space by both the Soviet Union and the United States that the original sense of wonder has been largely lost. Yet even in today's jaded atmosphere, it must be recognized that what the Soviet scientists accomplished with Venus 8 was one of the great technological accomplishments of the age. The scientists and technicians responsible for this great feat deserve the world's congratulations. It is self-evident that this work needs to be continued with respect to other planets . . . and that it can be done by international cooperation." (*NYT*, 9/13/72, 42)

September 13–15: First national conference on remotely manned systems was held at Jet Propulsion Laboratory. Dr. Stanley Deutsch of NASA Office of Life Sciences said in keynote address that usefulness of remotely controlled machines was "virtually limitless" both in space and on earth. "Using a remote system to scout Mars in advance of manned missions will enhance the safety of the man when we do send him to Mars." Remotely controlled systems also could be used in future to rescue and return an incapacitated astronaut and provide a method for capturing unstable satellites. Remotely controlled systems technology developed for space was also being applied to nonspace applications such as Navy undersea manipulators, Air Force and Army remotely piloted aircraft, handling material in radiation laboratories, industrial safety, mining, medicine and hospitals, police work, and fire fighting.

Dr. Ewald Heer of JPL's Advanced Technical Studies Div. said "intelligence" of automated systems would have to be increased greatly to ensure success of future earth orbital and planetary satellites. "Remotely manned systems are man-machine systems that augment and extend man's sensory, manipulative and intellectual capabilities to remote places. However while machines may have superior strength and endurance and resistance to hostile environments, they still depend on human intelligence for decision-making in the performance of complicated tasks. Advanced engineering automation is essential in space station and shuttle operations and scientific data gathering must be microminiaturized for sustained interplanetary operations of two or more years." (JPL Releases 625, 626; Swain, Pasadena *Star-News*, 9/14/72)

September 14: Senate approved U.S.–U.S.S.R. Interim Agreement on Limitation of Strategic Arms by vote of 88 to 2. Agreement was passed after adoption of amendment proposed by Sen. Henry M. Jackson (D-Wash.) that urged President Nixon to seek future treaty which would not limit U.S. to levels of intercontinental strategic forces inferior to limits provided to U.S.S.R. (*CR*, 9/14/72, S14859, S14868–14913)

- Senate passed with technical amendment H.R. 10243, bill to establish Office of Technology Assessment for Congress and to amend National Science Foundation Act of 1950. Bill had been passed by House Feb. 8. (*CR*, 9/14/72, S14915)
- NASA Management Issuance signed by Dr. James C. Fletcher, NASA Administrator, designated members of NASA Historical Advisory Committee: Dr. Louis Morton, Chairman of Dartmouth College Dept. of History, Committee Chairman; Dr. A. Hunter Dupree, Professor of History, Brown Univ.; Dr. Melvin Kranzberg, Callaway Professor of the History of Technology, Georgia Institute of Technology; and Dr. Rodman W. Paul, Professor of History, California Institute of Technology. Dr. Elting E. Morison, Institute Professor, Massachusetts

Institute of Technology, was appointed to committee in October. (NASA NMI 1156.2D; NASA Hist Off)

- Decision to place long-range orders for equipment and material to construct six Anglo-French Concorde supersonic transports was announced by British and French officials at Paris press conference. Final decision on whether to assemble aircraft would be deferred pending negotiations with potential buyers. French Transport Minister Robert Galley said first 16 Concordes were "practically sold." Decision "taken today should allow us to fix delivery dates in the course of negotiations with companies now in progress." (Agence France-Presse, *NYT*, 9/24/72, 14)

- Use of reconnaissance satellite under control of United Nations to spy out opium gardens throughout world was recommended by Gen. Lewis W. Walt (USMC, Ret.) in testimony before Senate Judiciary Committee's Subcommittee on Internal Security. Gen. Walt had made worldwide study of heroin addiction and smuggling. (UPI, *D News*, 9/15/72)

September 15: U.S.S.R. launched *Cosmos 518* from Plesetsk into orbit with 308-km (191.4-mi) apogee, 205-km (127.4-mi) perigee, 89.6-min period, and 72.8° inclination. Satellite reentered Sept. 24. (GSFC *SSR*, 9/30/72; *Sov Aero*, 9/25/72, 82)

- Findings to date of investigations into *Apollo 15* commercialism were released by Senate Committee on Aeronautical and Space Sciences and by NASA. Senate Committee reported on Aug. 3 closed hearing that received testimony from NASA management and *Apollo 15* astronauts on commercialization of postal covers and replicas of "Fallen Astronaut" sculpture that was carried to moon during mission. Hearing had revealed "no basic new facts" on incidents. NASA had disciplined astronauts for carrying unauthorized covers. There was "no indication at this time that the Apollo 15 crew either profited or intended to profit from" sales of sculpture. Sale of sculpture by artist Paul Van Hoeydonck "appears to be a direct contradiction of the intent of the Apollo 15 astronauts." Committee said Dept. of Justice had requested copy of hearing transcript and was reviewing situation for possible violation of Federal statutes.

  Committee said its continuing investigation had determined, after Aug. 3 hearing, "that several astronauts had accepted payment for autographing blocks of stamps. NASA was "continuing its investigation of this latest incident and is determining what management actions it will take as a result." Committee expressed concern "with the adequacy of NASA management procedures which would permit such incidents to occur or create an atmosphere or an environment which might encourage misconduct." It also was concerned with "awareness of responsible management personnel as to their responsibilities for the recognition of and need for initiation of timely and appropriate action on sensitive events coming to their attention."

  Dr. James C. Fletcher, NASA Administrator, released report on *Apollo 15* commercialism and disciplinary actions taken by NASA. Investigations and reviews had revealed weaknesses in NASA's administrative procedures, management communication channels, and internal operating relationships. Actions were under way to remedy weaknesses, including revision of policy and procedures on articles to be carried on manned flight missions. No more than 12 personal items weighing no more than 0.2 kg (0.5 lb) would be carried by each flight astronaut. All personal items would be approved by NASA Administrator. Items that could be commercially exploited by recipients were prohibited. List of items would be publicized no later than 30 days after flight's completion. Future missions would carry official flight kit containing items appropriate for official presentation by Government. Items would be approved by Administrator and announced before launch. Commercialism of flight kit was prohibited.

NASA PAO stated in response to inquiries about sale of autographs by astronauts, "NASA has determined . . . autographs were signed by the astronauts on their own time while not on duty, that no misuse of Government property or personnel was involved, but that a procedural infraction did occur in that no NASA management advice was sought prior to these actions." NASA employees had been "personally admonished . . . for this infraction of NASA procedures" and facts had been reported to Dept. of Justice. (Texts; NASA PAO)

- Library of Congress Congressional Research Service published *Cape Kennedy or Cape Canaveral? A Brief History, a Background, and an Analysis of S.J. Res. 193.* Report on Senate bill to return to Cape Kennedy, original name of Cape Canaveral, concluded that "relevant issues of fact are not in question in this legislative decision." Historical significance of Cape Canaveral was "genuine" and majority of Cape residents and Florida State Legislature favored return to old name. "Nor is the procedural legitimacy of the permanent change of name to 'Cape Kennedy' seriously questioned. The nature of the official act is generally accepted as a national gesture of tribute and respect to a martyred President. Evidence from historical precedent is scant; geographic names have often been applied and altered to honor persons, and these have subsequently been changed for a variety of reasons." Question was "whether sufficient cause exists . . . to reverse the act of November 27, 1963, changing the name of Cape Canaveral to Cape Kennedy." (Text)

- President Nixon transmitted Convention for the Suppression of Unlawful Acts Against the Safety of Civil Aviation to Senate for advice and consent to ratification. Convention had been signed at Montreal conference of International Civil Aviation Organization Sept. 23, 1971. In transmission message President said: "The problem of sabotage, armed terrorist attacks, and other criminal acts against aircraft and air travelers poses an increasingly grave threat to civil aviation around the world." Convention and hijacking convention adopted at The Hague in December 1970 were "vitally important to achieve safe and orderly air transportation for all the people of the world." (*PD*, 9/18/72, 1379)

- Communications Satellite Corp. announced new record volume of TV via satellite during 17-day coverage of Olympic Games in Munich, West Germany. Total of 1005 half-channel hours of satellite time—more than twice TV use of satellites for any event since commercial satellite service began in 1965—had been used for 144 different international telecasts via four Intelsat satellites during Olympics. (ComSatCorp Release 72-48)

- Release of first 24 in series of supersonic transport (SST) follow-on technology reports was announced by Federal Aviation Administration. Reports covered noise reduction, supersonic inlets, flight controls, fuel tank sealants, and titanium structures. (FAA Release 72-179)
- *Science* editorial said great development of generation had been "sharply increased realization of our mutual interdependence as citizens of this planet. The concept has been dramatized by the phrase 'spaceship earth' and given worldwide political recognition in the holding of the Stockholm Conference" [see June 5-16]. (Abelson, *Science*, 9/15/72)

September 16: U.S.S.R. launched *Cosmos 519* from Plesetsk into orbit with 313-km (194.5-mi) apogee, 204-km (126.8-mi) perigee, 89.7-min period, and 71.2° inclination. Satellite reentered Sept. 26. (GSFC *SSR*, 9/30/72; *Sov Aero*, 10/2/72, 6)

- North American Rockwell Corp. released statement that it was awarding $8 million in subcontracts to Grumman Corp. for engineering work on space shuttle. McDonnell Douglas Corp. would receive subcontracting worth $4 million, and $200 million worth of subcontracts would be parceled to companies throughout U.S. early in 1973. Grumman subcontracts would preserve jobs of some 300 top engineers and McDonnell Douglas subcontract, 140, *New York Times* reported. (NR Release; Andelman, *NYT*, 9/17/72, 23)

Air Force succesfully flight-demonstrated powered balloon system (POBAL) that had fly-and-float capability readily adaptable to military use in electronic fields. POBAL—coupling free helium-filled balloon to propulsion system—flew from White Sands Missile Range for three hours after reaching altitude, driven by 11.9-m (39-ft) propeller connected to 9-kw (12-hp) electric motor. Steering rudder, connected to lengthy tail boom, was 2.7 m (9 ft) high and 0.8 m (2.5 ft) wide. System was controlled by radio signals from the ground. POBAL would extend station-keeping capabilities by combining free balloon flight with propulsion and autopilot control system, with planned 12-hr flight under power at speed of 15 knots. It would float powerless remainder of time. (AF Release 129.72)

- NASA launched Aerobee 170 sounding rocket from White Sands Missile Range, N. Mex., carrying Massachusetts Institute of Technology astronomy experiment. Rocket and instrumentation performed satisfactorily. (SR list)

September 16-17: Air Force's 25th anniversary was celebrated at Andrews Air Force Base, Md., with aerial demonstrations of C-130 low-altitude parachute-extraction system (LAPES), HC-130 aircraft refueling HH-53 helicopter of Military Airlift Command, KC-135 aircraft, refueling B-52 bomber, and fire-suppression drill by H-42 helicopter. Static display included full-scale model for NASA launch rover, F-111 jet fighter aircraft, FB-111 swing-wing bomber, SR-71 photographic reconnaissance aircraft, T-38 supersonic trainer aircraft, and C-5 Galaxy, world's largest aircraft. [See Sept. 18.] (Program)

September 17-21: Air Force Assn. held 26th annual National Convention in Washington, D.C. Convention unanimously adopted policy resolutions including endorsement and support of space shuttle program, continuation and expansion of ballistic missile and military space systems technology, and increase in U.S. defense program research and development effort to level "at least equal to that of the Soviet Union."

Convention adopted 1972-1973 Statement of Policy Sept. 19. Statement said new environment created by Strategic Arms Limitation Talks (SALT) "places a high premium on added protection for our current strategic bomber force, on improved ballistic missile technology as an antidote to numerical inferiority, on more accurate and consistent surveillance capabilities, and on more secure and dependable communications." Nothing in SALT agreements impeded size and momentum of U.S.S.R.'s research and development effort. Soviet effort "for the past several years has surpassed that of the United States by a wide margin. The possibility of technological surprise has not diminished under SALT, unless the United States undertakes and maintains a military research and development program that is comprehensive, consistent, and competitive." If U.S. emphasized efforts on qualitative force improvement, enhanced forces survivability, and improved conventional weapon delivery, "risks involved in the SALT agreements can likely be kept within tolerable limits."

AFA awards presented during convention included Theodore von Kármán Award to L/C Donald G. Carpenter, Commander of 18th Surveillance Squadron, Air Defense Command, for "advancing the nation's space defense capability." David C. Schilling Award was presented to 1st Strategic Reconnaissance Squadron, Beale Air Force Base, Calif., for "pioneering operational long-range flight while performing vital defense missions with the SR-71 aircraft." Award was accepted by Col. Jerome F. O'Malley, Commander of 9th Strategic Reconnaissance Wing. (*AF Mag*, 11/72, 809, 54-5, 60-1)

September 18: New York Times published interview with Harris M. Schurmeier, Jet Propulsion Laboratory's manager of proposed 1977 Jupiter-Saturn mission. Two 680-kg (1500-lb) spacecraft launched by Titan IIIE-Centaur boosters would swing past Jupiter in 1979 and, gaining velocity from pull of Jupiter's gravity, speed past Saturn by 1981. Mission would cost between $250 million and $350 million—less than half cost of Grand Tour mission proposed earlier. Since mission would provide first close-up observation of Saturn, some 200 scientists had already submitted proposals for experiments. Proposals were being evaluated and spacecraft design would begin by end of 1972. (Wilford, *NYT*, 9/18/72, 15)

- Twenty-fifth anniversary of U.S. Air Force. First Secretary of the Air Force W. Stuart Symington had been sworn in Sept. 18, 1947, effective date of transfer of Army air activities to Dept. of the Air Force established by Armed Forces Unification Act of July 26, 1947.

Publications during September commemorated Silver Anniversary. *Airman* editorial noted it has been more than 50 yrs since advent of military aviation "and it is virtually impossible to divide the technological achievements and amazing advances over that span of years into categories like 'pre-autonomy' or 'post-autonomy.' " It was spirit that counted—"that indomitable, driving, probing, questioning, seeking perseverance, and sheer genius that has taken man from Kittyhawk to the moon in what is really an incredibly short period of time."

Air Force System Command's *Newsreview* traced historical highlights of its bases and organizations. Arnold Engineering Development Center (AEDC) near Tullahoma, Tenn., had been established in 1953 at recommendation of aeronautical scientist Dr. Theodore von Kármán. Center currently consisted of 40 test units that provided "simulated flight-test capability ranging in speed to more than 20 times the speed of sound and to altitudes around 1,000 miles [1600 kilometers]." Center's replacement value was $840 million. Patrick Air Force Base, Fla., had grown with space program. Launch of V-2 from Eastern Test Range at Patrick in 1950 "was the first step to the moon. That goal, now accomplished five times, leaves the future of space exploration on the threshold of a dream."

Future of the Air Force was discussed by Gen. John D. Ryan (USAF) in *Armed Forces Journal* issue dedicated to Air Force Silver Anniversary. "It would be very easy . . . to take off on rhetorical flights of overcoming the unlimited challenges of aerospace. The challenges do exist; and they could conceivably mean that the future of the Air Force will include the use of remotely piloted vehicles, hypersonic aircraft, and the laser. But if these, and anything else man can envision and produce, do exist in the future, they will only be the means for Air Force people to do their basic job of providing this country with the aerospace power required for defense and security. As in the past 25 years, success in the future depends heavily on the dedication and quality of Air Force people." (*Airman*, 9/72, 25; AFSC *Newsreview*, 9/72, 6, 9; *AFJ*, 9/72; Dept of AF, *A Chronology of American Air Space Events*, 1/3/61)

- Envelopes carried to moon aboard *Apollo 15* and later sold to stamp dealers had been ordered by Harold G. Collins, Chief of Kennedy Space Center Mission Support Office, Manager John H. Jacobs of Brevard Printing Co. told press. Collins had asked that bill be sent to Hughes Enterprises in Las Vegas, Nev. Associated Press said Collins and NASA officials and astronauts at Manned Spacecraft Center had declined to comment on Jacobs' statement because matter was under investigation by Dept. of Justice. (AP, *W Post*, 9/19/72, A3)
- U.S.S.R. had begun staging electronic-intelligence-gathering reconnaissance flights off U.S. East Coast, with aircraft returning to air base near Havana, Cuba, *Aviation Week & Space Technology* reported. Some U.S. officials believed U.S.S.R. was establishing precedent for using Cuba as base from which to launch and recover reconnaissance satellites. In first reported incident two naval reconnaissance versions of Tu-95 Bear turboprop aircraft made 12-hr flight from Cuba up route 80 km (50 mi) off U.S. coast to point north of Norfolk, Va. Flights had been detected by radar on U.S.S. *Forrestal* and had been monitored by F-4 and A-7 aircraft. Although Tu-95s had flown off U.S. East Coast for past several years, they had not taken off from Cuba, flown mission, and then returned to Cuba before. (*Av Wk*, 9/18/72, 11)

September 18–21: Twentieth International Congress of Aviation and Space Medicine was held in Nice, France. More than 500 scientists from 54 countries participated. Dr. Charles A. Berry, NASA Director for Life Sciences, discussed human problems in long-duration space flights: Long-term bedrest data offered some clues to potential effects of weightlessness on physiological systems. There were indications that some zero-g effects were self-limiting. Collection of good inflight data for longer periods, coupled with ground-based experiments, was vital in planning for longer duration space flight. Vestibular problems had not been great enough for U.S. crews to require preadaptation program. "Such a program must, however, be considered as the number of potential crewmen is increased, particularly as scientists with no test pilot experience are added to crews. However, I do not at the present time see any physiological requirement for the addition of artificial gravity for long duration space flight."

Soviet delegation was headed by Dr. Oleg G. Gazenko, Director of Institute of Space Biology of Soviet Academy of Sciences. (Text; FBIS–Sov, 10/3/72, A7)

September 19: Cosmos 520 was launched by U.S.S.R. into orbit with 39 319-km (24 431.7-mi) apogee, 652-km (405.1-mi) perigee, 710-min period, and 62.8° inclination. Western observers of Soviet space program noted that spacecraft reached orbit similar to that for communications satellites, although perigee was slightly higher and inclination slightly different from orbits of past Molniya satellites. U.S.S.R. later announced launch under name *Cosmos 520*. [See also Sept. 28.] (*Spacewarn*, 10/3/72; GSFC *SSR*, 9/30/72; *SBD*, 9/26/72, 106)

- House Committee on Public Works favorably reported H.R. 16645, bill to amend Public Buildings Act of 1959 to provide for construction of Dwight D. Eisenhower Memorial Bicentennial Civic Center in District of Columbia. Committee in Report recommended honoring certain members of Congress by naming public works buildings for them, including renaming Jet Propulsion Laboratory in Pasadena, Calif., "H. Allen Smith Jet Propulsion Laboratory." New name would honor Rep. H. Allen Smith (R-Calif.), who would be retiring after 16 yrs in House. Laboratory was operated by Cal Tech for NASA. (H Rpt 92–1410)

- NASA and Soviet Academy of Sciences had approved results of fourth meeting of joint working groups on Apollo-Soyuz Test Project (ASTP) at Manned Spacecraft Center July 6–18, NASA announced. Groups had agreed to base future work on three ASTP documents: Project Technical Proposal, Organization Plan, and Project Schedules. Number of working groups would be increased to five. Groups had agreed to launch Soyuz spacecraft before Apollo spacecraft and had decided that sufficient progress had been made to proceed with system design and development for 1975 joint mission. Areas requiring future study included possible crew transfer by extravehicular activity (EVA) in emergency situation, possibility of changing spacecraft pressure to shorten time required for crew transfer, TV system for docking, and choice of specific July 1975 date for mission. Next meeting of working groups was scheduled for October in Moscow. (NASA Release 72–190)

- Grumman Aerospace Corp. had said it would assume major portion of work it had been subcontracting on its F–14A jet fighter aircraft for Navy, *Wall Street Journal* reported. Grumman President Joseph G. Gavin, Jr., had said plan to reduce subcontracting was necessary to "increase program efficiency and reduce costs." First subcontractor to feel effect would be Republic Aviation Div. of Fairchild Industries Inc. (*WSJ*, 9/19/72, 13)

- Secretary of Transportation John A. Volpe announced award of $498 487 Federal Aviation Administration contract to Council for Airport Opportunity for training program to assist some 900 minority workers in New York area in qualifying for better jobs in aviation. (DOT Release 80-72)

September 19-21: U.S.-U.S.S.R. Joint Committee on Cooperation in the Field of Environmental Protection met in Moscow to discuss joint projects to benefit and protect environment of both countries. Committee had been established under treaty signed in Moscow by President Nixon and Soviet President Nikolay V. Podgorny May 23. Memorandum of Implementation of Environmental Agreement, signed at meeting, provided for 30 joint environmental projects to protect cities, farms, rivers, lakes, and air. Projects would include exchange of U.S. and Soviet scientists and would focus on air and water pollution, oil spills, seismic research, urban problems, pest-management to reduce use of chemical pesticides, atmospheric pollution by supersonic transports, and permafrost. (Smith, *NYT*, 9/22/72, 1; *Sci Pol Rev*, Vol 5, No 4)

September 19-27: Annual conference of Soviet and French scientists and specialists on cooperation and research in peaceful uses of space was held in Tbilisi, U.S.S.R. Conference examined results of space cooperation in 1971 and plans for joint work in space chemistry, meteorology, biology, medicine, and communications and in aeronautics. Also discussed were data on northern lights obtained by French satellite *Oreol* (launched by U.S.S.R. Dec. 27, 1971) and study of low-energy particles by *Prognoz 2* (launched June 29). (Tass, FBIS-Sov, 9/27/72, L8; 10/6/72, F5)

September 20: Astronauts Robert L. Crippen, Dr. William E. Thornton, and Karol J. Bobko emerged from Skylab simulator at Manned Spacecraft Center, where they had been isolated since July 26. They had conducted Skylab medical experiments altitude test (SMEAT) to gather medical data to compare with data from actual missions. Astronauts said that, although they were very glad to be out, confinement had not been as difficult as expected. Astronauts were slightly weaker and their muscles were slightly smaller, but they had experienced no changes that would affect Skylab plans. Crew had been on space food diet during test and would remain on same diet for two weeks while doctors evaluated effects of diet. (AP, B *Sun*, 9/21/72, A3; AP, *C Trib*, 9/21/72, 1)

- Dr. James C. Fletcher, NASA Administrator, denied reports that he had announced he might resign after Apollo 17 mission in December and return to Univ. of Utah. Reports had followed Administrator's dinner speech before Knife and Fork Club in Salt Lake City, Utah, Sept. 19. Dr. Fletcher—who had headed NASA since April 27, 1971—issued reply to queries received at NASA Hq. saying that "he had given no thought to resigning from NASA and could not possibly commit [himself] to going back to the University of Utah at this time." (NASA PAO; Off of Admin; *SBD*, 9/21/72, 84, 90)

- *Industrial Research* magazine's IR-100 award for 1972 had been awarded to Lewis Research Center and Arthur D. Little, Inc., for process for producing pure, very-high-strength, single-crystal, refractory ceramic fibers, LeRC announced. Award, given annually to developers of 100

September 20

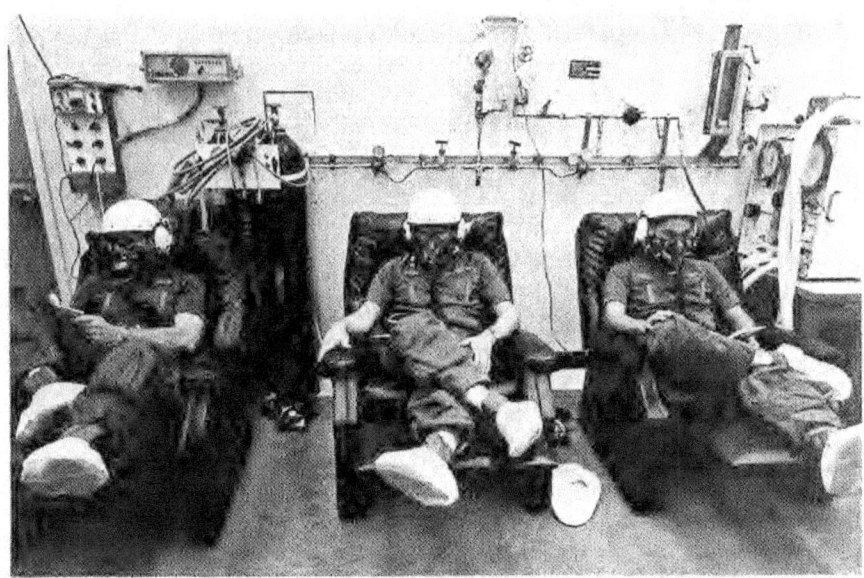

September 20: *Astronauts Dr. William E. Thornton, Karol J. Bobko, and Robert L. Crippen (shown left to right above in the altitude test chamber at Manned Spacecraft Center), emerged after isolation since July 26 in the Skylab medical experiments altitude test (SMEAT) to gather data for the 1973 Skylab Orbital Workshop mission. The Workshop had been delivered to Marshall Space Flight Center Director Eberhard F. M. Rees by McDonnell Douglas Astronautics Co. in a Huntington Beach, Calif., ceremony Sept. 7, for transfer to Cape Kennedy. The last components for the space station—the mated airlock module and multiple docking adapter—reached Kennedy Space Center Oct. 6. The airlock module flight version was photographed in the clean room at McDonnell Douglas in St. Louis, Mo., before mating.*

most significant new products selected from some 1000 entries throughout U.S., had been received by LeRC every year except one since Center first entered competition in 1966. (LeRC Release 72–82)

- First Lady Patricia Nixon visited Ames Research Center. She was received by ARC Director, Dr. Hans Mark; witnessed demonstration takeoff and landing by ARC augmentor-wing jet, short takeoff and landing (STOL), research aircraft; and was taken for "ride" in flight simulator for advanced aircraft. (ARC *Astrogram*, 9/28/72, 1)
- Plans for 1976 private space probe "Mankind 1" were described by spokesmen of Committee for the Future at Beverly Hills, Calif., press conference. Committee of private citizens proposed to buy U.S. or Soviet spacecraft and launch system for one of three proposed missions: orbiting laboratory, 14-day mission to place giant illuminator

with one sixth moon's reflective capacity in orbit 650 km (400 mi) above earth, or combination of both missions. Mankind 1 was successor to Project Harvest Moon which Committee had hoped to finance by selling moon rocks. Committee Chairman John J. Whiteside said first project had been scrapped after July 18 meeting with NASA officials who were skeptical about possibility of flying lunar mission after Apollo 17. (Sweeney, *LA Times*, 9/21/72)

- Tu-144 Soviet supersonic transport on flight from Moscow to Tashkent, Uzbekistan, flew 3000-km (1860-mi) distance in 1 hr 57 min, setting flight time record. (Tass, FBIS–Sov, 9/20/72, K1)
- National Oceanic and Atmospheric Administration announced appointment of Dr. George H. Ludwig as Director of System Integration. Dr. Ludwig, who had assumed position in early September, had been Associate Director for Data Operations at Goddard Space Flight Center since 1971. (NOAA Release 72–120)

September 21: North American Rockwell Corp. Downey Div. announced receipt of $5 500 000 contract from Air Force Systems Command's Space and Missile Organization (SAMSO) to build one-of-a-kind satellite for late 1973 launch. Spacecraft would be orbiting base from which quartet of instruments would gather scientific data for several Government agencies. (NR Release)

September 22–25: NASA launched *Explorer 47* (IMP–H) Interplanetary Monitoring Platform from Eastern Test Range at 9:20 pm EDT by three-stage, long-tank, thrust-augmented Thor-Delta booster. Satellite entered transfer orbit with 237 796-km (147 759.6-mi) apogee, 246-km (152.9-mi) perigee, 7365-min (5.11-day) period, and 28.6° inclination. Primary objective was to provide more detailed understanding of dynamics of regions discovered and broadly surveyed by previous earth-orbiting IMPs and lunar IMP by obtaining scientific data during period of decreasing solar activity. *Explorer 47* would continue studies of interplanetary radiation, solar wind and energetic particle emissions, and magnetic fields in earth's environment from orbit halfway to moon.

Apogee kick motor was fired Sept. 25 to place *Explorer 47* in circular orbit with 235 639-km (146 419.3-mi) apogee, 201 599-km (125 267.8-mi) perigee, 12.3-day period, and 17.2° inclination. By Sept. 25, 10 experiments had been commanded on and were operating satisfactorily. Three remaining experiments would be turned on about Oct. 5.

Explorer 47 was drum shaped, weighed 378 kg (833 lbs), and carried 13 experiments to measure energetic particles, plasmas, and magnetic and electric fields. Experiments were provided universities, industry, National Oceanic and Atmospheric Administration, Atomic Energy Commission, and Goddard Space Flight Center. *Explorer 47* was ninth spacecraft in IMP series. First—*Explorer 18* (IMP–A)—had been launched Nov. 26, 1963, and most recent—*Explorer 43* (IMP–I)—had been launched March 13, 1971. Four of the eight previously launched spacecraft were still operating. (NASA proj off; NASA Release 72–185)

September 22: Skylab Orbital Workshop, scheduled to carry three-man crews in earth orbit in 1973, arrived at Kennedy Space Center after two-week barge trip from McDonnell Douglas Co.'s Huntington Beach, Calif., plant. Apollo Telescope Mount (ATM) was flown to KSC from

Manned Spacecraft Center, where it had been undergoing thermal testing since mid-July. Airlock module (AM) and multiple docking adapter (MDA) were scheduled to arrive at KSC Oct. 9 and instrument unit Oct. 22. (UPI, *NY News*, 9/23/72; KSC Releases 268–72, 270–72)

- Soil sample collected by *Apollo 16* Astronaut Charles M. Duke, Jr., from natural hole beneath 3-m (10-ft)-high "Shadow Rock" on moon had been found to have characteristic resonance identical to that of sunlit soil near by, Naval Research Laboratory announced. Scientists were agreed that sample had not been in direct sunlight since being buried by rock propelled by ancient meteorite collision creating North Ray Crater on moon. NRL investigators were considering two explanations for similarity between soil under rock and exposed soil. Either North Ray Crater had been formed relatively recently or tens of millions of years of solar heat had not affected characteristic resonance. Either way, experiment had revealed that resonance of North Ray Crater soils probably contained fossil record of thermal and weathering histories of those soils during or immediately after North Ray collision. (NRL Release 53–72–9)

- Lewis Research Center scientists Dr. John V. Dugan, Jr., and Dr. Herman Mark told meeting of Northeast Ohio Congressional Council in Cleveland that space scientists and engineers dismissed because of aerospace recession should be hired for expanded programs to solve environmental and other societal problems. The LeRC scientists were invited by Rep. John F. Seiberling, Jr. (D-Ohio), to testify before House Committee on Science and Astronautics. (McCann, Cl *PD*, 9/23/72)

- National Aeronautic Assn. named Federal Aviation Administrator John H. Shaffer to receive Wright Brothers Trophy. Shaffer was cited for "outstanding leadership of the worldwide operations of the FAA, which has greatly enhanced all aspects of U.S. aviation to the benefit and safety of the general public and of all who fly." (AP, *W Post*, 9/23/72, A22)

- U.S.S.R. announced selection of Soviet aircraft designer Andrey N. Tupolev to receive "Sickle and Hammer," his third gold medal for exceptional service in development of air science and Soviet aircraft construction. Tupolev had designed some 150 aircraft, including Tu-144 supersonic airliner. (Tass, FBIS-Sov, 9/28/72, K8)

- Federal support to universities and colleges had reached $3.5 billion in FY 1971, increase of $253 million (8%), National Science Foundation reported. Total academic science activities were up $148 million, or 7%, to $2336 million. Federal funding of academic research and development increased by 7%, to $1544 million. Support of life sciences projects constituted 48% of total Federal obligations for academic R&D and 53% of all Federal support for fellowships, traineeships, and training grants. Total Federal obligations (exclusive of loans) to institutions of higher education increased 2% in constant dollars for first time since 1967. (NSF *Highlights*, 9/22/72)

- Need for national technology policy was urged in *Science* article by Dr. J. E. Goldman, Senior Vice President for Research and Development of Xerox Corp. As U.S. began to channel technical efforts toward programs of social significance, "we must remember one of the important lessons of our recent past; the history of the space program . . . is a lesson in the mastery of the institutional technique necessary to bring together the segments of the intellectual, industrial, and technological community needed to fulfill goals in a timely fashion. If we choose to ignore this, if we set aside the space program's experience as nothing more than a $20-billion waste, then by our irrationality we will endanger the ultimate achievement of such important societal objectives as better housing and the renewal of our inner cities." (*Science*, 9/22/72, 1078–80)

September 23: *Washington Post* reported that U.S.–U.S.S.R. environmental protection committee established by May 23 treaty would permit placement by both nations of seismic devices on each other's territories. Purpose would be to predict earthquakes but *Post* said devices also could detect underground nuclear tests and distinguish them from earthquakes. If used for test detection, "devices could close the last possible gap in an extremely controversial field." U.S. had maintained that underground tests could not be banned because of inability to monitor adherence to ban by both sides. *Post* said cursory check of Dept. of Defense and Arms Control and Disarmament Agency sources had yielded "no one who knew that the pact contained this provision." (Berger, *W Post*, 9/23/72)

September 25: NASA was instrumenting areas in five national parks to monitor any adverse effects of sonic boom generated by test flights of two YF–12 aircraft being operated by NASA. (FRC, Flight Operations; *Av Wk*, 9/25/72, 11)

- NASA launched Black Brant VC from Churchill Research Range, Canada, carrying Univ. of Minnesota fields and particles experiment. Rocket and instrumentation performed satisfactorily. (SR list)

September 26: President Nixon approved H.R. 15495, $20.9-billion authorization bill for Dept. of Defense procurement of aircraft, missiles, naval vessels, tracked combat vehicles, torpedoes, and other weapons. Authorization also covered research, development, test, and evaluation for Armed Forces and construction at certain Safeguard antiballistic missile system installations. (*PD*, 10/2/72, 1476; Sen Armed Services Com)

- Team of Naval Research Laboratory and Cornell Aeronautical Laboratory scientists had discovered chemical technique for stabilizing meteorological mists, NRL announced. Discovery had been made during attempts to eliminate fog. Release of certain organic compounds into atmosphere had been expected to diminish for intensity; instead, chemicals reduced fog evaporation rates. Findings might be used to change intensity and distribution of rainfall. (NRL Release 48–72–9)

- Siegfried O. Auer of West Germany was granted, on behalf of NASA, Patent No. 3 694 655 for particle-impact location detector. Detector could be carried on spacecraft to identify sources in interstellar space, including sun and comets, from which particles came. It could show angle, speed, and chemical composition of particles and could telemeter data to ground. (Pat Off PIO; Jones, *NYT*, 9/30/72, 39)

- NASA launched Astrobee F sounding rocket from Wallops Station, carrying Goddard Space Flight Center performance test payload. Rocket and instrumentation performed satisfactorily. (SR list)

September 27: NASA's M2–F3 lifting body, piloted by William H. Dana, completed 18th flight after air-launch from B–52 aircraft over Flight Research Center. Objectives—to expand flight envelope, obtain stability

September 27

and control data at mach 0.95, and check out pitch reaction control—were achieved. M2–F3 reached 21 000-m (68 000-ft) altitude and exceeded planned mach 1.34 speed. (NASA proj off)

- Dept. of Defense announced that two squadrons of F–111A fighter-bomber aircraft were being sent to Southeast Asia for first time in 4½ yrs. Aircraft, with low-level terrain-following radars, would replace four squadrons of F–4 Phantom fighter-bombers to increase U.S. all-weather and low-altitude capabilities during monsoon season in Vietnam war. (DOD PIO)

September 28: Claims that Government, news media, and public had pessimistic view of space program's future were refuted by Dr. George M. Low, NASA Deputy Administrator, in speech before National Space Club in Washington, D.C. With decision to build space shuttle, President Nixon had "committed his administration and the Nation to a continuing viable space program" that "meets the needs of the United States' future." President had taken another "great initiative in space" in signing U.S.–U.S.S.R. space cooperation agreement. Senate had supported space shuttle by vote of 61 to 21. "And for the first time in NASA's history, the Congress appropriated every last penny that we asked for." Editorial comment in U.S. had run 9 to 1 in favor of shuttle, 18 to 1 in support of *Apollo 16*, and "we have yet to encounter serious editorial criticism of the Apollo-Soyuz Test Project." While some segments of public were apathetic to space program, less than one percent of the 2200 letters NASA received from public daily was against space program. Most NASA mail was "in outright support of what we are doing" or sought additional information about space. One reason for pessimism "syndrome," Dr. Low said, was that "so many people in Washington . . . tend to equate 'program' with 'budget,' or . . . 'accomplishment' with 'dollars available.' " Space program challenges existed, "and whether or not we can meet them is relatively independent of whether NASA's budget is $3.4 billion, or $3.6 billion." But it "does depend strongly on the results we continue to achieve; . . . on our ability to create and innovate, on our skill and our dedication." (Text)

- Tenth anniversary of Canada's *Alouette 1* satellite launched by NASA from Vandenberg Air Force Base by Thor-Agena B booster. Satellite, designed and built by Canada to study ionosphere, was still transmitting data from 960-km (600-mi) circular orbit.

  Canada celebrated anniversary with special ceremonies Sept. 29, highlighted by official opening of new Spacecraft Assembly and Test Facility of Dept. of Communications Research Centre at Shirley Bay near Ottawa. Canadian Communications Minister Robert Stanbury said *Alouette 1*, as the "oldest satellite still sending back useful information from space, symbolizes the high standards of technology which Canada has maintained . . . and . . . has helped to meet two important national needs—expertise in space-age technology, and improved communications." (*SBD*, 9/27/72, 118; *A&A 1962*)

- U.S.S.R. apparently was testing spaceborne satellite intercept system at high altitudes, *Aerospace Daily* reported. "In five earlier tests of the system . . . Russian satellites had shown they could disable target satellites flying at relatively low altitudes." But *Cosmos 520*, launched Sept. 19 "into practically the same inclination as the earlier . . .

vehicles, has gone into a highly elliptical orbit with an apogee of more than 39,000 kilometers [24 000 miles]. This could mean that the Russians are showing that their interceptor can reach to synchronous orbit . . . where more than half of future U.S. military satellites will orbit." Orbit of *Cosmos 520* was not equatorial—as was that of U.S. vehicles in Air Force and TRW Program 647 ballistic-missile early-warning satellite system—and its high point was apparently over Northern Hemisphere and its intermediate altitude rather more than the 32 000-km (20 000-mi) altitude of U.S. satellites. Military implications of *Cosmos 520* mission were strong "since all SS-9 launched satellites are military vehicles. SS-9s have launched not only the interceptor series but all Fractional Orbital Bombardment System (FOBS) satellites and another class of satellites . . . strongly related to weapons development." (*Aero Daily*, 9/28/72)

- Roll-out ceremony for multinational European airbus and France's second Anglo-French Concorde supersonic transport was held in Toulouse, France. Aircraft were scheduled for maiden flights in few weeks. Concorde 2, similar to three service models entering production, was said to have eliminated smoke and noise that made Concorde 1 unwelcome in U.S. Airbus was standard preproduction model of 330-passenger, short- to medium-haul transport for 1975 operation. During ceremony British Aerospace Minister Michael Heseltine said British government was ready to confer with other European governments on possibility of integrated European aviation industry. (Lewis, *NYT*, 9/29/72, 8)

- U.S.S.R. was dismantling huge booster rocket at Tyuratam launch site [Baykonur Cosmodrome] near Aral Sea, unidentified U.S. intelligence sources told *New York Times*. Ground support equipment also had been removed. Booster had not been used for manned launch, but it was not known whether it had been used for unmanned mission. (Lyons, *NYT*, 10/1/72, 9)

- Soviet scientists had suggested system of time registration for moon in lunar stellar days, hours, minutes, and seconds, Tass announced. They had also calculated positions of 72 stars and of Jupiter for every 10 days of lunar year. Work was outlined in book *Formulas and Ephemerides for Field Observations on the Moon*. (FBIS–Sov, 10/2/72, L1)

- Air Force F-111 swing-wing jet bomber aircraft was lost over Vietnam on first day of F-111s' return to combat in Indochina since 1968. North Vietnam later claimed it had shot aircraft down but Dept. of Defense said it was unable to explain disappearance. (*W Post*, 10/3/72, A25)

- House Committee on Science and Astronautics, at closing session of term, elected to defer action on S.R. 193, bill passed by Senate July 21 to return Cape Kennedy to original name of Cape Canaveral, and on similar House bill originated by Rep. Lou Frey, Jr. (R-Fla.), in 1971. (*O Sen*, 10/1/72)

- Swing-wing Fitter-B fighter-bomber aircraft never before seen outside U.S.S.R. had been delivered to Egypt by U.S.S.R. shortly before Egyptian government ordered Soviet pilots and military advisers to leave Egypt in July, *New York Times* reported. Some U.S. Government analysts believed aircraft, version of fixed-wing Sukhoi aircraft which U.S.S.R. had supplied to Egypt for several years, had heightened Egyptian disillusionment with U.S.S.R. "The Egyptians wanted an advanced fighter-bomber and thought they had Russian promises to get some. But what was delivered is barely better than the Sukhoi-7, except that the swing-wings allow it to take off from a shorter field." (Beecher, *NYT*, 9/29/72, 19)

or structural problem with stand pipe in tank. Additional tests were

September 28–29: Oxidizer tank of Apollo 17 command and service module (CSM) reaction control system was overpressurized during test at Kennedy Space Center. Preliminary data indicated no bladder leak scheduled but problem was not expected to delay Apollo launch scheduled Dec. 6. (NASA PAO; *SBD*, 10/2/72, 139)

September 28–30: Symposium on High Energy Phenomena on the Sun was held at Goddard Space Flight Center to present recent findings to scientists from U.S., Australia, U.K., France, and Japan. Detection of first direct evidence of nuclear reactions on solar surface, during Aug. 4 and 7 by NASA's *Oso 7* Orbiting Solar Observatory (launched Sept. 29, 1971), was reported by Univ. of New Hampshire physicist Dr. Edward L. Chupp and team. Evidence had been obtained from two gamma-ray emission lines—radiation indicators of nuclear processes—during first observation of these radiations from sun. Findings proved that nuclear reactions did occur in large solar flares, and might contribute to goal of generating pollution-free electrical power from sustained thermonuclear reactions in laboratory. (NASA Release 72-193)

September 29: U.S.S.R. launched *Cosmos 521* from Plesetsk into orbit with 1014-km (630.1-mi) apogee, 976-km (606.5-mi) perigee, 105.0-min period, and 65.9° inclination. (GSFC *SSR*, 9/30/72; *SF*, 3/73, 114)

• Dr. James C. Fletcher, NASA Administrator, spoke at dedication ceremony for Univ. of Kansas Space Technology Center and at joint University and Chamber of Commerce luncheon in Lawrence, Kan., celebrating dedication. At dedication he said ceremony marked final step in NASA-sponsored program to build space-oriented research facilities at U.S. universities under Sustaining University Program started in 1960s. Univ. of Kansas Center was 37th and last facility to be completed. Dr. Fletcher said program had accomplished its mission "extremely well. These splendid facilities give our Nation the means of performing research that would be virtually impossible without them." Program's projects had been "invaluable in the successful accomplishment of many of our greatest achievements in space." NASA had found in universities "a concentrated source of one of the irreplaceable ingredients of advanced research programs—highly trained intellects. And now that they can work in facilities to match, we are expecting even more productive results."

Dr. Fletcher discussed universities' role in NASA activities in luncheon speech: "We work with large and small universities in aeronautics and space programs supporting projects . . . relevant to NASA's mission and . . . compatible with university interests and capabilities. We encourage the use of NASA laboratory facilities and the exchange of information between university faculty and NASA scientists." More than 900 university experimenters had used space data, more than 300 investigators from 42 universities had analyzed lunar samples, 25 universities had conceived Apollo mission experiments, and 2500 active research grants and contracts had been issued to universities by NASA. "As we look beyond this decade we can be confident that we have a mechanism, through the space shuttle, to provide quick and easy access to space and at a cost that should encourage new opportunities and initiatives, particularly from the university community." (Texts)

- U.S.S.R. ratified U.S.-U.S.S.R. Treaty on the Limitation of Anti-Ballistic Missile Systems and endorsed Interim Agreement With Respect to the Limitation of Strategic Offensive Arms. (Kaiser, *W Post*, 9/30/72, A24)
- Appointment of William H. Rock, Assistant Program Manager for Apollo-Skylab Programs, as Manager of new Sciences and Applications Office at Kennedy Space Center was announced by KSC Director, Dr. Kurt H. Debus. New office would become KSC's interface with NASA's Office of Applications. (KSC Release 279–72)
- U.S. Atomic Energy Commission and U.S.S.R. State Committee for the Utilization of Atomic Energy had agreed to renew their technical cooperation in peaceful uses of atomic energy, AEC announced. Fifth in series of memoranda of cooperation initiated in 1959 had been signed in Moscow. Memorandum called for expanded cooperation in controlled thermonuclear fusion reactions and breeder reactors. (AEC Release P-307)
- Laser fusion approach to thermonuclear power was discussed in *Science* article. Most U.S. fusion research for past 20 yrs had been directed toward containing fusion reaction with magnetic forces, but many scientists now thought fusion could first be achieved with high-powered lasers. "By bypassing the need for a magnetic field, the laser approach has also bypassed a long and depressing catalog of plasma instabilities that have thwarted the enthusiastic early hopes of physicists for proof of feasibility. But more than one major breakthrough per year in later development may be necessary before scientists can make hydrogen in a reactor hotter and denser than the sun." (Metz, *Science*, 9/29/72, 1180–1)
- Boeing Co. Board made executive changes: President T. A. Wilson was elected Board Chairman, succeeding William A. Allen, whose retirement would become effective Sept. 30. Senior Vice President Malcolm T. Stamper was elected President and member of Board. Wilson would retain responsibilities of Chief Executive. Charles M. Pigott was elected to Board. (*WSJ*, 10/3/72, 13)

September 30: U.S.S.R. launched *Molniya II-3* communications satellite to transmit Soviet TV programs and provide multichannel radio communications throughout Orbita network. Orbital parameters: 39 200-km (24 357.8-mi) apogee, 480-km (298.3-mi) perigee, 11-hr 43-min period, and 65.3° inclination. (FBIS–Sov, 10/2/72, L1; *Spacewarn*, 10/17/72)

- President Nixon signed H.J.R. 1227, ratifying U.S.-U.S.S.R. Interim Agreement with Respect to the Limitation of Strategic Offensive Arms. He said treaty was "beginning of a process that is enormously important that will limit now, and, we hope, later reduce the burden of arms, and thereby reduce the danger of war." (*PD*, 10/2/72, 1474)

During September: Four rare stellar explosions within Milky Way galaxy were recorded by observations in various parts of the world, beginning with Univ. of Toronto discovery by Dr. Philip C. Gregory [see Sept. 2]. Observations were made by radiotelescopes and other instruments and by x-ray scanners aboard *Oao 3* (*Copernicus*) Orbiting Astronomical Observatory (launched by NASA Aug. 21), *Explorer 42* (*Uhuru*) Small Astronomy Satellite (launched by Italy for NASA Dec. 12, 1970), and Dept. of Defense Vela satellite. Observations and speculations as to meanings were so extensive, Walter S. Sullivan said in *New York Times* article, that British journal *Nature Physical Science* devoted entire Oct. 23 issue to subject. Explosions were radio emissions from gas cloud expanding at about half the speed of light, Sullivan said, but were "clearly not those of a supernova, since they were observed only at radio wave lengths and they recurred several times." If, as some astronomers suspected, explosion source was also source of x-rays being emitted in 288-min cycle, object exploding probably was spinning once every 288 min and was "about the size of a star." (*NYT*, 11/1/72, 1; *Nature Physical Science*, 10/23/72)

- Space technology had become "forcing function to bring nations together in mutual undertakings with a common purpose," Dr. Allen E. Puckett, American Institute of Aeronautics and Astronautics President, said in *Astronautics & Aeronautics* editorial. U.S.–U.S.S.R. space agreement "expresses our intention to cooperate in various fields of space activity, and in particular proposes a joint experimental flight of U.S. and Soviet manned spacecraft in 1975. We may debate the immediate technical values of this experiment, but the overriding consideration is the fact that we will be working together. Regardless of the technological outcome we will be communicating and collaborating, getting to know each other better, in a common enterprise. On such small beginnings, a new era of mutual understanding and cooperation might evolve." (*A&A*, 9/72, 19)

- Engine concepts for space applications were described in *Air University Review*. Government and industry had spent more than $100 million on high-pressure-engine technology since 1961. Air Force XLR–129 engine program had provided strong technical base for space shuttle engine. Shuttle's orbiter stage would use advanced-concept staged combustion engine "very similar to that of a turbojet equipped with an afterburner; . . . there are two different stages of combustion." While first turbojet combustion occurred in main chamber, shuttle engine's first burning occurred in gas generator or preburner. Preferred engine concept for future spacecraft would be "single engine that could operate as a rocket, a ramjet, a scramjet, and a turbofan." At liftoff, rocket would be firing and "turbofan might also be operating to supply additional air to improve the performance of the rocket. After the vehicle is moving at a greater speed, the ramjet would start operating, and the turbofan would be removed from the airstream. The pure rocket would continue to operate briefly to aid the ramjet. As the vehicle reaches the outer fringes of the atmosphere, the inlets would be closed, and the pure rocket would be used alone for reaching, maneuvering in, and leaving orbit. To return to base after re-entry into the atmosphere, the ramjet and/or the turbofan might be used separately or they might be used like an afterburning turbofan." To improve spacecraft performance further, "we would like to get away from having to carry an oxidizer for the rocket portion of the flight."

Future space engines would "mate the tremendous energy available from nuclear explosions with the ability of a rocket to operate at high thrust levels." Performance of nuclear rocket "more than makes up for its being heavier than a normal rocket." Nuclear rocket's specific impulse was about twice that of "even the best chemical rocket." Nuclear stage also would be useful for seeding space with unmanned applications satellites. (Holder, Siuru, *Air Univ Rev*, 9/72)

- Europe's dilemma over altered conditions for participation in post-Apollo space program was described by Kenneth W. Gatland in *Spaceflight*. Europe had been invited to develop research and applications module (RAM) for space shuttle, rather than space tug as originally discussed. Elimination from Europe of any shuttle propulsion technology had "left little room for manoeuvre." Messerschmitt-Boelkow-Blohm would be "denied the opportunity to contribute their Attitude Control Propulsion System (ACPS) engine" which "is considered to be fully competitive with U.S. equivalents" because NASA had rejected ACPS as "too advanced and too complex" a European contribution. Space tug—with attributes of both launch vehicle and spacecraft—would have pushed propulsion technique "well beyond what is currently envisaged in Europe." Tug covered "very extensive range of the basic techniques of interest to the Europeans and might engender solutions to the most classic problems." Tug's "far-reaching integration with the shuttle and with the payload during operations would have afforded Europe effective participation in most American missions, i.e., the majority of space activities in the 1980s." (*SF*, 9/72, 322-3)

October 1972

October 1: Joint U.S.–U.S.S.R. space mission scheduled for 1975 would include 48-hr linkup of Apollo and Soyuz spacecraft and crew exchange, Soviet Cosmonaut Vladimir A. Shatalov said in report released by Tass. Coyuz spacecraft, manned by two cosmonauts, would be launched first. "In seven and a half hours, Apollo spaceship manned by three astronauts will be launched from Cape Kennedy. In 24 hours, Apollo will approach the Soyuz ship and link up will be made. The system that will be formed will be operated as a single spacecraft. Within two days, spacemen will move from one ship to another, scientific and technical experiments will be made and there will be television transmissions to the earth. At the end of the third day, the spaceships will be undocked and will soon land on their respective territories." Shatalov said he hoped manned Mars mission before end of 20th century would be international flight—"Possibly it will be the Soviet-American flight." (FBIS–Sov, 10/2/72, L2)

- Twenty-fifth anniversary of Office of Naval Research Project Skyhook was celebrated with launch from Cape Girardeau, Mo., of 940 000-cu-m (33.1-million-cu-ft) thin-walled plastic balloon. Balloon carried 1360-kg (3000-lb) experiment designed by Goddard Space Flight Center scientist Dr. Jonathan F. Ormes to measure high-energy cosmic rays. It reached altitude of nearly 40 000 m (more than 130 000 ft) and on-station time of more than 25 hrs. Payload was recovered 25 mi east of Nashville, Tenn. Project Skyhook, gathering information for future space flights, launched its first balloon Sept. 25, 1947. Since then more than 3000 flights had been launched throughout the world, including launches from the high seas. (ONR, *Naval Research Reviews*, 2/73, 1–10)

- U.S.S.R. had tested stellar inertial guidance system for missiles, according to series of interviews with unidentified Administration officials quoted in *New York Times*. System corrected course of 5600-km (3500-mi) Sawfly missile during flight by taking bearing from certain stars. Also reported tested were streamlined new warhead for intercontinental ballistic missiles that entered atmosphere more rapidly than current warheads, improving accuracy, and ground-scanning device for course correction. Improvements were among number cited in offensive and defensive missiles tested by U.S.S.R. since May 26 arms limitation accords between U.S. and U.S.S.R. (Beecher, *NYT*, 10/1/72, 17)

- *Apollo 14* Astronaut Edgar D. Mitchell retired from NASA and Navy to form Edgar D. Mitchell & Associates, Inc., organization to study psychic potential of man and other forms of life. (NASA PAO; AP, *NYT*, 10/3/72, 45)

October 1–7: Fédération Aéronautique Internationale held annual general conference in Paris. FAI Gold Space Medal—Federation's highest honor —was presented to *Apollo 15* Astronaut David R. Scott. Scott also accepted V. M. Komarov Diploma on behalf of *Apollo 15* crew. (*NASA Activities*, 11/15/72, 235; NASA Int Aff)

October 2: Air Force launched two satellites in Dept. of Defense Space Test Program (STP) from Vandenberg Air Force Base by single two-stage Atlas-Burner II booster.

Stp 72-1 entered orbit with 749-km (465-mi) apogee, 729-km (453-mi) perigee, 99.5-min period, and 98.4° inclination. Objectives were to measure background gamma radiation over entire earth; measure flux and spectra of low-altitude charged particles; determine effects of space environment on thermal control coatings; measure ultraviolet radiation and hydrogen and helium atoms and ions; and observe extreme- and far-ultraviolet radiation originating in interaction of solar wind with interplanetary medium or from galactic sources. Spacecraft was 566-kg (1248-lb) cylinder 2.1 m long and 1.4 m in diameter (7 by 4.5 ft), carrying antenna booms that extended 2.7 m (9 ft) from each end, gamma ray spectrometer, and low-altitude particle-measuring sensor.

Radcat entered orbit with 751-km (467-mi) apogee, 729-km (453-mi) perigee, 99.5-min period, and 98.4° inclination. Spacecraft—cylinder 3 m long with 1.2-m diameter (10 by 4 ft) and weighing 220 kg (485 lbs)—would provide passive optical and radar calibration target.

STP program was managed by Air Force Space and Missile Systems Organization (SAMSO). (Pres Rpt 73; *LA Times*, 10/3/72)

- Significance of NASA and Soviet Academy of Sciences joint Apollo-Soyuz test project (ASTP) was discussed by Dr. James C. Fletcher, NASA Administrator, in letter to NASA staff. Eyes of world would be on NASA and Soviet manned space flight teams as never before. There "must be a near certainty that the mission will be successful, or it should not be attempted." Not just NASA reputation but "future of manned space flight and international cooperation in space will be significantly influenced by the success or failure." Mission also was "important first step to facilitate rescue missions in space and to enable manned spacecraft from two or more countries to participate in future cooperative activities. And because several years of friendly cooperation are required just to plan and fly this one test mission, it takes on an added symbolic significance and gives us the feeling that even now, long before the flight, we may be making a contribution to the cause of peace and better understanding between our two countries." Decision to cooperate in project was "optimistic vote of confidence in the future of manned space flight by the two leading powers." But project had practical limits. Neither side would transfer its technology as such. "So neither side has to give away anything, but each has much to gain." Biggest concession to be made by each country was "recognition that close and visible cooperation between the two countries on an important aspect of manned space flight has become desirable." (*NASA Activities*, 10/15/72, 200-2)
- Flight Research Center announced successful remote-control landing of 1600-kg (3500-lb) modified Piper Comanche (PA-30) aircraft at FRC. Safety pilot was on board aircraft. Landing test demonstrated new technique to provide safer and more economical means of testing advanced aircraft and spacecraft when risk to aircraft or pilot was high or when costs precluded manned full-scale flight test vehicle.

PA-30 was equipped with preliminary version of system that used telemetry to send control commands to aircraft and flight information to pilot displays in ground "cockpit" several miles away. Ground pilot used TV and radar to navigate aircraft. In later tests, large-scale models up to 9 m (30 ft) long would be launched from carrier aircraft. Test pilot on ground would "fly" remotely piloted research vehicles (RPRV) using TV and telemetry. (FRC Release 16-72)

- Use of M518 multipurpose electric furnace developed by Marshall Space Flight Center for Skylab would permit increase in number of metals and metal-processing experiments to be carried on earth-orbiting station in 1973, NASA announced. New furnace would accommodate original composite casting experiments and 10 others to explore possibility of manufacturing and processing materials in zero g. Furnace would be delivered to NASA in December by principal contractor Westinghouse Corp. It would facilitate experimentation in solidification, crystal growth, composite structures, alloy structural characteristics, and other thermal processes that altered materials in weightlessness. (NASA Release 72-196)
- StarQuest Ltd., nonprofit firm to promote values of space exploration to U.S. public, had rented $350 000 ocean liner for cruise off Cape Kennedy during Apollo 17 launch Dec. 6, Chicago *Sun-Times* reported. Firm's honorary board included science fiction writer Arthur C. Clarke and former astronaut Edgar D. Mitchell. Firm intended to make motion picture of Apollo 17 launch. (Ziomek, C *Sun-Times*, 10/2/72)
- Air Force announced award of $2 471 953 fixed-price-incentive-fee contract to Lockheed Missiles & Space Co., Inc., for operations, technical and civil engineering support at Satellite Test Center, and scientific-technical support at Vandenberg Tracking Station. (DOD Release 692-72)
- Gerhard B. Heller, Director of Space Science Laboratory at Marshall Space Flight Center, died in Nashville, Tenn., at age 58 of injuries received in automobile accident. Heller had begun career in 1940 at Peenemuende Rocket Center in Germany. He had come to U.S. with Dr. Wernher von Braun's rocket group after World War II and worked with Army Ordnance Corps until transfer of group to MSFC in 1960. During NASA career Heller had pioneered in thermal control of earth satellites. He had served as director of Space Sciences Laboratory's Thermophysics Div. before becoming Laboratory director in 1969. (MSFC Release 72-128)

October 3: Tass reported *Luna 19* achievements as Soviet lunar probe launched Sept. 28, 1971, neared end of its activity. During more than 4000 lunar orbits spacecraft had transmitted data in more than 1000 communication sessions. Experiment data had indicated presence of plasma produced in near-lunar space by interaction of "space irradiation and the lunar surface." *Luna 19* had photographed lunar area "to study individual parts of the lunar surface and possibilities for using orbital panoramas for the navigation of space apparatus." Probe had recorded more than 10 increases in flux of solar cosmic ray protons, resulting from powerful solar flareups. During lifetime, *Luna 19* onboard systems had functioned normally. Tass said two-year monitoring of space radiation by *Luna 19* and predecessors *Lunokhod 1* (lunar roving vehicle landed on moon by *Luna 17*), *Venus 7* and *8*, and *Mars*

2 and *3* would "make it possible to ascertain the energy spectrum, the charge components of cosmic rays, and the conditions of their propagation in the interplanetary medium." (FBIS–Sov, 10/4/72, L1)

- President Nixon and Soviet Foreign Minister Andrey A. Gromyko spoke at White House ceremony marking entry into force of U.S.–U.S.S.R. Treaty on the Limitation of Anti-Ballistic Missile Systems and of Interim Agreement on the Limitation of Strategic Offensive Arms. Gromyko said: "Practical steps to limit rocket nuclear armaments rightfully hold an important place among the very real political changes taking place in relations between our two countries and this signifies a success for the policy of peaceful coexistence and it has a positive effect on the entire international scene as a whole." President said: "I think all of us are aware of the fact that the signing of these documents today, the signing of the documents that occurred earlier this year in the Kremlin, raise the hopes of all the people of the world for . . . a world of peace . . . in which peoples with different governments and different philosophies could live in peace together." (*PD*, 10/9/72, 1483–4)

- Communications Satellite Corp., Lockheed Aircraft Corp., and Microwave Communications, Inc., announced agreement on terms for jointly owned company to provide nationwide satellite communications services. Each corporation would have one-third ownership. Intention to establish jointly owned company had been announced Sept. 8. Agreement was subject to approval by Federal Communications Commission. (Joint ComSatCorp, Lockheed, MCI Release 72–49)

- House passed H.R. 16645, bill for construction of District of Columbia civic center that contained amendment to change name of Jet Propulsion Laboratory to H. Allen Smith Jet Propulsion Laboratory [see Sept. 19]. House then substituted Senate-passed bill, S. 3943, after amending it to contain language of House bill as passed. (*CR*, 10/3/72, H9027–51)

October 4: U.S.S.R. launched *Cosmos 522* from Plesetsk into orbit with 317-km (197-mi) apogee, 198-km (123.0-m) perigee, 89.6-min period, and 72.8° inclination. Satellite reentered Oct. 17. (GSFC *SSR*, 10/31/72; *SBD*, 10/6/72, 170)

- Fifteenth anniversary of space age inaugurated by 1957 launch of U.S.S.R.'s *Sputnik 1*, first man-made satellite. Soviet cosmonauts commemorated anniversary with interviews published in Soviet press. Cosmonaut Vladimir A. Shatalov said: "Mankind's entry into space has opened up . . . enormous prospects for the study not only of extraterrestrial space but also of the Earth. Modern space technology can be used to solve mankind's vitally important tasks, such as the protection of the natural environment and the rational utilization of the Earth's resources." Cosmonaut Konstantin P. Feoktistov discussed future in space in *Komsomolskaya Pravda* interview. In applied space research, "practical cosmonautics will move from purely informational tasks to exerting an active influence on . . . natural phenomena." It would be possible "to destroy embryonic hurricanes, influence weather . . . . And both automatic and manned craft can serve as instruments in resolving these tasks." Other space tasks of future would be study of Earth's environment and study of universe, but "I do not believe that the exploitation of the planets' natural resources for terrestrial needs will begin in our century." (FBIS–Sov, 10/17/72, L2–4)

- NASA's *Pioneer 10* Jupiter probe (launched March 2) was 424 million km (263 million mi) from earth and 377 million km (234 million mi) from Jupiter. Spacecraft, which entered Asteroid Belt July 15, had traversed radial distance of 102 million km (63 million mi) into belt in 79 days. Spacecraft performance remained satisfactory except that anomalies with stellar reference assembly had necessitated use of sun sensor assembly for primary roll reference until December; scientific instruments were providing good data. (NASA proj off)

- NASA released official summary of reasons for selection of North American Rockwell Corp. for $2.6-billion prime contract to build space shuttle. Summary said NR had been chosen over competitors—Grumman Aerospace Corp.; Lockheed Missiles & Space Co., Inc., Space Systems Div.; and McDonnell Douglas Corp.—because "North American Rockwell attained the highest score from a mission suitability standpoint, because its cost proposal was lowest and credible, and because its approaches to program performance gave high confidence to us, to the [Source Evaluation] Board, and to the Manned Space Flight center directors, that it will indeed produce the Shuttle at the lowest cost." Summary—signed Sept. 18 by Dr. James C. Fletcher, NASA Administrator—said NR's "greatest advantages" within mission suitability area were in management, with efficient control of program and best overall top project management team. NR design provided lightest dry weight. NR's "good understanding of all electrical power subsystems reflected the very thorough studies . . . made following the Apollo 13 accident." Analysis of maintainability was excellent. Further, NR management techniques "should provide earlier identification of cost problems" and planned constrained buildup of resources in beginning of program lent confidence in ability to control costs.

  Grumman Corp. had been "very close behind" but its plan to build up its work force to early manpower peak had been seen by NASA as threat of "premature commitment of resources during the course of the program." Grumman's greatest strength had been in its technical design; McDonnell Douglas design had strong features but some technical drawbacks; and Lockheed's proposal had lacked "consistent technical depth." NASA had originally made summary available only to General Accounting Office and to competing bidders because of potential embarrassment to losers. Agency had later released summary, *Wall Street Journal* said, "to demonstrate that politics didn't play any role in the decision and that the most painstaking selection procedures in the agency's history were observed." (Text; *WSJ*, 10/5/72, 12)

- NASA Assistant Administrator for International Affairs Arnold W. Frutkin received Bronze Medal from British Interplanetary Society (BIS) in London for outstanding contributions to international collaboration. He told BIS that time was "running out very rapidly" for Europe to work with NASA on post-Apollo program, "which today essentially means the sortie module." Europe faced dilemma in space. It could decide to work on regional basis of self-sufficiency, or become dependent in some measure on U.S. (*SF*, 1/73, 37)

- Semiannual survey released by Aerospace Industries Assn. said sharp three-year decline in aerospace industry employment had leveled off at slightly more than 900 000. Between June 1972 and June 1973, aerospace employment was expected to decline from 923 000 to 914 000, or one per cent. During previous three years more than 500 000 employees—one third of total aerospace work force—had been dropped from payroll. Survey projected December 1972 employment at 917 000, somewhat above earlier forecast of 887 000. Between June 1972 and June 1973, production workers were expected to decline by 1.8%, scientists and engineers by 0.6%, and technicians to increase by 1.6%. By June 1973 estimated 157 000 scientists and engineers would be employed; 1967 peak had been 235 000. (Text)

- Large 1793 penny secreted aboard *Gemini 7* earth-orbiting mission in 1965 recently had been sold for $15 000 although its numismatic value was about $2000, Associated Press reported. NASA spokesman said coin had been slipped into mission's inflight medical kit by flight surgeon who had left NASA several years ago. No one connected with NASA had profited from recent transaction. (NASA PAO; AP, *NYT*, 10/5/72, 38)

- Personal telephone call from Mrs. Rose P. Kennedy, mother of late President John F. Kennedy, to Rep. Thomas P. O'Neill, Jr. (D-Mass.), had helped kill bill to change name of Cape Kennedy back to Cape Canaveral, *Miami Herald* reported. Paper said Rep. O'Neill had asked Rep. George P. Miller (D-Calif.), Chairman of House Committee on Science and Astronautics, to block bill. O'Neill aide had admitted congressman had asked Miller to block bill, but had denied that O'Neill had acted on Mrs. Kennedy's behalf. (*M Her*, 10/4/72)

- Rep. Lou Frey, Jr. (R-Fla.), said in Washington, D.C., that he would carry his efforts to restore original name of Cape Canaveral to Cape Kennedy to courts if Congress refused to cooperate. If courts also refused, he would try to persuade Florida Gov. Reubin Askew and Florida residents to "ignore Federal name." (UPI, *NYT*, 10/6/72)

- Establishment in Vienna of International Institute for Applied Systems Analysis was announced by Dr. Philip Handler, President of National Academy of Sciences. Institute would specialize in complex global problems that resulted from industrialization, particularly those that would benefit from systematic comparison of remedies being attempted by various nations. Projects being considered included analytical study of short- and long-range projections of world supply of energy resources and demands for energy, dynamic substitutions among energy resources, future technologies, and hazards of each source. Projects would be in categories of environment, health care, municipal services, and engineering. Academician Dzherman M. Gvishiani of U.S.S.R. had been appointed Chairman of Institute Council. Dr. Howard Raiffa, Harvard Univ. mathematical statistician, would be Director. NAS would be represented on Council by its Foreign Secretary Dr. Harrison Brown. (NASA Release)

- Dept. of Defense spokesmen told press in Washington, D.C., that swing-wing F-111 jet fighter-bombers had been withdrawn from combat service in North Vietnam after Sept. 28 loss of one F-111 but aircraft had since returned to action. (AP, *W Post*, 10/5/72, A29)

- Air Force announced award of $1.7-million letter contract to North American Rockwell Corp. for repair and modification of attack radar and inertial navigation components of Mark II avionics system of F-111 aircraft. (DOD Release 697-72)

October 5: U.S.S.R. launched *Cosmos 523* from Plesetsk into orbit with 457-km (284-mi) apogee, 268-km (166.5-mi) perigee, 91.7-min period, and 70.9° inclination. Satellite reentered March 7, 1973. (GSFC *SSR,* 10/31/72; 3/31/73; *SBD,* 10/10/72, 174)
- NASA's M2-F3 lifting body, piloted by William H. Dana, completed 100th flight of lifting-body program and 19th flight of M2–F3 from Flight Research Center. Objectives—to obtain stability and control data at mach 0.95 and to check out pitch-reaction augmentation system—were achieved during "outstanding flight." M2–F3 reached mach 1.4 and 20 000-m (65 000-ft) altitude. First manned lifting-body flight had been made July 12, 1966, in M–2 piloted by Milton O. Thompson. (NASA proj off; FRC Release 17–72)
- House and Senate signed H.R. 10243, bill to establish Office of Technology Assessment for Congress and to amend National Science Foundation Act of 1950. Bill was presented to President for signature Oct. 10. (*CR,* 10/4/72, S16962, H9289; 10/10/72, H9454)
- U.K., France, and U.S.S.R. were negotiating for pooled supersonic airline service between Western Europe and Far East, British Airways Board Chairman David Nicholson said at New York press conference. He predicted that Japan "will want to be in on the deal" and would want to fly supersonic transports to U.S. "With these airlines knocking on the doors of America from both sides—with a supersonic girdle pretty well all around the world—I find it hard to see how the United States operators will be able to afford to stand aside." (Witkin, *NYT,* 10/6/72, 5)

October 6: Combined airlock module (AM) and multiple docking adapter (MDA) for 1973 Skylab space station were turned over to NASA at McDonnell Douglas facility in St. Louis and flown to Kennedy Space Center aboard Super Guppy aircraft. With arrival of shipment, weighing 28 000 kg (62 000 lbs), all major Skylab elements had been delivered to KSC. (MSFC Release 72–129; NASA Special Release; KSC PAO)
- Senate unanimously approved ratification of Convention on International Liability for Damage Caused by Space Objects. (*CR,* 10/6/72, S17097–17100, 17165)
- U.S. and U.S.S.R. signed protocol in Moscow calling for exchange of experience in planning, construction, and operation of electric power stations. Protocol also called for exchange of data on electric power transmission, magnetohydrodynamics, and use of solar and geothermal energy. (Tass, FBIS–Sov, 10/10/72, G4)
- President Nixon announced appointments to General Advisory Committee to Atomic Energy Commission: National Bureau of Standards physicist Dr. Evans V. Hayward; Stanford Univ. physicist Dr. Hubert Heffner; Dr. Michael M. May, Director of Lawrence Radiation Laboratory of Univ. of California at Livermore; and physicist Dr. Walter H. Zinn, former director of Argonne National Laboratory. (*PD,* 10/9/72, 1492)

October 8: Earth passed within 113 000 km (70 000 mi) of comet Giacobini-Zinner's orbit as U.S. and Canadian scientists on NASA's Convair 990 observatory aircraft *Galileo* studied comet fragments. Aircraft was flying at 12 000-m (40 000-ft) altitude over Bering Sea. Experiment to use light spectra to identify percentages of sodium, calcium, magnesium, and iron in Giacobini meteors was part of 12-day scientific mission during which *Galileo* was flown from Alaska as experimental station. Main objectives of flight series were to observe meteor showers, measure polar winds in upper atmosphere, observe barium ion cloud experiment, and measure auroras.

Nike-Apache sounding rocket launched by NASA from Kiruna, Sweden, carried Dudley Observatory experiment to 109.4-km (67.9-mi) altitude to collect and identify cosmic dust particles from Giacobini meteor shower. Launch was first in series of three. Rocket and instruments performed satisfactorily and good data were obtained. (ARC Release 72–98; FRC X–Press, 10/27/72, 2; NASA Rpt SRL)

- *Mariner 9* project scientists were pondering past and future of planet Mars as spacecraft (launched May 30, 1971) began final weeks of Mars probe, *New York Times* said. Questions of existence of water on Mars, Martian atmosphere, and possible future use of planet as refuge for man when earth became uninhabitable had "no clear answers." Dr. Bruce C. Murray of Cal Tech believed Mars was currently evolving, had never had any vast bodies of water, and might be facing atmospheric dead end." Harold Masursky of U.S. Geological Survey had said presence of river-like channels on Mars indicated existence of liquid water in recent past. Dr. Carl E. Sagán of Cornell Univ. had suggested that periodic changes in Mars' angle of rotation could make Martian climate unstable, alternating between present ice age and warm, earthlike areas that could support "thriving biology." Mars controversy centered on *Mariner 9* photos that showed many channels on Mars seemingly carved out by flowing water. (Wilford, *NYT*, 10/8/72, 68)

- Jet Propulsion Laboratory announced award of $4.8-million contract to Ball Brothers Research Corp. to design, build, and test high-spectral-resolution gamma ray spectrometer for NASA's High Energy Astronomy Observatory (HEAO) in 1977. Spectrometer would make exploratory search of universe for gamma-ray-line structures of low and medium energy. Observation of lines could unlock many secrets of supernovae and pinpoint regions where chemical elements—basic building blocks of universe—were synthesized. Spectrometer also would monitor intensity, position, time variation, and spectrum of x-rays and gamma rays as emitted from cosmic sources such as pulsars and x-ray stars. (JPL Release 630)

October 8–14: McDonnell Douglas DC–10 Series 40 jet transport aircraft set three point-to-point nonstop records during 48 942-km (30 411-mi) demonstration flight to seven cities in Asia, U.S., and South America. Aircraft flew from Los Angeles to Hong Kong (12 355 km; 7677 mi) in 14 hrs 44 min, from Honolulu to Buenos Aires (12 553 km; 7800 mi) in 14 hrs 18 min, and from Rio de Janeiro to Los Angeles (10 139 km; 6300 mi) in 11 hrs 52 min. (McDonnell Douglas 1973 calendar; McDonnell Douglas PIO)

October 9: President Nixon announced new policy making U.S. launch assistance available to interested countries and international organizations for satellites intended for peaceful purposes consistent with relevant international agreements. Launch assistance from U.S. sites would be on cooperative or reimbursable basis similar to that available to non-Government U.S. users. Assistance with launches from foreign sites would be available with purchase of U.S. launch vehicle. With foreign launches, U.S. would require assurance that launch vehicles would not be made available to third parties without its permission.

Under policy terms U.S. would provide launch assistance for telecommunications satellites recommended by International Telecommunications Satellite Consortium. In absence of INTELSAT recommendation, U.S. would assist systems it had supported within INTELSAT. For communications satellite system not recommended by INTELSAT and not previously supported by U.S., U.S. would decide on request for launch assistance after considering degree to which proposed system would be modified to meet INTELSAT approval. Future operational satellite applications without broad international acceptance would be considered for U.S. launch assistance only after international acceptance had been obtained. (White House Release; *PD*, 10/16/72, 1508)

- Nike-Apache sounding rocket launched by NASA from Kiruna, Sweden, carried Dudley Observatory experiment to 108.7-km (67.5-mi) altitude. Launch was second in series of three to collect and identify cosmic dust particles from Giacobini meteor shower [see Oct. 8]. Rocket and instruments performed satisfactorily and good data were obtained. (NASA Rpt SRL)

- Col. Thomas G. Lamphier (USA, Ret.), veteran pilot and former Deputy Administrator of Veterans Administration, died in San Diego, Calif., at 82. He had helped Charles A. Lindbergh plan routes for first U.S. passenger airline in 1928 and had served as Vice President of first airline—Transcontinental Air Transport—after retirement from Army in 1929. Col. Lamphier had testified in defense of Gen. William (Billy) Mitchell at general's court martial for insubordination in 1925. (AP, *NYT*, 10/12/72, 46)

October 9–15: International Astronautical Federation held 23rd Congress in Vienna. Dr. H. Guyford Stever, Director of National Science Foundation, stressed Congress theme "Space for World Development" in opening address: "This is the year of ever-increasing cooperative interests in space ventures, including the joint, manned space effort by the Soviet Union and the United States. Thanks to the genius of our early pioneers, . . . results of an era of experimentation now show us clearly the promise which the orderly and determined use of space and space vehicles has to offer." (Text; Congress Release, 10/9/72)

Soviet attendance at Congress was limited to 21 including Cosmonauts Anatoly V. Filipchenko and Valery N. Kubasov. More than half of scheduled Soviet participants failed to arrive or to send papers. Proxy readings of Soviet papers that were included eliminated most opportunities for clarification and discussion by panelists. Lack of Soviet participation had been attributed by Western scientists to U.S.S.R.'s downgrading of IAF in wake of increased space cooperation with U.S. and France and to Soviet budgetary hold-down. (*Av Wk*, 10/16/72, 16)

Dr. George E. Mueller, President of System Development Corp. and former NASA Associate Administrator for Manned Space Flight, said in interview following opening session that it would be technically feasible to put crews of mixed nationalities on Mars within 20 to 30 yrs. "Even the money is there. It's getting it allocated for the project that's difficult." He envisioned crew of 12 or more in giant spacecraft possibly powered by nuclear energy taking 12 to 18 mos to reach Mars. Special shuttle would be needed to carry supplies. "Such an expedition will require lots of resources . . . best . . . supplied by the world as a whole." (Reuters, B *Sun,* 10/10/72)

Dr. James H. Bredt of Advanced Missions Program Office in NASA Office of Manned Space Flight described new space processing experiments for Skylab missions in 1973. Experiments would broaden scope of experiments to correspond to program's increased area of interest, to gain experience with new techniques in order to develop advanced equipment for future missions, and to enable materials scientists and engineers to demonstrate value of space laboratories for applied materials research. Apparatus and experimental samples would be delivered to Kennedy Space Center for installation in spacecraft during first week of December. Experiments were to be performed by third and final Skylab crew in November and December 1973. Analysis of samples would begin immediately after their return to earth, and principal investigators probably would present preliminary reports of results at 25th IAF Congress. (Text)

Frank J. Malina, Chairman of International Academy of Astronautics' Manned Research on Celestial Bodies (MARECEBO) Committee, told 5th Lunar International Laboratory Symposium Oct. 10 that 1964 estimate of 1975–1985 decade for initiation of permanent lunar laboratory might have been "too optimistic." Men and machines had provided "great deal of hard working knowledge on the Moon and on men working there," but people and governments needed to be convinced that necessary funding should be provided. (Text)

Manned Spacecraft Center Skylab Program Manager Kenneth S. Kleinknecht discussed mission's rescue capability during 5th International Space Rescue Symposium. Special kit to be installed in Skylab command and service module (CSM) would permit CSM to rendezvous and dock with orbital workshop, using two-man crew. CSM then would be able to accept three stranded astronauts from workshop, undock, and reorbit with five persons on board. Rescue could be effected 48 days after alarm early in mission and as quickly as 10 days late in mission. Reason for variation in time required was that 22 days were needed to refurbish launch umbilical tower after previous launch. (Text)

Apollo-Soyuz Test Project (ASTP)—joint U.S.–U.S.S.R. rendezvous and docking mission scheduled for 1975—was discussed at Oct. 10 press conference. Participants were Cosmonauts Filipchenko and Kubasov; Soviet academician Dr. Leonid I. Sedov; Dr. Wernher von Braun, Corporate Vice President for Engineering and Development with Fairchild Industries and former NASA Deputy Associate Administrator for Planning; and Dr. Krafft A. Ehricke, North American Rockwell Corp. scientist. Dr. von Braun said transition from lower pressured Apollo spacecraft to higher pressured Soyuz would be easy for astronauts, while Soviet cosmonauts would have to spend about two hours in airlock to adjust to Apollo system. Dr. Sedov said U.S.S.R. had not accepted invitation from U.S. for joint mission when first proposed because of sensitive political situation. With successful Moscow summit meetings in May, situation had become "favorable." (IAF Release, 10/10/72)

Scout launch vehicle system was discussed in paper at Congress prepared by R. D. English of Langley Research Center and M. Green of LTV Aerospace Corp., presented by Green. Since inception of program in 1959, Scout vehicle had successfully launched 67 payloads. Studies had been made to ensure that system would provide flexibility for "unique missions to supplement the space shuttle." (Pamphlet reprint of Text)

Dale D. Myers, NASA Associate Administrator for Manned Space Flight, said European Launcher Development Organization (ELDO) studies of space tug had been discontinued but NASA studies were continuing. "Our present belief is that the required technical advances combined with the budgetary demands that the space shuttle will make . . . will make it impossible to develop within this decade the 'full capability' that we feel will eventually be necessary." NASA was investigating alternate approaches "to providing a Space Shuttle third stage which will give us a reduced but acceptable capability for the delivery of payloads during the early years of shuttle operation." Alternatives being considered were: to modify existing stage for use as expendable third stage; to make major modifications to existing cryogenic- or storable-propellant stage to enable it to carry payload to geosynchronous orbit and return to shuttle orbit for reuse; or to build new vehicle with less capability than needed, as part of evolutionary program leading to vehicle with required capability. Studies of first alternative were under way; contracted studies within next year were expected to examine latter two alternatives. (Text)

Sixth International History of Astronautics Symposium, held Oct. 13, was chaired by Dr. Eugene M. Emme, NASA Historian. In introduction Dr. Emme pointed out that 57 men had flown in space since October 1957, 10 had walked on lunar surface, 22 had flown around moon, and 2 (Astronauts James A. Lovell, Jr., and John W. Young) had traveled to moon twice and made two earth-orbiting flights. Papers presented included "Hungarian Rocketry in the 19th Century" by István G. Nagy, "Origins of Astronautics in Switzerland" by A. Waldis of Swiss Transport Museum, and "Development of the Technology of Rocketry and Space Research in Poland" by M. Subotowicz. Papers read in absentia were "Astronautic Pioneers" (analysis of rocket achievements of F. A. Tsander) by L. S. Dushkin and Ye. K. Moshkin and "Basic Stages of the Development of the Theory of Ramjet Engines" by I. A. Merkulov of Soviet Academy of Sciences. Memoir papers presented included "From GALCIT to Explorer 1, 1944–57" by Dr. William H. Pickering, Director of Jet Propulsion Laboratory; "From Wallops Island to Mercury, 1945–58" by Dr. Robert R. Gilruth, MSC Director; and "The Viking Rocket" by Milton W. Rosen, Senior Scientist in NASA Office of Dept. of Defense and Interagency Affairs. (Program; IAF Release)

Daniel and Florence Guggenheim International Astronautics Award for 1972 was presented to Dr. Reimar Luest, Director of Max Planck Institute and astrophysicist known for work on solar winds and interplanetary matter. (Program; IAA Annual Rpt, 8/31/72)

October 9–19: U.S. and U.S.S.R. representatives met in Moscow on Apollo-Soyuz Test Project (ASTP), joint rendezvous and docking mission scheduled for 1975. Meetings agreed on July 15, 1975, target launch date for Soyuz, with five launch opportunities for Apollo beginning 7½ hrs after Soviet launch—an addition of two opportunities. Atmospheric pressures of cabins would be adjusted while spacecraft were docked, to avoid necessity of prebreathing pure oxygen before transfer of crewmen from Soyuz to Apollo cabins. Soyuz pressure would be lowered from its normal 101 kilonewtons per sq m to 69 (from 14.7 psi to 10); Apollo pressure would remain at 35 kilonewtons per sq m (5 psi).

First joint crew training session was scheduled for summer of 1973, with cosmonauts visiting U.S. for several weeks. Astronauts would train equal time in U.S.S.R. in fall of 1973. Soviet two-fifths scale model of docking mechanism would be tested with U.S. model during Moscow meeting in December 1972. Further consideration of extravehicular crew transfer would be deleted; if any emergency prevented return of crewmen to their own ships, they would land in spacecraft they were visiting. Sufficient propellant would be budgeted for Apollo spacecraft to allow it to maintain attitude control while docked.

Exchange of working groups would be more frequent, with meetings at MSC in November, in Moscow in December, and at MSC in March 1973.

U.S. delegation included representatives of prime contractor North American Rockwell Corp. and three of five working groups established in July 6–18 meetings at MSC to define mission elements. Working group members included Astronaut Thomas P. Stafford. NASA delegation was headed by ASTP Director Glynn S. Lunney. Senior member of Soviet delegation was Prof. Konstantin D. Bushuyev, U.S.S.R.'s ASTP Director. (NASA Releases 72-198, 72-207, 72-211; AP, *C Trib*, 10/20/72)

October 10: Air Force launched unidentified reconnaissance satellite from Vandenberg Air Force Base by Titan IIID booster at 11:03 am PDT (2:03 pm EDT). Satellite entered orbit with 268-km (167-mi) apogee, 159-km (99-mi) perigee, 88.7-min period, 96.4° inclination and reentered Jan. 8, 1973. (Pres Rpt 73; *Av Wk*, 10/16/72, 19; GSFC *SSR*, 1/31/73)

October 10–12: Flat Conductor Symposium was held at Marshall Space Flight Center to exchange information on flat conductor cables and promote their wider use in commerce and industry and building trades. Cables were thin, parallel, metal conductors, embedded in plastics or other insulating materials. New fabricating methods had produced cable for conducting electricity that was ribbon-thin and backed with adhesives which adhered to wall surfaces. Cables required little space and were easily folded to conform to unusual shapes or patterns. (*Marshall Star*, 10/11/72, 1)

October 11: U.S.S.R. launched *Cosmos 524* from Plesetsk into orbit with 490-km (304.5-mi) apogee, 264-km (164-mi) perigee, 92-min period, and 71° inclination. Satellite reentered March 25, 1973. (GSFC *SSR*, 10/31/72; 3/31/73; *SBD*, 10/13/72, 193)

• NASA released specifications of Apollo 17 lunar roving vehicle (LRV) to be carried to moon on final Apollo mission Dec. 6. Following scheduled Dec. 11 lunar landing, LRV would travel about 37 km (23 mi), about 10 km (6 mi) farther than previous LRVs, and would have loaded earth weight of 725 kg (1600 lbs), increase of about 27 kg (60 lbs) over LRVs used during *Apollo 14* and *15*. Apollo 17 LRV would carry traverse gravimeter to measure variations in subsurface structure and provide data on whether Taurus mountains had deep roots or were deposits on uniform subsurface. Surface electrical properties experiment would measure physical properties of lunar interior to 1-km (0.6-mi) depth. Experiment, connected to LRV navigational system by cable, also would indicate subsurface electrical and mechanical properties, extent of subsurface layering, and degree of energy scattering at landing site, and possible existence of underground water. First direct LRV linking with experiment would enable scientists on earth to correlate its specific location with experiment data. (NASA Special Release)

- Apollo 17 crew had selected mission emblem to emphasize beginning of golden age of space flight that final Apollo flight would usher in Dec. 6, NASA announced. Emblem showed Apollo gazing toward planet Saturn and a galaxy, symbolizing that man's goal in space would someday include planets and even stars. Suspended behind Apollo image was American eagle with four red bars on wing, representing U.S. flag. Three white stars represented Astronauts Eugene A. Cernan, Ronald E. Evans, and Dr. Harrison H. Schmitt. Eagle's wing overlying moon suggested moon had been conquered by man. (NASA Special Release)

- X-24B experimental lifting body was delivered to Air Force and NASA by builder, Martin Marietta Corp., during ceremonies at Littleton, Colo. Lifting body could be used in joint NASA–Air Force research program at Flight Research Center and at Air Force System Command's Air Force Flight Test Center to test its handling qualities for conventional runway approach and landing and for extended near-earth flight. About 30 flights were planned in program starting April 1973. X-24B would be flown to maximum speed of mach 1.5 to obtain performance data from 1450 km per hr (900 mph) down to landing speed.

  Lifting body arrived at FRC Oct. 22. (FRC PAO; UPI, *W Post*, 10/13/72; ASFC *Newsreview*, 11/72, 3)

- Establishment of Institute for Computer Applications in Science and Engineering (ICASE) at Langley Research Center was announced by NASA. Institute would be operated by Universities Space Research Assn., consortium of 51 U.S. universities, as cooperative research effort. Purpose was to solve aeronautical and space-related problems, develop efficient use of newest-generation computers, and help universities use LaRC advanced computer facilities for research. (NASA Release 72-202)

- Manned Spacecraft Center's Cellular Analytical Laboratory had developed system to improve early detection of illness and assess flight crew health before and after Skylab missions, NASA reported. Unique medical research effort under direction of Dr. Stephen L. Kimzey used sophisticated instrumentation in spectrophotometry, cytofluorometry, and electron microscopy to supplement existing laboratory analyses in detection and identification of cellular elements of blood by structural and chemical analysis. Laboratory would coordinate Skylab M110 hematology experiment series of five experiments and seven scientific investigations outside NASA and would complete one experiment, M115 (special hematologic effects). Experiment would evaluate influence of long-duration space flight on man's immunological and hematological systems. Program would use complete electron microscopy facility to examine red blood cells for structural changes and x-ray analysis to detect alterations of internal electrolyte composition. Red cell represented model system to evaluate functional state of other body tissues and processes.

Laboratory would improve medical care available to public by cooperating with medical schools in MSC area and with Government agencies throughout U.S. Studies with Univ. of Texas Medical Branch had applied Laboratory's procedures to evaluation of lymphocytes of burn patients to determine their cellular resistance to infection; early detection of disease program was under way with Baylor Univ. College of Medicine; and work with Texas Children's Hospital would evaluate changes in muscle potassium and sodium concentrates associated with malnutrition. Joint effort with Environmental Protection Agency and Oak Ridge Laboratory would identify and quantitate early response of sputum cells exposed to agents having carcinogenic potential. (NASA Special Release)

- *New York Times* editorial praised President's launch assistance policy announced Oct. 9: "Only a few days ago the fifteenth anniversary of the orbiting of Sputnik I—the beginning of the space age—passed into history. President Nixon's announcement provides welcome evidence that the initial stage of nationalistic and irrational space competition may now be succeeded by a second stage of widening international space cooperation for the benefit of all peoples." (*NYT*, 10/11/72, 42)

October 12: U.S.S.R. proposed global ban on unrestricted use of satellites for direct TV broadcasting during debate on peaceful uses of outer space by United Nations First (main political) Committee. Soviet delegate Yakov A. Malik submitted draft of proposed convention as presented to U.N. Secretary General Kurt M. Waldheim Aug. 8. Signatories would agree to transmit to other countries by satellite only if recipient countries agreed. U.S. delegate Ambassador George Bush, speaking first, told Committee U.S. was concerned over limiting "what promises to become an important new means of making information widely and immediately available to the people of the world." (Astrachan, *W Post*, 10/13/72, A23; Worldwide Press Service, *Intl Her Trib*, 10/14–15/72, 5)

- Nobel Prize for medicine for 1972 was awarded jointly to Dr. Gerald M. Edelman of Rockefeller Univ. and Dr. Rodney R. Porter of Oxford Univ. for separate research on chemical structure of antibodies. In 1959 both scientists had presented first results of investigations that had since led to almost complete clarification of essential questions on blood proteins that played important part in human body's defense against infection and development of several diseases. Dr. Edelman and Dr. Porter would share $101 000 prize money. (*NYT*, 10/13/72, 1)
- Nike-Apache sounding rocket launched by NASA from Kiruna, Sweden, carried Dudley Observatory experiment to 108.0-km (67.1-mi) altitude. Launch was third and last in series to collect and identify cosmic dust particles from Giacobini meteor shower [see Oct. 9]. Rocket and instruments performed satisfactorily and good data were obtained. (NASA Rpt SRL)
- *Cleveland Press* editorial commented on NASA catalog of inventions available for commercial licensing [see Sept. 8]. "We already know of such

NASA-originated benefits as satellite communication, land surveys and crop inspection. The catalog simply is another illustration of the spinoff importance of the space program." (*Cleveland Press,* 10/12/72)

October 13: Three Nike-Apache sounding rockets launched by NASA from Poker Flat near Fairbanks, Alaska, carried GCA Corp. experiment to study thermospheric winds using vapor-trail technique. Experiment data would be compared and contrasted with data from several different ground-based and airborne measuring systems. Data were acquired by high-speed cameras aboard NASA's Convair 990 optical aircraft and by ground-based instruments—incoherent scatter radar, radio meteor winds system, ionosonde, interferometer, spectrographs, and high-speed cameras. Vapor trails for two launches were produced by sodium-lithium ejected from payload. Rockets performed satisfactorily and good data were obtained. Launches carried no instrumentation. Vapor trails for third launch were to have been produced by liquid trimethylaluminum ejected from payload in flight. Launch was adjudged failure when Apache 2nd stage failed to ignite. Payload did not reach experiment altitude of 80 to 200 km (50 to 125 mi) and no data were obtained. (NASA Rpt SRL; Wallops Release 72–10)

- Langley Research Center's award of contracts to study technology requirements for future supersonic commercial aircraft was announced by NASA. One-year, cost-plus-fixed-fee contracts—$316 415 to Boeing Co., $259 000 to McDonnell Douglas Aircraft Corp., and $231 015 to Lockheed Aircraft Corp.—were for independent and systematic assessment of existing aeronautical technology to determine state of readiness of supersonic transport and identify areas for additional research. Studies would seek ways to use advances in aerodynamics, propulsion, structures, materials, flight controls, and configurations to reduce SST noise and air pollution. Studies were part of NASA Office of Aeronautics and Space Technology advanced planning to ensure existence of technology to maintain U.S. leadership in world aircraft market. Parallel studies in advanced propulsion technology would be managed by Lewis Research Center. (NASA Release 72–203)

- President Nixon signed H.R. 10243, Technology Assessment Act of 1972 [see Oct. 5 and Oct. 26]. Act, which became Public Law 92–484, established Office of Technology Assessment to aid Congress in identifying existing and probable impacts of technological application. OTA would ascertain cause-and-effect relationships; identify alternative programs to achieve requisite goals, estimate and compare impacts of alternate methods and programs, present analyses to appropriate legislative authorities, identify areas that required additional research and data, and undertake associated activities. Bill also amended National Science Foundation Act of 1950 to authorize NSF to initiate and support specific scientific activities. (PL 92–484; *PD,* 10/23/72, 1538)

- Total of 176 persons were reported killed when Soviet Il-62 airliner crashed near Moscow on Leningrad-Moscow sector of flight that originated in Paris. If figure was correct, it was worst disaster in history of commercial aviation, *Washington Post* said later. (*W Post,* 10/16/72, A20)

- Senate passed and cleared for President's signature H.R. 16593, $74.4 billion FY 1973 Dept. of Defense appropriations bill. (*CR,* 10/13/72, S17963–67)

- Society of Air Safety Engineers presented Distinguished Service Award to Dr. James J. Ryan, Jr., professor emeritus of Univ. of Minnesota, for pioneering development of flight data recorder. Plaque was awarded during Society seminar in Washington, D.C. (Soc Air Safety Engineers; *Av Wk*, 10/23/72, 13)

October 13–18: U.S.S.R. conducted series of carrier rocket tests in Pacific, during which it launched two improved versions of SS-11 Savage intercontinental ballistic missile with three-warhead multiple reentry vehicle (MRV). (*Sov Aero*, 10/16/72, 18; 10/23/72, 23; *Av Wk*, 10/23/72, 15)

October 14: U.S.S.R. launched *Molniya I–21* communications satellite to transmit Soviet TV programs and provide multichannel radio communications throughout Orbita network. Orbital parameters: 39 300-km (24 419.9-mi) apogee, 480-km (298.3-mi) perigee, 11-hr 45-min period, and 65.3° inclination. (GSFC *SSR*, 10/31/72; *Spacewarn*, 10/31/72)

- Rep. Ken Hechler (D-W. Va.) criticized Sept. 13 action of Office of Management and Budget in impounding $44.9 million of $50.5-million FY 1973 NASA appropriation for aircraft noise suppression program, engine retrofit studies, and quiet, experimental, short takeoff and landing (QUESTOL) aircraft program. As Chairman of House Committee on Science and Astronautics' Subcommittee on Aeronautics and Space Technology, he signed telegram to Vice President Spiro T. Agnew, Chairman of National Aeronautics and Space Council, protesting action. (*CR*, 10/14/72, H10111)

- Twenty-fifth anniversary of supersonic flight. On Oct. 14, 1947, Capt. Charles E. Yeager (USAAF) flew Bell Aircraft Corp. X–1 experimental jet aircraft—launched from B–29 bomber from Muroc Flight Test Base, Calif.—to speed of 1078 km per hr (670 mph) in first flight to break sound barrier. Aircraft, designed by Bell chief designer Robert J. Woods, had been developed under Army Air Force contract with cooperation of National Advisory Committee for Aeronautics (NACA). (*CR*, 10/17/72, E8806–7)

- Harmon International Aviation Trust named winners of awards. Harmon Aviation Trophy for 1971 would be awarded jointly to chief test pilots of Anglo-French Concorde supersonic airliner, Brian Trubshaw of England and André Turcat of France. Trophy for 1972 would go to U.S. Air Force pilots L/C Thomas B. Estes and L/C Dewain C. Vick for unprecedented long-distance nonstop flight at three times speed of sound in USAF SR–71 strategic reconnaissance aircraft. Geraldine Cobb was named winner of Harmon Aviatrix Trophy for series of humanitarian flights over Amazon Basin in 1972. (AP, *NYT*, 10/16/72)

- President Nixon met with delegation of physicians from People's Republic of China at White House. Delegation was beginning three-week tour of U.S. as guests of National Academy of Sciences Institute of Medicine and American Medical Assn. (*PD*, 10/16/72, 1518)

October 15: *Noaa 2* (ITOS–D) National Oceanic and Atmospheric Administration meteorological satellite was successfully launched by NASA from Western Test Range at 10:19 am PDT by two-stage, long-tank, thrust-augmented Thor-Delta 0300 booster with three strap-on rockets. Orbital parameters: apogee, 1453.9 km (903.4 mi); perigee, 1448.2 km (899.9 mi); period, 114.9 min; and inclination, 101.8°.

October 14: *Twenty-five years ago—Oct. 14, 1947—Capt. Charles E. Yeager (USAAF) flew the Bell X-1 experimental jet aircraft (at top above), launched from a B-29 bomber from Muroc Flight Test Base, to the speed of 1078 kilometers (670 miles) per hour to break the sound barrier in level flight. The X-1 had been developed under an Army Air Force contract with National Advisory Committee for Aeronautics cooperation. In 1972, research plans included hypersonic flight, with flights in 1973 of the X-24B lifting body in a NASA–USAF program to develop technology for future aircraft that could cruise at hypersonic speeds at the edge of space. The X-24B was delivered to the USAF and NASA by Martin Marietta Corp. in Littleton, Calif., ceremonies Oct. 11 and unloaded at Flight Research Center Oct. 22 (lower photo). The B version of the X-24A, flown in 1971, had an extended nose, flattened underside, and blended wings.*

Primary objectives were to place 345-kg (760-lb) spacecraft in sunsynchronous orbit with local equator-crossing time between 9:00 and 9:20 am and conduct in-orbit engineering evaluation, before turning operational control of spacecraft over to NOAA's National Environment Satellite Service (NESS). Orbit would permit regular, dependable daytime and nighttime temperature soundings of earth's atmosphere and cloud-cover observations by both direct readout and onboard storage in support of National Operational Meteorological Satellite System (NOMSS). First flight to implement operational atmospheric sounding and very-high-resolution infrared cloud-cover viewing, *Noaa 2* carried two new sensors: redundant very-high-resolution radiometer (VHRR) and vertical-temperature-profile radiometer (VTPR). Equipment also included scanning radiometers (SRS) and single solar proton monitor (SPM) as secondary sensor.

Oscar 6 (Oscar C) Orbital Satellite Carrying Amateur Radio, carried pickaback on 2nd stage for Radio Amateur Satellite Corp. (AMSAT), was successfully ejected and entered orbit with 1456-km (904.7-mi) apogee, 1451-km (901.6-mi) perigee, 114.9-min period, and 101.7° inclination. *Oscar 6* was sixth in Oscar series; previous spacecraft was launched by NASA as secondary payload with *Itos 1* (Tiros-M) mission Jan. 23, 1970. Satellite, to conduct experimental program of multi-access communication techniques using low-powered earth terminals, had been developed by U.S., Australian, and German amateur groups working through AMSAT. The 18-kg (40-lb) satellite would receive at 145.9 mhz and transmit at 29.5 mhz.

Tiros Operational Satellite (TOS) program was joint effort of NASA and Dept. of Commerce under Jan. 30, 1964, agreement. Goddard Space Flight Center was responsible for design and development of spacecraft and ground systems, for launch, and for checkout and evaluation. NESS was responsible for operating spacecraft and processing data. First series of spacecraft included nine satellites—*Essa 1* through *Essa 9*—of which 8 and 9 were still in routine use and 2, 6, and 7 still functional and available for backup coverage. In second series—Improved TOS (ITOS)—prototype had been *Itos 1* and first NOAA-funded spacecraft, *Noaa 1*, had been launched Dec. 11, 1970, to observe globe twice daily. (NASA proj off; GSFC *SSR*, 10/31/72; NOAA Release 72–135; *A&A 1970*)

- Delegation of Soviet scientists headed by Soviet Academy of Sciences President Mstislav V. Keldysh flew from Moscow to New York. Scientists would participate in general meeting of U.S. National Academy of Sciences in Washington, D.C., and visit U.S. scientific institutes. (Tass, FBIS–Sov, 10/17/72, L1)

- Prediction that deep-space probes would be launched by combination rockets propelled by chemical, nuclear, and electric engines by 1980 had been made by Soviet aerospace designer Valentin P. Glushko in interview by *Moskovsky Komsomolets*, *New York Times* reported. Rockets would be launched with liquid or solid chemical fuel. Nuclear engine would take over when spacecraft penetrated earth's atmosphere.

Times also reported that color film "The Taming of the Fire" had been released in Moscow. Film, based on life of Sergey P. Korolev, head of Soviet space program until his death in 1966, raised philosophical questions of use of scientific advances for military purposes. Film showed Soviet dictator Josef Stalin at end of World War II calling for Soviet development of military rockets to counter weapons he expected Dr. Wernher von Braun to develop in U.S. Dr. von Braun, while not actually portrayed in film, was "frequently mentioned by name." (Shabad, *NYT*, 10/16/72, C17)

- NASA launched Strypi 4 sounding rocket from Kauai, Hawaii, carrying National Oceanic and Atmospheric Administration artificial aurora experiment. Rocket and instrumentation performed satisfactorily. (SR list)

October 16: Establishment of Office of Supply and Equipment Management under Assistant Administrator for Administration was announced by NASA. Office was responsible for receipt, use, and disposal of all NASA personal property, in-house and contractor held. It absorbed Property and Supply Div. and related functions of offices of Facilities and Procurement. William P. Risso, Director of Resources Management in Office of Applications, had been designated Director of new office, whose top priority was to ensure optimum use, reuse, and disposition of equipment throughout NASA. (NASA Ann)

- Preliminary design efforts were under way on two of three multimillion-dollar wind tunnels that had been declared essential by NASA and Dept. of Defense Aeronautics and Astronautics Coordinating Board (AACB), *Aviation Week & Space Technology* reported. Tunnels, to overcome serious deficiencies in U.S. aeronautical test facilities, were large vertical or short takeoff and landing (V/STOL) tunnel, High Reynolds Number Tunnel (HIRT), and Aeropropulsion System Test Facility (ASTF). NASA had been assigned responsibility for V/STOL tunnel; Air Force was responsible for HIRT and ASTF, which would be at NASA's Ames Research Center or Langley Research Center. Construction had been approved by AACB in July, and could be completed by decade's end. Announcement of ASTF design contractor was expected imminently; selection of HIRT contractor was under way. NASA was still analyzing requirements and alternate ways of meeting them, according to Chief Mark W. Kelly of ARC's Large-Scale Aerodynamics Branch. (*Av Wk*, 10/16/72, 42–5)

- General Telephone & Electronics Corp. announced formation of new subsidiary, GTE Satellite Corp., to provide domestic satellite communications system. System would include multipurpose satellite earth stations in Hawaii, California, Florida, Indiana, and Pennsylvania and could be operational within 24 mos after receipt of Federal Communications Commission approval. (GTE Release)

- Soviet academician Dr. Leonid I. Sedov had attributed 15-mo hiatus in Soviet manned space flight to major changes being made in Soyuz spacecraft, *Aviation Week & Space Technology* reported. Changes to improve cosmonauts' living conditions were believed to include retrofit of miniaturized control and communications equipment for more space in command module. Dr. Sedov had said faulty hatch system had been corrected and spacecraft was ready for 30-day orbital laboratory mission. (*Av Wk*, 10/16/72, 11)

- Aviation pioneers to be enshrined in Aviation Hall of Fame at Dayton, Ohio, Dec. 15 were listed by *Aviation Week & Space Technology*: L/G Claire L. Chennault of World War II Flying Tigers; Leroy R. Grumman, aircraft designer and manufacturer; James H. Kindelberger, aircraft executive; and Gen. Curtis E. LeMay, former Air Force Chief of Staff. (*Av Wk*, 10/16/72, 11)

- Dr. John E. Condon, Director of Reliability and Quality Assurance and Acting Director of Safety, left NASA to enter private industry. (*NASA Activities*, 11/15/72, 236)

October 17: Ariel 4, cooperative United Kingdom and U.S. satellite (launched by NASA Dec. 11, 1971), was adjudged successful. During more than 10 mos in orbit, spacecraft had operated satisfactorily except for tape recorder failures and minor experiment problems. Primary mission objective—to investigate interaction of electromagnetic waves, plasmas, and energetic particles in upper ionosphere—had been achieved.

High-frequency noise experiment had shown several apparently distinct kinds of radio noise, which were under study. Extremely-low- and very-low-frequency experiment and impulse experiment showed ELF noise spectrum depended on latitude and therefore on shape of magnetosphere. Univ. of Iowa charged-particle detector was observing auroral charged particles, large field-aligned currents over auroral zone, and higher-energy protons of ring current. Observations, when combined with those from other satellites, were expected to delineate dissipation mechanism for ring current responsible for magnetic storms. Electron density experiment was providing worldwide coverage for morphological study of topside ionosphere. (NASA proj off)

- Air Force Space Technology Program satellite *Sesp 1971–2* (launched Oct. 17, 1971) was still transmitting data after year in orbit. Satellite, designed to operate for 6 mos, had traveled more than 214 million km (133 million mi) in 5233 earth orbits. STP program manager at Lockheed Aircraft Corp., Robert T. Johnson, had said some experiments aboard satellite would be extended six more months. (Lockheed Release 72–73)

- Astronaut Thomas P. Stafford made joint "flight" with Soviet Cosmonaut Andrian G. Nikolayev in Soyuz spacecraft simulator at Zvezdny Gorodok (Star City), cosmonaut training center near Moscow. Stafford was member of NASA delegation to Oct. 9–19 Moscow meetings on Apollo-Soyuz Test Project (ASTP), joint rendezvous and docking mission scheduled for 1975. (UPI, W *Star & News*, 10/18/72)

- Soviet Academy of Sciences President Mstislav V. Keldysh and five members of Soviet Academy began 21-day coast-to-coast tour of U.S. scientific and space installations. Academician Keldysh addressed unpublicized session of U.S. National Academy of Sciences in Washington, D.C., from which even NAS staff members were banned. NAS had withheld announcement of Keldysh visit to discourage anti-Soviet demonstrations. He had been invited to U.S. by NAS President, Dr. Philip Handler. (*W Post*, 10/18/72)

- World's largest radiotelescope array was dedicated in Cambridge, England. Eight dish antennas, each 13 m (42 ft) in diameter, produced effect of radiotelescope 5 km (3 mi) in diameter. Dishes moved through 360° daily with earth's rotation to permit research into structure of stars and origin and evolution of universe. Major task of $5.2-million installation built for U.K.'s Science Research Council and operated by Cambridge Univ. would be to map galaxies and quasars by pinpointing their radio signals and producing computerized charts of their locations. (Reuters, B *Sun*, 10/18/72)

- Canadian government had said it would spend $80 million in three-year program to develop two prototypes of short takeoff and landing (STOL) aircraft, *Wall Street Journal* reported. (*WSJ*, 10/17/72)

- *Houston Chronicle* editorial commented on Soviet attempt to restrict satellite TV broadcasts as threat to world peace: "We should be very slow to accept any limitations upon the use of satellites. We have just begun to tap the benefits that can flow from the exploration of space. Our international space program stands in contrast to Russia's secret effort. While we are opening new avenues of communications, Russia typically is acting to keep its citizens in ignorance." (*H Chron*, 10/17/72)
- Second F-111 swing-wing fighter-bomber was lost over North Vietnam. First had been lost Sept. 28. U.S. Command in Saigon had no comment on North Vietnamese report that aircraft had been shot down. (AP, *NYT*, 10/19/72, 18)

October 17-19: Results of NASA research on powered-lift short takeoff and landing (STOL) aircraft were presented at Ames Research Center Conference. Papers were presented on short-haul transportation systems, aerodynamics, loads, flight dynamics, operation aspects, and quiet STOL propulsion by authors from ARC, Flight Research Center, Langley Research Center, Lewis Research Center, Federal Aviation Administration, and Army Air Mobility Research and Development Laboratory. (NASA Release 72-201; NASA SP-320)

October 18: U.S.S.R. launched *Cosmos 525* from Plesetsk into orbit with 292-km (181.4-mi) apogee, 206-km (128-mi) perigee, 89.5-min period, and 65.4° inclination. Satellite reentered Oct. 29. (GSFC *SSR*, 10/31/72; *SBD*, 10/23/72, 26)

- American Satellite Corp., Maryland-based company formed by Fairchild Industries and Western Union International, had told Federal Communications Commission it was ready to set up domestic communications satellite network, Fairchild announced. Company had said service was planned for late 1973 or early 1974. ASC President Emanuel Fthenakis had said proposed network would be developed in phases to provide point-to-point private line transmission for voice, data, and facsimile teletype. (Fairchild Release FI-9-229)
- Dr. Nicholas C. Costes of Marshall Space Flight Center Space Sciences Laboratory received Norman Medal, highest award of American Society of Civil Engineers, at ASCE annual meeting in Houston. Award was for paper "Apollo 11: Soil Mechanics Results," published by ASCE in 1970. (*Marshall Star*, 10/18/72, 1)
- Flight Safety Foundation, Inc., presented 1972 Admiral Luis de Florez Flight Safety Award to Dr. John T. Dailey of Federal Aviation Administration on final day of three-day seminar. Recognition was for "outstanding individual contribution through his development of a behavioral profile designed to identify potential hijackers." (Flight Safety Foundation; *Av Wk*, 10/23/72, 13)
- Award of $46-million contract to North American Rockwell Corp. for vertical or short takeoff and landing aircraft prototype was announced by Secretary of the Navy John W. Warner. Aircraft would use components of existing aircraft to reduce time and expense before first flight. If prototype was successful, Navy might develop operational V/STOL aircraft. (DOD Release 723-72)
- Dept. of Defense spokesman Jerry W. Friedheim said in Washington that U.S.S.R. had completed its latest missile tests in Pacific and might be ready to deploy "improved" SS-11 rockets armed with three nuclear warheads each. (Corddry, B *Sun*, 10/19/72, A5)

October 18–20: Group of 16 Nebraska officials visited Marshall Space Flight Center for orientation in telecommunications technology. Group also received briefings on simulation techniques and equipment that might be used in training disabled veterans as communications controllers. Veterans employment program was part of Nebraska's effort to use modern communications techniques in state's emergency medical service, law enforcement, and highway service. Nebraska Governor J. James Exon had requested NASA's aid. (MSFC Release 72–138; MSFC PAO)

October 19: NASA's M2–F3 lifting body, piloted by John A. Manke, completed 20th flight from Flight Research Center. Objectives—to check out pilot and to obtain stability and control data at mach 0.70—were achieved. M2–F3 reached mach 0.70 and 13 700-m (45 000-ft) altitude. (NASA proj off)

- Federal Bureau of Investigation and Dept. of Justice were investigating sale by Lamarque, Tex., flea market of technical items related to 1965 *Gemini 5* and *7* flights. Items included test card used in Gemini experiment, technical books, and tapes, NASA Regional Inspector at Manned Spacecraft Center Glenn L. McAvoy said. Market owner E. E. White had said he acquired material in 1970 from man who had identified himself as NASA employee. Investigation was completed in November with no charges brought. NASA later requested U.S. Attorney General efforts to retrieve Government property. (NASA Regional Inspector at MSC; AP, *NYT*, 10/20/72)

- Publication of *NASA Earth Resources Survey Program Weekly Abstracts* was announced by NASA. NASA-funded bulletin, prepared and distributed by National Technical Information Service of Dept. of Commerce, contained abstracts of technical reports on imagery received from Earth Resources Satellite *Erts 1* (launched July 23). (NASA Release 72–205)

- White House anounced that U.S.–U.S.S.R. Strategic Arms Limitations Talks (SALT) would resume in Geneva Nov. 21. (*PD*, 10/23/72, 1534)

- Retirement of Dr. Frank M. Branley as Chairman of American Museum-Hayden Planetarium was announced by museum. He would be succeeded by Dr. Kenneth L. Franklin, Assistant Chairman, codiscoverer of high-frequency emissions from Jupiter, and innovator of first lunar watch to tell time on moon. Dr. Branley would devote time to science writing and communications. (Museum-Planetarium Release)

October 20: United Nations General Assembly's First Committee concluded debate on peaceful uses of outer space and unanimously adopted resolution approving Committee report. Resolution noted progress in draft treaty on moon, welcomed progress in international cooperation in exploration of outer space, and approved of agreement between U.S. and U.S.S.R. on development of compatible docking equipment for carrying out joint scientific experiments.

Draft resolution on preparation of principles for use of satellites for direct broadcasting was endorsed by 68 votes with 12 against and 18 abstentions. Among those against were U.S.S.R. and other nations that favored defeated Soviet draft convention that would ban unrestricted use of satellites for direct TV broadcasting to other states [see Oct. 12 and Nov. 9]. (Tass, FBIS–Sov, 10/25/72, A6; UN Gen Assembly Documents A/8863, A/8864, 11/1/72; *NYT*, 11/10/72, 16)

- Swedish Royal Academy of Sciences announced award of 1972 Nobel Prizes in chemistry and physics to six U.S. citizens. Chemistry prize of more than $100 000 would be shared by National Institutes of Health scientist Dr. Christian B. Anfinsen and Rockefeller Univ. scientists Dr. Stanford Moore and Dr. William H. Stein for "pioneering studies" in enzyme ribonuclease. Dr. Anfinsen's investigations had "provided the answer to an important question concerning the way in which the active enzyme is formed in living cells; Moore and Stein have elucidated important principles related to the biological activity of the enzyme. These properties we generally associate with the concept of life and with living organisms."

  Physics prize was awarded for theory of superconductivity developed jointly by Dr. John Bardeen of Univ. of Illinois at Urbana, Dr. Leon N. Cooper of Brown Univ., and Dr. John R. Schrieffer of Univ. of Pennsylvania. Superconductivity was phenomenon in which electrical resistance in certain metals vanished when metals were cooled to temperatures near absolute zero (0 kelvins; $-459.7°F$). (Weinraub, *NYT*, 10/21/72, 1)

- National Academy of Sciences announced that seven-member Chinese scientific delegation would visit U.S. in late November. Delegation, second group of professionals from mainland China to visit U.S. since 1950s, would include specialists in high-energy physics and computer technology. (NASA Release)

- Claim that Central Intelligence Agency had "stolen" Soviet Sputnik spacecraft for three hours to examine it while it was on tour in 1958 was included in new book critical of CIA by former CIA agent, Baltimore *Sun* reported. Book was *CIA—The Myth & the Madness*, by Patrick J. McGarvey. (B *Sun*, 10/20/72, A1)

- Chen Chu, People's Republic of China delegate to United Nations General Assembly, issued statement denouncing use of name "China" by Formosa government of Chiang Kai-shek during U.N. debate on peaceful uses of outer space: "As from October 1, 1949, the day of the founding of the People's Republic of China, the Chiang Kai-shek clique has no right at all to represent China." Formosa government had "usurped" name of China to become party to agreement, treaty, and convention on peaceful uses of outer space. (FBIS–PRC Intl Aff, 10/24/72, A10)

- Dr. Harlow Shapley, one of world's leading astronomers, died in Boulder, Colo., at age 86 after long illness. He had been director of Harvard College Observatory from 1921 to 1952 and astronomer at Mount Wilson Observatory, Calif., from 1914 to 1921. At Mount Wilson Dr. Shapley had used variable stars and globular star clusters to fix position of solar system at some 30 000 light years from center of Milky Way galaxy. Previously earth and sun had been supposed near center of Milky Way galaxy. Finding had had immense scientific value and import for religious thought and philosophy. After World War II, Dr. Shapley shifted attention from scientific pursuits to public affairs. He had condemned cold war of 1940s and championed peaceful coexistence among nations of world. His advocacy of U.S.–U.S.S.R. friendship had led to his censure by House Un-American Activities Committee. Before retirement in 1952, Dr. Shapley had received 18 honorary degrees, many awards—including Pope Pius XI Prize in

1941—and had published more than 500 writings. (Weil, *W Post*, 10/21/72, B6; *NYT*, 10/21/72, 1)

October 21: President Nixon signed S. 3943, Senate version of H.R. 16645. Bill, for construction of civic center as memorial to President Dwight D. Eisenhower, carried rider that changed name of Jet Propulsion Laboratory to H. Allen Smith Jet Propulsion Laboratory, effective Jan. 4, 1973. Bill became Public Law 92–520. (*PD*, 10/30/72, 1548–9; PL 92–520)

- NASA launched Nike-Tomahawk sounding rocket from Poker Flat Rocket Range, Alaska, carrying Univ. of Minnesota auroral studies experiment. Rocket performed satisfactorily, but scientific objectives were not met. (SR list)
- President Nixon issued memorandum disapproving H.R. 56, bill to establish National Environmental Data System and create environmental centers in each state: "While both these titles sound desirable in theory, they would in reality lead to the duplication of information or would produce results unrelated to real needs and wasteful of talent, resources, and the taxpayers' money." (*PD*, 10/30/72, 1549)

October 21–23: Thirteenth annual Space Fair at Pacific Missile Range, Point Mugu, Calif., featured exhibition of *Apollo 12* moon rock, Apollo space capsule, aerospace equipment, and static display of aircraft. Fair was sponsored by naval commands and units at PMR and by Naval Construction Battalion Center at Port Hueneme, Calif. (PMR Release 933–72; PMR *Missile*, 10/13, 10/20, 10/27, 1)

October 22: Construction at United Nations Hq. in New York of "Hall of Fame for all spacemen—those who already have voyaged, or someday will voyage, in outer space" was proposed by *Parade* magazine. Hall would "underscore our need for international cooperation in space." Magazine said U.N. Secretary General Kurt M. Waldheim had endorsed proposal because "it is a practice of most of the astronauts to visit the U.N. when they return from their moon flights." (*Parade*, 10/22/72, 7)

- U.S., U.S.S.R., and several other countries were engaged in "multimillion-dollar efforts" to test feasibility of new approach to controlled fusion that would produce unlimited, pollution-free energy, *New York Times* reported. New method was to heat and implode, to superdense state, a hollow pellet of fusion fuel by smashing it from all sides with simultaneous pulses of laser beam. Concept had thus far been checked only in computer simulations. (Sullivan, *NYT*, 10/22/72, 1)
- Federal Aviation Administration announced appointment of Robert F. Bacon as Director of Office of Aviation Policy. He had been Acting Deputy Director of System Planning Div. in FAA's Airports Service. (FAA Release 72–200)

October 23: *Chemical & Engineering News* published views and plans of presidential candidates for science and technology beyond general pledges in Democratic and Republican Party platforms. Same set of questions had been submitted to campaign organizations of Sen. George S. McGovern (D–S.D.) and of President Nixon. President Nixon said he would continue general pattern of previous years on Federal research and development funding; its division among defense, space, and civilian sectors; its possible link to gross national product (GNP); and appropriate balance for basic research, applied research, and development. Pattern would include increasingly effective program in civilian sector R&D, "prudent space program, and a sufficient defense R&D budget." Administration believed R&D funding should be based on opportunities for effective programs and necessity for creating new opportunities through fundamental studies and that R&D resources, excepting basic research, should be allocated to areas that provided "greatest opportunity for leverage or that satisfy an immediate future need."

Sen. McGovern recommended increased Federal R&D funding. Principal problem was "government spending that has masqueraded as R&D, yet has little technical content. For example, when an Apollo is launched . . . all of the kerosene and liquid oxygen that it uses comes out of the federal R&D budget. Only a very small percentage of the federal R&D budget actually represents laboratory investigations, instrument design, feasibility studies, and other technical efforts that involve scientists and engineers." Sen. McGovern would emphasize civilian sector R&D and strong military R&D program as "good investment for long-term national security. My space program would emphasize unmanned planetary exploration and scientific satellites and also earth resource and communications satellites." Nixon Administration would continue to maintain "strong and up-to-date defense force" and to capitalize upon past space investments. Both candidates recommended increased priority on developments of new energy sources.

Nixon Administration was opposed to S. 32, reconversion bill introduced by Sen. Edward M. Kennedy to amend National Science Foundation Act of 1950 to establish framework of national science policy. Bill would focus policy and U.S. scientific talent and resources on Nation's priority problems. Administration believed segment of S. 32 that would establish independent agency within NSF was "particularly unwise" because it would "divorce the R&D from the departments of Government responsible for using its results" and "would divert attention from the primary mission of NSF—to support fundamental and basic research." Sen. McGovern would continue to support S. 32. "In its original form, before compromise to get Republican support, it would have provided $1.2 billion for a new NASA-type Civil Science Systems Administration that would fund genuine R&D across a broad range of civilian technology needs."

Administration did not believe in Cabinet-level science department or "supra" agency, nor that "existing science apparatus is totally adequate. As with all human and governmental institutions, it too should be dynamic and subject to change as conditions and events dictate." Sen. McGovern felt "more centralized federal science management and funding apparatus might be useful, but the key need is not to change the institutions but rather to give them more power in the Federal Government." He favored establishment of Office of Technology Assessment [see October 13] which would "assess each new technical program with an eye toward cost efficiency and effectiveness, and domestic betterment." Asked if federally owned laboratories should be expanded to deal with problems of health and environment, Nixon Administration pointed out that NASA Centers were "closely involved with the R&D components of other federal agencies in an effort to solve pressing national problems." Sen. McGovern believed Federal laboratories and research facilities had "plenty of problems in their immediate areas" with which to cope. "I have no intention of taking the easy route and putting them on safe subjects." (*Chemical & Engineering News*, 10/23/72, 14–19)

October 23–27: Conference on high-energy astrophysics was held at Cal Tech by division of American Astronomical Society. Dr. Kenneth I. Kellerman of National Radio Astronomy Observatory said most baffling phenomenon observed by radiotelescopes was speed at which objects seemed to be moving. If measurements taken in past two years were accepted at face value, objects associated with quasars were speeding away from each other faster than speed of light. Such speed had been considered impossible. Cal Tech astronomer Dr. W. L. W. Sargent said strange blue objects sighted in sky by Hale Observatory astronomers might be newly formed galaxies "about 100 million years old, and they seem to be made up of very young stars." Discovery indicated that galaxies might still be forming. Dr. Herbert Gursky of American Science and Engineering, Inc., said astronomers using *Uhuru* (*Explorer 42* Small Astronomy Satellite launched by NASA Dec. 12, 1970) had spotted several x-ray objects that could not be detected by wavelengths other than x-ray. Impact of x-ray astronomy might be that "this new wavelength range may show some different aspects of stars and galaxies." Through x-ray astronomy scientists "hope to . . . learn something about these objects that radio and optical observations can't tell us."

Evidence of black hole—star whose gravity field was so strong that light could not escape from it—in constellation Cygnus was reported by Dr. Kip Thorne of Cal Tech, Dr. J. P. Ostriker of Princeton Univ., and Dr. Riccardo Giacconi of American Science and Engineering. Evidence had come from photos taken at Hale Observatory that showed bright blue line near source Cygnus X–1 and from data gathered by *Uhuru* (see also Sept. 2). Scientists thought line in photos might be result of superhot gas being drawn from neighboring star in black hole. If finding was accurate, Cygnus X–1 would be first black hole discovered. Existence of black holes had been predicted by Einstein as part of his theory of relativity. (Cooke, Pasadena *Star-News*, 10/24/72; West, *LA Times*, 10/27/72; O'Toole, *W Post*, 10/28/72, A3; *W Star & News*, 10/28/72)

October 24: Soviet news agency Tass announced opening of Soviet National Conference on Space Biology and Medicine at Kaluga, U.S.S.R. Conference would hear papers on use of "terrestrial" atmosphere in spacecraft cabins, food and water regeneration, and establishment of bacterial medium in which future cosmonauts could live and work on extended space missions. Conference also would discuss weightlessness, crew selection and training, radiobiological aspects of flight, and psychophysiology of space crew work. Soviet Academician Oleg Gazenko had presented paper on topical problems of space biology and medicine at opening session. (FBIS–Sov, 10/25/72, L1)

- NASA launched Black Brant VB sounding rocket from White Sands Missile Range, N. Mex., carrying Goddard Space Flight Center payload to test rocket configuration that would be used in support of Skylab mission. Rocket and instrumentation performed satisfactorily. (SR list)

- Washington *Evening Star and Daily News* editorial commented on possible revival of U.S. supersonic transport aircraft program: "Last year the Senate shot down the supersonic transport program in a famous victory for environmentalism and good fiscal sense. We hoped that would end the noisy debate, at least for a few years. But new rumblings already are heard from the rubble of that costly, abortive endeavor, amid rumors that the Nixon administration may start revving up a new SST campaign after the election. We hope the President, assuming he wins the election, will let well enough alone. But the hints are fairly strong." (W *Star & News*, 10/24/72)

October 25: U.S.S.R. launched *Cosmos 526* from Plesetsk into orbit with 480-km (298.3-mi) apogee, 272-km (169.0-mi) perigee, 92-min period, and 70.9° inclination. Satellite reentered April 8, 1973. (GSFC *SSR*, 10/31/72; 4/30/73; *SBD*, 10/26/72, 250)

- *Oscar 6*, amateur radio satellite launched Oct. 15 as pickaback payload on *Noaa 2*, was operating satisfactorily and providing worldwide communications relay. Traffic through satellite's transponder had been particularly heavy over U.S. and Europe and use by stations in U.S.S.R. and Czechoslovakia had been reported. Anomaly in switching of experiment control logic had been overcome by use of ground command. Satellite's temperatures were higher than had been predicted. (NASA proj off)
- Plans for illuminating night launch of Apollo 17, scheduled for 9:53 pm EST Dec. 6, were released by NASA for photographic fans. Spacecraft would be surrounded on launch pad by 72, 20-kw xenon searchlights and 2, 60-kw xenon searchlight banks that would provide 225 foot-candles of light. Light from Saturn V 1st-stage booster engines would not increase illumination until liftoff, when some 7500 foot-candles emitted and reflected would create almost daylight exposure. Searchlight illumination for inflight coverage would not be available beyond first 18 m (60 ft) of flight. (NASA Release 72–208)
- Launch, by Italy, of NASA Small Astronomy Satellite B (SAS–B) had been postponed from Nov. 2 to Nov. 16, at earliest, because of malfunction of Scout launch-rocket gyro package, NASA announced. Package was being replaced with spare system and telemetry encoder had been returned to U.S. for repairs. (NASA Note to Editors)
- French Cabinet agreed in principle to mass production of Anglo-French Concorde supersonic transport aircraft and medium-range European Airbus. (Agence France Presse, FBIS–France, 10/26/72, T3)

October 26: Spartan antiballistic missile launched from Meck Island in Kwajalein Atoll successfully intercepted simulated intercontinental ballistic missile nosecone over mid-Pacific in 16th successful firing in Dept. of Defense second series of 18 firings. (*W Post*, 10/28/72, 36)

- Chronology of H.R. 10243, bill to establish Office of Technology Assessment for Congress (signed by President Nixon Oct. 13), was traced in *Washington Post* article. Bill was result of six-year effort begun by Rep. Emilio Q. Daddario (D-Conn.), who had since left Congress. House had passed earlier version Feb. 8; Senate had approved its version Sept. 14; and conference report worked out Sept. 21 had been approved by both houses. Bill had been supported in House by members of Committee on Science and Astronautics' Subcommittee on Science, Research, and Development and by Rep. Jack Brooks (D-Tex.)

October 26

and in Senate by Sen. Edward M. Kennedy (D-Mass.), Sen. B. Everett Jordan (D-N.C.), and Sen. Gordon Allott (R-Colo.). *Post* quoted Senate aide as saying recent renewed interest in bill had been prompted by debates over supersonic transport (SST) program and antiballistic missile systems. In both, Congress had had "no really authoritative expertise of its own." New Office of Technology Assessment would have authority and independence of General Accounting Office and was expected to depend mainly on contracts with outside study groups. (Cohn, *W Post*, 10/26/72, G3)

- Article on U.S. space reconnaissance program, written by Soviet engineers Yu. Safronov and Ya. Sukanov, was published in Moscow by *Krasnaya Zvezda* (Red Star). Authors quoted U.S. press reports and U.S. journal *Aerospace Daily* in describing Dept. of Defense space reconnaissance activities since 1962. Article concluded that DOD had "dispatched dozens, hundreds of automatic apparatuses into circumterrestrial space. Of 605 satellites launched . . . in the period 1958 through 1970, 366 were of military designation . . . two out of every three." It was "beyond doubt" that DOD's "increasingly extensive use of space technology for purposes very far removed from scientific research in no way accords with the tendencies which have begun to appear toward limiting the arms race and improving the international situation." (FBIS-Sov, 11/2/72, G1-4)

- President Nixon signed H.R. 16593, $74.4-billion FY 1973 Dept. of Defense appropriations act. Bill became Public Law 92-570. (*PD*, 11/6/72, 1619)

- European Space Conference failed to meet as tentatively scheduled. Formal meeting of conference ministers to decide fate of European launcher development program and European post-Apollo participation might not take place until late December, *Aviation Week & Space Technology* later reported. (*Av Wk*, 10/30/72, 14)

- Russian-born aviation pioneer Igor I. Sikorsky, who invented first successful helicopter in U.S., died at age 83 in Easton, Conn., after heart attack. Born in Kiev, he had come to U.S. in 1919 with only $600. In 1922 he established Sikorsky Aero Engineering Corp. in small shed. Firm produced amphibian aircraft in 1928, four-engine clipper in 1931 with which Pan American World Airways inaugurated first commercial route to Hawaii and to Asia, and first successful helicopter, which Sikorsky flew in 1939. Firm later became Sikorsky Aircraft Div. of United Aircraft Corp. Sikorsky retired as engineering manager in 1957 but had remained consultant. His honors included National Defense Transportation Award in 1953, first Dr. Alexander Klemin Award of American Helicopter Society, and National Medal of Science. (Hailey, *W Post*, 10/27/72)

- M/G Don R. Ostrander (USAF, Ret.), former NASA Director of Launch Vehicles, died at Andrews Air Force Base hospital, Md., of cancer at age 58. He had retired from Air Force as Vice-Commander of Ballistics System Division in 1965 and became Vice President for Planning, Bell Aero Systems Corp., until 1971. Gen. Ostrander had been appointed to NASA position in 1959. (*W Post*, 10/28/72, B10; Rosholt, *Administrative History of NASA*)

October 26–27: California newspapers reported reaction at Jet Propulsion Laboratory to news of renaming facility for retiring California

Congressman H. Allen Smith (D-Calif.) [see Oct. 3 and Oct. 21]. News had come as surprise to officials, who had refused to comment. JPL engineer Charles D. Hepburn had said that several engineers deplored renaming of facility for political office holder when there were many great local scientists who had contributed to space program. Hepburn had said it was "insult to technology.... Our work has to do with the physical universe, not with the meanderings of a political career." (*LA Times*, 10/26/72; Pasadena *Star-News*, 10/27/72)

October 27: U.S.S.R. launched *Meteor 13* weather satellite from Plesetsk to obtain "meteorological information needed for the operative weather forecast service." Orbital parameters: 891-km (553.6-mi) apogee, 866-km (538.1-mi) perigee, 102.5-min period, and 81.2° inclination. (GSFC *SSR*, 10/31/72; FBIS–Sov, 10/30/72, L1; *SBD*, 10/30/72, 266)

• Lunar science press conference at Manned Spacecraft Center discussed *Apollo 16* data and portent for Apollo 17 mission, scheduled for launch Dec. 6. Dr. Oliver A. Schaeffer of State Univ. of New York at Stony Brook reported that analysis of pea-size *Apollo 16* rock fragments had shown them ranging in age from 3.98 billion to 4.25 billion yrs. Rocks had crystallized some 300 million yrs after moon's formation. It was possible that Apollo 17 site would yield "very young rocks..., young volcanics. So one may be able to extend the mare type ages... and see just how long it took the surface of the moon... to crystallize before it was a rather hard crust... with very little, if any, igneous activity."

Dr. Gerald J. Wasserburg of Cal Tech reported *Apollo 16* rock No. 65015 seemed to have undergone two distinct melting steps. Bulk of this "two-history rock" had crystallized 3.93 billion yrs ago but had retained some crystals that had been formed between 4.40 and 4.48 billion yrs ago. First crystallization probably had been associated with great heat during moon's formation but scientists had no clue to cause of second melting period and subsequent recrystallization. Dr. Paul W. Gast, Chief of MSC Planetary and Earth Sciences Div., said "fundamental conclusion" was evolving from new understanding of moon gained "largely out of the Apollo program." Scientists "have to now look at each individual planet and somehow or another infer its chemical composition." They were beginning to see chemically heterogeneous solar system.

Dr. Harrison H. Schmitt, Apollo 17 lunar module pilot, described Apollo 17 landing site in moon's Taurus-Littrow area on rim of Serenitatis Basin: "It's in that general geographic ring that resembles the Apennine front near where Apollo 15 landed. It has, however, some very distinctive characteristics.... There is no clear evidence that we are close to mare. The nearest obvious mare surface is several hundred kilometers [100 km equals 62 mi] to the... northwest—in the Mare Serenitatis. We are landing on a plain that seems to overlie a slightly elevated flat surface that predates the mare. And it seems when you look at all of the basins on the moon, that that surface, a correlative surface in other basins, is somehow related to the formation of the basin. There are comparable plains around all of the big circular basins, and there is a flat floor that predates the mare that is cut by graben faulting.... Also in the site we have the... north and south massifs. These are large aggregates of peaks and mountains that . . . are part of the range structure of the Orientale basin."

Dark cover over valley floor and portions of highlands was younger than mare and was as little as 0.5 m (1.6 ft) thick. Radar data had shown cover was "finely fragmented mantle material" that was "ubiquitous in all of the low areas and locally is covering portions of the highlands." Dr. Schmitt said mantle, although of not as high a priority as the crustal examination in the south massif, had "potential of giving us our most fundamental evidence for studying this evolutionary model." Potential was high that "dark mantle material has sampled some significant portion of the crust. At least it has sampled its source area, and that source area is very probably deep within the moon." Dr. Schmitt told press that to be able to say that man was "on the verge, if not already there, of understanding the evolution of another planet is something that is really unprecedented in the history of man. And I think that's the thing that's the news of Apollo. That is what we have done." (Transcript)

- *Mariner 9* mission ended as attitude-control gas supply was depleted and engineering telemetry signal ceased during 698th Martian orbit of spacecraft launched May 30, 1971. *Mariner 9* had reached Mars Nov. 13, 1971, and was 383 675 000 km (238 416 000 mi) from earth when engineers at Jet Propulsion Laboratory sent 45 960th, and last, command—to turn off radio transmitter. Spacecraft was expected to remain in Mars orbit some 50 yrs.

  During 349 days (more than one half Martian year) of orbital activity, *Mariner 9* had maintained instrumented surveillance of Mars and made first closeup photos of Martian moons Phobos and Deimos. Major findings had included: geologically active planet with volcanic mountains and calderas larger than any on earth, equatorial crevasses more than 3000 km (2000 mi) long and three to four times deeper than Grand Canyon, indications that free-flowing water might have existed in Mars' geologic history, evolution of monumental dust storms that raged to altitudes of 50 to 60 km (31 to 37 mi) above Mars surface, and realization that dust storms and cloudiness accounted for much of variability of Mars' appearance over years.

  Mariner 9 was first man-made object to orbit another planet. Mariner project, managed for NASA's Office of Space Science by JPL, had been planned as two-spacecraft mission to Mars but Mariner 8, launched May 8, 1971, failed to enter orbit when Centaur stage of Atlas-Centaur booster malfunctioned after normal countdown and liftoff from Eastern Test Range. Spacecraft had reentered earth's atmosphere. Plans had been revised and *Mariner 9* had been launched successfully to conduct missions of both Mariners. Data from *Mariner 9* experiments would provide groundwork for 1975 Viking lander mission to search for evidence of life on Mars. (JPL Release 634; *A&A 1971*)

- Air Force System Command successfully launched 1.4-million-cu-m (47.8-million-cu-ft) research balloon, world's largest, from Chico, Calif. Balloon, developed by Air Force Cambridge Research Laboratories, carried 113-kg (250-lb) payload to unofficial record altitude of 51 800 m (170 000 ft). Previous record of 49 400 m (162 000 ft) had been established in 1969 by AFCRL balloon. Purpose of flight was to determine feasibility of flying large thin-film balloons and to study very-high-altitude parachute deployment characteristics. (AFSC Release 121.72)

- Naval Research Laboratory announced discovery of huge clouds of carbon monoxide in galaxy M-33 by NRL scientist Dr. Philip R. Schwartz and Aerospace Corp. scientists Dr. William J. Wilson and Dr. Eugene E. Epstein. Team had used National Radio Astronomy Observatory radiotelescope at Kitt Peak, Ariz., in first detection of extragalactic molecular emission line by radioastronomy. Finding indicated that presence of complex molecules in interstellar medium was not unique to earth's galaxy. Galaxy M-33 was 1.5 million light-years from center of Milky Way galaxy. (NRL Release 60-72-10)
- Community Action Agency officials announced Office of Economic Opportunity approval of $80 000 grant to start program of welfare aid for jobless aerospace workers and families in Brevard County, Fla. Families included those of 1500 workers laid off at Kennedy Space Center. (M Her, 10/28/72)
- Soviet delegation headed home after meeting with officials of North American Rockwell Corp. Aviation Services Div. at Bethany, Okla. Purpose of meeting was to discuss Soviet Yak-40 commuter aircraft. NR was considering purchase of Yak-40s for U.S. distribution. (AP, W Post, 10/28/72, A7)
- President Nixon signed S. 4022, authorizing U.S. participation in International Exposition on the Environment to be held in Spokane, Wash., in 1974. Bill became Public Law 92-598. (PD, 10/6/72, 1620)
- *New Haven Register* editorial commented on death of Igor I. Sikorsky [see Oct. 26]: "His death . . . removes from the scene perhaps one of the last links between aviation in its primitive stages and as it is known today, transcending even the boundaries of the earth to include space travel." (N Hav Reg, 10/27/72)

October 27-28: Academician Mstislav V. Keldysh, President of Soviet Academy of Sciences, met with Manned Spacecraft Center officials, toured MSC, and was briefed on NASA programs. Keldysh was visiting U.S. as guest of National Academy of Sciences President, Dr. Philip Handler. (NASA Int Aff; SBD, 10/30/72, 33)

October 28: European Airbus Consortium chief pilot Max Fischl flew 257- to 331-passenger prototype of A-300B Airbus on 1-hr 25-min flight from St. Martin du Touch Airfield near Toulouse, France. (AP, NYT, 10/29/72, 30; AF Mag, 1/73/29)

- Communications Commission of United Nations Educational, Cultural and Scientific Organization (UNESCO) voted 47 to 9 with 13 abstentions to approve declaration favoring ban of direct satellite broadcasting to other nations without their permission. U.S. Deputy Secretary of State for Educational and Cultural Affairs William B. Jones said that ban was "in direct opposition to our own national traditions, and to the basic principles of the United Nations." (Randal, W Post, 11/5/72, A14)

October 30: Pioneer 10 Jupiter probe (launched March 2) was 434 million km (270 million mi) from sun, at midpoint in Asteroid Belt, and nearly halfway to Jupiter. Spacecraft, which entered Asteroid Belt July 15, had encountered no difficulty although asteroid particles observed to date had been more numerous than expected. Finding led experimenters to preliminary conclusion that belt, to midway point,

offered no serious hazard to spacecraft on future outer planets missions. Spacecraft instruments had detected no decrease in rate of particle penetrations between end of earth orbit and edge of Asteroid Belt. Constant penetration rates perceived indicated particles probably were cometary debris. If penetration remained constant, *Pioneer 10* gas cell experiment to measure particle penetration rate was expected to provide good data far beyond Jupiter. (NASA Release 72-208)
- European Space Research Organization would use preliminary definition study of sortie laboratory for space shuttle application as basis for future negotiations with NASA, *Aviation Week & Space Technology* reported. Study, to be performed by Dornier, was part of European effort to participate in post-Apollo program. (*Av Wk*, 10/30/72, 9)
- NASA launched Aerobee 170 sounding rocket from White Sands Missile Range, carrying Univ. of Chicago solar physics experiment. Rocket performed satisfactorily but scientific objectives were not met. (SR list)

October 31: U.S.S.R. launched *Cosmos 527* from Plesetsk into orbit with 295-km (183.3-mi) apogee, 177-km (110-mi) perigee, 89.1-min period, and 65.4° inclination. Satellite reentered Nov. 13. (GSFC *SSR*, 10/31/72; 11/31[30]/72; *SBD*, 11/2/72, 2)
- Overview of Apollo 17 mission, scheduled for Dec. 6 launch, was given by Mission Director Chester M. Lee at NASA Hq. press briefing. Mission experiments would include lunar surface gravimeter to detect possible lunar surface deformation from pull of earth and sun and to confirm existence of gravity waves; lunar ejecta and meteorite experiment to determine mass, velocity, and frequency of meteorite impacts on moon and nature of their ejecta; lunar seismic profiling experiment to acquire data on lunar near-surface structure and layering; lunar atmospheric composition experiment (MASSPEC); experiment to measure solar flares; and lunar traverse gravimeter with receiver carried on lunar roving vehicle Rover to survey anomalies of gravity at all science stops.

Among major new experiments on Apollo 17 mission would be surface electrical properties experiment to determine lunar surface layering and measure electrical properties of material; lunar neutron probe to measure capture rate of low-energy cosmic ray secondary neutrons and neutron energy spectrum as function of depth; and medical experiment BIOCORE to be carried in command module (CM). BIOCORE would carry six pocket mice in environmentally controlled situation. On return to earth, mice would be returned to principal investigators for examination for effects of high-Z particles. New experiments to be carried in scientific instrument module (SIM) bay were far-ultraviolet spectrometer to determine composition density of lunar atmosphere; infrared scan radiometer to obtain data to correlate and confirm earth-produced thermal maps; and lunar sounder module to probe lunar near subsurface to depths of 1 to 1.05 km (0.6 to 1 mi).

First Apollo 17 launch window would open at 9:53 pm EST Dec. 6 and close at 1:31 am EST Dec. 7. Alternate launch windows would occur in January and February. To reach moon at optimum sun angle, Apollo 17 would be launched at night; entry into trajectory toward moon would be made over Atlantic rather than over Pacific as on previous Apollo missions. "This means the TLI [translunar insertion] will occur about 45 minutes later than . . . in the past." Countdown preparations were proceeding normally.

Astronauts Eugene A. Cernan and Dr. Harrison H. Schmitt were scheduled to land Dec. 11 on flat valley 9.7 km (6 mi) wide near southeastern edge of Sea of Serenity on northeastern side of moon's front face. Landing site was more difficult than that of previous Apollo missions, Lee said, because "here we're faced with a scarp that is a number of kilometers down." Site was surrounded with "sculptured hills at the east, the north massif and the south massif; you are coming into a little tighter valley, but we have . . . plenty of clearance." Astronauts would remain on moon for record 75 hrs, would make three traverses in Taurus-Littrow valley during three 7-hr extravehicular activity periods, and would drive electrified, four-wheeled Rover record 33.8 km (21 mi), including one drive that would take them almost 8 km (5 mi) from lunar module. Astronaut Ronald E. Evans would remain in lunar orbit in command module. Cernan and Schmitt were to rejoin CM at 5:56 pm EST Dec. 16. Splashdown was scheduled for 2:24 pm EST Dec. 19 in Pacific 560 km (350 mi) southeast of Samoa. Recovery ship would be U.S.S. *Ticonderoga.* (Transcript)

- Dr. George M. Low, NASA Deputy Administrator, described functions of *Erts 1* Earth Resources Technology Satellite (launched July 23) in speech before First International Conference on Computer Satellites in Agriculture at Ohio State Univ. Satellite "now circles the earth 14 times a day in near-polar orbit. Each pass covers a region 185 km (115 miles) wide. After 18 days, the satellite returns to the same position. The ERTS Multispectral Scanner scans the earth in four wavelength bands." Green band enhanced features within water, since they were transparent in band. Red band showed contrast between vegetation and man-made structures and was good for land use mapping. Infrared band was invisible to human eye but "vegetation strongly reflects this energy, and thus appears bright on film. Water absorbs it, and appears dark. This band is often used to identify plants or crops, since the brightness of the image depends greatly on the type of vegetation. It is also used to determine the health of plants, because healthy crops will appear to be much brighter than those which are diseased." Data to date had shown earth resources satellites like *Erts 1* would be able to take large-scale inventories of crop type and crop health. ERTS data "lend themselves to rapid automated area calculation procedures for all kinds of agricultural production parameters." (Text)

- NASA authorized construction of two additional buildings to augment visitor facilities at Kennedy Space Center. Buildings, designed by Charles Luckman Associates, would add some 3580 sq m (38 600 sq ft) of air-conditioned space for exhibits, films, lectures, snack bar, and ticket sale area. Extension of facilities, necessitated by increased public attendance, had been authorized by Congress. (KSC Release 298–72)

- Lewis Research Center announced award of two-year, $2-million-plus extension to contract with Martin Marietta Corp. for integration of Centaur upper-stage rocket and Titan III booster. Contract included proof test-flight of new launch vehicle combination in early 1974 and launch of Helios spacecraft later that year. (LeRC Release 72–93)

- Boston Museum of Science named *New York Times* Science Editor Walter S. Sullivan to receive 1972 Bradford Washburn Award for

October 31

increasing public understanding of science. Award, gold medal and $5000 honorarium, would be presented Nov. 19. (*NYT*, 11/1/72, 9)
- Soviet aircraft designer Andrey N. Tupolev received Sickle and Hammer gold medal and title "Hero of Socialist Labor" from Soviet President Nikolay V. Podgorny in Kremlin ceremony. Award was for exceptional services in development of Soviet aircraft construction. Tupolev had designed Soviet Tu-144 supersonic transport aircraft. (Tass, FBIS–Sov, 11/2/72, J1; 11/6/72, J1)
- Prominent Soviet astrophysicist Kronid A. Lubrasky had been sentenced to five years in labor camp for anti-Soviet activities, dissident sources in Moscow said. (AP, *H Post*, 11/1/72)

During October: Lunar Science Institute released *Post-Apollo Lunar Science*, report of July study by LSI scientists who met at Univ. of California at San Diego. Study had examined progress in lunar science, identified objectives for future work and drafted preliminary plan for long-term lunar science. Major recommendations were that NASA continue to "preserve and describe the samples, data, and photographs, and to make them available to the scientific community" through Lunar Receiving Laboratory at Manned Spacecraft Center; that Apollo lunar surface experiment package (ALSEP) network and subsatellite be operated continuously as long as they provided significant new findings; that selected raw photographic observations be processed to obtain usable scientific product; that lunar science program be continued at "substantial" level; that advisory panels be continued and augmented to emphasize geophysical and photographic data analysis; and that "small, selective program of mission planning be maintained" concurrently with instrument definition program "to provide the desired continuity between the present and the future." LSI also recommended high priority for high-inclination orbital mission to determine geophysical, geochemical, and geological variables on planetwide scale, and joint and multinational lunar missions. (Text)
- Tests of resistant characteristics of new space-age materials for furnishings in buildings and homes were conducted for NASA by Battelle Corp. Study included controlled experimental burning of prefabricated rooms containing furnishings treated with fire-resistant and fire-retardant materials. Tests would determine ignition time, fire-spread rate, smoke density, heat movement, chemical analyses, combustion products, and fire-fighting problems that arose. Purpose of program was to provide training for metropolitan firemen and new fire-safety data. (NASA Release 72–200)
- House Committee on Science and Astronautics released *For the Benefit of All Mankind: The Practical Returns from Space Investment*. Updated report of spinoff dividends from space exploration covered application of space technology to communications, weather forecasting, business and industry, management, agriculture, environment, aeronautics, education, medicine, home and marketplace, and municipal and urban affairs and in foreign countries. Committee concluded that dividends were "fallout of ingenious application of space flight experience." Dividends already paid, "coupled with those in sight for the near-term future, affect practically every facet of human convenience and concern. They promise continuing and increasing return on our space investment for the benefit of mankind." (H Rpt 92–1452)

- Results of U.S.S.R. research of Mars as released by Soviet press were summarized by *Interavia*. Initial analysis of data collected by *Mars 2* and *3* (launched May 19 and May 28, 1971) spacecraft during interplanetary flight had indicated large variations in composition of interplanetary plasma ions at 20 million km (12.5 million mi) from earth. Data had also revealed new information on distribution of solar plasma around solar system and its relationship with interplanetary and terrestrial magnetic fields. No magnetic field had been detected near Mars, indicating possible lack of liquid core.

  Orbits of spacecraft had been selected to cover areas of Martian surface difficult to investigate from earth-central latitudes of southern hemisphere to 65° south, equatorial zone, and northern hemisphere to 29° north. During Martian orbit infrared radiometer data had shown certain areas of surface to be definitely warmer than immediate surroundings, indicating possible volcanic activity.

  Mars 2 and *3* had found steam content in Martian atmosphere to be 2000 times less than that existing in terrestrial atmosphere. At times precipitated water could be 50 micrometers; at others, less than 10. Since photos taken by NASA's *Mariner 9* revealed impressive valleys which could have been formed by erosion, Soviet scientists had theorized that higher temperatures and pressures, sufficient to produce water in liquid form, existed on Mars in previous era. Very sharp differences in planet's brightness, never before observed from earth, were observed by visible range photometer. Data were collected from measurement of radio-wave reflection.

  After landing, scientific results from *Mars 2* and *3* had been meager. *Mars 2* had hardlanded a capsule on Mars Nov. 27, 1971. Photos and measurements by *Mars 3* (which softlanded a capsule Dec. 2, 1971) had indicated a dust cloud 6 to 8 km (4 to 5 mi) thick covering landing area, leading to speculation that either dust storms had raged over area and caused capsule to land incorrectly or capsule had sunk into dust layer.

  Soviet Academician Anatoly A. Blagonravov had said U.S.S.R. planned to use 1973 launch windows if program of Martian exploration continued as planned. (*Interavia*, 10/72, 1139–41; GSFC *SSR*, 1/31/73)

- *Erts 1* Earth Resources Technology Satellite, launched by NASA July 23, was regarded by scientists as "most significant unmanned satellite ever put into space," *Aerospace Facts* reported. More than 300 independent investigators were examining photos and data being returned by *Erts 1*. Investigators were from 43 states, District of Columbia, 31 foreign countries, and 2 international organizations. (*Aerospace Facts*, 10–12/72, 1)

- First complete monitoring by Navy satellite of effects on earth's atmosphere of major, long-duration solar flare was reported by Office of Naval Research's *Naval Research Reviews*. Preliminary data from experiment aboard Dept. of Defense Space Test Program satellite *Sesp 1971–2* (launched by Air Force Oct. 17, 1971) had shown Aug. 2–7 solar flare had produced more intense bombardment of earth by plasma showers than any previously watched by satellites. Data would aid Navy scientists in understanding effects of solar storms on polar radio communications and auroral and polar cap phenomena. (ONR *Naval Research Reviews*, 10/72, 30–31)

- Panel of Geophysics Research Board of National Research Council released *Summary Report of the Ad Hoc Panel on* (NO_x) *and the Ozone Layer*. National Oceanic and Atmospheric Administration had asked Panel to review paper by Univ. of California at Berkeley physicist Dr. Harold S. Johnston [see Aug. 1] and to consider research to resolve uncertainties that existed in Dr. Johnston's hypothesis that nitrogen oxides from exhausts of supersonic transport aircraft could have destructive effects on ozone concentration. Uncertainties existed because dynamic processes might distort chemical equilibrium model based on static atmosphere, rates of some reactions in scheme of chemical kinetics might not be known, and estimations of SST engine exhaust products seemed uncertain to order of magnitude, at least. Panel had concluded: "There appears to be general agreement that Harold Johnston's conclusions . . . are credible, but with the reservations cited." Possibility of serious effects on the normal ozone could not be dismissed. "To the extent that existing information permits, Johnston has done a very careful and detailed analysis." It was evident "that we suffer from serious ignorance of many fundamental aspects of the chemistry and dynamics of the stratosphere when we attempt to make quantitative assessments of the pollution consequences of high-altitude aircraft operations." (NAS–NRC–NAE *News Rpt*, 10/72, 2)
- National Research Council announced formation, at Dept. of Transportation request, of NRC Committee on the Department of Transportation Climatic Assessment Program. Committee, chaired by Dr. Henry G. Booker of Univ. of California at San Diego, would examine need for studies of effects of aircraft nitrogen-oxide emissions on earth's ozone layer; atmospheric dynamics, physics, and chemistry; nature of prospective emissions into stratosphere; and biological effects on earth of stratospheric activities. (NAS–NRC–NAE *News Rpt*, 10/72, 2)
- Communications Satellite Corp. released *At the 10th Anniversary of the Communications Satellite Act of 1962: Comsat Report to the President and the Congress.* Transmittal message dated Aug. 27—10th anniversary of Senate passage of Act—by ComSatCorp President Joseph V. Charyk and Board Chairman Joseph H. McConnell said 83-member International Telecommunications Satellite Consortium (INTELSAT) was operating successfully under ComSatCorp management. Satellite services had been provided to developing and developed countries, efficient and economical use was being made of electromagnetic frequency spectrum, and improved communications services were being provided at reduced charges. ComSatCorp believed "it is fulfilling its mandate successfully, and . . . progress is beyond what many expected 10 years ago." Report said regular full-time commercial satellite services were being provided globally by four satellites of Intelsat IV series and by network of 71 earth stations administered by 44 countries. Some 80 countries, territories, and possessions were being served directly, full time. Service was provided via more than 200 communications paths. (Text)
- TRW Inc. Systems Group Div., Hughes Aircraft Co., and Philco-Ford Corp. submitted proposals in competition conducted for North Atlantic Treaty Organization (NATO) by Air Force Space and Missiles Systems Organization (SAMSO). Competition was for development of NATO Phase 3 communications satellite program. (*Av Wk*, 10/30/72, 17)

November 1972

November 1: U.S.S.R. launched eight Cosmos satellites from Plesetsk with single booster.

Cosmos 528 entered orbit with 1469-km (912.8-mi) apogee, 1368-km (850.1-mi) perigee, 114.1-min period, and 74.0° inclination.

Cosmos 529 entered orbit with 1469-km (912.8-mi) apogee, 1404-km (872.4-mi) perigee, 114.5-min period, and 74.0° inclination.

Cosmos 530 entered orbit with 1469-km (912.8-mi) apogee, 1335-km (829.5-mi) perigee, 113.8-min period, and 74.0° inclination.

Cosmos 531 entered orbit with 1471-km (914.0-mi) apogee, 1423-km (884.2-mi) perigee, 114.7-min period, and 74.0° inclination.

Cosmos 532 entered orbit with 1469-km (912.8-mi) apogee, 1302-km (809.0-mi) perigee, 113.4-min period, and 74.0° inclination.

Cosmos 533 entered orbit with 1470-km (913.4-mi) apogee, 1318-km (819.0-mi) perigee, 113.6-min period, and 74.0° inclination.

Cosmos 534 entered orbit with 1469-km (912.8-mi) apogee, 1351-km (839.5-mi) perigee, 113.9-min period, and 74.0° inclination.

Cosmos 535 entered orbit with 1470-km (913.4-mi) apogee, 1386-km (861.2-mi) perigee, 114.3-min period, and 74.0° inclination.

Last U.S.S.R. eight-payload launch had been on July 20. Western specialists speculated that launches were part of global military communications satellite program. (GSFC *SSR*, 11/31/72; *SBD*, 11/2/72, 2; Shabad, *NYT*, 11/2/72, 11)

- NASA's M2-F3 lifting body completed 21st flight from Flight Research Center after air-launch from B-52 aircraft. Flight objectives were to check out pilot, John A. Manke, and to obtain stability and control data at mach 1.1 and mach 0.9 for stability augmentation systems only. Flight objectives were achieved. Vehicle reached 21 600-m (71 000-ft) altitude and mach 1.24. (NASA proj off)

- Doomed civilization theory was disputed by Dr. James C. Fletcher, NASA Administrator, in speech at Carnegie Institute Museum of Art in Pittsburgh. Speech was part of Man and Idea Series sponsored by museum's Women's Committee. "I think these views overlook the very center of civilization—man himself, his intellect and his capacity to learn and act, and thereby retain in some great measure control over his own destiny." Goal of space program was to "place the tools of planetary management into the hands of man." Rather than barrier or constraint, space had become to man "a natural extension of his sphere of activity. I do not see this as ephemeral or temporary; I believe that, both physically and intellectually, our operating environment has been permanently expanded. This is a real contribution to the human spirit." Human spirit was difficult to define; "it is measured by attitudes and aspirations and mature expectations. Our space experience, in the fourteen years past, has given us all a different—and better—view of the future. In a very real sense, the future is already here; we must learn to recognize it and live in it responsibly." (Text)

November 1

- NASA launched two sounding rockets from Wallops Station. Nike-Apache carried Univ. of Illinois payload containing Geiger counter, Langmuir probe, magnetic aspect sensor, and radio propagation experiment to 195-km (121.2-mi) altitude. Super Arcas carried Pennsylvania State Univ. payload containing electrometer to 78.9-km (49.3-mi) altitude. Primary purpose of flights was to study ionosphere near local midnight under disturbed magnetic conditions, with emphasis on role of energetic electrons as ionization source. Rockets and instrumentation performed satisfactorily. (NASA Rpt SRL)
- NASA announced selection of Xerox Data Systems for contract negotiations to provide central data-handling facility and remote terminals for Atmosphere Explorer (AE) C, D, and E program. Atmosphere Explorers, 450-kg (100-lb) spacecraft, were to be launched by Thor-Delta rockets into elliptical earth orbit between autumn 1973 and summer 1976 to study chemical processes and energy transfer mechanisms that controlled structure and behavior of earth's atmosphere and ionosphere. Project was managed by Goddard Space Flight Center under direction of Office of Space Science. (NASA Release 72-212)
- Retirement of Dr. Fred L. Whipple as Director of Smithsonian Astrophysical Observatory—effective July 1, 1973—was announced by Dr. S. Dillon Ripley II, Secretary of Smithsonian Institution. Dr. Whipple would be succeeded by Dr. George B. Field, professor of astronomy at Harvard College Observatory. Director of SAO since 1955, Dr. Whipple received President's Distinguished Federal Civilian Service Medal in 1963 for leadership in developing Baker-Nunn satellite tracking camera and optical tracking network that had gathered data since launch of world's first satellite in 1957. Known internationally for studies of moon, meteors, and comets, he would become a senior scientist at Smithsonian. (Smithsonian Release; *NYT*, 11/5/72, 12; 6/13/63, 21; Green, Lomask, *Vanguard—A History*, SP-4202)
- President Nixon signed ratification of Convention for the Suppression of Unlawful Acts Against the Safety of Civil Aviation and urged prompt action on other measures to combat international terrorism. Convention required extradition or prosecution of persons committing sabotage or violence against international civil aviation. Other measures were draft convention for prosecution or extradition of persons who attacked foreign officials, convention suspending air service to countries who failed to punish or extradite aircraft hijackers or saboteurs, and convention requiring prosecution or extradition of persons who injured, kidnaped, or killed innocent civilians to blackmail any state or international organization. (*PD*, 11/6/72, 1608)

November 2: American Astronautical Society presented its 1971 awards and honored newly elected fellows at Washington, D.C., luncheon. Flight Achievement Award was presented to *Apollo 14* Astronauts Alan B. Shepard, Jr., Stuart A. Roosa, and Edgar B. Mitchell for "achievements in flight testing which have contributed significantly to the advancement of manned space flight." Lloyd V. Berkner Award was presented to Willard F. Rockwell, Chairman and Chief Executive Officer of North American Rockwell Corp., for "outstanding contributions to the commercial utilization of space technology." Space Flight Award went to Joseph G. Gavin, Jr., Grumman Aerospace Corp.

President, for "outstanding efforts and achievements which have contributed significantly to the advancement of space flight and space science."

Victor A. Prather Award was presented jointly to Robert E. Breeding of United Aircraft Corp. Space Systems Dept. and Leonard F. Shepard of ILC Industries, Inc., for "outstanding contributions in . . . extravehicular protection in space." Randolph Lovelace II Award went to Dr. Maxime A. Faget, Manned Spacecraft Center Director of Engineering and Development, for "outstanding contributions in space technology." Melbourne W. Boynton Award for "significant research contributions to space flight safety" was presented to Dr. Willard R. Hawkins, Deputy Director for Medical Operations at MSC, and Dr. Harald J. von Beckh, Naval Air Development Center Director for Medical Research. Dr. James C. Fletcher, NASA Administrator, and Astronaut Eugene A. Cernan were among newly elected Fellows. (AAS Release)

- British Interplanetary Society statement expressing urgent need for United Kingdom space authority was forwarded by BIS to British Prime Minister. Space responsibilities in U.K. traditionally had been divided among several ministries. "There is not a British space programme as such, but only an assemblage of user projects co-ordinated by some undisclosed Cabinet machinery, which does not lead to any publicly discernible positive line of policy in relation to space as a whole." Statement urged immediate establishment of U.K. National Space Authority to plan and coordinate scientific, commercial, and social aspects of space development and act as liaison point with government departments responsible for telecommunications, meteorology, environment, aeronautics, marine affairs, and education and with international agencies. (Text)

- Boosted Arcas I launched by NASA from Churchill Research Range carried Univ. of Houston payload to 57-km (35.4-mi) altitude. Primary objective was to supplement comprehensive investigation of auroral zone disturbances conducted by Black Brant carrying Univ. of Texas payload and launched 60 sec later. Secondary objective was to study auroral phenomena in region above altitude already investigated, using balloons, and below altitude investigated, using rockets. Rocket and instrumentation performed nominally although one of two telemetry receivers provided marginal response. Recovery could not be effected because winds blew payload over Hudson Bay.

    NASA Black Brant, launched from Churchill Research Range 60 sec after Boosted Arcas, reached 240-km (149.1-mi) altitude. Primary objective of Univ. of Texas payload was to provide complete data on auroral precipitation events, analyze particle spectrum over wide energy range and observe spectrum in and above aurora, and observe optical emissions and ionospheric parameters. Rocket performed nominally but only one of nine experiments was successful because of faulty nosecone ejection mechanism. Nosecone did not deploy until descent portion of trajectory. Payload was recovered. (NASA Rpt SRL; NASA proj off)

November 3: U.S.S.R. launched *Cosmos 536* from Plesetsk into orbit with 543-km (337.4-mi) apogee, 517-km (321.3-mi) perigee, 95.2-min period, and 74.0° inclination. (GSFC *SSR*, 11/31/72; *SBD*, 11/15/72, 1)

- NASA achievements in 1972 were reviewed by Dr. James C. Fletcher, NASA Administrator, in letter to staff. NASA had received "great vote of confidence from the President, the Congress, and the country" in assignment to develop space shuttle. Assignment had been "most appropriate" in year when Apollo program would be concluded and was "a new challenge . . . worthy of the one we are about to complete." Year 1972 had also been one of "great decision" because in NASA FY 1973 budget "we have also won approval for the concept of a well-rounded space program, with funding for significant progress in all major fields of space activity during this decade. We will stress practical benefits from spacecraft in Earth orbit in this decade and the next. But we will also build greatly improved spacecraft to study the universe from Earth orbit and to explore throughout the solar system." NASA might be "moving faster with a bigger budget, but I don't see how we could be doing a better job." (*NASA Activities*, 11/15/72, 224)
- Large Space Telescope (LST) status briefing was held at Marshall Space Flight Center for representatives of potential industry participants in program to develop multipurpose telescope for launch by space shuttle in 1980s. LST feasibility studies would end in December. NASA expected to invite industry proposals on preliminary design and definition phase in early 1973. LST would operate from earth orbit to pick up galaxies 100 times fainter than those observed by most powerful ground-based optical telescopes. It would weigh maximum 11 000 kg (25 000 lbs) and have diffraction-limited mirror with 3-m (10-ft) diameter. (MSFC Release 72-142)
- Nike-Apache sounding rocket launched by NASA from White Sands Missile Range carried Univ. of Colorado payload containing six ultraviolet photometers and telemetry transmitter to 123-km (77-mi) altitude. Flight objective was to measure variation with altitude of atomic oxygen at night and to determine spectral distribution of night-glow continuum in altitude region between 85 and 110 km (53 and 68 mi). Rocket and instrumentation performed satisfactorily. (NASA Rpt SRL)
- Academician Mstislav V. Keldysh, President of Soviet Academy of Sciences, met with President Nixon at White House during visit to U.S. sponsored by National Academy of Sciences President, Dr. Philip Handler. (*PD*, 11/6/72, 1617)
- Selection of Dr. Manson Benedict, Massachusetts Institute of Technology nuclear physicist, to receive Enrico Fermi Award for 1972 was announced by Dr. James R. Schlesinger, Chairman of Atomic Energy Commission. President Nixon had approved nomination. President had informed Dr. Benedict that award was being given "in recognition of your pioneering leadership in the development of the Nation's first gaseous diffusion plant, while paying equal tribute to your imaginative contributions in the development of the nuclear reactor and its safe use for generation of electrical power." (AEC Release P-366)

November 4: President Nixon, in foreign policy speech over nationwide radio, referred to U.S.-U.S.S.R. agreements signed during May 22-29 Moscow summit meetings: "We agreed . . . to cooperate in protecting the environment, explore in space, fight disease. This means the day is fast approaching when a Russian cosmonaut and an American astronaut will shake hands in space, when a Russian chemist and an American biologist will work side by side to find a cure for cancer, and each time our nations join hands in the works of peace, we advance the day when nations will no longer raise their hands in warfare." (*PD*, 11/13/72, 1639–41)

- Protocol on agreement between National Academy of Sciences and Soviet Academy of Sciences for further development of scientific cooperation was signed in New York by NAS President, Dr. Philip Handler, and Soviet Academy President Mstislav V. Keldysh. Protocol agreed on planning 50% increase in interacademy exchange program and increasing interacademy cooperative research program. Parties had agreed to expand joint research after noting "that their efforts had already encouraged the development of the US-Soviet cooperation toward compatible docking mechanisms for space vehicles and other cooperation in space activities, organization of the International Institute for Applied Systems Analysis, as well as cooperation in very long base interferometry in radio, astronomy, oceanography, and other fields." (Text)

- NASA launched Nike-Apache sounding rocket from White Sands Missile Range, carrying Univ. of Colorado aeronomy experiment. Rocket and instrumentation performed satisfactorily. (SR list)

- Suspension in near-earth orbit of huge cosmic reflector to reflect sun's rays onto area of earth's surface with radius of more than 300 km (200 mi) had been discussed by Soviet Cosmonaut Vladimir A. Shatalov in *Krasnaya Zvezda* (Red Star) article, Tass reported. By concentrating sun's rays on definite sector of earth's surface, Shatalov had said, local temperature could be raised in various sections of earth. "Thus, an opportunity is opening up to control the climate." (FBIS–Sov, 11/4/72, L1)

- Eight nations had reserved satellite TV coverage time for 53 separate broadcasts on U.S. elections, according to Communications Satellite Corp. (AP, *NYT*, 11/5/72, 83)

November 5: NASA announced it had asked Dept. of Health, Education, and Welfare to help select community for 1975 field-testing of NASA's integrated medical and behavioral laboratory measurement system (IMBLMS). Test would appraise technical requirements for remote health care during long-duration space missions. If successful, it would also provide information and techniques to improve health care and medical services to remote earth areas. Data received at control center would enable physicians to diagnose conditions and prescribe treatment to trained paramedics at remote site. Joint NASA–HEW Site Selection Working Group had developed requirements for test site. Group would evaluate community applications received by Dec. 4. Final decision on site would be made by about Dec. 5. (NASA Release 72–210)

- Legislation to rename Jet Propulsion Laboratory as H. Allen Smith Jet Propulsion Laboratory effective Jan. 4 [see Oct. 21] had set off negative reaction, *Los Angeles Times* reported. "General feeling" among JPL scientists and engineers was that Laboratory and its work should not be involved with politics. Protest petition had been signed by 210 students of California Institute of Technology, which operated Laboratory for NASA. Cal Tech officials were surprised and concerned at not having been consulted about change and planned to talk to Government officials in Washington, D.C. (*LA Times*, 11/5/72)

- Spokesman for Cape Kennedy local of International Alliance of Theatrical and Stage Employees said Boeing Co. officials had defied union demands to submit wage dispute to Federal arbitrators. Attorney Bernard Mamet said 60 members of local had vowed to walk out by Dec. 5 if their request was denied. He said Boeing had halted negotiations in eight-month dispute and had refused to resume talks before Nov. 14. Boeing technical writers were asking at least partial restoration of April 1971 wage cut. Dispute had raised "slim threat" of walkout on eve of Apollo 17 launch, *Orlando Sentinel* said later. Sympathy strike by other locals of union could cause "serious problems." (*O Sen*, 11/8/72)

November 6: NASA selection of five potential landing sites on Mars for two unmanned Viking spacecraft in summer 1976 was announced by Dr. S. Ichtiaque Rasool, Deputy Director of Planetary Programs, in telephone interview. NASA also released *Mariner 9* photos of sites—two at edge of Nix Olympica slopes, another near slopes of three smaller volcanoes north of canyons along Martian equator, fourth on desert plain, and fifth in region of flattened craters at edge of highland terrain. Observations from *Mariner 9* (launched May 30, 1971) had shown sites were low, relatively warm regions, possibly moist, where life form was most likely to be discovered. Prime objective of Viking missions was to determine if Mars harbored life or potential for life. In choosing sites, project scientists had eliminated nearly one third of Mars because of high elevations. In mountainous regions, Viking parachutes would not have sufficient atmosphere to brake spacecraft descent. Scientists also had rejected areas where surface appeared too soft for spacecraft landing, too hard for retrieval of soil samples, or so steep that spacecraft might tip over on landing. Attention had centered

November 6: *Selection of five potential landing sites on Mars for NASA's two unmanned Viking spacecraft in 1976 was announced. The map showing the sites was based on photos televised from Mariner 9, launched May 30, 1971, and in orbit around Mars since Nov. 13, 1972. All the sites were fairly smooth areas near scientifically interesting geological features and were in regions of the planet where warmer temperatures and greater likelihood of water increased chances of finding evidence of life. The first map of the entire globe of Mars, prepared by the U.S. Geological Survey from 1500 of 7000 photos taken by Mariner 9, was published Nov. 27.*

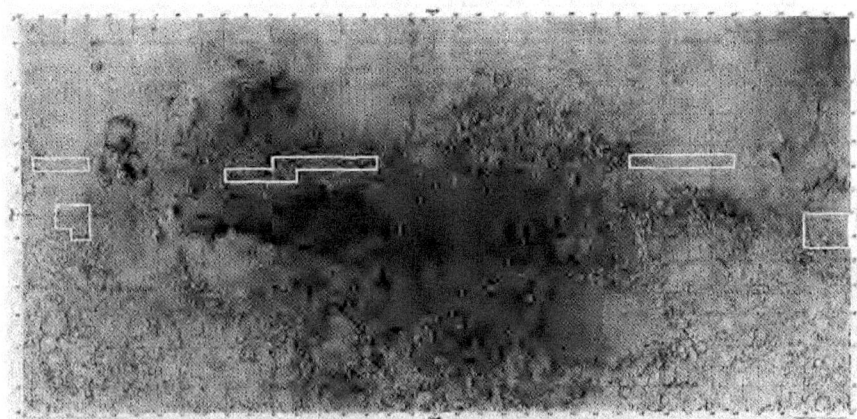

on broad Mars equatorial band with brightest sunlight and highest temperatures, which increased chance of finding water vapor in atmosphere and thus of finding biological activity. Landing zones were about 65 km (40 mi) wide and 650 km (400 mi) long. Precise sites would be selected after further analysis of *Mariner 9* photos. U.S. and Soviet scientists 'would meet during winter to compare Mars landing targets. There was "no use in landing on top of each other. We would also like to optimize the coverage of Mars." (Wilford, *NYT*, 11/7/72, 61; NASA PIO)

- U.S.S.R. award of $1-million contract to ITT Space Communications for earth station equipment for new Kremlin-White House satellite communications link was announced by parent firm International Telephone & Telegraph Corp. Existing hotline was direct Washington-Moscow teleprinter link for transmission of top-priority government messages. ITT Space Communications would provide orientation and training for Soviet personnel. U.S.S.R. would provide antenna structure. (ITT Release)

- European space activity was reported by *Aviation Week & Space Technology*: Dornier Aeros satellite had been cleared for December launch by NASA. Integration of Helios solar research satellite prototype would begin in early 1973 at Messerschmitt-Boelkow-Blohm's Ottobrunn facility. First Helios was scheduled for mid-1974 launch by NASA, to fly within 37 million km (23 million mi) of sun. Second launch was planned for 1975. (*Av Wk*, 11/6/72, 9)

- Air Force Systems Command successfully completed two subsystem tests of new high-altitude supersonic target (HAST) missile at its Armament Development and Test Center at Eglin Air Force Base, Fla. Tests over Gulf of Mexico used lightweight prototype target. First test was aerial recovery of HAST recovery test vehicle; payload test vehicle was towed by F-4 Phantom aircraft in second test. HAST was highly maneuverable target missile with mach 4 speed and 10 000- to 30 000-m (35 000- to 100 000-ft) altitude. First powered missile would be delivered to ADTC in December. (AFSC *Newsreview*, 12/72, 3)

November 7: Annual Revolution Day parade in Moscow featured only muted display of military power. One of leading floats reproduced *Pravda* headline requesting public support for U.S.-U.S.S.R. agreements signed during year. Politburo members, led by Communist Party Secretary Leonid I. Brezhnev, reviewed display but only speaker was Defense Minister Andrey Grechko, who called for "peaceful coexistence of states with different social systems." No new weapons were shown. (Seeger, *LA Times*, 11/8/72, 19)

- Unidentified Administration officials told *New York Times* there was evidence that People's Republic of China had deployed strategic missiles capable of reaching Moscow. Missiles were said to have 5600-km (3500-mi) range, carry 3-megaton warheads, and be installed in well protected launching sites. *Times* said officials had declined to reveal source of evidence but paper surmised source to be reconnaissance satellite. (Beecher, *NYT*, 11/8/72, 67)

November 7-8: Goddard Space Flight Center sponsored symposium "Significant Accomplishments in Technology" at Greenbelt, Md. Fifty-two papers covered spacecraft and vehicle technology, sensor technology, ground operations, and communications and navigation. Robert

J. Goss of GSFC reported design changes had increased Thor-Delta launch vehicle performance while maintaining reliability and low cost. Payload that could be placed in synchronous transfer orbit would have increased from 45 kg (100 lbs) in 1960 to 680 kg (1500 lbs) when H-1 booster was operational in 1974. Joseph Arlauskas of GSFC described multispectral scanner in orbit on *Erts 1*, launched July 23, that permitted simultaneous imaging in three visible bands and one near-infrared, to provide data for management of environmental resources.

GSFC also sponsored symposium "Significant Accomplishments in Science" on high-energy and solar astronomy; optical and ultraviolet astronomy; planetary, lunar, and cometary studies; earth observations; and earth physics. Dr. John A. Philpotts of GSFC reported that three years study of lunar samples from Apollo and U.S.S.R. Luna missions showed essentially three kinds of rocks: iron-rich basalts in mare basins, KREEP basalts in Mare Imbrium and Oceanus Procellarum, and anorthosites in highlands. He thought there was enough material in hand "to work out the origin of these . . . rocks, and from that, the evolution of the moon and its relation to other solar system materials." (Proceedings, NASA SP-326, SP-331)

November 8: Operational control of *Noaa 2* Improved Tiros operational weather satellite, launched by NASA Oct. 15, was officially transferred to National Oceanic and Atmospheric Administration's Environmental Satellite Service. (NASA Release 72-221)

- NASA announced signing of $64-million, cost-plus-fixed-fee award-fee contract with North American Rockwell Corp. for design, development, and test of docking module, docking system, and modification of Apollo command and service module (CSM) for Apollo-Soyuz Test Project. Hardware would be used in joint U.S.-U.S.S.R. rendezvous and docking mission scheduled for summer of 1975. (NASA Release 72-218)

- Designation of two new, lightweight, fighter prototype aircraft as YF-16 and YF-17 was announced by Secretary of the Air Force, Dr. Robert C. Seamans, Jr. YF-16, powered by one Pratt & Whitney F100 turbofan engine, had swept-wing, under-fuselage inlet, single vertical tail, and forebody strakes. YF-17, powered by two General Electric YJ101 turbojet engines, had twin tail, highly-swept-wing leading-edge extensions, and underwing side fuselage inlets. Both aircraft were expected to fly in early 1974. Air Force had no commitment to production. (DOD Release 763-72)

- United Kingdom Minister for Aerospace Michael R. D. Heseltine, at Paris meeting of science ministers of European Space Conference countries, urged creation of European space agency to achieve common policy as prerequisite for participation in post-Apollo space program. (*SF*, 1/73, 1)

- NASA launched Aerobee 170 sounding rocket from White Sands Missile Range, carrying Goddard Space Flight Center payload to measure spectrum of planetary atmosphere. Rocket and instrumentation performed satisfactorily. (SR list)

- Appointment of Dr. Robert L. Hirsch, Acting Director of Atomic Energy Commission's Div. of Controlled Thermonuclear Research, to position of Director was announced by AEC. Dr. Hirsch would succeed Dr. Roy W. Gould, who had returned to Massachusetts Institute of Technology in August. Dr. Hirsch would direct program to develop major new and environmentally attractive source of energy from nuclear fusion reactions. (AEC Release P-368)

- *Los Angeles Times* editorial criticized Jet Propulsion Laboratory name change: "While the alteration may meet with some approval in parts of the 20th Congressional District, it has gone over with a thud at the laboratory and on the Pasadena campus of Caltech." Scientists and teachers were asking "why the nonpolitical space lab has to be saddled with the name of a retiring congressman, especially one with no distinction in the technology of space . . . . Congress could reconsider; it should." (*LA Times*, 11/8/72)

November 9–15: Canadian Domestic Communications Satellite (*Telesat-A*)—named *Anik 1*, Eskimo for brother—was launched successfully at 8:14 pm EST by NASA from Eastern Test Range on three-stage, long-tank, thrust-augmented Thor-Delta launch vehicle. *Anik 1* was placed in highly elliptical transfer orbit with apogee of 36 470 km (22 661.4 mi); perigee, 188.9 km (117.4 mi); inclination, 26.9°. After checkout and reorientation of spacecraft, apogee kick motor on board *Anik 1* was fired at 3:55 pm EST Nov. 13, bringing spacecraft into near synchronous circular orbit with 36 508-km (22 658-mi) apogee, 35 822-km (22 258.8-mi) perigee, 1455-min period, and 0.4° inclination and drifting to final operating position in synchronous equatorial orbit off west coast of South America.

Under contract between NASA and Telesat Canada, NASA mission objective was to place *Anik 1* in orbit of sufficient accuracy to allow onboard propulsion systems to place spacecraft in stationary synchronous orbit while retaining sufficient station-keeping propulsion to meet mission lifetime requirements. Objective was met; mission was adjudged successful Nov. 15.

Launch was first on Thor-Delta "straight eight" launch vehicle, so called because diameter was eight feet for all three stages including fairing. Previous Thor-Deltas had been tapered at joint of first and second stage. New design allowed Delta to launch new generation of larger volume spacecraft. Flight was also first with nine solid-fuel thrust-augmentation rockets from Cape Kennedy. Third stage was TE-364-4 solid-fueled motor.

Anik 1—first of two—was built by Hughes Aircraft Co. under contract with Telesat and designed to provide transmission of TV, voice, and data throughout Canada for seven years. It would provide 10 color TV channels or up to 9600 telephone circuits. System could distribute analog or digital signals with overall performance comparable to that of terrestrial methods. Spacecraft, 1.8 m in diameter and 3.4 m high (6 by 11 ft), weighed 540 kg (1200 lbs) at launch and 270 kg (600 lbs) in orbit. Electronics system was powered by 23 000 solar cells, with onboard battery capacity to maintain service during sun eclipse.

Telesat Canada would reimburse NASA for cost of launch vehicle and services. Goddard Space Flight Center managed project for NASA Office of Space Science. (NASA proj off; NASA Release 72–206; GSFC Operations Center Br)

November 9: Air Force launched unidentified satellite from Vandenberg Air Force Base on Thor-Burner II into orbit with 879-km (546-mi)

November 9

- apogee, 824-km (512-mi) perigee, 101.5-min period, and 98.6° inclination. (Pres Rpt 73; *SBD*, 11/16/70, 70)
- NASA's M2–F3 lifting body completed 22nd flight at Flight Research Center after air-launch from B–52 aircraft. Flight objectives were to check out pilot, Cecil Powell, and to obtain stability and control data at mach 0.70. Both objectives were achieved. Vehicle reached 14 300-m (47 000-ft) altitude and mach 0.70 on flight using two chambers of four-chamber rocket engine. (NASA proj off)
- *Erts 1* Earth Resources Technology Satellite, launched by NASA July 23, was adjudged a success. Mission had accomplished primary objective, acquisition of synoptic, multispectral repetitive images for three-month period. By Oct. 24 *Erts 1* had obtained data which equaled or exceeded quantity and quality required. Data would be used for investigations in agriculture, forestry resources, mineral and land resources, and environment and for mapping and charting. (NASA proj off)
- Impact of large meteorite on moon's far side had provided first evidence suggesting moon still had molten core, Dr. Gary V. Latham, principal investigator for Apollo seismic experiments, said in press interview at Univ. of Texas. Data were preliminary and fragmentary but "I hope we'll be able to confirm or reject this core hypothesis in a month or so." Impact had been recorded July 17 by all four seismometers left on moon by Apollo astronauts but had not been discovered until recently because of delays in examining telemetry reports from instruments. Meteorite's size had not yet been determined but it "probably was larger than the 'whopper' recorded in May." Meteorite which impacted moon May 13 had been largest since seismometers began functioning in 1969. Seismic signals from July 17 impact had had "funny focusing effect that could be explained by a core." (Rossiter, W *Star & News*, 11/10/72, A3)
- United Nations General Assembly adopted, by vote of 102 to 1 with 7 abstentions, resolution to prepare international convention on principles governing use of artificial earth satellites for direct TV broadcasting. U.S. cast opposing vote; U.S.S.R. and 11 other nations that voted against similar measure in U.N. First Committee Oct. 20 voted for General Assembly's resolution. Resolution noted "need to prevent the conversion of direct television broadcasting into a source of international conflict and of aggravation of the relations among States and to protect the sovereignty of States from any external interference."

    U.S. Representative Robert C. Tyson said main reasons for U.S. opposition were that decision overlooked potential in direct satellite broadcasting for "furthering understanding among peoples" and importance of that potential to U.N. and did not sufficiently emphasize central importance of the "free flow of information and ideas in the modern world." (*Dept of State Bull*, 12/11/72, 691–2; *NYT*, 11/10/72, 16; *SF*, 3/73, 93)
- NASA held 14th Annual Awards Ceremony in Washington, D.C. Guest speaker Caspar W. Weinberger, Director of Office of Management and Budget, said: "A country needs—even a country as great as America—to be reminded of its greatness from time to time. And that is the thing of overwhelming importance that NASA has done . . . . They have reminded us, at a time when we desperately need it, of our greatness. They have taken us out of the aura of self-criticism and of doubt . . .

and given us a set of accomplishments that will be not only very hard to match, but a set of accomplishments in which every one of us . . . should take enormous pride."

Dr. James C. Fletcher, NASA Administrator, presented Outstanding Leadership Medal to Leonard Jaffe, Acting Director of Earth Observation Programs, for management of Earth Resources Survey program. He presented NASA Distinguished Service Medal to Dr. William R. Lucas, Marshall Space Flight Center Deputy Director, for exceptional contributions to Apollo program; to Richard C. McCurdy, Associate Administrator for Organization and Management, and Dr. Hans Mark, Ames Research Center Director, for management achievements; and to Dr. Paul W. Gast, Chief of Manned Spacecraft Center's Planetary and Earth Sciences Div., for Apollo science achievements. Secretary of the Interior Rogers C. B. Morton presented posthumous medal for Dr. William T. Pecora, late Under Secretary of the Interior, to Mrs. Pecora. Award was in recognition of work on Earth Resources Observation Systems (EROS) program.

Distinguished Public Service Medal was presented to Dr. Riccardo Giacconi of American Science and Engineering, Inc., for work on cosmic rays and to Dr. Gerald J. Wasserburg, Cal Tech physicist, for work on lunar samples. Third awardee, Dr. Brian O'Brien, physicist on NASA Space Program Advisory Council, was unable to be present.

Group Achievement Award was presented to Atlas Centaur Project Team, Hq., Lewis Research Center, and Kennedy Space Center; Deep Space Maser Development Group, Jet Propulsion Laboratory; Digital Fly-by-Wire Project Team, Flight Research Center; Earth Resources Technology Satellite Project Team, Goddard Space Flight Center; *Oao 3* Project Team, ARC, GSFC, KSC, and JPL; *Pioneer 6–9* Project Team, Hq.; Quiet Engine Team, LeRC; and Space Shuttle Technology Project Team, Hq. Other awards included Exceptional Service Medal to 41 persons and Exceptional Scientific Achievement Medal to 20. (Program; *NASA Activities*, 12/15/72, 245; NASA Hist Off; *Huntsville Times*, 11/9/72)

- National Public Affairs Center for TV began poll of 223 public TV station managers at request of Corp. for Public Broadcasting, to determine response to CPB proposal to show up to 21 hrs of live coverage from moon during Apollo 17 moonwalks Dec. 11, 12, and 13. *Washington Post* said later that NASA had offered public broadcasting opportunity to cover past Apollo missions but CPB had declined, to avoid conflict with commercial networks. TV coverage of astronauts' extravehicular activities (EVA), however, had been declining. During *Apollo 16* mission (April 16–27), American Broadcasting Co. had shown 6 of estimated 21 hrs of EVA, National Broadcasting Co. had shown 8 hrs, and Columbia Broadcasting System, 6½ hrs. (Carmody, *W Post*, 11/10/72, B1)

- Foreword to *Jane's Surface Skimmers, 1972–73* predicted intercontinental hovercraft could travel between U.K. and Japan 10 times faster than ocean liner, Associated Press reported. *Jane's* editor Roy McLeavy had said in annual survey of hovercraft and hydrofoil technology that hovercraft offered tremendous opportunity for Arctic travel. Vehicles would move people and cargo across ice cap and also "support oil and mineral extractions and scientific observations." U.S.S.R. was building similar craft called "ekranoplanes" for high-speed service in areas normally impassable in winter. (W *Star & News*, 10/9/72, C8)

November 10: Apollo 17 commander Eugene A. Cernan announced that command module had been named "America" and lunar module "Challenger," during Apollo 17 press briefing at Manned Spacecraft Center. "We have never turned our backs on a challenge and the name The Challenger fits into the theme of what made America what it is today." Mission scientist, astronaut Dr. Harrison H. Schmitt, said he was confident that Apollo 17 findings would "help us understand that moon so it can start telling us what happened in the early history of the earth." (*W Post*, 11/11/72, A12; Wilford, *NYT*, 11/12/72, 24)

- Interview with Glynn S. Lunney, U.S. Director of joint U.S.-U.S.S.R. Apollo-Soyuz Test Project (ASTP) scheduled for 1975, was published by *Detroit News*. Lunney had described agreements made during Oct. 9-19 ASTP meeting in Moscow. There had been complete cooperation between U.S. and U.S.S.R. during ASTP meetings to date, with both sides averaging about 50-50 on contributions. Biggest problem had been "reaching an understanding of what words mean. We have our jargon and they have theirs and there are different meanings for the same word. For instance, the word 'revise' seems to have a bad connotation and yet the word 'amend' can be used as a substitute." (Pipp, *D News*, 11/10/72)

- Families of Astronauts Edward H. White II and Roger B. Chaffee—who lost lives in Jan. 27, 1967, Apollo fire—received awards totaling $300 000 from spacecraft manufacturers. North American Rockwell Corp. and former North American Aviation, Inc., Rockwell-Standard Corp., and Rockwell Standard Co. had been charged with negligence in suits filed by Chaffee's widow Nov. 6 and White's widow Nov. 7. In out-of-court settlement, Mrs. Patricia White Davis would receive $50 000 for herself and $50 000 for each of two children; Mrs. Martha Chaffee Canfield would receive $70 000 and $40 000 for each of two children. Widow of Virgil I. Grissom, third astronaut killed in fire, had won $350 000 settlement from NR March 8. (W *Star & News*, 11/8/72, A4; AP, *NYT*, 11/11/72, 7)

- Award of $1.5-million contract to United Air Lines, Inc., for flight-test program to evaluate operational procedures and avionic system to reduce noise level near airport approach patterns was announced by Ames Research Center. Contract was part of ARC program in cooperation with Federal Aviation Administration. Avionics system would be developed by Collins Radio Co. under separate $233 000 NASA contract. (ARC Release 72-136)

- Award of $35 913 123 cost-plus-incentive-fee contract to TRW Systems Inc. for Fleet Satellite Communications (FLTSATCOM) System satellite was announced by Dept. of Defense. Contract called for design, development, construction, and test of qualification-model spacecraft, software, and support equipment. Separate contract for production of five flight-model spacecraft would be awarded later. When operational, FLTSATCOM would provide satellite communication link among mobile users including major Navy ships, selected Air Force and Navy aircraft, and ground stations. (DOD Release 775-72)

- Award of $95 219 456 contract to Boeing Co. and $85 901 510 contract to McDonnell Douglas Corp. to begin design and development of advanced, medium, short takeoff and landing transport (AMST) aircraft was announced by Dr. Robert C. Seamans, Jr., Secretary of the Air Force. (DOD Release 773-72)

- Inauguration of Traffic Control and Landing System (TRACALS, 404L Program) at Air Force System Command's Electronic Systems Div. at Hanscom Field, Mass., was announced by AFSC. Program would update management and control of Air Force aircraft takeoffs, en route flights, and landings in U.S. and overseas. Effort included development and acquisition of solid-state, low-cost, reliable electronic components and systems to replace outdated tube equipment in use for more than two decades. (AFSC Release 112.72)
- Directors of North American Rockwell Corp. proposed to change firm's name to Rockwell International Corp. at meeting in El Segundo, Calif. If change was approved by shareholders at Feb. 15, 1973, annual meeting all trace of name North American Aviation, Inc., founded in 1928, would disappear, *Los Angeles Times* said later. NR had been formed in 1967 through merger of North American Aviation and Rockwell-Standard Corp. (*LA Times*, 11/12/72)

November 11: Four hundredth anniversary of discovery of Tycho's Star by Danish astronomer Tycho Brahe. Supernova that increased in brightness by millions of times, Tycho's Star was first unusual star ever recorded by European astronomers and first definite proof that sky was not immutable. (American Museum–Hayden Planetarium Release, 11/6/72)

Dr. Wernher von Braun, Fairchild Industries, Inc., Vice President and former NASA Deputy Assistant Administrator for Planning, told press in Rio de Janeiro he did not believe in flying saucers. "Such things do not exist. I gather the whole business is a hallucination." Dr. von Braun was in Brazil to promote sale of communications satellite equipment to government. (AP, W *Star & News*, 11/12/72, A5)

November 12: Tristate Regional Planning Commission had reached agreement with NASA for astronauts aboard Skylab to photograph New York metropolitan area after launch from Cape Kennedy in April 1973, *New York Times* reported. Planning Agency would use infrared and standard photographs taken from height of 435 km (270 mi) for planning in three-state region. Photographs of housing trends and streets within 100-km (60-mi) radius of Times Square would complement those already taken by *Erts 1* Earth Resources Technology Satellite (launched July 23) and would be used to prepare computer-generated maps and charts available to any state, county, or local planning agency. (*NYT*, 11/12/72, 32)

November 13: Space shuttle program requirement review was held by prime contractor North American Rockwell Corp. at Downey, Calif. Work on shuttle orbiter stage under $2.6-billion contract had been accelerated and design of system had been altered to trim rising weight and cost of both orbiter and overall system. Orbiter weight had increased from 114 800 kg (253 000 lbs) to 125 600 kg (277 000 lbs); gross weight of overall system was up from 2.2 million kg (4.8 million lbs) to 2.4 million kg (5.3 million lbs). NR Space Div. Vice President and Program Manager Bastian Hello said orbiter currently planned would be about 38 m (126 ft) long, with 25.6-m (84-ft) wingspan. Manned Space Flight Center Space Shuttle Program Manager Robert

November 13

Thompson announced schedule for shuttle flights: first horizontal flight (orbiter only), 1976; unmanned vertical flight, 1977; first manned orbital flight, early 1978; first operational flight, early 1979; regular operations (60 flights per year), end of 1982. Program officials said program cost was estimated at $5.15 billion. (Miles, *LA Times*, 11/27/72; NR PIO)

- NASA policy on autographs was outlined by Dr. James C. Fletcher, NASA Administrator, in memo to officials of Hq. program and staff offices and field installations: "1. No autograph should be treated as a commercial commodity; nothing of value should be requested or accepted in exchange for an autograph. 2. Multiple autographs for the same individual or concern are discouraged; multiple autographs should not be provided when it is apparent that they will be used commercially by the requesting individual or concern." Policy applied to all NASA employees on and off duty but did not affect official signatures for official purposes or signing of autographs for non-commercial purposes. (Text)

- Apollo 17 Astronaut Eugene A. Cernan piloted lunar landing training vehicle on last flight before vehicle's retirement. First flight had been Oct. 8, 1968. All prime and backup commanders of lunar landing missions had practiced lunar module landings in LLTV at Ellington Air Force Base, Tex. (MSC *Roundup*, 12/8/72, 1)

- Dr. Charles A. Berry, NASA Director of Life Sciences, was chosen President of American College of Preventive Medicine at annual meeting in Atlantic City, N.J. (*NASA Activities*, 12/15/72; NASA PIO)

November 14: Dr. James C. Fletcher, NASA Administrator, visited Kennedy Space Center for briefings on KSC plans for 1970s. Programs included Apollo 17, Skylab, U.S.-U.S.S.R. Apollo-Soyuz Test Project, space shuttle, applications programs, and Viking unmanned Mars missions for 1975 launch. Dr. Fletcher's visit included Vehicle Assembly Building, to see Saturn V Orbital Workshop and Saturn IB launch vehicle for Skylab. (KSC Release 320-72)

- More than 70 students from six continents would witness Apollo 17 launch and visit major U.S. science centers during Dec. 4-17 tour sponsored by NASA in cooperation with Dept. of State, NASA announced. Students 15 to 17 yrs old with top academic ratings had been selected by their governments on NASA invitation. U.S. tour would give students broad view of work in science and space. Tour would include Washington, D.C.; Tennessee Valley Authority; Oak Ridge National Laboratory; Marshall Space Flight Center; Manned Spacecraft Center; Ames Research Center; Jet Propulsion Laboratory; and facilities of National Bureau of Standards, National Oceanic and Atmospheric Administration, and United Nations. Project also was supported by U.S. Information Agency and National Science Teachers Assn. (NASA Release 72-219)

- Successful "freezing" of orbiting solar research satellite during first space hibernation operation undertaken in Europe had been reported by European Space Research Organization (ESRO), *Christian Science Monitor* said. ESRO center at Darmstadt, West Germany, had rendered *TD-1A* satellite, launched for ESRO by NASA March 11, nonoperational during partial solar eclipse. Hibernation enabled satellite to preserve

full power intact until it emerged from eclipse. (Reuters, *CSM*, 11/14/72)
- NASA launched Aerobee 170 sounding rocket from White Sands Missile Range, carrying Goddard Space Flight Center solar physics experiment. Rocket and instrumentation performed satisfactorily. (SR list)
- Florida Governor Reubin Askew had been asked by citizens group to restore name of Cape Canaveral to Cape Kennedy by executive order, *Miami Herald* reported. (Markham, *M Her*, 11/14/72)
- Documentary film "Moonwalk One" was reviewed by *New York Post*. Film traced *Apollo 11* lunar landing and first moonwalk by Astronaut Neil A. Armstrong "from Stonehenge to Africa, Cape Kennedy to Houston, to the sun and its mighty flame-storms to camper-sightseers on the Florida shore, to the moon itself and the return." It was "documentary so vast and so beautiful and fascinating that its opening outside the big commercial theaters is a shocking sin of omission." Film, made by Francis Thompson Inc., was highlight of "New American Filmmakers Series" at Whitney Museum in New York. (Winsten, *NY Post*, 11/14/72)

November 14–17: Dept. of Transportation held conference at its Systems center in Cambridge, Mass., to discuss plans for stratospheric observations, laboratory experiments, and computer simulations in program to assess impact of supersonic air traffic on climate and life. Emphasis was on argument by Dr. Harold S. Johnston of Univ. of California at Berkeley and Dr. Paul J. Crutzen of Univ. of Stockholm that nitrogen oxides from supersonic transport exhaust could seriously deplete stratosphere ozone that absorbed much of harmful ultraviolet rays. Some scientists and engineers attending conference had agreed with theory but many had said it was essential to prove theory wrong. Dr. Harold F. Blum of State Univ. of New York at Albany had said that if additional UV got through stratosphere increase in skin cancer would occur, but he could not estimate amount of increase. Dr. Alan J. Grobecker, SST assessment program manager, said if chain reaction in which nitrogen oxides continuously removed ozone according to Johnston-Crutzen theory was very slow effect might escape notice. SST program studies would include use of high-flying U-2 and Concorde supersonic transport aircraft to make stratospheric observations and quick-freeze samples of air in stratosphere for laboratory analysis. (Sullivan, *NYT*, 11/19/72, 50; DOT PIO)

November 15: Apollo 17 astronauts began three-week preflight quarantine at Cape Kennedy while launch crews started final major rocket and spacecraft tests for scheduled Dec. 6 launch. Isolation was to minimize exposure to disease or illness that could delay mission. Week-long rocket and spacecraft tests would duplicate every phase of final countdown. Countdown would start Nov. 30. (AP, *B Sun*, 10/16/72, A9)
- Soviet space scientists were using laser beams and miniature artificial volcanoes to create lunite, material almost identical to moondust, Reuters reported. Man-made regolith was being produced at Soviet Academy of Sciences' Institute of Space Research in experiment to test whether regolith was product of bombardment over millions of years by micrometeorites or of volcanic activity. Pressure chambers provided temperature extremes and vacuum that simulated lunar conditions; laser beam created micrometeorite effect; and crucible of

tungsten filament containing terrestrial basalt was heated electrically to simulate vulcanism. Experiment had been described in newspaper *Soviet Industry*. (Reuters, *C Trib*, 11/16/72)

- GTE International, Inc., subsidiary of General Electronics Corp., announced it had received contract valued at about $4 million from Western Union International, Inc., for construction of satellite communications earth station in People's Republic of China. Station would be ground link for first direct communications satellite between China and Europe. (GTE Release)

November 16–27: Explorer 48 (SAS-B) Small Astronomy Satellite, second in series of three, was successfully launched for NASA by Italian crew at 1:14 am local time (5:14 pm Nov. 15 EST) from San Marco Facility off coast of Kenya. Four-stage solid-propellant Scout launch vehicle boosted 186-kg (410-lb) payload into circular orbit with 632.5-km (393-mi) apogee, 443.7-km (275.7-mi) perigee, 95.4-min period, and 1.9° inclination. Launch was 26th consecutive successful Scout launch, record for NASA vehicle.

Primary objective of *Explorer 48* was to measure spatial and energy distribution of primary galactic and extragalactic gamma radiation—within SAS program to make investigations of celestial sphere from above earth's atmosphere and search for sources inside and outside our galaxy, radiating in x-ray, gamma ray, ultraviolet, infrared, and other regions of electromagnetic spectrum. Gamma ray telescope was first satellite version of evolution of digitized spark-chamber telescopes that had begun with balloon-borne instruments. Advanced detector could establish gamma ray event and reject events that simpler instruments might identify erroneously. It could study gamma rays exceeding 20 mev in energy, in region of electromagnetic spectrum able to penetrate galactic and extragalactic matter.

On Nov. 19 half of experiment's high-voltage supplies were turned on and experiment was pointed toward earth. Real-time data acquisition indicated proper functioning of experiment, with detection of earth albedo gamma rays. Remaining high-voltage power was turned on Nov. 27. Experiment and spacecraft control section were functioning normally.

Spacecraft was cylinder 59 cm in diameter and 51 cm high (23-in dia by 20). When deployed, four solar paddles—27 cm wide and 135 cm long (10.5 by 53 in)—provided raw power. Unique was separate development of spacecraft control section and experiment package, which had permitted flexible concurrent development of experiments and minimized costs by fabrication of standardized control section.

NASA provided booster and satellite; Italian team was responsible for cost-reimbursable launch preparation. Goddard Space Flight Center and Applied Physics Laboratory crew performed ground test and checkout. Satellite was designed and built at GSFC, which managed project. Langley Research Center managed Scout. (NASA proj off; NASA Release 72–204; GSFC proj off)

November 16: Briefing on results of low-cost-design concept completed by Lockheed Missiles & Space Co. under NASA contract was held at Marshall Space Flight Center. Study had been conducted under NASA Low Cost Evaluation Project begun in May by Dr. George M. Low, NASA Deputy Administrator. Information from panels was to be

correlated and incorporated into program planning and execution. (MSFC Release 72-147)
- Five Soviet aviation officials toured Kennedy Space Center and were briefed on current and future NASA programs. They were among first Soviet citizens to visit KSC. Poet Yevgeny Yevtushenko had visited KSC and witnessed *Apollo 16* launch as NASA guest April 16. (KSC PAO)

November 17: Successful radio tracking of solar energetic particles through interplanetary space was reported in *Science* by Goddard Space Flight Center scientists Dr. Joseph Fainberg, Larry G. Evans, and Robert G. Stone. Satellite observations by *Explorer 6* (launched Aug. 7, 1959) of ejection of energetic sun particles at long radio wavelengths had provided means of investigating gross magnetic field configuration over distances of one astronomical unit. (*Science*, 11/17/72, 743-5)
- Selection of Lockheed-Georgia Co. Div. of Lockheed Aircraft Corp. for negotiations leading to contract for design and option for fabrication of quiet, fan-jet, propulsive-lift research aircraft (QUESTOL) was announced by NASA. Agency also authorized $600 000, four-month, cost-reimbursable design-refinement phase. If construction was authorized, Lockheed was to deliver research vehicle in two years. Project was managed by Ames Research Center with technical support from Langley, Lewis, and Flight Research Centers. (NASA Release 72-224)
- Effect of 1972 elections on congressional committees dealing with science and environment was discussed in *Science* article. Chairmen of both House and Senate space committees had been replaced. Sen. Clinton P. Anderson (D-N. Mex.), Chairman of Senate Committee on Aeronautical and Space Sciences, was retiring at age 77 after 31-year career in Congress. Rep. George P. Miller (D-Calif.) had been defeated in Democratic primary and would leave Congress at age 81 after serving in House since 1945. Rep. Earle Cabell (D-Tex.), member of Science and Astronautics Committee, was defeated in the general election. He was to have been one of six House members of congressional board to oversee new Office of Technology Assessment. Board also lost prospective member with defeat of Sen. Gordon Allott (R-Colo.). (Walsh, *Science*, 11/17/72, 25-6)

November 18: NASA and Boston Univ. Dept. of Astronomy sponsored symposium "Life Beyond Earth and the Human Mind" at Boston Univ. Panel Chairman Richard Berendzen, Boston Univ. astronomer, had said symposium was being held because recent findings in astronomy, biology, chemistry, engineering, and physics had indicated high probability of existence of extraterrestrial life. Many scientists believed that "within the next few decades some form of communication will be achieved with intelligent life elsewhere in the Universe." Panelists were anthropologist Dr. Ashley Montagu, Massachusetts Institute of Technology physicist Dr. Philip Morrison, Cornell Univ. astronomer Dr. Carl E. Sagan, Dean Krister Stendahl of Harvard Divinity School, and Dr. George Wald of Harvard Univ., winner of 1967 Nobel Prize for medicine and researcher in biochemical evolution.

Dr. Sagan said human technology had reached stage where man could detect another civilization as advanced as his. "Most optimistic estimate of civilizations in our galaxy is one million, which means one per many hundred thousand stars." Searching stars one by one could take hundreds of years. Dr. Montagu said that "some other forms of life are probably more intelligent than we, which can explain why they haven't contacted us." Dean Stendahl said discovery of life elsewhere would teach man that God's domain is larger and would give man better idea of his place in it. Contacting other beings might help man to stop viewing God in his own image. Dr. Wald said he was convinced that life existed elsewhere in galaxy but he doubted man would ever contact it. Dr. Morrison predicted man would receive and verify message from another civilization. Message would be technical, scientifically coded communication received over many months or years. NASA filmed symposium to produce educational film. (NASA Release 72-213; Leary, *W Post*, 11/24/72, C10)

November 19: SNAP-27 radioisotope thermoelectric generator completed third year of continuous operation on lunar surface. Atomic battery had been deployed on moon by *Apollo 12* astronauts Nov. 19, 1969, to power scientific experiments. SNAP-27, which had one-year life requirement of 63.5 w, was supplying more than 69 w. Generator, developed by Atomic Energy Commission, was fueled with plutonium 238 isotopes. (AEC Release P-372)

• Kennedy Space Center awarded $4.7 million contract to Morrison-Knudsen, Inc., to develop KSC facilities for Viking spacecraft encapsulation. Contract provided for modification of an addition to one structure and construction of new building in KSC Fluid Test Facility area. (KSC Release 346-72)

• Hall of Science, visited by some 5 million persons during 1964-65 World's Fair in Flushing, N.Y., was reopened with exhibits including NASA satellite and Apollo mockup. Hall had been revived and remodeled as free scientific exhibition for public under grants from Rockefeller and Ford Foundations. (Andelman, *NYT*, 11/2/72; NY Visitors Bur)

November 20: Apollo 17 crew underwent preliminary medical checks and men were reported in good shape, while final test countdown for mission, begun Nov. 14, proceeded smoothly at Kennedy Space Center. Crew was not present as propellants were pumped into Saturn V rocket, to avoid risk. They planned to participate in dry run Nov. 21 after flammable oxygen and hydrogen had been drained from rocket. (UPI, *NYT*, 11/21/72, 46)

• Apollo 17 experiments were discussed by Apollo Lunar Exploration Director William T. O'Bryant at NASA Hq. press briefing. Mission astronauts would perform "more scientific investigation than on any other Apollo mission." Experiments originally planned for Apollo 18, 19, and 20 would be aboard Apollo 17, last mission in Apollo series. (McGehan, *B Sun*, 11/21/72, A8)

• Fragments of Apollo 17 moon rock would be distributed to other nations for display in scientific and educational institutions, NASA announced. (NASA Release 72-225)

Ground-breaking ceremony for Smithsonian Institution's new National Air and Space Museum was held on Mall in Washington, D.C. Museum was scheduled to open July 4, 1976. Chief Justice Warren Burger, Chancellor of Smithsonian, was principal speaker. (Program)

• NASA launched Aerobee 200 sounding rocket from White Sands Missile Range, carrying Goddard Space Flight Center payload to test flight performance. Rocket and instrumentation performed satisfactorily. (SR list)

- Air-transportable satellite communications earth station for use in disasters would be designed by Nippon Electric Co. of Tokyo and Technology Resources of Berne, Switzerland, under United Nations contract, *Aviation Week & Space Technology* reported. (*Av Wk*, 11/20/72, 11)
- Army Aviation Systems Command issuance of requests for proposals to 10 aerospace companies for prototype development of advanced armed helicopter was reported by *Aviation Week & Space Technology*. Helicopter would be equipped with advanced air-to-surface missiles. Airframe contractors were expected to be selected in June 1973. (*Av Wk*, 11/20/72, 19)
- *Houston Chronicle* editorial commented on Apollo 17 mission and crew: "To some, the Moon trip is already looked upon as routine and there is less public anticipation than ever before. Nonetheless, there is no way to diminish our pride that these are Americans making this flight, that our nation has bridged space and that these men call the Houston area home." (*H Chron*, 11/20/72)

November 21–30: NASA successfully launched European Space Research Organization's *Esro 4* by four-stage Scout-D launch vehicle from Western Test Range at 4:17 PST. Satellite entered orbit with 1187-km (737.6-mi) apogee, 253-km (157.2-mi) perigee, 99-min period, and 91.1° inclination.

Primary NASA objectives were to place spacecraft—eighth in ESRO flight program on U.S. vehicles—in earth orbit that would permit investigations of phenomena in polar atmosphere and to provide tracking and telemetry support. By Nov. 30, flight data indicated objectives were met and mission was adjudged success.

Esro 4 carried six experiments from four countries: study of positive ions in ionosphere (United Kingdom); study of composition and total mass density of natural gas in upper atmosphere and exosphere (West Germany); study of low-energy-particle precipitation in auroral zones (Sweden); investigation of polar-cap-absorption events (Netherlands); measurement of flux and energy spectrum of solar flare particles, trapped particles in lower radiation belt of earth, and galactic and nonsolar energetic particles (West Germany); space-flight qualification of infrared horizon-sensing instrument for altitude measurements of spinning satellites (Netherlands).

European-designed and -built spacecraft weighed 114 kg (251 lbs) and was cylindrical, with body-mounted solar cells and four booms. ESRO would reimburse NASA for launch under December 1966 Memorandum of Understanding. NASA Office of Space Science directed project for NASA; Langley Research Center managed Scout program. (NASA proj off; NASA Release 72–214)

November 21: Apollo 17 Astronauts Eugene A. Cernan, Dr. Harrison H. Schmitt, and Ronald E. Evans participated in final prelaunch test aboard spacecraft. Astronauts wore spacesuits for simulated liftoff of Saturn V rocket. Following test, launch crew began to ready command module "America" and lunar module "Challenger" for start of actual countdown scheduled for Nov. 30. (AP, *B Sun*, 11/22/72, A5)

- Crowd estimated at more than half million would witness nighttime launch of Apollo 17, which experts believed would be visible within 800-km (500-mi) radius, Associated Press reported. Lester T. Keene of Kennedy Space Center Data Systems Div. had said exhaust from Saturn V rocket would be equivalent to brilliance of sunlight. Peak visibility would be reached about 2 min 30 sec after launch, when Saturn V was at 68-km (42-mi) altitude and 90 km (56 mi) northeast of Cape Kennedy, and before shutdown of 1st-stage engines that burned fuel producing orange flame. Pale blue flame of 2nd stage would not be seen for as great distance. (B *Sun*, 11/21/72, A8)

- NASA's M2–F3 lifting body, piloted by John A. Manke, completed 23rd flight from Flight Research Center after air-launch from B-52 aircraft to evaluate command augmentation system. Additional flight objectives were to obtain stability and control data at mach 1.1, mach 0.8, and mach 0.7 and to check out pitch-reaction augmentation system. M2–F3 reached 19 800-m (65 000-ft) altitude and mach 1.3. All objectives were achieved. (NASA proj off)

- U.S. and U.S.S.R. opened second round of Strategic Arms Limitation Talks (SALT II) in Geneva in effort toward comprehensive and permanent agreement to limit strategic nuclear weapons. SALT I, begun Nov. 17, 1969, had brought limitation accords signed May 26 during President Nixon's visit to Moscow. (Tuohy, *W Post*, 11/22/72, A10)

- Marshall Space Flight Center announced issuance of requests for proposals on space tug systems. Tug—space shuttle 3rd stage—would extend capability of shuttle booster and orbital stages by adding propulsion for higher orbit. Propulsion would be provided by temporary use of existing rocket stage adapted for launch from shuttle cargo bay or by use of interim tug. Studies of rocket stage use were under way. RFPs had been issued for possible interim tug development. Three $750 000 contracts would be awarded—two for work on cryogenic tug using liquid-hydrogen and liquid-oxygen propellants and one for study of tug using storable propellants. Proposals were due Dec. 22, with contracts to be awarded during first quarter of 1973. Air Force would assist in funding cryogenic tug studies and in selection of contractors and study management. Interim tug would be developed for use in space shuttle's first flight scheduled for late 1970s. (MSFC Release 72–155)

- Langley Research Center team working with Environmental Protection Agency had developed telescope-mounted laser to detect air-polluting gases from industrial smokestacks, NASA announced. Laser was beamed at smoke emitted and detected reflected light from pollutants through optical filters. EPA was interested in laser for remote sensing of industrial plants. Tests were being conducted at Asheville, N.C., plant of Carolina Power and Light Co., which had offered to cooperate with Government agencies in experiments. (NASA Release 72–222)

- Concorde 001, British prototype of Anglo-French supersonic transport, completed 300th and final flight in assigned test program. Flight brought combined total of Concorde 001 and 002 and preproduction aircraft 01 flights to total 654, representing 1354 hrs 50 min with 348 hrs 24 min at supersonic speeds. (BAC *Newsletter*, 12/9/72, 2)

- Measurement of speed of light "more accurately than ever before" with infrared laser was announced by National Bureau of Standards. NBS scientists using laser that generated extremely pure light beam, uncluttered by jumbled frequencies and wavelengths, had arrived at new light speed figure of 299 792.472 km per sec (186 282.396 mps). New value was 100 times more accurate than value that had been

November 21: *A telescope-mounted laser instrument was beamed at exhaust stacks of an industrial plant in tests proving the laser's ability to detect such air pollutants as sulfur dioxide. Remote detection from as far as one kilometer (3300 feet) was possible by the laser, developed at Langley Research Center. Tests were being conducted at the Carolina Power and Light Co. plant in Asheville, N.C., which had offered to cooperate in the LaRC and Environmental Protection Agency program.*

accepted for past 15 yrs and had opened up "new possibilities for an increase of 1,000 times in the number of frequency bands" used in communications. (NBS Release)

- Results of analyses of data collected by City of Cleveland's Air Pollution Control Div. from 1968 through 1971 were reported by Lewis Research Center scientists. State of Ohio standards for total suspended particulate and nitrogen dioxide had not been met in Cleveland; sulfur dioxide standards had been met only in limited residential areas. Air quality had improved in industrial valley but only sulfur dioxide pollution showed consistent decrease in remainder of city, and suspended particulate levels had increased to west of Cuyahoga River with decrease to east. Trends had been determined by LeRC Environmental Research Office from data obtained during two-year joint LeRC and City of Cleveland project. (LeRC Release 72–102)

- Officials of Corp. for Public Broadcasting (CPB), Public Broadcasting Service (PBS), and National Public Affairs Center for TV (NPACT) met to discuss NPACT feasibility study of possible special scientific coverage of Apollo 17 EVAs by PBS [see Nov. 9]. Following discussion CPB President Henry Loomis issued message to PBS stations. Decision had been made that "live coverage would be impractical for public television and that alternate idea for space project programming would be considered." Decision was based on study conclusions that time was too short to gather required staff, mission delays might disrupt TV schedules, and coverage by commercial networks would be more extensive than first expected. There was agreement that "scientific findings of the entire Apollo project, which concludes with Apollo, merited serious treatment on public television." (Text)

- Second delegation of scientists from People's Republic of China—seven specialists in physical and biological sciences—arrived in Washington, D.C., to begin four-week tour of major U.S. research centers and universities. Delegation included Chien Wei-chang, professor of Engineering at Chinghua Univ. in Peking, who had worked at Jet Propulsion Laboratory from 1940 to 1946. At National Academy of Sciences press conference Chien said: "We welcome and congratulate the work on space done by American scientists. We also congratulate you on the success of this problem; this is why we pay so much attention to work on space." In answer to question, Chien said his country was seeking to develop communications satellites but had no plans to put men in orbit. Visit was sponsored by National Academy of Sciences, Social Science Research Council, and American Council of Learned Societies. Delegation of physicians from PRC had toured U.S. in October. (McGehan, B *Sun*, 11/22/72; Schmeck, *NYT*, 11/22/72, 43)

- First successful experimental demonstration in U.S. of magnetic levitation system (MAGLEV) was announced by Secretary of Transportation John A. Volpe. MAGLEV—system for suspending vehicle above guideway with power supplied by supercooled magnets—would permit trains to travel at speeds up to 480 km per hr (300 mph) without physical contact with rails or guideway. Test was performed in Menlo Park, Calif., by Stanford Research Institute under contract to Dept. of Transportation. (DOT Release 100–72)

November 22: Deputy Director James A. Lovell, Jr., of Manned Spacecraft Center's Science and Applications Directorate announced intention to retire from space program to press at MSC. As astronaut he had spent more time in space than any other U.S. astronaut. Lovell had been *Gemini 4* backup pilot, *Gemini 7* pilot, *Gemini 9* backup command pilot, *Gemini 12* command pilot, *Apollo 8* command module pilot, *Apollo 11* backup command pilot, and *Apollo 13* commander. Lovell told press he was examining offers from New York and Chicago business firms. "This is not something that I would choose to do under other circumstances but . . . there are not enough slots for us guys who want to remain in the operations end of space flight." Lovell said he would remain with NASA until completion of Apollo 17 mission. (UPI, *W Post*, 10/24/72, A5)

- Instruments to be used globally to monitor air pollution in upper atmospheric airways were being flight-tested on Convair 990 at NASA's Ames Research Center, Lewis Research Center announced. Instruments—candidates for NASA Global Air Sampling Program (GASP) under direction of LeRC—would be installed on 10 to 15 foreign and domestic commercial aircraft to monitor gaseous and particulate pollutants. Particulates would also be collected by sampling device and later would be analyzed by electron microscopy and other methods. (LeRC Release 72–106)

- Huntsville, Ala., police department was using "elapsed time speed computer," developed by Chief Peter H. Broussard, Jr., of Marshall Space Flight Center Astrionics Laboratory's Sensors Branch, to identify traffic speed violators. System consisted of low-cost electronic computer and two roadside markers. Timing switch was turned on when vehicle passed first marker and off when vehicle reached second marker. Elapsed time was measured by internal computer clock and compared with distance between markers. System would perform basic functions of radar units more economically. (NASA Release 72-223)
- U.S. Command in Saigon reported loss of fourth F-111 swing-wing fighter-bomber aircraft in combat over Indochina since Sept. 28. Both crewmen were missing. (UPI, W Star & News, 11/22/72, A7)
- President Nixon had directed that regulations on travel of U.S. ships and aircraft (Transportation Order T-2) be changed to permit valid U.S. aircraft and vessels to visit People's Republic of China, White House announced. (PD, 11/22/72, 1688)

November 23: Dr. James C. Fletcher, NASA Administrator, during visit to Lewis Research Center told staff NASA's direction in coming years would include increased emphasis on sharing expertise with other Federal, local, and state government agencies and continued high performance in space-flight exploitation with an eye on the purse strings. He said, despite Federal budgetary restrictions, "I don't see a great change at NASA and feel we can continue in the direction we set forth this past year." (*Lewis News*, 12/15/72, 3)
- Leading figures in Federal Government monopoly suit against International Business Machines Corp. would visit Kennedy Space Center during Apollo 17 mission to examine computer operations, *Washington Post* reported. U.S. District Judge David N. Edelstein, with lawyers from IBM and Dept. of Justice, would fly from New York. Similar trip had been proposed for April 16–27 *Apollo 16* mission but NASA had opposed visit on grounds that it could lead to breach or relaxation of security and safety regulations. Visit was to educate court on nature of IBM equipment used during launch. (Mintz, *W Post*, 11/23/72, A29)
- Extremely sensitive low-light-level image-intensification tube was being used by Hale Observatory astronomers to aim 500-cm (200-in) telescope atop Mt. Palomar at most distant objects in universe. Tube, developed to detect enemy troops in Vietnam, amplified faint light from remote celestial objects 2000 or more times. Tube had been adapted to astronomical use by Dr. Edwin W. Dennison of Hale Observatory and Cal Tech. (LATNS, *W Post*, 11/23/72, H9)
- Electrostatic autopilot invented by Johns Hopkins Univ. physicist Maynard L. Hill stabilized small aircraft flight by harnessing horizontal voltage lines of earth's atmosphere, Associated Press reported. Device, simple enough to be made for $50, could provide small aircraft and helicopters with flying ability of airliners using costly autopilots. Hill had said instruments should open new techniques for studying cloud physics on board cheap and rugged remotely piloted vehicles (RPVs). (Haugland, AP, *W Post*, 11/23/72, H13)
- U.S.S.R. said it would conduct missile-carrier tests in Pacific through Dec. 31. It warned ships and aircraft to steer clear of area about 2400 km (1500 mi) east of Japan. (Tass, FBIS–Sov, 11/24/72, A1; AP, B *Sun*, 10/24/72, 6)

November 24: Twenty-fifth anniversary of first launch of Aerobee sounding rocket on Nov. 24, 1947, from White Sands Missile Range. Launch had been unsuccessful; rocket carrying Johns Hopkins Univ. payload had veered off course and had been destroyed. But since then arrow-shaped Aerobee had been busiest booster in U.S. space program, with launch rate of almost one per week and few failures out of 900 launches to date. Aerobees carrying scientific payloads to altitudes as high as 480 km (300 mi) had been launched by NASA, Air Force, Navy, and Federal scientific groups. Launches were mainly from NASA's Wallops Station, White Sands Missile Range, and Fort Churchill Range in Manitoba, Canada.

Aerobee—most widely used rocket in astronomy—had carried delicate instruments through blanket of air and dirt surrounding earth to area of greater visibility. Instruments launched June 13, 1969, by NASA on Aerobee 150 from Natal, Brazil, had detected x-ray star only 16 km (10 mi) wide that weighed as much as sun. Scientists at Univ. of California who conducted experiment had said stellar object might be first direct evidence of neutron star and first time size of x-ray source from deep space could be determined.

Other scientific payloads aboard Aerobees had measured Crab Nebula, collected micrometeoroids, taken temperatures aloft, and recorded air densities. In 1952 Air Force Aerobee carried two monkeys and two mice in biomedical experiments.

Rocket originally weighed 700 kg (1600 lbs), was 5.8 m (19 ft) long, and carried payloads under 45 kg (100 lbs). Over years Aerobee had grown to be 15 m (50 ft) long, powered by four clustered, liquid-fuel engines and solid-fuel booster. Largest Aerobee could blast 400-kg (900-lb) payload to altitude of 200 km (125 mi). (Lyons, *NYT*, 11/26/72, 39; *A&A 1915–1950*)

- Dr. Christopher C. Kraft, Jr., Manned Spacecraft Center Director and former MSC Director of Flight Operations, reviewed Apollo program in Associated Press interview published in Baltimore *Sun. Apollo 8* (launched Dec. 21, 1968) had been program's pivotal mission. "It proved so many things that had a bearing on the progress of the program—things that it might have disproved. The navigation to and from the moon, the ability of the spacecraft systems to perform, the ability of man to survive the deep space environment, all hinged on the Apollo 8 mission." Mission also had changed competitive position of U.S. and U.S.S.R. in space. "I really believed the Russians planned to fly a circumlunar mission, sending a manned spacecraft looping around and returning it without orbiting the moon. That way they could say they sent the first man to the vicinity of the moon. And when we . . . decided to fly Apollo 8 as a moon-orbit mission, there was nothing left for them to do." There was no major technological field that "hasn't been materially changed or had some reflection of not only Apollo but of the whole space program." Space technology had been important to maintenance of U.S. position as leading nation in world. Five manned lunar landing flights had produced greater scientific return than envisioned. "We have done just about all we should have done with the Apollo program in terms of lunar exploration. This is a good time to stop, evaluate what we have and use our resources to do other things in the space program." (B *Sun*, 11/24/72, A10)

- At least seven major U.S. companies had asked to lease part of Canadian telecommunications satellite system, Telesat, *Washington Post* reported. Negotiations were not final, but it was probable that at least one company would decide to use Canadian system until comparable U.S. systems were ready. Canadian system used U.S. designed and manufactured satellites launched by NASA under contract with Telesat Canada, first of which, *Anik 1*, had been launched Nov. 9. (Samuelson, *W Post*, 11/24/72)
- Former Congressman H. Allen Smith of California was quoted as having said in letters to Jet Propulsion Laboratory and Cal Tech that he would not seek to enforce law passed Oct. 21 to change JPL name to H. Allen Smith Jet Propulsion Laboratory. Smith had written: "I have no intention of bringing any lawsuit or any other action to force this name change. There are many laws on the federal books which are not enforced." *Wall Street Journal* said it was "doubtful" that Congress would reverse itself on measure. NASA officials had already found House Committee on Science and Astronautics "unwilling to help provide relief." (*WSJ*, 11/24/72)

Nov. 24–Dec. 8: U.S.–U.S.S.R. Working Groups 2 and 4, meeting at Manned Space Center on Apollo-Soyuz Test Project (ASTP), defined antenna locations on Soyuz and confirmed that U.S.S.R. would build antennas; negotiated interface signal characteristics for radio communication and ranging systems and confirmed 428-km (266-mi) ranging capability; agreed on end-to-end compatibility tests of radio and hardline communications; resolved U.S. very-high-frequency amplitude-modulation implementation requirements for Soyuz systems, defined design of cable communications systems including cable lengths, junction boxes, and Soviet supply responsibilities; and discussed intercenter control communications, establishing basis for March negotiations. (NASA proj off)

November 25: *Cosmos 537* was launched by U.S.S.R. from Baykonur Cosmodrome into orbit with 302-km (187.7-mi) apogee, 202-km (125.5-mi) perigee, 89.5-min period, and 64.9° inclination. Satellite reentered Dec. 7. (GSFC *SSR*, 11/31/72; 12/31/72; *SBD*, 11/29/72, 113)

- First three of 749 F-15 Eagle air superiority fighter aircraft in Air Force program had done well in initial flight tests, according to Air Force sources quoted by *New York Times*. Two were undergoing tests at Edwards Air Force Base, Calif., and third at McDonnell Douglas facility in St. Louis. Aircraft was U.S. answer to Soviet MiG–23 Foxbat and Mikoyan Fearless, new Soviet air superiority fighter forecast for 1974 deployment. (Middleton, *NYT*, 11/25/72, 17)

November 26: Sign posted by Grumman Aerospace construction personnel on gantry work platform near Apollo 17 spacecraft said, "This may be our last but it will be our best," United Press International reported. Grumman foreman had said sign was to inform Apollo crew that "they don't have to worry about our letting down." Jobs of 600 Grumman workers at Kennedy Space Center would terminate with completion of Apollo 17 mission. (*W Post*, 11/27/72, E6)

- Milton W. Rosen, Senior Scientist in NASA Office of DOD and Interagency Affairs, became Deputy Associate Administrator (Engineering) in Office of Space Science. Rosen had joined NASA in 1958 after 18 yrs with Naval Research Laboratory, where from 1947 to 1955 he was head of Rocket Development Branch. Rosen had proposed and developed Viking Upper-Air-Research Rocket and in 1955 proposed Vanguard Earth Satellite Project. He served as Technical Director during its development. (NASA Ann; NASA Release 72-237)

- Trailer in which NASA formerly quarantined astronauts returning from moon was being used by Air Force Military Airlift Command, *St. Louis Post-Dispatch* reported. Trailer was positioned at McGuire Air Force Base, N.J., and could be airlifted by Air Force C-141 cargo aircraft to transport patients with communicable diseases. Previously aircraft had to be decontaminated after carrying contagious patients. (*St. Louis P-D*, 11/26/72)

- *Washington Post* published first of series of articles on U.S. energy crisis. Population had doubled in 50 yrs, while energy use had almost quadrupled. Per capita electricity consumption had doubled five times—twice in last 15 yrs. Nationwide environmental movement had halted Government's Plowshare program to release trapped oil and gas by nuclear explosions, delayed Alaska pipeline, and forced near-moratorium on dam building. Ecologists had "forced power companies to abandon scenic river and lake sites in 10 states and have caused more than 20 delays in the construction of nuclear power plants." Disagreement was "rampant" over changes wrought by environmental movement. But ecologists had caused no shortage of energy in U.S. thus far. Many energy experts believed technology would "bail us out by letting us tap new, clean fuels and by cleaning up existing fuels." Mitre Corp. had recently completed exhaustive energy study. Its Senior Vice President Charles Zraket had said, "We believe the physical resources and technical options exist to get us out of this crisis in 10 years." (O'Toole, *W Post*, 11/26/72, A1)

November 27: NASA's TF-8A aircraft, equipped with supercritical wing and new fuselage fairings, was flight-tested at constant dynamic pressure of 9580 newtons per sq m (200 psf). Pressure distribution and wake survey measurements were made with instrumentation at outboard station. All tests with new fuselage fairings had been made with vortex generator off. Generators would be reinstalled and drag measurements made for direct comparison with previous tests with original fairings and vortex generators on and with wind tunnel results. Some measurement of shock wave dynamics was being made with microphones installed on wings. (NASA proj off)

- Apollo 17 astronauts began final week of training at Kennedy Space Center for Dec. 6 mission. NASA spokesman said countdown test begun Nov. 15 had proceeded so smoothly that launch pad technicians had "only to monitor the systems." Other technicians installed batteries and water in lunar landing vehicle and checked backup flight systems. (Wilford, *NYT*, 11/28/72)

- Map of entire globe of Mars—first detailed map of another planet, prepared by U.S. Geological Survey from 1500 of 7000 photos taken by *Mariner 9* (launched May 30, 1971) and processed by Jet Propulsion Laboratory—was published by *New York Times*. Map showed entire Martian surface in Mercator projection at scale of 1 to 25 million at equator and in two special polar projections. Other maps were being prepared for even more detailed study of Mars. Map reprinted in *Times* would be released by U.S. Geological Survey early in January.

Times commented that map was produced in Flagstaff, Ariz., few miles from mountain-top observatory where Dr. Percival Lowell in 1877 observed presumed canals on Mars that "inspired so many tales of Martians," canals that were "nowhere to be seen on the map of Mariner 9." (Wilford, *NYT*, 11/27/72, 1, 30; USGS Map Info Off)

- Flight Research Center began test series to evaluate method of simulating space shuttle orbiter in flight at subsonic speeds. Tests would be conducted on FRC's general-purpose airborne simulator (GPAS)—converted subsonic jet transport equipped with electronic variable stability and control system. Flights would determine aircraft's ability to match orbiter's final approach path in steep descent from 9 km (5.6 mi) down to landing. (FRC Release 21–72; FRC PAO)

- Laser Doppler airspeed indicator 10 times more precise than best systems in commercial aircraft was being tested by Ames Research Center in Global Atmospheric Research Program, NASA announced. System—proved accurate to one tenth of one per cent—focused carbon dioxide laser beam 18 m (60 ft) ahead of aircraft where air was undisturbed by bow wave, detected backscatter from small aerosols, and measured Doppler shift in frequency proportional to aircraft's airspeed. Doppler shift was then measured digitally and recorded by digital data system. (NASA Release 72–228)

- President Nixon exchanged remarks with President Felix Houphouet-Boigny in telephone call that inaugurated satellite communication service for Republic of Ivory Coast via ground station at Abidjan. President Nixon said he was pleased to talk "through this historic new satellite communication." (*PD*, 12/4/72, 1700)

- U.S.S.R. had offered to launch a Highly Eccentric Orbit Satellite (HEOS) for European Space Research Organization (ESRO) free of charge as part of general agreement on satellite research, *Aviation Week & Space Technology* reported. First two HEOS satellites had been launched Dec. 5, 1968, and Jan. 31, 1972, by NASA with Thor-Delta booster for charges of $4.5 million and $6 million. ESRO officials had estimated cost of planned launch at $7.5 million. (*Av Wk*, 11/27/72, 28)

- U.S.S.R. test-fired SSNX-8 Sawfly submarine-launched ballistic missile during series of tests announced Nov. 24. Secretary of Defense Melvin R. Laird said later that missile had traveled 7400 km (4000 nm). (*Av Wk*, 12/4/72, 5; *Aerospace Daily*, 12/15/72)

- Air Force was sponsoring "fly-off" of new attack aircraft prototypes at Edwards Air Force Base, Calif., to determine best aircraft, *Wall Street Journal* reported. Competition was "essential element" of Air Force A–X (attack-aircraft, experimental) program designed to produce new aircraft for supporting ground troops in battle at lowest possible cost. (Levine, *WSJ*, 11/27/72, 10)

- Resignation of Clarence H. Linder as first full-time President of National Academy of Engineering at close of May 1973 annual NAE meeting was announced by NAE. Linder was resigning for personal reasons. (NAE Release)

November 28: President Nixon announced resignation of Melvin R. Laird as Secretary of Defense and intention to nominate Elliot L. Richardson

—Secretary of Health, Education, and Welfare—as successor. (*PD*, 12/4/72, 1705-6)

November 29: NASA's M2–F3 lifting body, piloted by Cecil Powell, completed 24th flight from Flight Research Center after air-launch from B-52 aircraft. Flight objectives were to check out pilot and to obtain stability and control data at mach 1.1 and mach 0.9. All objectives were met. Vehicle reached 21 000-m (69 000-ft) altitude and mach 1.25. (NASA proj off)

- NASA announced organizational changes within Office of Manned Space Flight to fit post-Apollo/Skylab activities. New Mission and Payload Integration Office had been established, Space Station Task Force became Sortie Lab Task Force, Advanced Missions Office was redesignated Advanced Programs Office, and Apollo Program Office became Apollo/ASTP Program Office. Functions of Engineering and Operations Directorate in OMSF had been reassigned to Advanced Programs Office. Mission and Payload Integration Office would plan and coordinate requirements between payload sponsors and shuttle and sortie lab organizations. Sorties Lab Task Force would manage NASA planning and definition activities to produce manned module for research and applications activities with space shuttle. Current project emphasis was on technical definition and planning for hardware development phase. Parallel definition studies were being made by NASA and European Space Research Organization. If Europe undertook development of sortie lab, NASA Associate Administrator for Manned Space Flight Dale D. Myers had said, NASA Task Force would coordinate NASA and ESRO activities.

    Philip E. Culbertson was Director of Mission and Payload Integration Office and Acting Director of Advanced Programs Office. Douglas R. Lord was Director of Sortie Lab Task Force, and Dr. Rocco A. Petrone was Director of Apollo/ASTP Project for joint U.S.–U.S.S.R. rendezvous and docking mission in 1975. (NASA Release 72–232)

- Aerospace Industries Assn. of America, Inc., released *The National Technology Program: Utilization of Industry*, study of how Government and high-technology industry could best use technological resources available and under development. Study found that "positive policy actions by Government can restore lost technological momentum in the United States. Regained, this momentum will ensure that both Government and high-technology industry, working together, can move effectively toward the solution of many of the priority problems confronting the nation." AIAA recommended formulation of national technology strategy responsive to Nation's needs and supported both by "a mechanism that sets forth research and development goals, objectives, priorities and programs, and by adequate funding"; national program to identify major domestic problems having potential solution through technology and initiation and funding of demonstration programs to find solutions; national goal of maintaining positive trade balance supported by export policies that promoted internationally competitive high-technology products to increase international sales by U.S.; and Government procurement regulations that supported successful attainment of national needs. (Text)

- Spokesman for Friends of the Earth, coalition of environmentalists who lobbied against supersonic transport aircraft in 1971, said organization had set up office on Capitol Hill to lobby against expected Administration attempts to revive project in 93rd Congress. First priority was to present facts to 66 new representatives and 9 senators who had never voted on SST. Senate had voted 58 to 37 to defeat project funding in 1971; House had voted 215 to 204 against funding. (UPI, *W Post*, 11/30/72, A4)

- *New York Times* editorial commented on first global map of Mars produced by U.S. Geological Survey from *Mariner 9* photos [see Nov. 27]: "Ever since telescopes were invented people have wondered to what extent Mars resembles Earth. Now, for the first time, the existence of a reasonably comprehensive picture of Mars provides a basis for comparison." Despite differences from earth, "increased familiarity with the forces that have shaped its surface is bound to give man a more cosmopolitan view of how planets evolve—a view that should provide us with better understanding of the planet we call home." (*NYT*, 11/29/72)

November 30: Six-day countdown for Apollo 17 began at 8:30 am EST at Kennedy Space Center. Spacecraft was scheduled for launch at 9:53 pm EST Dec. 6 (KSC Hist Off; AP, *W Post*, 12/1/72, A22)

- Spokesman for Columbia Broadcasting System said in New York that CBS, American Broadcasting Co., and National Broadcasting Co. had dissolved plans for three-network-pool coverage of Dec. 6 Apollo 17 launch. Action would avoid problems that might develop because of 28-day-old strike against CBS by 1200 cameramen, technicians, and engineers. CBS would cover launch with cameras manned by supervisory personnel. ABC and NBC spokesmen said they would restore two-network pooled coverage. (AP, B *Sun*, 12/1/72, A5)

- Pan American World Airways' waiting list for reservations on first commercial flight to moon contained names of *Apollo 11* Astronauts Neil A. Armstrong—first man to walk on moon during July 16–24, 1969, mission—Edwin A. Aldrin, Jr., and Michael Collins. List had been closed after 90 000 reservations had been accepted. Pan Am spokesman had said flight would not be possible before 21st century, but most applicants had paid $10 or $25 as reservation fee. (*M Her*, 11/30/72)

- Dr. Karl Klager, Vice President and Director of Science for Aerojet Solid Propulsion Co., was named winner of 1972 James H. Wyld Propulsion Award by American Institute of Aeronautics and Astronautics for his development of hybrid and solid propellants. (*Av Wk*, 11/20/72, 11; AIAA PIO)

- Neil H. McElroy—Secretary of Defense from 1957 to 1959, during President Eisenhower's second term—died of cancer in Cincinnati, Ohio, at age 68. He had taken office five days after U.S.S.R. launched *Sputnik 1* (Oct. 4, 1957), in midst of controversy between branches of U.S. armed services over responsibility for missile and satellite programs. McElroy, former executive of Procter & Gamble Co., ordered no cuts to be made in $1-billion budget that had been approved by Congress for basic research. Research led to successful launchings in 1958 of Navy and Army satellites that helped U.S. to recover some prestige in space field. McElroy resigned in 1958 to return to private industry after receiving Medal of Freedom from President Eisenhower. (W *Star*, 12/1/72, B7)

During November: Space shuttle impact was described by Dr. Wernher von Braun, Vice President for Engineering and Development with Fair-

child Industries, Inc., and former NASA Deputy Administrator for Planning, in *American Scientist* magazine article: Shuttle was "key to future exploration and use of space" that represented "significant new operational capability rather than a discrete mission or missions. Its ultimate impact will be to expand space activity while at the same time reducing the high costs of the space effort." Shuttle combined in one transport "capability to launch or carry out unmanned or manned missions repeatedly and routinely. It will be the first true 'aerospace' vehicle in that it can fly and maneuver in both space and the atmosphere." Shuttle would contribute more than any previous rocket vehicle to NASA's progress in advancing technologies and capabilities for space flight and exploration, space science and study of aerospace phenomena, and expanded use of space and space technology for human benefit. "To achieve the smoothest, most economical progress and benefits at lowest cost to the taxpayers, all three objectives should go forward in fairly close concert, not piecemeal or with fluctuating budgetary peaks and valleys." (*Am Sci*, 11–12/1972, 730–8)
- DOD's *Tacsat 1* Tactical Communications Satellite ceased operations, after having served military service requirements for mobile communications in the Pacific since its launch Feb. 9, 1969. (*Av Wk*, 8/20/73, 21; *A&A 1969*)
- Space Science Awareness Month was celebrated by City of St. Louis, Mo., at instigation of Mayor Alphonso Cervantes. More than 100 space science lectures and demonstrations were presented in area to provide opportunity for teachers, students, and general public to understand space technology and its benefits to society. Eugene Hanses, education director of McDonnell Planetarium, was chairman of activities which included talks by Marshall Space Flight Center officials and display of moon rocks. (MSFC Release 72–145; *NASA Activities*, 12/15/72, 253)
- Aerospace opportunities for dirigibles existed in 1970s and 1980s, *Astronautics & Aeronautics* article said. Changes in economic and aviation worlds and progress in materials and engineering had renewed potential of rigid airship as passenger and cargo transport after 30-yr hiatus. Committee appointed in 1935 to examine overall airship record after loss of dirigible *Macon* had concluded: "On the whole . . . and with special reference to airships of the larger sizes, we believe that it is practicable to design, construct, and operate such airships with a reasonable assurance of safety." Authors saw use of short takeoff and landing (STOL) engineering in airship revival. "Innovations should probably first undergo test on a small experimental airstrip." Airship revival would "give impetus to technological innovation, one of the Federal government's highest priorities, and would offer opportunities for creative and original engineering to the industrial and academic worlds." (Morse *et al.*, *A&A*, 11/72, 32–40)
- *Pensée*, journal of Student Academic Freedom Forum, published first in series of 10 special issues on "Immanuel Velikovsky Reconsidered." Revolution in science and humanities inspired by Russian-born scientist-philosopher was "burning issue on campuses," *Pensée* said, but "most professional journals and learned societies continue their silence." *Pensée* publishers had seen "series of separate developments which reveal a gathering momentum behind the new assessment of Velikovsky's work." Among developments had been success of Velikovsky's Aug. 14

lecture at NASA's Ames Research Center and his participation in three-day symposium at Lewis and Clark College in Portland, Ore., Aug. 16–18. (*Pensée*, Fall, 1972)
- Mrs. Esther H. Goddard, widow of rocket pioneer Dr. Robert H. Goddard, received honorary Doctor of Humane Letters degree from Clark Univ. at inaugural meeting of Friends of the Goddard Library. (*Clark Now*, Winter 1972)

December 1972

December 1: *Intercosmos 8* was launched by U.S.S.R. from Plesetsk into orbit with 590-km (366.6-mi) apogee, 199-km (123.7-mi) perigee, 92.4-min period, and 70.9° inclination. Tass announced satellite would continue global research of ionosphere, including temperature and concentration of electrons, in cooperative program of socialist countries to study different cosmic ray particles and micrometeoroids which influenced weather, climate, radio communications, and biological processes on earth. Tass reported onboard scientific equipment and systems were functioning normally. *Intercosmos 8* carried equipment designed and built in Bulgaria, German Democratic Republic, Czechoslovakia, and U.S.S.R. Satellite reentered March 2, 1973. (GSFC *SSR*, 12/31/72; 3/31/73; FBIS–Sov, 12/1–2/72, L1; *SBD*, 12/2/72, 58)

- Apollo 17 Astronauts Eugene A. Cernan, Dr. Harrison H. Schmitt, and Ronald E. Evans passed final four-hour physical examination at Kennedy Space Center. Manned Spacecraft Center Deputy Director of Life Sciences for Medical Operations, Dr. W. Royce Hawkins, told press astronauts were "well rested, in good spirits, and physically in excellent shape." (Wilford, *NYT*, 12/2/72, 58)

- American Institute of Aeronautics and Astronautics announced election of 19 new Fellows, including Dr. James C. Fletcher, NASA Administrator; Bruce T. Lundin, Lewis Research Center Director; and Harris M. Schurmeir, Jet Propulsion Laboratory Project Manager. Fellows would be honored at annual banquet during Jan. 8–10, 1973, annual meeting and technical display. Dr. Fletcher had been cited for "pioneering work in missile guidance systems and for his farsighted administrative leadership of this nation's aeronautical and space research and development programs." Lundin citation was for "contribution to research in turbojet and ramjet engine technology and his management of many aircraft, missile and spacecraft research programs." Schurmeir was cited for "basic work in systems engineering for lunar and planetary spacecraft and in particular for his leadership of the Ranger photographic missions to the moon and the Mariner missions to Mars." Assistant Secretary of the Air Force Grant L. Hansen had also been elected AIAA Fellow, for "contributions to the electronic design of the Nike, Sparrow, Thor and Skybolt Systems, management and direction of the Centaur program and his perceptive leadership in military research and development." (AIAA Release)

- First results of new radar astronomy technique for accurately measuring lunar topography from earth were described in *Science* by Haystack Observatory astronomer Stanley H. Zisk. Earlier techniques had measured radar backscattering properties by analyzing lunar echo simultaneously in both time and delay Doppler frequency. New technique extended earlier technique to three-dimensional measurement using radio-interferometer receiver for lunar echo signals. Technique had been used to measure lunar area that included craters Ptolemaeus,

December 1

Alphonsus, and Arzachel and portions of Mare Nubium. Results had shown evidence of late episode of volcanism that had partially filled two craters through crustal fault of Imbrian origin. Other features coinciding with local gravitational anomalies could be correlated with flow events. *Science* cover photo of lunar topography showed floor of crater Ptolemaeus to be higher than large mare to its left. Lunar elevations were indicated by lighter shades of gray on black-to-white photo. (*Science*, 12/1/72, cover, 977–80)

- Major advance in research to control nuclear fusion for commercial energy was announced by Atomic Energy Commission. Experiments using adiabatic toroidal compressor (ATC) at AEC's Princeton Plasma Physics Laboratory had achieved, for first time, plasma density in range in which tokamak fusion reactor was expected to operate and had demonstrated new technique for heating tokamak plasma to thermonuclear temperatures. ATC was device for containing and heating plasma fuel to temperatures that would melt any other container. It was improved version of Soviet-developed tokamak containing and heating device that had been hailed internationally as promising approach to production of fusion power. Soviet results from T–3 tokamak experiment, published in 1969, had been confirmed by Princeton tokamak experiments in 1970. Dr. James R. Schlesinger, AEC Chairman, had said new ATC results "further brighten the prospect of realizing commercial fusion power in a tokamak confinement system." But AEC had "many years of hard work ahead" to develop concept into "practical power system." (AEC Release P–404; AEC PIO)

- Astronaut Thomas P. Stafford, Deputy Director of Flight Crew Operations at Manned Spacecraft Center, was promoted by Air Force to rank of brigadier general. At 42 he was youngest officer of flag rank in any U.S. service and third astronaut to attain flag rank. Alan B. Shepard, Jr., had been promoted to rear admiral by Navy in December 1971 and James A. McDivitt to brigadier general by Air Force in March. Stafford had been *Apollo 10* commander and *Gemini 6* pilot. (NASA Release 72–235)

December 2: U.S.S.R. launched *Molniya 1–22* communications satellite from Baykonur Cosmodrome to transmit radio communications and broadcast TV programs on the Orbita network to areas in north, Siberia, Soviet Far East, and Central Asia. Orbital parameters: 39 797-km (24 728.7-mi) apogee, 555-km (344.9-mi) perigee, 65° inclination and 11-hr 57-min period. (GSFC *SSR*, 12/31/72; Tass, FBIS–Sov, 12/4/72, L1; *SBD*, 12/5/72, 143; 12/14/72, 197)

- Apollo 17 astronauts flew proficiency runs in NASA T–38 jet aircraft from Patrick Air Force Base, Fla., while at Kennedy Space Center five mice were selected to fly on mission experiment. Mice had been selected from 40 small rodents brought from California desert as candidates for mouse-to-the-moon mission. They would remain in orbit around moon when lunar module descended to surface. (AP, B *Sun*, 12/3/72)

December 2–3: Chicago Daily News editorial commented on impact of Apollo program: "The costly moon race may have helped distort American thinking in a damaging way. If we can go to the moon, so the slogan ran, we can do anything—wipe out poverty, eradicate racism, guide the economy, cure every evil of society. The euphoria created by the burgeoning space program and the crowning successes of the landing may have contributed to the rise of expectations in other fields beyond any solid hope of their realization. Science and technology can do wonders, but social problems do not yield so readily to electronics or telemetry." (*C Daily News*, 12/2–3/72)

December 3: Kennedy Space Center technical writers, members of International Alliance of Theatrical and Stage Employees, agreed to accept 44% pay increase, ending threat of strike that might have delayed Apollo 17 launch [see Nov. 5]. (Auerbach, *W Post*, 12/4/72, A20)

- NASA controllers expected *Pioneer 10* (launched March 2) to pass behind Jupiter satellite Io because of Sept. 19 course correction. Brief firing of spacecraft's thrusters had increased velocity by 0.227 mps (0.745 fps) so that *Pioneer 10* would arrive at Jupiter 17.2 min earlier than originally planned. Spacecraft was expected to make its closest approach to planet at 9:23.5 pm EST Dec. 3, 1973, when Io would pass between spacecraft and earth. Io would then be about 531 700 km (330 400 mi) from spacecraft. Scientists could measure Io's atmosphere, if any, as *Pioneer 10*'s radio signals passed through it. (NASA Release 72–231)

- Apollo 17 launch date postmarks would be available from Kennedy Space Center and Manned Spacecraft Center, Washington *Sunday Star & Daily News* reported. Deadline for obtaining splashdown postmarks from recovery ship U.S.S. *Ticonderoga* had passed. Titusville, Fla., Post Office would accept covers to be postmarked at KSC for 18-hr period beginning 8 am Dec. 6. (Fairies, *W Star & News*, 12/3/72)

- *New York Times* said it had asked more than 20 scientists and scholars with no part in space program to assess probable place of space exploration in "broad sweep of history and in the evolution of man and man's perception of himself and his universe." British historian Arnold J. Toynbee and American anthropologist Dr. Margaret Mead had seen space experience as "turning point in human history." Univ. of Michigan biophysicist Dr. John R. Platt had said that "the great picture of earth taken from the moon is one of the most powerful images in the minds of men today and may be worth the cost of the whole Apollo project." He saw photo as "a great landmark in exploration—to get away from the earth to see it whole." Soviet poet Robert Rozhdestvensky had said: "Man has become . . . more perceptive to what worries other people. He has no possibility of hiding away . . . of escaping reality and shutting himself up in the narrow world of his own." French anthropologist Dr. Claude Levi-Strauss had said: "I never look at TV except when there's a moon shot, and then I am glued to my set." Dr. Garrett Hardin, ecologist of Univ. of California at Santa Barbara, had said it was time to pull back from further manned space exploration for several centuries. Space program "can only be called 'spiritual' in both the good and bad senses." Technological spinoffs of space program "could surely have been achieved with an expenditure of much less."

Dr. Daniel J. Boorstin, Director of Smithsonian Institution's National Museum of History and Technology, had said greatest thing about space exploration was "that we don't know what its payoff will be. This symbolizes the American civilization. The people who settled America had no idea what the payoff would be. They settled it before they explored it." Nobel Prize winning biologist Dr. Albert Szent-Gyorgyi had said Apollo flights demanded that word "impossible"

December 3

be struck from scientific dictionary. "They are the greatest encouragement for the human spirit." Oak Ridge National Laboratory Director, Dr. Alvin M. Weinberg, said successes in space and nuclear energy had given people "the feeling that essentially any technological problem can be solved. This has created within the technological community and within the public at large a kind of technological euphoria." (Wilford, *NYT*, 12/3/72, 1)

- Apollo program accomplishments were praised by Dr. Wernher von Braun, Vice President for Engineering and Development with Fairchild Industries, Inc., and former NASA Deputy Associate Administrator for Planning, in *New York Times* article: "Two developments stand out in my view: (1) The remarkable advance in sophistication of flight techniques and capabilities of the crews, and (2) . . . the eye opening increase in both the extent and results of the scientific activities of manned lunar exploration." After completion of Apollo 17 mission would come realization "that we have only begun the exploration of the moon. As the analysis of Apollo lunar data and specimens proceeds, it will dawn on us what an enormous treasure trove awaits mankind just offshore from our planet. It will also occur to the scientific community that a base on the moon is just as important as a base in the Antarctic." (*NYT*, 12/3/72, 68)

- Astronauts' image was discussed in *New York Times Magazine* article. Of 73 pilots and scientists selected for astronauts corps only 39 remained active, and corps faced future cuts. Three astronauts had died in Jan. 27, 1967, Apollo spacecraft fire; five others in aircraft and auto accidents. Of 25 who had left corps, "many have followed the conventional paths of glory into business and Government." But "in the behavior of others there is more than a touch of the eccentric, and a large dose of trouble, almost a mythological element: the wandering hero back among his tribe, after stealing the sacred fire and grappling with terrifying demons, condemned to ask tough questions." World might regard some astronauts as "Byrds and Lindberghs of our time." As space program emphasis shifted, "they may be an endangered species. For the hotshot pilot seems destined to be replaced in the space stations of the near-future with scientists given training of shorter duration for their flight duties." Astronauts had been "wandering heroes who tried to unite us with their vision of earth as a small ball floating in an oceanic universe, with all men as passengers. But they weren't really poets, and they flew at a time of bitter earthly quarrels, with many resentful of the price tag on their odyssey." Like most heroes, "they were magnified versions of all of us, men of great daring and courage whose competitiveness was no doubt fierce, whose outlook was often corporate and even corny, whose ingenuousness was at times profound. But their image as a comforting symbol to Middle America did them an injustice." (Muson, *NYT Magazine*, 12/3/72, 37, 142)

- Economists agreed that end of Vietnam war would have no major impact on U.S. economy, *New York Times* reported. Defense funds expected to be freed by coming of peace had already been spent as war wound down slowly. Many defense contractors in aerospace and communications expected to pick up business with war's end. Some of top 10 still had war-related business, but amount was small in relation to total sales. Many contractors were "seriously seeking to diversify into non-defense-related operations." (Shabecoff, *NYT*, 12/3/72)

- Columbia Univ. sociologist Dr. Amitai W. Etzioni suggested "most hopeful epitaph for Project Apollo" in *New York Times* article: "This was the last gasp of a technologically addicted, public-relations-minded society, the last escapade engineered by an industrial-military coalition seeking conquests in outer space, while avoiding swelling needs on earth." (*NYT*, 12/3/72, 64)

- London *Sunday Times* article said man was withdrawing from lunar exploration, "not because of the hostility of the lunar surface or of interplanetary space, but because of indifference back at home." U.S. had "vindicated her technological honor following the deep felt humiliation of being beaten into space by the Russians: the payoff in terms of international prestige has not come up to expectations; the hope that space had vast military potential has proved an illusion." (Silcock, London *Sunday Times*, 12/3/72, 33)

- Newspaper editorials commented on U.S. space program.

  Baltimore *Sun* on post-Apollo space program: "While it is tempting to contemplate what improvements could be financed here on earth from the billions being spent in space, it is well to bear several factors in mind. One is the decline in space spending from a $5.25 billion peak in 1965 to the present plateau of $3.35 billion a year. Another is the Nixon administration's decision to proceed cautiously on the rush-rush manned Mars expedition advocated by Vice President Agnew three years ago. Still another is the American-Soviet agreement on a rendezvous-and-docking mission for mid-1975, a concept once relegated to political science fiction. Finally, there is the very real prospect that experiments and programs conducted from space shuttles will be of use in improving the environment on earth. More accurate weather forecasting, air pollution studies, better communications—these are some of the more obvious opportunities. These trends surely justify reasonable financial support to continue a desired momentum in space technology." (*B Sun*, 12/3/72)

  New York News on last Apollo mission: "The Apollo program is being phased out before its time, a victim of what we consider false economy, and the horizon is barren of any prospect for manned exploration beyond the boundaries of Earth orbit. Even unmanned probes will be few and far between as things look now." (*NY News*, 12/3/72)

  Chicago Tribune on future space program: Cost of Skylab, Viking, and space shuttle programs would be far less than amount put into space since 1958. "The national priorities have been reordered and, we believe, properly so. Future space exploration should be a part of, but should not dominate, our national effort to improve the lot of Americans." (*C Trib*, 12/3/72)

December 4: Technicians at Kennedy Space Center's Apollo 17 launch site loaded helium used to pressurize lunar craft's descent rocket system. Liquid hydrogen and oxygen were pumped into command module fuel cell system, source of spaceship's electricity and water supply. Countdown, which began Nov. 30, continued on schedule with liftoff scheduled for 9:53 pm EST Dec. 6. (Wilford, *NYT*, 12/5/72, 1)

- Mariner Mars '71 extended mission of 259 days beyond primary mission was adjudged success. Objectives of extended mission of *Mariner 9* (launched by NASA May 30, 1971, and inserted in Mars orbit No. 13, 1971) had been to map 70% of planet with 1-km (0.6-mi) visual resolution and to study dynamic characteristics of Mars at reduced data-acquisition rate. During extended mission *Mariner 9* had mapped 100% of planet and had observed large areas with infrared and ultraviolet instruments. S-band occultations and celestial mechanics measurements had proved Mars to be geologically and meteorologically active. (NASA proj off)
- Dr. James C. Fletcher, NASA Administrator, described major alterations in "nature of the space effort" to be expected with 1978 introduction of space shuttle, in signed *New York Times* article. National space program had "turned an important corner in 1972—a year that, in a very real sense, was a year of decision" in choosing to proceed with shuttle. U.S. had "turned from a period of space exploration to a period of space exploitation for practical purposes."

    By 1978 it would be "realistic to consider a variety of missions either not practicable at present cost levels or that were beyond the state of the art as it was in the pre-shuttle era." Missions would include advanced, space-based communications systems that, when used for education, would "have the potential for eliminating illiteracy from the face of the earth." Shuttle-launched advanced-version Nimbus satellites, combined with weather-modeling ground computers, would permit experiments in weather modification to dampen violent storms. Earth resources program data would provide precise picture of crop and timber conditions and extent of attacks by disease, drought, and fire. By 1980s, pilot space manufacturing operations might be in progress. "Super pure vaccine could thus be produced. Perfectly homogeneous alloys, fiber composites and perfectly round ball bearings are other possibilities." During 1980s, "program of planetary exploration should record the completion of a close look at all the planets in our solar system as well as a detailed study of Mars, Jupiter and Venus. Large orbiting telescopes will be aimed at the outer reaches of the universe, examining such phenomena as quasars and pulsars." (*NYT*, 12/4/72, 50)
- Term "post-Apollo" had "outlived its usefulness," Dr. James C. Fletcher, NASA Administrator, said in message to staff. "When plans for this decade were still uncertain it made sense to refer to them as post-Apollo plans. But today we are advanced with new programs which stand on their own merits." Skylab, space shuttle, Viking, High Energy Astronomy Observatory (HEAO), and Pioneer and Mariner missions to planets drew heavily on technology of 1960s, "but they take this technology giant steps into the future." Dr. Fletcher suggested names for future years: 1973 would be "Year of Skylab" and of *Pioneer 10*, 1975 would be year of Apollo-Soyuz rendezvous, in 1976 Viking would land on Mars, and in 1978 first space shuttle orbital flight was scheduled. "Each of these events will make a memorable page in the story of human progress. Each of them will contribute importantly to mankind's knowledge of our planetary environment." (*NASA Activities*, 12/15/72, 244–5)

- Columbia Broadcasting System said it had rejoined original TV pool arrangement to cover Apollo 17 launch Dec. 6. Under arrangement, National Broadcasting Co. would act as pool agent for itself, CBS, and American Broadcasting Co. in relaying close-up TV photos of launch from NASA cameras. (AP, *W Post*, 12/5/72, A3)
- S.S. *Statendam* of Holland-American Line sailed from New York with about 100 passengers for cruise advertised as "Voyage Beyond Apollo." Itinerary included Cape Kennedy for close-up view of Apollo 17 launch, scheduled for Dec. 6. Passengers included novelists Norman Mailer and Katherine Anne Porter, science writer Isaac Asimov, and Northeastern Univ. physicist Dr. Robert D. Enzmann. Dr. Wernher von Braun—Fairchild Industries, Inc., Vice President and former NASA Deputy Associate Administrator—and former astronaut Edgar D. Mitchell and science fiction writer Arthur C. Clarke, who had been advertised as participants in space seminars aboard ship, did not appear. (Buckley, *NYT*, 12/12/72, 43)
- Environmental Protection Agency issued report on effect of aircraft emissions and proposed standards to limit emissions from aircraft and aircraft engines. Report, *Aircraft Emissions: Impact on Air Quality and Feasibility of Control*, said aircraft would continue to be major source of air pollution until 1980 even though 90% of U.S. jet transport aircraft had been fitted with engines that produced less emission fumes. New, more powerful engines to be used in future would increase nitrogen oxide level in air by 33% in next 10 yrs. Proposed regulations would limit exhaust emission from new and in-use gas turbine engines and aircraft, prohibit fuel venting from gas turbine engines, set exhaust and crankcase emission standards for new piston engines, and establish test procedures to determine compliance. (Text; EPA Release)
- Environmental Protection Agency was considering borrowing 12 Army helicopters to evaluate effectiveness of rotary wing aircraft in gathering evidence of waterway pollution, *Aviation Week & Space Technology* reported. (*Av Wk*, 12/4/72, 9)

December 4–15: Apollo-Soyuz Test Project (ASTP) Working Group 3 met in Moscow to work out details of July 1975 U.S.–U.S.S.R. joint mission. Group completed tests of U.S. and Soviet two-fifths-scale model, defined location of interface sensors and force and stroke necessary for actuation, defined surfaces requiring dry film lubrication, exchanged interface seal samples for unilateral testing, and defined load and thermal interaction requirements for docking systems. (NASA proj off)

December 5: President Nixon telephoned Apollo 17 crew at Kennedy Space Center to extend best wishes on eve of Apollo 17 mission. (*PD*, 12/11/72, 1735)
- Washington *Evening Star and Daily News* editorial commented on the last Apollo mission: A lot was riding on Apollo 17 "in its attempt to unlock the moon's most ancient secrets and thereby perhaps the earth's. Those three astronauts, with their record-size cargo of instruments, have the trickiest landing and the hardest lunar tasks in this whole program ahead of them." (*W Star & News*, 12/5/72, A12)
- NASA released report of *Explorer 46* Meteoroid Technology Satellite (MTS) Review Committee on failure of panels in bumper penetration experi-

December 5

ment to deploy following Aug. 13 launch. Partial instead of full extension of one pair of MTS bumper wings had been caused by desynchronization between deployment actuators for that wing pair. Experiment continued to provide good data on effectiveness of bumpers to protect against meteoroids in space, with 17 penetrations recorded to date. One secondary experiment, to measure impact flux of small mass meteoroids, had accomplished objective in detecting about 2000 micrometeoroids in near-earth space. Remaining secondary experiment, to measure velocities of meteoroids striking space vehicles, would require several months of successful operation to complete objective. (NASA Release 72–238)

- Further space budget cuts were forecast by Senate Finance Committee Chairman, Sen. Russell B. Long (D-La.), during Lafayette, La., press conference. He said President was trying to pare Federal budget to "reasonable limits." There was "a lot of appeal to making some reductions in the space program." (Sen Long Off; UPI, *W Post*, 12/7/72)
- U.S. communications firms announced agreements with Telesat Canada for use of Canada's satellite system to provide communications service in U.S. Chairman Robert W. Sarnoff of RCA said RCA would lease one full-time channel and occasional time on second channel. Subject to approval by U.S. and Canadian governments, RCA Global Communications would install earth stations in Washington–New York corridor, California, and Alaska to relay voice, message, and TV traffic between East and West Coasts and between both coasts and Alaska at reduced costs. American Satellite Corp. announced preliminary agreement with Telesat Canada for lease of up to three satellite channels to provide U.S. domestic communications market with private-line channels for voice, data facsimile, TV programming, and interconnections of cable TV systems and private and industrial video service. (RCA Release 12/5/72; AmSatCorp Release FL–9–240)
- NASA announced personnel actions: L/G Marvin L. McNickle (USAF, Ret.) was named Special Assistant to Associate Administrator for Manned Space Flight at NASA Hq. effective Dec. 3. L/G Carroll H. Dunn (USA) was elected Chairman of Aerospace Safety Advisory Panel to succeed Dr. Charles O. Harrington, who was retiring. (NASA Release 72–237; NASA Ann)
- Two sounding rockets were launched by NASA from Wallops Station.

  Nike-Apache carried Univ. of Illinois payload to 185-km (115-mi) altitude in first of three launches to measure ionospheric properties on three anomalous days: (1) "L," day with very low electron densities in 70- to 80-km (40- to 50-mi) region with 10-day minimal magnetic storm activity proceeding; (2) "H_1," day with very high electron densities with same altitude and magnetic storm prerequisites; and (3) "H_2," day with very high electron densities in same region with two or three days of major magnetic disturbance and magnetic storm aftereffects. Nike-Apache was fired on L day and measured electron densities, temperatures, and collision frequencies. Rocket and instrumentation performed satisfactorily.

  Super Arcas launched 44 min after Nike-Apache carried Pennsylvania State Univ. payload to 85-km (52.8-mi) altitude to collect data on positive and negative ion conductivities. Rocket and instrumentation performed satisfactorily. (NASA Rpts SRL)

- Secretary of Transportation John A. Volpe announced new air transport security program to protect air passengers against hijackers. Program was being implemented at President Nixon's direction. Security procedures required stationing armed local law enforcement officers at airport passenger check points during boarding periods, electronic screening of all passengers by airlines before boarding, and inspection by airlines of all carry-on items accessible to passengers during flight. (DOT Release 103-72)
- Sprint antiballistic missile successfully intercepted target cone over Pacific in DOD test of capability of missile site radar. Radar would be installed at U.S. ABM sites to launch and guide Sprint from site some distance from radar to long-range, low-altitude interception. (Reuters, B *Sun,* 12/7/72, A11; DOD PIO)
- International aspects of future space programs were discussed by Crosby S. Noyes in Washington *Evening Star and Daily News* column. Scientific and engineering aspects of 1975 joint U.S.-U.S.S.R. Apollo-Soyuz mission were "minimal compared to the psychological switch which the effort represents." Project had "most profound political implications for the future." U.S.S.R. participation in space shuttle development might raise "touchy questions" because of "obvious military aspects of the program." In distant future efforts, including possible manned flights to other planets, "it is . . . evident that international cooperation on a much wider scale will be essential. What the Apollo program has demonstrated is that manned space flight is an impractical proposition for any single nation to undertake. Which may turn out to be the most valuable discovery of all." (W *Star & News,* 12/5/72)

December 6: NASA's M2-F3 lifting body, piloted by Cecil Powell, completed 25th flight from Flight Research Center after air-launch from B-52 aircraft. Objectives of flight were to evaluate command augmentation system; to obtain stability and control data at mach 1.1, mach 0.8, and mach 0.7; and to check out pitch-reaction augmentation system. All objectives were achieved. M2-F3 reached 19 800-m (65 000-ft) altitude and mach 1.3. (NASA proj off)

- International Business Machines Corp. outlined variety of computers used in launching and tracking manned spacecraft from Kennedy Space Center for Judge David N. Edelstein, who was hearing Justice Dept. antitrust suit against IBM [see Nov. 23]. Judge and party would leave KSC for Goddard Space Flight Center Dec. 7 to examine computer operations there. (*Today,* 12/7/72)
- Newspaper editorials commented on close of Apollo program with Apollo 17 launch.

    Los Angeles Times: "Thanks to the marvels of satellite communications, 3,000 European doctors were able recently to 'attend' a medical meeting in San Antonio via television. A computer designed for the space agency is helping Detroit design safer steering linkages in cars and trucks. A technique for clarifying spacecraft photos of the moon is being used by hospitals to provide much sharper x rays. The thousands of men and women who have participated in the Apollo program, and the American people who supported it, have all cause for great pride in Apollo. It has been one of the best investments this country ever made, one of the most rewarding adventures it ever undertook." (*LA Times*, 12/6/72)

Christian Science Monitor: "To the layman, what has been learned about the moon and space has been an extension of our knowledge of the universe, not a radical departure. In this light, it is comforting to think that the Apollo program, remarkable though it has been historically, may illustrate that even the farthest reaches of today's unknown may prove to be no more than the captive of tomorrow's familiar." (*CSM*, 12/6/72)

New York Times: "Even as the world awaits this final launch, debate continues about the wisdom of the entire Apollo project. Some consider the moon landings the most brilliant scientific achievement in history; others still look at the whole venture as a waste of resources needed for urgent requirements here on earth. Yet one fact about the venture is beyond dispute: in the years 1969-72 men landed on another celestial body for the first time and showed they could live and work in the bizarre and literally inhuman conditions on that foreign planet. Long after most other developments of the twentieth century are forgotten, future generations will recall this as the century in which men broke the bonds of terrestrial gravity and began their cosmic destiny." (*NYT*, 12/6/72, 42)

December 7-19: Apollo 17 *lifted Astronauts Eugene A. Cernan, Ronald E. Evans, and Dr. Harrison H. Schmitt toward the moon on the last lunar landing mission in* NASA's *Apollo program. Light from the Saturn V launch vehicle's first stage was visible from Kennedy Space Center Launch Complex 39, Pad A, as far as South Carolina and Cuba in the first nighttime Apollo launch, starting off the 12½-day mission.*

December 7–19: Apollo 17 (AS–512), sixth and last successful NASA manned lunar landing mission, carried three-man crew—including first scientist-astronaut in space—to moon's Taurus-Littrow region after first nighttime launch. Lunar module *Challenger* (LM–12) landed on moon Dec. 11 and two astronauts deployed experiments, rode lunar roving vehicle (LRV), and explored lunar surface during 75-hr stay. *Challenger* rejoined orbiting command and service module *America* (CSM–114) in lunar orbit Dec. 14 for safe return to earth with record 115 kg (250 lbs) of lunar samples Dec. 19.

December 7–10: Launch from Kennedy Space Center Launch Complex 39, Pad A, by Saturn V launch vehicle was delayed 2 hrs 40 min. Automatic cutoff at T minus 30 sec in countdown required recycle and hold at T minus 22 min; additional hold was called at T minus 8 min. Holds were caused when terminal countdown sequencer failed to command pressurization of S–IVB liquid-oxygen tank. Tank was pressurized manually, but failure to arm tank's pressurized interlock prevented continuation of launch events until interlock was bypassed by jumper. Workaround was first thoroughly analyzed at Marshall Space Flight Center. Investigation later indicated cause of failure was defective diode on printed circuit card in sequencer.

Spacecraft—carrying Astronauts Eugene A. Cernan (commander), Ronald E. Evans (command module pilot), and Dr. Harrison H. Schmitt (lunar module pilot and geologist)—was launched at 12:33 am EST Dec. 7, with illumination from S–IC 1st stage creating brilliance of morning sun. Liftoff, witnessed by crowd estimated at 500 000 and by nationwide and foreign TV viewers, was visible to naked eye as far as South Carolina to north and Cuba to south.

Spacecraft and S–IVB combination entered parking orbit with 170-km (105.6-mi) apogee and 168-km (104.4-mi) perigee. Following checkout, spacecraft was inserted into trajectory for moon at 3:13 ground elapsed time (GET) as planned. CSM separation from LM/S–IVB/IU at 3:42 GET and docking with LM at 3:57 GET were shown on TV. During docking possible ring latch malfunction was indicated. Troubleshooting revealed latches 7, 9, and 10 were unlocked. Latch 10 was locked by pushing on handle; latches 7 and 9 were locked and manually fired to lock handles. Following hatch replacement, CSM/LM combination was ejected from S–IVB stage at 4:45 GET. S–IVB lunar impact maneuver targeted stage to strike moon's surface Dec. 10.

Midcourse correction (MCC–1) was not necessary because of nominal trajectory. MCC–2 maneuver was performed on time at 35:30 GET, changing velocity 3 m per sec (9.9 fps). Cernan and Schmitt entered LM at 40:10 GET and discovered docking latch 4 improperly latched. Evans moved latch handle 30°–45°, disengaging hook from docking ring. Mission Control and flight crew decided to wait until second LM entering at 59:59 GET for further action on latch. Remainder of LM housekeeping was nominal and LM was closed out at 42:11 GET.

Heat flow and convection demonstrations were conducted as planned; results were satisfactory. MCC–3 and –4 were not required because of near-nominal trajectory. LM housekeeping was resumed at 59:59 GET and completed at 62:16 GET with all systems nominal. Evans, following ground instructions, stroked docking latch 4 handle and succeeded in cocking latch. Latch was left in cocked position for CSM–LM rendezvous.

December 7–19

December 7–19: *Astronaut Eugene A. Cernan checked out the lunar roving vehicle in the first Apollo 17 extravehicular activity Dec. 11, after touchdown at the moon's Taurus-Littrow site. The stripped-down Rover was photographed by Scientist-Astronaut Harrison H. Schmitt before loadup with television assembly, communications equipment, lunar tools, and scientific gear. The South Massif was in the right background.*

Planned trajectory was modified continually because of late liftoff; coast toward moon was accelerated to ensure arrival of spacecraft at lunar orbit insertion (LOI) at scheduled GMT time. GET clock was updated 2 hrs 40 min at 65:00 GET to place all events on original flight plan schedule. Crew began one-hour visual-light-flash phenomenon observation at 68:19 GET and reported seeing bright to dull light flashes. Cernan reported continuing gas pains during private consultation via separate radio link with Dr. W. Royce Hawkins, Manned Spacecraft Center Deputy Director of Life Sciences for Medical Operations. He was told to continue taking antigas pills and to alter diet. Crew jettisoned scientific instrument module (SIM) door at 84:12 GET and said SIM bay looked good.

December 10–16: Apollo 17 entered lunar orbit with 315-km (195.7-mi) apolune and 97-km (60.3-mi) perilune after 398-sec LOI burn at 88:54 GET (2:47 pm EST Dec. 10). Cernan announced: "America has arrived on station for the challenge ahead."

S–IVB 3rd stage impacted on lunar surface at 89:39 GET (3:32 pm EST Dec. 10) at 4° 12′ south latitude and 12° 18′ west longitude. Impact was recorded by *Apollo 12, 14, 15,* and *16* seismometers.

Evans described crew's excitement at reaching vicinity of moon: "We're breathing so hard, the windows are fogging up." Schmitt interrupted running commentary on lunar topography with shout: "Hey I just saw a flash on the lunar surface!" When assured by Mission Control that impact of possible small meteor would have been masked

December 7-19: *Geologist-Astronaut Schmitt worked beside a huge split boulder at the base of North Massif on the lunar surface during EVA-3, photographed by Cernan. Above, Schmitt positioned a gnomon (a local vertical and color standard). Below, he chipped a sample from the other end of the rock, with the front of the Rover appearing at the left. Samples on this last Apollo EVA included blue-gray breccias, fine-grained basalts, crushed anorthositic rock, and soils. Meanwhile Ronald E. Evans—continuing to orbit the moon in the command and service module America—identified a series of volcanic domes in Aitken Crater on the moon's far side during EVA-3.*

on seismometers by S-IVB 3rd-stage impact, Schmitt groaned: "Just my luck." Cernan—who had orbited moon during May 18–26, 1969, *Apollo 10* mission—described sight as "still just as impressive."

Descent orbit insertion (DOI–1) was performed at 90:31 GET. CSM and LM undocked and separated on schedule at 107:48 GET. Circularization maneuver placed CSM in orbit with 129-km (80.2-mi) apolune and 100-km (62.1-mi) perilune.

DOI–2, at 109:50 GET, inserted LM into orbit with 111-km (69.0-mi) apolune and 11-km (6.8-mi) perilune. Powered descent was initiated and *Challenger* touched down on Taurus-Littrow site at 113:02 GET (2:55 pm EST Dec. 11).

Cernan descended LM ladder four hours later, at 117:02 GET, and said: "As I step off at the surface at Taurus-Littrow, I'd like to dedicate the first step of *Apollo 17* to all those who made it possible." Schmitt, first geologist to walk on moon, followed and asked, "Hey, who's been tracking up my lunar surface?" Cernan said surrounding hills looked like "wrinkled skin of an old, old, 100-year-old man."

Astronauts deployed lunar roving vehicle. Cernan made short test drive and said, "Hallelujah, Houston! *Challenger*'s baby is on the roll."

As Cernan and Schmitt set up TV equipment, Cernan called to Schmitt: "Hey, Jack, just stop. You owe yourself 30 seconds to look up over the South Massif and look at the Earth." Schmitt replied: "What? The Earth? You seen one earth, you've seen them all." Cernan unfurled U.S. flag on lunar surface while Schmitt took his picture. Cernan said, "It's got to be one of the most proud moments of my life." Flag had "flown in the MOCR [Mission Operations Control Room] since *Apollo 11*. And we very proudly deploy it on the Moon, to stay for as long as it can, in honor of all those people who have worked so hard to put us here and to put every other crew here and to make the country, the United States and mankind, something different than it was."

While preparing to traverse to Apollo lunar surface experiment package site, Cernan inadvertently knocked extension off of LRV right rear fender and repairs were made with tape. Astronauts crossed to ALSEP site, where they deployed ALSEP. Color TV camera beamed live telecast to earth.

LRV fender again fell off during traverse to Steno Crater, lunar station 1A. Crew deployed surface electrical properties experiment and explosive packages during 7-hr 12-min lunar surface exploration. First EVA ended at 124:13 GET.

Second EVA began 1 hr 20 min late, at 140:35 GET (6:28 pm EST Dec. 12). Cernan and Schmitt, on instructions from Mission Control, improvised replacement for lost LRV fender extension. Four chronopaque maps were taped together and held in place by two portable utility lamp clamps. "Call me the little old fender maker," Cernan said.

Cernan and Schmitt visited lunar stations 2, 2A, 3, 4, and 5. They deployed explosive packages, obtained photos, and collected and documented soil samples. At station 4, Shorty Crater, Schmitt shouted: "There is orange soil! It's all over! Orange!" Cernan confirmed finding: "Jack that is really orange. It's been oxidized." Schmitt dug trench and extracted sample which he described to Mission Control: "An essential portion of the zone . . . actually has a crimson hue, or

December 7–19: *Astronaut Evans photographed the crescent earth from lunar orbit as it rose over the moon, after Astronauts Cernan and Schmitt had returned to the Apollo 17 command and service module America from their explorations of the lunar surface and before insertion into trajectory for home. The lunar module Challenger's ascent stage, which had lifted the astronauts from the surface to dock with the CSM in orbit, had been separated and nudged into its flight path for a crash on the moon. The lunar far side is in the foreground.*

red hue. Outside of that it's orange. And outside of that, it's gray." Scientists on ground immediately speculated that soil might have originated from volcanism on moon as recently as 100 million yrs earlier. If so, it would contradict "dead moon" theory that moon had always been cold and inert. Dr. Robin P. Brett, head of MSC Geochemistry Branch, told press, "What we have been witnessing may be one of the most significant finds of Apollo geology."

Schmitt revisited ALSEP site to verify that lunar surface gravimeter had been leveled properly. EVA-2 ended at 148:35 GET (2:05 am EST Dec. 13), after 7 hrs 37 min. Total distance covered during second EVA was 19 km (12 mi).

Third EVA began 50 min late at 163:35 GET (5:26 pm EST Dec. 13). Cosmic ray detector was retrieved before start of traverse to prevent exposure to low-energy solar protons from small solar flare. Cernan and Schmitt took black and white and color photos and collected 66 kg (145.9 lbs) of samples. Samples included blue-gray breccias, fine-grained vesicular basalts, crushed anorthositic rocks, and soils. EVA-3 ended at 170:48 GET (0:41 am EST Dec. 14), after 7 hrs 16 min. During EVA-3 Evans, orbiting in CSM 113 km (70 mi) above lunar surface, identified series of volcanic domes in Aitken Crater, on moon's far side.

Before entering *Challenger* for last time, Cernan and Schmitt stepped before TV cameras. Cernan said: "To commemorate not just *Apollo 17*'s visit to the Valley of Taurus-Littrow but as an everlasting commemoration of what the real meaning of Apollo is to the world, we'd like to uncover a plaque that has been on the leg of our spacecraft." He read inscription: "Here man completed his first exploration of the Moon, December 1972 A.D. May the spirit of peace in which we came be reflected in the lives of all mankind." Dr. James C. Fletcher, NASA Administrator, congratulated astronauts from Mission Control at MSC and conveyed Godspeed message from President Nixon. As astronauts boarded LM, Cernan said: "I believe history will record that America's challenge of today has forged man's destiny of tomorrow. And, as we leave the moon at Taurus-Littrow, we leave as we came and, God willing, as we shall return, with peace and hope for all mankind. God speed the crew of *Apollo 17*."

Total time for three EVAs was 22 hrs 5 min 4 sec, during which Cernan and Schmitt traveled 35 km (22 mi) in LRV, collected 115 kg (250 lbs) of samples, and took 2120 photos. Good-quality TV transmission was received throughout EVAs.

Astronauts depressurized LM and discarded excess equipment. LM lifted off lunar surface at 188:02 GET (5:55 pm EST Dec. 14). TV

December 7–19: *Astronaut Evans retrieved film cassettes from the lunar sounder and mapping and panoramic cameras in three spacewalks to the scientific instrument module, as the Apollo 17 spacecraft coasted toward the earth on its return trip from the moon. The cylindrical object at Evans' left side is the mapping camera cassette.*

cameras in CSM provided excellent picture in Mission Control as LM approached. Cernan in LM said, "Good to see you . . . . It's been a good trip." Evans replied, "Good to have you all back up here." Observing that *"America* and *Challenger* are . . . good tight Navy formation," Cernan told Evans, "Command module looks as good as the day they put it on the pad." Evans answered, "And, you know, so does *Challenger."*

CSM and LM docked in lunar orbit at 190:17 GET and Cernan and Schmitt transferred to CSM with samples and equipment. LM ascent stage was jettisoned as planned. Impact of stage on lunar surface at 195:57 GET was recorded by four *Apollo 17* geophones and *Apollo 12, 14, 15,* and *16* ALSEPs. Explosive packages were detonated and events were picked up by lunar seismic profiling geophones. Flash and dust from EP 7 explosion were seen on TV.

Insertion into trajectory for earth began at 236:42 GET, after CSM's 147 hrs 48 min in lunar orbit.

December 17-19: Spacecraft left moon's sphere of influence at 250:40 GET, traveling at 1173 m per sec (3851 fps). Evans left CSM at 257:34 GET (3:27 pm EST Dec. 17) for 1-hr 7-min inflight EVA to retrieve lunar sounder film and panoramic and mapping camera cassettes in three trips to SIM bay. TV pictures relayed to Mission Control showed him cavorting in weightlessness in stiff white pressure suit and attached to 7.6-m (25-ft) line. Following space walk, astronauts settled down for final two days of mission.

CM separated from SM at 304:04 GET, 15 min before entry interface at 121 920 m (400 000 ft). Drogue and main parachutes deployed normally and CM splashed down in mid-Pacific 6.4 km (4.0 mi) from prime recovery ship U.S.S. *Ticonderoga* at 304:31 GET (2:25 pm EST Dec. 19).

Recovery helicopter dropped swimmers, who installed flotation collar and attached life raft. Astronauts were transported to recovery ship for postflight examination. They would be flown to Houston via Samoa the following day. CSM was retrieved by recovery ship for transport to San Diego, Calif.

Apollo 17 achieved primary objectives—to make selenological survey and sampling of materials and surface features in preselected area of Taurus-Littrow region, emplace and activate surface experiments, and conduct inflight experiments and photography. Launch vehicle and spacecraft system performances were near nominal throughout mission. Only minor discrepancies occurred, with no effect on safety or mission objectives. Flight crew performance was good. Mission was judged officially to be "a great success."

Accomplishments included sixth manned lunar landing and return, first geologist-astronaut on lunar surface, longest lunar surface stay time (74 hrs 59 min 38 sec), longest single lunar surface EVA (7 hrs 37 min 22 sec), longest total lunar surface EVA time (22 hrs 5 min 4 sec), longest total lunar distance traversed with LRV (35 km; 22 mi), longest Apollo mission (12 days 13 hrs 51 min), most samples returned (115 kg; 250 lb), and longest time in lunar orbit (147 hrs 48 min).

Apollo 17 was 14th and last scheduled mission in Apollo series and 11th manned Apollo mission. *Apollo 16* had flown April 16-27. Highlights of Apollo program had been first manned orbit of moon by *Apollo 8* (Dec. 21–27, 1968), first landing of men on moon during *Apollo 11* (July 16–24, 1969), and first use of lunar roving vehicle on moon during *Apollo 15* (July 26–Aug 7, 1971). Apollo program was directed by NASA Office of Manned Space Flight. Manned Spacecraft Center was responsible for Apollo spacecraft development, Marshall Space Flight Center for Saturn V launch vehicle, and Kennedy Space Center for launch operations. Tracking and data acquisition was managed by Goddard Space Flight Center under overall direction of Office of Tracking and Data Acquisition. (NASA proj off; NASA Release 72–220K; MSC Transcript 07629; *NYT*, 12/8–20/72; *W Post*, 12/8–20/72; NASA PIO)

December 7: President Nixon sent message to *Apollo 17* crew following successful launch: "With the final mission of the Apollo lunar exploration series man completes another step in his quest for knowledge of his universe and of himself. Those who come after will stand on the shoulders of the men of Apollo and their dedicated support team. I wish you good luck and Godspeed." (*PD*, 12/11/72, 1734–5)

- Scientist-Astronaut Harrison H. Schmitt was calmest of *Apollo 17* astronauts at launch time, United Press International reported. His heartbeat at liftoff was 115 beats per minute, with 130 for Astronauts Eugene A. Cernan and Ronald E. Evans. *Apollo 16* rates had been 108 for Astronaut John W. Young, 115 for Astronaut Thomas K. Mattingly II, and 130 for Astronaut Charles M. Duke. Launch Director Walter J. Kapryan had said *Apollo 17* crew was more tense because of liftoff delay. (*C Trib*, 12/8/72)

- Dr. James C. Fletcher, NASA Administrator, assessed Apollo program's meaning to U.S. in Associated Press interview published by *Today*: "Scientifically, it has allowed us to start obtaining a clear picture of what the moon is like, how it functions, its dynamics, how it has evolved and its relationship to the sun and our earth." After all Apollo data had been evaluated, "it may help us understand how our planet evolved from an uninhabited place." Apollo had "allowed us to see our earth as it really is" and "made a lot of people start talking about ecology and the need to preserve this fragile planet." When man first walked on moon "it gave Americans renewed confidence in themselves. They knew their country could really put together a complex program and do what it says it will do within the price it said it would cost." (Benedict, AP, *Today*, 12/7/72)

- Newspaper editorials commented on successful launch of *Apollo 17*:
  New York Times: "It is still a breathtaking concept, though five teams of astronauts have successfully accomplished similar feats since Apollo 11 made its historic breakthrough in July 1969. The near-catastrophe which forced abortion of the Apollo 13 mission provides a useful reminder of the dangers involved despite all the exquisitely painstaking care taken before blast-off." Astronauts had "special qualities which set them apart. Some day . . . taking a rocket to the moon will be as common as taking a plane to London is today. But the Apollo pioneers who have blazed the way had to have a special degree of competence and courage to embark on this extraordinary journey." (*NYT*, 12/7/72)

- President Nixon announced intention to nominate John A. Volpe to be U.S. Ambassador to Italy and Claude S. Brinegar to succeed Volpe as Secretary of Transportation. Nominations were sent to Senate Jan. 4. (*PD*, 12/11/72, 1735; 1/8/72, 12)

December 8: Soviet newspapers *Pravda* and *Komsomolskaya* carried detailed reports of *Apollo 17* progress and photos of astronauts on second day of moon-landing mission. *Pravda* story ended: "We wish them success." *Komsomolskaya* said mission was "one of the most daring measures ever carried out by mankind, and undoubtedly the most expensive experiment ever in the area of science and technology." *Washington Post* Foreign Service said "new tone in newspaper coverage" reflected high-level decision to cover some U.S. news more sympathetically in "period of Soviet-American detente." (Kaiser, *W Post*, 12/9/72, A3)

- United Kingdom and West Germany agreed to abandon plans for European launcher and to buy U.S. launch vehicles instead. Decision was made at meeting in Bonn between U.K. Aerospace Minister Michael R. D. Heseltine and West German Aerospace Minister Klaus von Dohnanyi. (Reuters, *NYT*, 12/9/72, C7)

- Proposed regulations to give industry access to Government technology for building plants to produce fissionable uranium were published in *Federal Register* by Atomic Energy Commission. *Washington Post* said next day that policy change was one of AEC's most important "since the dawn of the nuclear age." Government had previously kept monopoly on uranium plant building because of uranium's bomb-making potential. Dr. James R. Schlesinger, AEC Chairman, had said action was prompted by urgent need to plan atomic plants worth $15 billion or more within a few years because of U.S. energy shortage. (*Fed Reg*, 12/8/72, 26145; Cohn, *W Post*, 12/9/72, A3)

- Conclusive identification of absorptions caused by water frost in infrared reflectivities of Jupiter's Galilean satellites J II and J III was reported in *Science* by astronomers from Massachusetts Institute of Technology and State Univ. of New York at Stony Brook. Measurements had been made by rapid-scanning Fourier spectrometer on 150-cm (60-in) McMath solar telescope at Kitt Peak National Observatory. Percentage of frost-covered surface area had been determined at 50% to 100% for J II, 20% to 65% for J III, and possibly 5% to 25% for J IV. Leading side of J III had 20% more frost cover than trailing side, which explained visible geometric albedo differences between the two sides. Reflectivity of material underlying frost on J II, J III, and J IV resembled that of silicates. Surface of J I might be covered by frost particles much smaller than those of J II and J III. (Pilcher *et al.*, *Science*, 12/8/72, 1087–9)

- Vice Chairman Kuo Mo-jo of Standing Committee of National People's Congress of People's Republic of China told visiting French delegation to Peking that U.S.S.R. had begun underground tests of explosive devices in 250- to 300-megaton range Nov. 25. Tests, southeast of Ural Mountains, would continue to Dec. 30. He noted concomitance of European security talks in Helsinki, Finland, and of Dec. 6 *Apollo 17* launch. Kuo said both *Apollo 17* and Soviet nuclear tests showed technological supremacy with equal military potentialities. (Comparet, Agence France-Presse, *Atlanta JC*, 12/10/72, B1)

- Public attitude toward space program in late 1960s was described by Washington communications consultant Julian Scheer, former NASA

Assistant Administrator for Public Affairs, in *Washington Post* article. "There was a mixed feeling of pride and guilt . . . during those years." Program had moved forward as "almost a solitary symbol of something that was working. But the question of its rightful place in the scheme of things always hung over it." While "a cynic would conclude that the American people had turned from social problems," realist "might conclude that what we were hearing in the late 1960s were the first real rumblings against large federal programs. The space program, meanwhile, had gained its momentum while this disillusionment was setting in. We got to the moon, it seemed, almost without anyone knowing it. The program had the velocity—and there was no turning back." (*W Post*, 12/8/72, A26)

- *Houston Post* editorial commented on international aspects of space program: "The history of the world has provided few efforts more conducive to worldwide participation than space exploration. Its benefits are mutual. The space shuttle is emerging as a practical, self-supporting project and needs wider participation. Meanwhile, space investigations need to be continued. An international space agency would seem to be a practical way to perpetuate what the U.S. and Russia have started." (*H Post*, 12/8/72)
- NASA launched Aerobee 170 sounding rocket from White Sands Missile Range carrying Univ. of Wisconsin soft x-ray experiment. Rocket and instrumentation performed satisfactorily. (SR list)
- General Accounting Office released report to Congress that said Lockheed Aircraft Corp. would have to sell 275 TriStar transport aircraft to recover its investment in airliner program. As of Oct. 31, Lockheed had received firm order for 117 TriStars and purchase options for additional 67. (Reuters, *B Sun*, 12/9/72, B7)

December 9: Indian Space Research Organization's experimental satellite communication earth station at Ahmedabad began tracking *Apollo 17* spacecraft. It was first time India had tracked a space mission. (Delhi radio, FBIS–India, Bhutan, Sikim, 12/13/72, O1)

- President Nixon issued Proclamation 4174 calling upon public to observe December 17, 1972, as Wright Brothers Day, "both to recall the accomplishments of the Wright Brothers and to provide a stimulus to aviation in this country and throughout the world." (*PD*, 12/11/72, 1738–9)

December 10: *Nimbus 5* (Nimbus-E) meteorological satellite was successfully launched by NASA from Western Test Range at 11:56 pm EST by two-stage, thrust-augmented Thor-Delta booster. Satellite entered near polar orbit with 1101.3-km (684.3-mi) apogee, 1089.5-km (677.0-mi) perigee, 107.2-min period, and 99.95° inclination. Solar array lock-on and three-axis stabilization were achieved as planned. Activation was proceeding.

Nimbus 5 carried new and improved instruments to extend observation into cloudy areas and higher altitudes, making first vertical temperature and water vapor measurements through clouds of earth's atmosphere.

Assigned primary objectives were to improve and extend capability for vertical sounding of temperature and moisture in atmosphere with particular regard to altitude coverage and cloud interference, providing nearly total global coverage for first time, and to demonstrate improved thermal mapping for 10 wks. As secondary objectives spacecraft would determine feasibility of using passive microwave remote sensing techniques to obtain earth observational data and would demonstrate capabilities of improved selective chopper radiometer (SCR), provided by United Kingdom, to obtain atmospheric soundings.

Butterfly-shaped 769-kg (1695-lb) satellite carried six meteorological and earth resources experiments: SCR; infrared temperature profile radiometer (ITPR); Nimbus E microwave spectrometer (NEMS); electrically scanning microwave radiometer (ESMR); surface-composition-mapping radiometer (SCMR); and temperature-humidity infrared radiometer (THIR).

Nimbus 5 was sixth of seven spacecraft designed to provide atmospheric data for improved weather forecasting. With added sophistication program had grown to include wide range of earth sciences. Sensors aboard *Nimbus 5* would map Gulf Stream off east coast of U.S. and Humboldt Current off west coast of South America. By knowing locations of streams, shipping interests could make substantial saving. *Nimbus 5* would also provide information about El Nino—change in Humboldt Current. El Nino had been causing great economic damage by warming coastal waters, killing nutrients and fish.

Nimbus 2 (launched May 15, 1966, with design life of 6 mos) had returned data for 32 mos, for detailed study of effect of water, carbon dioxide, and ozone on earth's heat balance. Vertical temperature measurements of *Nimbus 3* (launched April 14, 1969) had been termed by meteorologists as one of most significant events in meteorological history, providing worldwide atmospheric information. Six of nine experiments on *Nimbus 4* (launched April 8, 1970) could still provide atmospheric data.

Nimbus program was managed by Goddard Space Flight Center under NASA Office of Applications direction. (NASA proj off; GSFC proj off; NASA Release 72-234)

- Ninety U.S. and foreign scientists would participate in Mariner Jupiter/Saturn 1977 (MJS77) mission, NASA announced. Group, chosen by NASA from more than 200 scientists who had submitted proposals in response to April invitation, represented 32 institutions in U.S. and abroad. France, Sweden, Germany, and United Kingdom were represented. Scientists had been grouped into 11 investigation areas. Each area would be represented in MJS77 Science Steering Group responsible for overall science program. MJS77 would launch two Mariner spacecraft in 1977 to fly by Jupiter and Saturn. (NASA Release 72-239)
- Walter S. Sullivan, *New York Times* science editor, commented in article on conclusion of Apollo program as *Apollo 17*, final mission launched Dec. 7, neared lunar orbit insertion: "In the spaceflight community one senses sadness at the finality of this mission, but resignation to the inevitable, rather than bitterness. The scientists are already so glutted with data and specimens that some say it will take years of study to formulate the questions that should be asked by experiments on future landings, be they manned or unmanned." (*NYT*, 12/10/72)
- Apollo program had been "project of peace, producing knowledge that has been and will be made available to all nations and to all peoples," *New York Times* editorial said. "No doubt men and women of many other nations will some day rocket across the cosmic tracks that Apollo

pioneered. Americans can be proud that this country led the way, and did so with the goal of making space a zone of peace, not a new arena for fratricidal war among nations." (*NYT*, 12/10/72, 10)

- *Houston Post* editorial described *Apollo 17* nighttime launch: "The star-studded sky exploded so violently and in such huge dimensions that it seemed as though a vast reflecting curtain, blended with the night, had been hung directly behind the Apollo 17 launch rocket with the specific role of casting all that fire and light right back over the audience. The climax was reached at the beginning. The drama dwindled swiftly to the view of a bright star fading farther and farther away as it headed toward the moon. An author could write tomes about the space program, but he couldn't write one line about the nighttime launch of an Apollo and get away with it." (*H Post*, 12/10/72)

December 11: Soviet news agency Tass announced "the perfect execution" of *Apollo 17* lunar landing "at 22 hours 55 minutes Moscow time in the north-eastern part of the visible side." Announcement was made five minutes after landing. [See Dec. 7–19.] (FBIS–Sov, 12/12/72, G1; Agence France-Presse, *NYT*, 12/12/72)

- Space program costs were recounted for press by unidentified NASA spokesman at Manned Spacecraft Center. U.S. had spent more than $26.657 billion on manned space program in past 15 yrs and planned to spend $8.85 billion in next 5 yrs. Apollo series—most expensive program—would cost $25 billion through completion of *Apollo 17*. Gemini program had cost $1283.4 million and Mercury, $392.1 million. Skylab was expected to cost $2.6 billion, and early estimates for space shuttle program were $6.1 billion. NASA had saved $1.5 billion by eliminating Apollo 18, 19, and 20. (UPI, *NYT*, 12/12/72, 38)

- Historians would have difficult time explaining decision to abandon Apollo program, *Time* magazine article said. "Having trained the men, perfected the techniques and designed the equipment to explore the earth's own satellite, having achieved the ability to learn more about man's place in the universe, Americans lost the will and the vision to press on. Barely three years after the first lunar landing, the nation that made it all possible has turned its thoughts inward and away from space." Though U.S. spent $5.9 billion to develop Apollo rocket system, "production of Saturn boosters has been halted. The painstakingly assembled team of skilled technicians, engineers and scientists that made Apollo possible is slowly being disbanded." Those who had branded lunar landings propaganda ploys or technological stunts were "prisoners of limited vision who cannot comprehend, or do not care, that Neil Armstrong's step in the lunar dust will be well remembered when most of today's burning issues have become mere footnotes to history." (*Time*, 12/11/72, 35)

- U.S. would continue to use most of Kennedy Space Center's $1 billion worth of launch facilities despite windup of Apollo program and reduced manned space-flight effort, *Aviation Week & Space Technology* reported. KSC Deputy Director Miles Ross had said current 13 500 contractor and Civil Service employees would grow slightly until April 30, 1973, when Skylab would be launched, and taper to about 10 000 by launch of Apollo-Soyuz mission July 15, 1975. At October 1968 peak KSC had employed about 23 000 contract workers and 3000 civil servants. December roster was 11 000 contract employees and 2400 Government employees. NASA anticipated that brunt of future cuts would be borne by contractors and that civil service employment would remain steady. (*Av Wk*, 12/11/72, 17)

- Navy announced it had exercised option to procure 48 F-14 aircraft from Grumman Aerospace Corp. in accordance with terms of original contract. Grumman issued statement same day saying costs had risen to point where it would go bankrupt producing aircraft at originally agreed-upon price of $16.8 million each. Its Board of Directors had decided against making any further deliveries. *Washington Post* reported later that original contract had called for delivery by Grumman of 313 F-14s at total cost of $5.3 billion, or $16.8 million each. Question was "whether the Nixon Administration will stand behind the Navy or give Grumman special aid." (Wilson, *W Post*, 12/12/72, A1; DOD PIO; Grumman release)

- NASA launched Aerobee 170 sounding rocket from White Sands Missile Range carrying Johns Hopkins Univ. payload to measure brightness of sun. Rocket performed satisfactorily but scientific objectives were not met. (SR list)

- Any future development program for U.S. supersonic transport was expected to be given to NASA rather than Federal Aviation Administration, *Aviation Week & Space Technology* reported. FAA, under retiring Dept. of Transportation Secretary John A. Volpe, had "lost so much of its authority" that "it no longer asserts itself" within Dept. of Transportation. (*Av Wk*, 12/11/72, 11)

- Indian Space Research Organization hoped to launch communications satellite in 1982, *Aviation Week & Space Technology* reported. Launch would be part of national effort to establish communications network to reach greater portion of population, 80% of which was "incommunicado." (*Av Wk*, 12/11/72, 20)

- Dept. of Defense officials had approved long-range plans to purchase additional Defense Satellite Communications System Phase 2 (DSCS 2) satellites, *Aviation Week & Space Technology* reported. First two had been launched in 1971. Plan called for eventual simultaneous deployment of four DSCS 2 spacecraft in orbit with the fourth satellite as a spare to handle special emergencies. Plans also called for spare DSCS 2 satellites and Air Force Titan IIIC launch vehicle on ground for fast orbit of replacement satellites. (Klass, *Av Wk*, 12/11/72, 64-7)

- Future of France's Kourou, French Guiana, Space Center was uncertain because of uncertainty over European space programs, *Aviation Week & Space Technology* reported. Center plans had not materialized because of delays in European launcher program and 18- to 20-month hiatus in French Diamant B launcher program. Only 1972 launches from Kourou had been 8 to 10 sounding rockets. French officials had discussed possibility of shutting down major portions of center between launches. Maintenance cost about $20 million annually. (*Av Wk*, 12/11/72, 18)

- Soviet delegation had visited Boeing Co. second time and had revived discussions of possible U.S. certification and marketing of Soviet Yak-40 short-haul transport aircraft by Boeing, *Aviation Week & Space Technology* reported. U.S.S.R. also had discussed Yak-40 with North American Corp. and World Airways Inc. [see Oct. 27]. (*Av Wk*, 12/11/72, 9)

- There was "much reason to suspect that sooner than most Americans realize there will be Soviet cosmonauts striding and riding on the moon's surface," *New York Times* article said. But U.S.S.R. was "no more capable of fully exploring the moon and achieving monopoly control of the lunar surface than the United States." It would take "resources of all nations to finance, equip and staff the permanent manned bases on the moon which will be needed if the moon's scientific, economic and other potentials are to be fully exploited." (Schwartz, *NYT*, 12/11/72, 39)
- *Aviation Week & Space Technology* editorial commented on Apollo program: Six lunar landings had marked "historic watershed in the annals of man." They had been "boldest, most imaginative and technically complex achievements of man" and had "added a dimension to the human spirit that cannot be fully measured for decades." Among Apollo's most important benefits were: (1) "Development of new technology faster, and on a far broader front than is generally realized. The spearhead technology spawned . . . is not just confined to aerospace but has spilled over into other broader areas such as medicine, communications and education." (2) "Creation of a cooperative blend of engineering and scientific effort that appears to have begun bridging a divisive gulf and providing a pattern for more fruitful future work." *Apollo 17* "with its first scientist-astronaut crewmember, foreshadows the increasing opportunity for scientists to work in space as Skylab and the space shuttle provide sufficient capacity for non-flightcrew specialists." (3) "Creation of a management capacity in both government and industry for marshaling vast resources to focus on a specific goal to achieve results within a limited time. Hopefully, these techniques can be applied to other complex problems facing modern society." (Hotz, *Av Wk*, 12/11/72, 7)

December 12: *Molniya II-4* communications satellite was launched by U.S.S.R. from Plesetsk into orbit with 39 300-km (24 420-mi) apogee, 470-km (292-mi) perigee, 11-hr 45-min period, and 65.3° inclination. Purpose of satellite was to provide system of distant telephone and telegraph radio communications in U.S.S.R. and transmission of Moscow TV programs to Orbita network. (Tass, FBIS–Sov, 12/13/72, L1; *SBD*, 12/5/73; 12/14/72; GSFC *SSR*, 12/31/72)

- Newspaper editorials praised *Apollo 17* performance as astronauts worked on moon's surface following Dec. 11 touchdown:

   Cleveland *Plain Dealer:* "On time and on target, the Apollo 17 mission has made it to the moon and provides further proof of excellence in American space technology. The temptation is to dismiss such performance as routine, but there is nothing at all routine about the assignments of the men and equipment on this far journey." *Apollo 17* was "establishing even more clearly the high degree of expertise and perfection that is built into the American space exploration effort. It is doing so, however, with Schmitt and Cernan continuing the very unroutine work begun by astronauts Neil A. Armstrong and Edwin E. Aldrin Jr. on the moon three years ago." (Cl *PD*, 12/12/72)

   Philadelphia *Evening Bulletin:* "With all the other firsts of this last Apollo mission to the moon, there is tinge of sorrow mingled with the superlatives of performance." To realize this was "last time humans will visit the moon in the foreseeable future is to feel a certain loss along with the sense of accomplishment." (P *Bull*, 12/12/72)

Miami Herald: "Knowledge gleaned from the Apollo program . . . is upsetting long-held theories. The venture thus is contributing to the demise of dogmatism, and encouraging open-mindedness which may lead mankind to many new truths." (*M Her*, 12/12/72)

- *New York Times* editorial criticized *Apollo 17* TV coverage: "One might have thought that the Apollo 17 moonwalks would be thoroughly covered by television as they took place. But the published television schedules indicate that the networks have chosen to dish out their usual commercial fare while giving only minimal live coverage to moonwalks that will make history. It is a poor choice from the point of view of public service." (*NYT*, 12/12/72, 46)

December 12–13: Review of plans for space shuttle external tank and solid-fueled rocket booster was held at Marshall Space Flight Center for 350 industry and Government representatives. Attendees heard presentations by NASA and North American Rockwell Dec. 12. NASA officials conducted separate briefings Dec. 13. (*Marshall Star*, 12/12/72, 1)

December 13: NASA's M2–F3 lifting body, piloted by William H. Dana, completed 26th flight from Flight Research Center after air-launch from B–52 aircraft. Objectives were to evaluate reaction augmentation system and to obtain stability and control data at maximum mach number. To obtain maximum speed, M2–F3 was launched from 14 300 m (47 000 ft) rather than usual 13 700 m (45 000 ft). Landing rockets were used during climb. M2–F3 reached 19 800-m (65 000-ft) altitude and mach 1.5. (NASA proj off)

- Library of Congress Congressional Research Service released *The Concorde SST* (72–257 SP). Report said Anglo-French supersonic transport aircraft remained "gamble" despite its technological successes, because of economic and environmental factors. Sonic boom and airport noise remained prime considerations although airlines had decided not to fly Concorde over populated areas. Aircraft's relatively high selling price—about $44 million—plus uncertainty over operating costs, recession in airline industry, and fear of environmental restrictions had slowed sales. Possible environmental hazards included upper-atmosphere and airport pollution from aircraft emissions. (Text)

- NASA launched Aerobee 170 sounding rocket from White Sands Missile Range carrying Univ. of Colorado experiment to measure solar ultraviolet light. Rocket and instrumentation performed satisfactorily. (SR list)

- National Science Foundation published *Company Funds Push Total Industrial R&D Spending to $18 Billion in 1971* (NSF 72–318). In 1971 total industrial research and development performance had amounted to $18.4 billion—equal to 1969 level. When measured in constant dollars, 1971 level was 3% below 1970 and 9% below 1969. Federal R&D funds in industry decreased by 1% between 1970 and 1971, to $7.7-billion level. Federal R&D funds to industry had decreased each year since 1968, when they had reached peak $8.6 billion. Company R&D funds totaled $10.7 billion in 1971, 3% above 1970 level of $10.4 billion, smallest year-to-year dollar gain in company R&D spending since NSF series began in 1953. But 1971 company R&D

funds accounted for 58% of total industrial R&D performance, highest ratio since 1953. (NSF *Highlights*, 12/13/72, 1)

- Hayden Planetarium in New York announced it was showing serigraphs created by photographer Len Gittleman from photos taken by panoramic camera on *Apollo 15* spacecraft. (Am Mus–Hayden Planetarium Release)

December 14: President Nixon issued statement following liftoff from moon of *Apollo 17* lunar module *Challenger:* "As the Challenger leaves the surface of the moon, we are conscious not of what we leave behind, but of what lies before us. The dreams that draw humanity forward seem always to be redeemed if we believe in them strongly enough, and pursue them with diligence and courage. Once we stood mystified by the stars; today we reach out to them. We do this not only because it is man's destiny to dream the impossible, to dare the impossible, and to do the impossible, but also because in space, as on earth, there are new answers and new opportunities for the improvement and the enlargement of human existence.

"This may be the last time this century that men will walk on the moon. But space exploration will continue, the benefits of space exploration will continue, the search for knowledge through the exploration of space will continue, and there will be new dreams to pursue based on what we have learned. So let us neither mistake the significance nor miss the majesty of what we have witnessed. Few events have ever marked so clearly the passage of history from one epoch to another. If we understand this about the last flight of Apollo, then truly we shall have touched a 'many-splendored thing.'" (*PD*, 12/18/72, 1758-9)

- *Cosmos 538* was launched by U.S.S.R. from Plesetsk into orbit with 280-km (174-mi) apogee, 204-km (126.8-mi) perigee, 65.4° inclination, and 89.3-min period. Satellite reentered Dec. 27. (GSFC *SSR*, 12/31/72; *SBD*, 12/18/72, 209)

- Air Force qualification of parachute decelerator system for NASA's Viking Mars lander, scheduled for 1975 launch, was announced by Air Force Systems Command. Qualification had followed July 11 and 26 and Aug. 13 and 19 tests by Martin Marietta Corp. at Roswell and White Sands, N. Mex. Tests had included supersonic, transonic, and free-fall environments. In Aug. 19 test, decelerator had been dropped from balloon at altitude of 27 000 m (88 000 ft). Previous tests had used rocket motors ignited after balloon had taken test system to desired altitude. In each test, decelerator deployed to slow down simulated softlanding instrument package. (AFSC Release 105.72; Martin Marietta Corp PIO)

- *New York Times* editorial commented on discovery by *Apollo 17* astronauts of orange dust on moon: Discovery had "astonished the scientific world. Nothing similar had been previously seen on the moon. It is premature, though tempting, to conclude that this is evidence of relatively recent—in geological terms—volcanic activity. Final judgments have to await analysis of this strange material . . . . Nevertheless, it is already evident that Apollo 17 is turning out to be the most productive of the moon visits; and the probability seems higher than ever that the crew will bring back both the oldest and youngest lunar rocks yet to be found. It is already indisputable that the Taurus-Littrow

site was an excellent choice for an Apollo landing since much of the terrain material there is quite different from that found on earlier trips." (*NYT*, 12/4/72, 46)
- Honorary lifetime membership in Auto Body Assn. of America was granted to *Apollo 17* Astronauts Eugene A. Cernan and Dr. Harrison H. Schmitt for fender repair work done on lunar rover on moon. Reg Predham, president of national organization of auto repairmen, said astronauts would be sent lapel pins and membership certificates. Boeing Co. would receive certificate for manufacturing "vehicle that can be repaired 250 000 miles [400 000 kilometers] from a part supplier." (AP, *C Trib*, 12/15/72)
- Tenth year of interplanetary exploration was celebrated by Jet Propulsion Laboratory on 10th anniversary of arrival at Venus of *Mariner 2*, launched by NASA Aug. 27, 1962.

  Five more Mariners—designed and built by JPL engineers—had been launched during decade. In 1965 *Mariner 4* (launched Nov. 28, 1964) flew by Mars, obtaining first close-up pictures. *Mariner 5* (launched June 14, 1967) helped determine atmosphere of Venus was 75 to 100 times more dense than earth's. *Mariner 6* (launched Feb. 24, 1969) and *Mariner 7* (launched Mar. 27, 1969) took photographs of Mars surface from as close as 3500 km (2200 mi). *Mariner 9* (launched May 30, 1971) gathered data during 698 revolutions of Mars, provided map of entire planet, and showed evidence of volcanic activity and free-flowing water in planet's geologic history. (JPL Release 640; *A&A 1962*; NASA Release 72-241)
- Highest energy output level ever reached by man-made machine—400 bev—was achieved by Atomic Energy Commission's nuclear particle accelerator near Batavia, Ill. Level was double original design level of 200 bev. (AEC Release P-433)
- Tass announced successful completion of Soviet carrier rocket launchings in Pacific and declared restricted area free for sea and air navigation. (FBIS-Sov, 12/14/72, A4)
- *New York Times* science editor Walter S. Sullivan had been named one of three winners of 1972 science writing awards sponsored by American Assn. for the Advancement of Science and Westinghouse Education Foundation, *New York Times* reported. Sullivan had won award for third time. Award was for his series "The Einstein Papers." Other winners were Dennis L. Meredith, science editor at Univ. of Rhode Island, and Eugene Kinkead, associate editor of *New Yorker*. (*NYT*, 12/14/72)
- *New York Times* editorial commented on Apollo program day after *Apollo 17* astronauts lifted off moon's surface: "Over the long term the chief product of Apollo is likely to be in man's consciousness, not only in scientific thinking. Man's entire perspective on the universe and on his place in it has been radically changed. Man evolved on the earth, but he is no longer chained to it. Man has walked on another planet and returned to tell the tale. The impact on the future must be enormous. Yesterday's farewell to the moon was certainly not a farewell to space." (*NYT*, 12/15/72, 46)
- *Apollo 17* Astronaut Ronald E. Evans would receive promotion from commander to captain in Navy despite Government freeze on promotions, Dept. of Defense spokesman Jerry W. Friedheim told press in

Washington, D.C. Exemption had been granted specifically for Evans. *Apollo 17* commander Eugene A. Cernan had received promotion to captain in Navy for previous space exploits. (*W Post*, 12/16/72, A4)

- President Nixon issued statement on receiving necessary ratifications of Definitive Agreements of the International Telecommunications Satellite Organization: "We can now look forward to the day when nations around the world will be linked together for instantaneous communications. The implications of this development are enormous, presaging improved international relations in the political, economic, cultural, and scientific spheres." (*PD*, 12/18/72, 1759)
- Rep. Olin E. Teague (D-Tex.), Chairman of House Committee on Veterans' Affairs, would become Chairman of House Committee on Science and Astronautics in January 1973, *Washington Post* reported. He would replace Rep. George P. Miller (D-Calif.), who was defeated for renomination to House. *Post* quoted Rep. Teague as saying he wanted to push U.S. space program forward because it "seems to be taking a back seat." (Lyons, *W Post*, 12/15/72)
- Federal Aviation Administrator John H. Shaffer received 1972 Wright Brothers Memorial Trophy at annual Wright Memorial Dinner in Washington, D.C. Citation was for "outstanding leadership of the worldwide operations of the Federal Aviation Administration which has greatly enhanced all aspects of U.S. aviation to the benefit and safety of the general public and all who fly." (NAA Release, 9/22/72; FAA PIO)
- Astronaut James A. Lovell—holder of record for hours logged in space [see Nov. 22]—had been elected to the board of American Bakeries Co., *Chicago Daily News* reported. (*C Daily News*, 12/15/72)
- *Science* article commented on renaming of Jet Propulsion Laboratory in honor of retiring California Congressman: Rep. H. Allen Smith's closest association with aerospace had come in late 1940s when he was security manager for Lockheed Aircraft Corp. "During his years in the House, Smith distinguished himself as a quiet, unwavering conservative dedicated to economy in government." He had voted against supersonic transport "and on at least three occasions opposed the space authorization bill, the ultimate wellspring of JPL's money." No one had asked JPL, California Institute of Technology, or NASA what they thought of redesignation. Rep. George P. Miller (D-Calif.) had approved but "was not in a position to object gracefully." Rep. Miller, too, was retiring and House Committee on Public Works had decided to affix his name to Federal building in Oakland, Calif. Article suggested solution might be "for JPL to acknowledge its new name with a small sign behind a fast-growing evergreen and let time take its course." (Gillette, *Science*, 12/15/72, 1178)

December 16–January 22, 1973: West Germany's *Aeros* Aeronomy Satellite was successfully launched by NASA from Western Test Range at 3:25 PST by four-stage, solid-fuel Scout D vehicle. Spacecraft was placed in planned orbit with apogee of 864.4 km (537.1 mi); perigee, 218.1 km (135.5 mi); period, 95.47 min; and inclination, 96.95°.

Primary objective was to place 127-kg (277-lb) payload in near-polar orbit to measure main aeronomic parameters of upper atmosphere and solar ultraviolet radiation in wavelength band of main absorption. By Jan. 22, 1973, all five experiments were performing satisfactorily and spacecraft housekeeping operations were normal, except that problem with synchronization of data rate to spin rate might degrade data from NASA's neutral atmosphere temperature experiment.

Satellite, cylinder 71 cm high and 91 cm in diameter (28 by 36 in) with four telemetry antennas, would contribute to understanding of thermal, chemical, and dynamic processes governing atmospheric behavior. Instruments included mass spectrometer, retarding potential analyzer, impedance probe, extreme-ultraviolet spectrometer, and NASA's neutral atmosphere experiment.

July 1965 agreement between West Germany and NASA had provided for series of cooperative satellite projects. *Aeros,* second in series, was agreed to in Memorandum of Understanding signed June 10, 1969. NASA was responsible for one experiment, Scout launch vehicle, launch, and limited tracking and data acquisition. West German Ministry for Education and Science (BMBW) was responsible for development of spacecraft and instrumentation and for operation and control of satellite after launch, as well as for four experiments. (NASA proj off; NASA Releases 69-91, 72-229)

December 16: Apollo 17 surface science press briefing was held at Manned Spacecraft Center as astronauts headed toward home from moon. Stanford Univ. seismologist Dr. Robert L. Kovach described results to date from lunar seismic profiling experiment. Experiment had recorded impact of lunar module (LM) on lunar surface 9 km (5.6 mi) southwest of landing site "right on target where we wanted it." Kinetic energy at impact was comparable to "1 ton of TNT going off." Signal was "very clearly recorded on the geophone array and we've acquired an extremely important data point; . . . we're effectively getting a depth sample down to 3 to 4 kilometers [1.9 to 2.5 miles] in the Moon. And . . . we should get some very important information about the underlying highland type materials." Data point was "in critical distance range to answer one of the mysteries . . . of the shallow Lunar interior. We've been very much concerned about the enormous change in physical properties by terrestrial standards, . . . velocity changes by enormous percentage over a very small pressure range." Data point should provide "answers as to just how these properties are changing." First two explosive charges deployed had not shown significant differences in subsurface of *Apollo 17* landing site and those of *Apollo 15* and *16* but *Apollo 17* site appeared to be "underlined by definitely more competent materials and they quite probably could be volcanic flows."

Dr. Gary V. Latham, principal investigator for passive seismic experiment, said *Apollo 17* data, when compared with previous data, suggested possibility of new interpretation of lunar structure. "We do find evidence for lunar crust as we did in the past, but we may have to thin it considerably." Velocity of seismic waves in crustal material and velocity of material underlying crust might also have to be lowered. "We're not dealing with pure anorthosite anymore. We're dealing perhaps with gabbros . . . and so we have . . . to review all of the data and to reconsider our interpretations some, not just the seismologist, but the geochemist and geologist, as well."

Dr. Marcus E. Langseth of Lamont-Doherty Geological Observatory, principal investigator for heat flow experiment, said *Apollo 17* experiment had "more than doubled" lunar heat flow data to date. "We've got a very successful emplacement in the operation of our instrument. It has been perfect." Lunar surface temperature recorded Dec. 14 had been about 360 K (189°F). At about 15 cm (6 in), it had dropped to 280 K (45°F). At about 65 cm (26 in) temperature dropped to 254 K (−2°F). "Below that depth, the temperature begins to increase again. It's about 257 degrees [3°F] at the bottom of the probes." Data indicated that heat flow at *Apollo 17* site would resemble that observed at 15 cm (6 in). If so, it would "give support to the kind of growing model of a warmer interior to the Moon" and "would require that there be a total gradient abundance of radioactive isotopes in the Moon as compared to Earth. So, there would be implication here for fundamental difference in composition, between the Moon and the Earth." Dr. Langseth also reported data from traverse gravimeter experiment had supported thesis that "high gravity anomalies associated with the mare basin—the so called mascons"—probably had been created by slab of basalt or lavas which had flooded into previously excavated basins.

Dr. William R. Muehlberger of Univ. of Texas, principal lunar geology investigator, said *Apollo 17* had provided "most exciting few days of my scientific career. This mission is truly the fitting climax to the whole Apollo program. The landing site held great promise that we were going to get a good chance to unravel much of the earlier lunar history as well as the sample and study of what appeared to be very young vulcanism . . . . And I think this promise was fulfilled." Orange materials at Shorty Crater had been "a spectacular plus." Scientists hoped "returned material will tell us that that was young vulcanism in volcanic alteration of Shorty Crater. There's still a possibility it could be an impact crater, however." Dr. Paul W. Gast, Chief of MSC Planetary and Earth Sciences Div., discussed Dr. Latham's suggestion that lunar crust might be thinner than originally expected: "I will certainly defend the proposition that the crust is made of a very aluminum plagioclase-rich rocks, . . . anorthosite gabbro or gabbroic anorthosite, . . . far from a basalt. I think that the chemistry of the surface rocks, . . . and the returned rocks bear this out and if the crust is thin, it reduces the likelihood that it is any different composition than what we see at the surface." (Transcript)

- Nearly half billion dollars worth of equipment, mostly junk, had been left on lunar surface during six Apollo landings, *Chicago Tribune* reported. List included rocket engines, electronic devices, golf ball driven by *Apollo 14* commander Alan B. Shepard, Jr., feather left by *Apollo 15* commander David R. Scott (who had dropped feather and hammer simultaneously during televised extravehicular activity to show that they both fell at same rate in vacuum), and 12 prs of boots costing $4000 a pair. Boots included those that *Apollo 11* Commander Neil A. Armstrong had used to make first footprints on moon. Technological items left on moon included 5 S–IVB rocket engines costing $20 million each, 6 $40-million lunar modules, 3 lunar rovers at $2 million each, TV and photography equipment costing $5 million, 12 backpacks costing $3.6 million, and assorted tools valued at $600 000.

Still operative on moon were five nuclear-powered scientific stations costing $25 million each and three laser reflectors costing total $2 million. (Kotulak, C Trib News Service, *C Trib*, 12/16/72)

December 17: Apollo 17 orbital science briefing was held at Manned Spacecraft Center while astronauts continued flight homeward from moon. Data from laser altimeter and S-band gravity experiments were described by Dr. Wilbur R. Wollenhaupt, Chief of MSC Experiments Support Section and S-band transponder coinvestigator. Profiles of near-side lunar basins appeared to be flat and depressed with surrounding terrain. Far side appeared mountainous. Copernicus Crater was deficiency region; Sinus Aestuum, Mare Serenitatis, and Mare Crisium were high regions; and Littrow landing site was low.

Harold Masursky of U.S. Geological Survey, member of Apollo Orbital Science Photographic Team, said panoramic camera aboard *Apollo 17* had "acted better than on the two previous flights [*Apollo 15* and *16*], and we essentially accomplished the tasks assigned to it." During *Apollo 17* mission "we saw the successful rectification of the 15 and 16 panoramic photographs, and they looked very good indeed. So, hopefully, we will have the rectification of 17 done shortly and put out a series of photomosaics, putting the high-resolution panoramic frames together into what we think will be the best series of maps that we have developed for the Moon." Correlation of photography with gravity tracking and laser altimetry had produced "a great deal of data" to confirm earlier interpretations from *Apollo 15* that lunar front side was low and that "largest area of mare material on the front side is an ocean basin floor, . . . not a result of the impact as the large gravity anomalies indicate in the impact basin."

Jet Propulsion Laboratory scientist Walter E. Brown reported lunar sounder experiment had "behaved about as we expected except that there were a couple of surprises." Experimenters had expected that "over the Mare, the smooth areas, . . . we'd have a reasonably smooth, well-behaved echo, and over the mountainous area it might bounce around a little bit." Sounder data had shown "it was fairly smooth over the mountains and it bounced around over the Mare." Until mission film was returned, "we won't know . . . just exactly what is going on, but it's . . . extremely interesting."

Dr. Frank J. Low of Univ. of Arizona and Rice Univ. said scanning radiometer experiment performance "was essentially perfect." It produced coverage of about one third of the surface area of the moon and made some 100 million independent temperature measurements. With "beautiful Atlas of thermal measurements generated by this magnificent flight," scientists had "foundation of a new technique for exploration of planetary surfaces without atmospheres."

Johns Hopkins Univ. scientist William E. Fastie said far-ultraviolet spectrometer had "performed perfectly . . . and all of the observations that were intended were accomplished." Experiment had detected neither major nor minor constituents of atmosphere. "And, pending further analysis, we have . . . identified only one trace constituent of the lunar atmosphere. The immediate conclusion . . . is, that the Moon is not outgassing, because the only other possible gases that could be in the lunar atmosphere are neon and argon, and neither of them are the primordial atmosphere of the Moon."

December 17

Mood of briefing was conveyed by Apollo orbital science program scientist Floyd I. Roberson: "I think we've got a lot of tools in our hands now and we sit up here and we're all tired and we talk about the results of our experiments and maybe we don't convey the excitement that's upon us. But, we really feel that this is a beginning as far as lunar science is concerned, as far as the Apollo data is concerned, and we're really going to get there now." (Transcript)

- *New York Times* editorial praised Apollo program in which U.S. had "attained one of its finest hours." Astronauts had gone to moon in "spirit of peace and selflessness." They had been "seeking knowledge, not gold or slaves or other wealth. They went there consciously as representatives of all mankind, not as imperialists seeking to bring the moon under the Stars and Stripes. What they learned has been made available to all peoples and all nations without any request for a quid pro quo. In the atmosphere created by Project Apollo—with its constant reminder that what united men is stronger than what divides them—it was easier to reduce cold war tensions, to end the original space race, and to begin genuine international collaboration in cosmic exploration. Project Apollo has helped consecrate the moon and space to the cause of peace and cooperation among all men. This has been a glorious adventure in whose successful outcome the people of America and of the world can properly take pride." (*NYT*, 12/17/72)

- Sixty-ninth anniversary of first flight by Wright brothers was celebrated in Kitty Hawk, N.C., as *Apollo 17* astronauts headed home from sixth manned landing on the moon. (AP, *NYT*, 12/17/72, 42)

- Dr. Robert C. Cameron—first director of NASA's Johannesburg, South Africa, tracking station from 1958 to 1959—died in Washington, D.C., at age 47. He had resigned earlier in year as astronomer at Goddard Space Flight Center. Dr. Cameron had made first successful photos of U.S. satellite (*Explorer 1*, launched Jan. 31, 1958) and had received special award from GSFC in 1966 for writings on astronomy. Before joining NASA, he had worked for Smithsonian Astrophysical Observatory. In 1952 he had discovered asteroid Winifred, named after his wife. (*W Post*, 12/17/72, D12)

December 18: NASA announced award of $16 794 556 contract extension to North American Rockwell Corp. for testing, checkout, and launch operations in Skylab program. Cost-plus-fixed-fee supplement agreement with Kennedy Space Center covered third year of work under Nov. 29, 1970, contract. (NASA Release 72–243)

- Retiring California Congressman H. Allen Smith said in joint letter to Dr. William H. Pickering, Jet Propulsion Laboratory Director, and Dr. Harold Brown, California Institute of Technology President, that he would request legislation to repeal law passed by Congress Oct. 3 that changed name of JPL to H. Allen Smith Jet Propulsion Laboratory. "I have never sought personal recognition. I think it could cause some confusion and possibly some misunderstanding" to rename JPL. (Pasadena *Star-News*, 12/18/72)

- American Institute of Aeronautics and Astronautics would sponsor U.S. aerospace industry trade exhibit in Moscow in early spring 1973, *Aviation Week & Space Technology* reported. Exhibit would center on air traffic control equipment but probably would include other aviation and avionics gear. U.S.S.R. was expected to purchase $10 million to $20 million worth of equipment but total orders might exceed $1 billion. (*Av Wk*, 12/18/72, 30)

December 19: Following successful splashdown—after Dec. 7 launch on longest lunar mission—*Apollo 17* astronauts were greeted aboard recovery ship U.S.S. *Ticonderoga* by 12 Congressmen. President Nixon telephoned his personal congratulations on successful completion of mission. Sen. Barry M. Goldwater (R-Ariz.) said Apollo program results would "give us such scientific values that we will be required to have another lunar program before the end of the century." (AP, *NYT*, 12/20/72, 30; *PD*, 12/25/72, 1789)

- NASA officials at Manned Spacecraft Center celebrated successful completion of last Apollo mission, following splashdown of *Apollo 17*. Dr. Rocco A. Petrone, Apollo Program Director since 1969, said mission "goes into the record books as the most perfect mission, the most sophisticated science mission ever conducted." *Apollo 17* "closes a golden chapter in the age of space exploration. In a way, it brings a close to what has been a very romantic era in space exploration. But . . . the book is still being written. The next chapter will open the age of exploiting and utilizing space for other benefits."

    Dr. George M. Low, NASA Deputy Administrator, said, "All great things must come to an end so we can move on to even greater things." Dr. Christopher C. Kraft, Jr., MSC Director, said he had "never seen a more perfect mission" than *Apollo 17*. Apollo program had been greatest engineering feat of all our lifetimes." Before Apollo, "my mind, like most of us, was bound to earth. It's certainly not that way anymore."

    NASA spokesman at MSC told press *Apollo 17* was "the last, longest, and most successful of seven manned lunar landing missions." (Wilford, *NYT*, 12/20/72, 1; AP, *B Sun*, 12/20/72, 1; UPI, *W Star & News*, 12/20/72, A4)

- President Nixon issued statement on U.S. space program following splashdown of *Apollo 17* command module: "The safe return of the command module America marks the end of one of the most significant chapters in the history of human endeavor. In October 1958, this Nation set about sending men into a hostile, unknown environment. We had little idea what lay before us, but there was new knowledge to be gained and there was a heritage of meeting historical challenge—the challenge of greatness—to be sustained. Project Mercury, begun in 1958, taught us that man could survive and work in space. In 1961, President Kennedy voiced the determination of the United States to place a man on the moon. We gained the understanding and the technology to embark on this great mission through Project Gemini and we accomplished it with the Apollo lunar exploration series. In 1969, for the first time, men from the planet Earth set foot on the moon.

    "Since the beginning of Apollo, nine manned flights have been made to the moon. Three circled that nearest neighbor in the universe, six landed and explored its surface. We have barely begun to evaluate the vast treasure store of extraterrestrial data and material from these voyages, but we have already learned much and we know that we are probing our very origins. We are taking another long step in man's ancient search for his own beginnings, pressing beyond knowledge of the means of human existence to find, perhaps, the meaning of human existence.

"Nor is this great work ending with the return of Gene Cernan, Jack Schmitt, and Ron Evans from the moon today. Rather it has barely begun. As Sir Isaac Newton attributed his accomplishments to the fact that he stood 'upon the shoulders of Giants,' so Newton himself is one of the giants upon whose shoulders we now stand as we reach for the stars. The great mathematician once wrote: 'I do not know what I may appear to the world; but to myself I seem to have been only like a boy playing on the seashore, and diverting myself in now and then finding a smoother pebble or a prettier shell than ordinary, whilst the great ocean of truth lay all undiscovered before me.' I believe we have finally moved into the great ocean and we are trying now to understand what surrounds us.

"The making of space history will continue, and this Nation means to play a major role in its making. Next spring, the Skylab will be put into orbit. It will be aimed not at advancing exploration of deep space, but at gaining in space new knowledge for the improvement of life here on earth. It will help develop new methods of learning about the earth's environment and the earth's resources, and new methods of evaluating programs aimed at preserving and enhancing the resources of all the world. It will seek new knowledge about our own star, the sun, and about its tremendous influence on our environment. Scientists aboard the Skylab will perform medical experiments aimed at a better knowledge of man's own physiology. Also, they will perform experiments aimed at developing new industrial processes utilizing the unique capabilities found in space. Skylab will be our first manned space station. It will be in use for the better part of a year, permitting the economy of extended usage, and laying the groundwork for further space stations.

"Economy in space will be further served by the Space Shuttle.... It will enable us to ferry space research hardware into orbit without requiring the full expenditure of a launch vehicle as is necessary today. It will permit us to place that hardware in space accurately, and to serve or retrieve it when necessary instead of simply writing it off in the event it malfunctions or fails. In addition, the Shuttle will provide such routine access to space that for the first time personnel other than trained astronauts will be able to participate and contribute in space as will nations once excluded for economic reasons.

"The near future will see joint space efforts by this Nation and the Soviet Union in an affirmation of our common belief that the hopes and the needs that unite our people and all people are of greater consequence than the differences in philosophy that divide us.

"Finally, we will continue to draw knowledge from the universe through the use of unmanned satellites and probes.

"We cannot help but pause today and remember and pay homage to those many men and women—including those who made the ultimate sacrifice—whose hopes, whose energies, skill, and courage enabled the first man to reach the moon and who now have seen with us perhaps the last men in this century leave the moon. But the more we look back the more we are reminded that our thrust has been forward and that our place is among the heavens where our dreams precede us, and where, in time, we shall surely follow.

"Though our ancestors would have called the deeds of Apollo miraculous, we do not see our age as an age of miracles. Rather, we deal in facts, we deal in scientific realities, we deal in industrial capacity, and technological expertise, and in the belief that men can do whatever they turn their hands to. For all this, however, can we look at the record of 24 men sent to circle the moon or to stand upon it, and 24 men returned to earth alive and well, and not see God's hand in it?

"Perhaps, in spite of ourselves, we do still live in an Age of Miracles. So if there is self-congratulation, let it be tempered with awe, and our pride with prayer, and as we enter this special time of spiritual significance, let us reserve a moment to wonder at what human beings have done in space and to be grateful." (PD, 12/25/72, 1788–9)

- Man's lunar theories were undergoing major revision because of *Apollo 17*, *Wall Street Journal* said. Observations by astronauts of orange soil on moon—clue to past volcanic activity—had promoted review of *Apollo 14* color photos. Review had disclosed brown and orange discolorations on dome in crater Langrenus. Orange soil meant there might be other relatively young volcanic features. Discovery could result in major revision of thermal theories of moon. (*WSJ*, 12/19/72)

December 20: Air Force launched unidentified satellite by Atlas-Agena booster from Eastern Test Range. Satellite entered orbit with 40 728-km (25 307-mi) apogee, 31 012-km (19 270-mi) perigee, 1440.4-min period, and 9.7° inclination. (Pres Rpt 1973; Sheldon, Sci Policy Research Div, Library Congress; *SBD*, 12/26/72, 245)

- *Apollo 17* astronauts were examined on board recovery ship U.S.S. *Ticonderoga* by Dr. Charles K. La Pinta, Apollo medical team leader. He pronounced them "as good as any crew of astronauts that I have ever seen. They appear to be one of the more well-rested crews to have returned from a space flight." (Reuters, *W Post*, 12/21/72, A17)

- Soviet President Nicolay V. Podgorny had sent congratulations to President Nixon on successful conclusion of *Apollo 17* mission, Moscow radio reported. Message had conveyed best wishes to brave crew, their flight, and explorations that "have contributed greatly to the study of space." (FBIS–Sov, 12/21/72, G1)

- NASA's M2–F3 lifting body, piloted by John A. Manke, completed 27th and last flight after air-launch from B–52 aircraft. Flight objective was to evaluate reaction augmentation system during boost. M2–F3 reached 19 800-m (65 000-ft) altitude and mach 1.4. (NASA proj off)

- Representatives of 12 nations at European Space Conference ministerial meeting in Brussels agreed in principle to participate in space shuttle phase of U.S. post-Apollo space program, support French plan to build L–35 launcher, and fuse European Launcher Development Organization (ELDO) and European Space Research Organization (ESRO) into one agency by Jan. 1, 1974. Space programs of individual European countries would be blended into common European program with each country maintaining independent choice of what projects to join. Conference President Theo Lefevre, Belgian Minister of Space, commented at close of meeting, "For the first time I can say I am satisfied

December 20

with the results of the space conference." United Kingdom and West Germany, who had agreed to abandon plans for European launcher [see Dec. 8], did not rule out participation (ESRO Release, 1/19/73; NYT, 12/21/72)

- Discovery of ordinary star and pulsar "waltzing" together in constellation Hercules was reported by Dr. Harvey Tananbaum of American Science & Engineering, Inc., during Sixth Texas Symposium on Relativistic Astrophysics in New York. Discovery had been made through x-ray observation from *Uhuru* (*Explorer 42*) Small Astronomy Satellite, launched by Italy for NASA Dec. 12, 1970. X-rays from pulsar caused nearest part of its companion star to glow brightly as pulsar flew its orbit. Star became brightest when pulsar was on earth side of it, then suddenly dimmed as pulsar eclipsed bright region briefly, creating waltzing effect. (Sullivan, NYT, 12/22/72, 8)
- Dr. James C. Fletcher, NASA Administrator, said in Christmas message to NASA staff: "We are nearing the end of a historic and decisive year for NASA. We have completed the first great era of space exploration and use and begun a promising new one, with more emphasis this time on use. Well done, Apollo! Well begun, Shuttle!" (Text)
- Australian Defense Minister Lance Barnard announced closing of deep space tracking station near Woomera, South Australia. He said station was no longer needed to meet NASA requirements. Station's last operation had been tracking of *Pioneer 10* Jupiter probe, launched March 2. (AP, NYT, 12/21/72)
- Newspaper editorials commented on successful conclusion of *Apollo 17*.

New York Times: "Astronauts Cernan, Schmitt and Evans returned safely to earth yesterday in a splashdown that marked the brilliantly successful end of one of the most successful series of scientific experiments in history. Less than four years ago men had never been even in the vicinity of the moon. Now about two dozen men have orbited the satellite, and a dozen have walked and ridden on its surface. Hundreds of pounds of lunar rock and soil are now . . . on earth, available for study, while instruments planted on the moon give daily reports of lunar conditions, and will do so for years. The moon is now a distant but accessible outpost of earth, better known now than the North Pole was before Admiral [Robert E.] Peary discovered it in 1909. Project Apollo may be ended, but man's interest in the moon is still very much alive." (NYT, 12/20/72, 42)

Baltimore *Sun:* "No doubt the success of the $25 billion Apollo program will encourage astronauts of the future to hurtle far past the moon to the solar planets, and perhaps beyond. But the immediate significance of Apollo is much closer to home. We are today much more aware of the fragility of our spaceship earth and of the unique chance mixture of elements that sustain life as we know it. If this vision of a rather small but infinitely precious planet can inspire us to take better care of it, Apollo will have produced its greatest dividend." (B Sun, 12/20/72, A12)

Pittsburgh Press: Apollo program demonstrated "that the U.S. is still a will-do, can-do nation rather than a won't-do, can't-do has-been. In the days ahead we must hold fast to this national legacy of pioneering—and continue to nurture among our people the spirit of adventure, the quest for knowledge, the drive to explore the unknown, whether it be in the vast vacuum of space or in the tiny test tube of a laboratory. For therein lies our future, and the future of all mankind." (*Pittsburgh Press*, 12/20/72)

St. Louis Post-Dispatch: "So let Apollo's epitaph be that the program was magnificent but that it is also done. The evils that Apollo was never intended to cure remain and deserve—as they have all along—attention. If money is to be spent on further space projects such as the shuttle and the skylab, let Congress insist first that funds go where they are needed most. And that, as 1972 ends, is not out among the stars but in America. The moon, after all, can wait for its next human visitor. But for the poor, the hungry and the dispossessed, time wasted is misery compounded. The measure of our greatness will be found in how quickly the nation turns to them." (*St. Louis P–D*, 12/20/72)

Washington Post: At outset of Apollo program "we spoke glowingly of the challenge but we were principally in a 'race'—one that the cynical conventional wisdom of the time had it we would lose." Apollo had ended "with the prospect of the joint Apollo-Soyuz testing mission. Somewhere in all this . . . mixed in with the great political trends and the particular political accidents that have contributed to the altered relations among the world's super powers, the technology of Apollo itself played a part. You do not have to be a sentimentalist or a hopeless romantic to acknowledge that manned space flight in general and the successful moon landings in particular have had at least a subliminally humbling and unifying impact upon people, that the first astronauts to land on the moon were seen to be proxies for humankind, that their bravery and their achievement were regarded as human triumphs and that the first photographs of earth taken from the moon conveyed a unique and invaluable sense of planetary vulnerability and openness." (*W Post*, 12/20/72)

Detroit News: "Today we know that the dream was not an impossible dream. And there will be more dreams to come and new worlds to conquer, all of which is, in essence, the true purpose of civilization, provided all men share the knowledge acquired and thus reap the harvest of benefits." (*D News*, 12/20/72)

Chicago Tribune: "Welcome home. . . . The trail you helped to blaze will be used again some day." (*C Trib*, 12/20/72)

December 21: Air Force launched unidentified satellite by Titan IIIB-Agena booster from Vandenberg Air Force Base into orbit with 391-km (243-mi) apogee, 138-km (86-mi) perigee, 90.0-min period, and 110.5° inclination. Satellite reentered Jan. 23, 1973. (Pres Rpt 73; *SBD*, 12/26/72, 245; GSFC *SSR*, 1/31/73)

- *Cosmos 539* was launched from Plesetsk by U.S.S.R. into orbit with 1381-km (858.1-mi) apogee, 1343-km (834.5-mi) perigee, 74° inclination, 112.9-min period. (GSFC *SSR*, 12/31/72; *SBD*, 1/3/73, 2)

- *Apollo 17* astronauts arrived at Ellington Air Force Base, Tex., after flight from Samoa, via Hawaii, aboard Air Force C-141 jet transport. After reception at airport by crowd of 5000, they went to Manned Spacecraft Center for debriefings that would continue through Dec. 23. (Wilford, *NYT*, 12/22/72, 15; *WSJ*, 12/22/72)

- Romanian President Nicolay Ceausescu had sent telegram to President Nixon on successful conclusion of *Apollo 17* mission, Romanian news

December 21

agency Agerpres announced in Bucharest. Telegram had said: "We express our conviction that the remarkable successes in outerspace exploration scored during the Apollo programme that concluded with this brilliant mission will serve the general progress of science and technique and will be put in the service of humanity, peace and understanding among people." (FBIS–Romania, 12/22/72, H1)

- Newspaper editorials continued comment on Apollo program.

    Christian Science Monitor: "Such technological feats as going to the moon do not absolve people of responsibilities on earth. It may be only an irony but the last Apollo splashdown, amazingly within a mile of its target, occurred while the United States again renewed bombing of North Vietnam in that excruciating war. Millions in America and elsewhere are hungry or unhappy. Great leaps in human relationships comparable to those in space are called for. Nonetheless, though we do not yet fully grasp the impulse that led to the moon venture, and though it but the more reveals what remains unresolved in the human lot, we are aware that Apollo 17 has written finis to something marvelous in mankind's annals." (*CSM*, 12/21/72)

    Pasadena, Calif., *Star-News* editorial praised Apollo Program: "The American knowhow and resources, which have made the Apollo missions and space travel so successful, have confirmed for the world our scientific leadership and perfection. This has resulted in more pride and esteem for our nation than all the foreign aid, diplomacy or pats on the back that Uncle Sam has bestowed in his lifetime." (Pasadena *Star-News*, 12/21/72)

December 22: Retirement of Dr. Eberhard F. M. Rees as Marshall Space Flight Center Director and appointment of Dr. Rocco A. Petrone, Apollo Program Director, to succeed Dr. Rees in January 1973 were announced by Dr. James C. Fletcher, NASA Administrator. Dr. Rees, with Dr. Wernher von Braun, had pioneered U.S. space and rocket activities. He had succeeded Dr. von Braun as MSFC Director in 1970. Dr. Petrone had been Apollo Program Manager at Kennedy Space Center before becoming Apollo Program Director at NASA Hq. in 1969. Dr. Fletcher commended Dr. Rees for "successful completion of the Marshall Center's role in the Apollo program. The performance of the Saturn launch vehicle . . . has been magnificent. The development of the Lunar Roving Vehicle and its performance . . . is a tribute to his leadership." (NASA Release 72–244)

- NASA released color photo of Shorty Crater area of *Apollo 17* Taurus-Littrow lunar landing site. Photo, taken Dec. 12 during second extravehicular activity of Dec. 7–19 mission, confirmed existence of orange soil discovered by astronaut-geologist Dr. Harrison H. Schmitt. (NASA photo 72–HC–933)

- Question of whether moon was hot or cold was discussed in *Science* by California Institute of Technology scientists Dr. Don L. Anderson and Dr. Thomas C. Hanks. High surface concentrations of uranium, thorium, and potassium found on moon and *Apollo 15* heat-flow value of 33 ergs per sq cm per sec had indicated high present-day temperatures in lunar interior. But recent interpretations of lunar conductivity profile—"non-hydrostatic shape of the moon, the existence of mascons, the remarkable aseismicity of the moon, and the absence of present-day volcanism"—had suggested lunar interior had always been cold. "We find that the basic observations do not demand a presently cold moon and are, in fact, consistent with a hot moon. We find that an iron-deficient, highly resistive, hot lunar interior, capped by a cool rigid lunar lithosphere with a thickness of several hundred kilometers [100 kilometers = 62 miles] can explain the relevant observations and is a reasonable model of the moon today." (*Science*, 12/22/72, 1245-9)

- Scientists were developing estimate of interactions between unvisited deep-space planets and solar wind in National Oceanic and Atmospheric Administration project to program future deep-space missions. Series of space environment models had been developed by Ames Research Center scientists Dr. Arthur W. Rizzi, NOAA physicist Dr. Murray Dryer, and Dr. Wen-we Shen of Texas Instruments, Inc. Scaled magnetopause or ionospheric shell had been made for each planet by using available data on earth, Mars, and Venus, and extending models that had worked well for these planets to cases of Jupiter, Saturn, Uranus, Neptune, and Pluto. (NOAA Release 72-161)

- *Boston Globe* editorial commented on *Apollo 17* splashdown. It was "little short of astounding. After a trip of about 500,000 miles [800 000 kilometers], the astronauts brought their command module safely home just a half mile [0.08 kilometer] from the planned splashdown point and within one second—repeat, one second—of the time scheduled when it blasted off two weeks ago. It would be unbelievable if it hadn't happened." (*B Globe*, 12/22/72)

December 23: Twenty-fifth anniversary of transistor—invented Dec. 23, 1947, by Dr. John Bardeen, Dr. Walter H. Brattain, and Dr. William Shockley at American Telephone & Telegraph Co.'s Bell Telephone Laboratories. Invention had introduced age of electronic miniaturization, advanced computer development, and simplified communications. *New York Times* said that, "without the tiny transistor, it is highly unlikely that man would yet have walked on the moon." (Smith, *NYT*, 12/22/72, 45)

- V/A Charles E. Weakley (USN, Ret.), NASA Assistant Administrator for Management Development, died in Bethesda, Md., at age 66. He had retired in 1967 as commander of Atlantic Fleet's antisubmarine warfare force and had held NASA post since 1968. Adm. Weakley had received Legion of Merit and Bronze Star. During World War II he had devised random-screen method of protecting merchant convoys in experiment with early sonar. (*W Star & News*, 12/26/72, A6)

- *Saturday Review* editorial appraised Apollo program's final significance: Militarily "little strategic value beyond increased sophistication in rocketry and global surveillance seems to have been realized." Scientific benefits were difficult to determine until data were analyzed. Program "signals the end—for the time being—of what must be regarded as a kind of dream . . . wherein human captains ply their ships through the reaches of the heavens. In a way we are all vicarious captains of space ships, and with the end of Apollo it will be infinitely more difficult to imagine ourselves aboard craft that are guided only by the circuitry of electronic crews. As machinery for exploration, the human captain has been superannuated, at least for the foreseeable future. The inescapable corollary is that we are earthbound once again, tightly contained in the atmosphere of our single spinning island by insurmountably vast distances and the laws of physics." *Apollo 11* Astronaut Neil A. Armstrong's "giant leap for mankind" had been, perhaps, "that the Apollo program, for all its costs and whatever the legitimacy of its birth, created a cosmic event—earth-rise played on the retinas of terrestrially evolved organisms." (Meyer, *Saturday Review*, 12/23/72, 22)

- Soviet aircraft designer Andrey N. Tupolev died in Moscow at age 84. During 50-yr career Tupolev and associates had pioneered design of 120 different aircraft from 35-hp ANT–1 in 1920s to Tu-144 Soviet supersonic transport scheduled for service in 1975. He had been regarded as originator of all-metal aircraft, which he first designed in early 1920s. In 1934 he had built huge eight-engine aircraft with 63-m (207-ft) wingspan. Aircraft, named *Maxim Gorky* after Russian author, actually flew, but was demolished in 1935 crash. Tupolev, one of most decorated Soviet citizens, had won three Stalin Prizes and one Lenin Prize and was three-time Hero of Socialist Labor—highest Soviet civilian honor. He had been imprisoned in 1936 for allegedly divulging Soviet aviation secrets to Germany. During three years of forced labor he designed Tu-2 dive bomber which was flown by U.S.S.R. in World War II. Among postwar airliners designed by Tupolev design bureau were Tu-134 medium-range, 80-passenger transport and Tu-154, three-engine, 150-passenger transport scheduled to replace Tu-104 on Soviet domestic routes. (*NYT*, 12/24/72, 42)

December 24: NASA released *Apollo 17* photos including one of best full-disc photos of earth taken from space. Photo, taken Dec. 7 as *Apollo 17* spacecraft headed toward moon, showed earth from North Africa and Arabia to Antarctic polar icecap. (Photo 12–H–1578; NASA PIO)

- *Erts 1* Earth Resources Technology Satellite (launched July 23) took photo from space of Nicaraguan capital of Managua that showed why city had been devastated by earthquake Dec. 23. Picture showed that Managua was built on ashes of volcanoes that stretched in straight line from Managua almost to Pacific shore. Quake had registered only 6.5 on Richter scale, but it had been close enough to surface to shake loose ash fill beneath city. (O'Toole, *W Post*, 12/31/72, A2)

- Dr. Rocco A. Petrone, NASA Apollo Program Director, reviewed program in *Los Angeles Times* article: "As one looks back over the 12 years of Apollo, the biggest effort . . . was the engineering effort—the understanding of the demands that would be placed on the flight hardware, the understanding of the laws of nature that we would have to work with in order to fly to the moon, the conversion of dreams into hardware, the manufacture and testing of this hardware, so that we could feel confident to fly our men to the moon and return.

    "But within that overall engineering program, we knew we would be exploring, we knew we would open up new vistas and for that reason we developed the experiments to be taken along on later flights." There had been "only one reason for flying Apollo after the first landing and that was to gather information for science, to continue the exploration of the Moon." Underlying drive of Apollo "—to reach into space, to seize these new challenges, to learn more about the moon and also about earth, to explore for new opportunities for man to use in his future development—is not well understood by the public."

Apollo had "lifted mankind from his cradle and started him on the pathway to the stars." (*LA Times*, 12/24/72)

- Discovery through infrared measurements that Titan, largest of Saturn's six moons, had minimum temperature of 205 K (−90°F), same as atmosphere in which life originated on earth 2 billion yrs ago, was discussed by Cornell Univ. astronomer Dr. Carl E. Sagan in interview published by *Washington Post*. Discovery had been made by Cornell Univ. and Univ. of Minnesota astronomers using infrared measurements. Titan was about 1.5 billion km (900 million mi) from sun and received only one percent of sunlight that earth received. "We have to ask ourselves what kind of a planet would make this atmosphere so far from sun. The only answer we have been able to come up with is something identical to our primeval earth. Possible life form on Titan would not be earthlike, "but it would be earth chauvinism to think that no form of life whatever could survive so far off in the solar system." Dr. Sagan believed Titan was made up of chemical ice, except for small molten core of radioactive rock. Core melted ice far below surface, forcing volcanic ice to surface, where it burst into brilliant red on contact with sun's ultraviolet light. "We see Titan as a red disc through the telescopes, which is exactly what we get in the laboratory when we react methane, hydrogen, water ice, and ammonia with ultraviolet rays." (O'Toole, *W Post*, 12/24/72, A1)

- *New York Times* editorial asked, "Which America?" In same week *Apollo 17* astronauts fired their spacecraft home from moon, "American pilots fired bombs that broke through the heavens over a peasant nation in Asia. America the ingenious and America the vengeful had both struck." U.S. and its people were being judged "for what our Government is doing with its mighty technology. Are we now the enemy—the new barbarians?" (*NYT*, 12/24/72)

December 26: *Cosmos 540* was launched from Plesetsk by U.S.S.R. into orbit with 805-km (500.2-mi) apogee, 785-km (487.8-mi) perigee, 74° inclination, and 100.7-min period. (GSFC *SSR*, 12/31/72; *SBD*, 1/3/73, 2)

- NASA announced immediate changes in key personnel. Director Joseph F. Malaga of Resources Analysis Div. became Assistant Administrator for Institutional Management. Assistant Administrator for Administration William E. Lilly became NASA Comptroller. Malaga had received William A. Jump Memorial Foundation Meritorious Award for exemplary achievement in public administration and NASA Exceptional Service Award in 1966. Lilly, former Director of Program Control in Office of Manned Space Flight, had received NASA Exceptional Service Award in 1965 and 1969. (NASA Ann)

- Phase II of prototype development program for advanced medium STOL (short takeoff and landing) transport (AMST) was authorized by Dr. Robert C. Seamans, Jr., Secretary of the Air Force. Contractors Boeing Co. and McDonnell Douglas Corp. would design, build, and test two prototype aircraft, using existing engines. (DOD Release 863–72)

December 26–31: American Assn. for the Advancement of Science held 139th Annual Meeting in Washington, D.C. Fifteenth Annual Meeting of Society for the History of Technology (SHOT) was held Dec. 27–29 in conjunction with AAAS meeting.

Dissident group, Scientists and Engineers for Social and Political Action, unsuccessfully attempted to disrupt opening AAAS meeting and later held press conference to protest alleged AAAS attempts to "exclude political dissent from scientific and community groups." (Lyons, *NYT*, 12/27/72, 14)

Space Shuttle Payloads Symposium Dec. 27–28 considered space shuttle system and its capability, science payloads, space operation roles, and other aspects of space shuttle system.

Dr. S. Ichtiaque Rasool, Deputy Director of Planetary Programs in NASA Office of Space Science, discussed effect of data derived from Mars and Venus probes on understanding of solar system, during Dec. 29 session on NASA planetary research. (*AAAS Bull*, 11/72; AAAS PAO)

Dr. Raymond L. Bisplinghoff, Deputy Director of National Science Foundation, presented progress report on NSF's Research Applied to National Needs (RANN) Program Dec. 29. RANN, originated in August 1968, was "gaining in strength and effectiveness." It had been receiving "excellent proposals in great numbers from good people" and had in hand "proposals which would cost over $400 million to support—which unfortunately, would somewhat overstrain our present budget." (Text)

William A. Fischer of U.S. Geological Survey said *Erts 1* Earth Resources Technology Satellite (launched by NASA July 23) had discovered at least two major deposits that might contain large quantities of copper. "And if I were a geologist looking for oil and natural gas, I'd buy up every ERTS photograph of Northern Alaska and Northern Canada that I could get my hands on." (Kirkman, *Cl Press*, 12/28/72)

During SHOT meeting Dr. John B. Rae, Chairman of Dept. of Humanities and Social Sciences at Harvey Mudd College, was elected SHOT President for term expiring Dec. 31, 1974; Dr. Eugene M. Emme, NASA Historian, was elected to Advisory Council. (NASA Hist Off)

December 27: U.S.S.R. launched *Cosmos 541* from Plesetsk into orbit with 348-km (216.2-mi) apogee, 218-km (135.5-mi) perigee, 81.3° inclination, and 90.2-min period. Satellite reentered Jan. 8, 1973. (GSFC *SSR*, 12/31/72; 1/31/73; *SBD*, 1/3/73, 2)

- Houston, Tex., public TV station KUHT–TV announced it had been only TV station to carry entire coverage of *Apollo 17* extravehicular activity. On Dec. 11, 12, and 13 station had shown uninterrupted coverage of Astronauts Eugene A. Cernan and Dr. Harrison H. Schmitt on moon. On Dec. 17 station had carried Astronaut Ronald E. Evans' space walk live from earth orbit. KUHT EVA coverage had totaled 25 hrs. British Broadcasting Corp. had sent 13 hrs of *Apollo 17* programming through KUHT's facilities to United Kingdom via satellite. (KUHT Release)

- NASA announced appointment of Raymond J. Sumser as Director of Personnel, effective Jan. 21, 1973. Sumser had been Personnel Director at Goddard Space Flight Center since July 1967 and was awarded NASA's Exceptional Service Medal in 1972. (NASA Ann)

December 27–28: Joint Congressional Committee on Economics held hearings to probe possible Nixon Administration intention to revive U.S. supersonic transport (SST) program that was killed by Congress in 1971. Economist Milton Friedman said he favored building SST in U.S.

"if private enterprise finds it profitable to do so after paying all costs, including any environmental costs imposed by third parties." He opposed governmental subsidizing of SST because "a governmental decision to produce an SST largely at its own expense is a step toward socialism and away from free enterprise." Committee Chairman, Sen. William Proxmire (D-Wis.), criticized failure of Administration witnesses to testify on possible SST plans. "If nothing is planned, why don't they come forward and say so?" (Testimony; UPI, *W Post*, 12/28/72, A7)

December 28: Cosmos 542 was launched from Plesetsk by U.S.S.R. into orbit with 640-km (397.7-mi) apogee, 527-km (327.5-mi) perigee, 81.2° inclination, 96.3-min period. (GSFC *SSR*, 12/31/72; *SBD*, 1/3/73, 2)

December 29: Lunar scientists at Manned Spacecraft Center told press they had not yet been able to determine from study of *Apollo 17* lunar samples whether orange soil found on lunar surface proved existence of lunar volcanic activity. Dr. Wallace C. Phinney, Chief of MSC Geologic Branch, said soil had no "distinct evidence of alteration" as from steam or sulfur gas in volcanic action. "As to what caused the orange coloration, we must wait until we have more chemistry. The most interesting thing is it has very fine grain soil samples, greater than 90 percent in glass." Dr. Paul Gast, Chief of Planetary and Earth Sciences, said glass could indicate volcanic or other crust-related activities. Soil was not particularly rich in water or sulfur, properties associated with volcanic action. (UPI, W *Star & News*, 12/30/72, A3)

- Ames Research Center announced successful completion of $30-million contract with Univ. of Illinois and Burroughs Corp. for design, development, and manufacture of ILLIAC IV computer system.

    ILLIAC IV, most powerful computer in existence, would be integrated into large remote-access system at ARC as computation resource on nationwide Advanced Research Projects Agency (ARPA) network. System would be accessible to universities and Government.

    ILLIAC IV would solve problems conventional computers could not solve economically, in global climate dynamics, distant seismic event detection, multisensor processing, fluid dynamics, and computational aerodynamics. (ARC Release 72-145)

- U.S.S.R. successfully flight-tested new intercontinental missile over 6300-km (3900-mi) distance between Tyuratam in central U.S.S.R. to Kamchatka Peninsula. U.S. intelligence experts later said flight was first successful test of improved missile about as large as Soviet SS-9. (Hoffman, AP, W *Star & News*, 1/7/73, A5)

December 31: Use of solar energy to combat encroaching energy crisis was being examined seriously, Oliver Bell reported in *Washington Post* article. Solar energy panel had been established in Office of Science and Technology's Energy Research and Development Goals Study and NASA was "actively researching whether it can become involved in the solar energy picture." Solar satellite power station concept developed by Peter E. Glaser of Arthur D. Little, Inc., was being considered. Satellite would circle earth in synchronous orbit at 35 900-km (22 300-mi) altitude exposed to sun 24 hrs a day. Solar collector—lightweight panel of solar cells—would convert sun's light into electricity, which would then be changed to microwave energy, beamed to earth, collected on antenna, and reconverted into electricity.

Other proposals included earth-based system to reflect sun's radiation onto solar furnace and boiler atop 460-m (1500-ft) tower, and system to use steel collecting surfaces with "greenhouse" effect to produce temperatures to 813 K (1004°F). Congress would decide whether Government would fund research into unconventional sources of power. (*W Post*, 12/31/72)

During December: House Committee on Science and Astronautics released *Solar Energy Research: A Multidisciplinary Approach*, responses of Federal agencies to requests for information on solar energy research. NASA and National Science Foundation—agencies primarily responsible for Government's solar energy efforts—had recommended Federal program of research and development to apply solar energy to U.S. heat and power needs. NASA had recommended development of economical systems for heating and cooling buildings; economical methods for producing and converting organic materials to liquid, solid, and gaseous fuels; and economical methods for generating electricity. If programs were successful, NASA had said, building heating could reach public use in 5 yrs and building cooling in 6 to 10 yrs, synthetic fuels could be produced from organic materials in 5 to 8 yrs, and electricity could be generated from solar energy in 10 to 15 yrs. There were no significant environmental disadvantages to wide use of solar energy. (Text)

- Asteroid discovered in constellation Leo by Crimean Astrophysical Observatory scientists had been named after late Soviet Cosmonaut Yuri A. Gagarin—first man in space, on *Vostok 1* April 12, 1961—*Spaceflight* reported. (*SF*, 12/72, 457)
- *Armed Forces Journal* article described small tactical aerial mobility platform (STAMP), two-man version of individual flight system designed by Bell Aerospace Div. of Textron Corp. and powered by Williams Research fan-jet engine. STAMP could carry 227-kg (500-lb) payload plus driver and would fly for 30 min with 50-km (30-mi) range at 100–170 km per hr (60–90 knots). Marine Corps was experimenting with STAMP for infantry use. (*AFJ*, 12/72, 20, 34)
- Award by U.K. government of two contracts for definition phase of Geostationary Technology Satellite (GTS) was reported by *Spaceflight*. Hawker Siddeley Dynamics Ltd. would be responsible for project coordination and design of spacecraft and equipment; Marconi Co. Ltd. Space and Defence Systems Div. would design communications package and electronic subsystems. (*SF*, 12/72, 457)
- Task Force on Energy of House Committee on Science and Astronautics' Subcommittee on Science, Research, and Development published *Energy Research and Development*. Report called for immediate implementation of "greatly increased national energy research and development effort" with adequate funding and technical manpower; national energy R&D effort as part of overall policy with White House as focal point; establishment of operating agency to manage Government-supported R&D; increased R&D for environmental protection and energy conservation; and establishment of priorities in basic research, materials research, and solar, geothermal, and nuclear energy research. (Text)

Summary

During 1972: U.S. orbited 39 spacecraft in 30 launches—including *Apollo 16*'s 4 payloads and *Apollo 17*'s 3. U.S.S.R. orbited 89 payloads in 74 launches. Italy launched one satellite for NASA on NASA booster. And Japan launched one spacecraft on its own booster. U.S. total included 16 payloads orbited by Dept. of Defense in 13 launches and 23 by NASA in 17 straight flawless launches—NASA's first perfect annual launch record.

NASA's Apollo program came to successful conclusion with *Apollo 17*, most productive of all NASA lunar landing missions and most scientifically rewarding. *Apollo 17*, sixth successful manned landing mission, set records for longest manned lunar landing flight (301 hrs 51 min), largest lunar sample return (115 kg; 250 lbs), longest total extravehicular activity time (22 hrs 5 min), and longest time in lunar orbit (147 hrs 48 min). Earlier in year *Apollo 16* also produced valuable data from surface and lunar orbit investigations.

Unmanned NASA program was highlighted by *Pioneer 10*, first spacecraft to fly beyond Mars orbit, first to penetrate Asteroid Belt, first intended to look at Jupiter close up, and first destined to escape solar system. At year's end *Pioneer 10* had passed safely through three fourths of Asteroid Belt. *Mariner 9*, launched in 1971, had circled Mars 698 times, photographing Martian moons and completely mapping planet's surface before Oct. 27, 1972, shutdown. Findings revised previous concepts of Mars, showing planet to be geologically active, with volcanic mountains higher than any mountain on earth, with a crevasse three to four times deeper than Grand Canyon, and with indications that free-flowing water might once have existed on Mars. *Erts 1*, first Earth Resources Technology Satellite, laid groundwork for global inventory of earth's resources and environment. *Copernicus* (*Oao 3*) Orbiting Astronomical Observatory, with largest telescope placed in space, set new standards of pointing accuracy and opened new channels of information on far reaches of universe. *Nimbus 5* meteorological satellite demonstrated new techniques for daytime and nighttime temperature readings through clouds at all levels in atmosphere.

Other unmanned missions included *Explorer 47* (IMP) to study earth-sun interactions from orbit halfway to moon, *Explorer 48* (SAS; launched for NASA by Italy) to survey gamma ray sources, *Noaa 2* weather satellite for National Oceanic and Atmospheric Administration, and *Triad OI–1X* Transit satellite for Navy. Two dozen spacecraft orbited in earlier years continued to return data on solar system and universe beyond.

All components of Skylab manned orbital research satellite were assembled at Kennedy Space Center for 1973 launch. Three flight crews and two backup crews began training. Final space shuttle design was selected and prime contract for shuttle orbiter development was awarded, as well as main engine contract. Contractors had progressed

U.S. AND U.S.S.R. SPACE PROGRAMS, 1957–1972

Selected Cumulative Totals Through December 1972

Activity	U.S.	U.S.S.R.	World Total
Successful launches into orbit or beyond.	575	621	1216 (incl. 20 by other nations)
Spacecraft launched (including multiple payloads on single boosters).	772	679	1472 (incl. 21 by other nations)
Spacecraft still in orbit.	361*	205	600 (incl. 34 belonging to other nations)
Lunar missions (that reached or passed moon).	30	25	55
Unmanned.	21	25	46
Manned.	9	0	9
Planetary and interplanetary probes.	12	13	25
Manned space flights.	27	18	45
Number of crewmen who have flown in space.	34	25	59
Total manned space-flight duration.	3526 hrs	2097 hrs	5 623 hrs
Total man-hours in space.	9500 hrs	4403 hrs	13 903 hrs
Total extravehicular activity (EVA)			
Duration.	96 hrs 2 min	1 hr 12 min	97 hrs 14 min
Man-hours.	177 hrs 7 min	1 hr 12 min	178 hrs 19 min
Duration on moon.	80 hrs 39 min	0	80 hrs 39 min
Man-hours on moon.	161 hrs 18 min	0	161 hrs 18 min
Total time lived on moon.	301 hrs	0	301 hrs
International launches**	50	10	60
Launched for other nations.	10	2	12
Launched for international organizations or programs.	33	8	41
Launched by other nations with U.S. aid.	4		4
Launched by other nations for U.S.	3		3

* Does not include spacecraft launched for other nations.
** In addition to experiments carried for other nations and to launches of cooperative sounding rockets.

well into system definition, design, and early development. Kennedy Space Center and Vandenberg Air Force Base were selected as shuttle launch and landing sites.

M2-F3 lifting body completed 27th and last scheduled flight in joint NASA and Air Force program begun July 12, 1966. During 14 lifting-body flights in 1972, maximum speed of 1718 km per hr (1066 mph) and maximum altitude of 21 800 m (17 500 ft) were attained.

In aeronautics, Flight Research Center celebrated 25th anniversary of its creation in 1947 as Muroc Flight Test Unit of National Advisory Committee for Aeronautics. NASA TF-8A jet aircraft with supercritical wing was flown in configuration planned for future flights. Digital fly-by-wire computerized control system was demonstrated in series of flights from FRC. System, for potential use in space shuttle, enhanced aircraft handling and reduced pilot workload. NASA awarded contract to refine design of quiet, jet, propulsive-lift, experimental aircraft QUESTOL in major effort to develop technology for short takeoff and landing (STOL) aircraft to relieve airport congestion. Progress was made in quiet jet engine, techniques to reduce aircraft emissions, composite materials, avionics, aerodynamics including supercritical and antisymmetrical wings, general-aviation safety, supersonic and hypersonic research, and research in basic materials and structures.

In sounding rocket program, more than 80 successful flights were made to study atmosphere, ionosphere, auroras and airglow, geomagnetic storms, meteor streams, and trapped radiation fluctuations; to make astronomical observations in x-ray, ultraviolet, and radio regions of electromagnetic spectrum; for special projects; and for support. NASA flew 61 balloon flights in its program to study near-earth phenomena, including balloon that lifted 545 kg (1200 lbs) of scientific equipment to 45 000-m (148 000-ft) altitude.

U.S.S.R. launches decreased from record 83 in 1971 to 74. The 89 payloads included 72 Cosmos satellites, 3 Intercosmos, 1 Luna, 3 Meteor, 2 Prognoz, 3 Molniya I, 3 Molniya II, 1 Venus, and secondary payload *Sret I* (launched for France).

Japan launched *Denpa* Radio Explorer Satellite. (NASA Releases 73-48 and unnumbered of 12/27/72; NASA Lifting Body Off; GSFC Sounding Rocket Br.; GSFC SSR, 11/31/72; 12/31/72; *A&A 1972*; Myers testimony, 2/27/73)

- President Nixon's January decision to proceed with development of reusable space shuttle keyed transition in U.S. space program from space exploration to space exploitation for practical and immediate benefits to mankind. While budget restrictions and national emphasis of technology applications to improve conditions on earth brought early end to Apollo program, U.S.-U.S.S.R. space agreement opened new era of international space cooperation. Agreement inspired intensive planning for 1975 joint rendezvous and docking mission of Soviet Salyut and U.S. Apollo spacecraft in orbit.

Apollo program—major drive of U.S. space program since 1961, triumphantly concluded in 1972—had begun to give clear picture of moon, how it had evolved and its relationship to sun and earth, and would contribute to understanding of how earth had evolved. Apollo program had also given man new awareness of earth and had given Americans self-confidence in ability to achieve gigantic goal.

During 1972

Picture of a manned spaceship
... (courtesy of NASA)

During 1972: *Conclusion of the Apollo program was commemorated by Scott Long's cartoon, published in the* Minneapolis Tribune *Dec. 6. The manned lunar explorations had given man a new awareness of the earth and the need to preserve it, as well as a wealth of information on the moon and possible origins of the earth and solar system.*

New space program for 1970s was designed to be funded at constant budget level over several years—yet to maintain space flight capability, contribute to national security, increase opportunities for new scientific knowledge, encourage international cooperation, address problems of society and environment, and continue U.S. leadership in aeronautics. Reusable shuttle was expected to give "routine access to space" by cutting complexity and cost of launches and multiplying uses of satellites. It could be used for both manned and unmanned missions and, in military use, could serve national security.

Revision of NASA FY 1972 program to phase in new programs while holding expenditures to $3.4-billion budget level required basic changes in planning. Further stretch-out of period over which NASA long-term objectives in space and science could be achieved was necessary. Shuttle configuration costs were pared to about half of original cost estimate; plans for Grand Tour missions to explore all five outer planets in late 1970s were cut back to focus on Jupiter and Saturn with less expensive spacecraft; and NASA and Atomic Energy Commission's nuclear engine for rocket vehicle application (NERVA) program was terminated. Emphasis on aeronautics was increased at urging of congressional committees, but $44.9 million of $50.5-million NASA FY 1973 appropriation for aircraft noise suppression program, engine retrofit studies, and quiet, experimental, short takeoff and landing (QUESTOL) aircraft program was impounded by Office of Management and Budget. Tight NASA funding and end of Apollo program brought reductions in personnel, especially at manned space-flight Centers. NASA total in-house personnel decreased by 1181 from Dec. 31, 1971, to 27 801 persons Dec. 31, 1972, with further cuts scheduled for 1973. Marshall Space Flight Center employment was reduced by 258, to 5501 persons; Manned Space Center 159, to 3882; and Kennedy Space Center 123, to 2518.

Employment in aerospace industry throughout the Nation also declined slightly, from 924 000 workers at end of 1971 to 917 000 at end of 1972, despite infusion of space shuttle program. Aerospace industry remained largest manufacturing employer, however, and industry sales increased for first time since 1968.

To emphasize transition from Apollo to Skylab and shuttle activities, NASA established new Mission and Payload Integration Office within Office of Manned Space Flight. Space Station Task Force became Sortie Lab Task Force. Apollo Program Office became Apollo/ASTP Office to prepare for 1975 U.S.–Soviet mission. Office of Advanced Research and Technology (OART) was renamed Office of Aeronautical and Space Technology (OAST) to emphasize increasing NASA aeronautical activity. Office of Applications had been separated from Office of Space Science at end of 1971 to increase emphasis on use of space technology for benefits on earth. Charles W. Mathews, Deputy Associate Administrator for Manned Space Flight since 1968, headed new Office of Applications.

Dr. Wernher von Braun retired as NASA Deputy Associate Administrator for Planning to become Corporate Vice President for Engineering and Development at Fairchild Industries, Inc., after more than quarter of century as leader of space rocket development in the U.S. His NASA career had begun in 1960 when his Army Ballistic Missile Agency team—which had launched first U.S. satellite, *Explorer I* in 1958—was transferred to form nucleus of Marshall Space Flight Center. Dr. von Braun had served as MSFC Director from 1960 to 1970, where

Saturn V launch vehicle for Apollo moon missions was his major engineering achievement.

In technology utilization, NASA continued to distribute information about aerospace technology to private and public sectors of national economy at ever-increasing rate. NASA-developed technology was being used in medicine, nondestructive testing, and engineering design. NASTRAN computer program to analyze structures under stress was used by more than 70 industrial firms, universities, laboratories, and government agencies to solve structural engineering problems.

Technology and interpretive systems developed by NASA in cooperation with other agencies prepared way for worldwide environmental pollution and monitoring network.

NASA FY 1973 appropriation of $3.408 billion was equivalent to budget request and $110 million above FY 1972 appropriation of $3.298 billion. Increase reflected funds for design and development of space shuttle. (*A&A 1972;* Fletcher testimony, 2/28/73; AIA *Aerospace*, 3/73; NASA *Pocket Statistics*, 3/72, 1/73)

- In international cooperation program, NASA successfully launched European Space Research Organization's *Heos 2* to study interplanetary space and magnetosphere, *Esro 4* to investigate solar ionosphere, and *TD-1A* astronomical observatory satellite that carried seven experiments from six European universities and research organizations. NASA also launched *Anik 1* (*Telesat A*) domestic communications satellite for Telesat Canada, *Aeros* scientific satellite in joint U.S.–West German project, and *Oscar 6* (developed by U.S., German, and Australian radio amateurs) for Radio Amateur Satellite Corp.—in addition to two International Telecommunications Satellite Consortium comsats. Italian crew trained by NASA launched NASA's *Explorer 48* Small Scientific Satellite from Italian platform off coast of Kenya.

May 24 agreement signed in Moscow by President Nixon and Soviet Premier Aleksey N. Kosygin on cooperation in peaceful uses of outer space included Apollo Soyuz Test Project in which Apollo spacecraft would rendezvous and dock with Soviet Soyuz spacecraft in earth orbit. Mission was scheduled for July 1975 and meetings in U.S. and U.S.S.R. advanced project to point of testing models of spacecraft linking device. U.S. and U.S.S.R. exchanged data from 1971 *Mariner 9* and *Mars 2* and *3* probes of planet Mars.

After series of meetings in U.S. and abroad, ESRO decided to develop sortie laboratory to fly with NASA's space shuttle in 1980s.

President Nixon announced new policy making U.S. launch assistance available to interested countries and international organizations for satellites intended for peaceful purposes.

Third Annual Lunar Science Conference at Lunar Science Institute in January was attended by more than 600 researchers from U.S. and 17 foreign countries. Participating Soviet scientists consulted with U.S. scientists, examined lunar samples, and helped select *Apollo 14* samples to return to U.S.S.R. as part of U.S.–U.S.S.R. exchange agreement. *Apollo 15* lunar soil samples were exchanged for soil samples collected by U.S.S.R.'s *Luna 20*. *Apollo 16* materials were distributed to scientists in 15 foreign countries, and *Apollo 17* samples also would be distributed to other nations for analysis and display.

NASA launched 29 sounding rockets in cooperative international program to test weather data systems for France, Japan, and U.S. Program was sponsored by United Nations World Meteorological Organization's Commission for Instruments and Methods of Observation.

Since beginning of international program in 1962 with launch of U.K. satellite *Ariel 1* (April 26) and Canadian satellite *Alouette 1* (Sept. 29), number of foreign countries and international organizations with which NASA had participated in space activities had reached 134. Total of 55 foreign agreements included cooperative flight projects with 25 countries, earth resources survey with 39, lunar sample analysis with 20, tracking and data acquisition with 22, and reimbursable launchings with 4. Scientists from 85 countries and organizations had worked with NASA—80 in meteorological research, 41 in personnel exchanges, and 52 in other space projects. Total countries and organizations exchanging scientific and technical information with NASA reached 63; total sending visitors to NASA reached 126, and total cooperating in some form with NASA reached 94. (NASA Release, 12/27/72; *A&A 1972*; Fletcher, *NASA Activities*, 1/15/73, 2–3; NASA Int Aff)

- Aerospace industry sales increased for first time since 1968. Sales for 1972 were $23.5 billion, up 5.9% over 1971. Commercial aerospace sales increased 11.6%, to reach estimated $4.8 billion. Increase reflected increase in deliveries of helicopters and general-aviation aircraft. Major aerospace sales included $13.8 billion to Dept. of Defense, up from $12.6 billion to DOD in 1971; missile sales to DOD at $5.2 billion, up from $4.7 billion; and military aircraft sales at $8.1 billion, up from $7.4 billion.

  Space sales continued to decline in 1972, to $3.0 billion from $3.2 billion in 1971. Utility and executive aircraft sales increased from $321 million in 1971 to $500 million—up 55.8%. Units delivered increased 23.2%. Civilian helicopter sales increased from $69 million in 1971 to $95 million in 1972, gain of 37.7%

  Aerospace exports declined for first time since 1964—from $4.2 billion in 1971 to $3.9 billion, 6.6% decrease. Major reason for decline was drop in military aerospace exports—down 23.8% from $1.1 billion in 1971 to $854 million in 1972. Aerospace imports in 1972 were valued at $500 million, increase of 34.0% from $373 million in 1971. Aerospace industry profits were expected to increase from 1.8% in 1971 to 2.2%.

  Aerospace industry employment declined from 924 000 in December 1971 to estimated 917 000 in December 1972, but industry remained Nation's largest manufacturing employer. (AIA *Aerospace*, 3/73, 3)

- Worldwide scheduled airline industry increased 14% in scheduled traffic over 1971—double 1971 growth rate. Total 1972 traffic was estimated at 47 120 million ton-miles. Revenue-passenger-miles flown increased 13% over 1971, with estimated 450 million passengers flown. U.S. international carriers showed 22.5% increase in revenue-passenger-miles, with 12% increase in available seat-miles. Load factors rose 4.8% over 50% level of 1971. U.S. domestic carriers increased revenue-passenger-miles 11.2%, with only 1.9% increase in available seat-miles. Average load factor rose to 52.4% from 48% in 1971. (*Av Wk*, 1/8/73, 22)

- At least 1944 persons died in civilian aircraft accidents involving aircraft weighing more than 5700 kg (12 500 lbs). World accident rate in terms of facilities per 100 000 flight-hours was 0.126, up from 0.094 in 1971 and highest rate since 1969's 0.134. Number of actual deaths in U.S. airline accidents dropped from 203 in 1971 to 190 in 1972. (*Interavia*, 3/73, 201)
- NASA published *Physics of the Space Environment* (NASA SP-305). Series of lectures, presented originally in colloquia at Univ. of Alabama 1970–1971, had been sponsored by Marshall Space Flight Center's Aero-Astrodynamics Laboratory. Major topics covered were dynamics and transient state of upper atmosphere, chemical composition of upper atmosphere, and solar flare forecasting. Flare forecasting "fiasco" was described by Frederick W. Ward, Jr., Air Force Cambridge Laboratories physicist: Forecasters dreamed of discovering cause, energy source, and antecedent conditions for solar flares. "This 'tunnel vision' has had a profound and deleterious effect on the state-of-the-art. It has diverted attention from many promising approaches and relegated them to limbo. More importantly, it has actually inhibited the scientific search for the underlying physical mechanisms." Examples of neglected aspects ranged from observing techniques and equipment to availability and accuracy of archived data. Inspection of available data "leads to an obvious conclusion: there are at least two suns up there." (Text)
- Federal Aviation Administration completed installation of automated flight-data processing system in all 20 of its contiguous air route traffic control centers, rendered 34 automated radar terminal systems (ARTS III) operational, and installed 53 instrument landing systems. (DOT Release 99-S-72)
- U.S. helicopter manufacturers exported record total of 259 rotary-wing aircraft valued at $73 673 000. Shipment was 17% increase in dollar value and 6% increase over the 244 helicopters valued at $62 853 000 shipped in 1971. (AIAA Release 72-2)
- Kennedy Space Center minority and female employees increased steadily. During first 18 mos of KSC Equal Employment Opportunity—from April 1971 through September 1972—contractor employees increased by 249, to 11 382. Female employment increased by 128 persons, to 1240, an 11.5% increase. Minority personnel members increased from 609 to 847, 39.1%. During third quarter 1972, KSC contract personnel increased by 162 and minority employees increased by 70 to peak 7.44% of KSC contractor work force. Minority population in Brevard County, based on 1970 census, was 9.4%. (KSC Release 348-72)
- Smithsonian Institution's National Air and Space Museum completed first year under new Director, former astronaut Michael Collins. Major emphasis was on preparation for new museum to be opened in U.S. Bicentennial year, 1976. Domed planetarium chamber 9 m (30 ft) high was being constructed as laboratory for experimentation and design of programs and equipment for Spacearium in new museum. Museum library was being reorganized to standards of other Smithsonian Institution libraries.

    Smithsonian Astrophysical Observatory (SAO) progress and plans were being assessed against standards recommended by Astronomy

Survey Committee of National Academy of Sciences and National Research Council [see June 1]. (*Smithsonian Year 1972*, 61, 62)

- Swiss Transport Museum in Lucerne had record 608 000 visitors—increase of about one third over 1971. Museum was considered most active museum in Switzerland and Europe's most popular transport museum. Planetarium rated second to New York's American Museum–Hayden Planetarium in number of annual visitors. Swiss Transport Museum's Air and Space Wing was dedicated July 1. Dedication ceremonies were attended by Dr. James C. Fletcher, NASA Administrator; *Apollo 11* Astronaut Neil A. Armstrong; and Mercury Astronaut John H. Glenn. During year, James E. Webb, former NASA Administrator, officially presented model of Goddard Space Flight Center's computer facility to Museum. (Dir Swiss Trans Museum letter to NASA Hist Off, 2/19/72; *A&A 1972*)

Appendix A

SATELLITES, SPACE PROBES, AND MANNED SPACE FLIGHTS

A CHRONICLE FOR 1972

The following tabulation was compiled from open sources by Leonard C. Bruno of the Science and Technology Division of the Library of Congress. Sources included the United Nations Public Registry; the *Satellite Situation Report* compiled by the Operations Control Center at Goddard Space Flight Center; and public information releases of the Department of Defense, NASA, NOAA, and other agencies, as well as those of the Communications Satellite Corporation. Russian data are from the U.N. Public Registry, the *Satellite Situation Report*, translations from the Tass News Agency, statements in the Soviet press, and international news services reports. Data on satellites of other foreign nations are from the U.N. Public Registry, the *Satellite Situation Report*, governmental announcements, and international news services reports.

This tabulation lists payloads that have (a) orbited; (b) as probes, ascended to at least the 6500-kilometer (4000-mile) altitude that traditionally has distinguished probes from sounding rockets, etc.; or (c) conveyed one or more human beings into space, whether orbit was attained or not. Furthermore, only flights that have succeeded—or at least can be shown by tracking data to have fulfilled our definition of satellite or probe or manned flight—are listed. Date of launch is referenced to local time at the launch site. An asterisk by the date marks dates that are one day earlier in this tabulation than in listings which are referenced to Greenwich Mean Time. A double asterisk by the date marks dates of Soviet launches which are a day later in this compilation than in listings which are referenced to Greenwich Mean Time.

World space activity decreased for the first time in three years. There was a decrease in the total successful launches—106 against 120 in 1971, the lowest since 1964—and a sharp drop in total payloads orbited—126 against 151 in 1971. The difference between launches and payloads is of course accounted for by the multiple-payload launches (DOD, the principal user of this system in the past, made 3 multiple launches in 1972, orbiting 6 payloads; NASA made only 1 multiple launch, with 2 payloads; the U.S.S.R. made 3 multiple launches, orbiting 18 payloads).

Of the 1972 world total, the United States launched 30 boosters carrying 35 payloads (counting the *Apollo 16* and *17* payloads as 1 for each mission). In 1971 29 U.S. boosters launched 47 payloads. Although the U.S. 1972 launch total increased by 1, its payload total was the lowest since 1961. Of these 1972 totals, DOD was responsible for 13 launches and 16 payloads. Of NASA's 17 launches, 8 were non-NASA missions—*Intelsat-IV F-4* and *Intelsat-IV F-5* for ComSatCorp, *Heos 2*, *TD-1A*, and *Esro 4* for ESRO, *Anik 1* for Canada, *Aeros* for West Germany, and *Triad OI-1X* for the U.S. Navy. The Soviet Union once again more than doubled United States space totals, launching 89 payloads with 74 launches. It had launched 97 payloads with a record 83 launches in 1971.

The year was very successful for both manned and unmanned activity. The U.S. completed its Apollo program of manned lunar exploration with two immensely successful flights. *Apollo 16* explored the moon's Descartes region and returned with 96.6 kg (213 lbs) of lunar samples. *Apollo 17*, the sixth and last manned lunar landing, was the most productive and most scientifically rewarding. Among its many records, this longest of manned lunar landing flights spent the longest time in lunar orbit, conducted the longest total EVA on the moon, and returned with the largest sample of lunar material (115 kg; 250 lbs). *Pioneer 10*, launched early in the year, successfully streaked through the Asteroid Belt toward its December 1973 Jupiter flyby objective and would eventually become the first spacecraft to escape the solar system. NASA also launched *Erts 1*, the first of a series of experimental satellites to monitor the earth's resources and environment; *Oao 3*, the largest telescope ever placed in space; *Nimbus 5*, a more sophisticated experimental meteorological satellite; and *Anik 1* for Canada, which became the world's first operational domestic communications satellite system.

The Soviet Union scored a major interplanetary success with the soft-landing on Venus of a capsule from *Venus 8*, which conducted the first soil analysis of another planet. *Luna 20* continued a series of unmanned Soviet lunar accomplishments with its successful lunar landing and retrieval of lunar soil. The U.S.S.R., in a typically active year, also orbited six communications satellites, including an improved second-generation series, and three Meteor weather satellites.

Also during 1972, Japan launched its fourth satellite, the second one to carry scientific instruments. Italy launched the U.S.-built *Explorer 48* on a NASA vehicle from its launch facility off the coast of Kenya. And the Soviet Union launched a French-built scientific satellite, *Sret 1*, as a secondary payload. It was the first of three planned launches in a cooperative French-Soviet program.

As we have cautioned in previous years, the "Remarks" column of these appendixes is never complete, because of the inescapable lag behind each flight of the analysis and interpretation of results.

ASTRONAUTICS AND AERONAUTICS, 1972

Launch Date	Name, Country, International Designation, Vehicle	Payload Data	Apogee in Kilometers (and st mi)	Perigee in Kilometers (and st mi)	Period in Minutes	Inclination in Degrees	Remarks
Jan. 12	Cosmos 471 (U.S.S.R.) 1972-1A Not available	Total weight: Not available. Objective: "Continuation of Cosmos scientific satellite series." Payload: Not available.	317 (197)	194 (120.5)	89.5	64.9	Reentered 1/25/72.
Jan. 20	DOD Spacecraft (United States) 1972-2A Titan IIID and	Total weight: Not available. Objective: Develop space flight techniques and technology. Payload: Not available.	340 (211)	150 (93)	89.3	96.9	Two spacecraft launched with single booster. Reentered 2/29/72.
	DOD Spacecraft 1972-2D	Total weight: Not available. Objective: Develop space flight techniques and technology. Payload: Not available.	550 (342)	470 (292)	94.8	96.5	Still in orbit.
Jan. 22*	Intelsat-IV F-4 (United States) 1972-3A Atlas-Centaur	Total weight: 1402 kg (3090 lbs) at launch; 720 kg (1587 lbs) after apogee motor firing. Objective: Place satellite and apogee motor into proper transfer orbit; provide tracking and telemetry and backup calculations through transfer orbit so satellite can be injected into synchronous orbit for commercial communications. Payload: 528-cm-high x 238-cm-dia (208- x 93.7-in) cylindrical satellite capable of carrying 3000-9000 telephone circuits simultaneously or 12 color TV channels or a combination; spin-stabilized; 12 communications repeaters (transponders); 6 antennas (2 transmit horns, 2 receive horns, and 2 steerable 127 cm [50 in] dish spot-beam antennas); 42 240 solar cells.	36 523 (22 694.3) After apogee motor firing, 35 787 (22 237)	5984 (3718.3) 35 625 (22 136.4)	1432	28.2 0.7	Launched by NASA into good transfer orbit for ComSatCorp. Apogee kick motor fired 1/24/72, stationing satellite in synchronous orbit over Pacific Ocean at 174° east longitude. Used during President Nixon's China trip. Spacecraft operating normally. Still in orbit.
Jan. 25	Cosmos 472 (U.S.S.R.) 1972-4A Not available	Total weight: Not available. Objective: "Continuation of Cosmos scientific satellite series." Payload: Not available.	1536 (954.4)	193 (119.9)	102.2	82	Reentered 8/18/72.

Launch Date	Name, Country, International Designation, Vehicle	Payload Data	Apogee in Kilometers (and st mi)	Perigee in Kilometers (and st mi)	Period in Minutes	Inclination in Degrees	Remarks
Jan. 31	Heos 2 (U.S.-ESRO) 1972-5A DSV-3L	Total weight: 117 kg (257 lbs). Objective: Place satellite in earth orbit that will permit investigation of interplanetary space and of high-latitude magnetosphere and its boundary in region of northern neutral point. Payload: 16-sided cylindrical polyhedron 130 cm high x 70 cm in dia (51.2 x 27.6 in) contains central octagonal tube 74 cm (29.1 in) across. Tripod axial boom and experiment BLF loop antenna extend over length to 239 cm (94.2 in). Central tube holds most of spacecraft's experiment instrumentation, electronic controls, batteries, telemetry and command equipment. Contains seven experiments. About 70% of outer surface covered with solar cells.	244 011 (151 621.4)	416 (258.5)		90.2	ESRO-built Heos 2 launched into highly eccentric orbit by NASA three-stage Thor-Delta. All experiments turned on and spacecraft fully operational. Still in orbit.
Feb. 3	Cosmos 473 (U.S.S.R.) 1972-6A Not available	Total weight: Not available. Objective: "Continuation of Cosmos scientific satellite series." Payload: Not available.	356 (221.2)	176 (109.4)	89.6	65	Reentered 2/15/72.
Feb. 14	Luna 20 (U.S.S.R.) 1972-7A Not available	Total weight: Not available. Objective: Carry out scientific investigation of moon and circumlunar space; automatic delivery of lunar soil to earth. Payload: 2-stage (ascent and descent) spacecraft; 4 legs; telephotometer; mechanical arm with improved core drilling apparatus.	Lunar orbit, 100 (62)	100 (62)	118	65	Spacecraft entered lunar orbit 2/18/72 and landed between moon's Sea of Fertility and Sea of Crises 2:19 pm EST 2/21/72. Earth-operated drilling rig penetrated lunar surface to 35 cm (13.8 in). Samples obtained and transferred to container in return capsule and hermetically sealed. Spacecraft remained on moon 27 hrs 39 min; liftoff from moon 5:58 pm EST 2/22. Returned to earth 2:12 pm EST 2/24. Total flight time 11 days 16 hrs. Analysis of lunar samples indicate anorthosite. Findings contrast with Luna 16 Sea of Fertility samples, which were primarily basaltic rock.

ASTRONAUTICS AND AERONAUTICS, 1972

Date	Spacecraft	Description	Orbit	Weight kg (lbs)	Apogee km (mi)	Period (min)	Inclination (deg)	Remarks
Feb. 16	Cosmos 474 (U.S.S.R.) 1972-8A Not available	Total weight: Not available. Objective: "Continuation of Cosmos scientific satellite series." Payload: Not available.		365 (226.8)	179 (111.2)	90	65	Reentered 2/29/72.
Feb. 25	Cosmos 475 (U.S.S.R.) 1972-9A Not available	Total weight: Not available. Objective: "Continuation of Cosmos scientific satellite series." Payload: Not available.		1003 (623.2)	967 (600.9)	104.8	74.1	Still in orbit.
Mar. 1	DOD Spacecraft (United States) 1972-10A Titan IIIC	Total weight: Not available. Objective: Develop space flight techniques and technology. Payload: Not available.		35 964 (22 347)	35 418.5 (22 008)	1429.9	0.2	Still in orbit.
Mar. 1	Cosmos 476 (U.S.S.R.) 1972-11A Not available	Total weight: Not available. Objective: "Continuation of Cosmos scientific satellite series." Payload: Not available.		634 (394)	615 (382.1)	97.1	81.2	Still in orbit.
Mar. 2*	Pioneer 10 (United States) 1972-12A Atlas-Centaur-TE-M-364-4	Total weight: 258 kg (569 lbs). Objective: Obtain, during 1972 Jovian opportunity, precursory scientific information beyond orbit of Mars with emphasis on: (a) investigation of interplanetary medium, (b) investigation of nature of asteroid belt, (c) exploration of Jupiter and its environment. Payload: Hexagonal spacecraft with auxiliary offset hexagonal compartment for scientific instruments and 2.7-m (9-ft)-dia parabolic antenna reflector; 4 radioisotope thermoelectric generators (RTG) on trusses 120° apart extend 1.7 m (5.6 ft) radially beyond periphery of antenna reflector; 4-segment folding magnetometer boom extends radially 4.7 m (15.4 ft) beyond reflector and 120° from RTGs; from top of high-gain antenna feed to bottom of low-gain antenna measures 2.9 m (9.5 ft). Eleven scientific instruments measure magnetic fields; plasma; cosmic rays and charged particles; electromagnetic radiation in ultraviolet, visible and infrared ranges; and asteroid/meteoroid population. Six thrusters; spin-stabilized.	Heliocentric, later to become solar-escape trajectory					*Pioneer 10* first of new generation Pioneer series. Reached highest launch velocity ever attained at 51 500 km per hr (32 000 mph) with first use of Atlas-Centaur as 3-stage vehicle. First NASA spacecraft powered entirely by nuclear energy and first intended to escape solar system ultimately into interstellar space. By 2/15/73, spacecraft had emerged undamaged from 7-mo journey through Asteroid Belt and was about 550 million km (340 million mi) from sun. During flight, spacecraft experiments found elements sodium and aluminum among solar high-energy particles for first time, also measured helium atoms in interplanetary space for first time. Spacecraft operating normally and Jupiter encounter scheduled 12/3/73.

461

Launch Date	Name, Country, International Designation, Vehicle	Payload Data	Apogee in Kilometers (and st mi)	Perigee in Kilometers (and st mi)	Period in Minutes	Inclination in Degrees	Remarks
Mar. 4	Cosmos 477 (U.S.S.R.) 1972-13A Not available	Total weight: Not available. Objective: "Continuation of Cosmos scientific satellite series." Payload: Not available.	309 (192)	214 (133)	89.5	72.8	Reentered 3/16/72.
Mar. 11*	TD-1A (U.S.–ESRO) 1972-14A Thor-Delta N	Total weight: 472 kg (1038 lbs). Objective: Place satellite in earth orbit that will permit UV spectrometer measurements of celestial sphere in approximate 180-day cycle. Payload: Boxlike structure 98 x 216 cm (38.6 x 85 in) resembles large refrigerator. Divided into 2 sections, smaller bottom section contains all of spacecraft electronics and other subsystems, and top section contains experiment complement. Two large solar panels attached to sides deploy when orbit achieved. Fixed to spacecraft bottom is 2.7-m (106.3-in) omnidirectional antenna which deploys after orbital insertion; 3-axis stabilized; contains 7 experiments.	539 (334.9)	531 (329.9)	95.3	97.6	ESRO-built satellite launched into sun-synchronous orbit by NASA 2-stage Thor-Delta. Largest and most advanced scientific satellite built in Western Europe. Spacecraft 3-axis stabilized and all experiments turned on. Performed first European "hibernation" mode as satellite preserved intact its full power and systems during partial solar eclipse. Still in orbit.
Mar. 15	Cosmos 478 (U.S.S.R.) 1972-15A Not available	Total weight: Not available. Objective: "Continuation of Cosmos scientific satellite series." Payload: Not available.	282 (175.2)	176 (109.4)	89	65	Reentered 3/28/72.
Mar. 17	DOD Spacecraft (United States) 1972-16A Titan IIIB-Agena	Total weight: Not available. Objective: Develop space flight techniques and technology. Payload: Not available.	396 (246)	141 (88)	89.8	110.9	Reentered 4/11/72.
Mar. 22	Cosmos 479 (U.S.S.R.) 1972-17A Not available	Total weight: Not available. Objective: "Continuation of Cosmos scientific satellite series." Payload: Not available.	541 (336.2)	514 (319.4)	95.1	74	Still in orbit.

ASTRONAUTICS AND AERONAUTICS, 1972

Date	Spacecraft	Description	Perigee km (mi)	Apogee km (mi)	Period (min)	Inclination (deg)	Remarks
Mar. 24	DOD Spacecraft (United States) 1972–18A Thor-Burner II	Total weight: Not available. Objective: Develop space flight techniques and technology. Payload: Not available.	884 (549)	802 (498)	101.7	98.7	Still in orbit.
Mar. 25	Cosmos 480 (U.S.S.R.) 1972–19A Not available	Total weight: Not available. Objective: "Continuation of Cosmos scientific satellite series." Payload: Not available.	1202 (746.9)	1174 (729.5)	109.1	82.9	Still in orbit.
Mar. 25	Cosmos 481 (U.S.S.R.) 1972–20A Not available	Total weight: Not available. Objective: "Continuation of Cosmos scientific satellite series." Payload: Not available.	511 (317.5)	269 (167.2)	92.3	71	Reentered 9/2/72.
Mar. 27	Venus 8 (U.S.S.R.) 1972–21A Not available	Total weight: 1177.5 kg (2596 lbs). Objective: Continuation of investigation of planet Venus. Payload: Spacecraft consisted of service module and descent module; contained stellar tracking unit, orientation nozzles, cosmic particle counter, narrow-angle parabolic antenna, radiator for temperature control system, S-band antenna, and solar panels.	Softlanded on Venus.				Venus 8 softlanded 495-kg (1091-lb) spherical capsule on Venus 7/22/72. Capsule survived landing and transmitted data for 50 min. Performed first soil analysis of another planet. Surface layer density of Venus measured at 0.8 kg per cu m (0.05 lb per cu ft) and resembled earth's granite rock. Determined that, while Venusian atmosphere considerably weakens sunlight, sunlight does reach planet's surface. Dayside temperature measured at 738 K (869°F) and atmospheric pressure 93 times that of earth. Venusian atmosphere 97% carbon dioxide.
Mar. 30	Meteor 11 (U.S.S.R.) 1972–22A Not available	Total weight: Not available. Objective: Acquisition of meteorological information for use by weather service. Payload: Cylindrical body with 2 large solar paddles attached; 3 data collection systems (TV cameras, infrared sensors, and actinometric scanners); 3-axis attitude control system.	890 (553)	867 (538.7)	102.5	81.2	Meteor 11 meteorological satellite still in orbit.
Mar. 31	Cosmos 482 (U.S.S.R.) 1972–23A Not available	Total weight: Not available. Objective: "Continuation of Cosmos scientific satellite series." Payload: Not available.	9798 (6088.2)	208 (129.3)	201.4	52.1	Some observers speculate Cosmos 482 was Venus 8 follow-on which failed to get out of earth orbit. Still in orbit.
Apr. 3	Cosmos 483 (U.S.S.R.) 1972–24A Not available	Total weight: Not available. Objective: "Continuation of Cosmos scientific satellite series." Payload: Not available.	299 (180.2)	203 (126.1)	89.5	72.8	Reentered 4/15/72.

Launch Date	Name, Country, International Designation, Vehicle	Payload Data	Apogee in Kilometers (and st mi)	Perigee in Kilometers (and st mi)	Period in Minutes	Inclination in Degrees	Remarks
Apr. 4	Molniya I-20 (U.S.S.R.) 1972-25A Not available and	Total weight: Not available. Objective: Continue operation of long-range telephone, telegraph and radio communications system and transmission of U.S.S.R. central TV programs to stations in Orbita network in remote areas of U.S.S.R. Payload: Not available.	39 911 (24 799.6)	442 (274.7)	717.7	65.5	Molniya I-20 communications satellite launched successfully with French-built secondary payload, Sret 1. Still in orbit.
	Sret 1 (France) 1972-25B	Total weight: 15 kg (33 lbs). Objective: Study characteristics of solar batteries for space operations and degeneration of solar cells from cosmic ray exposure in Van Allen Belts. Payload: Octahedronal satellite covered with test solar cells of cadmium sulfide and cadmium telluride semiconductor materials; advanced battery components inside spacecraft.	39 250 (24 388.8)	458 (284.6)	704.6	65.6	Sret 1, first of series of three planned for launch in French-Soviet cooperative program, launched as secondary payload by U.S.S.R. Still in orbit.
Apr. 6	Cosmos 484 (U.S.S.R.) 1972-26A Not available	Total weight: Not available. Objective: "Continuation of Cosmos scientific satellite series." Payload: Not available.	236 (146.6)	177 (110)	88.6	81.5	Reentered 4/18/72.
Apr. 7	Intercosmos 6 (U.S.S.R.) 1972-27A Not available	Total weight: Not available. Objective: Study particles of primary radiation with energy range of 10^{12} to 10^{13} electron volts, chemical composition and energy spectrum of space beams in high-energy sphere, and meteoric particles in inner space. Payload: Primary instrument 1070-kg (2354-lb) photo-emulsionary block and ionization calorimeter.	334 (207.5)	197 (122.4)	90	51.8	First of new kind of Intercosmos capable of automatic reentry. Primary spacecraft instruments designed by U.S.S.R., Hungary, Mongolia, Poland, Romania, and Czechoslovakia. Satellite recovered 4/11/72. Recorded cosmic ray particles from deep space with highest energy yet detected in vicinity of earth.
Apr. 11	Cosmos 485 (U.S.S.R.) 1972-28A Not available	Total weight: Not available. Objective: "Continuation of Cosmos scientific satellite series." Payload: Not available.	462 (287.1)	267 (165.9)	91.8	70.9	Reentered 8/30/72.

ASTRONAUTICS AND AERONAUTICS, 1972

Apr. 14	Prognoz 1 (U.S.S.R.) 1972-29A Not available		199 667 (124 067.3)	1005 (624.5)	5782.1	65	Total weight: 845 kg (1863 lbs). Objective: Study processes of solar activity and their influence on interplanetary medium and earth's magnetosphere. Payload: Carried 15 instruments to study corpuscular gamma-ray, and x-ray solar radiation; solar plasma flows and their interaction with magnetosphere.	New interplanetary monitoring platform, Prognoz (Forecast), launched into highly eccentric orbit and functions as solar observatory. Still in orbit.
Apr. 14	Cosmos 486 (U.S.S.R.) 1972-30A Not available		253 (157.2)	218 (135.5)	89.1	81.3	Total weight: Not available. Objective: "Continuation of Cosmos scientific satellite series." Payload: Not available.	Reentered 4/27/72.
Apr. 16	Apollo 16 (United States) 1972-31A Saturn V		175.9 (109.3) Lunar orbit, 315.4 (196)	166.7 (103.6) 107.7 (66.9)			Total weight: 140 046.6 kg (308 750 lbs) at initial earth orbit insertion, including S-IVB stage, instrument unit, spacecraft LM adapter, LM, and CSM. Objective: Perform selenological inspection, survey, and sampling of materials and surface features in preselected area of Descartes region; emplace and activate surface experiments; conduct inflight experiments and photographic tasks. Payload: 'S-IVB/IU/LM adapter/LM/CSM, 34.8 m (114 ft) long, carrying lunar roving vehicle.	Apollo 16, 5th successful manned lunar landing mission, carried Astronauts John W. Young, Thomas K. Mattingly II, and Charles M. Duke, Jr. After lunar trajectory insertion and CSM separation from S-IVB, LM was extracted from S-IVB and docked with CSM. One midcourse correction during coast. Spacecraft entered lunar orbit 4/19 at 3:22 pm EST. LM undocked, initiated descent, and touched down about 230 m (755 ft) northwest of planned target of Descartes area at 9:23 pm EST 4/20. Crew performed 3 EVAs. During 1st EVA (7 hrs 11 min), Astronauts Young and Duke rode lunar roving vehicle, collected lunar samples, and deployed ALSEP; heat flow experiment accidentally broken by Young and canceled. Rover used during 2nd EVA (7 hrs 23 min). All activities accomplished during 3rd EVA (5 hrs 40 min). Total astronaut EVA time 20 hrs 15 min and total distance traversed 27.1 km (16.8 mi). Total weight of lunar samples 96.6 kg (213 lbs). LM liftoff from moon at 8:26 pm EST 4/24 covered with Rover TV camera. Total lunar stay time 71 hrs 14 min. After CSM-LM docking and crew transfer, LM ascent stage jettisoned, began tumbling and went into lunar orbit instead of impacting moon near landing site. Scientific subsatellite launched into lunar orbit by CSM. CSM entered trajectory for earth. Astronaut Mattingly left CSM for 1-hr 24-min inflight EVA to retrieve film from SM camera. Two midcourse corrections required during transearth coast. CM separated from SM and

and

Launch Date	Name, Country, International Designation, Vehicle	Payload Data	Apogee in Kilometers (and st mi)	Perigee in Kilometers (and st mi)	Period in Minutes	Inclination in Degrees	Remarks
	Apollo 16 Subsatellite 1972-31D	Total weight: 40.8 kg (90 lbs). Objective: Pursue extensive, continuing lunar scientific studies from lunar orbit. Payload: Hexagonal satellite 77 cm long, 63 cm in dia (30 x 14 in); three 1½-m (5-ft) booms deploy after launch; carries 3 experiments, S-band transponder, particle shadows and boundary layer experiment, and magnetometer experiment; solar-cell array and nickel-cadmium battery.	Lunar orbit, 123.3 (76.6)	97.9 (60.8)			splashed down in mid-Pacific 2:44 pm EST 4/27. Total flight time 11 days 1 hr 51 min 5 sec. Subsatellite stored in service module sim bay, and spring-ejected into lunar orbit 4/24. Shaping burn to optimize orbit not performed because of CSM engine problems. Subsatellite lifetime was decreased, Impacted moon 5/29, after 425 revolutions.
Apr. 19	DOD Spacecraft (United States) 1972-32A Thorad-Agena	Total weight: Not available. Objective: Develop space flight techniques and technology. Payload: Not available.	247 (153)	155 (96)	88.4	81.4	Reentered 5/12/72.
Apr. 21	Cosmos 487 (U.S.S.R.) 1972-33A Not available	Total weight: Not available. Objective: "Continuation of Cosmos scientific satellite series." Payload: Not available.	504 (313.2)	267 (165.9)	92.2	70.9	Reentered 9/24/72.
May 5	Cosmos 488 (U.S.S.R.) 1972-34A Not available	Total weight: Not available. Objective: "Continuation of Cosmos scientific satellite series." Payload: Not available.	317 (197)	185 (115)	89.5	65.2	Reentered 5/18/72.
May 6	Cosmos 489 (U.S.S.R.) 1972-35A Not available	Total weight: Not available. Objective: "Continuation of Cosmos scientific satellite series." Payload: Not available.	1004 (623.9)	967 (600.9)	104.7	74	Still in orbit.
May 17	Cosmos 490 (U.S.S.R.)	Total weight: Not available. Objective: "Continuation of Cosmos scien-	339 (210.6)	202 (125.5)	89.9	65.4	Reentered 5/29/72.

Date	Name/Designation/Source	Objective/Payload	Weight kg (lbs) / Perigee km (mi)	Apogee km (mi)	Period min	Inclination deg	Remarks
May 19	1972-36A Not available	tific satellite series." Payload: Not available.	39 295 (24 416.8)	433 (269.1)	705.1	65.3	Still in orbit.
	Molniya II-2 (U.S.S.R.) 1972-37A Not available	Total weight: Not available. Objective: Continue operation of long-range telephone and telegraph radio communications system within Soviet Union and transmission of U.S.S.R. central TV programs to stations in Orbita network and participating international networks. Payload: Not available.					
May 25	Cosmos 491 (U.S.S.R.) 1972-38A Not available	Total weight: Not available. Objective: "Continuation of Cosmos scientific satellite series." Payload: Not available.	369 (229.3)	173 (107.5)	89.9	64.9	Reentered 6/8/72.
May 25	DOD Spacecraft (United States) 1972-39A Thorad-Agena	Total weight: Not available. Objective: Develop space flight techniques and technology. Payload: Not available.	259 (161)	154 (96)	89.2	96.4	Reentered 6/4/72.
June 9	Cosmos 492 (U.S.S.R.) 1972-40A Not available	Total weight: Not available. Objective: "Continuation of Cosmos scientific satellite series." Payload: Not available.	316 (196.4)	202 (125.5)	89.7	65	Reentered 6/22/72.
June 13	Intelsat-IV F-5 (United States) 1972-41A Atlas-Centaur	Total weight:1402 kg (3090 lbs) at launch; 720 kg (1587 lbs) after apogee motor fire. Objective: Place satellite and apogee motor into proper transfer orbit; provide tracking and telemetry and backup calculations through transfer orbit so satellite can be injected into synchronous orbit for commercial communications. Payload: 528-cm-high x 238-cm-dia (208- x 93.7-in) cylindrical satellite capable of carrying 3000-9000 telephone circuits simultaneously or 12 color TV channels or combinations, spin-stabilized; 12 communications repeaters (transponders); 6 antennas (2 transmit horns, 2 receive horns, and 2 steerable 127-cm [50-in] dish spot-beam antennas); 42 240 solar cells.	35 963 (22 346.4)	35 777 (22 230.8)	1440.4	0.4	Launched by NASA into good transfer orbit for ComSatCorp. Apogee kick motor fired 6/14, stationing satellite in synchronous orbit over Indian Ocean at 61.4°E longitude. Used during Olympic games held in Munich, West Germany. Spacecraft operating normally. Still in orbit.

Launch Date	Name, Country, International Designation, Vehicle	Payload Data	Apogee in Kilometers (and st mi)	Perigee in Kilometers (and st mi)	Period in Minutes	Inclination in Degrees	Remarks
June 21	Cosmos 493 (U.S.S.R.) 1972–42A Not available	Total weight: Not available. Objective: "Continuation of Cosmos scientific satellite series." Payload: Not available.	266 (165.3)	201 (124.9)	89.1	64.9	Reentered 7/3/72.
June 23	Cosmos 494 (U.S.S.R.) 1972–43A Not available	Total weight: Not available. Objective: "Continuation of Cosmos scientific satellite series." Payload: Not available.	804 (499.6)	789 (490.3)	100.7	74	Still in orbit.
June 23	Cosmos 495 (U.S.S.R.) 1972–44A Not available	Total weight: Not available. Objective: "Continuation of Cosmos scientific satellite series." Payload: Not available.	320 (198.8)	171 (106.3)	89.4	65.4	Reentered 7/6/72.
June 26	Cosmos 496 (U.S.S.R.) 1972–45A Not available	Total weight: Not available. Objective: "Continuation of Cosmos scientific satellite series." Payload: Not available.	253 (157.2)	176 (109.4)	88.8	51.6	Reentered 7/2/72.
June 29	Prognoz 2 (U.S.S.R.) 1972–46A Not available	Total weight: 845 kg (1863 lbs). Objective: Study processes of solar activity and their influence on interplanetary medium and earth's magnetosphere. Payload: Carried 15 instruments to study corpuscular gamma ray and x-ray solar radiation; solar plasma flows and their interaction with magnetosphere. Carried French instruments for studying solar plasma and solar neutron flux.	200 000 (124 000)	550 (340)	5820	65	Prognoz 2 (Forecast), second interplanetary monitoring platform, launched into highly eccentric orbit and functions as solar observatory. Still in orbit.
June 30	Intercosmos 7 (U.S.S.R.) 1972–47A Not available	Total weight: Not available. Objective: Continue joint studies of sun's ultraviolet and x-radiation and its effects on structure of earth's upper atmosphere. Payload: Not available.	550 (341.8)	260 (161.6)	92.7	48.4	Intercosmos 7, similar to nonrecoverable, first-generation version, contained improved instruments designed and built by East Germany, Czechoslovakia, and U.S.S.R. Reentered 10/5/72.

Date	Name	Description	Weight kg (lb)	Orbit dimensions km (mi)	Period (min)	Inclination (deg)	Remarks
June 30	Cosmos 497 (U.S.S.R.) 1972-48A Not available	Total weight: Not available. Objective: "Continuation of Cosmos scientific satellite series." Payload: Not available.	787 (489)	271 (168.4)	95.2	71	Still in orbit.
June 30	Meteor 12 (U.S.S.R.) 1972-49A Not available	Total weight: Not available. Objective: Acquisition of meteorological information for use by weather service. Payload: Cylindrical body with 2 large solar paddles attached; 3 data-collection systems, TV cameras, infrared sensors, and actinometric scanner; 3-axis attitude control system.	904 (561.7)	888 (551.8)	102.8	81.2	Meteor 12 meteorological satellite still in orbit
July 5	Cosmos 498 (U.S.S.R.) 1972-50A Not available	Total weight: Not available. Objective: "Continuation of Cosmos scientific satellite series." Payload: Not available.	469 (291.4)	267 (165.9)	91.8	70.9	Reentered 11/25/72.
July 6	Cosmos 499 (U.S.S.R.) 1972-51A Not available	Total weight: Not available. Objective: "Continuation of Cosmos scientific satellite series." Payload: Not available.	284 (176.5)	205 (127.4)	89.4	51.9	Reentered 7/17/72.
July 7	non Spacecraft (United States) 1972-52A Titan IIID and non Spacecraft 1972-52C	Total weight: Not available. Objective: Develop space flight techniques and technology. Payload: Not available. Total weight: Not available. Objective: Develop space flight techniques and technology. Payload: Not available.	257 (160) 501 (311)	173 (108) 499 (310)	88.7 94.5	96.8 96.1	Dual payload launched with single booster. 52A reentered 9/13/72. 52C still in orbit.
July 10	Cosmos 500 (U.S.S.R.) 1972-53A Not available	Total weight: Not available. Objective: "Continuation of Cosmos scientific satellite series." Payload: Not available.	545 (338.7)	508 (315.7)	95.1	74	Still in orbit.
July 12	Cosmos 501 (U.S.S.R.) 1972-54A Not available	Total weight: Not available. Objective: "Continuation of Cosmos scientific satellite series." Payload: Not available.	2105 (1308)	215 (133.6)	108.5	48.4	Still in orbit.

Launch Date	Name, Country, International Designation, Vehicle	Payload Data	Apogee in Kilometers (and st mi)	Perigee in Kilometers (and st mi)	Period in Minutes	Inclination in Degrees	Remarks
July 13	Cosmos 502 (U.S.S.R.) 1972-55A Not available	Total weight: Not available. Objective: "Continuation of Cosmos scientific satellite series." Payload: Not available.	248 (154.1)	208 (129.3)	89.1	65.4	Reentered 7/25/72.
July 19	Cosmos 503 (U.S.S.R.) 1972-56A Not available	Total weight: Not available. Objective: "Continuation of Cosmos scientific satellite series." Payload: Not available.	308 (191.4)	170 (105.6)	89.2	65.4	Reentered 8/1/72.
July 20	Cosmos 504 (U.S.S.R.) 1972-57A Not available and	Total weight: Not available. Objective: "Continuation of Cosmos scientific satellite series." Payload: Not available.	1497 (930.2)	1323 (822.1)	113.9	74	Eight satellites launched with single booster. Still in orbit.
	Cosmos 505 1972-57B and	Total weight: Not available. Objective: "Continuation of Cosmos scientific satellite series." Payload: Not available.	1498 (930.8)	1354 (841.4)	114.3	74	Still in orbit.
	Cosmos 506 1972-57C and	Total weight: Not available. Objective: "Continuation of Cosmos scientific satellite series." Payload: Not available.	1498 (930.8)	1384 (860)	114.6	74	Still in orbit.
	Cosmos 507 1972-57D and	Total weight: Not available. Objective: "Continuation of Cosmos scientific satellite series." Payload: Not available.	1497 (930.2)	1414 (878.6)	114.9	74	Still in orbit.
	Cosmos 508 1972-57E and	Total weight: Not available. Objective: "Continuation of Cosmos scientific satellite series." Payload: Not available.	1497 (930.2)	1445 (897.9)	115.3	74	Still in orbit.

	Cosmos 509 1972–57F and	Total weight: Not available. Objective: "Continuation of Cosmos scientific satellite series." Payload: Not available.	1500 (932.1)	1475 (916.5)	74	Still in orbit.
	Cosmos 510 1972–57G and	Total weight: Not available. Objective: "Continuation of Cosmos scientific satellite series." Payload: Not available.	1512 (939.5)	1496 (929.6)	74	Still in orbit.
	Cosmos 511 1972–57H	Total weight: Not available. Objective: "Continuation of Cosmos scientific satellite series." Payload: Not available.	1547 (961.3)	1496 (929.6)	74	Still in orbit.
July 23	* Erts I* (United States) 1972–58A Thorad-Delta	Total weight: 941 kg (2075 lbs). Objective: Acquire synoptic, multispectral repetitive images for 3 mos, from which useful data can be obtained for investigations in such disciplines as agriculture and forestry resources, mineral and land resources, land use, water resources, marine resources, mapping and charting, and environment. Payload: 3.1-m-high × 3.4-m-wide (10- × 11-ft) spacecraft shares same basic structure and design as Nimbus satellite. Consists of 3 major elements: 142-cm (56-in) torus ring forms base; smaller hexagonal housing, connected to ring by truss, houses altitude stabilization and control system; two 0.9- × 1.5-m (3- × 5-ft) canted solar paddles. Carries 2 multispectral scanners, multispectral scanner subsystem (MSS), return beam vidicon (RBV) camera subsystem (3 cameras), data-collection system (DSC), and 2 wideband video tape recorders (WBVTR). Active 3-axis stabilization.	906.9 (563.6)	899.7 (559)	99.1	*Erts I* launched by 2-stage Thorad-Delta into polar orbit which allows spacecraft to photograph nearly entire planet during 18-day period. Two camera systems photograph identical scenes in different spectral bands. Spacecraft met all objectives and photographed all major land masses. Also monitored volcanoes; discovered unknown mountains in Antarctica, unknown earth fracture in Tennessee, and buried faults on Atlantic coast; used in water-pollution and land-use studies. Tape recorder developed problems and was turned off after exceeding 500-hr lifetime. Still in orbit.
July 28	*Cosmos 512* (U.S.S.R.) 1972–59A Not available	Total weight: Not available. Objective: "Continuation of Cosmos scientific satellite series." Payload: Not available.	273 (169.6)	202 (125.5)	65.3	Reentered 8/9/72.
Aug. 2	*Cosmos 513* (U.S.S.R.) 1972–60A Not available	Total weight: Not available. Objective: "Continuation of Cosmos scientific satellite series." Payload: Not available.	322 (200.1)	201 (124.9)	65	Reentered 8/15/72.

Launch Date	Name, Country, International Designation, Vehicle	Payload Data	Apogee in Kilometers (and st mi)	Perigee in Kilometers (and st mi)	Period in Minutes	Inclination in Degrees	Remarks
Aug. 13	Explorer 46 (United States) 1972-61A Scout	Total weight: 167.8 kg (370 lbs). Objective: Measure meteoroid-penetration rates in bumper-protected target in near-earth environment to evaluate effectiveness of bumpers. Payload: Cylindrical satellite 3.2 m (10.5 ft) long. Four bumper wings deployed after launch measure 7 m (23 ft) tip to tip; wings covered with 2 thin, stainless-steel bumper sheets. Three scientific experiments; four solar-cell arrays; two telemetry systems. Spin-stabilized.	814.5 (506.1)	495.8 (308.1)	98	37.7	First inflight test of bumper concept. Meteoroid Technology Satellite launched into proper orbit, but only 2 of 4 bumper wings deployed. Primary experiment functioning properly; impact-flux secondary experiment accomplished objective; meteoroid velocity secondary experiment requires several months of successful operation. Still in orbit.
Aug. 16	Cosmos 514 (U.S.S.R.) 1972-62A Not available	Total weight: Not available. Objective: "Continuation of Cosmos scientific satellite series." Payload: Not available.	975 (605.8)	957 (594.7)	104.3	82.9	Still in orbit.
Aug. 18	Cosmos 515 (U.S.S.R.) 1972-63A Not available	Total weight: Not available. Objective: "Continuation of Cosmos scientific satellite series." Payload: Not available.	286 (177.7)	179 (111.2)	89.1	73	Reentered 8/31/72.
Aug. 19	Denpa (Japan) 1972-64A Mu-4S	Total weight: 75 kg (165 lbs). Objective: Conduct scientific observations in ionosphere and magnetosphere. Payload: 68-cm (22.3-ft)-long satellite carried ionospheric plasma probes, electromagnetic and plasma wave receivers, cyclotron instability experiment, electron flux analyzer, and fuxtgate magnetometer.	6302 (3915.9)	239 (148.5)	157.2	31	Denpa (Radio Wave) fourth satellite launched by Japan. Unusually strong head winds caused satellite to enter lower orbit than planned. Subsequent encoder fault prevented transmission of experiment data. Still responds to ground commands and sends fragmentary data on plasma density in earth's magnetic field and ionosphere. Still in orbit.
Aug. 21	Oso 3 (United States) 1972-65A Atlas-Centaur	Total weight: 2200 kg (4900 lbs). Objective: Obtain high-resolution spectra of number of stars in ultraviolet range between 1000 A and 3000 A to investigate composition, density, and physical state of matter in interstellar space and stellar sources.	745 (462.9)	740 (459.8)	99.7	35	Copernicus Orbiting Astronomical Observatory heaviest and most complex scientific satellite launched by U.S. Contains largest telescope ever orbited. Spacecraft pointing accuracy 3 times better than planned. Princeton experiment package made 1780 observations of 47 unique objects, and

Date	Spacecraft	Payload/Objective					
Aug. 21	*Cosmos 516* (U.S.S.R.) 1972-66A Not available	Total weight: Not available. Objective: "Continuation of Cosmos scientific satellite series." Payload: Not available.	264 (164)	250 (155.3)	89.6	64.9	Still in orbit.
Aug. 30	*Cosmos 517* (U.S.S.R.) 1972-67A Not available	Total weight: Not available. Objective: "Continuation of Cosmos scientific satellite series." Payload: Not available.	285 (177.1)	203 (126.1)	89.3	64.9	Reentered 9/11/72.
Sept. 1	DOD Spacecraft (United States) 1972-68A Titan IIIB-Agena	Total weight: Not available. Objective: Develop space flight techniques and technology. Payload: Not available.	381 (237)	142 (88)	89.7	110.4	Reentered 9/30/72.
Sept. 2	*Triad OI-IX* (United States) 1972-69A Scout	Total weight: 93.9 kg (207 lbs). Objective: Correct long-term drift of satellite, forcing it to fly highly predictable orbit, and test environmental and component quality factors in performance; provide general navigation support in Navy Transit system. Payload: Overall length 7.3 m (24 ft); 3 gravity-stabilized components separated by 3.1-m (10-ft) booms. Earth-facing body contains Doppler system, computer system, and directional antenna. Central body contains DISCOS (disturbance com-	838 (520.7)	743 (461.7)	100.6	90.1	Launched by NASA for U.S. Navy. Spacecraft powered by Atomic Energy Commission radioisotope thermoelectric generator (RTG); which would provide 30 w power for 5 yrs. Still in orbit.

Payload: 8-sided structure with central tube which carries astronomical observing equipment; sun baffle mounted on viewing and of spacecraft; solar-cell arrays, containing about 55 000 solar cells, and inertial balance booms folded flat against sides during launch, are deployed in orbit; spacecraft measures 3.1 m long by 2.1 m wide (10 x 7 ft) in space; central tube 1.2 m (4 ft) in dia. Carries 2 experiments. Princeton experiment package (PEP), 81-cm (32-in.)-dia reflecting telescope housed in central tube; University College London (UCL) experiment consists of 3 small x-ray telescopes and collimated proportional counter. Spacecraft covered with thin aluminum skin; altitude control system; 2 transmitters; 4 receivers; thermal control system.

University College London experiment made 191 observations of 58 unique objects, as of 1/12/73. Spacecraft accomplished all objectives and was adjudged successful. Still in orbit.

Launch Date	Name, Country, International Designation, Vehicle	Payload Data	Apogee in Kilometers (and st mi)	Perigee in Kilometers (and st mi)	Period in Minutes	Inclination in Degrees	Remarks
		pensation system) for station-keeping. Spacecraft body is radioisotope thermal generator system with experimental solar cells.					
Sept. 15	Cosmos 518 (U.S.S.R.) 1972-70A Not available	Total weight: Not available. Objective: "Continuation of Cosmos scientific satellite series." Payload: Not available.	308 (191.4)	205 (127.4)	89.6	72.8	Reentered 9/24/72.
Sept. 16	Cosmos 519 (U.S.S.R.) 1972-71A Not available	Total weight: Not available. Objective: "Continuation of Cosmos scientific satellite series." Payload: Not available.	313 (194.5)	204 (126.8)	89.7	71.2	Reentered 9/26/72.
Sept. 19	Cosmos 520 (U.S.S.R.) 1972-72A Not available	Total weight: Not available. Objective: "Continuation of Cosmos scientific satellite series." Payload: Not available.	39 319 (24 431.7)	652 (405.1)	710	62.8	Still in orbit.
Sept. 22**	Explorer 47 (United States) 1972-73A Thorad-Delta	Total weight: 378 kg (833 lbs). Objective: Perform detailed and near continuous studies of interplanetary environment for orbital periods comparable to several rotations of active solar regions; study particle and field interactions in distance magnetotail including cross-sectional mapping of tail and neutral sheet. Payload: 16-sided drum-shaped structure 158 cm (62 in) high and 135 cm (53 in) in dia. Upper portion of spacecraft contains aluminum honeycomb shelf which supports experiments and spacecraft electronics. Lower portion has thrust tube 46 cm (18 in) dia to accommodate solid-propellant kick motor. Spacecraft structure consists of aluminum honeycomb sr shield panels and three bands of solar panels (16 panels per band) mounted on	237 796 (147 759.6) After apogee motor firing, 235 639 (146 419.3)	246 (152.9) 201 599 (125 267.8)	7365 17 760	28.6 17.2	Explorer 47 launched into highly elliptical transfer orbit. On 9/25/72 apogee kick motor was fired, injecting satellite into final circular orbit halfway between earth and moon. Provides detailed scientific information on solar-lunar-terrestrial relations; 12 of 13 scientific instruments operational. Still in orbit.

Date	Name (Country) Designation Launch vehicle	Description			Period (min)	Inclination (deg)	Remarks
Sept. 29	Cosmos 521 (U.S.S.R.) 1972-74A Not available	aluminum honeycomb substrate. Two diametrically opposed experiment booms (each 3.1 m [10 ft] long) and two attitude control system booms (each 1.2 m [4 ft] long) spaced 90° from experiment booms, appended to spacecraft exterior and deployed after launch. Spacecraft fitted with 8 equally spaced, hf antennas (4 active, 4 passive turnstile type) which extend radially. Contains 13 experiments; spin-stabilized. Total weight: Not available. Objective: "Continuation of Cosmos scientific satellite series." Payload: Not available.	1014 (630.1)	976 (606.5)	105	65.9	Still in orbit.
Sept. 30	Molniya II-3 (U.S.S.R.) 1972-75A Not available	Total weight: Not available. Objective: Continue operation of long-range telephone and telegraph radio communications system within Soviet Union and transmission of U.S.S.R. central TV programs to stations in Orbita network and participating international networks. Payload: Not available.	39 200 (24 357.8)	480 (298.3)	703	65.3	Still in orbit.
Oct. 2	Sts 72-1 (United States) 1972-76A Atlas-Burner II and	Total weight: 566 kg (1248 lbs). Objective: Measure background gamma radiation over whole earth in 100-300 kv and over 700 kv ranges; measure fluxes and spectra of low-altitude charged particles as function of time and magnetospheric position; determine effects of space environment on thermal control coatings; measure uv radiation which generate and maintain nighttime ionosphere, and measure hydrogen and helium atoms and ions; observe extreme and far uv originating in interaction of solar wind with interplanetary medium or from galactic sources. Payload: Cylinder 2.1 m long, 1.4 m in dia (7 x 4.5 ft); antenna booms extend 2.7 m (9 ft) from each end coincident with spin axis; carries gamma ray spectrometer and low-altitude particle-measuring sensor.	749 (465)	729 (453)	99.5	98.4	Dual payload launched with single booster in Dept. of Defense Space Test Program (STP). Still in orbit.
	Radcat 1972-76B	Total weight: 220 kg (485 lbs). Objective: Provide passive optical and radar	751 (467)	729 (453)	99.5	98.4	Still in orbit.

ASTRONAUTICS AND AERONAUTICS, 1972

Launch Date	Name, Country, International Designation, Vehicle	Payload Data	Apogee in Kilometers (and st mi)	Perigee in Kilometers (and st mi)	Period in Minutes	Inclination in Degrees	Remarks
		calibration target of about 5 sq m (53.8 sq ft) cross-section. Payload: Cylinder 3 m long with 1.2-m dia (10 ft by 4-ft dia).					
Oct. 4	Cosmos 522 (U.S.S.R.) 1972-77A Not available	Total weight: Not available. Objective: "Continuation of Cosmos scientific satellite series." Payload: Not available.	317 (197)	198 (123)	89.6	72.8	Reentered 10/17/72.
Oct. 5	Cosmos 523 (U.S.S.R.) 1972-78A Not available	Total weight: Not available. Objective: "Continuation of Cosmos scientific satellite series." Payload: Not available.	457 (284)	268 (166.5)	91.7	70.9	Reentered 3/7/73.
Oct. 10	DOD Spacecraft (United States) 1972-79A Titan IIID	Total weight: Not available. Objective: Develop space flight techniques and technology. Payload: Not available.	268 (167)	159 (99)	88.7	96.4	Reentered 1/8/73.
Oct. 11	Cosmos 524 (U.S.S.R.) 1972-80A Not available	Total weight: Not available. Objective: "Continuation of Cosmos scientific satellite series." Payload: Not available.	490 (304.5)	264 (164)	92	71	Reentered 3/25/73.
Oct. 14	Molniya I-21 (U.S.S.R.) 1972-81A Not available	Total weight: Not available. Objective: Continue operation of long-range telephone and telegraph radio communications system and transmission of U.S.S.R. central TV programs to stations in Orbita network and remote areas of U.S.S.R. Payload: Not available.	39 300 (24 419.9)	480 (298.3)	705	65.3	Molniya I-21 communications satellite still in orbit.
Oct. 15	Noaa 2 (United States) 1972-82A Thor-Delta 0300	Total weight: 345 kg (760 lbs). Objective: Launch spacecraft into sun-synchronous orbit of sufficient accuracy to enable spacecraft to accomplish opera-	1453.9 (903.4)	1448.2 (899.9)	114.9	101.8	Launched by NASA on 2-stage Thor-Delta. First operational spacecraft to provide temperature soundings of earth's atmosphere as well as direct readout and globally

tional mission requirements, conduct in orbit evaluation and checkout of spacecraft and, upon completion, turn operational control over to NOAA/NES. Orbit has local equator-crossing time between 9:00 am and 9:20 am to permit regular and dependable daytime and nighttime meteorological observations in both direct readout and stored modes of operation in support of National Operational Meteorological Satellite System.
Payload: Rectangular, box-shaped spacecraft with deployable 3-panel solar array. Base of main body about 102 x 102 cm (40 x 40 in), and overall height about 122 cm (48 in). Total area of array 4.5 sq m (48 sq ft) with each of 3 panels measuring 93 x 162 cm (36.4 x 63.8 in). Three-axis-stabilized earth-oriented satellite carries two very-high-resolution radiometer (VHRR) instruments and two scanning radiometer (SR) sensors for daytime and nighttime coverage, and two vertical-temperature-profile-radiometers (VTPR) which permit determination of earth's atmospheric vertical temperature profile over every part of earth's surface at least twice daily. Thermal control system; 4 antennas.

...recorded cloud-cover data. Spacecraft functioning normally and turned over to NOAA 11/8/72 for operational use. Employed with Erts I in Federal-State project to monitor water pollution of New Jersey. Still in orbit.

Oscar 6 1972-82B	Total weight: 18 kg (40 lbs). Objective: Conduct experimental program of multiple-access communication techniques using large number of relatively low-powered earth terminals. Payload: Powered by solar cells and battery, spacecraft carries 2- to 10-m (6.6- to 32.8-ft) linear translator with bandwidth of 100 khz; input frequency centered on 145.95 mhz and output centered at 29.5 mhz. Peak power output of transmitter about 1 w; also carries message storage device.	1456 (904.7)	1451 (901.6)	114.9	101.7	Launched by NASA as secondary payload. Built by U.S., Australian, and German amateur groups working through Radio Amateur Satellite Corporation (AMSAT). Spacecraft functioning normally; still in orbit.
Oct. 18	Cosmos 525 (U.S.S.R.) 1972-83A Not available	Total weight: Not available. Objective: "Continuation of Cosmos scientific satellite series." Payload: Not available.	292 (181.4)	206 (128)	89.5	65.4	Reentered 10/29/72.

Launch Date	Name, Country, International Designation, Vehicle	Payload Data	Apogee in Kilometers (and st mi)	Perigee in Kilometers (and st mi)	Period in Minutes	Inclination in Degrees	Remarks
Oct. 25	Cosmos 526 (U.S.S.R.) 1972-84A Not available	Total weight: Not available. Objective: "Continuation of Cosmos scientific satellite series." Payload: Not available.	480 (298.3)	272 (169)	92	70.9	Reentered 4/8/73.
Oct. 27	Meteor 13 (U.S.S.R.) 1972-85A Not available	Total weight: Not available. Objective: Acquisition of meteorological information for use by weather service. Payload: Cylindrical body with 2 large solar paddles attached; 3 data-collection systems, TV cameras, infrared sensors, and actinometric scanner; 3-axis attitude control system.	891 (553.6)	866 (538.1)	102.5	81.2	Still in orbit.
Oct. 31	Cosmos 527 (U.S.S.R.) 1972-86A Not available	Total weight: Not available. Objective: "Continuation of Cosmos scientific satellite series." Payload: Not available.	295 (183.3)	177 (110)	89.1	65.4	Reentered 11/13/72.
Nov. 1	Cosmos 528 (U.S.S.R.) 1972-87A Not available and	Total weight: Not available. Objective: "Continuation of Cosmos scientific satellite series." Payload: Not available.	1469 (912.8)	1368 (850.1)	114.1	74	Eight satellites launched with single booster. Still in orbit.
	Cosmos 529 1972-87B and	Total weight: Not available. Objective: "Continuation of Cosmos scientific satellite series." Payload: Not available.	1469 (912.8)	1404 (872.4)	114.5	74	Still in orbit.
	Cosmos 530 1972-87C and	Total weight: Not available. Objective: "Continuation of Cosmos scientific satellite series." Payload: Not available.	1469 (912.8)	1335 (829.5)	113.8	74	Still in orbit.
	Cosmos 531 1972-87D and	Total weight: Not available. Objective: "Continuation of Cosmos scientific satellite series." Payload: Not available.	1471 (914)	1423 (884.2)	114.7	74	Still in orbit.

Date	Name	Description	Weight kg (lbs)	Apogee/Perigee km (mi)	Period min	Inclination deg	Remarks
	Cosmos 532 1972-87E and	Total weight: Not available. Objective: "Continuation of Cosmos scientific satellite series." Payload: Not available.	1469 (912.8)	1302 (809)	113.4	74	Still in orbit.
	Cosmos 533 1972-87F and	Total weight: Not available. Objective: "Continuation of Cosmos scientific satellite series." Payload: Not available.	1470 (913.4)	1318 (819)	113.6	74	Still in orbit.
	Cosmos 534 1972-87G and	Total weight: Not available. Objective: "Continuation of Cosmos scientific satellite series." Payload: Not available.	1469 (912.8)	1351 (839.5)	113.9	74	Still in orbit.
	Cosmos 535 1972-87H	Total weight: Not available. Objective: "Continuation of Cosmos scientific satellite series." Payload: Not available.	1470 (913.4)	1386 (861.2)	114.3	74	Still in orbit.
Nov. 3	Cosmos 536 (U.S.S.R.) 1972-88A Not available	Total weight: Not available. Objective: "Continuation of Cosmos scientific satellite series." Payload: Not available.	543 (337.4)	517 (321.3)	95.2	74	Still in orbit.
Nov. 9	DOD Spacecraft (United States) 1972-89A Thor-Burner II	Total weight: Not available. Objective: Develop space flight techniques and technology. Payload: Not available.	879 (546)	824 (512)	101.5	98.6	Still in orbit.
Nov. 9*	Anik 1 (Canada–U.S.) 1972-90A Thorad-Delta	Total weight: 540 kg (1200 lbs) at launch; 270 kg (600 lbs) after apogee kick motor firing. Objective: Place satellite into orbit of sufficient accuracy to allow spacecraft propulsion systems to place spacecraft in stationary synchronous orbit while retaining sufficient station-keeping propulsion to meet mission lifetime requirements. Payload: Cylindrical spacecraft 1.8 m in dia and 3.4 m high (6 by 11 ft). Spin-stabilized; 1.5-m (5-ft) optically transparent antenna weighing 4.1 kg (9 lbs) affixed to top of spacecraft remains stationary, pointed toward Canada, as space-	36 470 (22 661.4) After apogee motor firing, 36 508 (22 685)	188.9 (117.4) 35 822 (22 258.8)	1455	26.9 0.4	Anik 1 (Telesat A) launched by NASA on 3-stage Thor-Delta for Canadian Domestic Communications Satellite System into elliptical transfer orbit. Apogee kick motor fired by Canada 11/13 and spacecraft placed in stationary equatorial orbit off west coast of South America. Operational service initiated 1/73. World's first operational domestic comsat. Use of system for limited interim domestic communications service in U.S. is planned. Still in orbit.

Launch Date	Name, Country, International Designation, Vehicle	Payload Data	Apogee in Kilometers (and st mi)	Perigee in Kilometers (and st mi)	Period in Minutes	Inclination in Degrees	Remarks
		craft revolves. Spacecraft provides 10 color TV channels or up to 9600 telephone circuits; 23 000 solar cells.					
Nov. 16**	Explorer 48 (United States-Italy) 1972-91A Scout	Total weight: 186 kg (410 lbs). Objective: Measure spatial and energy distribution of primary galactic and extragalactic gamma radiation. Payload: Dome-shaped cylinder 59 cm in dia and 51 cm high (23 by 20 in). Four solar paddles 27 cm wide and 135 cm long (10.5 by 53 in) hinged to outer shell are folded downward during launch and upon deployment are canted 30° from 2 axes and perpendicular to cylinder surface. Command and telemetry antennas attached to tips of paddles; spacecraft measures 396 cm (156 in) tip to tip with paddles deployed. Inside spacecraft shell, honeycomb deck contains basic spacecraft systems; nickel-cadmium battery, command receivers and decoders, telemetry system, and spin control system. Spacecraft experiment consists of 32-level digitized spark chamber gamma ray telescope to detect rare celestial gamma rays and determine intensity, energy, and direction of arrival.	632.5 (393)	443.7 (275.7)	95.4	1.9	NASA-built satellite launched into equatorial orbit from San Marco platform by Italian launch crew. Single experiment operating flawlessly and carrying out most comprehensive study of gamma rays ever undertaken. Still in orbit.
Nov. 21*	Esro 4 (ESRO-United States) 1972-92A Scout	Total weight: 114 kg (251 lbs). Objective: Place satellite in earth orbit which will enable successful achievement of scientific objectives and provide tracking and telemetry support for spacecraft. Payload: Cylindrical spacecraft about 137 cm high with 76-cm dia (54 x 30 in) with 6990 body-mounted solar cells. Three radical booms, hinged at bottom of craft, folded along sides during launch, deployed in orbit; 4th boom, mounted on spacecraft's bottom, stowed inside center of	1187 (737.6)	253 (157.2)	99	91.1	ESRO-built spacecraft launched successfully by NASA for ESRO into near-polar elliptical orbit. Satellite monitoring polar ionosphere and obtaining data to be correlated with ground-based observations. Still in orbit

Date	Name	Description	Weight kg (lbs)	Payload kg (lbs)	Period (min)	Remarks
Nov. 25	Cosmos 537 (U.S.S.R.) 1972-93A Not available	craft during launch and deployed immediately after radial booms. Radial booms contain sensors for part of ionospheric experiment. Spacecraft carries 6 experiments from U.K., Sweden, West Germany, and Netherlands and is spin-stabilized. Total weight: Not available. Objective: "Continuation of Cosmos scientific satellite series." Payload: Not available.	302 (187.7)	202 (125.5)	89.5	64.9 Reentered 12/7/72.
Dec. 1**	Intercosmos 8 (U.S.S.R.) 1972-94A Not available	Total weight: Not available. Objective: Continue global research of ionosphere, including temperature and concentration of electrons, in cooperative program to study different cosmic ray particles and micrometeoroids which influence weather, climate, radio communications, and biological processes on earth. Payload: Not available.	590 (366.6)	199 (123.7)	92.4	70.9 Intercosmos 8 carried equipment designed and built in Bulgaria, East Germany Czechoslovakia, and U.S.S.R. Reentered 3/2/73.
Dec. 2	Molniya 1-22 (U.S.S.R.) 1972-95A Not available	Total weight: Not available. Objective: Continue operation of long-range telephone and telegraph radio communications system and transmission of U.S.S.R. central TV programs to stations in Orbita network and remote areas of U.S.S.R. Payload: Not available.	39 797 (24 728.7)	555 (344.9)	717.7	65 Molniya 1-22 communications satellite still in orbit.
Dec. 7	Apollo 17 (United States) 1972-96A Saturn V	Total weight: 141 135.7 kg (311 151 lbs) at initial earth-orbit insertion. Objective: Perform selenological inspection, survey, and sampling of materials and surface features in preselected area of Taurus-Littrow region; emplace and activate surface experiments; conduct inflight experiments and photographic tasks. Payload: S-IVB/IU/LM adapter/LM/CSM, 34.8 m (114 ft) long, carrying lunar roving vehicle.	170 (105.6) Lunar orbit, 315 (195.7)	168 (104.4) 97 (60.3)		Apollo 17, 6th successful manned lunar landing mission, carried Astronauts Eugene A. Cernan, Ronald E. Evans, and Harrison H. Schmitt. After insertion into lunar trajectory and CSM separation from S-IVB, LM was extracted from S-IVB and docked with CSM. One midcourse correction during coast. Spacecraft entered lunar orbit 2:47 pm EST 12/10, LM undocked, initiated descent, and touched down at planned Taurus-Littrow site 2:55 pm EST 12/11. Crew performed 3 EVAS. During 1st EVA (7 hrs 12 min), Astronauts Cernan and Schmitt rode lunar roving vehicle to Steno Crater and deployed ALSEP. Rover used again during 2nd EVA and 5 stations explored (7 hrs 37 min).

Launch Date	Name, Country, International Designation, Vehicle	Payload Data	Apogee in Kilometers (and st mi)	Perigee in Kilometers (and st mi)	Period in Minutes	Inclination in Degrees	Remarks
							Remaining stations explored during 3rd EVA (7 hrs 16 min). Total astronaut EVA time 22 hrs 5 min, and total distance traveled in lunar hover about 35 km (22 mi). Total weight of lunar samples about 115 kg (250 lbs). LM liftoff from moon at 5:55 pm EST 12/14 covered with Rover TV camera. Total lunar stay time 74 hrs 59 min 38 sec. After CSM-LM docking and crew transfer, LM ascent stage was jettisoned, deorbited, and impacted moon. Impact recorded by four Apollo 17 geophones and Apollo 12, 14, 15, and 16 ALSEPS. Trajectory for earth entered 12/16. Astronaut Evans left CSM for 1 hr 7 min inflight EVA to retrieve film from SM camera. One midcourse correction required during transearth coast. CM separated from SM and splashed down in mid-Pacific 2:25 pm EST 12/19. Total flight time 12 days 13 hrs 51 min.
Dec. 10*	Nimbus 5 (United States) 1972-97A Thorad-Delta	Total weight: 769 kg (1695 lbs). Objective: Improve and extend capability for vertical sounding of temperature and moisture in atmosphere, particularly with regard to altitude coverage, and interfering effects of clouds, by acquisition of synoptic data for 10 wks from either infrared temperature profile radiometer (ITPR), or Nimbus E microwave spectrometer (NEMS); demonstrate improved thermal mapping of earth by obtaining data for 10 wks from either electrically scanning microwave radiometer (ESMR) or surface composition mapping radiometer (SCMR). Payload: Butterfly-shaped spacecraft 3.1 m high and 3.4 m wide (10 by 11 ft), consists of 3 major elements; 1.5-m (5-ft) torus ring forms base and houses major spacecraft electronics; smaller hexagonal housing, connected by truss, houses alti-	1101.3 (684.3)	1089.5 (677)	107.2	99.95	Nimbus 5 launched successfully by 2-stage Thorad-Delta into sun-synchronous, nearly circular polar orbit. Achieved all mission objectives and was adjudged successful. Still in orbit.

Date	Name/Designation	Description	Weight	Orbit param	Incl.	Remarks
Dec. 12	*Molniya II-4* (U.S.S.R.) 1972-98A Not available	tude stabilization and control system; and two solar paddles about 1 by 2.4 m (3 by 8 ft). Active 3-axis stabilization. Carries 6 experiments: selective chopper radiometer (SCR), infrared temperature profile radiometer (ITPR), Nimbus E microwave spectrometer (NEMS), electrically scanning microwave radiometer (ESMR), surface composition mapping radiometer (SCMR), and temperature humidity infrared radiometer (THIR). Solar cells provide 550 w of power and nickel-cadmium batteries average 277 w. Total weight: Not available. Objective: Continue operation of long-range telephone and telegraph radio communications system within Soviet Union and transmission of U.S.S.R. central TV programs to stations in Orbita network and participating international networks. Payload: Not available.	39 300 (24 420)	470 (292)	705	65.3	Still in orbit.
Dec. 14	*Cosmos 538* (U.S.S.R.) 1972-99A Not available	Total weight: Not available. Objective: "Continuation of Cosmos scientific satellite series." Payload: Not available.	280 (174)	204 (126.8)	89.3	65.4	Reentered 12/27/72.
Dec. 16	*Aeros* (West Germany–U.S.) 1972-100A Scout	Total weight: 125.7 kg (277 lbs). Objective: Measure main aeronomic parameters of upper atmosphere and solar ultraviolet radiation in wavelength band of main absorption. Payload: Circular cylindrical shell welded to bottom conical shell to form structural unit 91 cm in dia and 71 cm high (36 × 28 in). Flat honeycomb solar-cell-array lid attached to spacecraft cylinder top. Carries 5 scientific instruments: mass spectrometer (MS), retarding potential analyzer (RPA), impedance probe (IP), EUV-spectrometer, and neutral atmosphere temperature experiment (NATE). Spin-stabilized; 4 telemetry antennas; 2 batteries.	864.4 (537.1)	218.1 (135.5)	95.5	96.95	West-German-built satellite launched into polar orbit by NASA. All 5 science experiment instruments commanded on and performing satisfactorily. Still in orbit.
Dec. 20	DOD Spacecraft (United States) 1972-101A Atlas-Agena	Total weight: Not available. Objective: Develop space flight techniques and technology. Payload: Not available.	40 728 (25 307)	31 012 (19 270)	1440.4	9.7	Still in orbit.

Launch Date	Name, Country, International Designation, Vehicle	Payload Data	Apogee in Kilometers (and st mi)	Perigee in Kilometers (and st mi)	Period in Minutes	Inclination in Degrees	Remarks
Dec. 21	Cosmos 539 (U.S.S.R.) 1972-102A Not available	Total weight: Not available. Objective: "Continuation of Cosmos scientific satellite series." Payload: Not available.	1381 (858.1)	1343 (834.5)	112.9	74	Still in orbit.
Dec. 21	DOD Spacecraft (United States) 1972-103A Titan IIIB-Agena	Total weight: Not available. Objective: Develop space flight techniques and technology. Payload: Not available.	391 (243)	138 (86)	90	110.5	Reentered 1/23/73.
Dec. 26	Cosmos 540 (U.S.S.R.) 1972-104A Not available	Total weight: Not available. Objective: "Continuation of Cosmos scientific satellite series." Payload: Not available.	805 (500.2)	785 (487.8)	100.7	74	Still in orbit.
Dec. 27	Cosmos 541 (U.S.S.R.) 1972-105A Not available	Total weight: Not available. Objective: "Continuation of Cosmos scientific satellite series." Payload: Not available.	348 (216.2)	218 (135.5)	90.2	81.3	Reentered 1/8/73.
Dec. 28	Cosmos 542 (U.S.S.R.) 1972-106A Not available	Total weight: Not available. Objective: "Continuation of Cosmos scientific satellite series." Payload: Not available.	640 (397.7)	527 (327.5)	96.3	81.2	Still in orbit.

*Local time at site; 1 day later by Greenwich time.
**Local time at site; 1 day earlier by Greenwich time.

Appendix B

CHRONOLOGY OF MAJOR NASA LAUNCHES, 1972

This chronology of major NASA launches in 1972 is intended to provide an accurate and ready historical reference, compiling and verifying information previously scattered in several sources. It includes launches of all rocket vehicles larger than sounding rockets launched either by NASA or under "NASA direction" (e.g., in 1972 NASA provided vehicles and launch facilities and launched the Communications Satellite Corporation's two Intelsat IV satellites and the European Space Research Organization's *Heos 2*, *TD–1A*, and *Esro 4*, as well as *Anik 1* for Canada, *Aeros* for West Germany, *Triad OI–1X* for the U.S. Navy, and *Noaa 2* for the National Oceanic and Atmospheric Administration; also, under a NASA and University of Rome agreement, an Italian team launched a NASA satellite, *Explorer 48*, from Italy's San Marco platform off the coast of Kenya, on NASA's Scout booster). During 1972, NASA achieved its first perfect annual record, experiencing no launch or mission failures.

An attempt has been made to classify performance of both the launch vehicle and the payload and to summarize total results in terms of primary mission. Three categories have been used for evaluating vehicle performance and mission results—successful (S), partially successful (P), and unsuccessful (U). A fourth category, unknown (Unk), has been added for payloads when vehicle malfunctions did not give the payload a chance to exercise its main experiments. These divisions are necessarily arbitrary; many of the results cannot be neatly categorized. Also they ignore the fact that a great deal is learned from missions that may have been classified as unsuccessful.

Date of launch is referenced to local time at the launch site. Open sources were used, verified when in doubt with the project offices in NASA Headquarters and with NASA Centers. For further information on each item, see Appendix A of this volume and the entries in the main chronology as referenced in the index. The information was compiled in May 1973 by Leonard C. Bruno of the Science and Technology Division of the Library of Congress.

Date	Name (NASA Code)	General Mission	Launch Vehicle (Site)	Performance			Remarks
				Vehicle	Payload	Mission	
Jan. 22*	*Intelsat-IV F-4*	Operational communications satellite	Atlas-Centaur (ETR)	S	S	S	Launched into elliptical transfer orbit by NASA for ComSatCorp; on 1/24/72 ComSatCorp fired apogee kick motor to circularize synchronous orbit and put satellite over Pacific. Operating normally.
Jan. 31	Heos 2	Scientific satellite, magnetosphere	Thor-Delta DSJ-3L (WTR)	S	S	S	ESRO-built satellite launched into highly eccentric orbit by NASA. All 7 experiments turned on and spacecraft fully operational.
Mar. 2*	*Pioneer 10* (Pioneer-F)	Scientific interplanetary probe	Atlas-Centaur TE-M-364-4 (ETR)	S	S	S	Reached highest launch velocity ever attained at 51 500 km per hr (32 000 mph) with first use of Atlas-Centaur as 3-stage vehicle. First NASA spacecraft powered entirely by nuclear energy and first intended ultimately to escape solar system. By 2/15/73 spacecraft had passed through Asteroid Belt undamaged; Jupiter encounter scheduled 12/3/73.
Mar. 11*	TD-1A	Scientific satellite, high-energy emissions	Thor-Delta N (WTR)	S	S	S	ESRO-built satellite launched into sun-synchronous orbit by NASA. Largest and most advanced scientific satellite built in Western Europe. Operating normally.
Apr. 16	*Apollo 16* (AS-511, CSM-113, LM-11) and	Manned lunar landing flight	Saturn V (KSC)	S	S	S	Fifth successful lunar landing mission. Astronauts John W. Young, commander; Thomas K. Mattingly II, CM pilot; Charles M. Duke, Jr., LM pilot. Launched from KSC at 12:54 pm EST 4/20. Touched down at planned Descartes site at 9:23 pm EST 4/20. Astronauts Young and Duke performed three EVAs totaling 20 hrs 15 min. Crew deployed ALSEP, drove lunar roving vehicle, took photos, and obtained 96.6 kg (213 lbs) of lunar samples. LM lifted off moon at 8:26 pm EST 4/24. CM landed in Pacific near U.S.S. *Ticonderoga* at 2:44 pm EST 4/27. Total mission time 11 days 1 hr 51 min 5 sec.
	Apollo 16 Subsatellite	Scientific satellite, lunar orbit studies			S	P	Subsatellite spring-ejected in lunar orbit from service module SIM (scientific instrument module) bay 4/24. Shaping burn to optimize orbit not performed because of CSM engine problems. Subsatellite lifetime decreased. Impacted moon 5/29 after 425 revolutions.
June 13	*Intelsat-IV F-5*	Operational communications satellite	Atlas-Centaur (ETR)	S	S	S	Launched into elliptical transfer orbit by NASA for ComSatCorp; on 6/14 ComSatCorp fired apogee kick motor to circularize synchronous orbit and put satellite over Indian Ocean at 61.4°E longitude. Operating normally.

Date	Name	Purpose	Launch vehicle (site)				Remarks
July 23	*Erts 1* (ERTS-A)	Earth resources technology satellite	Thorad-Delta (WTR)	S	S	S	Launched into polar orbit which allows spacecraft to photograph nearly entire planet during 18-day period. Two camera systems photograph identical scenes in different spectral bands. Spacecraft photographed all major land masses, monitored volcanoes; was used in water pollution and land use studies. Tape recorder developed problems and was turned off after exceeding 500-hr lifetime.
Aug. 13	*Explorer 46* (MTS)	Meteoroid technology satellite	Scout (WS)	S	S	S	First inflight test of bumper concept for protection against meteoroids. Satellite was launched into proper orbit, but only 2 of 4 bumper panels deployed. Primary experiment functioning properly; impact flux secondary experiment accomplished objective; meteoroid velocity secondary experiment requires several months of successful operation.
Aug. 21	*Oao 3* (OAO-C)	Scientific satellite, astronomical observations	Atlas-Centaur (ETR)	S	S	S	Heaviest and most complex scientific satellite launched by U.S. Contains largest telescope ever orbited. Spacecraft pointing accuracy 3 times better than planned. Accomplished all objectives and adjudged successful.
Sept. 2	*Triad OI-1X*	Experimental navigation satellite	Scout (WTR)	S	S	S	Launched by NASA for U.S. Navy. Spacecraft powered by Atomic Energy Commission radioisotope thermoelectric generator (RTG); would provide 30 w of power for 5 yrs.
Sept. 22*	*Explorer 47* (IMP-H)	Interplanetary monitoring platform	Thorad-Delta (ETR)	S	S	S	Launched initially into highly elliptical orbit. Satellite apogee kick motor firing placed satellite in final circular orbit halfway between earth and moon. Spacecraft provides detailed scientific information on solar-lunar-terrestrial relationship; 12 of 13 scientific instruments operational.
Oct. 15	*Noaa 2* (ITOS-D) and	Meteorological satellite	Thor-Delta 0300 (WTR)	S	S	S	First operational spacecraft to provide temperature soundings of earth's atmosphere as well as direct readout and globally recorded cloud-cover data. Spacecraft functioning normally and turned over to NOAA 11/8/72 for operational use.
	Oscar 6 (Oscar C)	Radio transmitter		S			Launched by NASA as secondary payload. Built by American Australian, and German amateur groups working through Radio Amateur Satellite Corp. (AMSAT).
Nov. 9*	*Anik 1* (Telesat-A)	Operational communications satellite	Thorad-Delta (ETR)	S	S	S	Launched by NASA for Canadian Domestic Communications Satellite System into transfer orbit. Apogee kick motor fired by Canada 11/13 placed spacecraft in synchronous orbit off west coast of South America. Initiated operational service 1/73. World's first operational domestic ComSat.

Date	Name (NASA Code)	General Mission	Launch Vehicle (Site)	Performance Vehicle	Performance Payload	Performance Mission	Remarks
Nov. 16**	Explorer 48 (SAS-B)	Scientific satellite, astronomy	Scout (Kenya)	S	S	S	NASA-built satellite launched into equatorial orbit from San Marco range by Italian crew. Single experiment operating flawlessly and carrying out most comprehensive gamma ray study ever attempted.
Nov. 21*	Esro 4	Scientific satellite, ionosphere	Scout (WTR)	S	S	S	ESRO-built spacecraft launched successfully by NASA into near-polar elliptical orbit. Monitors polar ionosphere.
Dec. 7	Apollo 17 (AS-512, CSM-114, LM-12)	Manned lunar landing flight	Saturn V (KSC)	S	S	S	Sixth successful manned lunar landing mission. Astronauts Eugene A. Cernan, commander; Ronald E. Evans, CM pilot; Harrison H. Schmitt, LM pilot. Launched from KSC at 12:33 am EST. Touched down at planned Taurus-Littrow site at 2:55 pm EST 12/11. Astronauts Cernan and Schmitt performed 3 EVAs totaling 22 hrs 5 min. Crew deployed ALSEP, drove lunar roving vehicle, took photos, and obtained 115 kg (250 lbs) of lunar samples. LM lifted off moon at 5:55 pm EST 12/14. CM landed in Pacific near U.S.S. Ticonderoga at 2:25 pm EST 12/19. Total mission time 12 days 13 hrs 51 min.
Dec. 10*	Nimbus 5 (Nimbus-E)	Experimental meteorological satellite	Thorad-Delta (WTR)	S	S	S	Launched by 2-stage Thorad-Delta into sun-synchronous, nearly circular, polar orbit. Achieved all mission objectives and was adjudged successful.
Dec. 16	Aeros	Scientific satellite, ionospheric F-region	Scout (WTR)	S	S	S	West-German-built satellite launched into polar orbit by NASA. All 5 science experiment instruments commanded on and performing satisfactorily.

*Time at launch site; 1 day later by Greenwich time.
**Time at launch site; 1 day earlier by Greenwich time.

Appendix C

CHRONOLOGY OF MANNED SPACE FLIGHT, 1972

This chronology contains basic information on all manned space flights during 1972 and—taken with Appendix C to the 1965, 1966, 1968, 1969, 1970, and 1971 volumes of this publication—provides a summary record of manned exploration of the space environment through 1972. The information was complied by Leonard C. Bruno of the Science and Technology Division of the Library of Congress.

Compared with 1971, 1972 saw a halving of manned space flight, mainly attributable to U.S.S.R. inactivity. While two U.S. Apollo and two U.S.S.R. Soyuz flights placed a total of 12 men in space during 1971, 1972 saw two Apollo flights and no Soyuz mission, for a total of 6 men in space.

Apollo 16, launched in April, continued to improve and to extend U.S. manned lunar capability by spending more than 20 hours exploring the moon's surface and returning with 96.6 kg (213 lbs) of lunar samples. The Apollo program was brought to a highly successful conclusion with the near-perfect December flight of *Apollo 17*. Breaking nearly all previous manned lunar exploration records, the *Apollo 17* mission was the longest and most productive. The two-man crew spent some 22 hours exploring the moon's surface and returned with 115 kg (250 lbs) of lunar material. With this sixth and last manned lunar landing, 12 men had walked on the moon and spent nearly 12½ days on its surface. They had returned a total of about 380 kg (841 lbs) of lunar rocks and soil to earth for future study.

The Soviet Union made no manned flight attempts during 1972, apparently still recovering from the tragic June 1971 flight of *Soyuz 11* that resulted in the death of its three-man crew.

By the end of 1972, the United States had conducted a total of 27 manned space flights—2 suborbital, 16 in earth orbit, 3 in lunar orbit, and 6 lunar landings—with a total of 34 different crewmen. Of the 34 American astronauts, 9 had participated in two flights each, 5 had flown three times, and 2 had flown four times. The Soviet Union had made 18 manned flights, all in earth orbit, with 25 cosmonauts. Three had participated in two flights each and two had flown three times. Cumulative totals for manned spacecraft hours in flight had reached 3526 hours 27 minutes for the United States and remained at 2097 hours 16 minutes for the Soviet Union. Cumulative total man-hours in space were 9499 hours 30 minutes for the United States and remained 4403 hours 12 minutes for the U.S.S.R.

Data on U.S. flights are the latest available to date within NASA, although minor details are subject to modification as data are refined.

Date Launched	Date Recovered	Designation (NASA Code)	Crew	Weight in Kilograms (and in lbs)	Revolutions	Maximum Distance from Earth in Kilometers (and st mi)	Duration	Remarks
Apr. 16	Apr. 27	*Apollo 16* (AS-511, CSM-113, LM-11)	John W. Young, Thomas L. Mattingly II, Charles M. Duke, Jr.	140 046.6 (308 750)	65	383 000 (238 000)	265 hrs 51 min	Sixth manned lunar landing mission; fifth successful one. Spacecraft launched by Saturn V booster. After insertion into trajectory for moon and CSM separation, LM was extracted from S-IVB and docked with CSM. S-IVB propulsion system leak canceled 2nd lunar trajectory burn; untracked stage impacted moon 4/19. One midcourse correction during CSM/LM coast; live color TV transmitted. Spacecraft entered lunar orbit 4/19 at 3:22 pm EST. LM undocked, initiated descent, and touched down in Descartes region 9:23 pm EST 4/20. Young and Duke performed 3 EVAs. During 1st EVA 4/21 (7 hrs 11 min), astronauts drove lunar roving vehicle (LRV) and deployed Apollo lunar surface experiments package (ALSEP); heat flow experiment cable was accidentally broken by Young. LRV used during 2nd EVA 4/22 (7 hrs 23 min); crew explored Survey Ridge, Stone Mountain, and South Ray Crater area. On 3rd EVA 4/23 (5 hrs 40 min) crew explored North Ray Crater and 2 other stations. Live color TV transmitted during all 3 EVAs. Total weight of collected lunar samples 96.6 kg (213 lbs). Total astronaut EVA time 20 hrs 15 min. LM liftoff from moon at 8:26 pm EST 4/24 covered with live TV from LRV camera stationed nearby. Total lunar stay time 71 hrs 14 min. After CSM-LM docking and crew transfer, LM ascent stage was jettisoned, began tumbling, and went into lunar orbit instead of impacting moon. Scientific subsatellite launched into lunar orbit by CSM. CSM was inserted into trajectory for earth, and Astronaut Mattingly left CSM for 1-hr 24-min inflight EVA to retrieve film from SM camera. Two midcourse corrections were required during coast toward earth. CM separated from SM and splashed down in mid-Pacific at 2:44 pm EST 4/27. Recovery by U.S.S. *Ticonderoga*.

Dec. 7	Dec. 19	Apollo 17 (AS-512, CSM-114, LM-12)	Eugene A. Cernan Ronald E. Evans Harrison H. Schmitt	141 135.7 (311 151)	75	400 000 (248 500)	301 hrs 51 min	Seventh manned lunar landing mission; sixth successful one. Spacecraft launched by Saturn V booster. After insertion into trajectory for moon and CSM separation, CSM was extracted from S-IVB and docked with CSM. S-IVB was set on lunar impact trajectory and impacted moon 3:32 pm EST 12/10. One midcourse correction during CSM/LM coast; live color TV transmitted. Spacecraft entered lunar orbit 2:47 pm EST 12/10. LM undocked, initiated descent, and touched down at Taurus-Littrow site 2:55 pm EST 12/11. Cernan and Schmitt performed 3 EVAs. During 1st EVA 12/11 (7 hrs 12 min), astronauts drove lunar roving vehicle (LRV) to Steno Crater and deployed Apollo lunar surface experiments package (ALSEP). LRV used during 2nd EVA 12/12 (7 hrs 37 min) and explored 5 stations. Remaining stations explored during 3rd EVA 12/13 (7 hrs 16 min). Live color TV transmitted during all 3 EVAs. Total weight of collected lunar samples about 115 kg (250 lbs). Total astronaut EVA time 22 hrs 5 min and total distance traveled in lunar rover about 35 km (22 mi). LM liftoff from moon at 5:55 pm EST 12/14 covered with live TV from LRV camera stationed nearby. Total lunar stay time 74 hrs 59 min 38 sec. After CSM-LM docking and crew transfer, LM ascent stage was jettisoned, deorbited, and impacted moon. Impact recorded by 4 Apollo 17 geophones and Apollo 12, 14, 15, and 16 ALSEPs. CSM was inserted into trajectory for earth, and Astronaut Evans left CSM for 1-hr 7-min inflight EVA to retrieve film from SM camera. One midcourse correction required during transearth coast. CM separated from SM and splashed down in mid-Pacific at 2:25 pm EST 12/19. Recovery by U.S.S. *Ticonderoga*

Appendix D

ABBREVIATIONS OF REFERENCES

Listed here are abbreviations for sources cited in the text. This list does not include all sources provided in the chronology, for some of the references cited are not abbreviated; only references that appear in abbreviated form are listed below. Abbreviations used in the chronology entries themselves are cross-referenced in the Index.

A&A	American Institute of Aeronautics and Astronautics' magazine, *Astronautics & Aeronautics*
A&A 1972	NASA's *Astronautics and Aeronautics, 1972* [this publication]
ABC	American Broadcasting Company
AEC Release	Atomic Energy Commission News Release
Aero Daily	*Aerospace Daily* newsletter
Aero Med	*Aerospace Medicine* magazine
AF Mag	Air Force Association's *Air Force Magazine*
AFFTC Release	Air Force Flight Test Center News Release
AFHF Newsletter	*Air Force Historical Foundation Newsletter*
AFJ	*Armed Forces Journal* magazine
AFNS Release	Air Force News Service Release
AFOSR Release	Air Force Office of Scientific Research News Release
AFRPL Release	Air Force Rocket Propulsion Laboratory News Release
AFSC *Newsreview*	Air Force Systems Command's *Newsreview*
AFSC Release	Air Force Systems Command News Release
AF&SD	*Air Force and Space Digest* magazine
AFSSD Release	Air Force Space Systems Division News Release
AIA *Release*	Aerospace Industries Association News Release
AIAA *Facts*	American Institute of Aeronautics and Astronautics' *Facts*
AIAA *News*	American Institute of Aeronautics and Astronautics' *News*
AIAA Release	American Institute of Aeronautics and Astronautics News Release
AIP *Newsletter*	American Institute of Physics *Newsletter*
AP	Associated Press news service
ARC *Astrogram*	NASA Ames Research Center's *Astrogram*
Astro Journ	American Astronomical Society's *Astrophysical Journal*
Atlanta JC	*Atlanta Journal Constitution* newspaper
Av Daily	*Aviation Daily* newsletter
Av Wk	*Aviation Week & Space Technology* magazine
Battelle *Sci Pol Rev*	Battelle Memorial Institute's *Science Policy Reviews*
B News	*Birmingham News* newspaper
B *Sun*	Baltimore *Sun* newspaper
Bull Atom Sci	Education Foundation for Nuclear Science *Bulletin of the Atomic Scientists*
Bus Wk	*Business Week* magazine
C Daily News	*Chicago Daily News* newspaper
C Trib	*Chicago Tribune* newspaper

Can Press	Canadian Press news service
CBS	Columbia Broadcasting System
C&E News	*Chemical & Engineering News* magazine
ClPD	Cleveland *Plain Dealer* newspaper
Cl Press	*Cleveland Press* newspaper
Columbia J Rev	*Columbia Journalism Review* magazine
ComSatCorp Release	Communications Satellite Corporation News Release
CQ	*Congressional Quarterly*
CR	*Congressional Record*
CSM	*Christian Science Monitor* newspaper
CTNS	Chicago Tribune News Service
DASA Release	Defense Atomic Support Agency News Release
D News	*Detroit News* newspaper
D Post	*Denver Post* newspaper
DJ	Dow Jones news service
DOC PIO	Department of Commerce Public Information Office
DOD Release	Department of Defense News Release
DOT Release	Department of Transportation News Release
EOP Release	Executive Office of the President News Release
FAA Release	Federal Aviation Administration News Release
FBIS–Sov	Foreign Broadcast Information Service, Soviet number
FonF	*Facts on File*
FRC Release	NASA Flight Research Center News Release
FRC X-Press	NASA Flight Research Center's *FRC X-Press*
GE Forum	*General Electric Forum* magazine
Goddard News	NASA Goddard Space Flight Center's *Goddard News*
GSFC Release	NASA Goddard Space Flight Center News Release
GSFC *SSR*	NASA Goddard Space Flight Center's *Satellite Situation Report*
GT&E Release	General Telephone & Electronics News Release
H Chron	*Houston Chronicle* newspaper
H Post	*Houston Post* newspaper
JA	*Journal of Aircraft* magazine
JPL *Lab-Oratory*	Jet Propulsion Laboratory's *Lab-Oratory*
JPL Release	Jet Propulsion Laboratory News Release
JPRS	Department of Commerce Joint Publications Research Service
JSR	American Institute of Aeronautics and Astronautics' *Journal of Spacecraft and Rockets* magazine
KC Star	*Kansas City Star* newspaper
KC Times	*Kansas City Times* newspaper
KSC Release	NASA John F. Kennedy Space Center News Release
LA *Her-Exam*	Los Angeles *Herald-Examiner* newspaper
LA Times	*Los Angeles Times* newspaper
Langley Researcher	NASA Langley Research Center's *Langley Researcher*
LaRC Release	NASA Langley Research Center News Release
LATNS	Los Angeles Times News Service
LC *Info Bull*	Library of Congress *Information Bulletin*
LeRC Release	NASA Lewis Research Center News Release
Lewis News	NASA Lewis Research Center's *Lewis News*
M Her	*Miami Herald* newspaper
M News	*Miami News* newspaper
M Trib	*Minneapolis Tribune* newspaper
Marshall Star	NASA George C. Marshall Space Flight Center's *Marshall Star*
MJ	*Milwaukee Journal* newspaper
MSC Release	NASA Manned Spacecraft Center News Release
MSC *Roundup*	NASA Manned Spacecraft Center's *Space News Roundup*
MSFC Release	NASA George C. Marshall Space Flight Center News Release
N Hav Reg	*New Haven Register* newspaper
N News	*Newark News* newspaper
N Va Sun	*Northern Virginia Sun* newspaper
NAA *News*	National Aeronautic Association *News*

NAC Release	National Aviation Club News Release
NAE Release	National Academy of Engineering News Release
NANA	North American Newspaper Alliance
NAS Release	National Academy of Sciences News Release
NAS–NRC Release	National Academy of Sciences–National Research Council News Release
NAS–NRC–NAE *News Rpt*	National Academy of Sciences–National Research Council–National Academy of Engineering *News Report*
NASA Ann	NASA Announcement
NASA HHR–37	NASA Historical Report No. 37
NASA Hist Off	NASA Historical Office
NASA Hq *WB*	NASA Headquarters *Weekly Bulletin*
NASA Int Aff	NASA Office of International Affairs
NASA *LAR* X/8	NASA *Legislative Activities Report*, Vol. X, No. 8
NASA proj off	NASA project office (for the project reported)
NASA Release	NASA Headquarters News Release
NASA Rpt SRL	NASA Report of Sounding Rocket Launching
NASA SP–4014	NASA Special Publication #4014
NASC Release	National Aeronautics and Space Council News Release
Natl Obs	*National Observer* magazine
NBC	National Broadcasting Company
NGS Release	National Geographic Society News Release
NMI	NASA Management Instruction
NN	NASA Notice
NOAA Release	National Oceanic and Atmospheric Administration News Release
NR *News*	North American Rockwell Corp. *News*
NR Release	North American Rockwell Corp. News Release
NR *Skywriter*	North American Rockwell Corp. *Skywriter*
NSC *News Letter*	National Space Club *News Letter*
NSC Release	National Space Club News Release
NSF *Highlights*	National Science Foundation's *Science Resources Studies Highlights*
NSF Release	National Science Foundation News Release
NY News	*New York Daily News* newspaper
NYT, 5:4	*New York Times* newspaper, section 5 page 4
NYTNS	New York Times News Service
O Sen	*Orlando Sentinel* newspaper
Oakland Trib	*Oakland Tribune* newspaper
Omaha W–H	*Omaha World-Herald* newspaper
OST Release	Office of Science and Technology News Release
P *Bull*	Philadelphia *Evening* and *Sunday Bulletin* newspaper
P *Inq*	*Philadelphia Inquirer* newspaper
PAO	Public Affairs Office
PD	National Archives and Records Service's *Weekly Compilation of Presidential Documents*
PIO	Public Information Office
PMR *Missile*	USN Pacific Missile Range's *Missile*
PMR Release	USN Pacific Missile Range News Release
Pres Rpt 73	*Aeronautics and Space Report of the President: 1972 Activities*
SAO Release	Smithsonian Astrophysical Observatory News Release
SBD	*Space Business Daily* newsletter
Sci Amer	*Scientific American* magazine
SciServ	Science Service news service
SD	*Space Digest* magazine
SD Union	*San Diego Union* newspaper
SET *Manpower Comments*	Scientific Manpower Commission's *Scientific Engineering, Technical Manpower Comments*
SF	*Spaceflight* magazine
SF Chron	*San Francisco Chronicle* newspaper
SF Exam	*San Francisco Examiner* newspaper

Sov Aero	*Soviet Aerospace* newsletter
Sov Rpt	Center for Foreign Technology's *Soviet Report* (translations)
SP	*Space Propulsion* newsletter
Spaceport News	NASA John F. Kennedy Space Center's *Spaceport News*
Spacewarn	IUWDS World Data Center A for Rockets and Satellites' *Spacewarn Bulletin*
SR	*Saturday Review* magazine
SR list	NASA compendium of sounding rocket launches
SSN	*Soviet Sciences in the News*, publication of Electro-Optical Systems, Inc.
St Louis G–D	*St. Louis Globe-Democrat* newspaper
St Louis P–D	*St. Louis Post-Dispatch* newspaper
T Picayune	New Orleans *Times-Picayune* newspaper
Tech Rev	Massachusetts Institute of Technology's *Technology Review*
Testimony	Congressional testimony, prepared statement
Text	Prepared report or speech text
Transcript	Official transcript of news conference or congressional hearing
UPI	United Press International news service
USGS Release	U.S. Geological Survey News Release
US News	*U.S. News & World Report* magazine
W Post	*Washington Post* newspaper
W *Star & News*	Washington *Evening/Sunday Star and News* newspaper (as of July 13, 1972, merger)
WH Release	White House News Release
WJT	*World Journal Tribune* newspaper
WS Release	NASA Wallops Station News Release
WSJ	*Wall Street Journal* newspaper

INDEX AND LIST OF ABBREVIATIONS AND ACRONYMS

A

A-7 (fighter aircraft), 321
A-7G (Corsair), 314
A-300 (European consortium airbus), 94
A-300B, 365
AAAS. See American Assn. for the Advancement of Science.
AACB. See Aeronautics and Astronautics Coordinating Board.
AAS. See American Astronautical Society.
Abbot, Dr. Charles G., 168
Abbott Laboratories, 247
ABC. See American Broadcasting Co.
Abidjan, Ivory Coast, 397
ABM. See Antiballistic missile system.
ABMA. See Army Ballistic Missile Agency.
Abrahamson, Col. James A. (USAF), 16
Abzug, Rep. Bella S., 193
Academician Sergei Vavilov (U.S.S.R. research ship), 228
Academy of Sciences, U.S.S.R. See Soviet Academy of Sciences.
Accelerator, 79, 175
Accident (see also Fire; Explosion), 45-46, 73, 131, 174, 266
  aircraft
    F-14, 244, 271
    F-111, 230, 233, 236
    general aviation
      prevention, 204
      statistics, 454
    Lockheed Electra, 15
    Orville Wright, 26
    T-38, 25, 178, 179, 214
    U.S.S.R., 349
  astronaut, 91, 153, 169-170
  cosmonaut, 166
  lunar experiment, 146
  radioactive fallout, 80
  spacecraft
    Apollo 13, 339
    Apollo 16 Subsatellite, 208
    Apollo 17 LRV, 416
    Luna 18, 166
    Navy Transit satellite, 80
  TRANSPO '72, 206, 222
Achievement Award (SBA), 247
Adiabatic toroidal compressor (ATC), 404
Adler, Dr. Isidore, 14

Administration and Program Support (MSFC) (NASA), 243-244
Administration and Technical Services Directorate (MSFC) (NASA), 243-244
Advance Systems Construction, Inc., 60
Advanced medium STOL transport (AMST), 77, 382-383, 443
Advanced Missions Office, OMSF (NASA), 398
Advanced Programs Office, OMSF (NASA), 398
Advanced Research Projects Agency (ARPA), 18, 247, 445
Advanced Technology Applications program (RANN), 85
Advertising Club of Baltimore, 49
AE. See Atmosphere Explorer.
AEC. See Atomic Energy Commission.
AEC-NASA Space Nuclear Systems Office, 59, 79, 130-131
AEDC. See Arnold Engineering Development Center.
Aerial Phenomenon Research Organization, 241
Aerial photography, 280-281
Aero Club of Washington, 162, 263
Aerobee (sounding rocket)
  150
    aeronomy experiment, 156
    atmospheric composition experiment, 237
    auroral astronomy experiment, 109, 125
    x-ray star detection, 394
  170
    aeronomy experiment, 306
    astronomy experiment, 180, 259, 288, 319
    planetary atmosphere data, 378
    soft x-ray experiment, 422
    solar brightness measurement, 425
    solar physics experiment, 158, 225, 242, 255, 274, 276, 366
    solar ultraviolet light measurement, 427
    test flight performance, 388
    x-ray astronomy, 61, 127, 288
  design, 394
  history, 394
Aerodynamic Test Range (NASA), 264
Aerojet-General Corp., 35, 252
  Aerojet Nuclear Systems Co., 59
  Aerojet Solid Propulsion Co., 399

Aeronautical Propulsion Div., OAST (NASA), 176
Aeronautical Research Associates, 20
Aeronautics (see also Federal Aviation Administration)
  Aeronautics and Space Report of the President: 1971 Activities, 99
  aircraft. See Aircraft.
  anniversary, 311–312, 449
  awards and honors, 69, 114, 210, 403, 430
  cooperation, 6, 49–50, 67, 79, 204, 330, 347
  employment, 85, 233
  exhibit, 110, 136, 434–435
  facilities, 106, 110, 353
  funds
    appropriation, FY 1973, 187–188, 273–274
    authorization, FY 1973, 115–116, 160, 172, 178
    constant level budget, 450–451
    House hearings, 21–23, 66–67, 84–85, 109–110, 209
    impoundment, 315
    program costs, 93
    program revision, 451
    request, FY 1973, 30, 32
    Senate hearings, 99–100, 106–107, 116
  general aviation. See General aviation.
  hijacking. See Hijacking of aircraft.
  international competition, 53–54
  JT3D/JT8D Refan Program Office, 176
  meeting, 20–21, 49, 185–186, 221, 314–315, 385
  military (see also U.S. Air Force, aircraft), 67, 106, 389, 397
  NASA program. See National Aeronautics and Space Administration, programs.
  National Aviation System (NAS) Ten Year Plan and Policy Summary, 141–142
  research (see also individual aircraft and Lifting body)
    AWACS, 42
    DFBW, 200, 312, 449
    FRC programs, 311
    hypersonic, 153
    instrumentation, 42, 43–44, 235, 302, 312, 350, 392, 393
    landing methods, 56, 124, 186, 336–337, 397
    LaRC programs, 153
    quiet engine technology (see also Noise, aircraft), 66–67, 186, 231, 387, 449
    simulation and testing, 42, 191–192, 247, 396, 397, 449
    supersonic, 59, 190, 191–192, 349, 390, 395
    Terminal Configured Vehicles and Avionics Operating Experiments Program, 84
  wing studies
    antisymmetrical, 191–192
    augmentor-wing, 169, 172, 324
    delta, 272
    supercritical, 50–51, 72–73, 184, 186, 193, 214, 449
  safety, 49, 78, 113, 116, 178, 187–188, 204, 209, 242, 273–274
  space technology use in, 46–47, 106, 400
  technology assessment, 2, 319
Aeronautics and Astronautics Coordinating Board (AACB), 353
Aeronautics and Space Engineering Board. See National Academy of Engineering.
Aeronautics and Space Report of the President: 1971 Activities, 99
Aeronca, Inc., 293
Aeropropulsion System Test Facility (ASTF), 353
Aeros (scientific satellite), 377, 430–431, 452
Aerosat. See Air traffic control satellite program.
Aerosonic Corp., 123
Aerospace Div. Conference (ASME), 314
Aerospace Employment Project, 304
Aerospace Industries Assn. of America, Inc. (AIA), 118, 162, 188, 204, 339–340, 398
Aerospace industry
  achievements, 133
  business base, 151
  contracting and procurement, 15, 53, 55, 274, 319, 406–407
  cooperation, 23, 204
  defense, 39
  employment
    Aerospace Employment Project, 304
    McGovern, Sen. George S., 310, 314
    manpower increase, 4, 5, 9, 55, 156, 198, 204, 208, 451
    manpower reductions, 59, 133–134, 274, 280, 281, 297–298, 301, 310, 314, 326, 339–340, 365
    recruitment, 61
    wages, 6, 10, 16–17, 46
  Europe, 218, 239
  exhibit, 434
  international competition, 4, 10, 50, 193
  meeting, 60
  metric system conversion, 182
  NASA technology use, 92, 130–131, 452
  overproduction, 192–193
  public opinion, 2
  trade, 133–134, 451, 453
  West Germany, 159
Aerospace Mechanisms Symposium, Seventh, 310
Aerospace Primus Club, 313
Aerospace Research Pilot School, 71
Aerospace Safety Advisory Panel. See NASA Aerospace Safety Advisory Panel.
Aerospace Safety Research and Data Institute, 244

ASTRONAUTICS AND AERONAUTICS, 1972

Aérospatiale (Société Nationale Industrielle Aérospatiale), 271
AFCRL. See Air Force Cambridge Research Laboratories.
AFGE. See American Federation of Government Employees.
AFML. See Air Force Materials Laboratory.
Africa (see also South Africa), 63, 250-251
AFSC. See Air Force Systems Command.
Ag-Cat (agricultural aircraft), 309
Agena (booster upper stage), 108, 190, 200, 305, 328, 439
Agena (star), 255
Agency for International Development (AID), 302
Agnew, Vice President Spiro T., 37, 142, 197-198, 350, 407
Agreement. See International cooperation; International cooperation, space; Memorandum of Understanding; Treaty.
Agreement for Scientific and Technological Cooperation, 230
Agreement on Cooperation in the Field of Environmental Protection, 193-194, 245, 374-375
Agreement on Cooperation in the Field of Medical Science and Public Health, 194, 222, 245, 374-375
Agreement on Exchanges and Cooperation in Scientific, Technical, Educational, Cultural, and Other Fields, 195, 222, 245, 374-375
Agreement on the Rescue of Astronauts, the Return of Astronauts, and the Return of Objects Launched into Outer Space, 195
Agriculture, Dept. of
 cooperation, 68, 169-170, 301-302
 ERTS, 210, 235, 270, 280
 Federal earth resources survey program, 302
 funds, 51
 reorganization, 122
Agrippa Crater (moon), 46
AH-56A (Cheyenne) (helicopter), 288
Ahmedabad, India, 422
AIA. See Aerospace Industries Assn. of America, Inc.
AIAA. See American Institute of Aeronautics and Astronautics.
AICMA. See Association Internationale des Constructeurs de Materiél Aérospatial.
AID. See Agency for International Development.
AIDJEX. See Arctic Ice Dynamic Joint Experiment.
Aiken, S. C., 236
Air Force. See U.S. Air Force.
Air Force Aero Propulsion Laboratory, 20
Air Force Assn., 319-320
Air Force Cambridge Research Laboratories (AFCRL), 216, 275, 295, 302, 364
Air Force Materials Laboratory, 210-211, 308
Air Force/NASA STOL Coordinating Council, 77
Air Force Prototype Systems Program Office, 67
Air Force Space and Missile Systems Organization. See Space and Missile Systems Organization.
Air Force Systems Command (AFSC)
 Armament Development and Test Center, 377
 award, 313
 contract, 51, 122, 325
 cost control initiatives, 294
 Electronic Systems Div., 383
 history, 321
 launch, 364-365
 lifting body research, 347
 missile, 16, 377
 parachute decelerator system, Viking lander, 428
 particle identifier, 216
 personnel, 140, 189, 303, 308
 symposium, 293-294
Air Force Weather Service, 122-123
Air Industries Assn. of Canada, 204
Air Line Pilots Assn. (ALPA), 203-204
Air pollution
 air sample collection, 154
 aircraft, 66, 136, 280, 292, 349, 370, 409, 449
 airport standards, 203-204
 control, 206
 data analysis, 63
 Earth-Watch, 218
 Environmental Quality: The Third Annual Report of the Council on Environmental Quality-August 1972, 285
 monitoring, 21, 35, 84, 101, 216-217, 218, 285, 390, 392
 prevention, 2, 5, 31
 Republican National Platform, 299
 research, 50, 66, 73-74, 276-277, 407
 technology utilization, 101
 U.N. Conference on the Human Environment, 217
Air Products and Chemicals, Inc., 224
Air show, 157
Air superiority aircraft, 263
Air traffic control (ATC)
 AIAA exhibit, 434
 airport congestion, 84, 136, 182, 203-204
 automated flight-data processing system, 454
 FAA program, 104
 facilities, 141-142
 instrument landing system (ILS), 141-142
 international, 12-13
 microwave landing system (MLS), 20
 radar, 25, 38, 454
 satellite technology, 141-142
 space shuttle role, 25
Air traffic control satellite (aerosat) program, 12-13
Air turbulence. See Clear-air turbulence.

Airborne Instruments Laboratory, 35
Airborne science, 303
Airborne visible-laser optical-communications (AVLOC), 292-293
Airborne warning and control system (AWACS), 31, 42
Airbus (see also A-300; A-300B), 329, 361
Aircraft (see also individual aircraft, such as A-300, B-52, Convair 990, F-111, TF-8A, etc.)
 accident, 15, 25, 26, 178, 179, 204, 214, 230, 233, 236, 244, 271, 349, 454
 agricultural, 309
 air traffic control. See Air traffic control.
 airbus, 94, 361, 365
 amphibious, 362
 augmentor-wing. See Wing, aircraft, augmentor.
 avionics system, 42, 263, 382
 award, 442
 bomber, 31, 93, 201-202, 209, 329
 cargo, 341
 carrier, 31, 337
 civil, 20, 22-23, 93, 106
 collision avoidance, 116, 178
 commercial, 23, 50, 72, 312, 349, 393
 cost, 77, 93, 260, 263
 design, 20, 131, 191-192, 209, 231
 development
  Air Force Prototype Systems Program Office, 67
  AMST, 382-383, 443
  CTOL, 66-67
  NASA technology, 46-47, 154
  QUESTOL, 23, 66-67, 161, 350, 387, 451
  remotely piloted vehicle, 51, 316, 321, 336-337, 393
  STAMP, 446
  STOL, 2, 23, 31, 40, 53-54, 66-67, 77, 118, 124, 172, 182, 185-186, 268-269, 324, 355, 449
  V/STOL, 31, 67, 141-142, 161, 187, 353, 355
  VTOL, 30, 40, 56, 66-67
 engine. See Engine, aircraft.
 environmental effects, 136, 155, 203, 217
 exhibit, 206, 222, 307, 358
 fighter, 240, 263, 314, 328, 378
  supersonic, 26, 31, 151-152, 209, 220, 230, 233, 236, 241, 244, 263, 282, 322, 328, 340, 355, 377, 395, 425
 flight tests, 67, 73, 155, 163, 169, 177, 184, 186, 193, 200, 215
 foreign, 10, 94, 106, 177, 248
 U.S.S.R., 263, 285, 317, 329-330, 349, 354, 365, 425, 442
 funding, 30-31
 general aviation, 161, 311, 314, 453
 glider, 161, 275
 helicopter, 40, 56, 136, 148-149, 187, 190-191, 303, 362, 389, 393, 409
 hijacking. See Hijacking of aircraft.
 industry, 22-23, 62, 192-193, 309, 453
 instrumentation, 42, 43-44, 235, 302, 312, 350, 392, 393
 interceptor, 177
 jet, 178, 203, 309, 392, 397
 landing methods, 56, 124, 186, 336-337, 397
 light aircraft, 15, 78
 military, 20, 67, 122, 131, 133-134, 201-202, 206, 263, 312, 319, 321, 327, 329, 396, 397
  Military Aircraft Programs Office, 67
 multipurpose, 263
 NASA. See National Aeronautics and Space Administration.
 navigation, 56
 noise. See Noise, aircraft.
 performance, 192-193
 reconnaissance, 72, 118, 162, 216, 255-256, 263, 280-281, 350
 record, 61
 remote control (see also Remotely piloted vehicle), 316, 336-337
 rescue, 15
 research (see also Aeronautics, research; and CTOL, STOL, TF-8A, etc.), 21-23, 24, 40, 67, 128, 293, 324, 396
  clear-air turbulence, 43, 302
  combustor design, 66, 276-277
  fuselage fairings, 274-275
  subsonic flight characteristics, 155, 247
  tire testing, 47-48
 rocket, 233
 safety, 49, 78, 113, 116, 178, 187-188, 204, 209, 242, 273-274
 specialization, 263
 sport, 309
 supercritical wing. See Wing, aircraft, supercritical.
 supersonic. See Supersonic transport; Concorde; F-111; Tu-144.
 survey, 21, 68, 216, 221-222, 226-227, 235, 254, 270, 302, 314
  airborne experiment, 24, 136, 178, 341, 349
 takeoff technology, 177
 terminal-configured, 84
 tilt-rotor, 40
 traffic control. See Air traffic control.
 training, 25, 214, 309
 transport (see also Supersonic transport), 2, 23, 31, 93, 94, 107, 110, 120, 124, 131, 154, 157, 161, 329, 422, 425, 439
  jet, 4-5, 162, 173, 186, 192-193, 248, 251, 288, 292, 298, 314, 342, 409
 turboprop, 61
 U.S. leadership, 50, 193, 451
 utility, 177, 309, 397, 453
Aircraft carrier, 31
Aircraft Collision Avoidance Systems, 104
Aircraft Div. See Northrop Corp.
Aircraft Emissions: Impact on Air Quality and Feasibility of Control, 409

Aircraft Engine Research Laboratory, 244
Airlines (see also Air pollution; Air traffic control; Airports; Noise, aircraft; and Supersonic transport)
  bathroom commode system, 293
  computerized reservation systems, 180
  cooperation, 21, 23, 341
  cost, 118, 182, 427
  FAA airport-airways system plan, 141–142
  hijacking. See Hijacking of aircraft.
  industry, 118, 161, 162, 200, 427
  navigation, 33
  personnel, 72–73
  security, 93, 411
  statistics, 453, 454
  STOL importance, 118
  symposium, 72–73
  two-segment landing approach, 124
Airlock module (AM), 219, 262, 280, 282, 325, 341, 344
Airports (see also Air pollution; Air traffic control; Noise, aircraft; and individual airports such as Dulles International Airport)
  automation, 84
  CFD system, 123
  congestion, 67, 83–84, 118, 182, 203–204, 449
  Discrete Addressable Beacon and Microwave Landing Systems, 84
  environmental problems, 427
  FAA airport-airways system plan, 141–142
  facilities, 136, 139–140, 141–142
  navigation, 20
  planning, 31, 206
  Planning Grants Program (FAA), 272
  runways, 2, 84
  security, 67, 93, 411
  wake turbulence detection system, 92
Airship, 400
Airspeed indicator, 397
Aitken Crater (moon), 418
Alabama, 69, 74, 129–130, 206, 208, 393
Alabama Earth Resources Information Committee, 294
Alabama Space and Rocket Center, 117, 177, 277
Alabama, Univ. of, 95, 454
Alaska, 44, 70, 122–123, 262, 308, 341
  earth station, 173, 410
  AIDJEX test site, 128
  satellite data, 124, 131, 235, 270, 444
  solar eclipse viewing site, 254
  sounding rocket launch, 90, 91, 182–183, 186–187, 349, 358
  TV transmission via satellite, 27, 173, 262
  weather buoys, 77
Alaska Educational Television Broadcasting Commission, 173
Alaska pipeline, 396
Alaska, Univ. of, Geophysical Institute, 182–183

Albert, Rep. Carl, 286–287
Aldrin, Col. Edwin E., Jr. (USAF, Ret.), 18, 71, 265, 313, 399, 426
ALFMED. See Apollo light-flash moving-emulsion detector.
Alfven, Dr. Hannes O., 2, 141
Algol III (rocket engine), 290
Algonquin Park, Canada, 307
Alibrando, Alfred B., 52
Allen, J. H., 308
Allen, William A., 331
Allen, William M., 215
Allison, L/C Edgar L., Jr., (USAF), 61
Allott, Sen. Gordon, 362, 387
Alouette I (Canadian satellite), 83, 328, 453
ALPA. See Air Line Pilots Assn.
Alphonsus Crater (moon), 62, 404
ALSEP. See Apollo lunar surface experiment package.
Altimeter, 123
Aluminum, 14, 183–184
AM. See Airlock module.
Amazon Basin, 350
Ambler, Ernest, 216
Amchitka Island, Alaska, 308
America (Apollo 17 CSM). See Command and service module.
American Airlines, Inc., 186, 189, 216–217 308
American Association for the Advancement of Science (AAAS), 283–284, 429, 443–444
American Astronautical Society (AAS), 98, 314, 360, 373–374
American Astronomical Society, 112, 131
American Aviation Corp., 309
American Bakeries Co., 430
American Broadcasting Co. (ABC), 381, 399, 409
American Chemical Society, 225
American College of Preventive Medicine, 384
American Council of Learned Societies, 392
American Federation of Government Employees (AFGE), 242–243, 255
American Geophysical Union, 233
American Helicopter Society, 187, 362
American Institute of Aeronautics and Astronautics (AIAA), 36, 69, 83, 137, 314, 399, 403, 434
American Institute of Architects, California Council, 241
American Institute of Biological Sciences, 24
American Institute of Chemical Engineers, 225
American Institute of Nondestructive Testing, 225
American Medical Assn. (AMA), 350
American Mountains (moon), 237
American Museum–Hayden Planetarium, 185, 211, 356, 428, 455
American Physical Society, 159
American Satellite Corp., 355, 410

American Science and Engineering, Inc., 127, 255, 380–381
American Society for Metals, 225
American Society for Quality Control, 225
American Society for Testing Materials, 93, 225
American Society of Civil Engineers (ASCE), 355
American Society of Mechanical Engineers (ASME), 137, 314
  Man of the Year Award, 202–203
American Society of Public Administrators, 117
American Telephone & Telegraph Co. (AT&T), 230–231, 255, 261–262, 309, 441
American Topical Assn., 253
Ames Research Center (ARC) (NASA)
  acoustic–aerodynamic research, 67
  AIDJEX participation, 128
  air pollution research, 392
  Airborne Research Program, 68, 303, 314
  aircraft research, 21, 45, 124, 169, 191–192
  aircraft survey, 280–281, 314–315
  Arizona Land Use Experiment, 235
  automated visual sensitivity tester, 41
  cloud research, 24
  conference, 355
  contract, 382
  cooperation, 67–68, 79, 382
  electron microscope improvement, 101
  electrophysics research program, 79
  Environmental Problems Symposia, 101
  ERTS participation, 210
  evaluation studies, 303
  GARP, 397
  ILLIAC IV computer, 445
  laser-doppler CAT detection system test, 302
  lecture, 291, 401
  Life Scientist Program, 298
  personnel, 152, 158, 223, 264, 380–381
  QUESTOL program, 387
  request for proposals, 5, 94
  tour, 384
  visit, Mrs. Nixon, 324
  wind tunnel, 353
AMSAT. See Radio Amateur Satellite Corp.
AMST. See Advanced medium STOL transport.
Anaheim, Calif., 314
Anderson, Dr. Don L., 440
Anderson, Sen. Clinton P., 98, 132–133, 277, 282–283, 387
Andoeya, Norway, 15, 17, 53, 210
Andover, Me., 255
Andrews AFB, Md., 190, 303, 313, 319, 362
Andrews, Edward P., 159
Androgynous docking mechanism, 197–198

Anfinsen, Dr. Christian B., 357
Anglo-French supersonic transport. See Concorde.
Anik I (Telesat A) (Canadian comsat), 379, 395, 452
Animal experiments, space, 394, 404
Anniversary
  Aerobee, 394
  Alouette I, 328
  Annual Revolution Day, U.S.S.R., 377
  Apollo 11, 265
  Bicentennial, 131
  communications satellite, 370
  Copernicus, birth, 183, 257, 274
  Cosmos 1, 108
  discovery of Tycho's star, 383
  Flight Research Center, 311–312, 449
  Friendship 7 mission, 61
  Kepler, Johannes, birth, 60
  LaRC visitor center, 217
  Mariner 2 Venus arrival, 429
  Navy Astronautics Group, 134
  Oso 1, 90
  Project Skyhook, 335
  Saturn V, 33
  Sputnik I, 338, 348
  supersonic flight, 350
  Telstar I transoceanic TV, 255
  transistor, 441
  USAF, 258, 319, 320–321
  U.S.S.R. reactor, 14–15
  Vanguard I, 109
  von Braun, Dr. Wernher, 117
  Vostok I, 138
  Wright brothers flight, 434
Annual Federal Report on Earth Resources Survey Programs, 301–302
Annual Revolution Day, U.S.S.R., 377
Anorthosite (lunar surface sample), 14, 179, 377–378
ANT-1 (U.S.S.R. aircraft), 442
Antarctica, 171, 262
Antares (*Apollo 14* LM). See Lunar module.
Antenna, 31, 65, 82, 113, 146, 148, 208, 336, 354
Anthropology, 54
Antiballistic missile (ABM) system (see also Safeguard), 362
  budget, 30–31, 327
  Minuteman supplement, 72
  public attitude, 13
  test, 108, 174, 260, 300, 361, 411
  treaty, 201–202, 213, 283, 299, 331
Antigua, West Indies, 184, 195
Antimatter, 176
Antiproton, 176
Antisubmarine warfare, 441
Antisymmetrical wing, 191–192
Antitechnology, 135–136
APK. See Astronaut preference kit.
Apollo (program), 2, 100, 250–251, 380–381
  achievements, 164–165, 265, 345, 394, 406, 412, 426–427, 435, 438, 439, 440, 441, 442–443, 447, 449

502

astronaut. See Astronaut.
  benefits, 157, 394, 420, 426–427, 438–439, 440
  budget, 64, 300, 359
  conclusion, 58, 423, 424, 447, 449
  cost, 7, 30, 36, 71–72, 87, 424
  criticism, 162, 407
  evaluation, 142, 405–406
  experiments, 6, 102, 180–181, 300
  funds, 125, 176–177
  impact, 151, 204–205, 404–405, 411, 424
  launch, 169–170, 208
  launch vehicle, 169–170, 452
  lunar debris, 432
  lunar landing trainer, 311
  lunar model development, 54
  lunar samples, 52, 152, 377–378
  management, 248
  mission cancellations, 175–176
  Nixon, President Richard M., 435–437
  operational maturity, 57–58
  personnel, 9, 14, 27, 29, 33, 62, 63, 78, 117, 170, 197–198, 205, 440
  phase-out, 9, 18–19, 27, 157, 205
  press comment, 133, 151, 152–153, 160–161, 404–405, 411–412, 423, 424, 429, 434, 440, 443
  responsibilities, 420
  results, 2–4, 13–14, 63, 226–227, 441
  significance, 69, 420, 441–442, 449
  support services, 248
  surplus hardware use, 28
  symposium, 310
  television coverage, 391–392
  training, 23–24
  U.S.S.R. comparison, 140–141
Apollo (spacecraft)
  accident, 91, 266
  Apollo-Soyuz Test Project, 128, 130, 174–175, 195–196, 197–198, 247, 250, 257, 260, 335, 344, 346
  command and service module. See Command and service module.
  control system, 200
  development responsibility, MSC, 149–150, 273
  exhibit, 358
  international cooperation, 106
  lunar module. See Lunar module.
Apollo 5 mission, 169–170
Apollo 7 mission, 169–170, 192, 298
Apollo 8 mission, 137, 205, 237, 392, 394, 419–420
Apollo 9 mission, 33, 142, 197–198, 221
Apollo 10 mission, 39, 60, 115, 144–146, 404, 416
Apollo 11 mission, 119, 133, 167, 355, 392, 420
  anniversary, 265
  astronaut, 63, 71, 74, 96, 205, 313
  distance traveled, 63
  documentary film, 385
  lunar samples, 33, 179, 291
  significance, 192–193

"Apollo 11: Soil Mechanics Results," 355
Apollo 12 mission
  astronaut, 63, 178
  exhibit, 206, 358
  experiments, 142–144, 388, 414, 419
  launch, 141
  lunar samples, 179, 206, 358
Apollo 13 mission, 259, 302, 339, 392, 420
Apollo 14 mission, 259
  astronaut, 63, 188, 194, 280
  commercialism, 277, 280
  experiments, 54–55, 142–144, 180–181, 414
  landing site, 148, 180–181, 219–220
  lunar samples, 13–14, 54, 203, 452
  lunar water vapor, 83, 86
  photographs, 437
Apollo 15 mission, 57–58, 96–97, 149–150, 168
  astronaut, 1, 4, 20, 35, 41, 61, 63, 65–66, 152, 164, 194, 206, 210, 232, 256, 266–268, 270, 273, 277, 280, 282–283, 289, 300, 303, 317–318, 321, 336
  award, 152, 164, 206, 210, 267–268, 336
  commercialism, 232, 256, 266–268, 270, 273, 277, 280, 282–283, 289, 300, 317–318, 321
  press comment, 258, 259, 262, 276
  cost, 71–72
  exhibit, 153
  experiments, 142–144, 414, 419
  foreign reaction, 28
  landing site, 69, 146–148, 431
  lunar roving vehicle, 57–58, 99, 420
  lunar sample, 35–36, 62, 138, 146, 215–216, 257, 452
  lunar surface data, 13–14, 428, 433
  medical aspects, 97, 152, 164
  postage stamp commemoration, 157
  press comment, 258, 259, 262, 276, 289
Apollo 15 Preliminary Science Report (NASA SP-315), 168
Apollo 15 Subsatellite, 50, 152
Apollo 16 mission, 228–229, 259, 387, 393, 414, 447
  accident, 174
  achievements, 149–150
  astronaut, 2, 14, 62–63, 120, 165, 169, 170–171, 172, 178, 180, 184, 187, 193, 200, 228–229
  award, 187, 200, 238, 258
  commemorative medal, 264
  cost, 71–72
  criticism, 157, 162, 178
  docking, 142–144
  experiments, 62, 68, 142–144, 146–148, 149–150, 165, 193, 254
    ALSEP, 146, 419
    cosmic ray detection, 102, 149–150
    electrophoresis demonstration, 144
    far-ultraviolet camera and spectrograph, 102, 179–180, 271, 306
    lunar magnetism, 88, 148–149, 152

MEED, 96
  seismometer, 414
  extravehicular activity, 146–148, 149–150, 152, 157, 165, 170–171, 173
  landing site, 62, 165, 431
  launch, 142
  launch vehicle, 33
  lunar landing, 142, 144–146, 148
  lunar liftoff, 148
  lunar roving vehicle, 97, 142, 146–149, 169, 172, 264
  lunar samples, 142, 146–148, 149–150, 170–171, 238, 257, 326, 363, 452
  medical aspects, 2, 14, 97, 120, 135, 148–150, 152, 164, 166, 172, 279, 420
  mission objectives, 62
  mission preparations, 2, 8, 14, 30, 33, 41, 46, 96, 112, 120, 135, 151
    countdown, 113, 123, 134, 142
    flight readiness review, 94
    spacecraft repair, 46
  Nixon, President Richard M., 112–113, 157, 162–163, 163–164, 228–229
  On the Moon with Apollo 16, 71
  photographs, 62, 142–144, 148, 149–150, 158, 169, 170–171
  press comment, 142, 151, 152–153, 154, 157, 158, 160–161, 162, 164–165
  press conference, 6, 62, 96–97, 109, 142, 148–149, 164, 170–171, 172, 179–180, 183–184
  press support, 328
  quarantine, 120
  records, 149–150
  rendezvous, 148
  results, 149–150, 158, 179–180, 183–184
  splashdown, 148–149
  Subsatellite. See Apollo 16 Subsatellite.
  technical problems, 33, 142–144, 148, 149–150, 152–153, 157, 162, 208
  translunar injection, 142–144
  TV broadcasts, 142–144, 146–149, 151, 156–157, 158, 381
Apollo 16 Subsatellite, 148–149, 152, 208
Apollo 17 mission, 265, 301, 392, 393, 395, 447
  accident, 416
  achievements, 412, 427, 447
  astronaut, 54–55, 194, 297, 301, 366–367, 382, 384, 385, 388, 389, 396, 404, 409, 413, 419, 426, 437, 439
  cutbacks, 297–298
  docking, 413, 419
  emblem, 347
  experiments, 54–55, 68, 366, 388, 414, 416, 419, 433
    ALSEP, 54–55, 416–417
    convection demonstration, 413
    far-ultraviolet spectrography, 433
    gravimeter, 54–55, 432
    heat flow, 54–55, 413, 432
    laser, 54–55, 433
    lunar roving vehicle, 346–347
    mouse-to-the-moon, 404
    scanning radiometer, 54–55, 433
    seismometer, 54–55, 431

  extravehicular activity, 416–419, 444, 447
  flag unfurling, 416
  flight rules, 267
  foreign reaction, 421, 439
  launch, 337, 361, 384, 413
  lunar landing, 414
  lunar liftoff, 418–419, 428
  lunar orbit insertion, 413
  lunar roving vehicle, 54–55, 346–347, 413, 416–419, 429
  lunar samples, 388, 413, 416–417, 419, 445, 447, 452
  management, 142
  medical aspects, 385, 388, 403, 414, 420, 437
  midcourse corrections, 413
  mission objectives, 382
  mission preparations, 30, 182, 297, 301, 382, 403, 407
    countdown, 385, 388, 389, 396, 399, 407, 413
  mission profile, 366–367, 413–419
  Nixon, President Richard M., 409, 418, 420, 428, 435–437, 439–440
  photographs, 416–417, 418, 419, 440, 442
  postmarks, launch date, 405
  press comment, 389, 409, 420, 426–427, 428–429, 434, 438–439, 441, 442
  press conference, 363–364, 366–367, 382, 388, 428–429, 431–432, 433–434, 435
  public opinion, 389
  quarantine, 385
  records, 409, 413, 419, 447
  rendezvous, 413
  results, 363, 419, 426, 431–432, 433–434, 435, 437
  significance, 428
  splashdown, 297, 313, 419, 435, 441
  strike threat, 377, 405
  technical problems, 330, 413, 419
  tracking, 422
  TV broadcast, 381, 391–392, 399, 409, 413, 416–418, 419, 427, 444
  U.S.S.R. reaction to, 421, 424, 437
  "Voyage Beyond Apollo" cruise, 409
Apollo 18 mission, 388
Apollo 20 mission, 388
Apollo/ASTP Program Office, 398, 451
Apollo light-flash moving-emulsion detector (ALFMED), 6
Apollo lunar surface experiment package (ALSEP), 54–55, 62–63, 146–148, 368, 416–417
Apollo Program Office, 398, 451
Apollo Soil Mechanics Investigation Science Team, 13–14
Apollo-Soyuz Test Project (ASTP), 61, 247, 384, 439
  agreements, 185, 190, 197–198, 207, 244, 322, 395, 452
  astronaut, 174–175, 175–176, 185, 195–196, 197–198, 207, 335, 346, 354
  benefits, 197–198, 214, 222, 228
  Congress, 198–199, 209

cosmonaut, 175–176, 197
cost, 197–198, 206–207
crew requirement, 205
criticism, 328
employment impact, 208, 424
feasibility, 175–176
funds, 198–199, 213
language, 197, 382
management, 197, 297, 451
meetings, 252, 281, 310, 322, 346, 354, 409, 452
Nixon, President Richard M., 213, 228–229, 240–241
plans, 48–49, 174–175, 195–196, 197, 204–205
political implications, 411
press conference, 250, 257, 260–261, 344
rendezvous and docking, 182, 241, 322, 346, 378, 408
safety, 197
significance, 192–193, 197–198, 336
spacecraft, 197–198, 344, 346
training, 197, 346
working groups, 197, 322
Apollo Telescope Mount (ATM), 23–24, 44, 69, 70, 169, 239, 262, 280, 325
Apollonius Crater (moon), 179
Applications of Aerospace Technology in the Public Sector, 75
Applications satellite, 166, 185
Applications Technology Satellite (ATS) (see also individual ATS satellites, such as ATS–F, ATS–G, and *Ats 1*), 30, 241, 292–293
Applied Physics Laboratory, Johns Hopkins Univ., 301, 306, 386
APS. See Auxiliary propulsion system.
Arabia basin (moon), 237
Aral Sea, 329
ARC. See Ames Research Center.
Arcas (sounding rocket), 173, 184, 195
 Boosted Arcas I, 19, 373
 Super Arcas, 8, 39, 372, 410
Archeology, 298
Arctic, 21, 70–71, 186–187, 228, 262, 381–382
Arctic Ice Dynamic Joint Experiment (AIDJEX), 128
Arctic Ocean, 77
Arenosillo, Spain, 248, 251
Argentina, 173
Ariel 1 (U.K. satellite), 453
Ariel 4, 354
Aristillus Crater (moon), 35–36
Arizona, 148, 235, 396–397
Arizona Land Use Experiment, 235
Arizona, Univ. of, 24
Arlauskas, Joseph, 377–378
Armament Development and Test Center. See Eglin AFB, Fla.
Armed Forces Management Assn., 293
Armed Forces Unification Act, 258, 320
Arms. See Weapons.
Arms limitation. See Disarmament.
Arms race, 55–56

Armstrong, Neil A., 205, 247, 265, 385, 399, 424, 426, 432, 442, 447, 455
 award, 96, 315
Armstrong, Neil, Air and Space Museum, 265
Army. See U.S. Army.
Army Air Mobility Research and Development Laboratory, 355
Army Aviation Systems Command, 303, 389
Army Ballistic Missile Agency (ABMA), 202–203, 451
Army Corps of Engineers, 184, 216, 270, 301–302
Army Ordnance Corps, 337
Arnold Engineering Development Center (AEDC), 321
Arnold, Gen. Henry H. (USA), 275
Arnold, Dr. James R., 14
ARPA. See Advanced Research Projects Agency.
ARTS. See Automated radar terminal system.
Arzachel Crater (moon), 404
ASCE. See American Society of Civil Engineers.
Asheville, N.C., 390
Asia, 8, 263, 404, 443
Asimov, Isaac, 409
Askew, Gov. Reubin (Fla.), 340, 385
ASME. See American Society of Mechanical Engineers.
Aspin, Rep. Les, 155
Assembly of Croatia, Yugoslavia, 35
Associated Press Man of the Year in Science Award, 202–203
Association Internationale des Constructeurs de Materiél Aérospatial (AICMA), 204
Asteroid, 227, 259–260, 314, 446
Asteroid Belt, 81, 83, 200, 254, 259–260, 305, 339, 365–366, 447
ASTF. See Aeropropulsion System Test Facility.
ASTP. See Apollo-Soyuz Test Project.
Astrobee F (sounding rocket), 327
Astrodynamics Conference, 314
Astronaut (see also Cosmonaut; Extravehicular activity; Scientist–astronaut; and Space biology and medicine)
 accident, 91, 153, 169–170, 178, 179
 achievements, 420, 538
 anniversary, 265
 Apollo 10, 60
 Apollo 11, 71, 385, 399
 Apollo 14, 372–373
 Apollo 15, 1, 4, 20, 35, 41, 61, 63, 65–66, 152, 164, 194, 206, 210, 232, 256, 266–268, 270, 273, 277, 280, 282–283, 289, 300, 303, 317–318, 321, 336
 Apollo 16, 2, 14, 62–63, 120, 165, 169, 170–171, 172, 178, 180, 184, 187, 193, 200, 228–229

Apollo 17, 54–55, 194, 297, 301, 366–367, 382, 384, 385, 388, 389, 396, 404, 409, 413, 419, 426, 437, 439
Apollo-Soyuz Test Project, 174–175, 175–176, 185, 195–196, 197–198, 207, 335, 346, 354
appointment, 45, 273
art exhibit, 17–18, 153
awards and honors, 96, 187, 200, 206, 210, 265–266, 277, 303, 315, 372–373, 429
Chapel of the Astronauts, 35, 44–45, 55
commemorative stamp, 157
commercialism, 232, 256, 266–268, 270, 273, 277, 280, 282–283, 289, 300, 317–318, 321
Congress, visit to, 184
cooperation in space, 28, 198, 231–232, 374–375
corps status, 205
criticism, 178
damage suit award, 382
death, 406
debriefing, 439
domestic life, 153
equal employment opportunity, 11
flight status, 97–98
food rations, 71–72
former, 45, 74, 185, 192, 205, 406
Friendship 7 mission, 61
goodwill tour, 41
investigation board, 179
long-duration mission, 129, 291, 308
medical aspects, 52, 65–66, 67–68, 97–98, 120, 150, 172, 187, 248, 281, 285–286, 388, 403, 419
Memphis space information program, 72
Mercury (program), 205
performance, 57–58, 426
personal preference kit, 267
photographs, 421
Presidential message to, 162–163, 409, 418, 420
press comment, 160–161, 258, 259, 262, 276, 406, 409
press conference, 172, 228
promotion, 39, 60, 162–163, 172, 429–430
public appearance, 1, 20, 28, 35, 165, 247
public image, 406
quarantine, 120, 385, 396
reassignment, 194, 303, 314
rescue, 127, 174, 195–196, 197–198, 199, 229, 316, 344
  Agreement on the Rescue of Astronauts, the Return of Astronauts, and the Return of Objects Launched into Outer Space, 195
resignation, 194, 205, 259, 406
retirement, 18, 142, 192, 205, 303, 335, 392
safety, 68, 132, 174, 197

scientist-astronaut, 55, 185
Skylab, 23–24, 68, 165–166, 189, 209, 219, 229, 282, 285–286, 383, 426
SMEAT, 323
space shuttle, 175–176
splashdown, 419, 435, 441
Swiss Transport Museum inauguration, 247
time in space, statistics, 61
training, 23–24, 73, 128, 146–148, 197, 201, 204–205, 228–229, 235–236, 239–240, 282, 322
U.N. Hall of Fame (proposed), 358
White House visit, 41
woman, 15
Astronaut Day celebration, Chicago, Ill., 187
"Astronaut, Fallen" (sculpture), 266
Astronaut Office (MSC) (NASA), 194, 303
Astronaut preference kit (APK), 267
Astronaut wings (award), 265–266
Astronautics Engineer Award (NSC), 98
Astronomical unit, 387
Astronomy (see also individual celestial bodies, observatories, planets, probes, sounding rockets, stars, and telescopes, such as Comet, Large Space Telescope, Mars (planet), Nike-Apache, Pioneer, Smithsonian Astrophysical Observatory, Quasars, etc.), 17, 176, 372, 434
anniversary, 383, 429
Asteroid Belt. See Asteroid Belt.
award, 345
cosmic ray, 102, 159, 176, 189, 235, 264, 335, 337,
exhibit, 117
galactic, 102, 200, 222, 332, 357, 365, 374, 386
gamma ray, 56–57, 139, 181, 287, 342
international cooperation, 45, 109, 177, 187, 208, 237, 375
museum, 454
NASA program, 30, 31, 104, 107, 125, 176, 213–214, 227, 377–378, 393, 409–410, 441, 447, 451
  aircraft, 303
  satellite, 46, 69, 81, 83, 90, 107, 132, 208, 254, 257, 264, 273, 275, 276, 287–288, 295–296, 300, 384
  Skylab, 24, 63–64, 70
  sounding rocket, 125, 180, 189, 224, 242, 255, 276, 288, 319, 366, 394, 425, 427
personnel, 55, 233, 357, 387–388
press comment, 151, 233, 399
publication, 246
radar, 403
radio, 115, 213, 365
solar, 81, 83, 108, 224, 228, 239, 242, 255, 273, 276, 280, 284, 290–291, 315, 366, 377–378, 384–385, 421
  brightness, 425, 427
  corona, 254
  energy R&D, 85, 250, 251
  evolution, 252–253

occultation, 132
plasma, 107
radiation, 90, 95–96, 252–253, 330
Stein waves, 131
stellar, 69, 94–95, 108, 208, 213–214, 220, 233, 254, 257, 261, 270–271, 327, 342, 360, 365, 383, 394, 438, 446
   black hole, 233, 307, 360
   mapping, 354
   Milky Way, 104, 271, 332, 357, 365
   radio emissions, 104, 307
   supernova, 181, 342, 383
ultraviolet. See Ultraviolet.
universe, 18, 56–57, 165, 167, 176, 233, 271
U.S.S.R., 86, 113, 119, 123, 242, 255, 267, 329, 337
x-ray. See X-ray.
Astronomy and Astrophysics for the 1970's, 213–214
Astronomy Survey Committee, 213–214, 287–288
ASW-12 (glider), 161
At the 10th Anniversary of the Communications Satellite Act of 1962: Comsat Report to the President and the Congress, 370
AT&T. See American Telephone & Telegraph Co.
ATC. See Adiabatic toroidal compressor and Air traffic control.
Atlanta, Ga., 162
Atlantic City, N.J., 384
Atlantic Ocean, 26–27, 60–61, 104, 194, 250–251, 254, 275, 312–313, 366
Atlas (booster)
   Atlas-Agena, 437
   Atlas-Burner II, 336
   Atlas-Centaur, 19, 26–27, 51–52, 81, 224–225, 257, 295, 364
Atlas-Centaur Project Team, 380–381
ATM. See Apollo Telescope Mount.
Atmosphere
   cloud study, 118, 120, 127, 351–352
   fog stabilization, 327
   geophysical warfare, international control, 250
   international cooperation
      GARP, 84, 174, 306, 397
      U.S. World Weather Watch, 262–263
   planetary. See Individual planets, such as Mars (planet).
   radioactive fallout, 80
   solar. See Solar atmosphere.
   temperature sounding, 96, 228, 351–352, 422–423, 447
   upper (earth), 8, 51, 105, 108, 174, 182–183, 184, 186–187, 195, 203, 238, 243, 252–253, 255, 306, 307, 342, 349, 354, 389, 393, 430–431, 454
      aurora, 15, 19, 44, 46, 53, 54, 70, 182–183, 184, 227, 342, 353, 373, 389
      composition, 37, 51, 454
      density, 54, 354, 394
      ionosphere, 227, 244, 372, 389
      magnetosphere, 139, 354
      SST impact, 274, 370, 385, 427
      U.S.S.R. program, 87, 236–237
      x-ray penetration, 159
   weather. See individual meteorological satellites, such as Nimbus; and Meteorology.
Atmosphere Explorer C (AE–C) satellite, 372
Atmosphere Explorer D (AE–D) satellite, 372
Atmosphere Explorer E (AE–E) satellite, 372
Atoll. See Kwajalein Atoll.
Atomic energy, 53, 55, 186–187, 207, 243, 287
Atomic Energy Commission (AEC) (see also AEC–NASA Space Nuclear Systems Office; and NERVA and SNAP programs), 29, 78, 85, 89, 93, 169–170, 194–195, 248, 262–263, 315, 325, 374, 425
   atomic waste disposal, 53
   contract, 194–195
   cooperation, 24–25, 81, 94, 131, 155, 194–195, 256, 331, 451
   Div. of Applied Technology, 259
   Div. of Controlled Thermonuclear Research, 378–379
   environmental research park, 236
   funds, 31, 51, 231, 245
   General Advisory Committee, 341
   Global Inventory and Distribution of Pu-238 from SNAP-9A, 80
   Meson Physics Facility, 132–133, 222
   NERVA termination, 451
   nuclear accident report, 80
   nuclear research, 88, 308, 404, 429
   personnel, 288
   press comment, 88
   radioisotope thermoelectric generator, 307, 388
   technology utilization, 106
Atomic Energy Institute (U.S.S.R.), 14–15
Ats 1 (Applications Technology Satellite), 19, 126, 131
ATS–F, 27, 45, 110, 171
ATS–G, 66, 171, 241, 293, 294
Attitude control propulsion system (ACPS), 333
Auburn Univ., 294
Auer, Siegfried O., 327
Augmentor-wing, 169, 172, 324
Aurora, 15, 19, 44, 46, 53, 54, 70, 182–183, 184, 227, 342, 353, 373, 389
Aurora 7 mission, 97
Aurora borealis, 284
Australia, 63, 70, 120, 153, 233–234, 330, 352, 438
Austria, 189, 190, 191, 213, 340
Auto Body Assn. of America, 429
Autographs, 384
Autolycus Crater (moon), 35–36
Automated radar terminal systems (ARTS III), 38, 454

Automated visual sensitivity tester (AVST), 41
Automatic data processing, 34–35
Automatic Microbiological Laboratory, 96
Autopilot, electrostatic, 393
Auxiliary propulsion system (APS), 142–144, 144–146
Avco Corp.
  Research Laboratories, 259
  Systems Div., 294–295
Aviation. See Aeronautics; Commercial aviation; General aviation.
Aviation Hall of Fame, 353
Aviation/Space Writers Assn., 192–193
Avionics, 263, 382
AVLOC. See Airborne visible-laser optical-communications flight tests.
AVST. See Automated visual sensitivity tester.
AWACS. See Airborne warning and control system.
Awards
  civic, 202–203, 372–373
  foreign, 2, 26, 313, 339, 442
  government, 131–132, 160, 191, 202–203, 291–292, 368, 372, 444
    NASA. See National Aeronautics and Space Administration.
  institutions, 202, 206, 313, 350, 355, 362, 443
  literary, 119
  military, 258, 263, 268–269, 303, 313, 320
  scholastic, 401
  society
    achievement, 61, 202, 247, 357–358
    aeronautics, 69, 114, 139, 187, 210, 233, 315, 326, 350, 355, 362, 430
    astronautics, 69, 202, 238, 241, 313, 315, 335–336, 381, 403
    science, 20–21, 202, 345, 367–368, 374, 387
AX (close-support aircraft), 67, 397
Azores, 254

B

B–1 (advanced strategic bomber), 31, 67, 311
B–1A, 93
B–29 (bomber), 350
B–52 (Stratofortress), 319
  lifting body test, 271, 327, 371, 379, 380, 390, 398, 411, 427, 437
B–57 (bomber), 43, 163
BAC. See British Aircraft Corp.
Bacon, Robert F., 358
Baker-Nunn satellite tracking camera, 372
Bakrac, Vice President Boris (Yugoslavia), 35
Baldin, Aleksander, 10
Ball Brothers Research Corp., 163, 342
Ballentine, Dr. David, 259

Ballistic missile (see also Antiballistic missile; Intercontinental ballistic missile), 71, 202–203
Balloon
  anniversary, 335
  ATM flight feasibility, 70
  auroral phenomena study, 373
  cosmic ray detection, 176
  exhibit, 74
  experiment, 179, 449
  gamma ray study, 16
  military use, 319
  parachute test, 163, 256, 364–365
  powered, 319
  remote sensing, 101
  stratospheric experiments, 315
  supernova observation, 181
  Viking Mars lander test, 295
  weather, 77
Baltimore, Md., 49
Bangkok, Thailand, 192
Barashev, Pavel, 285
Barberton Mountain Land, South Africa, 158
Bardeen, Dr. John, 357, 441
Barium cloud experiment, 342
Barker, Dr. Ed, 112
Barnard, Defense Minister Lance (Australia), 438
Barnett, Henry C., 115
Barr, Joseph M., 89
Barrow, Alaska, 173, 262
Barstow, Calif., 113
Barth, Dr. Charles A., 43
Bartlett ground station, Alaska, 173
Basalt, 14, 179, 377–378
"Basic Principles of Relations" (U.S.–U.S.S.R. agreements outline), 207
Bastille Day (France), 259
Batavia, Ill., 55, 78, 88, 94, 314, 429
Battelle Corp., 368
Battelle Memorial Institute, 161
Battery, 70, 73, 178–179, 235–236
Baykonur Cosmodrome, U.S.S.R., 111, 329
  launch
    Cosmos, 15, 44, 200, 221, 236, 241, 251, 280, 297, 301, 395
    Intercosmos 6, 132
    Luna 20, 52
    Molniya 1–20, 128
    Molniya 1–22, 404
    Prognoz 1, 139
    Prognoz 2, 242
    Sret 1, 128
    Venus 8, 119
  mission control center, 174–175
Baylor Univ. College of Medicine, 348
BBC. See British Broadcasting Corp.
Beale AFB, Calif., 162, 320
Bean, Capt. Alan L. (USN), 23–24
Beaufort Sea, 128
Behrend, A. F., Jr., 293
Belgium, 60, 108, 157, 262, 266–267
Belgrade, Yugoslavia, 28

Bell Aerospace Co. Div. (Textron, Inc.), 254–255, 446
Bell Helicopter Co. Div., 187
Bell Aerosystems Corp., 362
Bell Aircraft Co., 350
Bell, Oliver, 445
Bell Telephone Laboratories, Inc., 441
Bellcomm, Inc., 83, 86, 141
Belyayev, Y. I., 13–14
Bendix Corp., 35, 282
Benedict, Howard, 99
Benedict, Dr. Manson, 374
Berendzen, Richard, 387–388
Bergland, Rep. Bob S., 35
Bergstrom AFB, Tex., 178, 179
Bering Sea, 341
Berkner, Lloyd V., Award, 372–373
Bermuda, 194, 242
Berne, Switzerland, 314, 389
Berry, Dr. Charles A., 50, 67–68, 279, 298, 321, 384
  Apollo 16 mission, 97, 135–136, 150, 152, 164
Bethany, Okla., 365
Bethel, Alaska, 173
Bethesda, Md., 86, 441
Bethpage, N.Y., 136, 274
Beverly Hills, Calif., 324–325
Biarritz, France, 161
Bicentennial, 45, 131, 454
Big Sur, Calif., 280–281
Biggs, Dr. Geoffrey A., 225–226
BIOCORE (*Apollo 17* medical experiment), 366
Biological warfare, 134
Biomedical and Public Sector Technology Application Team Program, 75
Biophysical acoustics, 287
Biosatellite 2, 170
Biostack (*Apollo 16* medical experiment), 6
Birmingham, Ala., 206
BIS. See British Interplanetary Society.
Bisplinghoff, Dr. Raymond L., 119, 444
Black Arrow (U.K. booster), 126
Black Brant (sounding rocket), 19, 327, 373
Black Brant VB, 360
Black Brant VC, 35, 92, 210, 222
Black hole (space phenomenon), 233, 307, 360
Blagonravov, Dr. Anatoly A., 229, 231–232, 369
Blue Book Project, 300–301
Blum, Dr. Harold F., 385
BMBW. See Germany, West, Ministry for Education and Science.
BOAC. See British Overseas Airways Corp.
Bobko, Maj. Karol J. (USAF), 273, 281, 285–286, 323
Boeing 707 (jet passenger transport), 42, 124, 186, 248, 251
Boeing 720 (jet transport), 186
Boeing 747 (jet passenger transport), 72, 94, 186

Boeing Co., 215
 AMST prototype, 443
 antitrust suit, 122
 award, 99, 429
 Boeing Aerospace Group, 227
 C–8A research aircraft, 169
 Commercial Airplane Group, 231
 Compass Cope, 72
 contract, 190, 216–217, 227, 266, 292, 349, 382–383
 export license grant, 251
 Field Operations and Support Div., 44
 personnel, 331
 Saturn V, 266
 SST, 166, 306
 STOL, 53–54
 U.S.S.R. aircraft marketing, 425
 Vertol Div., 303
 wage dispute, 376
Bogart, L/G Frank A. (USAF, Ret.), 29
Bolling AFB, Washington, D.C., 303
Bomber, strategic, 201–202
Bond, Donald S., 263
Bonestell, Chesley, 185, 211
Bong Bong 11 (Philippine rocket), 96
Bonn, West Germany, 421
Bonner, Dr. William A., 74–75
Booker, Dr. Henry G., 370
Boorstin, Dr. Daniel J., 405
Boosted Arcas I. See Arcas.
Booster. See individual boosters and stages such as Agena, Centaur, Europa, Saturn, Scout, Thor-Delta, Titan, etc.
Borisov, T., 108
Borisov, V., 240–241
Borman, Col. Frank (USAF, Ret.), 117–118, 205
Boston Museum of Science, 367
Boston Univ., Dept. of Astronomy, 387–388
Boulder, Colo., 36, 243, 357
Boyden Observatory, 181
Boynton, Melbourne W., Award, 372–373
Brady, Joseph L., 165
Brahe, Tycho, 383
Brand, Vance D., 23
Branley, Dr. Frank M., 356
Brasilia, Brazil, 42
Brattain, Dr. Walter H., 441
Brazil, 42, 92, 173, 214, 270, 383, 394
Breccia, grey (lunar surface sample), 238
Bredt, Dr. James H., 344
Breeding, Robert E., 372–373
Breguet Aviation, 177
Bremen, West Germany, 251
Brett, Dr. Robin P., 417
Brevard County, Fla., 308, 365, 454
Brevard Printing Co., 321
Brezhnev, Leonid I., 185, 191, 201–202, 207, 240, 243, 377
Brimelow, Sq. Ldr. Brian (RAF), 20–21
Brinegar, Claude S., 420
British Aircraft Corp. (BAC), 301, 307, 309
British Airways Board, 288, 341

British Broadcasting Corp. (BBC), 444
British Interplanetary Society (BIS), 339, 372–373
  Gold Medal, 202–203
British Overseas Airways Corp. (BOAC), 200
Brittany, France, 255
Broadcasting Satellite System, 155
Bromley, Dr. David A., 287
Brooks, David R., 314
Brooks, Rep. Jack, 361
Broussard, Peter H., Jr., 393
Brown, Dr. Harold, 310, 434
Brown, Dr. Harrison, 340
Brown & Root-Northrop, 73
Brown Engineering Co., 181
Brown, Gen. George S. (USAF), 294, 303, 308, 313
Brown, Walter E., 433
Bruckner, D. J. R., 157
Brussels, Belgium, 108, 226, 262, 437
Bucharest, Romania, 440
Buckley, Sen. James L., 66
Buffalo (augmentor-wing research aircraft), 169
Bulgaria, 403
Bull, L/Cdr John S. (USN, Ret.), 205
Bumerang Project (West Germany), 303–304
Buoy, 97
Burchard, Hank, 88
Bureau of Mines. See U.S. Bureau of Mines.
Burger, Chief Justice Warren, 388
Burke, Walter F., 309
Burroughs Corp., 445
Bush, Ambassador George, 348
Bushuyev, Konstantin D., 197, 252, 260–261, 346
Buster Crater (moon), 146–148
Butenandt, Dr. Adolf F. J., 251
Butz, Helmut F., 277

C

C-5 (Galaxy) (military cargo aircraft), 110, 120, 131, 319
C-5A, 93
C-8A (Buffalo) (augmentor-wing research aircraft), 169
C-130 (military transport), 31, 319
C-130E, 154
C-141 (troop carrier aircraft), 45, 439
CAB. See Civil Aeronautics Board.
Cabell, Rep. Earle, 387
Cajun. See Nike-Cajun.
Calcium 171, 183–184
Calhoun Technical College, 129–130
California, 51, 96, 113, 126, 208–209, 343
  accident, 266
  aerospace employment, 59, 61, 208, 281
  Aerospace Research Pilots School, 18
  aircraft unveiling, 240
  anniversary, 134
  Apollo animal experiments, 404
  dedication ceremony, 215
  earth station, 235, 270, 353, 410
  earthquake faults, 313
  environment, 35, 101
  explosion, 266
  Federal building naming, 430
  flight corridor, 264
  magnetic levitation system demonstration, 392
  medical center, 65–66, 74–75
  meeting, 12, 20–21, 173
  Navy Astronautics Group, 134
  Nixon, Pres. Richard M., announcement, 2–4
  political election, 200–201, 220, 286
  sonic boom, 80
  space shuttle, 139–140, 181–182, 214–215, 276, 383–384
  Viking parachute test, 163
California Crime Technological Research Foundation, 126
California Div. of Highways, 35
California Institute of Technology (Cal Tech)
  astronomy, 131, 222, 393
  contract, 264
  Division of Geology and Planetary Sciences, 300
  JPL name change, 322
  meeting, 360
  personnel, 14
California Museum of Science and Industry, 291
California, Univ. of, 165, 222, 394
  Berkeley, 101, 118, 170, 203, 280, 385
  Los Angeles, 152, 158, 247, 292
  Riverside, 315
  San Diego, 2, 14, 61, 141, 264, 287, 368
Calvin, Dr. Melvin, 170
Cambridge, Mass., 385
Cambridge, U.K., 354
Cambridge Univ., 354
Camera (see also Photography and Television)
  aircraft, 190–191, 216, 235
  Apollo 16 mission, 146–148, 148–149
  cold emulsion, 220
  ERTS, 210, 284
  far-ultraviolet, 149–150, 179–180, 270–271
  Grand Tour spacecraft, 65
  infrared, 47–48
  metric, 237
  multispectral scanner, 269, 283
  panoramic, 428, 433
  patent, 220
  reconnaissance, 190–191, 280–281
  return-beam-vidicon, 235, 269
  tracking, 372
  TV, 409, 416
  U.S.S.R. Mars probe, 90
Cameron, Dr. Robert C., 434
Cameron, Dr. Roy E., 171
Campbell, Dr. Wesley G., 281

Canada
  aircraft, 169, 354
  Alouette 1, 328, 453
  astronomy, 254, 307
  communications satellite, 379, 395
  Department of Communication Research Centre, 328
  earth resources, 444
  ground station, 270
  international cooperation, 128
  launch, 19, 92, 109, 315, 327
  meeting, 49, 204
  treaty, 174, 227
Canadian Centre for Remote Sensing, 184
Canadian Domestic Communication Satellite, 379
Cancer, 1, 385
Canfield, Mrs. Martha Chaffee, 382
Cannikin (nuclear explosion), 308
Cannon, Sen. Howard W., 178
Cape Canaveral, Fla. (see also Cape Kennedy, Fla.), 9, 41–42, 329
Cape Chat, Quebec, 254
Cape Girardeau, Mo., 335
Cape Girardeau Municipal Airport, Mo., 315
Cape Kennedy, Fla., 151
  anniversary, 255
  Apollo 17 cruise, 337, 409
  Apollo 17 launch site, 389–390
  ASTP launch plans, 335
  labor dispute, 375
  name change, 41, 91, 215, 216, 241, 266, 318, 329, 340, 384
  press conference, 52
  Skylab delivery, 262, 309
  Skylab launch site, 383
  solid-fuel thrust-augmentation rocket, 379
  space shuttle, 181–182
Cape Kennedy Air Force Station, 44
Cape Kennedy or Cape Canaveral? A Brief History, a Background, and an Analysis of S.J. Res. 193, 318
CARD. See Civil Aviation Research and Development.
Cardiac pacemaker, 248
Carnegie Institute Museum of Art, 371
Carolina Power and Light Co., 390
Carpenter, L/C Donald G. (USAF), 320
Carr, L/C Gerald P. (USMC), 23, 24
Carruthers, Dr. George R., 254, 270–271, 306
Cartography. See Mapping.
Casper (Apollo 16 CSM). See Command and service module.
Caspian Sea, 2
Castro, Premier Fidel (Cuba), 243
CAT. See Clear-air turbulence.
Cayley Plains (moon), 146–148, 183–184
CBS. See Columbia Broadcasting System.
CCI Aerospace Corp., Marquardt Co., 261
CDDT. See Countdown demonstration test.

Ceausescu, President Nicolay (Romania), 439–440
Celestial mechanics, 81
Centaur (booster upper stage) (see also Atlas and Titan), 367
Centaur program, 403
Central Intelligence Agency (CIA), 357
Central Switzerland Cultural Foundation, 313
Central Television Service (U.S.S.R.), 188
Centre National d'Études Spatiales (CNES) (France), 241–242
Cernan, Capt. Eugene A. (USN)
  Apollo 17 mission, 414, 436
    accident, 416
    command module naming, 382
    emblem, 347
    extravehicular activity, 416–419, 426, 444
    launch, 413
    lunar landing, 367
    lunar landing training vehicle, 384
    medical aspects, 403, 414, 420
    prelaunch test, 389
    preliminary timeline, 297
    press comment, 426, 438
  awards and honors, 373, 429
  promotion, 430
Cervantes, Mayor Alphonso (St. Louis, Mo.), 400
Cesium cloud experiment, 118, 120, 122, 127
CFD. See Cold fog dissipation system.
Chafee, Secretary of the Navy John H., 130
Chaffee, L/Cdr Roger B. (USN), 382
Challenger (Apollo 17 LM). See Lunar module.
Chamber of Commerce, St. Louis, Mo., 118
Changes in Graduate Programs in Science and Engineering, 1970–1972 and 1972–1974 (NSF 72-311), 267
Chantilly, Va., 136, 206
Chapel of the Astronauts (proposed) (KSC), 35, 44–45, 55
Chapelle, Emett W., 258
Chapman, Dr. Philip K., 205, 259
Charlesworth, Clifford E., 173
Charlie Brown (Apollo 10 CM). See Command module.
Charyk, Dr. Joseph V., 370
Chemical injection, 275
Chemical Propulsion Div. (LeRC), 244
Chen Chu, 357
Chennault, L/C Claire L. (USAF), 353
Cherry, Dr. George W., 22, 67, 110, 189
Chesapeake Bay, Md., 136, 216, 244, 255–256
Chiang Kai-shek, 357
Chicago, Ill., 135–136, 187, 188
Chicago, Univ. of, 302, 315, 366
Chico, Calif., 364
Chicom 1 (P.R.C. satellite), 54
Chicom 2, 54
Chien Wei-chang, 392

Childs, J. Howard, 115
Childs, Sen. Lawton M., Jr., 41–42
China, People's Republic of (P.R.C.)
 British Air Show attendance, 307
 denunciation of Formosa, 357
 earth station, 28–29, 86, 98, 294, 386
 foreign aircraft purchase, 251
 ICBM launch detection, 69
 launch
  Chicom 1, 54
  Chicom 2, 54
 missile program, 39, 54, 96, 310, 377
 Nixon visit, 2, 26–27, 28–29, 59, 72, 86, 98, 264
 nuclear test resolution, 219
 reconnaissance flight ban, 72
 Skylab fly-over, 63
 space program, 125
 U.N. Conference on the Human Environment, 225, 392
 U.S. visits to, 181, 350, 357, 393
 U.S.S.R. border, 177
 U.S.S.R. missile target, 87
 U.S.S.R. underground testing, 421
 visit to U.S., 392
 weaponry, 49
China, Republic of (Nationalist), 61, 357
Chincoteague Bay, Va., 136
Ching Chuan Kang Air Base, Taiwan, 61
Chinghua Univ., 392
Choctawhatchee Bay, Fla., 233
Christman, Albert B., 99
Chupankhin, M. S., 13–14
Chupp, Dr. Edward L., 330
Churchill, Manitoba, 254
Churchill Research Range, Canada, 19, 54, 109, 125, 283, 284, 315, 327, 373
CIA. See Central Intelligence Agency.
CIA—The Myth & the Madness, 357
CIAP. See Climatic Impact Assessment Program.
CIC. See International Coordinating Committee.
CICAR. See Cooperative Investigation of the Caribbean and Adjacent Region.
Ciepluch, Carl C., 186
Cincinnati, Ohio, 77, 399
Cincinnati, Univ. of, 96, 205
Civil Aeronautics Board (CAB), 22
Civil Air Patrol, 277
Civil aviation. See Aeronautics.
Civil Aviation Research and Development (CARD), 6, 22, 116, 160
Civil Service Commission, 12, 131–132, 255, 269
Clark Univ., 400–401
Clarke, Arthur C., 337, 409
Clarke, James, 9
Clarksburg, Md., 104
Classification Review Committee, 187
Clean Air Act of 1970, 285
Clear-air turbulence (CAT), 43–44, 302
Cleaver, A. V., 125
Cleveland, Ohio, 50, 154–155, 326, 391
Cleveland (Ohio) Air Pollution Control Div., 154

Clifton, K. Stuart, 254
Climatic Impact Assessment Program (CIAP), 155
Clock rate, 194
"Clock, universal," 81
Clouds, 70–71, 228, 293
CM. See Command module.
CNES. See Centre National d'Études Spatiales (France).
Cobb, Geraldine, 350
Cocoa Beach, Fla. 154
COCOM. See Coordinating Committee on Trade in Strategic Materials.
Cohn, Victor, 9
Cold fog dissipation (CFD) system, 122–123
Cold war, 10, 201
Coleman, Dr. Paul J., 152
Collective-effect accelerator, 175
College Park, Md., 131
Collier, Robert J., trophy, 114, 210, 233
Colliers (magazine), 211
Collimated proportional counter, 295–296
Collins, Harold G., 321
Collins, Col. Michael (USAF, Ret.), 45, 74, 454
Collins Radio Co., 382
Collision avoidance, aircraft, 116, 178
Colorado, 36, 92, 243, 259
Colorado Air National Guard, 118
Colorado, Univ. of, 43, 210, 224, 242, 259, 374, 375, 427
Columbia Broadcasting System (CBS), 134–135, 381, 399, 409
Columbia Univ., 14, 159, 219, 264
Colwell, Dr. Robert N., 118
Comet (see also individual comets such as Forbes and Halley's), 372, 377–378
Comision Nacional de Investigación del Espacio (CONIE) (Spanish Space Commission), 248, 251
Command and service module (CSM)
 Apollo 7, 169–170
 Apollo 16 (Casper) (CSM-113), 142–144, 144–146, 148–149, 150
 Apollo 17 (America) (CSM-114), 330, 413–420, 435, 441
 Apollo-Soyuz Test Project, 61, 197–198, 378
 reusable, 175–176
 Skylab, 262, 298, 344, 378
Command module (CM), 366, 382, 389
 Apollo 10 (Charlie Brown), 115
 Apollo 12 (Yankee Clipper), 206
 Apollo 15 (Endeavor), 61
 Apollo 16 (Casper)
  splashdown, 41, 62, 109, 142–144, 148–149, 152
  technical problems, 8, 46, 156–157, 174, 208
 Apollo 17 (America), 366–367, 382, 389, 413–420
 Apollo-Soyuz Test Project, 257
 Skylab, 298
Command Pilot Astronaut Badge (USAF), 258

Commerce, Dept. of (DOC) (see also Maritime Administration, National Oceanic and Atmospheric Administration, etc.)
  bulletin, 356
  conference, 314
  cooperation, 1, 22, 301, 302, 352
  ERTS, 270, 280, 302
  executive reorganization, 122
  export license issue, 98, 251
Commercial aviation, 50, 239
Commission of the European Community, 218
Commission on Government Procurement, 89
Committee for the Future, 324
Committee on Inventions and Discoveries (U.S.S.R.), 18
Committee on Space Research (COSPAR), 178
Committee on the Peaceful Uses of Outer Space. See United Nations.
Commonwealth Club of San Francisco, 51
Communications, 135–136, 258, 377–378, 390–391, 403
Communications satellite (see also individual satellites: *Molniya 1-20, Telstar 1, Telstar 2*), 27, 383
  agreement, 286, 430
  anniversary, 255
  benefits, 164–165
  Broadcasting Satellite System, 155
  cooperation, international, 19, 306, 352, 395, 410, 452
  cooperation, jointly owned system, 230, 309, 338
  cost, 102
  development, 309
  Development Test Satellite, 208
  domestic, 190–191, 283, 355, 379
  earth station, 173, 353, 386, 389, 410, 422
  FCC policy, 230, 261–262
  Hot Line, 20, 377
  implications, 24–25, 185, 430
  inauguration of, 397
  Israel, 91, 273
  launch
    Intelsat-IV F-4, 26
    Intelsat-IV F-5, 224
    Molniya 1-21, 350
    Molniya 1-22, 404
    Molniya II-3, 331
    Molniya II-4, 426
  launch assistance, 343
  military, 30–31, 116, 233, 371, 425
  NASA program, 425
  NATO, 370
  P.R.C., 392
  public right to access, 190–191
  space shuttle, 9, 136–137
  technology, 141–142
  technology utilization, 114
  television, 173, 188, 197–198, 273
  use
    anniversary message, 370
    Apollo 17 mission, 444
    capacity, 309
    India, 425
    Indian Ocean, 276
    limitation, 322, 348, 356, 365
    mail service, 32
    policy recommendations, 104
    press comment, 355
    White House review policy, 72
Communications Satellite Act of 1962, 370
Communications Satellite Corp. (ComSatCorp), 318
  aerosat program, 12–13
  anniversary, 370
  communications satellite demand, 309
  cooperation, 242, 309, 338
  earth station, 173
  FCC regulation, 230–231, 261–262
  Intelsat-IV F-2, 104
  Intelsat-IV F-3, 26-27 60–61
  Intelsat-IV F-4, 26, 47, 53, 61, 173, 224
  Intelsat-IV F-5, 224, 242, 276
  services, 104, 276
  U.S. election TV coverage, 375
Communications Satellite Corp. Laboratories, 104
Communications Technology Satellite (Canada), 188
Communist China. See China, People's Republic of.
Community Action Agency, 365
Community Development, Dept. of (proposed), 122
Company Funds Push Total Industrial R&D Spending to $18 Billion in 1971. (NSF 72-318), 427
Compass Cope (reconnaissance drone aircraft), 72
Computer
  advances, 63
  aeronautics, 200, 449
  air traffic control, 38
  aircraft astronomy observatory, 45
  automated ship-satellite communications and navigation system, 281
  award, 160
  booster inertial guidance, 182
  cooperation, 347, 452
  demographic study, 5–6
  docking dynamics use, 310
  DOD research, 190–191
  facility, 455
  Halley's Comet orbit computation, 165
  ILLIAC IV, 445
  industry, 151
  launch and tracking use, 411
  malfunction, 109, 117
  mapping, 354
  MSC inventory system, 21
  NASTRAN program, 71, 452
  national security classification system, 187
  National Symposium on Technology Transfer, 225
  nuclear fusion use, 358
  Oao 3, 295–296

pollution research use, 155
P.R.C., 357
production management use, 201
public impact, 24–25, 180
remote sensing, 34–35
Skylab, 261
space shuttle, 34, 79
telemetry data analysis, 50
TELOPS, 73–74
traffic use, 393
transistor, 441
transportation use, 411
UNIVAC 1108, 131–132
university programs, 267
weather use, 262–263
World Weather Program, 172–173
Computer sciences, 236–237
Computer Sciences Corp., 293
Comsat. See Communications satellite.
ComSatCorp. See Communications Satellite Corp.
Concorde (Anglo-French supersonic transport), 50, 94
  award, 350
  Congressional report, 427
  engine, 248, 307
  environmental problems, 307, 385
  exhibit, 157, 307, 329
  flight test, 390
  production, 317, 361
  sales, 200, 271, 301
  test, 297, 390
  U.S. SST comparison, 153
Condon, Dr. Edward U., 300–301
Condon, Dr. John E., 353
Conference on Aircraft Engine Noise Reduction, 185–186
Conference on High-Energy Astrophysics, 360
Conference on Survival and Growth of Small R&D Firms, 223
Congress, 99, 253, 395, 438
  Apollo 15 commercialism, 289
  Apollo 16 mission, 228
  Apollo-Soyuz Test Project, 198–199
  astronaut reception, 184
  astronomical research center, 183
  Bicentennial, 45
  briefing, 229–230
  communications satellite, 370
  hearings
    aeronautical research and development, 21–22
    domestic comsat system, 190–191
    space shuttle, 36–37
    U.S.–U.S.S.R. summit agreements, 203
  helicopter contract cancellation, 288
  International Exposition on the Environment, 292
  Joint Committee on Atomic Energy, 132–133
  Joint Committee on Defense Production, 62
  Joint Committee on Economics, 123, 444–445
  joint conference, 273–274
  JPL name change, 379, 434
  KSC facilities expansion, 367
  military aircraft sale restraint, 133–134
  Moscow agreements, 218
  NAS resolution, 159–160
  NASA budget, 231–232
  National Aerospace Museum of the West, 291
  nuclear breeder reactor, 162
  Office of Technology Assessment, 341, 349, 361
  Presidential message
    arms limitation, 224–225
    Bicentennial, 45
    environment, 47, 285
    European visit, 213
    executive reorganization, 122
    National Science Foundation, 139
    science and technology, 105–106
    State of the Union, 24–25
    United Nations, 311
    urban transportation, 52
    U.S. foreign policy, 48–49
    World Weather Program, 172–173
  R&D funds, 125, 399
  reports submitted to, 62, 64
  space program, 168, 219, 374
  space shuttle, 7, 8–9, 35, 156–157, 181–182, 185
  supersonic transport, 36–37, 190, 398–399
  TriStar sales, 422
  unconventional power sources, 446
  von Braun, Dr. Wernher, 220
Congress, House of Representatives
  bills defeated, 231, 398–399
  bills introduced, 41
  bills passed, 44–45, 48, 74, 115–116, 155, 161, 164, 193, 209, 338
  Committee on Appropriations, 187–188
    Subcommittee on DOD Appropriations, 64
    Subcommittee on Department of Housing and Urban Development–Space–Science–Veterans, 89
  Committee on Government Operations, 104
    Subcommittee on Foreign Operations and Government, 234
  Committee on Public Works, 322, 430
  Committee on Science and Astronautics, 53, 231, 265, 326, 387, 395
    Apollo 16 briefing, 184
    Cape Kennedy name change, 329
    committee members, 35, 41, 286–287, 430
    For the Benefit of All Mankind: The Practical Returns from Space Investment, 368
    meeting, 34–35
    NASA budget, 46–47, 109–110, 115–116, 135
    Solar Energy Research: A Multidisciplinary Approach, 446

Subcommittee on Advanced Research and Technology, 21-22
Subcommittee on Aeronautics and Space Technology, 21-22, 59, 66-67, 73-74, 79-80, 84-85
Subcommittee on International Cooperation in Science and Space, 203, 229, 303
Subcommittee on Manned Space Flight, 53, 57-58, 63, 67-68, 78, 166-167, 209
Subcommittee on NASA Oversight, 33, 287
Subcommittee on Science, Research, and Development, 361, 446
Subcommittee on Space Science and Applications, 56, 65, 78, 84, 91, 171
Committee on Veterans' Affairs, 430
joint conference, 273-274
nominations approved and confirmed, 89
resolutions introduced, 220
resolutions passed, 274
Un-American Activities Committee, 357
Congress, Senate, 338
advice and consent, 318
bills defeated, 398-399
bills introduced, 318
bills passed, 44-45, 66, 178, 295, 316, 329, 349, 358
NASA authorization, 189, 226
Committee on Aeronautical and Space Sciences
Apollo 15 commercialism, 277, 282-283, 300, 317
bills reported on, 35
committee members, 66, 387
NASA budget, 172
approval, 160
space shuttle, 102, 137, 140
testimony, 99-100, 106-107, 114, 116, 130-131
report, 7, 166
U.S.-U.S.S.R. space agreements, 239-240
Committee on Appropriations, 69, 209, 226, 247
Subcommittee on Housing and Urban Development, Space-Science-Veterans, 136-137, 289
Committee on Armed Services, 55-56, 59, 115, 151-152, 221, 235
Committee on Foreign Relations, 230, 265
Committee on Interior and Insular Affairs, 91
Committee on Judiciary, 216
Subcommittee on Internal Security, 317
conventions ratified, 201-202, 283, 316, 341
conventions submitted to, 230, 288-289
joint conference, 273-274

nominations approved and confirmed, 34, 45, 60, 130, 172
nominations submitted to, 39, 281-282, 288, 421
reports submitted to, 131, 273-274, 370
resolutions introduced, 41-42, 216, 220
resolutions passed, 266
space shuttle, 6-7, 37, 102, 137, 140, 214-215, 328
SST, 361
Congressional Research Service. See Library of Congress.
CONIE. See Comisión Nacional de Investigación del Espacio.
Connecticut, 362
Conquest of the Moon, 211
Conquest of Space, 211
Conrad, Capt. Charles, Jr. (USN), 23, 150, 178-179, 219, 282
Consumer Power Co., 221
Continental Drift, 250-251
Contract
cost-plus-award-fee, 17, 243, 292, 295
cost-plus-fixed-and-award-fee, 272, 378
cost-plus-fixed-fee, 45, 172, 191, 203, 231, 234, 254, 264, 277, 283, 349, 434
cost-plus-incentive-fee, 221-222, 303, 382
cost-reimbursable, 387
cost sharing, 266, 281
firm-fixed-price, 42, 47, 55, 113, 132, 236, 247, 254, 261, 266
fixed-fee, 240
fixed-price, 138, 256, 266
fixed-price-incentive-fee, 337
letter, 129, 292, 340
Contract Office (MSFC) (NASA), 243-244
Contractor Equality Opportunity Programs Office (MSC), 60
Controlled fusion, 287, 289
Convair 990 (*Galileo*) (jet research aircraft), 21, 24, 68, 128, 302, 341, 349, 392
Convair Aerospace Div. See General Dynamics Corp.
Convention for the Suppression of Unlawful Acts Against the Safety of Civil Aviation, 311, 318, 372
Convention on International Liability for Damage Caused by Space Objects, 121-122, 230, 311, 341
Convention on the Prohibition of the Development, Production and Stockpiling of Bacteriological (Biological) and Toxin Weapons and on Their Destruction, 134, 288-289, 311
Conventional takeoff and landing aircraft. See CTOL.
Cooper, Dr. Leon N., 357
Cooper, Dr. Theodore, 86
Cooperative Investigation of the Caribbean and Adjacent Regions (Project CICAR), 236
Coordinating Committee on Trade in Strategic Materials (COCOM), 86
Copernicus. See *Oao 3.*

Copernicus Crater (moon), 13–14, 69, 433
Copernicus, Nicolaus, 167, 183, 257, 274, 300, 305
Copernicus, Nicolaus, Day, 274
Cornell Aeronautical Laboratory, Inc., 45, 327
Cornell Univ., 6, 14, 252–253, 261, 387–388, 443
Cornwall, U.K., 255
"Corporate Continuity in Space: The Case for NASA's Future," 168
Corp. for Public Broadcasting (CPB), 45, 381, 391–392
Corps of Engineers. See Army Corps of Engineers.
Cortright, Edgar M., 40, 50, 153
Cosmic plasma, 141
Cosmic ray, 381
   Apollo 17 experiment, 366
   balloon experiment, 335
   biomedical effects, 129
   detection, 102, 148, 149, 176, 235
   earth surface effects, 403
   energy measurement, 163
   observation, 159, 337
   study, 189, 264
   Van Allen Belt, 128
   very-high-energy, 315
Cosmic reflector, 375
Cosmology, 283
Cosmonaut, 243, 374–375
   asteroid name, 440
   ASTP mission, 174–175, 175–176, 197–198, 257, 260, 335, 346
   award, 268
   death, 7, 166, 279
   lunar landing, 426
   lunar site nomenclature, 168
   meeting, 343
   professions, 111
   rendezvous and docking mission, 174–175, 197
   Soyuz living conditions, 353
   space biology, 360
   space cooperation, 231
   speech, 338
   stellar astronomy study, 255
   training, 110–111, 201, 204, 205, 228, 229, 239–240, 354
Cosmonauts Day, 138
Cosmos (U.S.S.R. satellite), 7, 108, 167–168, 449
Cosmos 1, 108
Cosmos 92, 108
Cosmos 149, 108
Cosmos 166, 108
Cosmos 215, 108
Cosmos 261, 108
Cosmos 348, 108
Cosmos 463, 9–10
Cosmos 464, 9–10
Cosmos 471, 15
Cosmos 472, 33
Cosmos 473, 44
Cosmos 474, 55
Cosmos 475, 69
Cosmos 476, 77, 118
Cosmos 477, 87
Cosmos 478, 102
Cosmos 479, 113
Cosmos 480, 119
Cosmos 481, 119
Cosmos 482, 123
Cosmos 483, 127
Cosmos 484, 131
Cosmos 485, 135
Cosmos 486, 139
Cosmos 487, 155
Cosmos 488, 174
Cosmos 489, 174
Cosmos 490, 186
Cosmos 491, 200
Cosmos 492, 221
Cosmos 493, 236
Cosmos 494, 239
Cosmos 495, 239
Cosmos 496, 241
Cosmos 497, 243
Cosmos 498, 250
Cosmos 499, 251
Cosmos 500, 253
Cosmos 501, 256
Cosmos 502, 257
Cosmos 503, 264
Cosmos 504, 264
Cosmos 505, 264
Cosmos 506, 264
Cosmos 507, 265
Cosmos 508, 265
Cosmos 509, 265
Cosmos 510, 265
Cosmos 511, 265
Cosmos 512, 274
Cosmos 513, 280
Cosmos 514, 292
Cosmos 515, 294
Cosmos 516, 297
Cosmos 517, 301
Cosmos 518, 317
Cosmos 519, 319
Cosmos 520, 322, 328–329
Cosmos 521, 330
Cosmos 522, 338
Cosmos 523, 340
Cosmos 524, 346
Cosmos 525, 355
Cosmos 526, 361
Cosmos 527, 366
Cosmos 528, 371
Cosmos 529, 371
Cosmos 530, 371
Cosmos 531, 371
Cosmos 532, 371
Cosmos 533, 371
Cosmos 534, 371
Cosmos 535, 371
Cosmos 536, 373
Cosmos 537, 395
Cosmos 538, 428
Cosmos 539, 439
Cosmos 540, 443

Cosmos 541, 444
Cosmos 542, 445
COSPAR. See Committee on Space Research.
Cost–A Principal System Design Parameter (symposium), 293
Cost Benefit Analysis Used in Support of the Space Shuttle Program (B-173677), 215
Costa Rica, 142
Costes, Dr. Nicholas C., 13-14, 355
Council for Airport Opportunity, 323
Council of Europe, 29
Council of Ministers (U.S.S.R.), 18
Council on Environmental Quality, 285, 289
Countdown demonstration test (CDDT), 113
Cox, Tricia Nixon, 265
CPB. See Corp. for Public Broadcasting.
Crab (U.S.S.R. mobile submarine apparatus), 228
Crab Nebula, 159, 394
Crater. See Moon, crater.
Crimean Astrophysical Observatory, 446
Crippen, L/Cdr Robert L. (USN), 281
Croatia, Yugoslavia, 35
Crutzen, Dr. Paul J., 385
CSM. See Command and service module.
CTOL (conventional takeoff and landing) aircraft, 66
Cuba, 236, 321, 413
Culbertson, Philip E., 68, 398
Cultural exchange, 195
Cunard Lines, Ltd., 104
A Current Index of Technical Briefs (C1-1), 311
Curtin, M/G Robert H. (USAF, Ret.), 100, 109
Cutts, Dr. James A., 178
Cuyohoga River, 391
CVAN-70 (aircraft carrier), 31
Cyclotron, 295
Cygnus (constellation), 307, 360
Cygnus X-1 (pulsar), 360
Cygnus X-3, 307
Czechoslovakia, 132, 243, 361, 403

D

D-558 II (Skyrocket) (rocket research aircraft), 311
DAD. See Dual Air Density Explorer satellite.
Daddario, Rep. Emilio Q., 361
Dailey, Dr. John T., 355
Daley, Mayor Richard J. (Chicago, Ill.), 187
Dallas, Tex., 39-40, 209, 313
Dana, William H., 271, 289, 299, 315, 327, 341, 427
Daniloff, Nicholas, 133
Darmstadt, West Germany, 384
Darwin, Charles, 167

Dassault-Breguet Mirage 3 (French aircraft), 177
Dassault Milans (French fighter-bomber aircraft), 314
Data Collection System (DCS), 269-270
Data processing (see also Computer), 34-35, 220, 372
Data-relay satellite, 233-234
Data Systems Test, 84
David, Dr. Edward E., Jr., 34-35, 60, 185, 197-198, 236-237, 257, 275
Davis, L/G Benjamin O., Jr. (USA, Ret.), 221
Davis, Mrs. Patricia White, 382
Day, LeRoy E., 36
Dayton, Ohio, 353
Daytona Beach, Fla., 37
DC-3 (jet transport), 192-193
DC-8, 124, 186, 248
DC-9, 4-5
DC-10, 94, 186, 298, 342
DCR. See Design certification review.
DCS. See Data Collection System.
Death Valley National Monument, 80
Debus, Dr. Kurt H., 86, 139-140, 205, 222-223, 331
Decatur, Ala., 129-130
Deep Space Maser Development Group, 380-381
Deep Space Network (DSN), 114, 305
Defense, Dept. of (DOD) 224, 347, 399
  aircraft. See Aircraft.
  award, 202-203
  budget, 29, 39, 53-54, 260
    Congressional consideration, 64, 69, 247, 349, 362
    Presidential action, 24-25, 30-31, 105
  contract, 382
  cooperation, 16, 22, 40, 67, 116, 139-140, 194-195, 301-302
  documents security classification, 93
  Joint Parachute Test Facility, 163
  launch, 108, 139-140, 447
  microwave landing systems, 67
  missile program, 300, 361, 453
  national security, 64, 218
  nuclear tests, underground, 327
  personnel, 45, 64-65, 69, 189
  R&D, 125, 134
  report to Congress, 64
  satellite, 116, 190-191, 192, 307, 362, 369, 400, 425
  space program, 102, 245, 336
  space shuttle, 55, 139-140
  technology transfer, GAO report, 267
  Telecommunications Office, 18
  weapon system, 68, 87, 221, 236, 246, 263, 327, 397
Defense Satellite Communications System (DSCS), 116, 425
Defense Transportation Award, 362
de Florez, Admiral Luis, Flight Safety Award, 355
de Havilland Aircraft Co., 275
de Havilland, Sir Geoffrey, 275

Deimos (Martian moon), 364
Del Rio, Tex., 308
Delaware Valley Council, 139
Delpit, Dale, 136
Delta (booster upper stage). See Thor-Delta.
Democratic Party, 358
Demography, 5–6
Denmark, 60
Dennison, Dr. Edwin W., 393
DeNoyer, Dr. John M., 21
Denpa (Japanese Radio Explorer Satellite), 295, 449
Denver, Colo., 92, 259
Deprit, Dr. André, 160
Descartes (lunar landing site), 14, 62–63, 71, 142, 144–146, 170–171
  surface sample, 62–63, 149–150, 165, 170–171, 183–184
"Description, Dissection and Subsampling of *Apollo 14* Core Sample 14230" (NASA Technical Memorandum), 54
Design certification review (DCR), 280
Detlein, Dr. Lawrence F., 132
Detroit, Mich., 16, 411
Deutsch, Dr. Stanley, 316
Development Test Satellite (DTS), 208
DFBW. See Digital-fly-by-wire program.
Diamant B (French booster), 425
Diepen, R. J. L., 204
Diesel-Allison Div. See General Motors Corp.
Digital Equipment Corp., 201
Digital fly-by-wire program (DFBW), 200, 312, 449
Digital Fly-By-Wire Project Team, 380–381
Dirigible, 400
Disarmament
  military satellites, 362
  national security, 236
  SALT, 90, 121, 190, 201–203, 214, 218, 224–225, 229–230, 232, 252, 265, 288, 289, 299, 310, 316, 331, 332, 338, 356, 390
  status, 24–25
  treaty, 188, 299, 327
  U.S.-U.S.S.R. cooperation, 189, 190, 203, 206, 207, 335
Disaster warning satellite, 154
DISCOS. See Disturbance compensation system.
Discrete Addressable Beacon and Microwave Landing Systems programs, 84
Disney World, Fla., 308
Distinguished Civilian Service Award (DOD), 202–203
Distinguished Service Medal (NASA), 29, 202–203, 310, 381
Distinguished Service Medal (USAF), 258, 303
Distinguished Service Medal (USN), 265–266
District of Columbia (see also Washington, D.C.), 45, 255, 270, 322, 338

Disturbance compensation system (DISCOS), 306–307
Djezkazgan, Kazakhstan, U.S.S.R., 53
Dobbins AFB, Ga., 178
DOC. See Commerce, Dept. of.
Docking
  Apollo 16 mission, 8, 142–144, 148, 228–229
  Apollo 17 mission, 413–414
  Apollo-Soyuz Test Project, 61, 174–175, 192–193, 196, 244, 335, 449
  budget, 198–199
  congressional briefing, 209, 229, 239–240
  contract, 378
  enzyme production experiment, 247
  joint communique, 207, 252
  launch hardware, 28
  Presidential message, 106
  press comment, 407
  press conference, 197–198, 228, 250, 257–258, 260–261, 344
  "rendock," 182, 205
  spacecraft compatibility, 120, 178, 196, 197–198, 239–240, 240–241, 344, 356, 378
  technical requirements, 136–137, 158
  training, 354
  U.N. resolution, 356
  working groups, U.S.-U.S.S.R., 120, 127, 178, 252, 282, 322, 409
  design, 180, 280, 282, 310
  dynamics, 310
  multiple docking adapter, 280, 282
  Skylab, 27, 344
  space shuttle, 15, 229
DOD. See Defense, Dept. of.
Domestic Council (White House), 53–54
Donnelly, John P., 72
Doomed civilization theory, 371
Doppler shift, 302, 397
Dornier Aeros (European research satellite), 377
DOT. See Transportation, Dept. of.
Douglas Aircraft Co. See McDonnell Douglas Astronautics Co.
Downey, Calif., 262, 383–384
Drake, Dr. Frank D., 261
Drifting platforms, 77
Dryden, Dr. Hugh L., 20–21, 155–156
Dryden, Hugh L., Memorial Fellowship, 98
Dryden, Hugh L., Memorial Fund, 155–156
Dryden, Hugh L., Research Lecture, 20–21
Dryer, Dr. Murray, 441
DSCS. See Defense Satellite Communications System.
DSN. See Deep Space Network.
DTS. See Development Test Satellite.
Dual Air Density (DAD) Explorer satellite, 294
Dubna, U.S.S.R., 10
Ducayet, Edwin J., 187

Dudley Observatory, Albany, N.Y., 341, 343, 349
Dungan, Dr. John V., Jr., 326
Duke, Col. Charles M., Jr. (USAF)
  Apollo 16 mission
    debriefing, 180
    extravehicular activity, 62–63, 142, 148–149, 169, 170–171, 326
    medical aspects, 120, 135–136, 150, 164, 166, 420
    press conference, 109, 172
    test, 123
    U.S.S.R. reaction to, 163–164
  Apollo 17 mission, 194
  awards and honors, 187, 258, 264
  illness, 2, 14
  press comment, 157, 178
  promotion, 162–163, 172
  public appearance, 165, 184, 228–229
Duke, Dr. Michael B., 138, 178
Dulles International Airport, Va., 38, 136, 157, 183, 206, 214, 222
Dunn, L/G Carroll H. (USA), 410
Dunseith, Lynwood C., 248
Dupree, Dr. A. Hunter, 316
Dupree, Kirby C., 73
Dushkin, L. S., 345
Dyal, Dr. Palmer, 152
Dyna-Soar (research aircraft), 79

E

Eagle (Apollo 11 LM) See Lunar module.
Early warning satellite system, 69, 233–234
Earth, 58, 184, 378
  atmosphere. See Atmosphere.
  climate control, 375
  evolution, 101
  magnetic field, 108, 148–149, 152, 174, 177–178, 184, 276, 308
  mantle, 219–220
  open-world concept, 125
  origin, 179
  photographs, 142–144, 405, 442
  rotation, 243
  Skylab experiments, 63–64
  "Spaceship," 319
  tides, 54–55
Earth Observations Satellite (EOS), 64, 116, 160
Earth resources, 338, 408
  management, 17
  photography, 118, 120, 221–222, 444
  remote sensing, 25, 34–35, 75, 229, 234, 301–302, 303, 314–315, 408
Earth resources experiment package (EREP), 24, 283, 310
Earth Resources Information Storage, Transformation, Analysis and Retrieval (ERISTAR) system, 294
Earth Resources Observation System (EROS), 234, 302, 380–381
Earth Resources Program Review, Fourth Annual, 21
Earth Resources Survey Data Center, 302
Earth resources survey program (ERSP), 30, 33, 34–35, 75, 266, 270, 302, 380–381, 408
Earth Resources Technology Satellite (see also Erts 1), 27, 56, 210, 380–381, 383, 447
  applications, 226–227, 234, 408, 444
  data, 34–35, 219
  funds, 91, 116
  international cooperation, 184, 226–227, 230
  TV transmission, 124–125, 184
Earth station, 130, 194, 208
  Apollo 17 tracking, 422
  domestic satellite communications system, 353, 410
  ERTS, 184
  India, 422
  installation, 91, 410
  Intelsat, 53, 60, 91, 173
  P.R.C. purchase, 28–29, 86, 98, 294
  statistics, 370
Earth-watch, 218–219
Earthquake, 1, 236–237, 308, 313, 327, 442
Eastern Airlines, 162, 205
Eastern Test Range (ETR), 139–140, 321
  launch
    Anik 1, 379
    delay, 71
    Explorer 47, 325
    failure, 364
    Intelsat-IV F-4, 26–27
    Oao 3, 295
    Pioneer 10, 81
    reconnaissance satellite, 77
    unidentified satellite, 437
  Mariner 10 launch site, 307
Easton, Conn., 362
Echeverria Alvarez, President Luis (Mexico), 162–163, 229–230
Eclipse, lunar, 38
ECM. See Electronic countermeasures.
Ecology (see also Air pollution; Noise, aircraft), 1, 6, 21, 138, 206, 223, 298
Economic Affairs, Dept. of (proposed), 122
Economic Analysis of the Space Shuttle System, 39
Edelman, Dr. Gerald M., 348
Edelstein, Judge David N., 393, 411
Editorial. See Press comment.
Edmonds, Thomas E., 169
Edwards AFB, Calif. 18, 71, 282, 395, 397
Edwards, Calif., 264, 311
Edwards Plateau, Tex., 308
EEO. See Equal Employment Opportunity.
Eglin AFB, Fla., 233, 377
  Armament Development and Test Center, 377
EGNG Inc., Bedford, Mass., 122
Egypt. See United Arab Republic.
Ehricke, Dr. Krafft A., 125, 344
Einstein, Albert, 194, 360

"*Einstein Papers,*" 429
Eisele, Col. Donn F. (USAF, Ret.), 192
Eisenhower, President Dwight D., 358, 399
Eisenhower, Dwight D., Memorial Bicentennial Civic Center, 322
Eisenhower, John S. D., 187
Ekranoplane (U.S.S.R. hovercraft), 381–382
El-Baz, Dr. Farouk, 83, 237
El Centro, Calif., 163
El Niño (current), 423
El Segundo, Calif., 383
ELDO. See European Launcher Development Organization.
Electrical energy, 162, 169, 188, 236–237, 284, 289, 374
Electrically scanning microwave radiometer (ESMR), 423
Electromagnetic radiation, 227, 253, 295, 386
Electrometer, 372
Electron flux analyzer, 295
Electron microscope, 101, 292
Electronic countermeasures (ECM), 61–62
Electrophoresis, 41
Electrophysics research, 79
Electrostatic autopilot, 393
Ellender, Sen. Allen J., 289
Ellington AFB, Tex., 148–149, 178, 292–293, 384, 439
Emeg Haela, Israel, 91, 273
Emme, Dr. Eugene M., 345, 444
Emmen, Switzerland, 177
Endeavor (*Apollo 15* CM). See Command module.
Energy
  coal, 162
  cosmic plasma, 141
  electron accelerators, 175
  nuclear, 2, 10, 14–15, 24–25, 130–131, 162, 175
  R&D, 47
  resources, 236–237, 298–299, 386
  solar, 160, 445, 446
    international cooperation, 138, 199, 237
  solar cell, 1, 38, 169
  solar propulsion, 39–40
Energy crisis, 225, 396
Energy Research and Development (Congressional study), 446
Enevoldson, Einar, K. 50

Engine (see also individual engines and motors, such as JT3D, JT8D, TE-M-364-4, etc.)
  aircraft, 78
    emissions, 31, 409, 449
    exhibit, 240, 307, 378
    jet, 93, 288, 378, 403, 449
    noise, 130–131, 185–186, 350
    SST, 297, 307
    STOL, 2, 40, 66, 124, 315, 443
  command module, 208

  nuclear, 333
  rocket, 19, 35, 39, 53, 90, 304, 352, 379
    space shuttle, 108, 129, 332, 333
  small tactical aerial mobility platform, 446
Engineers, 233
  employment, 61, 64–65, 67, 134, 319, 326, 340
  Eole transponder test, 242
  foreign, 8, 193, 205, 239
  helicopter test, 56
  lunar landing contribution, 133
  meeting, 16, 60, 72–73, 74, 135–136, 154–155, 201, 385
  NASTRAN, 71
  remote sensing, 302
  space shuttle, 129, 140–141
England. See United Kingdom.
England, Dr. Anthony W., 205, 259
English, R. D., 344
Environment (see also Air pollution; Noise, aircraft), 5–6, 101, 160–161, 232, 298, 380
  aircraft effects, 190
  Apollo data application, 63
  atomic waste disposal, 53
  funds, 47
  international cooperation, 47, 217, 225
    press comment, 201
    U.S.-U.S.S.R., 127, 190, 191, 203, 298, 323
      agreement, 193–194, 195–196, 203, 323
    remote sensing, 177, 229
  natural resource management, 17, 235, 378
  Nixon, President Richard M., statement, 47, 206, 236
  recycling, 195
  remote sensing, 21, 24, 34–35, 63–64, 172–173, 301–302
  space shuttle effects, 10
Environmental Action Group, 162
Environmental Control and Life Support Conference, 293
Environmental Control Co., 271
Environmental engineering. See Geophysical warfare.
Environmental Protection Agency (EPA), 73–74, 270, 301–302, 348, 390, 409
Environmental Quality: The Third Annual Report of the Council on Environmental Quality–August 1972, 285
Environmental Research Office (LeRC) (NASA), 94
Environmental satellite. See *Erts 1*; Meteorological satellite; Nimbus; *Noaa 1* and *2*; *Sret 1*.
Environmental systems resources program, 85
Enzmann, Dr. Robert D., 409
Enzyme, 247
Eole (French satellite), 66, 242
EOS. See Earth Observation Satellite.
EPA. See Environmental Protection Agency.

epndb: effective perceived noise in decibels.
Epstein, Dr. Eugene E., 365
Equal employment opportunity (EEO), 11, 60, 80–81, 188, 276
Eratosfyen Crater (moon), 46
EREP. See Earth resources experiment package.
Erikson, Leif, Day, 297
ERISTAR. See Earth Resources Information Storage, Transformation, Analysis and Retrieval System.
ERNO–Raumfahrttechnik GmbH (West Germany), 303–304
EROS. See Earth Resources Observation System.
ERSP. See Earth resources survey program.
ERTS. See Earth Resources Technology Satellite.
Erts 1 (ERTS–A)
  Canada, 184
  funding, 91
  global inventory, 383, 444, 447
  information dissemination, 91, 184, 210, 235, 269–270, 280, 294, 356
  launch, 269–270
  mission profile, 235, 270
  photographs, 210, 235, 269–270, 308, 313, 367, 378, 380, 442
  press comment, 272
  Return-beam-vidicon system shutdown, 284
  significance, 91, 279, 369
  tape recording systems, 235, 282, 288
ESC. See European Space Conference.
ESMR. See Electrically scanning microwave radiometer.
ESOC. See European Space Operations Center.
ESP. See Extrasensory perception.
ESRANGE (European Sounding Rocket Range), Kiruna, Sweden, 244
ESRO. See European Space Research Organization.
Esro 4 (scientific satellite), 389, 452
Essa 1 through *Essa 9* (Environmental Science Services Administration Meteorological satellite), 352
Esso Research & Engineering Co., 188
ESTEC. See European Space Technology Center.
Estes, M. F., 16
Estes, L/C Thomas B. (USAF), 162, 350, 363
Esthesiometer, 184–185
ESTRACK. See European Space Tracking Stations.
ETR. See Eastern Test Range.
Etzioni, Dr. Amitai W., 407
Europa I (ELDO booster), 59
Europa II, 59, 267
Europa III, 108, 304
Europe, 218–219, 250, 341
  aerospace industry, 192–193
  aviation industry, 329

  communications satellite, 255, 361, 386
  earth station, 208
  exhibit, 307
  Nixon, President Richard M., visit to, 189, 190, 191
  scientists, 64
  Skylab fly-over, 63
  space program, 48–49, 239, 362, 377, 425
  space shuttle participation, 11–12, 48, 251–252, 333, 339, 362
  U.S.S.R. missile target, 87
European Airbus Consortium, 365
European Broadcasting Union, 156–157
European Conference on Space and Youth, First, 75
European Launcher Development Organization (ELDO)
  American manned space program participation, 239
  booster, 59, 108, 267, 304
  contract, 267
  reorganization, 437
  space tug, 238, 345
  termination demand, 262
  U.K. withdrawal, 126
European Sounding Rocket Range. See ESRANGE.
European Space Conference (ESC), 48–49, 108, 226, 241–242, 362, 378, 437–438
European Space Operations Center (ESOC), 38, 96
European Space Research Organization (ESRO)
  international cooperation, 49–50, 177–178, 208, 188, 389
  launch, 95–96
  reorganization, 437
  satellite, 12–13, 38, 46, 95–96, 384–385, 389, 397
  space shuttle participation, 238, 241–242, 366, 398, 452
European Space Technology Center (ESTEC), 38, 241–242
European Space Tracking Stations (ESTRACK), 39, 96
EUV. See Ultraviolet, extreme.
ev: electron volt.
EVA. See Extravehicular activity.
Evans, Albert J., 67
Evans, J. V., 174
Evans, Larry G., 387
Evans, Llewellan J., 253
Evans, Capt. Ronald E. (USN)
  Apollo 17 mission, 367, 414–419
  emblem, 347
  extravehicular activity, 444
  launch, 413
  medical aspects, 403, 420
  Nixon message, 420
  prelaunch test, 389
  press comment, 438
  timetable, 297
  promotion, 429–430
Everett, Mass., 259
Evvard, Dr. John C., 115

Ewbank, Thomas, 253
Exceptional Scientific Achievement Medal (NASA), 270–271, 381
Exceptional Service Medal (NASA), 381, 443, 444
Exhibit
  astronomical art, 17–18, 153, 185, 211
  "Ballooning: 1782–1972," 74
  electrical energy uses, 112
  "Eyewitness to Space" (NASA art program), 17–18
  "Fallen Astronaut" (sculpture), 153
  "Research and Development in the U.S.A.," 115
  Space Fair, 358
  "Space Science and Awareness Month," 72
  technology utilization, 112
  TRANSPO '72, 136, 157, 183, 206, 214, 222, 224
  U.S. aerospace industry, 434–435
  von Braun, Dr. Wernher, personal papers, 117, 185
  World's Fair Hall of Science, 388
Exon, Gov. J. James, 356
Exosphere (geocorona), 389
The Experiments of Biosatellite II (NASA SP-204), 170
Exploration of Mars, 211
Explorer (spacecraft), 55–56, 177–178, 275
Explorer 1 (satellite), 202–203, 451
Explorer 6, 387
Explorer 18 (IMP-A) (Interplanetary Monitoring Platform), 325
Explorer 41, 280
Explorer 42 (*Uhuru*) (Small Astronomy Satellite), 56–57, 107, 261, 307, 332, 360, 438
Explorer 43 (IMP-I), 280, 325
Explorer 45 (Small Scientific Satellite), 233, 280
Explorer 46 (Meteoroid Technology Satellite), 290, 409–410
Explorer 47 (IMP-H), 325, 447
Explorer 48 (SAS-B) (Small Astronomy Satellite), 361, 386, 447, 452
Explosion, 266
Exposition. See Exhibit.
Extrasensory perception (ESP), 188
Extraterrestrial life, 56, 171, 258, 291, 298, 387–388
Extravehicular activity (EVA), 345
  Apollo 15 mission, 432
  Apollo 16 mission, 6, 14, 62–63, 96–97, 142, 144–146, 149–150, 170–171
  Apollo 17 mission, 297, 367, 381, 391–392, 416–419, 440, 444
  Apollo-Soyuz Test Project, 252, 322, 346
  medical aspects, 152
  protective equipment, 132, 372–373
  Skylab, 23–24
  spacesuit, 185, 262
"Eyewitness to Space" (NASA art program), 17–18

F

F-1 (rocket engine), 58
F-4 (Phantom) (fighter-bomber aircraft), 321, 328, 377
F-5E (international fighter aircraft), 263
F-8 (supersonic carrier fighter), 73, 200, 312
F-14 (supersonic fighter aircraft), 93, 151–152, 220, 244, 260, 271, 311, 425
F-14A, 26, 322
F-15 (supersonic fighter aircraft), 31, 67, 93, 241, 263, 282, 395
F-100 (turbofan engine), 378
F-111 (supersonic fighter aircraft), 311
  accident, 230, 233, 236
  exhibit, 319
  lawsuit, 209
  supercritical flight program, 50
  Vietnam war, 329, 340, 355, 393
F-111A, 328
F-111D, 263
FAA. See Federal Aviation Administration.
Facilites. See National Aeronautics and Space Administration, facilites.
Faget, Dr. Maxime A., 372–373
FAI. See Fédération Aéronautique Internationale.
Fainberg, Dr. Joseph, 387
Fairbanks, Alaska, 44, 70, 90, 91, 182–183, 235, 270, 349
Fairchild Hiller Corp. See Fairchild Industries, Inc.
Fairchild Industries, Inc., 110, 202–203, 233, 283, 383, 399, 451
  American Satellite Corp., 355
  Republic Aviation Div., 322
Falcon 20T (European commuter transport), 94
Falcon (*Apollo 15* LM). See Lunar module.
"Fallen Astronaut" (sculpture), 153, 266–267, 317
FAMOUS. See Franco-American Midocean Undersea Study.
Far-ultraviolet camera-spectrograph, 102, 254, 270–271
Faraday, Dr. Bruce J., 291–292
Farnborough Air Show, England, 307
FAS. See Federation of American Scientists.
Fast luciferace automated assay of specimens for hospitals (FLASH), 258
Fastie, William E., 433
FB-111 (supersonic fighter), 319
FBI. See Federal Bureau of Investigation.
FCC. See Federal Communications Commission.
Federal Aviation Administration (FAA)
  air traffic control, 20, 38, 67, 141
  airlines, 93
  airports, 203–204, 272
  astronaut flight status, 98
  award, 326

budget, 298
collision prevention, 104
contract, 35, 92, 161, 266, 268-269, 323
cooperation, 12-13, 16, 67, 74, 136, 176, 204, 382
cooperation, international, 49-50
landing systems, 67, 116, 124, 141, 454
meeting, 355
meteorology, 93
National Airspace System Program Office (NASPO), 301
noise, aircraft, 31, 124
personnel, 221, 301, 358, 430
Planning Grant Program, 272
satellite communications systems, 49-50 141-142
statistics, 161
Systems Research and Development Service (SRDS), 301
transport, supersonic (see also Supersonic transport), 166, 319, 425
Federal Bureau of Investigation (FBI), 221, 222, 356
Federal City College, Washington, D.C., 222
Federal Communications Commission (FCC)
  communications satellite, 32, 190, 261-262, 286, 309, 355
  corporation approval, 338, 353
  policy, 104, 230-231, 286
Federal Council on Science and Technology, 85
Federal District Court, Washington, D.C., 242-243, 255
Federal Electric Corp., 248
Federal Funds for Academic Science, Fiscal Year 1970 (NSF 72-301), 258
Federal Funds for Research, Development, and Other Scientific Activities, Fiscal Years 1970, 1971, 1972 (NSF 71-35), 245
Federal German Ministry for Education and Science (West Germany), 303-304
Federal Grand Jury, Dallas, Tex., 209
Federal Pay Board, 6, 10, 16-17, 46
Federal R&D Funding Continues to Rise (NSF 72-314), 289
Federal Scientific, Technical, and Health Personnel in 1970 (NSF 71-47), 64-65
Federal Support to Universities, Colleges, and Selected Nonprofit Institutions, Fiscal Year 1970 (NSF report), 51
Fédération Aéronautique Internationale (FAI), 268, 335
Federation of American Scientists (FAS), 181
Feoktistov, Konstantin P., 338
Fermi, Enrico, Award, 374
Ferrer, President Jose Figueres (Costa Rica), 142
Field, Dr. George B., 372
Filipchenko, Anatoly V., 343-344
Fingerprints, 126
Finland, 121, 421
Fiorio, Franco, 173

Fire, 45-46, 266, 280-281
Fischer, William A., 444
Fischl, Max, 365
Fishman, Dr. G. J., 181
FIT. See Florida Institute of Technology
Fitter-B (U.S.S.R. fighter-bomber aircraft), 329
Flag Crater (moon), 146-148
Flagstaff, Ariz., 148, 396-397
FLASH. See Fast luciferace automated assay of specimens for hospitals.
Flat Conductor Cable Symposium, 346
Fleet Satellite Communications (FLTSATCOM) System, 382
Fletcher, Dr. James C.
  agreement signing, 77
  Apollo 15 commercialism, 282-283, 317
  Apollo 17 congratulations, 418
  Apollo program assessment, 420
  Apollo-Soyuz Test Project, 136-137, 175-176, 185, 208, 228-229, 279, 336
  awards and honors, 238, 268, 310, 372-373, 380-381, 403
  ceremonies, 247, 286, 309, 311, 330, 455
  Equal Employment Opportunity, 80, 276
  FRC reorganization approval, 165
  LeRC visit, 393
  letters, 21-22, 36, 64, 88, 119, 214, 279, 305, 374, 438
  magazine articles, 2, 248, 408
  meeting, 2, 34, 51, 135-136, 192-193, 219, 223, 371
  NASA budget, 4-5, 32, 168
  testimony, 22, 46-47, 99-100, 102, 136-137
  personnel, 117-118, 202-203, 208, 316, 323, 440
  press conference, 4-5, 32, 102, 128, 139, 197-199
  space program, 27-28, 46-47, 135-136, 175-176, 192-193, 219, 279, 286-287
  space shuttle, 4-5, 12, 36, 51, 120, 123, 136-137, 139, 181-182, 339, 408
  U.S.S.R. space program, 248
Flight Achievement Award (AAS), 372-373
Flight Crew Operations (MSC) (NASA), 205
Flight data recorder, 350
Flight Research Center (FRC) (NASA)
  aircraft, 66, 67, 72-73, 311-312, 355, 387
    flight tests, 67, 72-73, 182, 200, 336-337
    TF-8A, 94, 104, 109, 113, 123, 155, 163, 184, 186, 193
  anniversary, 311-312, 449
  award, 311, 380-381
  clear-air turbulence, 43
  contract, 264
  lifting body, 312, 347
    test, 271, 289, 299, 315, 327, 341, 356, 371, 379, 390, 398, 411, 427
  reorganization, 165
Flight Safety Foundation, Inc., 355

Flight Test Center (USAF), Calif., 347
Florensky, K. P., 13-14
Florida
  airlines, 162
  Apollo 17 launch, 389-390, 405
  astronaut visit, 1
  Cape Kennedy name change, 41-42, 340, 385
  earth station, 353
  employment, 208
  fingerprint transmission, 126
  land use, 308
  legislature, 318
  meeting, 37, 154
  Nixon, President Richard M., statement, 162-163
  Presidential campaign, 98
  space shuttle, 139-140, 181-182
  student space experiment, 136
Florida Institute of Technology (FIT), 138
Florida Presbyterian College, 17
Florida Technical Univ. (FTU), 138
FLTSATCOM. See Fleet Satellite Communications System.
Fluid and Thermal Engineering Seminar, 74
Flushing, N.Y., 388
Flying Tigers, 353
FOBS. See Fractional orbital bombardment system.
Foolscap (American yawl), 242
Foote Mineral Co., 309
For the Benefit of All Mankind: The Practical Returns from Space Investment, 368
Forbes (comet), 314
Ford Foundation, 388
Forecast. See *Prognoz 1.*
Forestry, 209, 236-237, 380
Formosa, 357
Formulas and Ephemerides for Field Observations on the Moon, 329
Fort Churchill Range, Canada, 394
Fort Worth, Tex., 313
Fort Yukon, Alaska, 173
Foster, Dr. John S., Jr., 69, 115, 116, 245
Fra Mauro Crater (moon), 180-181, 203, 215-216 219-220
Fractional orbital bombardment system (FOBS), 252, 329
France
  aircraft (see also Concorde), 133-134, 161, 204, 314, 361, 365
  award, 350
  international, cooperation, 200, 204 250-251, 341
  international cooperation, space, 27, 28, 75, 128, 174, 208, 242, 323, 423
  KSC visit, 60
  long-duration experiment, 308
  meeting, 29, 48, 75, 168, 330, 335, 378
  meteorology, 105, 453
  missile, 259
  P.R.C. visit, 421
  space program, 70, 125-126, 128, 425
  treaty, 174, 250-251
Franco-American Agreement on Scientific Cooperation, 250-251
Franco-American Midocean Undersea Study (FAMOUS), 250-251
Frank, Herbert J., 123
Frankel, Max, 207
Franklin, Dr. Kenneth L., 356
Franklin Mint, Philadelphia, Pa., 280
FRC. See Flight Research Center.
Freedom of Information Act, 234
Freeman, Arthur B., 223
Freeman, honorary title (Scotland), 96
Freeman, Dr. John W., 83
French Speleological Institute, 308
Frequency measurement, 42
Frey, Rep. Louis, Jr., 329, 340
Friedheim, Jerry W., 87, 221, 355, 429
Friedman, Milton, 444
Friends of the Earth, 398-399
Friends of the Goddard Library, 400-401
Friendship 7 mission, 61
Frutkin, Arnold W., 173, 209, 257, 339
Fryxell, Dr. Roald, 54
Fthenakis, Emanuel, 355
Funds for Research, Development, R&D Plant and Scientific Technical Information, Fiscal Years 1971-1973: Annual Report to the National Science Foundation (NASA report), 125
Fusion technology, 199

G

G booster, 7, 166
Gabriel, David S., 79, 107
Gagarin, Col. Yuri A. (U.S.S.R.), 138, 243, 446
Galabert International Astronautical Prize, 202-203
Galaxy (see also Milky Way), 354, 357, 360, 365, 374
Galileo. See Convair 990.
Galley, Robert, 317
Gallup, George H., 241
Galveston, Tex., County Health District, 204
Gamma ray, 16, 56-57, 139, 148, 181, 287, 342
GAO. See General Accounting Office.
GARP. See Global Atmospheric Research Program.
Garrick, I. Edward, 80
Garriott, Dr. Owen K., 23
GASP. See Global Air Sampling Program.
Gast, Dr. Paul W., 13-14, 138, 183-184, 238, 363, 380-381, 432, 445
Gatland, Kenneth W., 239, 333
Gavin, Joseph G., Jr., 274, 275, 284-285, 322, 372-373
Gayler, Adm. Noel A. M. (USN), 163
Gazenko, Dr. Oleg G., 170, 279, 322, 360
GCA Corp., 349

GCEP. See United Nations, Governing Council for Environmental Programs.
GE. See General Electric Co.
Gegenschein experiment, 6
Geiger counter, 372
Gemini (program), 226–227, 424, 435
Gemini (spacecraft), 118
Gemini 4 mission, 142, 221, 392
Gemini 5 mission, 356
Gemini 6 mission, 404
Gemini 7 mission, 340, 356, 392
Gemini 9 mission, 392
Gemini 11 mission, 150
Gemini 12 mission, 392
General Accounting Office (GAO), 131, 179, 215, 263, 267, 362, 389, 422
 contract review, 45, 90, 110, 120, 123
General aviation, 161, 311–312
General Aviation Accident Prevention Industry Advisory Committee, 204
General Dynamics Corp., 203, 209, 247
 Convair Aerospace Div., 247, 283
 Convair Div., 247, 283

General Electric Co. (GE), 378
 aerodynamic test, 186
 Aerospace Group, 261
 Aircraft Engine Group, 172, 231
 award and honorarium, 20–21
 contract, 34, 110, 172, 194–195, 257, 281, 308, 386
 electric power production, 188
 metric conversion, 182
 Space Div., 34, 281
 spacecraft systems, 194–195
 turbojet engine, 20–21, 34, 182, 184–185, 186, 188, 194–195, 205, 281, 308

General Motors Corp. (GMC), Diesel-Allison Div., 172
General-purpose airborne simulator (GPAS), 397
General Services Administration (GSA), 287
General Telephone & Electronics Corp.
 GTE International, Inc., 386
 GTE Satellite Corp., 353
"Genesis bean," 14
"Genesis rock," 62, 146
Geneva, Switzerland, 134, 174, 356, 390
Geocorona, 179, 184, 254
Geology, 34–35, 124–125, 235, 298
Geophysical stations, 265
Geophysical warfare, 247, 250
George Washington Univ., 75
Georgia, 74, 162, 178
Geostationary Operational Environmental Satellite (GOES) system, 31, 262
Geostationary satellite, 154
Geostationary Technology Satellite (GTS), (U.K.), 126, 446
Geothermal energy, 236–237
German Research and Test Establishment for Aviation and Spaceflight, 304
Germany, 6, 202–203, 205
Germany, East, 218, 243, 403
Germany, West, 161, 224, 251
 Aeros satellite, 430, 431, 452
 aerospace development, 159
 Bumerang project, 303–304
 CFD participation, 123
 ELDO phaseout proposal, 262
 ESC participation, 437–438
 ESRO experiment, 389
 exhibit, 157
 interstellar research, 327
 Mariner Jupiter/Saturn mission participation, 423
 Max Planck Society for Advancement of the Sciences, 251
 Ministry for Education and Science, 303–304, 430–431, 452
 satellite communications, 352
 solar research, 384–385
 sounding rocket payload, 92
 U.S. launcher purchase proposal, 262
 visit to KSC, 60
GET: ground elapsed time.
ghz: gigahertz (1 billion cycles per second).
Giacconi, Dr. Riccardo, 360, 380–381
Giacobini meteor shower, 341, 343, 348
Giacobini-Zinner (comet), 341
Gibson, Dr. Edward G., 23
Gilbreath, Kenneth B., 80
Gillette, Robert, 36–37
Gilruth, Dr. Robert R., 17, 85, 114, 205, 210, 345
Ginter, R. D., 85
Gittleman, Len, 428
Glaser, Peter E., 445
Glasser, L/G Otto J. (USAF), 192–193
Glenn, Col. John H., Jr. (USMC, Ret.), 61, 205, 247, 455
Glennan, Dr. T. Keith, 288
Global Air Sampling Program (GASP), 216–217, 392
Global Atmospheric Research Program (GARP), 84, 174, 306, 397
Global Inventory and Distribution of Pu-238 from SNAP-9A (AEC report), 80
Glushko, Valentin P., 352
GMC. See General Motors Corp.
GMT. See Greenwich Mean Time.
Gnaegi, Defense Minister Rudolph (Switzerland), 314
Goddard, Mrs. Esther H., 400–401
Goddard Institute of Space Studies, 167, 246
Goddard Memorial Dinner, 98
Goddard, Dr. Robert H., 400–401
Goddard, Robert H., Award, 20–21
Goddard, Robert H., Historical Essay Award, 98–99
Goddard, Robert H., Memorial Trophy, 98, 202–203
Goddard Space Flight Center (GSFC) (NASA), 60, 138, 235, 242, 385, 387
 AE project management, 372
 Anik 1 project management, 379
 Apollo program, 149–150, 420
 balloon experiment, 315
 computer, 173, 270, 411, 455

contract, 155, 264
ERTS, 235, 313
Explorer 47 experiment, 325
meeting, 17, 98, 330, 377-378
meteorology experiment, 24
Nimbus program, 423
Oao 3, 297
patent, 258
personnel, 21, 160, 258-259, 325, 434, 444
Small Astronomy Satellite program, 386
sounding rocket experiments (see also Sounding rocket)
  atmospheric data, 51, 186-187, 195, 248, 378, 379
  cosmic radiation, 189
  fields and particles, 315
  ionospheric composition, 251
  parachute deployment, 173
  performance test, 327, 388
  polar cap absorption event, 283, 284
  rocket configuration, 361
  wind data, 90, 186-187
symposium, 379
TOS program, 352
Godfrey, Roy E., 110
Godynn Crater (moon), 46
GOES. See Geostationary Operational Environmental Satellite.
Gold Space Medal, 335
Gold, Dr. Thomas, 14
Goldman, Dr. J. E., 326
Goldsmith, Harry, 253
Goldstone Tracking Station, Calif., 113, 221, 235, 270
Goldwater, Sen. Barry M., 435
Goodman, David M., 16
Goodrich, B. F., Co., 47-48
Goonhilly Downs, Cornwall, U.K., 255
Gorman, Harry H., 64, 78
Goss, Robert J., 377-378
Gould, Dr. Roy W., 378-379
Government Accounting Office (GAO), 120
Government Printing Office (GPO), 71
GPAS. See General-purpose airborne simulator.
GPO. See Government Printing Office.
Grand Tour mission, 28, 32, 39-40, 56, 64, 65
Gravitational Redshift Space Probe Experiment, 194
Gravity (see also Weightlessness), 54-55, 194
Great Galactic Ghoul, 200
Great Lakes, 21
Great Miami River, 77
Grechko, Defense Minister Andrey (U.S.S.R.), 377
Green Bank, W. Va., 18, 307
Green, Constance M., 75
Green, M., 345
Greenbelt, Md., 235, 270, 284, 377-378
Greenwich Mean Time (GMT), 243
Gregory, Dr. Philip C., 307, 332

Grenade experiment, 186-187, 193
Grissom, Mrs. Betty, 91, 382
Grissom, L/C Virgil I. (USAF), 91
Grobecker, Dr. Alan J., 385
Gromyko, Foreign Minister Andrey A. (U.S.S.R.), 338
Grooms, Red, 17-18
Grosse, Hans-Werner, 161
Ground station. See Earth station.
Group Achievement Award (NASA), 381
Grumman Corp., 26, 123, 151-152, 253, 280, 309, 319
  Grumman Aerospace Corp.
    award, 372-373
    contract
      F-14 aircraft, 151-152, 425
      F-14A aircraft, 26, 322
      shuttle training aircraft, 247
      space shuttle, 180, 266, 274, 275-276, 308, 339
    personnel, 372-373, 395
    TACRV, 136
Grumman, Leroy R., 353
GSA. See General Services Administration.
GSFC. See Goddard Space Flight Center.
GTE International, Inc. See General Telephone & Electronics Corp.
GTE Satellite Corp. See General Telephone & Electronics Corp.
GTS. See Geostationary Technology Satellite.
Guam, 233-234
Guggenheim, Daniel and Florence, International Astronautics Award, 345
Guggenheim Museum, 17-18
Gulf of Maine, 250-251
Gulf of Mexico, 377
Gulf Stream, 423
Gulfstream II (jet aircraft), 309
Gurney, Sen. Edward J., 41-42
Gursky, Dr. Herbert, 360
Guyot Crater (moon), 148-149
Gvishiani, Dzherman M., 340

H

H-1 (booster), 377-378
H-42 (helicopter), 319
Hadley region (moon), 35-36, 45
The Hague, Netherlands, 49, 318
Hahn, Dr. T. Marshall, Jr., 281
Haifa, Israel, 80
Hale Observatory, 181, 360, 393
Hall of Science, 388
Halley's Comet, 165
Hamby, Herman G., 294
Hamilton, Jeffrey T., 271
Hamilton Standard Div. See United Aircraft Corp.
Hampton Technical Center. See LTV Aerospace Corp.
Handler, Dr. Philip, 127, 229, 299, 303, 340, 354, 365, 374, 375
Hanks, Dr. Thomas C., 440

Hanover Air Show, 157
Hanscom Field, Mass., 383
Hansen, Grant L., 77, 245, 403
Hanses, Eugene, 400
Hardin, Dr. Garrett, 405
Harford, James J., 121
Hargis, Dr. William J., 216
Harmon International Aviation Trust, 350
Harmon International Aviator's Trophy, 350
Harnett, Daniel J., 73–74, 100, 293
Harnish, R/A William M. (USN), 289
Harr, Dr. Karl G., Jr., 162, 204
Harrington, Dr. Charles O., 410
Harris, Mrs. Ruth Bates, 11
Harrison, Dr. Anna J., 281
Harvard College Observatory, 274, 276, 357, 372
Harvard Divinity School, 387–388
Harvard Univ., 13, 387–388
Harvest Moon Project, 325
Hasbrouck, Dr. W. P., 308
HAST. See High-altitude supersonic target missile.
Haughton, Daniel J., 173, 288
Havana, Cuba, 321
Hawaii, 42, 112, 148–149, 165, 302, 306, 353, 439
Hawaii, Univ. of, 19
Hawker Siddley Dynamics Ltd., 267, 446
Hawkins, Dr. Willard R., 132, 166, 372–373, 403, 414
Haystack Observatory, 403
Hayward, Dr. Evans V., 341
Hazeltine Corp., 35
HC–130 (tanker aircraft), 319
HC–130H, 61
Heacock, Raymond L., 266
Health, Education, and Welfare, Dept. of (HEW) 27, 45, 51, 67–68, 73–74, 122, 169–170, 245, 375
HEAO. See High Energy Astronomy Observatory.
Heart, artificial, 86
Heat flow experiment (HFE), 146–148
Heath, Mark C., 25, 214
Hechler, Rep. Ken, 21–22, 350
Heer, Dr. Ewald, 316
Heffner, Dr. Hubert, 281, 341
Heiss, Klaus P., 39, 137
Helicopter
  airborne laser system, 136
  American Helicopter Society, 187
  astronaut pickup, 148–149, 419
  development, 40, 288, 362
  electrostatic autopilot, 393
  exhibit, 319
  manned automatic landings, 56
  noise reduction, 67
  prototypes, 303, 389
  sales, 453, 454
  surveillance camera, 190–191
  water pollution study, 409
Heligoland (island) (West Germany), 303–304

Helios (solar research satellite), 367, 377
Helium, 272
Heller, Gerhard B., 337
Hello, Bastian, 383–384
Helsinki, Finland, 121, 421
HEOS. See Highly Eccentric Orbit Satellite.
Heos 1, 38
Heos 2, 38, 46, 60, 78, 96, 452
Hepburn, Charles D., 363
Hercules (constellation), 261, 438
Hero of Socialist Labor (U.S.S.R. honor), 368, 442
Heseltine, Aerospace Minister Michael R. D. (U.K.), 288, 329, 378, 421
Hess, Dr. Wilmot N., 227
HEW. See Health, Education, and Welfare, Dept. of.
HFE. See Heat flow experiment.
HH–53 (helicopter), 319
Hickam AFB, Hawaii, 148–149, 165
High-altitude supersonic target (HAST) missile, 377
High Energy Astronomy Observatory (HEAO), 69, 287–288, 408
  experiments, 107, 163, 213–214, 243, 264, 342
  funding, 30, 64, 116, 125, 160
  program management, 243–244
High Reynolds Number Tunnel (HIRT), 353
High-speed interferometer (HSI), 35
Highly Eccentric Orbit Satellite (see also *Heos 1; Heos 2*), 397
Hijacking of aircraft, 49, 221, 227, 242, 318, 355, 372, 411
Hill, Maynard L., 393
Hills, Dr. H. Kent, 83
Hines, William, 16, 206–207, 266
Hinners, Dr. Noel W., 141
Hinz, Leslie F., 244
Hirsch, Dr. Robert L., 378
Hjellming, Dr. Robert, 307
HL–10 (lifting body), 312
Ho Chi Minh trails, 247
Holland. See Netherlands.
Holland-American Line, 409
Holloman AFB, N. Mex., 272
Holloway, Mike, 136
Holmdel, N.J., 255
Holmquest, Dr. Donald L., 205
Holography, 53
Holzman, Arthur D., 222
Honeywell, Inc., 236, 308
Hong Kong, 19
Honolulu, Hawaii, 312–313
Hoover Institute on War, Revolution, and Peace, 281
Hospitals, 65–66, 180
Houbolt, Dr. John C., 20–21
Houphouet-Boigny, President Felix (Ivory Coast), 397
Housing and Urban Development, Dept. of (HUD), 122
  aerospace industry employment transfer, 304

appropriations, 187, 193, 209, 273–274, 291
cooperation, 74, 131
technology utilization, 1, 74, 131
Houston, Tex., 20, 60, 111, 156, 175, 385, 419, 444
Houston, Univ. of, 19, 323
Hovercraft, 381–382
Hovis, Dr. Warren A., 24
Howard, Jean Ross, 118
Howard, Needles, Tammen and Bergendorff, 240
Howardite (meteorite), 14
HSI. See High-speed interferometer.
HST. See Hypersonic transport.
HUD. See Housing and Urban Development, Dept. of.
Hudson Bay, 373
Hughes Aircraft Co., 16, 20–21, 286, 294–295, 309, 370, 379
Hughes Enterprises, 321
Human Factors in Long-Duration Spaceflight (SSB report), 129, 153
Human Resources, Dept. of (proposed), 122
Humboldt Crater (moon), 144–146
Humboldt Current, 423
Humphrey, Sen. Hubert H., 98
Hungary, 132, 345
Hunn, Spencer S., 301
Huntington Beach, Calif., 209, 309, 325
Huntoon, Harrison H., 204
Huntsville, Ala., 69, 393
Huntsville Hospital, Ala., 95
Hurricane Agnes, 216
Hussein, King (Jordan), 142
Hydrogen, 58, 224, 270–271, 272, 304, 407
Hydrogen bomb, 289
Hydrology, 34–35, 210, 298
Hydroponics, 68
Hynek, Dr. J. Allen, 300–301
Hypersonic aircraft, 321
Hypersonic research, 153
Hypersonic transport (HST), 40

I

IAEA. See International Atomic Energy Agency.
IAF. See International Astronautical Federation.
Iapigya region (Mars), 113
Ibilisi, Georgia (U.S.S.R.), 115
IBM. See International Business Machines Corp.
ICAO. See International Civil Aviation Organization.
ICASE. See Institute for Computer Applications in Science and Engineering.
ICBM. See Intercontinental ballistic missile.
ICC:ERSP. See Interagency Coordination Committee for Earth Resources Survey Programs.

ICCAIA. See International Coordinating Council of Aerospace Industries Associations.
Ice, 21, 24, 128, 178
Ice Age, 17, 101
ICSU. See International Council of Scientific Unions.
Idaho, 230
IEEE. See Institute of Electrical and Electronics Engineers.
Il-62 (U.S.S.R. jet airliner), 349
ILC Industries, Inc., 373
ILLIAC IV (computer), 445
Illinois, 55, 61, 78, 88, 94, 314, 429
Illinois, Univ. of, 39, 372, 410, 445
ILS. See Instrument landing system.
Image dissector camera system (IDCS), 124
Image Transforms, Inc., 156
IMBLMS. See Integrated medical and behavioral laboratory measurement system.
Imbrium Basin (moon), 69
IMP (Interplanetary Monitoring Platform). See *Explorer 18, 43, 47.*
Improved Tiros Operational Satellite. See ITOS, *Itos 1*, etc.
India
  Apollo 17 mission tracking, 422
  communications satellite, 27
  international cooperation, 75, 105, 132, 223
  scientists and engineers, 8, 208
  space program, 125, 138, 223
  sounding rocket experiment, 105, 132
India-Pakistan war, 9–10
Indian Ocean, 38, 224–225, 233–234, 241, 276
Indian Space Research Organization (ISRO), 75, 422, 425
Indiana, 353
Indiana Univ., 74–75
Indochina, 247, 250, 329, 393
Industrial College of the Armed Forces, 19
Industrial Research (magazine) IR-100 Award, 323–324
Infrared horizon-sensing instrument, 389
Infrared radiation, 386
Infrared temperature profile radiometer (ITPR), 423
Injun-F (spacecraft), 115
Inlet choking, 231
Institute for Computer Applications in Science and Engineering (ICASE), 347
Institute of Electrical and Electronics Engineers (IEEE), 112
Institute of Experimental Physics, Warsaw, Poland, 20
Institute of High Energy Physics (U.S.S.R.), 221
Institute of Medicine (NAS), 299
Institute of Space Research (U.S.S.R.), 385–386
Institute of Texas Cultures, 233
Instrument landing system (ILS), 141–142
Instrument unit (IU), 142–143

Insulation, 261
Integrated medical and behavioral laboratory measurement system (IMBLMS), 15, 283, 375
INTELSAT. See International Telecommunications Satellite Consortium.
Intelsat IV (communications satellite series), 309, 370
Intelsat-IV F-2, 104
Intelsat-IV F-3, 26-27, 60-61
Intelsat-IV F-4, 26, 47, 53, 61, 173, 224
Intelsat-IV F-5, 223, 224, 242, 276
Interagency Coordination Committee for Earth Resources Survey Programs (ICC:ERSP), 301-302
Intercontinental ballistic missile (ICBM)
  early warning system, 69
  electronic countermeasures, 61-62
  People's Republic of China, 54, 310
  U.S.S.R., 49, 87, 89, 115, 335, 350
Intercosmos Council. See Soviet Academy of Sciences.
Intercosmos (U.S.S.R. satellite), 449
Intercosmos 6, 132, 235
Intercosmos 7, 243
Intercosmos 8, 403
Intergovernmental Personnel Act of 1970, 204, 222
Interim Agreement on Limitation of Strategic Arms, 316
Interim Agreement on Strategic Offensive Arms, 224-225, 230, 338
Interim Agreement with Respect to the Limitation of Strategic Offensive Arms, 201-202
Interior, Dept. of
  aerospace technology applications, 314
  Apollo launch equipment transfer, 169
  Cape Kennedy wildlife refuge, 216
  cooperation, 1, 73-74, 235
  earth resources, 235, 270, 280, 301-302
  executive reorganization, 122
  solar energy research, 85
International Aerospace Hall of Fame, 275
International Alliance of Theatrical and Stage Employees, 376, 405
International Astronautical Federation (IAF), 343-345
International Astronomical Union, 168, 237
International Atomic Energy Agency (IAEA), 256, 288
International Biophysics Conference, 291
International Business Machines Corp. (IBM), 250, 308, 393, 411
International Civil Aviation Organization (ICAO), 49, 204, 227, 311, 318
International Club of Rome, 5-6
International Conference on Computer Satellites in Agriculture, First, 367
International Congress of Aviation and Space Medicine, Twentieth, 321
International cooperation (see also Disarmament; Treaty)
  agreement, 9, 20, 136, 189, 190,191, 193-194, 199, 201-202, 203, 208, 213, 216, 218, 219, 222, 224, 225, 229-230, 236-237, 239-240, 241, 250, 265, 275, 283, 285, 288, 289, 299, 331, 338, 342, 374-375, 377, 410
  AIDJEX program, 128
  aircraft, 53-54, 157, 169, 200, 204, 297, 329, 365
  hijacking resolutions, 49, 227
  atomic energy, 94, 259, 331
  award, 339
  computer sciences, 236-237, 357
  congressional hearings, 203
  cultural exchange, 195
  defense, 240
  earth resources, 48-49, 177, 217, 218-219, 230, 236-237, 239-240, 285, 289, 292, 323, 327
  Hot Line, 20, 377
  meteorology, 84, 174, 216, 239-240, 263, 306, 397
  oceanography, 250-251, 256, 375
  Peace Corps, 192
  power stations, 341
  press conference, 90
  radio communications, 375
  reconnaissance flights, 72
  science, 90, 127, 136, 170, 181, 199, 213, 215-216, 222, 230, 236-237, 239-240, 241, 275, 291, 308, 357, 374-375
  solar energy, 138
  summit accord, 185, 189, 191, 201, 203, 204-205, 206, 207
  tourism and visitation, 60, 354, 365, 374, 387, 392
  trade, 90
International cooperation, space (see also Apollo-Soyuz Test Project; European Launcher Development Organization; European Space Research Organization; Global Atmospheric Research Program; International Telecommunications Satellite Consortium; *Isis 2*, etc.), 48, 173, 238, 245, 387, 449, 452-453
  aerosat program, 12-13
  aerospace industry, 159, 204, 307
  agreements, 49, 241
    Agreement Concerning Cooperation in the Exploration and Use of Outer Space for Peaceful Purposes, U.S.-U.S.S.R., 128, 185, 195-197, 201, 203, 204, 206-207, 208, 213, 214, 239-240, 374, 452
    Agreement on the Rescue of Astronauts, the Return of Astronauts, and the Return of Objects Launched into Outer Space, U.S.-U.S.S.R., 195
  arms control, 9, 190, 203
  peaceful use of space, 203
  press comment, 201
  publications, 107, 303
Apollo program, 151, 228-229, 434
ASTP. See Apollo-Soyuz Test Project.

astronomy, 45, 151, 187, 341, 375
benefits, 214, 219, 222, 239–240, 240–241
dictionary, 239–240
earth resources, 323, 452, 453
European Conference on Space and Youth, First, 75
International Space Agency (proposed), 422
joint mission, U.S.–U.S.S.R., 10, 15, 19, 27, 28, 48–49, 100, 117, 127, 128, 129, 130, 133, 134, 136–137, 151, 158, 174–175, 178, 182, 185, 190, 195, 199, 201, 203, 204–205, 206–207, 208, 213, 214, 219, 222, 228–229, 231–232, 239–240, 240–241, 276, 281, 282, 298, 310–311, 322, 335, 336, 344, 345–346, 354, 375, 377, 382, 395, 407, 422, 436, 449
joint working groups, U.S.–U.S.S.R., 158, 177, 178, 180, 189–190, 279–280, 282, 305–306, 322, 346, 395
launch assistance, 48, 342–343, 348, 421, 452
lunar programs, 48, 100, 134, 138, 174, 189, 215–216, 229, 231–232, 299–300, 305–306, 368, 452
Lunar Science Conference, Third Annual, 13–14, 215–216, 452
Memorandum of Understanding, space biology and medicine, U.S.–U.S.S.R., 170
meteorology, 48–49, 172–173, 453
NASA
 -Canada, 188, 379, 395
 -ESRO, 46, 177–178, 208, 366, 384, 389, 398, 452
 -Europe, 48, 60, 173
 -France, 27
 -India, 132
 -INTELSAT, 26–27, 47, 60–61, 224
 -Italy, 361, 386
 -Norway, 17, 53, 210
 -Soviet Academy of Sciences, 48–49
 -Sweden, 46
 -U.K., 208
Outer Space Liability Convention, 49, 230, 341
peaceful use of space, (see also agreements), 173, 174, 323
planetary exploaration, 28, 178, 187, 189–190, 201, 228–229, 273, 343
 NASA-U.S.S.R., 177, 178, 187, 189–190, 229, 335, 336–337
Presidential report, 106, 119
press comment, 117, 133, 273, 276, 316
rendezvous and docking. See joint mission.
satellite(see also Intelsat IV satellites)
 Anik 1, 379, 395, 452
 communications. See Communications satellite.
 Eole, 242
 Esro 4, 389, 452
 Explorer, 177–178, 208, 261, 290, 386, 409–410, 447, 452
 France–U.S.S.R., 28, 128, 208, 449
 Helios, 377
 Heos 2, 38, 46, 60, 78, 96, 452
 India–U.S.S.R., 75
 International Ultraviolet Explorer, 208
 launch assistance, 342–343
 Mariner, 423
 NASA
  -Canada, 83, 188
  -Europe, 38, 48–49, 95–96
  -Italy, 361
  -Japan, 306
  -U.K., 354
  -U.S.S.R., 100, 252, 298, 411
  -West Germany, 431
 Oscar 6, 352, 361, 452
 SAS-B, 361
 TD-1A, 95–96, 384–385, 452
 tracking, 242
science, 170, 180, 198–199, 201, 208, 213, 222, 352
Skylab program, 27–28, 63–64, 131
sounding rocket. See Sounding rocket, international programs.
space biology and medicine, 49, 170, 279
space object damage liability, 49, 230, 341
space research, 24–25, 32, 92, 105, 107, 151, 228, 229, 247, 254, 257, 423, 453
 U.K.–ESRO, 208
 U.S.
  -Europe, 11–12, 48, 239, 241–242, 251–252, 333, 366, 437
  -Norway, 17
  -U.S.S.R., 10, 15, 18–19, 48–49, 199, 203, 228, 239–240, 240–241, 257, 310–311, 449, 451, 452
  -West Germany, 303–304
 U.S.S.R.
  -East Germany, 243
  -France, 128, 174, 208, 242
  -Hungary, 132
  -India, 75, 208, 223, 239
  -Mongolia, 132
  -Poland, 132
  -Romania, 132
space shuttle, 4–5, 121
space transportation system, 48–49
space tug, 11–12, 238, 239, 333
Spacelab, 366, 398
Summary of Results of Discussions on Space Cooperation, U.S.–U.S.S.R., 195–196
Summary of Results of the Meeting Between Representatives of NASA and the Academy of Sciences on the Question of Developing Compatible Systems for Rendezvous and Docking of Manned Spacecraft and Space Stations of the U.S.A. and the U.S.S.R., 195–196
technology, 332

Treaty on Principles Governing the Activities of States in the Exploration and Use of Outer Space, Including the Moon and Other Celestial Bodies, 195
  unmanned space flight, 201
International Coordinating Committee (CIC) (France), 75
International Coordinating Council of Aerospace Industries Associations (ICCAIA), 204
International Council of Scientific Unions (ICSU), 84
International Exposition on the Environment, 292, 365
International History of Astronautics Symposium, Sixth, 345
International Institute for Applied Systems Analysis, 340, 375
International Institute for Strategic Studies, 288, 310
International R&D Trends and Policies: An Analysis of Implications for the U.S. (AIAA study), 29
International rendezvous and docking mission (IRDM), 310
International Satellite for Ionospheric Studies. See *Isis 2.*
International Space Log (U.N.), 252
International Space Rescue Symposium, Fifth, 344
International Telecommunications Satellite Consortium (INTELSAT) (see also Intelsat IV Satellites), 48–49, 318
  communications satellite, 309, 318, 343, 452
  launch, 26–27, 47, 60–61, 224, 276
  Nixon, President Richard M., statement, 430
  operations, 370
International transportation exhibition. See TRANSPO '72; TRANSPO '74.
International Ultraviolet Explorer (IUE) (satellite), 208
Interplanetary Monitoring Platform. See *Explorer 18, 43, 47.*
Interrogation, recording, and location system (IRLS), 77
Inventors' Premium Fund (proposed), 253
Io (Jupiter moon), 83, 405
Ioffe Physical-Technical Institute. U.S.S.R., 176
Ionosphere
  Alouette 1 experiments, 328
  Atmosphere Explorer Program, 372
  Denpa experiments, 295
  Esro 4 experiments, 373
  Intercosmos 8 experiments, 403
  Isis 2 experiments, 83
  solar, 452
  sounding rocket experiments, 39, 248, 251, 373, 410
IOTA Engineering Co., 60
Iowa, Univ. of, 115, 354
IR-100 award, 323–324

Iran, 189, 190, 213
IRDM. See International rendezvous and docking mission.
Iris 1 (International Radiation Investigation Satellite), 96
IRLS. See Interrogation, recording, and location system.
Iron, 33, 179, 183–184, 291
Irwin, Col. James B. (USAF, Ret.)
  Apollo 15 mission, 41, 61, 232, 256, 270, 282–283
  awards and honors, 114, 206, 210, 267–268
  press conference, 13–14, 20–21
  public appearance, 1, 28, 35
  resignation, 194, 205
Isayev, Boris, 215–216
Isis 2 (International Satellite for Ionospheric Studies), 83
Islander (light aircraft), 15
Isotope power systems, 80, 85
Israel, 80, 91, 227, 273
Israel Annual Conference on Aviation and Astronautics, Fourteenth, 80
ISRO. See Indian Space Research Organization.
Italian Research Organization (SORIS), 218
Italy
  aeronautical equipment sales, 133–134
  aircraft, 53–54
  international cooperation, 53–54
  KSC visit, 60
  social sciences study, 5–6
  space program, 173, 218, 361, 386, 438, 447, 452
  U.S. ambassador appointment, 420
Itek Corp.
  Optical Systems Division, 221–222
ITOS (Improved Tiros Operational Satellite) program, 31
Itos 1 (TIROS-M), 352
ITOS-B, 70–71
ITOS-C, 78, 182
ITOS-D, 182
ITPR. See Infrared temperature profile radiometer.
ITT Federal Electric Corp., 238–239
ITT Gilfillan, Inc., 35, 362
ITT Space Communications, 377
IU. See Instrument unit.
IUE. See International Ultraviolet Explorer.
Ivory Coast, 397

J

Jackass Flats, Nev., 119
Jackson, Sen. Henry M., 98, 316
Jackson, Mich., 221
Jackson, Nelson P. Aerospace Award, 99
Jackson, Roy, P., 83
  aircraft, 23, 59, 66, 77, 130–131, 185–186
  NASA budget testimony, 79, 85, 106

Jacobs, John H., 321
Jaffe, Leonard, 177, 380–381
Jane's Surface Skimmers 1972-73, 381–382
Japan, 19, 29, 86, 254
  aircraft, 133–134, 204, 298
  international cooperation, 62, 128, 306, 341
  launch, 295
  meeting, 302–303, 330
  science and technology, 105, 453
  space program, 26, 126, 447, 449
Japan Meteorological Agency, 105
Jastrow, Dr. Robert, 167
Javelin (sounding rocket), 51, 315
Jaycees (Fla.), 37
Jerusalem, Israel, 91
Jet Combustor Exhaust Study, 277
Jet Propulsion Laboratory (JPL) (Cal Tech), 79
  anniversary, 429
  contract, 342
  funds, 430
  Goldstone Tracking Station, 113, 221
  high-speed interferometer, 35
  Mariner 9 mission, 90–91, 225–226, 310, 364, 396
  Mariner 10 mission, 307
  Mariner Jupiter/Saturn 1977 project, 65, 266, 320
  name change, 322, 338, 358, 362–363, 375, 379, 430, 434
  personnel, 171, 178, 233, 266, 274, 380–381, 403
Johannesburg, South Africa, 434
Johns Hopkins Univ., 109, 254, 306, 393–394, 425
  Applied Physics Laboratory, 301, 306, 386
Johnsen, Irving, 244
Johnson, Alexis, 239–240
Johnson, Dr. Francis S., 39–40
Johnson, President Lyndon B., 91, 132–133
Johnson, Robert T., 354
Johnson, Theodore, 201
Johnson, Vincent L., 78
Johnston, Dr. Harold S., 280, 370, 385
Johnston, Richard S., 50, 132, 281
Joint Chiefs of Staff, 55–56, 71, 236
Joint Commission on Scientific and Technical Cooperation (U.S.-U.S.S.R.), 199, 275
Joint Committee on Cooperation in the Field of Environmental Protection (U.S.-U.S.S.R.), 193–194, 289, 323, 327
Joint Editorial Board (U.S.-U.S.S.R.), 170
Joint Hail Research Experiment, 118
Joint Parachute Test Facility (DOD), El Centro, Calif., 163
Joint Working Group on Compatibility of Docking Systems and Tunnels (U.S.-U.S.S.R., 282
  Working Group II, 178
Joint Working Group on Guidance and Control Systems (U.S.-U.S.S.R.), 282
Joint Working Group on Space Biology and Medicine (U.S.-U.S.S.R.), 180, 279
Joint Working Group on the Exploration of Near-Earth Space, the Moon and the Planets (U.S.-U.S.S.R.), 177, 306, 311
Joint Working Group on the Natural Environment (U.S.-U.S.S.R.), 177, 298
Jonash, Edmund R., 115
Jones, R/A Don A. (USN), 130
Jones, J. Lloyd, Jr., 264
Jones, Jesse C., 80
Jones, Dr. Robert T., 191–192
Jones, U.S. Deputy Secretary of State, William B., 365
Jordan, 142
Jordan, Sen. B. Everett, 362
JPL. See Jet Propulsion Laboratory.
JT3D (turbofan engine), 176, 292
JT8D, 176, 292
JT9D, 72
Jump, William A., Memorial Foundation Meritorious Award, 443
Juneau, Alaska, 173
Jupiter (missile), 202–203
Jupiter (planet), 291, 329, 441
  atmosphere, 176
  exploration, 27–28, 39–40, 56, 64, 408
    cost, 90, 451
    international cooperation, 178, 311, 423
  Grand Tour mission, 32
  gravity, 65
  high-frequency emissions, 356
  Mariner Jupiter/Saturn 1977 project, 65, 266, 320, 324, 423
  moons, 227, 421
  Pioneer 10 mission, 19, 71, 81, 140–141, 155, 200, 226, 253–254, 259–260, 305, 339, 365, 405, 438, 447
  U.S.S.R. probe, 7
Jupiter-C (booster), 202–203
Justice, Dept. of, 73–74, 120, 122, 300, 317, 318, 321, 356, 393, 411

K

Kagoshima Prefecture, Japan, 264
Kagoshima Space Center, Japan, 295
Kaiser Wilhelm Society, 251
Kaluga, U.S.S.R., 360
Kamchatka Peninsula, U.S.S.R., 445
Kansas, Univ. of, 330
Kapryan, Walter J., 420
Kapustin Yar, U.S.S.R., 243, 256
Karnow, Stanley, 232
Kash, Dr. Don. E., 95
Kauai, Hawaii, 353
Kayten, Gerald G., 67
Kazakhstan, U.S.S.R., 53, 111
KC–135 (Stratotanker), 319
Keene, Lester T., 389–390

Keldysh, Prof. Mstislav V., 138, 229, 352, 354, 365, 374, 375
Kellerman, Dr. Kenneth L., 360
Kelly, Mark W., 353
Kennedy, Sen. Edward M., 20, 359, 362
Kennedy, President John F., 41–42, 91, 259, 279, 340, 435
Kennedy, Mrs. Rose P., 340
Kennedy Space Center (KSC) (NASA), 19, 41–42
 accident, 169–170
 anniversary, 61
 Apollo 16 mission, 94, 113, 120, 134, 135, 142, 150, 200
 Apollo 17 mission, 330, 388, 396, 399, 403, 404, 405, 409
 award, 247, 265, 380–381
 Chapel of the Astronauts, 35, 44–45, 55
 computers, 411
 contract, 44, 238–239, 240, 248, 388, 434
 ecological studies, 138
 employment, 80–81, 301, 395, 405, 424
 Equal Employment Opportunity Program, 454
 facilities, 78, 139–140, 367
 Fluid Test Facility, 388
 launch, 169–170, 177–178, 297, 413
 launch operations (see also Apollo missions, Launch Complex 14, 34, 37, and 39), 8, 12, 33, 41, 407, 420
 Launch Site Medical Operations, 52
 meetings, 16, 24, 222–223, 273
 personnel, 78, 86, 205, 278–279, 331, 440, 451
 programs, 384
 Skylab hardware delivery, 209, 266, 298, 325, 341, 344
 space shuttle, 139–140, 153, 449
 Vehicle Assembly Building, 46, 139–140, 150
 visitors, 55, 60, 169–170, 217, 384, 387, 393
Kenya, 386, 452
Kepler, Johannes, 60
Kerwin, Cdr. Joseph P. (USN), 23, 24, 219, 282
Key Biscayne, Fla., 162–163
Key Personnel Development. See NASA Key Personnel Development.
Khodarev, Yu. K., 177
Kiev, U.S.S.R., 362
Kilgore, Edwin C., 85
Kimzey, Dr. Stephen L., 347
Kindelberger, James H., 353
Kinkead, Eugene, 429
Kirillin, Vladimir A., 199, 275
Kiruna, Sweden, 46, 244, 342, 343, 348
Kissinger, Dr. Henry A., 49–50, 229
Kitt Peak National Observatory, 365, 421
Kitty Hawk, N.C., 45, 434
Klager, Dr. Karl, 399
Klein, Milton, 130–131
Kleinknecht, Kenneth S., 344
Klemin, Dr. Alexander, Award, 187, 362

Knife and Fork Club, Salt Lake City, Utah, 323
Knott, Judge James R., 91
Kodiak, Alaska, 173
Kolin, Alexander, 247
Komarov, V. M., Diploma, 336
Korolev, Sergey P. 352
Kosygin, Premier Aleksey N. (U.S.S.R.), 185, 195, 203, 452
Kotelnikov, Vladimir A., 130
Kourou Space Center, French Guiana, 425
Kovach, Dr. Robert L., 431
Kowal, Charles, 181
Kraft, Dr. Christopher C., Jr., 17, 129, 170, 174–175, 252, 257, 260, 394, 435
Kramer, James J., 124, 176
Kranzberg, Dr. Melvin, 316
KREEP (lunar material), 14, 377–378
Kremlin, U.S.S.R., 20, 185, 191, 206, 368, 377
The Kremlin & The Cosmos, 133
Krenkel, E. T., Hydrometeorological Observatory, 228
Krier, Gary E., 163, 200
Krym Astrophysical Observatory (U.S.S.R.), 123
Ksamfomaliti, L., 113
KSC. See Kennedy Space Center.
Kubasov, Valery, N., 343–344
Kubicki, Ronald W., 273
KUHT (TV station), 444
Kvenvolden, Dr. Keith A., 158
Kwajalein Atoll, 108, 174, 361

L

L-35 (French launcher), 437
L-1011 (TriStar) (jet transport), 93, 162, 173, 288, 298
Labor, Dept. of, 122, 304
Laboratory for Extraterrestrial Physics (GSFC) (NASA), 20–21
Lafayette, La., 410
Laird, Secretary of Defense Melvin R., 18, 54, 64, 89, 110, 163, 185, 221, 397–398
Lake Erie, 77
Lake Ontario, 77
Lake Superior, 21
Lake Tahoe, Calif., 313
Lamarque, Tex., 356
Lamont-Doherty Geological Observatory, 62, 219–220
Lamphier, Col. Thomas G. (USA, Ret.), 343
Langholm, Scotland, 96
Langley Memorial Aeronautical Laboratory (NACA), 223, 233, 244
Langley Research Center (LaRC) (NASA), 223, 287, 386, 389
 airborne laser system, 136
 aircraft research, 66–67, 387
 contract, 115, 210, 234, 349
 electrophysics research program, 79
 Explorer 46 satellite, 290

Institute for Computer Applications in Science and Engineering (ICASE), 347
Life Scientist Program, 298
  patent, 101
  personnel, 40, 50, 56, 80, 248
  planetary program, 314
  pollution detection, 101, 390
  technology utilization, 5, 38
  Visitor Center, 217
  wind tunnel, 73, 233, 353
Langmuir probe, 132, 372
Langrenus Crater (moon), 437
Langseth, Dr. Marcus E., 62, 432
LAPES. See Low-altitude parachute-extraction system.
La Pinta, Dr. Charles K., 437
Lapp, Dr. Ralph E., 86–87, 136, 181
LaRC. See Langley Research Center.
Large Space Telescope (LST), 69, 176, 243, 302, 374
Las Vegas, Nev., 93, 119, 321
Laser, 287, 321
  AIDJEX project, 128
  development, 42, 79, 138, 190–191, 331
  Laser-Doppler system, 302, 397
  use, 136, 248, 385–386, 390–391
    communications, 66, 292, 241, 258, 295
      energy source, 188, 331
      nuclear fusion control, 289, 358
      tracking, 130–131, 258
LASL. See Los Alamos Scientific Laboratory.
Latham, Dr. Gary V., 14, 180–181, 219–220, 237, 380, 431
Lathram, Ernest H., 124
Launch Complex 14 (KSC), 61
Launch Complex 34, 169–170
Launch Complex 37, 169–170
Launch Complex 39, 46, 139–140, 142, 240, 301
Launch vehicle. See individual launch vehicles and stages, such as Agena, Centaur, Europa, Saturn, Scout, Thor-Delta, Titan, etc.
Law Enforcement Assistance Administration, 126
Lawrence, Kan., 330
Lawrence Livermore Laboratory, 155, 165
Lawsuit, 91, 122, 411
Lear, William P., Sr., 214
Lee, Chester M., 96, 366
Leesburg, Fla., 136
Lefevre, Minister of Science Theo (Belgium), 437–438
Lehn & Fink Products Co., 220
LeMay, Gen. Curtis E. (USAF, Ret.), 353
Lenin Prize, 442
Leningrad, U.S.S.R., 349
Lenoir, Dr. William B., 23, 219
Leo (constellation), 446
Leonidov, L., 253
LeRC. See Lewis Research Center.
Levi-Strauss, Dr. Claude, 405
Levine, Joseph H., 225

Lewis and Clark College, 400–401
Lewis Research Center (LeRC) (NASA)
  air pollution, 50, 154, 266–267, 391, 392
  aircraft research, 66–67, 78, 349, 355
  awards and honors, 16, 323–324, 380–381, 403
  contract, 66, 367
  electrophysics research program, 79
  energy conversion, 251
  Environmental Research Office, 94, 391
  meeting, 154–155, 185–186
  personnel, 115, 186, 244
  quiet engine program, 124, 130–131, 186, 387
Ley, Willy, 211
Library of Congress
  Congressional Research Service, 28, 318, 427
  Science Policy Research Div., 7, 166
Licensing, 248, 251
Life Beyond Earth and the Human Mind (symposium), 387–388
Life, origin of, 158
Life Scientist Program. See NASA Life Scientist Program.
Life support system, 129, 187, 194–195, 293, 298
Lifting body
  HL–10, 312
  M2, 312
  M2–F3, 271, 289, 299, 315, 327–328, 341, 356, 371, 380, 390, 398, 411, 427, 437, 449
  X–24A, 312
  X–24B, 347
Light, speed of, 390–391
Light baffle, 295–296
Lighthill, Sir James M., 299–300
Lilienthal, Otto, 275
Lilly, William E., 443
Lima, Agriculture Minister Cirne (Brazil), 42
Lind, Dr. Don L., 23
Lindbergh, B/G Charles A. (USAF, Ret.), 274, 343
Linder, Clarence H., 397
Ling-Temco-Vought, Inc. See LTV Aerospace Corp.
Lipari Islands, 228
Liquid hydrogen, 58, 224, 272, 304, 407
Liquid oxygen, 272, 304, 407
Little, Arthur D., Inc., 123, 251, 276–277, 323
Littleton, Colo., 347
Littrow Crater (moon), 54
LLRV. See Lunar landing research vehicle.
LLTV. See Lunar landing training vehicle.
LM. See Lunar module.
Lockheed Aircraft Corp.
  aircraft, 15, 110, 288, 298, 307, 422
  communications satellite system, 309, 338

contract, 247, 284–285, 288, 349
meeting, 173
overpayment, 120
personnel, 430
Lockheed-Georgia Co., 154, 251, 387
Lockheed Missiles & Space Co., 53, 70, 158, 208, 310
    contract, 91, 180, 261, 283, 337, 339, 386–387
Lockheed Propulsion Co., 35
LOI. See Lunar orbit insertion.
Lok Sabha (Indian parliament), 138
Loki-Datasonde (sounding rocket), 105
Lomask, Milton, 75
Lombardo Mint, 264
Lomonsov Gold Medal, 2
London *Financial Times*, 60
London, U.K., 101, 121, 134, 188, 288, 310, 339
London, Univ. of, University College, 257, 296
Long Island Assn. of Commerce and Industry, 220
Long-range laser traversing system, 138
Long, Sen. Russell B., 410
Loomis, Henry, 391–392
Lord, Douglas R., 398
Los Alamos, N. Mex., 132–133
Los Alamos Scientific Laboratory (LASL), N. Mex., 182–183, 254
Los Angeles, Calif., 152, 158, 247, 291
Los Angeles Air Pollution Control District, 35
Losey, Robert M., Award, 20–21
Loudon, James A., 121, 185
Louisiana State Univ., 282
Lousma, Maj. Jack R. (USMC), 23
Lovelace, Dr. Alan M., 308, 372–373
Lovell, Capt. James A., Jr. (USN), 137, 345, 392, 430
Low-altitude parachute-extraction system (LAPES), 319
Low-energy proton electron differential energy analyzer, 115
Low, Dr. Frank J., 433
Low, Dr. George M.
    astronaut commercialism, 256, 282–283
    NASA budget testimony, 100
    meeting, 2–4, 130
    press conference, 257, 435
    space program, 188, 293, 328, 386–387
    space shuttle, 5, 32, 59, 102, 136–137, 139–140
    U.S.–U.S.S.R. cooperation, 229, 239–240
Low-light-level TV, 254, 393
Lowell, Dr. Percival, 396–397
Lowey/Snaith, Inc., 42
Lowman, Dr. Paul D., 313
Lowry, John, 156
LRV. See Lunar roving vehicle.
LSI. See Lunar Science Institute.
LST. See Large Space Telescope.
LTV Aerospace Corp., 132, 214, 281
    Hampton Technical Center, 210
    Service Technology Corp., 179

Vought Missiles & Space Co., 210
Lubbock, Tex., 181
Lubrasky, Kronid A., 368
Lucas, Dr. William R., 380–381
Lucerne, Switzerland, 247, 313, 455
Luckman, Charles, Associates, 367
Ludwig, Dr. George H., 325
Luebeck Airport, West Germany, 161
Luest, Dr. Reimar, 251, 345
Luftwaffe, 6
Lukasik, Stephen J., 247
Luna (U.S.S.R. lunar probe), 7, 108, 449
Luna 3, 237
Luna 16, 13–14, 33, 52–53, 71, 179, 291, 299–300
Luna 17, 53
Luna 18, 52–53, 166
Luna 19, 38, 46, 52–53, 94, 337
Luna 20, 52–53, 71, 86, 138, 179, 257, 452
Lunar. See Moon.
Lunar atmospheric composition experiment, 366
Lunar International Laboratory Symposium, 5th, 344
Lunar landing research vehicle (LLRV), 311
Lunar landing training vehicle (LLTV), 311, 384
Lunar module (LM), 54, 170, 200, 253, 266, 432
    Apollo 11 (Eagle), 313
    Apollo 14 (Antares), 194
    Apollo 15 (Falcon), 194
    Apollo 16 (Orion), 2, 8, 14, 41, 62–63, 142–144, 146–148
    Apollo 17 (Challenger)(LM-12), 54, 297, 367, 382, 389, 404, 413–420, 428, 431, 432
    EVA, 418–419
Lunar Nomenclature Committee, International Astronomical Union, 168
Lunar orbit insertion (LOI), 144–146, 413
Lunar roving vehicle (LRV) (Rover) (see also *Lunokhod 1*), 432
    Apollo 15, 57–58, 99, 420
    Apollo 16, 97, 142, 146–149, 169, 172, 264
    Apollo 17, 54–55, 346–347, 413, 416–419, 429
    exhibit, 17–18, 72, 115, 319
    stamp, 157
Lunar sample. See Moon, surface sample.
Lunar Samples Analysis Planning Team, 215–216
Lunar Science Conference, First Annual, 13–14
Lunar Science Conference, Third Annual, 13–14, 26, 69, 215–216, 452
Lunar Science Institute, 13–14, 368, 452
Lundin, Bruce T., 115, 403
Lunite, 385
Lunney, Glynn S., 197–199, 209, 250, 252, 260, 346, 382

Lunokhod 1 (U.S.S.R. lunar surface explorer), 7, 13–14, 53, 128, 166, 228, 256, 337
Lyman-alpha light, 179–180

M

M2 (lifting body), 312
M2-F3, 449
  test, 271, 289, 299, 327, 341, 356, 371, 379, 390, 398, 411, 427, 437
  failure, 315
M33 (galaxy), 365
M-100 (meteorological rocket), 238
McAvoy, Glenn L., 356
McCandless, L/Cdr Bruce, II (USN), 23, 219
McConnell, Joseph H., 370
McCulloch Electronics Corp., 178–179
McCurdy, Richard C., 88, 100, 109, 139, 380–381
McDivitt, B/G James A. (USAF, Ret.), 33, 142, 170, 197–198, 221, 303, 404
Macdonald, J. Ross, 277
MacDonald Observatory, 112
McDonnell Douglas Corp., 219, 248, 284–285, 298, 395, 443
  contract, 44, 72, 174, 180, 261, 292, 382–383
  employment, 281, 319
  McDonnell Douglas Aircraft Co., 241, 266, 341, 342, 349
  McDonnell Douglas Astronautics Co., 39, 180, 219, 266, 281, 309
  Skylab, 39, 209, 325
  space shuttle, 303–304, 319, 339
McDonnell Planetarium, 400
McDonnell, Sanford N., 281
McDowell, Edwin, 156
McElroy, Neil H., 399
McElroy, Dr. William D., 37
McGarvey, Patrick J., 357
McGovern, Sen. George S., 200–201, 310, 314, 358–359
McGuire AFB, N.J., 396
Mackay Trophy, 263
McKee, Col. Daniel D., 20
McLeavy, Roy, 381–382
MacLeod, Dr. Norman H., 17
McMurtry, Thomas C., 73, 94, 104, 109, 113, 123, 155, 163, 184, 186, 193, 274
McNamara, Robert S., 39, 218–219
McNickle, L/G Marvin L. (USAF), 410
Macomber, Frank, 181–182
Macon (airship), 400
Madrid, Spain, 178, 189–190, 310
MAGLEV. See Magnetic levitation system.
Magnetic aspect sensor, 372
Magnetic field, 174, 177–178, 182–183, 183–184, 387
Magnetic levitation system (MAGLEV), 392
Magnetic shells, 291
Magnetohydrodynamics, 199, 236–237
Magnetometer, 50, 152
Magnetosphere (earth), 46, 184, 233, 242
Magruder, William M., 93–94, 192–193, 271
Maickel, Dr. Roger P., 74
Mailer, Norman, 119, 409
Maine, 255
Maine, Gulf of, 250–251
Malaga, Joseph E., 443
Malik, Yakov A., 348
Malina, Dr. Frank J., 233, 344
Mamet, Bernard, 376
Managua, Nicaragua, 442
Manganiello, Eugene J., 16
Manhattan, N.Y., 122
Manitoba, Canada, 394
Manke, John A., 356, 371, 390, 437
Mankind 1 (private space probe), 324–325
Manned space flight (see also Apollo, Apollo-Soyuz Test Project, Astronaut, Cosmonaut, Gemini, Mercury, Salyut, Skylab, Soyuz, Space biology and medicine, Space shuttle, etc.)
  achievements, 85–86, 133, 265, 345, 420, 426–427, 428, 435, 438, 439
  anniversary, 265
  art display, 17–18
  benefits, 151, 231–232
  capability, 32
  cooperation, 195–196, 198–199, 201, 207, 239, 343–344
  cost, 71–72, 424
  credibility, 271
  earth orbit emphasis, 58
  employment, 297–298, 301
  funding, 30
  impact, 338, 429
  justification, 166–167
  long duration, 68, 248, 308, 322, 347–348
  lunar landing. See Moon, exploration, manned.
  nuclear power, 130–131
  policy and plans, 2–4, 28, 36, 56–57, 111–112, 162, 205, 262, 266–267, 335, 420
  press conference, 285–286
  R&D, 64
  remotely manned systems, 316
  safety, 195–196, 207, 227, 372–373
  significance, 192–193
  statistics, 33, 102
  technology utilization, 244–245
  training, 239–240, 322
  U.S.S.R., 138, 166, 167, 353
  world leadership, 175–176
Manned Spacecraft Center (MSC) (NASA), 228–229, 239, 326, 382, 383, 384, 394, 395
  accident, 73, 179, 235–236
  aircraft optical communications system, 292–293
  altitude test chamber, 301
  anniversary, 265
  Apollo 15 mission, 50, 321

Apollo 16 mission, 96, 109, 164, 166, 170–171, 172, 179–180, 183–184, 237
Apollo 17 mission, 182, 403, 405, 431, 433–434, 435
Apollo spacecraft, 149–150, 170, 420
Apollo-Soyuz Test Project, 250, 252, 346
Apollo Telescope Mount, 44
Astronaut Office, 194, 303, 314
astronaut training, 235–236
astronauts at, 321
AVLOC flight tests operation, 292–293
award, 114
Bioengineering Systems Div., 132
Biomedical Research Div., 132
capabilities, 78
Cellular Analytical Laboratory, 347
computerized inventory system, 21
contract, 60, 132, 221–222, 256, 281, 283
  space shuttle, 34, 180, 220, 254–255, 261
    cargo handling system, 301
    crew compartment, 271–272
    engine, 90
    heat protection, 203
    materials, 55, 91, 214
    orbiter, 42, 45, 47, 191, 214, 236, 266
    safety, 220
    simulator, 174, 247, 266
    waste collection system, 277
Contractor Equality Opportunity Programs Office, 60
cooperation, 261
Earth Resources Program Office, 173
Earth Resources Program Review. Fourth Annual, 21
employment, 78, 87, 297–298, 301, 451
flight crew operations, 172, 205, 232, 384
Geophysics Branch, 152, 183–184
Health Services Div., 132
Life Sciences Directorate, 132
Life Scientist Program, 298
Lunar Receiving Laboratory, 368
Management Analysis Office, 204
Medical Research and Operations Directorate, 132, 372–373
meeting, 120, 180, 282, 310, 322
Mission Control, 187
personnel
  appointments, 80, 141, 172, 173, 204, 248, 273
  ASTP management, 197
  award, 131–132, 210
  former, 71
  reassignment, 302
  reorganization, 132
  resignation, 205
  retirement, 29
Planetary and Earth Sciences Div., 170–171, 183–184, 238, 380–381
press conference
  Apollo 16 mission, 109, 164, 170–171, 172, 179–180, 183–184, 237
  Apollo 17 mission, 403, 431, 433–434, 435
Apollo-Soyuz Test Project, 257–258, 260
lunar science, 152, 363–364, 445
Skylab, 23–24
space program, 424
proposals, 5, 15, 37, 262
reorganization, 132
Science and Applications Directorate, 205, 392
Skylab, 23–24, 44, 173, 261, 262, 273
sounding rocket experiments, 237
space biology and medicine, 65–66, 187, 279, 298, 347–348
space shuttle, 34, 42, 45, 47, 55, 90, 91, 174, 180, 191, 194, 203, 214, 220, 236, 247, 254–255, 261, 266, 271–272, 277, 281, 283, 293, 301, 383–384
spacesuit, 262
Urban Systems Project Office, 131
visit to, 60, 310, 365
Manned Spacecraft Recovery Forces (USN), 312–313
Manpower and Financial Resources Allocated to Academic Science and Engineering Activities, 1965–71 (NSF 72–302), 112
Manufacturers Aircraft Assn., 122
Mapping, 1, 14, 21, 39, 235, 380
Marconi Co. Ltd.
  Space and Defence Systems Div., 446
Mare Crisium (Sea of Crises) (moon), 179, 433
Mare Fecunditatis (Sea of Fertility) (moon), 53, 179
Mare Imbrium (Sea of Rains) (moon), 14, 377–378
Mare Nubium (Sea of Clouds) (moon), 404
Mare Serenitatis (Sea of Serenity) (moon), 54, 363, 367, 433
Mare Tranquillitatis (Sea of Tranquility) (moon), 265
MARECEBO. See International Academy of Astronautics, Manned Research on Celestial Bodies Committee.
Mariner (program), 65, 112, 266, 403, 408
Mariner (spacecraft), 56
Mariner 2 (Venus probe), 429
Mariner 4 (Mars probe), 225–226, 429
Mariner 5 (Venus probe), 429
Mariner 6 (Mars probe), 429
Mariner 7, 43, 429
Mariner 8 (Mariner-H), 51–52, 364
Mariner 9 (Mariner-I), 117, 364
  achievements, 56, 447
  award, 310
  computer malfunction, 109, 117
  data exchange, 452
  Deep Space Network support, 114
  dust particle composition, 113
  mission profile, 51–52
  photographs, 6, 15, 42–43, 90–91, 117, 221, 225–226, 309, 396–397
  canals, 225–226, 244, 342

mapping, 1, 253, 399
  possible landing sites, 376–377
  water ice, 178
press conference, 42–43
radio signal test, 253
results, 342, 369, 408, 429, 447
solar occultation, 132, 199–200
Mariner 10 (Venus/Mercury probe), 307–308
Mariner Jupiter/Saturn 1977 (MJS 77) mission, 65, 266, 320, 324, 423
Mariner Mars 1971 Project, 274
Mariner Mars mission, 51–52, 408
Mariner Venus/Mercury (spacecraft), 30
Maritime Administration, 281
Mark, Dr. Hans, 324, 380–381
Mark, Dr. Herman, 326
Mark II (quiet experimental engine), 186
"Marmes Man," 54
Marquardt Co. See CCI Aerospace Corp.
Mars (planet) (see also Mariner; *Mars 2, and 3*; Viking), 99, 221, 289
  asteroid belt proximity, 259
  atmosphere, 43, 56, 113, 176, 225–226, 244, 299, 369
  canals, 42–43
  crater, 42–43, 244
  data, 444
  dust, 42–43, 51–52, 113, 252–253, 273, 342, 364, 369
  experiments, 101
  exploration, 151, 291, 407, 408
    manned, 27, 335, 343
    remotely manned systems, 316
    U.S.S.R., 369
  extraterrestrial life, 56, 112, 376–377
  gravity, 43
  Great Galactic Ghoul, 200
  international cooperation, 178, 189, 228, 310–311
  landing site, 376–377
  life-detection robot, 96
  mapping, 1, 90–91, 117, 253, 396–397, 399
  moon, 56
  orbit, 291
  oxygen, 112
  photographs, 6, 15, 56, 90–91, 114, 117, 253
  polar cap, 43, 273, 309
  softlanding, 56, 86
  space environment model, 441
  space shuttle, 121
  spectrometer data, 56
  surface, 56, 178, 225–226, 244, 313, 364, 369
  unmanned landing, 131, 163
  upper mantle, 273
  volcanism, 42–43, 369
  water, 43
  wind, 42–43
Mars (U.S.S.R. Mars probe), 108
Mars 2, 60, 166, 299, 337–338
  achievements, 86
  data exchange, 452

objectives, 28
photographs, 28, 90
results, 10–11, 38, 113, 369
Mars 3, 60, 166, 338
  data exchange, 452
  landing, 167
  objectives, 28
  photographic equipment, 90
  photographs, 28
  results, 10–11, 38, 113, 369
  softlanding, 86
Marshall Space Flight Center (MSFC) (NASA), 223–224, 239, 374, 380–381, 384, 400
  Aero-Astrodynamics Laboratory, 454
  Apollo 17 mission, 413
  Apollo Telescope Mount, 44
  Astrionics Laboratory, 54–46, 393
  Astronautics Laboratory, 44, 68
  awards, 238
  budget, 16
  capabilities, 78
  CAT detection system, 302
  contract, 66, 95, 227
    HEAO, 163, 264
    Skylab, 39, 288
    space shuttle, 35, 45, 90
    space tug, 129
  Contracts Office, 243–244
  cooperation, 184, 261
  earth resources, 294
  employment, 129, 182, 242–243, 255, 297–298, 451
  Environmental Applications Office, 294
  facilities, 184
  flight tests, 292–293
  HEAO, 163, 243, 264
  LST, 243, 302
  materials technology experiment, 337
  meeting, 74, 135, 177, 292, 346
  organization, 451
  personnel, 94, 181, 205, 451
    appointment, 302, 440
    death, 337
    retirement, 440
  press conference, 386–387
  Quality and Reliability Assurance Laboratory, 44
  reorganization, 243–244, 284
  request for proposals, 129, 241, 390
  Saturn V launch vehicle, 33, 149–150, 420
  Shuttle Construction Office, 302
  Shuttle Program Office, 110
  Skylab, 39, 60, 68, 166, 280, 288, 309
  solar-cell-array system, 169
  solar eclipse observations, 254
  Sortie Can Task Team, 129
  "Space and Science Awareness Month," 72
  space flight orientation course, 277
  Space Science Projects Office, 243–244
  space shuttle, 35, 45, 90, 309, 427
  technology utilization, 45, 309
  telecommunications technology, 356
  visits to, 60

Martin Marietta Corp., 153, 261, 271–272, 290, 295, 301, 347, 367, 428
Martin Stamping and Stove Co., 309
Maryland, 49, 131, 136, 190, 235, 244, 355, 377–378
Maryland, Univ. of, 210
Maser, 287
Mass spectrometer, 294
Mass transit, 1, 31, 206
Massachusetts, 259, 385
Massachusetts Institute of Technology (MIT), 5–6, 313, 374, 378–379, 421
   contract, 264
   Measurement Systems Laboratory, 259
   personnel, 174, 180, 205
   sounding rocket experiment, 189, 288, 319
MASSPEC (lunar atmospheric composition experiment), 366
Masursky, Harold, 15, 42–43, 225–226, 342, 433
Matagorda Island, 25, 214
Materials technology, 179, 210–211, 309, 323–324, 337, 368
   aircraft, 59, 154
   space shuttle, 55, 79, 203, 294
Mathematica, Inc., 39, 58, 137, 181, 215
Mathews, Charles W., 21, 84, 91, 114, 171, 226–227, 451
Mattingly, Cdr. Thomas K., II (USN), 228–229
   Apollo 16 mission, 144–146, 148–149
   countdown demonstration test, 123
   debriefing, 180
   experiments, 6
   launch, 142
   medical aspects, 120, 135–136, 150, 166, 420
   press conference, 6, 109, 172
   U.S.S.R. reaction, 163–164
   assignment, 194
   awards and honors, 187, 264, 265–266
   congressional reception, 184
   promotion, 162–163, 172
   public appearance, 165
Maverick (missile), 116
Max Planck Institute of Nuclear Physics, 345
Max Planck Society for Advancement of the Sciences, 251
Maxim Gorky (U.S.S.R. transport aircraft), 442
May, Dr. Michael M., 341
Mayer, Dr. Jean, 13
Mayer, Dr. Maria G., 61
Mazelspoort, South Africa, 181
MCC: midcourse correction.
MCI–Lockheed Satellite Corp., 309
MDA. See Multiple docking adapter.
Mead, Dr. Margaret, 405
Meck Island, 108, 361
Meckling, Dr. William H., 282
Medal for Outstanding Leadership (NASA), 202–203
Medal of Freedom, 399
Mediterranean Sea, 228

MEED. See Microbial ecological evaluation device.
Meir, Prime Minister Golda (Israel), 273
Melbourne, Fla., 138
Memorandum of Agreement, 248, 251
Memorandum of Implementation of Environmental Agreement, 323
Memorandum of Understanding, 49–50, 77, 170, 389, 431
Memphis, Tenn., 72
Mendonca, Dr. Fernando, 173
Menlo Park, Calif., 392
Menon, M. G. K., 75
Mercator projection, 396–397
Mercure (European short-range transport), 94
Mercury (planet) (see also Mariner-Venus/Mercury spacecraft), 56, 65, 307–308
Mercury (program), 61, 205, 345, 424, 435
Meredith, Dennis L., 429
Merkulov, I. A., 345
Merritt Island, Fla., 41
Merritt Island National Wildlife Refuge, 216
Meson Physics Facility (AEC), 132–133
Messerschmitt-Boelkow-Blohm, 304, 333, 377
Meteor, 372
Meteor (U.S.S.R. meteorological satellite), 87, 449
Meteor 11, 123
Meteor 12, 243
Meteor 13, 363
Meteor shower (Giacobini), 341, 343, 348
Meteorite, 14, 63, 132, 152, 158, 203
   detection, 38
   lunar impact, 54–55, 62, 180–181, 219–220, 237
Meteoroid, 81–82, 155, 200, 226, 254, 260, 290, 409–410
Meteoroid Technology Satellite (MTS). See *Explorer 46*.
Meteorological probe, 113
Meteorological rockets, 105
Meteorological satellite (see also individual meteorological satellites, such as *Nimbus 5*, *Noaa 1*, etc.), 24–25, 154, 164–165, 172–173, 229, 243, 306, 378
   achievements, 70
   cost, 136–137
   data output, 34–35
   funding, 31
   instrumentation, 27
   technology transfer, 24
   tracking station, 42
   U.S.S.R., 87, 123
Meteorology (see also National Weather Service), 25, 403
   DOT–NASA–NOAA cooperation, 203
   experiments, 24, 77, 101, 229
   fog stabilization, 327
   international cooperation, 105, 195–196, 216
   Joint Hail Research Experiment, 118
   lunar composition, 215–216

Nixon, President Richard M., 172–173
solar effects, 243
television monitoring, 93
U.S.S.R. program, 87, 363, 403
weather forecasting, 84, 174, 407, 423
weather modification, 84, 236–237, 247, 250, 375, 408
Metric system, 93, 182, 295, 297
mev: million electron volts.
Mexico, 162–163, 230, 270
MHD. See Magnetohydrodynamics.
mhz: megahertz (one million cycles per second).
Miami, Fla., 1, 162, 242
Michael Reese Hospital, Chicago, Ill., 187
Michigan, 16
Michigan, Univ. of, 125, 185, 186–187, 220
Michlovic, Joseph, 181
Michoud Assembly Facility (MSFC) (NASA) 209, 227, 266, 292
Microbial ecology evaluation device (MEED), 6, 96–97, 148–149
Micrometeorite, 385–386
Micrometeoroid, 68, 394, 403, 410
Microwave Communications, Inc., 309, 338
Microwave instrument landing system (MILS), 20, 141–142
Microwave landing system (MLS) (see also MLS Interagency Advisory Group; MLS Project Office, FAA), 20, 35, 67
Middle East, 310
MiG-23 (Foxbat) (U.S.S.R. fighter aircraft), 395
Mikoyan Fearless (U.S.S.R. fighter aircraft), 395
Milch, Erhard, 6
Military Aircraft Programs Office, OAST, 67
Military Airlift Command, 319, 396
Military Assistance Service, 240
Military Balance 1972-73, 310
Milky Way, 104, 271, 332, 357, 365
Miller, Rep. George P., 41, 219–220, 231, 286–287, 340, 387, 430
Miller, William H., 244
Millionschikov, M. D., 86
Mills, Dr. Ralph F., 266
MILS. See Microwave instrument landing system.
Minnesota, Univ. of, 37, 264, 294, 327, 358, 443
Minority Business Development Program, 247
Minority Business Enterprise Program, 60
Minuteman (missile), 72, 108, 115, 191
Mirage 3 (French supersonic fighter aircraft), 177
MIRV. See Multiple independently targetable reentry vehicle.
Missile, 16, 174, 185, 191, 202–203, 230
air-to-surface (ASM), 389
antiballistic (ABM), 13, 30–31, 174, 201–202, 300, 361, 411
Safeguard, 31, 72, 108
ballistic, 39, 71, 202–203, 205, 319–320, 397
budget, 327
cost, 263
evasion technology, 65
foreign
France, 259
People's Republic of China, 39, 310, 377
U.S.S.R., 3, 35, 85, 87, 89, 115, 177, 210, 328–329, 350, 355, 445
guidance systems, 403
intercontinental ballistic (ICBM), 61–62, 69, 87, 89, 115, 361
MIRV, 221
sales, 453
submarine-launched (SLM), 24–25, 210
test, 174, 300, 361, 377, 397, 411, 445
treaty, 213, 224–225, 331
undersea long-range missile system (ULMS), 70, 185
warning, 233–234
Mission and Payload Integration Office (OMSF), 398
Mission Control (MSC), 144–146, 148, 174–175, 187, 265
Mississippi, 74
Mississippi Test Facility (MTF) (NASA), 302
Missouri, 118, 241, 335, 341, 395, 400
MIT. See Massachusetts Institute of Technology.
Mitchell, Capt. Edgar D. (USN, Ret.), 188, 194, 205, 335, 337, 372–373, 409
Mitchell, Edgar D., & Associates, Inc., 335
Mitchell, B/G William (Billy) (USA), 343
Mitre Corp., 396
Mittauer, Richard T., 52
MIUS. See Modular-sized integrated utility system.
MLS. See Microwave landing system.
MLS Interagency Advisory Group, 67
MLS Project Office (FAA), 67
Mobile launcher (KSC), 298
Modular-sized integrated utility system (MIUS), 131
Moffett Field Laboratory (ARC), 223
Molecular hydrogen, 270–271
Molniya I (U.S.S.R. comsat series), 449
Molniya I-20, 128
Molniya I-21, 350
Molniya I-22, 404
Molniya II, 449
Molniya II-2, 188
Molniya II-3, 331
Molniya II-4, 426
Mondale, Sen. Walter F., 6–7, 177, 178, 215
Mongolia, 132
Monitoring Earth Resources from Aircraft and Spacecraft (NASA SP-275), 118
Montagu, Dr. Ashley, 387–388

Montana State Low Income Organization, 157
Monterey, Calif., 280–281, 313
Montreal, Canada, 49, 318
Moon, 88, 99, 112, 113, 123, 372, 394
  age, 203
  atmosphere, 52, 54–55, 306, 433
  base, 63, 406
  colonization, 26
  composition, 219–220, 238, 432
  Conquest of the Moon, 211
  cosmic ray detector, 50, 417
  crater (see also specific craters such as Apollonius, Copernicus, Littrow, etc.), 144–146, 148, 165, 168, 215–216, 432
    landing site, 142
    magnetism survey, 152
    photographs, 46, 169
    soil sample, 416
    topography, 403, 404, 433
    traverse, 416
    volcanic domes, 417
  crust, 219–220
  debris, 533
  earth tidal pull, 54–55
  eclipse, 38
  evolution, 152, 377–378, 385–386, 420
  experiments, 54–55, 101, 102, 346–347, 366, 380, 388, 432
    ALSEP, 54–55, 163, 416–417, 419
    electrical properties, 416
    geocorona, 306
    gravimeter, 54–55, 417, 432
    grenade, 193
    heat flow, 54–55, 163, 417, 440
    infrared scanning radiometer, 54–55, 433
    laser sounder, 54–55
    magnetometer, 50, 88, 152
    orbital science, 54–55, 62
    seismometer, 54–55, 146–148, 180–181, 219–220, 237, 431
    sounder, 433
    spectrograph, 306, 433
    spectrometer, 54–55
  exploration
    astronomical contribution, 151
    manned
      Apollo program
        experimental objectives, 442
        results, 69, 363–364
      Apollo 11, 265, 313, 385
      Apollo 14, 219–220
      Apollo 15, 57–58, 168, 215–216, 232, 279, 282–283
      Apollo 16, 41, 71, 112, 142, 148, 165, 172, 184
        experiments, 306
        photographs, 169
        results, 184
        schedule, 62
        soil sample, 326
      Apollo 17, 194, 366, 404, 413–419, 428
        achievements, 447
        experiments, 346–347, 388
        press comment, 404–405, 409, 426–427, 438
        schedule, 297
        surface samples, 428–429
        television coverage, 381, 391–392
        U.S.S.R. reaction, 421, 423
    commercialism, 317–318
    cooperation, international, 134, 174, 177, 189, 195–196, 231–232, 306
    credibility, 271
    future, 406, 407
    impact, 404–405, 429
    monopolization, 426
    Nixon, President Richard M., statement, 435–437
    priority, 162
    significance, 192–193, 406, 420
    statistics, 265, 345
    U.S.S.R., 175–176
    Vanguard satellite, 75
  unmanned, 423
    Apollo 16 Subsatellite, 208
    international cooperation, 134
    laser beam transmission, 181
    Luna 16, 179, 291
    Luna 19, 38, 46, 94, 237, 337
    Luna 20, 52–53, 179
    Ranger 9, 62
    SNAP-27, 388
    U.S.S.R., 7, 86
  "Genesis bean," 14
  gravity, 94, 148–149, 404
  interior, 417, 431, 432, 440–441
  laboratory, 344
  landing, 416
    crash, 53, 62, 166
  landing site
    Apollo 14, 180–181, 203
    Apollo 15, 69, 168
    Apollo 16, 14, 62–63, 71, 142, 144–146, 170–171, 183–184
    Apollo 17, 54, 297, 363, 416, 428–429
    Luna 20, 53, 138, 179
  lunar roving vehicle. See Lunar roving vehicle.
  lunar science, 368
  magnetism, 13–14, 69, 152, 165, 183–184
  mantle, 219–220
  mapping, 14, 39, 53, 54–55, 177, 229, 305–306, 433
  mascon, 440
  meeting, 13–14, 26, 69, 215–216, 344, 452
  meteorite, 180–181, 380
  mountains, 62–63, 146–148
  origin of, 62, 69, 158, 165, 170–171, 215–216
  photographs
    Apollo 15, 168
    Apollo 16, 169, 170–171, 172, 179–180, 184, 237
    lunar orbit, 148, 149–150
    lunar surface, 62, 142–144, 148–149, 149–150

    Apollo 17, 416–417, 421, 433
    Haystack Observatory, 404
    Luna 3, 237
    Luna 19, 46, 337
    Luna 20, 53, 138
  plaque, 59
  radiation, 148
  radioactivity, 14, 69
  "Shadow Rock," 326
  shape, 62
  surface, 13–14, 39, 101, 219–220, 237, 337, 403, 431–432
  surface sample
    analysis, 152, 203, 330, 377–378
    Apollo 11, 33, 179, 291, 355
    Apollo 12, 179, 206
    Apollo 14, 13–14, 54–55, 203, 452
    Apollo 15
      analysis, 14, 35–36, 69, 168, 215–216
      exchange of, 138, 452
      "Genesis rock," 62, 146
    Apollo 16, 63, 142, 156
      analysis, 170–171, 179, 183–184, 237–238, 326
      collection. 62–63, 146–148, 149–150
      core sampling, 62
      exchange, 452
      press conference, 170–171, 183–184
      significance, 165
    Apollo 17, 54, 413, 419, 447
      analysis, 432, 445
      exchange, 388, 452
      significance, 428–429
      soil, 416, 417, 437, 440, 445
    exchange, 48–49, 452
    exhibit, 72, 142, 301, 400
    Luna 16, 13–14, 33, 53, 71, 179, 291, 299–300
    Luna 20, 53, 71, 138, 179
    soil, 52, 63
  temperature, 433
  tides, 237
  time, 329, 356
  thermal theories, 437
  treaty, 134, 356
  volcanic domes, 418
  volcanism, 14, 54–55, 62, 404, 417, 428, 431–432, 440, 445
  water, 83, 86
Moonquake, 237
"Moonwalk One" (documentary film), 385
Moonwalker. See *Lunokhod 1*.
Moore, Dr. H. J., 62
Moore, Dr. Stanford, 357
Moorer, Adm. Thomas H. (USN), 55–56, 71, 236
Morelli, Frank A., 171
Morgenstern, Dr. Oskar, 39, 137
Morison, Dr. Elting E., 316
Moriss, Frank, 95
Moritz, Bernard, 109, 188, 289
Morocco, 271
Moroz, V., 113

Morris, Owen G., 170
Morrison, Dr. Philip, 387–388
Morrison-Knudsen, Inc., 388
Morton, Dr. Louis, 316
Morton, Secretary of the Interior Rogers C. B., 380–381
Moscow, U.S.S.R., 18, 133, 185, 221, 243, 349, 352, 354
  ABM radar complex, 201–202
  agreements, 218, 224, 235, 265, 452
  Annual Revolution Day parade, 377
  Apollo 16 launch coverage, 151
  atomic energy reactor, 14–15
  ceremony, 134, 138, 299–300
  convention signing, 121
  exhibit, 115, 434
  Mars photographs, 28
  meeting, 130, 199, 236–237
    ASTP, 178, 192–193, 240, 252, 322, 346, 409
    biophysics, 291
    electronic industry, 201
    environment, 177, 193–194, 289, 298
    peaceful use of space, 323
    space biology and medicine, 180, 279
    space cooperation, 127, 177, 197–198
    summit, 190, 201–202, 203, 241
      accord, 193–194, 206, 207
      agreements, 195, 199, 201–202, 303, 374–375
  memorandum signing, 331
  Nixon, President Richard M., visit to, 191
  press conference, 199
  protocol signing, 216, 275, 341
  tour, 259
  treaty, 188, 224
  TV transmission, 426
  Venus 8 data published, 275
Moscow Univ., 291
Moshkin, Ye. K., 345
Mt. Holyoke College, 281
Mt. Kilauea, Hawaii, 77
Mt. Palomar Observatory, 208–209, 393
Mt. Wilson Observatory, Calif., 357
Mountain Home AFB, Idaho, 230
Mouse, 404
MSC. See Manned Spacecraft Center.
MSFC. See Marshall Space Flight Center.
MSS. See Multispectral scanner subsystem.
MT-135 (Japanese sounding rocket), 105
MTF. See Mississippi Test Facility.
MTS. See *Explorer 46* (Meteoroid Technology Satellite).
Mu-4S (Japanese booster), 264, 295
Muehlberger, Dr. William R., 170–171, 432
Muelhause, Carl M., 191
Mueller, Dr. George E., 343
Mulholland, Donald R., 314
Mullen, Dr. George, 252–253
Multiple docking adapter (MDA), 219, 262, 280, 282, 325, 341
Multiple independently targetable reentry vehicle (MIRV), 221, 229–230, 429

Multipurpose spacecraft (proposed), 256
Multispectral scanner subsystem (MSS), 235, 269–270, 284, 367, 377–378
Mundt, Sen. Karl E., 66
Munich, West Germany, 224–225, 318
Munk, Dr. Max M., 20–21
Muroc Flight Test Base, Calif., 311, 350, 449
Murray, Dr. Bruce C., 300–301, 342
Museum. See individual museums, such as American Museum–Hayden Planetarium.
Musgrave, Dr. Franklin Story, 23, 219
Myers, Dale D., 11–12, 57–58, 102, 198–199, 209, 241–242, 345, 398

N

NAA. See National Aeronautic Assn.
NACA. See National Advisory Committee for Aeronautics.
NAE. See National Academy of Engineering.
Nagy, Istvan G., 345
NAS. See National Academy of Sciences.
NAS–NRC Space Science Board. See National Academy of Sciences.
NASA. See National Aeronautics and Space Administration.
NASA Aerospace Safety Advisory Panel, 135, 234
NASA Authorization Act of 1968, 234
NASA Earth Resources Survey Program Weekly Abstracts, 356
NASA–HEW Site Selection Working Group, 375
NASA Historical Advisory Committee, 234, 316
NASA Key Personnel Development program, 17
NASA Life Scientist program, 74–75, 298
NASA Low Cost Evaluation Project, 386–387
NASA Office of Advanced Research and Technology (OART), 17, 21–22, 130–131
NASA Office of Aeronautics and Space Technology (OAST), 17, 21–22, 66–67, 84–85, 302, 349
  Aeronautical Operating Systems Office, 189
  JT3D/JT8D Refan Program Office, 176
  Military Aircraft Programs Office, 67
  personnel, 85, 185–186, 264
  Transport Experimental Program Office, 67
NASA Office of Applications (OA), 32, 177, 270, 331, 451
NASA Office of DOD and Interagency Affairs, 395–396
NASA Office of Equal Opportunity (OEO), 81, 188
NASA Office of Industry Affairs and Technology Utilization (OIATU), 270, 293
NASA Office of Life Sciences, 316

NASA Office of Manned Space Flight (OMSF), 64, 68, 141, 149–150, 398, 420
  Advanced Programs Office, 398
  Apollo/ASTP Program Office, 398
  Mission and Payload Integration Office, 398, 451
  Sortie Lab Task Force, 398
NASA Office of Public Affairs, 17, 52
NASA Office of Space Science (OSS), 24, 297, 307, 364, 379, 389, 395–396, 451
NASA Office of Supply and Equipment Management, 353
NASA Office of Tracking and Data Acquisition (OTDA), 130–131, 149–150, 420
NASA Patent Abstracts Bibliography (SP-7039), 299
NASA Quiet Engine Program, 185–186
NASA Regional Dissemination Centers, 100
NASA Research and Technology Advisory Council (RTAC), 234
NASA Site Review Board, 12, 139–140
NASA Space Program Advisory Council (SPAC), 140, 234, 381
NASA Space Shuttle Technology Conference, 137
NASA Space Systems Committee, 140
NASA Summer Seminar on History and Space, 246
NASA Tracking and Data Acquisition Panel, 234
NASC. See National Aeronautics and Space Council.
Nashville, Tenn., 335, 337
NASPO. See Federal Aviation Administration, National Airspace System Program Office.
NASTRAN (NASA's structural analysis computer program), 71, 452
Natal, Brazil, 92, 394
Natelson, Dr. Samuel, 187
National Academy of Engineering (NAE), 119, 299, 397
National Academy of Sciences (NAS), 39–40, 95, 127, 136, 155–156, 159–160, 183, 287, 299, 352, 354, 357, 374, 375, 392
  Advisory Committee to Air Force Systems Command, 140
  Astronomy Survey Committee, 213–214, 233, 287–288, 455
  Board on Science and Technology for International Development, 138
  Institute of Medicine, 299, 350
  Physics Survey Committee, 287
  Space Science Board, 39–40, 64, 65, 129, 153, 178
National Accelerator Laboratory, 94, 287, 314
National Advisory Committee for Aeronautics (NACA), 155–156, 223, 311, 350, 449
National Aeronautic Assn. (NAA), 114, 210, 326

National Aeronautics and Space Administration (NASA), 17–18, 18–19, 30, 32, 46–47, 53, 55, 59, 160, 168, 175–176, 176–177, 182, 187, 262, 292–293, 317–318, 384
   accident, 45–46, 73, 91, 153, 174, 178, 179
   accomplishments (see also Space program, national), 219, 305, 311–312, 345, 374, 406, 412, 447–449
   aircraft (see also individual aircraft, such as Convair 990; and Aircraft), 43–44, 45, 50–51, 128, 155, 163, 169, 184, 186, 193, 200, 210, 215–216, 235, 280, 302, 303, 311, 312, 336–337, 341–342, 349, 392, 396
   anniversary, 33, 61, 90, 217, 265, 311, 328, 394, 429, 449
   Apollo 14 commemorative medal, 277
   Apollo 15 postal covers, 256, 258, 259, 262, 267, 270, 273, 276, 277
   astronaut. See Astronaut.
   autograph policy, 384
   awards and honors, 69, 88, 139, 202–203, 238, 241, 247, 310, 311, 323, 384, 443, 444
      Annual Awards Ceremony, 380–381
      astronaut, 187, 200, 206, 210, 315, 372–373
      Distinguished Public Service Medal, 381
      Distinguished Service Medal, 29, 202–203, 310, 381
      Exceptional Scientific Achievement Medal, 270–271, 381
      Exceptional Service Medal, 381, 443, 444
      Goddard Memorial Trophy, 98
      Group Achievement Awards, 381
      Outstanding Handicapped Federal Employee, 131–132
      Outstanding Leadership Medal, 202–203, 381
      Space Science Award, 20–21
   budget, FY 1973, 5, 11, 32, 70, 86, 125, 134, 160, 167–168, 171, 173, 185, 187–188, 198–199, 209, 214–215, 219, 245, 297–298, 314, 315, 328, 374, 393, 410, 450, 451, 452
   bill introduced, 30
   bills passed, 115–116, 164, 189, 207, 226
   bills signed, 189, 207, 231, 291
   House consideration
      appropriations, 21–22, 89–90, 187–188, 193
      authorization, 41, 46–47, 56–58, 59, 63–64, 66–68, 73–74, 78, 84–85, 91, 109–110, 135, 155, 160, 164–165
   impoundment, 350
   joint conference consideration, 273–274
   Senate consideration
      appropriations, 136–137, 209
      authorization, 99–100, 102, 106, 107, 114, 130–131, 137, 140, 160, 172, 178
   contract, 63–64, 179, 224, 234, 274, 378
   Aerodynamic Test Range sites support, 264
   aeronautics, 154, 216–217, 231, 292, 382
   Apollo 16 mission, 156
   Apollo-Soyuz Test Project study, 61
   ATS-E, 110
   ATS-G, 66, 295
   base support services, 44
   building-ground maintenance, 179
   communications, 238–239, 394
   construction of facilities, 60
   data processing, 372
   docking system, 378
   Dual Air Density Explorer, 294
   esthesiometer, 184–185
   HEAO, 243, 264, 342
   IMBLMS, 15, 283
   International Ultraviolet Explorer, 208
   long-range laser traversing system, 138
   low-cost-design concept, 386–387
   multipurpose spacecraft, 256
   multispectral camera systems, 221–222
   performance practices, 179
   photographic support, 17
   public sector problems, 73–74
   publications, 179
   quadraplegic aids, 95
   safety criteria study, 220
   satellite energy applications, 251
   Saturn V launch vehicle, 227
   Scout launch vehicle, 210
   ship communication and navigation system, 281
   Skylab, 39, 44, 288, 434
   solar cell, 1
   space shuttle, 280
      checkout system, 33
      crew compartment, 271–272
      design criteria and study, 254–255, 284–285, 383–384
      economic analysis report, 39
      GAO review, 123
      hardware, 12
      heating system, 37, 266
      helium regulator systems, 261
      landing site survey, 240
      mission simulator, 174
      orbital maneuvering system (OMS), 266
      orbiter, 42, 45, 47, 180, 191, 214, 220, 247, 266, 272, 283
      passenger couch, 261
      payload systems, 220
      polymer seal materials, 55
      propulsion system, 35, 45, 90, 129, 180, 292
      RCS, 191
      RSI, 261

safety criteria, 220
simulator requirements, 266
subsonic flight, 247
technical proposals, 180
thermal protection, 203, 220
waste collection system, 277, 293
weld-bonding materials, 91
space tug, 390
space-walk studies, 132
spacecraft systems, 294-295
SST, 349
STOL aircraft, 2, 172, 231
support services, 17, 44, 248, 293
Synchronous Meteorological Satellite, 171
technology utilization, 5
television transmission, 156
Viking spacecraft encapsulation, 388
waste-incinerator system, 194-195
water-recovery systems, 1, 194-195
cooperation, 23, 25, 71, 226-227, 265, 270, 393, 452
AEC, 79, 131, 169-170, 194-195, 451
Agriculture, Dept. of, 67-68, 169-170, 301-302
Air Force Military Airlift Command, 396
Arizona, 235
Army Corps of Engineers, 184, 301-302
Boston Univ., 387-388
Bureau of Mines, 67-68
CAB, 22
Cleveland, Ohio, 391
DOC, 22, 169-170, 301-302, 352
DOD, 8, 16, 22, 40, 66-67, 116, 139-140, 194-195, 301-302, 306-307
DOT, 6, 22, 29, 43, 73-74, 160, 176, 203
EPA, 73-74, 301-302, 390
FAA, 66-67, 74, 104, 136, 176
Geological Survey, 235
HEW, 27, 45, 67-68, 73-74, 169-170, 375
HUD, 73-74, 131
Interior, Dept. of, 1, 73-74, 169-170, 216, 235, 301-302
Justice, Dept. of, 73-74
NASC, 6, 22
NCAR, 156
NOAA, 73-74, 203
NSF, 85
NSTA, 265
Postal Service, 73-74
State, Dept. of, 301-302, 384
Tristate Regional Planning Commission, 383
U.N., 105
USA, 40
USAF, 50-51, 67, 77, 102, 258, 295, 347, 449
VA, 73-74
cooperation, industrial, 226-227
ComSatCorp, 224, 242
CPB, 45
INTELSAT, 224
Martin Marietta Corp., 295
small R&D firms, 223
Telesat Canada, 379
cooperation, international. See Apollo-Soyuz Test Project; International cooperation; International cooperation, space; Sounding rocket, international programs.
criticism, 178, 214-215, 231
aerospace employment, 310, 314
Apollo program, 407
Apollo 16, 157, 162
astronaut, 289, 300, 317
budget allocation, 177, 314
national priorities, 6-7, 20, 55, 185, 439
Pioneer 10 plaque, 88, 95
press comment, 10, 314
research and development, 358-359
space benefits, 200-201
space shuttle
budgetary factors, 10, 177, 314
developmental costs, 16, 51, 137
national priorities, 6-7, 20, 55, 185
payloads, 181
press comment, 10, 314
reusability, 16
West Coast site, 181-182
technological progress, 37
Current Index of Technical Briefs (C1-1), 311
"Description, Dissection and Subsampling of *Apollo 14* Core Sample 14230," 54
document security classification, 93
employment, 242-243
aeronautical research, 21-22
budget considerations, 78, 100, 109
EEO program, 11, 80-81, 276, 454
industrial base, 27
manned space flight, 9, 301, 424-425
reduction in force, 22, 78, 106, 109, 297-298, 451
astronauts, 205
KSC, 86
manned space flight, 301
Nuclear Rocket Development Station, 119
restraining order, 255
salaries, 12, 405
termination notices, 182
exhibit, 136, 142, 206, 271, 388, 400
facilities
aeronautics testing, 353
budget and funding, 109-110, 116, 160, 187-188, 209
construction of, 100, 160
dismantling of, 119, 169-170
maintenance, 128
modifications, 139-140
post-Apollo use, 424
public sector projects, 73-74
visitation, 367
grant, 45, 138
launch
Apollo 16 (AS-511), 142

Apollo 17 (AS-512), 413
postponed, 8, 57, 71, 182
probe, *Pioneer 10*, 81
reentry vehicle, 275
satellite
  Aeros, 430
  Anik 1, 379
  Erts 1 (ERTS-A), 269–270
  Esro 4, 389
  Explorer 46 (MTS), 290
  Explorer 47 (IMP-H), 325
  Explorer 48 (SAS-B), 386
  Heos 2, 38
  Intelsat-IV F-4, 26
  Intelsat-IV F-5, 224
  Nimbus 5 (NIMBUS-E), 422
  Noaa 2 (ITOS-D), 350
  Oao 3 (OAO-C) (*Copernicus*), 295–296
  TD-1A, 95
  Triad 01-1X, 306
site selection, 139–140
sounding rocket
  Aerobee 150, 109, 125, 156
  Aerobee 170, 61, 127, 158, 180, 189, 255, 259, 274, 276, 288, 306, 319, 366, 378, 385, 422, 425, 427
  Aerobee 200, 388
  Arcas, 173, 184, 195
    Boosted Arcas I, 19, 373
    Super Arcas, 8, 39, 105, 372, 410
  Astrobee F, 327
  Black Brant, 373
  Black Brant VB, 360
  Black Brant VC, 210, 222, 327
  Javelin, 51, 315
  Loki-Datasonde, 105
  MT-135 (Japanese), 105
  Nike, 73
  Nike-Apache, 15, 37, 39, 54, 132, 210, 222, 248, 283, 284, 342, 343, 348, 349, 372, 374, 375, 410
  Nike-Cajun, 186–187, 210, 251
  Nike-Javelin III, 119, 120, 122, 127
  Nike-Tomahawk, 17, 44, 46, 53, 70, 90, 91, 358
  Strypi 4, 353
Trailblazer II, 275
management, 117, 287
meeting, 13–14, 16, 21, 24, 26, 48, 69, 72–73, 74, 80–81, 101, 135, 137, 177, 185–186, 222–223, 270, 292, 293, 305–306, 310, 314, 316, 321, 325, 330, 387–388, 427, 454
  U.S.-U.S.S.R., 170, 178, 180, 250, 252, 257, 261–262, 279–280, 282, 298, 322, 345–346
organization, 17, 32, 115, 165, 176, 234, 243–244, 284, 353, 397, 451

patents, 16, 258, 299, 327
personnel, 9, 15, 17, 21–22, 64–65, 67, 78, 80–81, 100, 129–130, 131–132, 134–135, 150, 155–156, 182, 185, 188, 189, 192–193, 197–198, 208, 255, 276, 282–283, 313, 323, 403, 438, 451

  appointment, 50, 52, 64, 67, 80, 115, 159, 170, 171, 172, 173, 204, 248, 264, 273, 297, 302, 303, 331, 395–396, 410, 440, 443, 444

  death, 19, 20, 25, 223, 233, 258–259, 269, 337, 362, 434, 441

  promotion, 39, 404

  resignation, 52, 89, 194, 205, 221, 259, 293

  retirement, 18, 29, 202–203, 205, 217, 220, 244, 303, 335, 353, 392, 410, 451

press conference. See Press conference.
programs

  aeronautics, 2, 17, 21, 22–23, 24, 30, 31, 32, 40, 50–51, 56, 59, 66–67, 69, 73, 77, 90, 94, 100, 102, 106–107, 153, 155, 160, 163, 169, 172, 176, 178, 182, 184, 185–186, 187–188, 191–192, 193, 200, 209, 214, 216–217, 231, 233, 235, 247, 251, 254–255, 277, 289, 299, 302, 306, 311, 315, 324, 327, 337, 349, 350, 355, 387, 396, 397, 403, 425, 449, 451

  airborne science, 303

  astronomy, 30, 39–40, 45, 56–57, 63–64, 65, 69, 81, 83, 90, 104, 107, 115, 116, 125, 132, 160, 176, 180, 189, 213–214, 246, 250, 254, 255, 257, 259–260, 261, 264, 273, 274, 276, 279, 287, 288, 295–296, 300, 303, 305, 306, 307, 314, 319, 320, 325, 327, 330, 332, 342, 376–377, 384, 394, 396–397, 399, 405, 407, 408, 409–410, 423, 425, 427, 429, 438, 444, 447

  communications, 255, 258, 292–293

  computer, 71, 261, 452

  earth resources, 21, 30, 56–57, 84, 85, 91, 101, 105–106, 116, 118, 124–125, 184, 210, 219, 221–222, 230, 234, 235, 266, 269–270, 279, 280, 282, 283, 284, 294, 301–302, 303, 308, 310, 313, 356, 367, 369, 380–381, 392, 397, 408, 423, 431, 442, 445–446, 447

  education, 51, 74–75, 154–155, 288, 298, 330, 347–348

  lifting body, 271, 289, 299, 312, 315, 327, 341, 347, 356, 371, 380, 390, 398, 411, 427, 437, 449

manned space flight (see also *Apollo 14* through *18*; Apollo-Soyuz Test Project; *Gemini 4* through *12*; Skylab; *Soyuz 11* and *12*; Space shuttle), 17-18, 30, 32, 33, 58, 64, 71-72, 102, 130-131, 138, 151, 166-167, 175-176, 192-193, 195-196, 198-199, 201, 207, 239-240, 244-245, 265, 271, 297-298, 301, 316, 322, 338, 343-344, 353, 424, 429
   achievements, 85-86, 133, 164-165, 265, 345, 412, 420, 426-427, 428, 435, 438-439, 440, 441, 442-443, 447, 449
   benefits, 157, 394, 412, 420, 426-427, 438-439, 440, 441, 442-443, 449
   long duration, 68, 248, 308, 322, 347-348
   policy and plans, 2-4, 28, 36, 56-57, 111-112, 162, 205, 262, 266-267, 335, 420
   press conference, 285-286
   safety, 195-196, 207, 227, 372-373
meteorology, 17, 77, 124, 184, 186-187, 195, 262-263, 422, 423, 447
nuclear propulsion, 30, 32, 59, 85
space biology and medicine, 6, 15, 24, 30, 41, 50, 52, 63-64, 65-66, 67-68, 71-72, 74-75, 86, 92, 95, 96-97, 119-120, 127, 129, 132, 135, 148-149, 150, 152, 153, 164, 166, 170, 171, 172, 180, 187, 189, 194-195, 201, 203, 207, 227, 229, 239-240, 240-241, 279-280, 281, 283, 285-286, 288, 291, 293, 298, 308, 321-322, 323, 347-348, 360, 375, 385, 388, 394, 403, 404, 414, 419-420, 437, 464
space station (see also Salyut; Skylab), 1, 9, 27, 28, 30, 35, 63, 127, 128, 137, 141, 166, 173, 175-176, 181, 198, 206, 207, 229, 231, 241, 265, 316, 406
space tug, 11-12, 48, 129, 226, 238, 239, 333, 345, 390
technology utilization. See Technology utilization, space.
tracking and data acquisition (see also Tracking), 30, 149-150, 155
property, GSA disposal of, 287
"Space and Science Awareness Month," 72
suit against, 255
test, 67, 191-192, 223, 239, 253
   aircraft. See Aircraft, flight tests.
   AVLOC, 292-293
   digital-fly-by-wire control system, 200
   ESP, 188
   Gravitational Redshift Space Probe, 194
   Heos 2, 46
   missile, 260
   NERVA, 59
   nondestructive, laser holography, 53

quiet engine nacelles, 186
remote health care, 375
reusable rocket, 309
Skylab, 209, 273, 281, 285-286, 323
SMEAT, 273, 281, 285-286, 323
supercritical wing. See Supercritical wing.
Viking Mars Lander parachute system, 163, 256, 272-273, 290, 295
Youth Science Congress, 154-155
National Aeronautics and Space Council (NASC), 6, 22, 350
National Aerospace Museum of the West (proposed), 291
National Air and Space Museum, 45, 74, 388, 455
National Association of Aerospace Clubs of France, 75
National Assn. of Broadcasters, 135-136
National Assn. of Student Councils, 222-223
National Aviation System (NAS) Ten Year Plan and Policy Summary, 141-142
National Book Awards, 119
National Book Committee, 119
National Broadcasting Co. (NBC), 381, 399, 409
National Bureau of Standards (NBS), 42, 106, 155-156, 215-216, 243, 384, 390-391
National Center for Atmospheric Research (NCAR), 16, 118, 156
National Defense Transportation Award, 362
National Environmental Satellite Service (NESS) (NOAA), 351-352, 378
National Environmental Data System (proposed), 358
National Gallery of Art, 17-18
National Geographic Society, 313
National Heart and Lung Institute, 86, 247
National Institutes of Health (NIH), 86
National League of Cities, 304
National Machinery Import and Export Corp. (P.R.C.), 28-29
National Medal of Science, 362
National Microwave Landing System (MLS) program, 67
National Moon Walk Day (proposed), 216
National Ocean Survey (NOAA), 130
National Oceanic and Atmospheric Administration (NOAA)
   cooperation, 73-74, 93, 203, 302
   National Environmental Satellite Service, (NESS), 351-352, 378
   personnel, 21, 130, 325
   research, 15, 31, 118, 236, 308, 353, 370, 441
   satellite, 70-71, 154, 171, 182, 235, 325, 350, 447
   sounding rocket experiment, 51, 186-187
   tour, 384

World Weather Program for Fiscal Year 1973, 262
National Operational Meteorological Satellite System (NOMSS), 352
National Patterns of R&D Resources: Funds and Manpower in the United States 1953–1972 (NSF 72-300), 134
National Portrait Gallery, 17–18
National Press Club, 20
National Public Affairs Center for TV (NPACT), 381, 391–392
National Radio Astronomy Observatory, 18, 307, 365
National Research Council (NRC), 39–40, 159–160, 299, 455
   Astronomy Survey Committee, 213–214, 233
   Committee on the Department of Transportation Climatic Assessment Program, 370
   Geophysics Research Board, 370
   Space Science Board, 39–40, 64, 65, 129, 153, 178
National Research Council of Canada, 19, 307
National Science Board, 281–282
National Science Foundation (NSF), 25, 104, 106, 118, 139, 283–284, 289, 299, 314, 326, 359, 446
   Changes in Graduate Programs in Science and Engineering, 1970–1972 and 1972–1974 (NSF 72-311), 267
   Company Funds Push Total Industrial R&D Spending to $18 Billion in 1971 (NSF 72-318), 427
   Federal Funds for Academic Science, Fiscal Year 1970 (NSF 72-301), 258
   Federal Funds for Research, Development, and Other Scientific Activities, Fiscal Years 1970, 1971, 1972 (NSF 71-35), 245
   Federal R&D Funding Continues To Rise (NSF 72-314), 289
   Federal Scientific, Technical, and Health Personnel in 1970 (NSF 71-47), 64–65
   Federal Support to Universities, Colleges, and Selected Nonprofit Institutions, Fiscal Year 1970 (NSF report), 51
   funding and budget, 31, 112, 125, 161–162, 191, 245, 262
   Manpower and Financial Resources Allocated to Academic Science and Engineering Activities, 1965–1971 (NSF 72-302), 112
   National Patterns of R&D Resources: Funds and Manpower in the United States 1953–1972 (NSF 72-300), 134
   personnel, 34–35, 37, 343
   RANN program, 1, 85, 191, 444
   Scientists, Engineers, and Physicians from Abroad: Trends Through Fiscal Year 1970 (NSF report), 8
   Solar Energy in Developing Countries: Perspectives and Prospects, 138
   Technology Assessment Act of 1972, 349
National Science Foundation Act of 1950, 316, 341, 349, 359
National Science Teachers Assn. (NSTA), 1, 61, 131, 154–155, 265, 384
National Secondary Principals Assn., 222–223
National security, 29, 168
   classification system, 92–93, 159–160, 187
   Convention on the Prohibition of the Development, Production, and Stockpiling of Bacteriological (Biological) and Toxin Weapons, and on Their Destruction, 134, 288–289, 311
   defense, 46–47, 55–56, 64, 69, 162
   National Security Strategy of Realistic Deterrence (DOD report to Congress), 64
   space program use, 32, 33, 167
National Security Agency (NSA), 163
National Security Council, 187
National Security Industrial Assn., 50, 93, 293
National Security Strategy of Realistic Deterrence (DOD report to Congress), 64
National Space Club (NSC), 59, 98–99, 153, 226–227, 286–287, 328
   Press Award, 99
National Space Development Agency (Japan), 26
National Symposium on Technology Transfer, 225
National Technical Information Service (NTIS), 291, 356
The National Technology Program: Utilization of Industry (AIAA study), 398
National Weather Service, 93, 154
NATO. See North Atlantic Treaty Organization.
Natural Resources, Dept. of (proposed), 122
Naugle, Dr. John E., 56, 65, 107, 189–190
Naumann, Dr. Robert J., 254
Naval Air Development Center, 372–373
Naval Air Station, San Diego, Calif., 174
Naval Construction Battalion Center, 358
Naval Missile Center, 215
Naval Observatory, 243
Naval Research Laboratory (NRL), 52, 127, 254, 270, 306, 326, 327, 365, 395–396
   awards, 291–292
   contract, 264
   personnel, 179–180
   satellite, 291–292
   sounding rocket experiment, 119, 120, 122, 180, 255
   Space Systems Div., 291–292
Naval Weapons Center
   Aerothermochemistry Div., 21
Navigation (see also Air traffic control), 33, 141–142, 243, 377–378
   aircraft, 20, 178, 182, 200

landing systems, 20, 35, 56, 67
  ship, 134, 301
  spacecraft, 182
Navigation satellite, 134, 192, 306
Navy. See U.S. Navy.
Navy Astronautics Group, 134
Navy Navigation Satellite System, 134
NBC. See National Broadcasting Co.
NBS. See National Bureau of Standards.
NCAR. See National Center for Atmospheric Research.
Nebraska, 356
NEMS. See Nimbus E microwave spectrometer.
Neptune (planet), 165, 441
NERVA. See Nuclear engine for rocket vehicle application.
Ness, Dr. Norman F., 21
NESS. See National Environmental Satellite Service (NOAA).
Netherlands, 60, 123, 241-242, 389
Network Project, 190-191
Neuilly, France, 48
Nevada, 59, 118, 146-148, 230, 321
New Hampshire, Univ. of, 91, 315
New Jersey, 255, 313, 384, 396
New Mexico, 127, 132-133, 154, 156, 180, 189, 202-203, 210
New Mexico State Univ., 42-43
New York, 136, 220, 248, 266, 275, 388, 393, 399
New York, N.Y., 93, 162, 313, 323, 383
  announcement, 286
  exhibition, 17-18, 185, 211
  media, 156-157, 399
  meeting, 112, 117, 159, 192-193, 352, 375
  museum, 385, 428, 455
  press conference, 341
  "Voyage Beyond Apollo," 409
  Yankee Stadium, 6
New York Public Library, 2, 13
New York Univ., 16, 421
Newell, Dr. Homer E., 167, 176-177, 192-193, 301-302
Newhall, Calif., 266
Newport-Bermuda Race, 242
Newton, Sir Isaac, 167, 436
Nicaragua, 442
Nice, France, 321
Nicholson, David, 341
"Nicolaus Copernicus Day," 274
Nierenberg, Dr. William A., 282
NIH. See National Institutes of Health.
Nike (asteroid), 260
Nike (missile), 403
Nike (rocket), 73
Nike-Apache (sounding rocket)
  atomic oxygen study, 374
  auroral study, 15
  cosmic dust particle collection, 242, 343, 348
  electron and proton spectra, 210, 283, 284
  galactic astronomy experiment, 222
  ionosphere study, 248, 372, 410

  langmuir probe, 132
  launch failure, 37
  night-glow study, 374
  nitric oxide, 54
  polar cap absorption (PCA) data, 283, 284
  thermospheric winds experiment, 349
  winter anomaly study, 39
Nike-Cajun (sounding rocket)
  atomic oxygen distribution, 186-187
  grenade payload, 186-187
  ion mass spectrometer, 210
  ionosphere study, 251
  ozone measurement, 184, 186-187
Nike-Javelin III (sounding rocket), 119, 120, 122, 127, 147
Nike-Tomahawk (sounding rocket)
  auroral aeronomy, 44, 46, 53, 70, 358
  energetic particle experiment, 91
  fields and neutral winds, 90, 91
  particles and fields, 17
Nikolayev, Andrian G., 354
Nile Blue (DOD report), 247
Niles, Ohio, 62
Nimbus (meteorological satellite), 30, 408
Nimbus 2, 423
Nimbus 3, 17, 423
Nimbus 4, 77, 124, 184, 186-187, 195, 423
Nimbus 5, 422, 423, 447
Nimbus E microwave spectrometer (NEMS), 423
Nippon Electric Co., Tokyo, Japan, 389
Nitrogen oxide, 385, 391
Nix Olympica (Mars), 42-43, 376-377
Nix Olympica-Tharis (Mars), 225-226
Nixon, Mrs. Richard M., 324
Nixon, President Richard M.
  aeronautical satellite program, 49-50
  Aeronautics and Space Report of the President: 1971 Activities, 99
  airline security regulations, 93
  Apollo 13 mission tribute, 302
  Apollo 16 mission, messages and statements, 112-113, 157, 162-163, 163-164, 228-229
  Apollo 17 mission, messages and statements, 409, 418, 420, 428, 435-437, 439-440
  Apollo-Soyuz Test Project (ASTP), 213, 228-229, 240-241
  appointments and nominations by, 34, 110, 172, 187, 236, 264, 277, 281-282, 288, 289, 341, 374, 397-398, 420
  assignment approval, 189
  astronaut visit, 41
  bills signed
    appropriations, 189, 207, 231, 289, 291, 298, 327, 362
    Chapel of the Astronauts, 55
    Convention for the Suppression of Unlawful Acts Against the Safety of Civil Aviation, 372

Convention on the Prohibition of the Development, Production and Stockpiling of Bacteriological (Biological) and Toxin Weapons and on Their Destruction, 134
Eisenhower memorial, 358
international aeronautical exposition, 110
International Exposition on the Environment, 365
Technology Assessment Act of 1972, 349
bills submitted to, 341, 349
budget message, 30–31
China visit, 2, 28–29, 59, 72, 86, 96, 98, 350, 393
criticism, 310
Domestic Council programs, 53–54
Environmental Quality: The Third Annual Report of the Council of Environmental Quality–August 1972, 285
Erickson, Leif, Day, 297
European visit, 189, 190, 191, 213
Executive Order, 92–93, 217–218
Executive reorganization message, 122
family, 265
Gallup poll, 241
GCEP, 218–219
Government document security, 92–93
hijacking of aircraft, 372, 411
Hot Line approval, 20
Intelsat Israel connection, 273
international cooperation, 224–225, 241, 306
international cooperation, space, 213–214, 228–229, 232, 240–241, 430, 436
International Exposition on the Environment, 292, 365
Iran visit, 189, 190
launch assistance policy, 342–343, 348, 452
memorandum, 358
messages, 311, 318
messages to Congress, 24–25, 45, 47, 48–49, 52, 172–173, 285, 292
Minority Business Development Program, 247
National Science Foundation (NSF), 18, 51
"Nicolaus Copernicus Day," 274
protocol signing ceremony, 275
reports transmitted to Congress, 139
resignations accepted by, 130, 221, 280, 397–398
resolutions, 159–160, 216
science and technology, 105–106, 358–360
space program, national, 219, 328, 374, 407, 410, 435–437
space shuttle, 46–48, 449
 endorsement, 2–4, 5, 6–7, 8, 9, 10, 11, 12, 16, 18–19, 32, 36–37, 253
 funding, 34
speech, 374–375
staff, 192–193

supersonic transport, 361, 444
super-secret satellite spy system, 214
treaty, 201–202, 224–225, 323, 338
United Nations Conference on the Human Environment, 236
U.S.S.R. agreements
 ABM treaty, 201–202, 224–225
 Agreement Concerning Cooperation in the Exploration and Use of Outer Space for Peaceful Purposes, 195
 Agreement on Cooperation in the Field of Environmental Protection, 193–194
 Apollo-Soyuz Test Project, 213, 240–241
 "Basic Principles of Relations," 207
 Limitation of Strategic Offensive Arms, 331
 peaceful use of space, 452
 SALT accord, 201–202, 214
 summit accord, 206
U.S.S.R. visit, 185, 189, 190, 191, 197–198, 203, 207, 232, 241, 390
Wright Brothers Day, 422
NOA: new obligational authority.
NOAA. See National Oceanic and Atmospheric Administration.
Noaa 1 (National Oceanic and Atmospheric Administration meteorological satellite), 70, 352
Noaa 2, 350, 351–352, 361, 378, 447
Nobel Peace Prize, 264
Nobel Prize, 95, 162
 chemistry, 357
 medicine, 348, 387–388
 physics, 2, 61, 357
Noise, aircraft
 abatement, 74, 106, 130–131, 136
 antisymmetrical wing effect, 191–192
 DOT program, 203–204
 funds, 31, 47, 110, 116, 178, 185–186, 187–188, 209, 273–274, 350, 451
 JT3D/JT8D Refan Program, 176, 292
 landing systems, 20, 67, 182
 Quiet Engine Program, 124
 sonic boom research, 80, 327
 CARD policy study, 21–22
 Conference on Aircraft Engine Noise Reduction, 185–186
 contract, 231, 248, 251, 266, 292, 382
 CTOL, 66
 QUESTOL, 66, 449
 STOL, 2, 40, 66, 124–125, 182
 supersonic transport, 248, 306, 307, 319, 329, 349, 427
 terminal-configured aircraft, 83–84
 TRANSPO '72 exhibition, 206
 TriStar, 307
 V/STOL, 67
 VTOL, 66–67
Nome, Alaska, 173
NOMSS. See National Operational Meteorological Satellite System.
Nondestructive testing, 47–48, 53
Noordwijk, Netherlands, 241–242

NORAD. See North American Air Defense Command.
Norfolk, Va., 312–313, 321
Norman Medal, 355
North America, 38, 254
North American Air Defense Command (NORAD), Space Defense Center, 109
North American Aviation, Inc., 382
North American Rockwell Corp. (NR), 383, 425
 ASTP Working Group, 346
 Aviation Services Div., 365
 contract, 274
  ASTP, 378
  F-111 components, 340
  RPV man-machine study, 51
  Skylab, 262, 434
  space shuttle, 45, 90, 129, 180, 254, 272, 276, 279, 280, 284–285, 292, 308, 319, 339, 383–384
  STOL prototype, 355
  subcontract, 275, 308, 319
 Downey Div., 325
 lawsuit, 91, 382
 meeting, 427
 personnel, 372–373
 Rocketdyne Div., 45, 90, 91, 123, 272, 292
 Space Div., 61, 180, 254, 256, 272, 383–384
North Atlantic Treaty Organization (NATO), 86, 370
North Carolina, 390
North Hollywood, Calif., 156
North Ray Crater (moon), 62–63, 148, 326
North Vietnam, 340, 355, 440
Northeast Corridor Report, 67
Northeast Ohio Congressional Council, 326
Northrop Corp., Northrop Aircraft Corp., 240, 275
Northrop, John K., 275
Northwestern Univ., 300–301
Norway, 15, 17, 53, 210
Novak, Dr. Grga, 35
Novosti News Agency (U.S.S.R.), 110
Noyes, Crosby S., 411
NPACT. See National Public Affairs Center for TV.
NR. See North American Rockwell Corp.
NRC. See National Research Council.
NRL. See Naval Research Laboratory.
NSA. See National Security Agency.
NSC. See National Space Club.
NSF. See National Science Foundation.
NSTA. See National Science Teachers Assn.
NTIS. See National Technical Information Service.
NTNF. See Royal Norwegian Council for Scientific and Industrial Research.
Nuclear engine for rocket vehicle application (NERVA), 30, 32, 59, 90, 106, 119, 160, 451

Nuclear explosion, 95, 308, 327
Nuclear fusion, 331, 358, 404
Nuclear particle accelerator, 88, 429
Nuclear physics, 79, 132–133
Nuclear power, 39–40, 79–80, 85, 130–131, 248, 330, 378–379
Nuclear reactor, 2, 10, 14–15, 24–25, 79–80, 130–131, 162, 474
Nuclear Rocket Development Station (NASA–AEC), 119
Nuclear test, 421
Nuclear weapons, 9, 13, 49, 54, 55–56, 63, 71–72, 188, 192, 201–202, 207, 218–219, 259

O

OA. See NASA Office of Applications.
Oak Ridge National Laboratory, 348, 384
Oakland, Calif., 430
Oao 1 (Orbiting Astronomical Observatory), 297
Oao 2, 254, 297
Oao 3 (*Copernicus*), 254, 257, 295–296, 300, 305, 332, 447
Oao 3 Project Team, 381
OAO-B, 296
OAO-C. See *Oao 3*.
OART. See NASA Office of Advanced Research and Technology.
OAST. See NASA Office of Aeronautical and Space Technology.
Oberth, Hermann, Award, 69, 202–203
O'Brian, Hugh, 134–135
O'Brian, Hugh, Foundation, 222–223
O'Brien, Dr. Brian, 140, 380–381
O'Bryant, William T., 388
Observatory (see also individual observatories, such as Boyden Observatory, Hale Observatory, High Energy Astronomy Observatory, Mt. Palomar Observatory, Upper Atmosphere Observatory, etc.), 95, 276
Ocean. See Atlantic Ocean, Indian Ocean, Pacific Ocean.
Ocean of Storms (Oceanus Procellarum) (moon), 14, 148, 377–378
Oceanography
 remote sensing, 302
 research, 34–35, 66, 128, 136, 228, 229, 252, 256
 U.S.–U.S.S.R. cooperation, 177, 229, 298, 375
O'Connor, John J., 190–191
O'Connor, L/G Edmund F., 277, 303
O'Dell, Dr. Charles R., 302
Odenthal, Col. H. J. (USAF, Ret.), 26
Oehler, Dorothy, 158
OEO. See NASA Office of Equal Opportunity and Office of Economic Opportunity.
Of a Fire on the Moon, 119

Office of Aeronautics and Space Technology. See NASA Office of Aeronautics and Space Technology.
Office of Economic Opportunity (OEO), 365
Office of Management and Budget (OMB), 12, 75, 267, 315, 350, 380–381, 451
Office of Naval Research (ONR), 244, 315, 335, 369
Office of Science and Technology (President's) (OST), 191, 445
Office of Space Science (GSFC) (NASA), 372
Office of Technology Assessment (OTA), 48, 316, 341, 349, 359, 361–362, 387
Office of Telecommunications Policy (White House), 72
Ogo 5 (Orbiting Geophysical Observatory), 179–180
Ohio, 50, 62, 77, 154–155, 186, 391, 399
Ohio State Univ., 367
Ohsumi (Japanese satellite), 295
OIATU. See NASA Office of Industry Affairs and Technology Utilization.
O'Leary, Dr. Brian T., 55, 121, 185
Olivine, 179
Olympic Games, 224–225, 318
"Olympics of Science," 253
O'Malley, Col. Jerome F. (USAF), 320
OMB. See Office of Management and Budget.
OME. See Orbit maneuver engine.
OMS. See Orbital maneuvering system.
OMSF. See NASA Office of Manned Space Flight.
O'Neal, Dr. Russell D., 282
O'Neil, L/G John W. (USAF), 277, 303
O'Neill, Rep. Thomas P., Jr., 340
On the Moon with Apollo 16, 71
ONR. See Office of Naval Research.
Onverwacht Strata, South Africa, 158
"Open skies" agreement, 190
Ophir–Eos region (Mars), 225–226
Ophiuchus (constellation), 296
Opium detection, 317
Optokinetic Nystagmus Reflex, 41
Orange Bowl Classic (football game), 1
Orbit maneuver engine (OME), 47
Orbita network (U.S.S.R. communications satellite system), 128, 188, 331, 350, 404, 426
Orbital maneuvering system (OMS), 42, 266
Orbital Workshop (OWS) (Skylab component), 23–24, 169, 209, 239, 262, 280, 309, 325, 384
Orbiter. See Space shuttle.
Orbiting Astronomical Observatory (see also *Oao 1*, OAO–B, etc.), 305
Orbiting Geophysical Observatory (OGO) (satellite) (see also *Ogo 5*), 275
Orbiting laboratory. See Orbital Workshop and Salyut.
Orbiting Solar Observatory (OSO) (see also *Oso 1*, *Oso 5*, *Oso 7*, etc.), 213–214, 273, 275, 276
Ordnance Missile Command, Redstone Arsenal, 19
Oreol (French satellite), 323
Orientale basin (moon), 364
Orion (*Apollo 16* LM). See Lunar module.
Orion (NASA barge), 266, 298
Orion (U.S.S.R. astronomical system), 255
Orlando, Fla., 98, 138
Ormes, Dr. Jonathan F., 335
Orthophotoquads, 235
Oscar 6 (OSCAR–C) (amateur radio satellite), 352, 361, 452
OSO. See Orbiting Solar Observatory.
Oso 1, 90
Oso 5, 280
Oso 7, 107, 273, 276, 280, 284, 330
OSO–L, 213–214
OSO–M, 213–214
OSO–N, 213–214
OSS. See NASA Office of Space Science.
OST. See Office of Science and Technology.
Ostrander, M/G Don R. (USAF, Ret.), 362
Ostricker, Dr. J. P., 360
OTDA. See NASA Office of Tracking and Data Acquisition.
Ottawa, Canada, 184, 328
Ottobrunn, West Germany, 304, 377
Outer Planets Exploration, 1972–1985 (NAS–NRC Space Science Board study), 39–40
Outer Space Liability Convention, 49
Outer Space Treaty, 48–49
Outstanding Handicapped Federal Employee of the Year Award, 131–132
Outstanding Leadership Medal (NASA), 380–381
Owens, Roy, Interests, Inc., 60
OWS. See Orbital Workshop.
Oxygen, 254, 272, 304, 374
Ozone, 43, 173, 184, 186–187, 195, 370, 476

P

P–51 (aircraft), 192–193
Pacific Missile Range (PMR), 289, 358
Pacific Ocean
  ABM test, 174, 300, 361, 411
  Apollo 16 splashdown, 148–149
  Apollo 17 philatelic mail cancellation, 312–313
  Apollo 17 splashdown, 297
  earthquake potential, 442
  Intelsat–IV satellite activities, 26–27, 47, 53, 60–61, 173, 224
  PEACESAT broadcasting, 19
  Safeguard test, 108, 300
  Sprint missile test, 260, 411
  Tacsat 1 activity, 400
  U.S.S.R. missile launches, 350, 355, 393, 429
Packard, David M., 34

Page, Dr. Thorton L., 101, 179, 229, 231–232, 306
Paine, Dr. Thomas O., 205
Pakistan-India war, 9–10
Palaora, Hans R., 129
Palmetto Crater (moon), 152
Palo Alto, Calif., 314
Palomar-Lyden (asteroid), 260
Pan American World Airways, Inc., 362, 399
Pan-Pacific Education and Communications Experiments Using Satellites (PEACESAT), 19
Panel on Science and Technology, 34–35
Pant, Minister of State K. C. (India), 138
Parachute
  booster recovery, 309
  decelerator system, 223–224, 290, 295, 428
  deployment test, 173, 187, 195, 364–365
  sounding rocket, 105, 173, 184, 195
  U.S.S.R. Venus probe, 267
  Viking, 163, 256, 272–273, 290, 295
Paralytic control device, 38
Parapsychology, 188
Paris, France, 75, 168, 268, 271, 317, 335, 378
Park, Dr. Arch B., 235
Parker, Marshall, 88–89
Parnell, Dr. Thomas A., 16, 181
Parsons, Brinckeroff, Quade and Douglas, Inc., 268–269
Particle accelerator, 55, 78, 88, 429
Particle identifier, 216
Particle-impact location detector, 327
Pasadena, Calif., 322, 379
Patent
  astronaut backpack harness, 101
  cold camera, 220
  esthesiometer, 184–185
  FLASH, 258
  mail delivery system, 263
  particle-impact location detector, 327
  Rasquin-Estes diamond-maker, 16
  "Smokey," 258
  solar cell, 250
  television, 156
Patrick AFB, Fla., 2, 14, 321, 404
Patuxent River Naval Air Station, Md., 244
Paul, Dr. Rodman W., 316
Pauling, Dr. Linus C., 162
Pavlov, Vladimir N., 256
PBS. See Public Broadcasting Service.
Peace Corps, 192
PEACESAT. See Pan-Pacific Education and Communications Experiments Using Satellites.
Pearl Harbor, Hawaii, 96
Peary, Adm. Robert E. (USN), 438
Pecora, Dr. William T., 380–381
Pecora, Mrs. William T., 380–381
Peebles, Ohio, 186
Peenemuende Rocket Center, Germany, 202–203, 205, 337
Peking, P.R.C., 294, 301, 392, 421

Pelehach, Michael, 220
Pendray, G. Edward, Award, 20–21
Pennino, Walter A., 124
Pennsylvania, 54, 59, 139, 353, 371
Pennsylvania State Univ., 8, 39, 372, 410
Pentagon papers, 247
People's Republic of China (P.R.C.). See China, People's Republic of.
PEP. See Princeton experiment package.
Perkins, Dr. Courtland D., 140, 244–245
Pershing (missile), 202–203
Peru, 15
Petrone, Dr. Rocco A., 62, 63, 88, 112–113, 297, 398, 435, 440, 442
Petrov, Dr. Boris N., 127, 140–141, 199, 208, 252, 257–258, 260
Petrov, Dr. Georgy I., 189–190, 311
Petrovsky, Minister of Health Boris V. (U.S.S.R.), 194
Philadelphia, Pa., 139
Philco-Ford Corp., 370
Philippines, 19, 96
Phillips, L/C Samuel C. (USAF), 163, 313
Philpotts, Dr. John A., 377–378
Phinney, Dr. Wallace C., 445
Phobos (Mars moon), 364
Phosphorus, 14
Photogeology, 237
Phototelevision, 90
Photoemulsion plate, 235
Photography (see also Mars, photographs and Moon, photographs)
  Apollo 16 mission, 142–144, 148–149, 149–150, 158, 165, 169, 254
    press conference, 170–171, 172
  Apollo 17 mission, 148–149, 361, 433, 442
  atmospheric, 184
  atomic reactor, 14–15
  battery applications, 178–179
  contract, 17
  earth environment and resources, 118, 120, 142–144, 179–180, 184, 254, 269, 313, 383, 405, 438, 442, 444
  ERTS, 210, 269–270, 313, 369
  fingerprint transmission, 126
  high-altitude, 120
  infrared, 383
  land use, 118, 269, 383
  lunar, 138, 148–149, 368, 404, 428, 432, 433, 440
  Mariner Mars probe, 6, 15, 42–43, 56, 132, 178, 342, 364, 429
  meteorological, 184, 262–263, 269
  multispectral, 269
  RBV, 269
  reconnaissance satellite, 201–202
  Skylab missions, 23–24, 60
  spectrographic, 254, 270–271
  U.K. sounding rocket, 120
  ultraviolet, 142–144, 165
  U.S.S.R. lunar probe, 138
  U.S.S.R. Mars probe, 10–11, 28, 90
  x-ray, 411
Photometer, 254
Photopolarimeter, 254, 260

Photosynthesis, 158, 254
Physicians, 8, 24, 131
Physicist, 152
Physics and astronomy program (NASA), 160
Physics in Perspective (NAS study), 287
Physics of the Solar System (NASA SP-300), 246
Physics of the Space Environment (NASA SP-305), 454
Picciolo, Dr. Grace, 258
Pickering, Dr. William H., 310, 345, 434
Pigott, Charles M., 331
Pigpen (comic strip character), 24
Pilatus Porter (Swiss aircraft), 177
Pilots (see also Aerospace Research Pilots School)
  Apollo 16 CM, 6
  Asian bombing, 443
  award, 258, 263
  cardiac response, 150
  death, 25
  Gemini 4 mission, 142
  glider aircraft, 161
  human time cycle study, 308
  lifting body test, 271
  lunar module, 258
  noise reduction technique, 124
  press comment, 443
  remote-control landing, 336
  SR-71 aircraft, 162
  STOL, 269
  TF-8A flight test, 155, 163
  U.S.S.R., 138, 329
  weather monitoring, 93
Pink Palace Museum, 72
Pinkel, Irving I., 244
Pioneer (program), 19, 30, 39-40, 408
Pioneer (spacecraft), 64, 65, 294-295
Pioneer 5 (interplanetary probe), 83
Pioneer 6, 83, 280, 300, 314
Pioneer 6-9 Project Team, 380-381
Pioneer 7, 83, 300, 305
Pioneer 8, 83, 300
Pioneer 9, 83, 300
Pioneer 10 (Pioneer-F), 408
  achievements and progress, 81, 88, 140-141, 155, 200, 226, 253-254, 305, 339, 447
  Asteroid Belt penetration, 365-366
  communications, 82
  data transmission, 305
  experiments, 81-83, 226, 305, 366
  Jupiter observations, 81, 405
  plans, 39-40, 56, 71, 314
  plaque, 81, 87-88, 94, 95
  press comment, 87-88, 89, 94
  reconnaissance, 259-260
  solar observations, 81, 280, 300
  spacecraft checkout, 19
  tracking, 438
Pioneer-F. See *Pioneer 10*.
Pioneer-G, 39-40, 314
Pioneer-H, 39-40
Piper Comanche (PA-30) (research aircraft), 336-337

Pittsburgh, Pa., 371
Pittsburgh, Univ. of, 54
Pius XI, Pope, Prize, 359
Planck, Max, Society for Advancement of the Sciences, 251
Planetarium, 428, 454
Planetary and Earth Sciences Div., MSC (NASA), 13-14
Planetary exploration. See specific planets and probes.
Plankton, 136
Planning Grant Program (FAA), 272
Plasma, 79, 108, 141, 275, 337, 404
Plasma physics, 83
Plastic, 214, 256
Platt, Dr. John R., 405
Plemeur-Bodou, Brittany, 255
Plesetsk, U.S.S.R.
  Cosmos launch, 33, 69, 77, 87, 102, 113, 119, 127, 131, 135, 139, 155, 174, 186, 239, 243, 250, 253, 257, 264-265, 274, 292, 294, 317, 319, 330, 338, 340, 346, 355, 361, 366, 371, 373, 428, 439, 443, 444, 445
  Intercosmos 8 launch, 403
  Meteor 12 launch, 243
  Meteor 13 launch, 363
  Molniya II-2 launch, 188
  Molniya II-4 launch, 426
Plourde, Gary A., 20-21
Plowshare program, 396
PLSS. See Portable life support system.
Plum Brook Station, Ohio, 94, 287
Plum Crater (moon), 146-148
Pluto (planet), 165, 200, 441
Plutonium
  238 isotope, 80, 194-195, 388
  244 isotope, 13-14, 203
PMR. See Pacific Missile Range.
pndb: perceived noise in decibels.
POBAL. See Balloon, powered.
Podgorny, President Nikolay V. (U.S.S.R.), 163-164, 191, 193-194, 323, 368, 437
Poetry Society of Texas, 61
Pogue, L/C William R. (USAF), 23
Pohwaro (pulsated, overheated, water rocket), 177
Point Barrow, Alaska, 128, 186-187
Point Mugu, Calif., 134, 215, 355, 358
Poker Flat Rocket Range, Fairbanks, Alaska, 44, 70, 90, 91, 182-183, 349, 358
Poland
  astronaut visit, 20, 28, 41
  astronomical research center, 183
  Copernicus, Nicolaus, birthday, 257
  Intercosmos 6 instrumentation, 132
  Nixon, President Richard M., visit, 189, 190, 213
  space technology, 345
Polar cap, 273, 283, 389
Polar Morning, 228
Polaris (missile), 31, 241
Politics, 133, 181-182, 185, 201
Pollock, Howard W., 154

Pollution control (see also Air pollution; Noise, aircraft; Recycling; Water pollution), 2–4, 5–6, 193–194, 194–195, 217, 248, 276, 302
Population, world, 5–6
Port Hueneme, Calif., 358
Portable life support system (PLSS), 146–148
Porter (utility aircraft), 177
Porter, Katherine Anne, 409
Porter, Dr. Rodney R., 348
Portland, Ore., 400–401
Poseidon (missile), 31
Poseidon (NASA barge), 261
Post-Apollo Lunar Science, 368
Postage Day (Belgium), 157
Postal covers (*Apollo 15* mission), 256, 258, 259, 262, 267–268, 270, 273, 276, 277, 317, 321
Potassium, 14, 65–66, 148, 172
Poverty, 133, 135–136, 161, 164
Powell, R/A Allen L. (USN), 130
Powell, Cecil, 379, 398, 411
Prather, Victor A., Award, 372–373
Pratt & Whitney Div. See United Aircraft Corp.
P.R.C. See China, People's Republic of.
Predham, Reg, 429
Preliminary Examination Team. See *Apollo 15* Preliminary Examination Team.
Present, Stuart M., 25, 214
Presidential Prizes for Innovation (proposed), 191, 253
President's Distinguished Federal Civilian Service Medal, 372
Press comment
 aerospace industry, 2, 10
 Apollo 15 mission, 258, 259, 262, 276, 289
 Apollo 16 mission
  commemorative medal, 264
  Congressional support, 328
  criticism, 162
  future, 151
  priority, 142
  public complacency, 154, 157
  scientific contribution, 158
  technical problems, 152–153, 160–161
  technological success, 164–165
  U.S.S.R. coverage, 151
 Apollo 17 mission
  achievements, 389, 420, 428–429, 438–439
  contributions, 427
  impact, 423
  mission profile, 409, 441–442
  nighttime launch, 424
  nostalgia, 426–427
  peaceful use of space, 434
  splashdown, 441
  TV coverage, 427
 Apollo program
  achievements, 152–153, 426, 434, 440
  communications, 411–412, 423
  conclusion, 423, 424
  cost, 133
  criticism, 162, 440
  impact, 404–405, 429
  international cooperation, 151
 Apollo-Soyuz Test Project, 328
 astronaut, 160–161
 Copernicus (Oao 3), 300
 Defense, Dept. of, 218
 Earth Resources Observation System, 234
 earth resources survey program, 266
 electronic countermeasures, 61–62
 Erts 1, 272
 geophysical warfare, 250
 Human Factors in Long-Duration Spaceflight (SSB report), 153
 international cooperation, space, 151, 276
 Jet Propulsion Laboratory, 362–363, 379
 launch assistance policy, 348
 Lunar Science Conference, 13–14, 26
 McGovern, Sen. George S., 314
 Mars (planet), 399
 metric system, 297
 NASA
  budget, 164–165
  inventions and licensing, 348–349
 New York Public Library, 13
 noise, aircraft, 80
 nuclear reactor, 88
 Pioneer 10, 87–88, 89, 94
 pollution control, 217
 reconnaissance aircraft, 255–256
 reconnaissance satellite (USAF), 96
 science, 37
 Sikorsky, Igor I., 365
 sonic boom, 80
 space program, national, 11, 16, 156, 162, 167, 178, 272, 314, 348–349, 407, 411–412, 422, 423, 426, 440
 space shuttle, 215, 231, 295
  benefits, 35, 244–245
  congressional consideration, 7
  cost and funding, 11, 16, 156
  economic consequences, 275
  endorsement impact, 8–9, 10, 11, 18–19
  international cooperation, 15, 251–252
  launch site, 8
  public support, 328
 space trends, 176–177
 "Spaceship Earth," 319
 supersonic transport, 361
 TRANSPO '72, 222, 224
 TRANSPO '74, 224
 TV broadcast via satellite, 355
 United Nations Conference on the Human Environment, 217
 U.S.–U.S.S.R. space cooperation agreement, 201, 204–205, 206–207
 U.S.–U.S.S.R. summit meeting, 207
 U.S.S.R. space program, 120
 Venus 8 (U.S.S.R. space probe), 273, 294, 315
 Vietnam war, 406, 443

von Braun, Dr. Wernher, 217
Zvezdny Gorodok (Star City), U.S.S.R., 111, 117
Press conference
  Apollo program, 394
  Apollo 16 mission, 6, 62, 96, 237–238
    astronaut, 164
    code names, 109
    inflight briefing, 148–149
    international cooperation, 228–229
    lunar science, 6, 170–171, 363
    MEED experiment, 6, 96–97
    mission review, 172
    photographic results, 170–171, 179–180
    preflight briefing, 142
    preliminary science results, 179–180, 183–184
    U.S.S.R. reaction, 150
  Apollo 17 mission, 382, 388
    astronaut, 403, 429–430
    completion, 435
    lunar science, 363–364, 431–432, 433–434, 445
    overview, 366–367
  Apollo-Soyuz Test Project, 228, 229, 250, 257–258, 260–261, 344–345, 382
  astronomy, international program, 151
  communications satellite, 286
  Concorde, 271, 317
  Earth Resources Technology Satellite, 235, 313
  F-111 aircraft, 340
  hijacking of aircraft, 221
  Jupiter–Saturn mission, 320
  Lovell, Capt. James A., Jr. (USN), retirement, 392
  lunar science, 152, 380
  Mankind 1 (space probe), 324–325
  Mariner 9 (interplanetary probe), 42–43, 225–226
  NASA
    budget, 32, 410
    programs, 219
  noise, aircraft, 124
  pollution monitoring, 21
  Skylab medical experiment altitude test (SMEAT), 285–286
  Skylab program, 23–24
  space program, national, 310, 410, 424
  space shuttle, 5, 11–12, 102, 139–140
  supercritical wing, 214
  supersonic airline service, 341
  Teegan, Dr. John, resignation, 52
  Trident, 185
  unidentified flying objects (UFO), 383
  U.S.–U.S.S.R.
    bilateral cooperation, 90, 236–237
    joint mission, 127, 128, 129
    science and technology agreement, 199
    space cooperation agreement, 197–198, 198–199
  Venus 8 (U.S.S.R. space probe), 284
Price, Edward W., 20–21
Price, Shirley, 131–132

Prince Albert, Saskatchewan, Canada, 184, 270
Princeton experiment package (PEP), 296
Princeton Plasma Physics Laboratory (AEC), 404
Princeton Univ., 254, 257
  Dept. of Aerospace and Mechanical Sciences, 140, 244–245
Probe (see also individual probes and experiments, such as Gravitational Redshift Space Probe Experiment, *Luna 19*, *Luna 20*, *Mariner 8*, *Mariner 9*, *Mars 2*, *Mars 3*, *Pioneer 10*, *Ranger 9*, and *Venus 8*)
  balloon,' 16
  Jupiter, 56, 65, 71, 140–141, 200
  lunar. See *Luna 16*, *Luna 17*, *Luna 18*, *Luna 19*, *Luna 20*, and *Ranger 9*.
  Mars, 1, 6, 10–11, 15, 28, 38, 42–43, 51–52, 56, 60, 65, 132, 166, 167, 178, 199–200
  Mercury, 65
  meteorological, 24
  Saturn, 56, 65
  Venus, 65, 140–141
Procter & Gamble Co., 399
Procurement Office (MSFC) (NASA), 243–244
Prognoz (U.S.S.R. satellite), 449
Prognoz 1 (Forecast), 139
Prognoz 2, 242, 323
Progress Report on Flight Investigations of Supercritical Wing Technology, 72–73
Project Blue Book, 300–301
Project Harvest Moon, 325
Project Search (System for Electronic Analysis and Retrieval of Criminal Histories), 126
Propulsion
  electric, 208
  hydrazine, 208
  launch vehicle, 352
  Minuteman III, 191
  nuclear, 106, 343
  solar-and-nuclear-electric, 107
  space shuttle, 79, 333
  supersonic transport, 349
  systems development, 90, 107
Proton accelerator, 221
Proxmire, Sen. William, 10, 120, 131, 179, 214, 260, 289, 445
Ptolemaeus Crater (moon), 403
Public Broadcasting Corp. See Corp. for Public Broadcasting.
Public Broadcasting Service (PBS), 391–392
Public Buildings Act of 1959, 322
Public health, 64–65, 194
Public Health Service. See U.S. Public Health Service.
Public relations, 169–170
Public sector problems. See Technology utilization, space, social problems.
Public Technology Inc., 270
Puckett, Dr. Allen E., 20–21, 332
Puerto Rico, 277

Puget Sound, Wash., 252
Pulsar, 81, 107, 123, 159, 233, 261, 342, 408, 438
Pulsed laser radar acquisition-and-ranging system, 292-293
Purchasing Office (MSFC) (NASA), 243-244

Q

Quadraplegics, 95
Quality control, 47-48
Quantum optics, 287
Quarantine, 120, 385
Quasar (quasi-stellar object), 107, 123, 125, 187, 213, 354, 360, 408
Queen Elizabeth 2 (ship), 104
QUESTOL. See Quiet, experimental, short takeoff and landing aircraft.
Quiet Engine Project, 124, 130, 131, 292
Quiet Engine Team (LERC) (NASA), 469
Quiet, experimental, short takeoff and landing (QUESTOL) aircraft, 23, 66, 77, 161, 315, 350, 387, 449, 451
Quiet sun, 315

R

Radar
　air traffic control, 25, 38, 141-142, 454
　aircraft, 240, 336-337, 340
　　detection, 321
　ionosphere experiment, 244
　lunar exploration, 364, 403
　missiles, 174, 201-202, 411
　Salyut detection, 277
Radcat (Space Test Program satellite), 336
Radiation
　belts, 115, 227, 389
　corpuscular, 139, 228, 242
　fields, 85, 139
　gamma ray, 90, 139, 242, 384
　infrared, 386
　Jupiter (planet), 39-40
　lunar, 14
　nuclear, 94, 130-131
　nuclear weapons, 192
　particles, 129, 132
　solar, 6, 17, 90, 102, 108, 130, 139, 155, 184, 242, 243
　ultraviolet, 6, 90, 155, 184, 243, 385, 386
　x-ray. See X-ray.
Radio
　astronomy, 115, 213, 365
　experiment, 244, 275, 319
　noise, 354
　stellar emissions, 332
　tracking, 387
　transmission, 174, 182-183, 253, 284
　　U.S.S.R. communications satellite, 128, 331, 350, 404, 426
　transmitter, 364
　waves, 83, 307

Radio Amateur Satellite Corp. (AMSAT), 352, 452
Radio Corp. of America. See RCA.
Radio explorer satellite, 264
Radio Physics Institute (U.S.S.R.), 273
Radio propagation experiment, 372
Radioactivity, 53, 80, 256
Radioastronomy, 365
Radio-interferometer receiver, 403
Radioisotope thermoelectric generator (RTG), 80, 85, 226, 306, 388
Radiometer, 128, 305, 352, 366, 369, 423, 433
Radiotelescope, 18, 123, 187, 213-214, 233, 307, 354, 360, 365
Rae, Dr. John B., 444
Raiffa, Dr. Howard, 340
RAM. See Research and applications modules.
Ramey, James T., 132-133
Ramjet engine, 345
RAND Corp., 123, 155
R&D. See Research and development.
R&PM. See Research and program management.
Rangel, Rep. Charles D., 155
Ranger (program), 403
Ranger 9 (lunar probe), 62
RANN. See Research Applied to National Needs.
Ransone, Robert K., 189
Rasool, Dr. S. Ichtiaque, 17, 246, 311, 376-377, 444
Rasquin, John R., 16
Raven Industries, Inc., 315
Raytheon Co., 35, 123, 302
RB-211 (jet engine), 288, 307
RBV. See Return-beam-vidicon camera.
RCA (Radio Corp. of America), 184, 220, 221, 263, 410
　Aerospace Systems Div., 138
　Electronics Div., 263
　Government Communications Systems, 288
　RCA Global Communications, Inc., 28-29, 98, 294, 410
　RCA Service Co., 264
RCS. See Reaction control system.
RD-107B (U.S.S.R. booster), 128
RDC. See NASA Regional Dissemination Centers.
RDT&E (research, development, test, and engineering). See Research and development.
Reaction control system (RCS), 33, 144-146, 191
Reactive Metals, Inc., 62
Reagan, Gov. Ronald, 101
Rechtin, Dr. Eberhardt, 18, 34, 45
Reconnaissance drone, 72
Reconnaissance satellite, 377
　criticism, 190
　exhibit, 214
　India-Pakistan war monitoring, 9-10
　launch, 77, 108, 154, 200, 305, 346
　opium surveillance, 317

peaceful use, 9–10
press comment, 96
SALT compliance monitoring, 201–202, 288
submarine launch monitoring, 96
U.S.S.R.–Cuban launch site, 321
U.S.S.R. reaction, 362
Redondo Beach, Calif., 200–201
Redstone Arsenal, Ala., 19, 45–46, 292
Reed, Nathaniel P., 216
Reed, Sylvanus Albert, Award, 20–21
Rees, Dr. Eberhard F. M., 45–46, 69, 205, 309, 440
Regolith, 385–386
Reinecke, Lt. Gov. Ed (Calif.), 101
Reis, Herbert K., 134
Relativity, theory of, 132, 194, 360
Relay (communications satellite), 255
Remote sensing (see also Sensor)
  aircraft, 101, 314–315
  cost-effectiveness, 302
  environment, 21–24, 101, 128, 173, 177, 229, 314–315
  ERTS
    cooperation, 301–302
    Interagency Coordination Committee for Earth Resources Survey Program, 75
    programs, 75, 302
  international cooperation, 173, 177, 184, 229, 302
  technology, 34–35, 235, 314–315
Remote Sensing Agreement, 230
Remote terminals, 372
Remotely manned systems, 316
Remotely piloted research vehicle (RPRV), 336–337
Remotely piloted vehicle (RPV), 51, 316, 321, 336–337, 393
Rendezvous (see also Docking)
  Apollo 16 mission, 148, 228
  Apollo-Soyuz Test Project, 61, 174–175, 192–193, 196, 244, 335, 449
    budget, 198–199
    congressional briefing, 209, 229, 239–240
    contract, 378
    joint communique, 207, 252
    launch hardware, 28
    Presidential message, 106
    press comment, 407
    press conference, 197–198, 228, 250, 257–258, 260–261, 344
    "rendock," 182, 205
    spacecraft compatibility, 120, 178, 196, 197–198, 239–240, 240–241, 344, 356, 378
    technical requirements, 136–137, 158
    training, 354
    U.N. resolution, 356
    working groups, U.S-U.S.S.R., 120, 127, 178, 252, 282, 322, 409
  computer simulation, 310
  rescue operations, 137, 240, 344
  Skylab, 344
  space shuttle, 229
"Rendock," 182, 205
Renton, Wash., 42
Republic Aviation Div. See Fairchild Industries, Inc.
Republican National Platform, 298–299
Rescue. See Space rescue.
Research and applications modules (RAM). 69, 333
Research and development (R&D)
  aircraft industry, 22, 122
  awards, 403
  budget, 30–31, 160, 161–162, 178, 187–188, 209, 427
  domestic problems, 24–25, 106, 277, 399, 446
  funding
    congressional recommendation, 274, 289
    environmental technology, 47, 116
    NASA, 64, 78, 125
    presidential campaign issue, 358–359
    scholastic, 258, 326
  military, 65, 233–234, 293–294, 319–320
  national trends, 29, 153
  publications, 134, 311
  small businesses, 223
  solar energy, 85, 445, 446
  space shuttle, 231, 261
  transportation, 52
"Research and Development in the U.S.A.," 115
Research and engineering (DOD), 245
Research and program management (R&PM), 160
Research Applied to National Needs (RANN), 1, 85, 191, 444
Research Council, St. Louis, Mo., 118
Return-beam-vidicon (RBV) camera, 235, 269–270, 284
Reusable surface insulation (RSI), 261
Reynolds, Dr. Joseph M., 282
rf: radio frequency.
Rice, Dr. Donald Blessing, 123
Rice Univ., 44, 70, 83, 86, 315
Richard, Paul, 17–18
Richardson, Robert W., 184–185
Richardson, Secretary of Health, Education, and Welfare Elliot L., 397–398
Rio de Janeiro, Brazil, 42, 383
Ripley, Dr. S. Dillon, II, 372
Risso, William P., 352
Riverside, Calif., 315
Rizzi, Dr. Arthur W., 441
RO&AS Joint Venture (see also Advance Systems Construction, Inc.), 60
Roberson, Floyd I., 434
Roberts, R. Danning, 241
Rochester International Conference on High Energy Physics, 16th, 314
Rochester, Univ. of, 188, 282
Rock, William H., 331
Rockefeller Foundation, 388
Rocket (see also Launch vehicle and Sounding rocket)
  auroral expedition, 373
  cost, 90

design, 256, 373
hydrogen detection, 270–271
launcher, 269
meteorological program, 118, 395–396
monitoring test, 96
P.R.C., 96
propulsion systems, 107, 234, 333
solid-fueled, 309
space shuttle, 272, 332
U.S.S.R., 96, 352, 429
water-entry simulation, 308
Rocket engine. See Engine.
Rocketdyne Div. See North American Rockwell Corp.
Rockwell International Corp., 383
Rockwell Standard Corp., 382, 383
Rockwell, Willard F., 372–373
Rocky Mountains, 27, 45
Rodgers, Donald R., 286–287
Rogers, Secretary of State William P., 49–50, 90, 122, 194, 199
Rohr Industries, Inc., 52
Rolls-Royce (1971) Ltd., 288, 297
Romania, 132, 439
Rome, Italy, 5–6
Roosa, L/C Stuart A. (USAF), 179, 194, 372–373
Rosen, Milton W., 345, 395–396
Ross, Miles, 424
Roswell, N. Mex., 272, 428
Rotary Club (Salt Lake City), 219
Rover. See Lunar roving vehicle.
Royal Greenwich Observatory, 243
Royal Norwegian Council for Scientific and Industrial Research (NTNF), 210
Royal Society of London, U.K., 299–300
Rozhdestvensky, Robert, 405
RPRV. See Remotely piloted research vehicle.
RPV. See Remotely piloted vehicle.
RSI. See Reusable surface insulation.
RTAC. See NASA Research and Technology Advisory Council.
RTG. See Radioisotope thermoelectric generator.
Rubin, Andrey, 291
Rush, Kenneth, 34, 35
Ryan, Dr. James J., Jr., 258, 263, 321, 350

S

S–IC. See Saturn V (booster), stage, 1st.
S–IVB. See Saturn V (booster), stage, 3rd.
S–Band occultation, 81
Sacramento, Calif., 126
SAE. See Society of Automotive Engineers.
Safeguard (antiballistic missile system), 31, 49, 72, 108, 260, 300, 327
Safety (see also NASA Aerospace Safety Advisory Panel; Space rescue)
aircraft, 25, 31, 67, 178, 186, 187–188, 190, 204, 209
astronaut, 68, 129, 132, 298
experimental vehicles, 206
facilities, 139
Life Scientist Program, 298
mine, 73–74
nuclear reactor, 162
space shuttle sortie payload, 220
structural engineering, 71, 225
Safronov, Yu., 362
Sagan, Dr. Carl E., 6, 252–253, 342, 387–388, 443
Saigon, South Vietnam, 393
St. John, Jeffrey, 133
St. Louis, Mo., 118, 219, 241, 281, 282, 341, 395, 400
St. Luke's Hospital, Denver, Colo., 92
St. Martin du Touch Airfield, France, 365
St. Petersburg, Fla., 17
Sakhorov, Andrey D., 240
SALT. See Strategic Arms Limitation Talks.
SALT II (agreement), 218
Salt Lake City, Utah, 323
Salt Lake City, Utah, Rotary Club, 219
Salyut (U.S.S.R. space station), 58
achievement, 86, 166
docking, 61, 108, 128, 174–175, 197–198, 206–207
failure, 277
plans, 166
simulator, 111
Salyut 1–Soyuz 11 (U.S.S.R. orbital station), 255
Salzburg, Austria, 190, 191
Samford Univ., 206
Samoa, 367, 419, 439
SAMSO. See Space and Missiles Systems Organization.
San Antonio, Tex., 38, 137, 233, 411
San Clemente, Calif., 2–4, 12
San Diego, Calif., 20–21, 96, 141, 148–149, 174, 343
San Francisco, Calif., 65–66, 270, 271, 293
San Francisco Medical Center Cardiovascular Research Institute, 74–75
San Francisco State College, 55
San Marco (Italian launch site), Indian Ocean, 386
Sanders, Joseph F., 170
Sanders, Newell D., 186, 244
Sandhawk Tomahawk (sounding rocket), 182–183
Santa Monica Freeway, 35
SAO. See Smithsonian Astrophysical Observatory.
Sargent, Dr. W. L. V., 360
Sarnoff, Robert W., 410
SAS. See Small Astronomy Satellite.
SAS–B (Small Astronomy Satellite). See *Explorer 48*.
Saskatchewan, Canada, 184
Satellite for Environmental and Technical Research. See *Sret 1, 2, 3*.
Satellite Test Center, 337

Saturn (booster), 33, 45, 175–176, 280, 309, 424, 440
  Saturn I, 169–170
  Saturn IB, 28, 33, 169–170, 197–198, 298, 384
  Saturn V
    Apollo missions, 150, 432
    Apollo 16 mission, 28, 46, 112, 142–144, 151
    Apollo 17 mission, 301, 361, 413, 420
    contract, 227
    launch, 142, 413
    lifting power, 140
    magnetomotive hammer, 16
    Skylab mission, 189
    stage
      1st (S-IC), 112, 227, 266, 361, 413
      3rd (S-IVB), 180–181, 237, 413–415, 432
    von Braun influence, 202–203
  Saturn V Orbital Workshop. See Skylab.
Saturn (planet)
  Apollo 17 emblem, 347
  atmosphere, LST monitoring, 176
  exploration
    cost, 90, 451
    Mariner Jupiter/Saturn 1977 project, 266, 320, 423
    plans, 28, 39–40, 56, 64, 65
    U.S.–U.S.S.R. cooperation, 178, 311
  extraterrestrial life, 291
  size, 165
  space environment model, 441
  Titan (moon), 65
Saturn Program Office (MSFC) (NASA), 243–244
Sawfly (U.S.S.R. missle), 71, 335, 397
SBA. See Small Business Administration.
SCAD. See Strategic bomber penetration decoy.
Scandinavia, 133–134
Scanning radiometer (SR), 352
Schaeffer, Dr. Oliver A., 363
Scheer, Julian, 421–422
Schilling, David C., Award, 320
Schirra, Capt. Walter M., Jr. (USN, Ret.), 271
Schlesinger, Dr. James R., 53, 288, 374, 404, 421
Schliesing, John A., 310
Schmitt, Dr. Harrison H.
  Apollo 17 mission, 436
    emblem, 347
    EVA, 414–419, 440, 444
    launch, 413
    lunar landing, 367
    medical aspects, 403, 420
    prelaunch test, 389
    preliminary timeline, 297
    press comment, 438
    press conference, 363–364, 382
    spacecraft naming, 382
  awards and honors, 429
Schneider, William C., 63
Schneiderman, Dan, 98, 274, 310

Scholastic cooperation, 85, 254, 264, 265, 330
Schomburg, L/G August (USA, Ret.), 19
Schopf, Dr. J. William, 158
Schrieffer, Dr. John R., 357
Schull, Wilfred E., 235
Schultz, M/G Kenneth W., 189
Schumacher, Howard E., 20–21
Schurmeier, Harris M., 266, 320, 403
Schwartz, Dr. Philip R., 365
Schweickart, Russell L., 23, 72, 219
Science
  awards and honors, 95, 191, 202–203, 357, 429
  education, 1, 154–155, 201, 283–284, 299
  funds, 112, 359
  human needs, 25, 29, 159–160, 230
  international cooperation
    agreements, 195, 199, 201, 203, 206, 207, 303
    basic research, 94
    environment, 298
  national policy and goals, 1, 37, 60, 160–161, 200–201, 287, 298–299
  Presidential message, 105–106
  space program, 57, 63, 101, 102, 152–153, 167, 176–177, 183–184
Science and Applications Directorate (MSC) (NASA), 205
Science and Technology Agency (Japan), 26
Science Research Council (U.K.) 208, 354
Scientific instrument module (SIM), 142–144, 148–149, 366, 414, 419
Scientist-astronaut, 23–24, 55, 121, 185, 322, 406, 413, 426
Scientists
  American Physical Society, 159
  awards, 21, 253
  credibility, 13
  employment, 64–65, 134, 326, 339
  Environmental Action Group, 162
  Federation of American Scientists, 181
  foreign, 8, 71, 202–203, 239, 392
  international cooperation, 127, 128, 136, 173, 299–300
    U.S.–P.R.C., 181
    U.S.–U.S.S.R., 136, 187, 189–190, 195–196, 199, 208, 352
  life sciences, 68, 158, 298
  lunar science, 52, 54, 71, 133, 181
  manned space flight, 11, 41, 64, 70, 121, 140, 204–205, 344, 423, 426
  meetings, 17, 21, 60, 72–73, 154–155, 159, 162, 178, 291, 321, 330, 385
  NASA programs, 135, 176, 302, 331, 405
  patent, 184–185
  planet discovery, 165
  request for proposals, 177–178, 275
  research and development, 42, 70, 134, 175, 359
  satellite programs, 28, 48, 174, 208, 254, 320, 369, 423

Scientists and Engineers for Social and Political Action, 444
Scientists, Engineers, and Physicians from Abroad: Trends Through Fiscal Year 1970, 8
SCMR. See Surface-composition-mapping radiometer.
Scotland, 96
Scott, Col. David R. (USAF), 150
  Apollo 15
    EVA, 432
    lunar observations, 13-14
    appointment, 194, 273
    awards and honors, 99, 114, 157, 210, 267-268, 335-336
    commercialism, 232, 256, 276, 282-283
    press comment, 276
    public appearance, 1, 20, 28, 35, 61
    White House visit, 41
Scott, Sen. Hugh, 226, 264
Scott, James E., 73
Scott, Sheila, 77
Scott, Wallace, 161
Scott AFB, Ill., 61
"Scott and Erwin on the Moon" (sculpture), 17-18
Scout (booster), 115, 210, 290, 306, 344-345, 361, 386
Scout-D, 194, 381, 430
SCR. See Selective chopper radiometer.
Scripps Institute of Oceanography, 282
Sea of Clouds (Mare Nubium) (moon), 404
Sea of Crises (Mare Crisium)(moon), 179, 433
Sea of Fertility (Mare Fecunditatis) (moon), 53, 179
Sea of Rains (Mare Imbrium) (moon), 14, 377-378
Sea of Serenity (Mare Serenitatis) (moon), 54, 363, 367, 433
Sea of Tranquility (Mare Tranquillitatis) (moon), 265
Seal Beach, Calif., 309
Seamans, Dr. Robert C., Jr., Secretary of the Air Force, 8, 77, 258, 263, 286-287, 378, 382-383, 443
Search, Project (System for Electronic Analysis and Retrieval of Criminal Histories), 126
Seattle, Wash., 93, 166, 169, 256
Second Supplemental Appropriations Act, 1972, 207
Secretariat for Electronic Test Equipment (SETE), 16
Sedov, Dr. Leonid I., 137, 344, 353
Seiberling, Rep. John F., Jr., 326
Seismology, 95
Seismometer experiment, lunar, 142-144, 219-220, 237, 380
Selb Manufacturing Co., 209
Selective chopper radiometer (SCR), 423
Senate. See Congress, Senate.
Sensors, 43, 63-64, 69, 377-378
Serenitatis Basin (moon), 363
Serpukhov, U.S.S.R., 10, 55, 221, 314

Service propulsion system (SPS), 142-144, 144-146, 148
Service Technology Corp. See LTV Aerospace Corp.
SES-100A (surface effect ship), 252
Sesp 1971-2 (Space Test Program satellite), 354, 369
SETE. See Secretariat for Electronic Test Equipment.
Shadow bands, 254
Shafer, Robert J., 52
Shaffer, John H., 93, 326, 430
Shanghai Airport, 86, 98
Shanghai, People's Republic of China, 28-29, 294
Shapley, Dr. Harlow, 357
Shapley, Willis H., 216, 234
Shatalov, Vladimir A., 110-111, 335, 338, 375
Shelby County, Tenn., 72
Sheldon, Dr. Charles S., II, 7, 28, 166
Shen, Dr. Wen-we, 441
Shepard, R/A Alan B., Jr. (USN), 122, 272-273, 404, 432
Shepard, Leonard E., 372-373
Sherman Antitrust Act, 122
Sherrod, Robert, 231-232
Shevchenko, U.S.S.R., 2
Shinsei (Japanese satellite), 295
Ships
  military, 134, 301, 327
  passenger, 104, 409
  recovery, 148-149, 162, 313, 366, 405, 419, 435
  research, 21, 128, 136
  surface effect, 252
  yawl, 242
Shirley Bay, Canada, 328
Shklovskiy, Josif, 187
Shobert, Erle I., II, 93
Shockley, Dr. William, 441
Short takeoff and landing aircraft. See STOL.
Shorty Crater (moon), 416, 432, 440
SHOT. See Society for the History of Technology.
Shuttle Task Team, 110
Shuttle training aircraft (STA), 247
Siberia, 254, 404
Sickle and Hammer (U.S.S.R. gold medal), 326, 368
Sieger, Herman E., 232
Siffre, Michel, 308
Significant Accomplishments in Science (symposium), 378
Significant Accomplishments in Technology (symposium), 377-378
Sikorsky Aero Engineering Corp., 362
Sikorsky Aircraft Div. See United Aircraft Corp.
Sikorsky, Igor I., 362, 365
Silicon cell, 250
Silver, Dr. Leon T., 14
SIM. See Scientific instrument module.
Simmons, Dr. Gene M., 71
Singer Co., 266

Sinus Aestuum (moon), 433
Site Review Board. See NASA Site Review Board.
Sjoberg, Sigurd A, 172
Skybolt (missile), 403
Skyhook Project, 335
Skylab (program), 23–24, 28, 58, 68, 69, 128, 205, 288, 298, 408
  ATM, 44, 70, 262
  contract, 39, 44, 434
  cost factors, 27, 30, 64, 125, 198–199, 245, 297–298, 407, 424, 439
  crew, 23, 281, 282, 426, 447
  development and testing, 261, 262, 282, 341, 360, 447
  employment, 78, 301, 424
  international cooperation, 27
  management, 63, 159
  medical aspects, 24, 68, 97, 184–185, 279, 282, 347–348
  Nixon comment, 436
  plans
    experiments, 24, 41, 44, 68, 70, 96–97, 219
      astronomy, 24, 70, 239
      earth observation, 24, 63–64, 221–222, 283, 383
      materials science, 27, 41, 210–211, 337
  press conference, 23–24, 285–286
  safety considerations, 135, 344
  Saturn IB booster, 266, 280, 298
  shower assembly development, 165–166
  status 33, 63–64
  support services, 44, 248
  technology utilization, 309
Skylab (spacecraft), 58, 189
  airlock module (AM), 219, 262, 280, 282, 325, 341, 344
  multiple docking adapter (MDA), 219, 262, 280, 282, 325, 341
  Workshop, 60, 68, 169, 209, 262, 298
Skylab A, 33
Skylab B, 33
Skylab Educational Conference, 1, 265
Skylab medical experiments altitude test (SMEAT), 273, 281, 285–286, 323
Skylab Student Project, 1, 35, 61, 131, 177, 222–223, 265, 288
Skylark (U.K. sounding rocket), 120
Slayton, Maj. Donald K. (USAF, Ret.), 97, 128, 205, 232, 280
Sloan, Alfred P., Foundation, 299
SM (service module). See Command and service module.
Small Astronomy Satellite (SAS), 361, 386
  Explorer 42 (Uhuru), 56–57, 107, 261, 307, 332, 360, 438
  Explorer 48 (SAS-B), 361, 386, 447, 452
Small Business Administration (SBA), 88–89, 311
Small Business Assn. Achievement Award, 247
Small Business Investment Companies (SBICS), 106
Small Scientific Satellite (SSS), 233, 280

Small tactical aerial mobility platform (STAMP), 446
SMEAT. See Skylab medical experiments altitude test.
Smith, Arthur, 167
Smith, Dr. Bradford, 42–43
Smith, Dr. Edward J., 266
Smith, Rep. H. Allen, 322, 430, 434
Smith, H. Allen, Jet Propulsion Laboratory, 322, 338, 358, 363, 375, 395
Smith, Kenneth M., 221
Smith, Dr. Malcolm C., 71–72
Smithsonian Astrophysical Observatory, 14, 38, 130–131, 194, 372, 434, 454
Smithsonian Institution, 153, 291, 372
  Aerospace Art Hall, 211
  Arts and Industries Building, 74
  National Air and Space Museum, 45, 388, 454
Smog, 35, 101
Smokey (long-range laser traversing system), 138, 258
Smoky Mountain (moon), 146–148
SMS. See Synchronous Meteorological Satellite.
SMS-A, 171
SMS-B, 171
SNAP-9A (radioisotope thermoelectric generator), 80
SNAP-27, 388
Snyder, C. Thomas, 98
Social problems, 46–47, 73–74, 133, 135–136
Social Science Research Council, 392
Society for the History of Technology (SHOT), 443–444
Society of Air Safety Engineers, 350
Society of Automotive Engineers (SAE), 16, 137
Society of Engineering Science, 227
Society of Japanese Aircraft Constructors, 204
Socorro, N. Mex., 104
Soil Mechanics Investigation Science Team. See Apollo Soil Mechanics Investigation Science Team.
Soil studies, 298
Solar array, 208, 270
Solar astronomy. See Astronomy, solar.
Solar atmosphere, 300
Solar cell, 1, 38, 85, 128, 169, 188, 250, 290, 291–292, 379, 389
Solar corona, 254
Solar eclipse, 254
Solar emissions, 242, 387
Solar energy, 123, 128, 250, 252–253, 330, 387
  data exchange, U.S.–U.S.S.R., 199, 236–237, 341
  development and use, 85, 138, 160
  satellite system, 123, 251, 291–292
Solar Energy: A Multidisciplinary Approach, 446
Solar Energy in Developing Countries: Perspectives and Prospects, 138
Solar evolution, 253–254, 420

Solar flare, 290–291, 315, 330, 337, 366, 369, 389, 454
Solar gravity, 132
Solar neutrons, 242
Solar occultation, 132
Solar paddle, 386
Solar physics (see also Apollo Telescope Mount)
  Apollo program, 69
  atmospheric models, 70
  Prognoz 1, 139
  Skylab program, 63–64, 70
  sounding rocket experiment, 158, 210, 224, 242, 255, 276, 385
  Stein wave discovery, 131
Solar plasma, 107, 139, 242, 369
Solar propulsion, 39–40
Solar proton monitor (SPM), 352
Solar radiation, 6, 17, 90, 102, 108, 139, 155, 184, 243, 375
Solar satellite, 123, 251, 291–292, 384–385, 445
Solar storm, 273, 276, 280, 284, 300
  exploration, 28, 39–40, 166, 176–177, 374
  formation, 176, 246
  lunar programs, 57–58, 377–378
  magnetic shells, 291
  tenth planet, 165
Solar wind, 51, 441
  measurement, 177–178, 300
  spacecraft observations, 28, 38, 242, 308, 325
  spacecraft propulsion, 141
Solid-fuel motor, 35, 39
Solid-waste management, 1, 5
Sonar, 441
Sonic boom, 74, 80, 191–192, 306, 327
Soo Canal, 21
SORIS. See Italian Research Organization.
Sorokhov, Gen. Simha (Israel), 91
Sortie can (space shuttle payload carrier), 11–12, 68, 129, 239, 366, 452
Sortie Can Task Team, 129
Sortie Lab Task Force (OMSF), 398, 451
Sortie Laboratory, 452
Sortie mission, 68, 69
Sounding rocket (see also individual sounding rockets such as, Aerobee, Arcas, Javelin, Nike-Apache, Skylark, etc.), 245, 373, 374, 378, 449
  experiments
    analysis of, 178
    astronomy, 259
    atmospheric, 8, 51, 315, 410
    auroral studies, 182–183, 353, 358, 373
    cosmic dust collection, 342, 343, 348
    earth resources, 120
    field and particles, 19, 109, 283, 284, 327
    ionospheric study, 248, 251
    meteorology, 105, 228, 238, 395–396
    ozone measurement, 184, 186–187, 210
    Polar Morning, 228
    solar physics, 255, 274, 276, 366
    thermospheric winds, 349
    winter anomaly study, 39
  history of, 394
  international programs, 248, 251, 453
    NASA
      -Canada, 19, 109, 283, 284, 327
      -Norway, 15, 17
      -Spain, 248, 251
      -U.K., 120
    U.N. World Meteorological Org., 105
    U.S.S.R.-France, 228
  launch
    Aerobee 150, 156, 287
    Aerobee 170, 61, 127, 158, 180, 189, 222, 242, 288, 422, 427
    Arcas, 173, 184, 195
    Black Brant VC, 210, 222
    Nike-Apache, 54, 132, 210, 222, 283, 284
    Nike-Cajun, 184, 186–187, 210
    Nike-Javelin III, 119, 127
    Nike-Tomahawk, 44, 46, 53
    Sandhawk Tomahawk, 182–183
  launch failure, 37
  meridional networks, 229
  test, 327, 360, 388
South Africa, 155, 158, 181
South America, 38, 63, 133–134, 263, 379
South Atlantic Anomaly, 216
South Carolina, 236, 413
South China Sea, 96
South Massif (moon), 416
South Ray Crater (moon), 146–148, 172
Southern Contractors Service, 169–170
Southwest Research Institute, 38
Soviet Academy of Sciences, 2, 14, 71, 85, 86, 138, 179, 299–300, 374, 385–386
  Council for International Cooperation in Investigation and Utilization of Outer Space, 127
  Institute of Space Studies, 10–11
  Intercosmos Council, 140–141, 208, 252
  international cooperation
    Joint Working Group on the Natural Environment, 298
    scientist exchange, 136
    visitation, 354
  international cooperation, space, 14, 75, 136, 195–196, 208, 282, 305, 310–311, 375
    information exchange, 208
    joint mission, U.S.-U.S.S.R., 48–49, 130, 157, 207, 322, 336
    Joint Working Group on Compatibility of Docking Systems and Tunnels, 282
    Joint Working Group on Space Biology and Medicine, 170, 279
    lunar science, 14, 305
    planetary exploration, 310–311
    science and applications, 239–240
  Ioff Physical-Technical Institute, 176

Soviet Institute of Outer Space Research, 177, 189–190
Soviet Ministry of Health, 170
Soviet Mountains (moon), 237
Soviet National Conference on Space Biology and Medicine, 360
Soviet Politburo, 299
Soviet Space Program, 1971, 166
Soviet Space Programs, 1966–70, 7
Soviet Standards Committee, 215–216
Soyuz (program), 7, 28, 166, 279
Soyuz (U.S.S.R. spacecraft) (see also Apollo-Soyuz Test Project), 61, 108, 130, 353, 354
Soyuz 11 mission, 167, 180, 279
Soyuz 12 mission, 166
Soyuz 13 mission, 166
SPAC. See NASA Space Program Advisory Council.
Space and Defence Systems Div. See Marconi Co. Ltd.
Space and Missile Systems Organization (SAMSO) (USAF), 163, 189, 313, 336, 370
Space and Science Awareness Month, 72
Space biology and medicine, 24, 30, 50, 63–64, 67–68, 75, 97, 101, 132, 227, 248, 273
  ALFMED experiment, 6
  animal experiments, 394, 404
  Applications of Aerospace Technology in the Public Sector, 75
  astronaut, 52, 65–66, 68, 119–120, 153, 279–280, 420
  AVST, 41
  award, 348
  Biosatellite II, 170
  cardiology, 86
  enzyme research, 247, 357
  extraterrestrial life, 171, 291
  Human Factors in Long-Duration Spaceflight, 129, 153
  IMBLMS, 283
  international cooperation, 15, 127, 170, 180, 194, 195–196, 201, 203, 207, 229, 239, 240–241, 279–280
  Life Scientist Program, 74–75, 298
  life support system, 129, 187, 194–195, 293, 298
  lung tissue study, 67–68
  medical glossary, 279
  NASA program
    Apollo 8 mission, 394
    Apollo 15 mission, 65–66
    Apollo 16 mission, 68, 119–120, 135, 148–149, 150, 152, 164, 166, 172, 279–280
    Apollo 17 mission, 68, 385, 388, 404, 414, 419–420, 437
    Skylab, 41, 63–64, 189, 281, 285–286, 323, 347–348
    space shuttle, 293
  nutrition, 71–72
  remote health care, 464
  spacecraft sterilization, 24
  technology utilization. See Technology utilization, space, medicine.
  time cycle study, 308
  U.S.S.R. program, 360, 403
  weightlessness effects, 15, 24, 41, 65–66, 73, 129, 136, 241, 279–280, 321–322, 337, 360, 419
Space Congress, Ninth, 154
Space Data Corp., 113
Space debris, 109
Space Environmental Simulation Laboratory, 301
Space Fair, Point Mugu, Calif., 358
Space Flight Award (AAS), 372–373
Space hibernation, 384
"Space in the Age of Aquarius" (TV program), 134–135
Space law, 197
Space, military use of
  Apollo program, 441
  communications, 116, 190, 371, 425
  moon, 134
  powered balloon system, 319
  press comment, 441
  reconnaissance, 25, 96, 100, 154, 190, 200, 201–202, 305, 346
  space shuttle, 8, 9, 18–19, 34, 139–140, 157, 451
  "super-secret satellite spy system," 214
  Transit navigation satellite, 192
  U.S., 130–131, 245, 319, 407
    DOD, 8, 31, 190, 362
  U.S.S.R., 7, 120, 167–168, 252, 328–329, 352, 411
  weather control, 247, 250
  West Germany, 159
Space Nuclear Rocket Development Station, 59
Space Nuclear Systems Office. See AEC–NASA Space Nuclear Systems Office.
Space Ordnance Systems, Inc., 266
Space, peaceful use of, 35, 342–343, 371, 452
  international cooperation
    Agreement Concerning Cooperation in the Exploration and Use of Outer Space for Peaceful Purposes, 195–197
    Outer Space Liability Convention, 49
    U.N. activity, 348, 356, 357
    U.S.–U.S.S.R., 127, 191, 197, 203, 204–205, 207, 336, 338, 452
    U.S.S.R.–France, 323
  press comment, 423, 434
  Treaty on Principles Governing the Activities of States in the Exploration and Use of Outer Space, Including the Moon and Other Celestial Bodies, 195, 203
  United Nations Committee on the Peaceful Uses of Outer Space, 173
  U.S. foreign policy, 48–49

Space program, national (see also individual programs, such as Apollo program; and National Aeronautics and Space Administration), 72, 202–203, 214–215, 218, 219, 330, 371, 384, 394, 403
  achievements, 112–113, 135–136, 286, 374, 394, 442–443, 447
  Aeronautics and Space Report of the President: 1971 Activities, 99
  benefits (see also Space results), 24–25, 34–35, 46–47, 105–106, 134–135, 164–165, 301–302, 408, 428, 449, 452
  budget (see also National Aeronautics and Space Administration, budget), 64, 86–87, 89–90, 99–100, 116, 133, 161–162, 164–165, 219, 297, 410, 450
  Conquest of Space, 211
  cost, 86–87, 89–90, 99–100, 137, 188, 190–191, 293, 407, 424
  criticism, 55, 200–201, 223, 328, 407, 424
  goals and priorities, 27–28, 32, 36, 121, 142, 279, 326, 450–451
  history, 132–133, 405
  international aspects (see also International cooperation, space; Space race), 48–49, 125, 126, 128, 167, 239, 251–252, 333, 342–343, 407
    U.S.–U.S.S.R., 7, 136–137, 140–141, 150, 166, 167–168, 175–176, 185, 197–198, 362
  Nixon, President Richard M., 2–4, 49–50, 112–113, 428, 435–437
  opinion, 134–135, 231, 314, 359, 422, 430
  outlook, 11, 58, 140, 175–176, 177, 181–182, 219, 328, 393, 408, 441
  press comment, 11, 16, 156, 162, 167, 178, 272, 314, 348–349, 407, 411–412, 422, 423, 426, 440
  significance, 57, 135–136, 162, 167, 175–176, 184, 192–193, 204–205, 233, 287, 404–405, 406
  United States Space Science Program: Report to COSPAR, 178
Space race, 9, 10, 64, 98, 100, 175–176, 404–405, 407, 434, 439
  press comment, 348, 434
Space rescue, 128, 136–137, 174, 195–196, 197–198, 199, 207, 239–240, 336, 344
  Agreement on the Rescue of Astronauts, the Return of Astronauts, and the Return of Objects Launched into Outer Space, 195
Space results (see also Research Applied to National Needs), 52, 74, 112, 153, 164–165, 223, 228, 271, 309, 311, 394
  Applications of Aerospace Technology in the Public Sector, 75
  astronomy, 179–180, 220, 300, 447
  communications, 84, 114, 131, 242, 407, 408
  computer systems, 63, 71, 452
  earth resources, 2–4, 24, 84, 122, 226–227, 271, 272, 314–315, 408, 447
  education, 114
  engineering, 45, 47–48, 56, 71, 84, 178–179, 271, 452
  environment (see also Earth Resources Technology Satellite; *Erts I*), 53, 63, 68, 122
  forest fire reconnaissance, 280–281
  industry, 45, 84, 223, 239–240, 248, 271, 348–349
  law enforcement, 74, 75, 114, 271
  medicine, 38, 41, 68, 74, 75, 84, 92, 95, 101, 106, 112, 114, 248, 258, 270, 271, 316, 375, 396, 408, 411, 452
  meteorology, 407–408
  National Symposium on Technology Transfer, 225
  navigation, 281
  Nixon, President Richard M., 435–437
  nuclear fission, 203
  oceanography, 136, 250–251, 316
  pollution control, 35, 68, 74, 105, 390, 407
  press comment, 156, 272, 348–349, 411–412
  safety, 74, 75, 101, 219, 270, 271, 316, 368
  social problems, 1, 2–4, 5, 32, 314, 356
  space applications, 314, 316, 374, 408, 452
  submarine, 228
  technology application team, 75
  transportation, 75, 84, 105, 411, 423
  urban problems, 75, 84, 114, 126, 131, 270
Space Science Award (AIAA), 20–21
Space Science Awareness Month, 400
Space Science Board (SSB) (NAS–NRC), 39–49, 64 65, 129, 153, 178
Space Science Projects Office (MSFC) (NASA), 243
Space shuttle, 28, 98, 175–176, 198–199, 204, 219, 238, 345, 438
  award, 69
  benefits, 55, 58, 59, 166–167, 244–245, 399–400
  booster, 58
  cost, 39, 102
  development, 345, 427
  recovery test, 309
  reusable, 102
  composite structural materials, 292
  contract
    booster, 35
    checkout system, 34
    crew compartment, 271
    employment repercussions, 274, 276
    engine, 42, 90, 254–255, 272, 292
    GAO review, 123
    heat protection, 203, 220
    heating system, 37
    operational processing, 153
    orbital maneuvering system, 219, 266
    orbiter, 42, 129, 261, 266, 272, 281, 283
      cargo handling system, 301
    orbiter propulsion, 180

orbiter RCS, 191
orbiter simulator, 45, 174, 247, 281
payload systems compatability, 220
press comment, 276
prime contract, 279, 284–285, 292, 339
   loss impact, 280, 292
   technical proposals, 180
subcontracts, 275–276, 319
waste collection system, 277
weld-bonding materials, 91
cooperation, 8, 116, 331
cost, 36–37, 51, 55, 69, 86–87, 219, 399–400, 407, 451
  booster, 39, 102
  congressional testimony, 89, 102, 136, 140
  orbiter, 136, 383–384
  payload value, 181, 188
  press conference, 102, 424
crew compartment, 271
criticism, 6–7, 16, 20, 51, 55, 177, 181, 185, 214–215, 231
design, 58, 71, 79, 90, 261, 308
development, 33, 79, 108, 155, 253, 272, 447
DFBW, 200, 312, 449
earth resources applications, 407
employment, 5, 9, 27, 34, 81, 214, 272, 276, 319, 451
endorsements, 32, 36, 37, 119, 319, 374
engine, 45, 292, 332
facilities, 109–110, 184, 187–188, 209, 273–274
feasibility studies, 58
flight schedule, 383–384
funds
  congressional approval, 160, 178, 185, 193, 209
  development, 452
  NASA request, 30, 32, 64
  press comment, 156, 157, 439
  press conference, 32
  R&D, 30, 125, 209, 245
  space program impact, 11
impact, 175–176
  employment, 276, 319, 451
  scientist-astronaut, 121
  space exploration, 11, 328, 399–400
  space program, 120–121, 140, 175–176, 408, 436
  U.S. technology leadership, 49
international cooperation, 120, 219
  NASA
    -Europe, 48, 239, 251–252, 303–304, 366, 437, 452
    -U.S.S.R., 136–137, 229, 253, 411
LaRC technology development, 153
Large Space Telescope (LST), 176
launch and landing sites, 12, 102, 139–140, 181–182, 187–188, 240, 449
Mars mission use, 343
Mathematica, Inc., report, 58
mechanisms, 310
meeting, 137, 292, 310, 444

military applications, 219
NASA Site Review Board, 139–140
NASA Space Shuttle Technology Conference, 137
Nixon, President Richard M., 2–4, 4–5, 7, 8–9, 10, 11, 16, 18, 19, 25, 32, 34, 36–37, 46–47, 48, 105
orbital maneuvering system, 42, 47, 219, 266
orbiter
  approach-landing simulation, 45
  cargo handling system, 301
  development, 272
  interior, 42
  liquid-hydrogen fuel, 58
  simulator, 45, 174, 247, 281, 397
  statistics, 383–384
  thermal protection, 214, 281
  winged model flight, 303–304
parachute deceleration system, 223–224
payloads symposium, 444
press comment
  contract, 276
  endorsement, 8–9, 10, 11, 18–19
  funds, 7, 11, 16, 156, 157, 295, 424, 439
  impact, 35, 276
  launch site, 8
  scientist-astronaut, 121
  U.S.S.R. cooperation, 15
press conference, 4–5, 11–12, 32, 139–140
propulsion, 390
reentry, 283
remotely manned systems, 316
Republican National Platform, 298
request for proposals, 55, 102, 108
rescue, 137
reusable, 140–141, 219, 449
safety, 229
scientist-astronaut, 121, 426
simulation study, ARC, 303
Sortie can, 11–12, 68, 129, 239, 366, 452
specifications, 272, 383–384
technology, 79, 106, 121, 244–245, 408
technology utilization, 105, 408
waste collection system, 277, 293
Space Shuttle Payloads Symposium, 444
Space Shuttle–Skylab: Manned Space Flight in the 1970's, 33
Space Shuttle Technology Project Team, 380–381
Space stamps, 253
Space station (see also Salyut; Skylab)
  award, 241
  design, 241
  funding, 30
  international cooperation, 28, 127, 137, 173, 198, 207, 229
  military applications, 9
  remotely manned systems, 316
  rendezvous and docking, 198, 206, 229
  rescue operations, 229
  scientist-astronaut, 406
  significance, 231

space shuttle application, 271
student participation, 1, 35
U.S.S.R., 28, 128, 141, 166, 175-176
Space Station Task Force (OMSF), 398, 451
Space Systems Committee. See NASA Space Systems Committee.
Space Systems Div. See Naval Research Laboratory.
Space Test Program (STP) (USAF), 336, 354, 369
Space Tracking and Data Acquisition Network (STADAN), 130-131
Space tug, 11-12, 48, 129, 226, 238, 239, 333, 345, 390
Spacecraft (see also individual spacecraft)
 accident, 91
 compatible docking systems, 240-241
 experiments analysis, 178
 Jupiter-Saturn mission, 320
 Mars mission, 343
 onboard data-management systems, 220
 power systems, 79-80, 291-292
 resource management, 235
 solar panel, 291-292
 space biology, 92, 360
 systems engineering, 403
 technology, 377-378
Spacecraft Sterilization Seminar, KSC, 24
"Spaceship Earth," 319
Spacesuit, 8, 142, 146-148, 185, 189, 262, 282
Spain, 178, 189-190, 248, 251
Sparrow (missile), 403
Spartan (antiballistic missile), 361
Special Achievement Award (NRL), 291 292
Special Astrophysical Observatory U.S.S.R.), 208-209
Spectrograph, 254, 255, 270-271, 306, 433
Spectrometer
 Aeros, 431
 Apollo 16, 144
 Apollo 17, 54-55, 366
 extreme-ultraviolet, 431
 far-ultraviolet, 366
 gamma ray, 148, 336, 342
 HEAO, 342
 mass, 54-55, 210, 431
 McMath solar telescope, 421
 microwave, 423
 Nimbus 5, 422
 Oao 3, 296
 photoelectric, 296
 rapid-scanning Fourier, 421
 solar ultraviolet, 210
 sounding rocket, 210
 stellar dust composition, 291
 Stp 72-1, 336
Spectrophotometer, 254
Sperry Rand Corp., 51
Spirit of St. Louis Award, 315
Spiritual Frontiers Fellowship, 188

SPM. See Solar proton monitor.
Spokane, Wash., 292, 365
Spook Crater (moon), 146-148
Sprint (missile), 108, 174, 260, 300, 411
SPS. See Service propulsion system.
Sputnik (U.S.S.R. satellite), 87, 167, 315, 357
Sputnik 1, 348, 399
SR-71 (reconnaissance aircraft), 162, 263, 319, 320, 350
SRDS. See Systems Research and Development Service (FAA).
Sret 1 (Satellite de Recherches et d'Environment Technique) (French Satellite for Environmental and Technical Research), 128, 449
SRET 2, 128
SRET 3, 128
Sriharikota, India, 138
S.S. *Statendam* (ship), 409
SS-4 (U.S.S.R. ICBM), 87
SS-5, 87
SS-9, 115, 252, 329, 445
SS-11 (Savage) (U.S.S.R. ICBM), 87, 350, 355
SSB. See Space Science Board.
SSN-8 (U.S.S.R. submarine-launched ballistic missile), 210
SSNX-8 (Sawfly), 71, 335, 397
SSS. See Small Scientific Satellite.
SST. See Supersonic transport.
STA. See Shuttle training aircraft.
Staats, Elmer B., 123
Stack, John, 233
STADAN. See Space Tracking and Data Acquisition Network.
Stafford, B/G Thomas P. (USAF), 39, 60, 128, 346, 354, 404
Stalin, Josef, 352
Stalin Prize, 442
STAMP. See Small tactical aerial mobility platform.
Stamper, Malcolm T., 331
Stanberry, Communications Minister Robert (Canada), 328
Stanford Research Institute, 392
Stanford Univ., 74-75, 152, 281, 306
Star, 108, 181, 187, 394, 438, 486
Star City. See Zvezdny Gorodok.
Star tracker, 295-296
StarQuest Ltd., 337
State, Dept. of, 226, 227, 301-302, 384
State Univ. of New York
 Albany, 385
 Stony Brook, 246, 421
Statler, Richard L., 291-292
STC. See Service Technology Corp.
Stein, Alan, 131
Stein, Dr. William H., 357
Stein waves (sun), 131
Steinbacher, Dr. Robert H., 225-226
Stellar reference assembly, 253
Stendahl, Dr. Krister, 387-388
Steno Crater (moon), 416
Sterilization, spacecraft, 24, 65
Sterling, Claire, 225

Stern, Robert, 86
Stever, Dr. H. Guyford, 34–35, 343
Stockholm, Sweden, 217, 218–219, 226–227, 232, 319
Stockholm, Univ. of, 385
Stockton Metropolitan Airport, Calif., 186
Stoeckenius, Dr. Walther, 74–75
STOL (short takeoff and landing) aircraft, 324, 355, 449
  augmentor wing, 172
  cooperative research, 23, 77
  design and development, 40, 53–54, 67, 118, 400
  foreign programs, 94, 354
  funding, 31, 66–67
  landing systems, 67, 182
  noise reduction, 66–67, 185–186
  propulsion systems, 2, 124, 172, 315
STOLport, 268–269
Stone Mountain (moon), 62–63, 146–148
Stone, Robert G., 387
Stony Brook, N.Y., 246, 421
Storms, 118, 255–256, 273, 275, 276, 354
STP. See Space Test Program.
Stp 72–1 (Space Test Program satellite), 336
Strangway, Dr. David W., 152, 183–184
Strasbourg, France, 29
Strategic Arms Limitation Talks (SALT), 49, 121, 201–202, 214, 218, 236, 252, 265, 288, 310, 320, 356, 390
Strategic bomber penetration decoy (SCAD), 31
Stratoscope (balloon-borne telescope), 69
Stratoscope I, 64
Stratoscope II, 64
Stratosphere (see also Atmosphere, upper), 80, 262, 280, 315, 385
Strauss, R/A Lewis L. (USN), 132–133
Structures, Structural Dynamics and Materials Conference, 13th, 137
Strypi 4 (sounding rocket), 353
Student Academic Freedom Forum, 400–401
Submarine, 25, 71, 96, 201–202, 210, 228, 308, 397
Subotowicz, M., 345
Subsatellite. See *Apollo 15 Subsatellite*; *Apollo 16 Subsatellite*.
Subsonic aircraft, 155
Suitland, Md., 302
Sukanov, Ya., 362
Sukhoi-7 (U.S.S.R. fighter-bomber aircraft), 330
Sulfur dioxide, 391
Sullivan, Francis J., 79–80
Sullivan, Walter S., 290–291, 332, 367–368, 423, 429
Sulzberger, Cyrus L., 223
Summary of Results of Discussion on Space Cooperation Between the U.S. National Aeronautics and Space Administration and the Academy of Sciences of the U.S.S.R., 195–196
Summary of Results of the Meeting Between Representatives of the U.S. National Aeronautics and Space Administration and the U.S.S.R. Academy of Sciences on the Question of Developing Compatible Systems for Rendezvous and Docking of Manned Spacecraft and Space Stations of the U.S.A. and U.S.S.R., 195–196
Summary Report of the Ad Hoc Panel on (NO_x) and the Ozone Layer, 370
Summer Institute for Astronomy and Astrophysics, Fourth, 246
Summer Seminar on History and Space, Tenth, 246
Summit meeting, U.S.–U.S.S.R., 189, 190, 191, 193–194, 195–197, 203, 206, 207, 303
Sumser, Raymond J., 444
Sun (see also Solar headings), 85, 165, 394, 420
  quiet sun, 315
  satellite data, 90, 108, 132, 199–200, 300
  sounding rocket data, 425
Sunspots, 131
Super Arcas. See Arcas.
Super Guppy (NASA cargo aircraft), 341
Superconductivity, 79, 357
Supercritical wing (see also Wing, aircraft), 50–51, 72–73, 155, 163, 184, 186, 193, 214, 274–275, 449
Supernova, 181, 332, 342, 383
Supersonic transport (SST) (see also Concorde and Tu-144)
  air show display, 157
  anniversary, 350
  benefits, 190
  Boeing Co. model, 166
  congressional considerations, 6–7, 20, 37, 362, 398–399, 430, 444–445
  contract, 349
  cost, 16
  criticism, 20, 274
  development revival, 306
  engine specifications, 248, 307
  environmental effects, 155, 280, 370, 427
  foreign
    award, 326, 350, 362, 368
    competition, 50
    economic factors, 427
    engine design, 248, 307
    environmental factors, 427
    equipment and materials, 317
    exhibit, 157, 307
    flight time record, 325
    mass production, 361
    patent, 263
    personnel, 442
    pooled airline service, 341
    purchase, 200, 271, 301
    roll-out ceremony, 329
    U.S.S.R., 157, 183, 263, 285, 325, 326, 442
  funding, 445
  future capability, 40

hazards, 385
justification, 50
NASA technology, 59, 153, 306, 311, 425
postponement, 29
press comment, 10, 80, 306, 361
public attitude, 13, 37
sonic boom, 80, 427
space shuttle comparison, 10
stratospheric observation, 385
technology report, 319
trade competition, 193
wing conversion, 191–192
Supersonic Transport Development Program, 291
Surface-composition-mapping radiometer (SCMR), 423
Surface effect ships, 252
Surface Effect Ships Project Office (USN), 252
Surveillance aircraft, 21, 24
Surveillance satellite, 9–10, 21, 190–191
Survey Ridge (moon), 146–148
Suslov, Mikhail A., 299
Sustaining University Program (NASA), 330–331
Swaziland, 158
Sweden, 46, 60, 173, 217, 232, 244, 342, 343, 349, 389, 423
Swedish Royal Academy of Sciences, 357
Swider, J. E., Jr., 293
Swigert, John L., Jr., 128
Swiss Transport Museum, 313
Air and Space Wing, 247, 455
Switzerland, 134, 174, 177, 247, 313, 314, 345, 389, 390, 455
Symington, Rep. James W., 121
Symington, Secretary of Air Force Stuart W., 320
Symposium on High Energy Phenomena on the Sun, 330
Symposium on Significant Accomplishments in Science, 377–378
Symposium on Significant Accomplishments in Technology, 377–378
Synchronous Meteorological Satellite (SMS), 171, 262
Syncom (communications satellite), 255
Syrtis Major (Mars crater), 28
Systems Center (DOT), 385
Systems Research and Development Service (SRDS) (FAA), 301
Syvertson, Clarence A., 22, 314
Szent-Georgyi, Dr. Albert, 405–406

T

T3 (Arctic island), 77
T-3 (U.S.S.R. tokamac experiment), 404
T-34 (high-bypass-ratio engine), 66
T-38 (jet trainer), 25, 178, 179, 319, 404
TACRV. See Tracked air-cushion research vehicle.
Tacsat I (tactical communications satellite), 400

TACV. See Tracked air-cushion vehicle.
Taiwan. See China, Nationalist.
Talkeetna, Alaska, 173
Tallahassee, Fla., 126
The Taming of the Fire (U.S.S.R. film), 352
Tananbaum, Dr. Harvey, 438
Tanka, Prime Minister Kakuei (Japan), 302–303, 306
Tansei (MST-1) (Japanese satellite), 295
TAOS. See Thrust-assisted orbiter shuttle.
Tashkent, U.S.S.R., 325
Taurus-Littrow (lunar landing site), 54, 297, 363, 367, 413, 416, 418, 419, 428–429, 440
Taurus Mountains (moon), 54, 347
Tbilsi, U.S.S.R., 323
TD-1A (ESRO scientific satellite), 95–96, 384–385, 452
TE-M-364-4 (solid-fueled rocket motor), 19
Teague, Rep. Olin E., 33, 166–167, 430
Technicolor Inc., 17
Technology, 1, 46–47, 63–64, 72, 139, 159–160, 174–175, 191, 192–193, 230, 389
  aeronautics, 50, 59, 67, 72–73, 77, 93, 106, 233
  Apollo contributions, 152–153, 157, 426
  benefits, 47, 117, 134–135, 154–155, 162, 173, 179, 206, 233, 396, 398, 452
  funding, 1, 29, 116, 162, 213–214
  goals and public opinion, 2, 24–25, 48, 105–106, 134–135, 135–136, 298–299, 359–360, 426
  international aspects, 29, 49, 53–54, 98, 199, 201, 203, 206, 207, 236–237, 443
  military, 65, 162, 190–191
  space program contribution, 98, 101, 162, 164–165, 165–167, 452
Technology Assessment Act of 1972, 349
Technology utilization, space (see also Research Applied to National Needs), 52, 74, 112, 153, 164–165, 223, 228, 271, 309, 311, 394
  Applications of Aerospace Technology in the Public Sector, 75
  astronomy, 179–180, 220, 300, 447
  communications, 84, 114, 131, 242, 407, 408
  computer systems, 63, 71, 452
  earth resources, 2–4, 24, 84, 122, 226–227, 271, 272, 314–315, 408, 447
  education, 114
  engineering, 45, 47–48, 56, 71, 84, 178–179, 271, 452
  environment (see also Earth Resources Technology Satellite; *Erts I*), 53, 63, 68, 122
  forest fire reconnaissance, 280–281
  industry, 45, 84, 223, 239–240, 248, 271, 348–349
  law enforcement, 74, 75, 114, 271

medicine, 38, 41, 68, 74, 75, 84, 92, 95, 101, 106, 112, 114, 248, 258, 270, 271, 316, 375, 396, 408, 411, 452
meteorology, 407–408
National Symposium on Technology Transfer, 225
navigation, 281
Nixon, President Richard M., 436
nuclear fission, 203
oceanography, 136, 250–251, 316
policy and opinion, 105, 187–188, 200–201, 225, 244–245, 326–327, 368
pollution control, 35, 68, 74, 105, 390, 407
press comment, 156, 272, 348–349, 411–412
program funding, 116, 405
safety, 74, 75, 101, 219, 270, 271, 316, 368
submarine, 228
technology application team, 75
transportation, 75, 84, 105, 411, 423
urban problems, 75, 84, 114, 126, 131, 270
Teegan, Dr. John, 52
Tel Aviv, Israel, 80, 227
Telecommunications Office (DOD), 18
Telegraph, 426
Telemetry, 290, 373, 380
Telemetry on-line processing system (TELOPS), 73–74
Telephone, 284
Telephone-telegraph radio communication system, 128
Teleprinter, 377
Telesat Canada, 379, 395, 410, 452
Telesat-A (Canadian comsat), 379
Telescope (see also Apollo Telescope Mount; Large Space Telescope Radiotelescope), 123, 131, 136, 181, 286, 332, 408
  balloon, 64
  development, 125, 138, 213–214, 255, 386, 393
  gamma ray, 16, 386
  infrared, 45, 213–214
  land surveying system, 248
  OAO observation, 254, 257, 296, 308, 447
  observatories, 181, 208–209, 446
  optical, 213–214, 374
  planetary investigation, 112, 421, 443
  plankton measurement, 136
  reflecting, 208–209
  supernova observation, 181
  x-ray, 108
Television
  Alaska Educational Broadcasting Commission, 173
  Apollo 16 coverage, 142–144, 146–149, 151, 156–157, 158, 381
  Apollo 17 coverage, 381, 391–392, 399, 409, 413, 416–418, 419, 427, 444
  Apollo-Soyuz Test Project, 197–198, 252, 322
  applications, 93, 114, 254, 336–337
  Crab submarine, 228
  educational, 45, 72
  ERTS, 27, 125, 184
  European Broadcasting Union, 156–157
  Grand Tour spacecraft, 65
  lunar debris, 432
  Mariner 10 mission, 307
  NASA public affairs, 52
  Nixon, Kremlin broadcast, 206
  Orbita TV network (U.S.S.R.), 128, 331, 426
  PEACESAT program, 19
  "Space in the Age of Aquarius" program, 134–135
  U.N. resolution, 286, 380
  U.S.S.R., 28, 90, 128, 188, 189, 350, 404
  via satellite, 17, 26–27, 45, 126, 173, 255, 262, 273, 318, 355, 375, 410, 411
Teller, Dr. Edward, 17, 354
TELOPS. See Telemetry on-line processing system.
Telstar 1 (communications satellite), 255
Telstar 2, 255
Temperature-humidity infrared radiometer (THIR), 423
Tennessee, 72, 74, 335, 337
Tennessee Valley Authority (TVA), 384
TERLS. See Thumba Equatorial Rocket Launching Station.
Terminal Configured Vehicles and Avionics Operating Experiments program, 84
Texas, 19, 20, 38, 60, 61, 148–149, 384, 444
  accident, 25, 179, 214
  balloon launch, 181
  exhibit, 233
  Federal Grand Jury, Dallas, 209
  Galveston County Health District, 204
  Matagorda Island, 214
  NASA Space Shuttle Technology Conference, 137
Texas Children's Hospital, 348
Texas Instruments Inc., 35
Texas Southern Univ., 131–132
Texas Symposium on Relativistic Astrophysics, Sixth, 438
Texas, Univ. of, 39–40, 170–171, 348, 373, 380
Textron, Inc., 123
  Bell Aerospace Co. Div., 254–255, 446
  Bell Helicopter Co. Div., 187
TF-8A (jet research aircraft), 73, 94, 104, 109, 113, 123, 155, 163, 184, 186, 193, 274–275, 396, 449
Thailand, 192
Thermal protection, 79, 220, 281
Thermoelectric outer planets spacecraft (TOPS), 39–40
Thermoelectric power conversion, 80
Thermonuclear energy, 31, 331
Thiokol Chemical Co., 35
THIR. See Temperature-humidity infrared radiometer.

Thole, John M., 110
Thomas, David F., 101
Thompson, Francis, Inc., 385
Thompson, Milton O., 341
Thompson, Robert F., 383–384
Thor (missile), 403
Thor-Agena (booster)
  Thorad-Agena, 154, 200
  Thor-Agena B, 328
Thor-Burner II (booster), 117
Thor-Delta (booster), 38, 78, 95, 177–178, 182, 269, 350, 372, 377–378, 379, 397, 422
Thorium, 148
Thorne, Dr. Kip, 360
Thornton, Dr. William E., 273, 285–286, 323, 343
Thrust-assisted orbiter shuttle (TAOS), 39
Thumba Equatorial Rocket Launching Station (TERLS), 132, 138, 238
Thumba, India, 75
Thurmond, Sen. Strom, 184
Ticonderoga. See U.S.S. *Ticonderoga*.
Tilt rotor aircraft, 40
Tindall, Howard W., Jr., 172
Tiros Operational Satellite (TOS) (see also ITOS-B; *Noaa 1*), 262, 352
Tischler, Adelbert O., 79
Titan (booster)
  Titan III-Centaur, 367
  Titan IIIB-Agena, 108, 190, 305, 439
  Titan IIIC, 77, 425
  Titan IIID, 25, 252, 346
  Titan IIIE-Centaur, 320
  Titan-Centaur, 65
  Titan-Centaur-Burner III, 65
Titan (Saturn moon), 65, 443
Titanium, 62
Titusville, Fla., 98
Titusville, Fla., Post Office, 405
Tizard Memorial Lecture, 101
Tokamak fusion reactor, 404
Tokyo, Japan, 389
Tokyo, Univ. of, 264
Tolubko, Gen. Vladimir F. (U.S.S.R.), 177
Tomahawk. See Nike-Tomahawk; Sandhawk Tomahawk.
TOPS. See Thermoelectric outer planets spacecraft.
Toronto, Univ. of, 332
Toros (earth moon), 287
TOS. See Tiros Operational Satellite.
Toulouse, France, 292, 329, 365
Tower, Sen. John G., 216
Towl, E. Clinton, 151–152
Townes, Dr. Charles H., 64
Toynbee, Arnold J., 405
TRACALS. See Traffic Control and Landing System.
Tracked air-cushion research vehicle (TACRV), 52, 136
Tracking (see also Earth station), 30, 38, 39, 42, 70, 73–74, 130–131, 149–150, 422, 438

Trade, 49, 62, 93, 189, 190, 192–193, 200, 207
Trade and Industry, Dept. of (U.K.), 120
Traffic control and landing system (TRACALS), 383
Trailblazer II (reentry vehicle), 275
Train, Russell E., 289
Trains, 392
Trans World Airlines, Inc. (TWA), 216–217
Transcontinental Air Transport, 343
"Transfer of Space Technology to Community and Industrial Activities," 98
Transim (navigation aid), 301
Transistor, 441
Transit (navigation satellite), 80, 192, 301, 306–307
TRANSPO '72, 136, 157, 183, 206, 214, 222, 224,
TRANSPO '74, 224
Transport Experimental Program Office, OAST (NASA), 67
Transportation, 1, 2–4, 31, 73–74, 140, 162, 298–299
Transportation, Dept. of (DOT) (see also Climatic Impact Assessment Program)
  aircraft, 40, 203–204, 425
  airport security regulations, 93, 227
  contract, 52, 136, 190, 250, 392
  cooperation, 1, 6, 22–23, 24–25, 29, 43, 73–74, 155, 160, 176, 203, 204
  executive reorganization, 122
  funding, 31, 116, 298
  noise abatement, 251
  Northeast Corridor Report, 67
  prototype radiometric sensor, 43
  research grant, 280
  SST environmental effects, 280, 370, 385
  technology utilization conference, 314
  tracked air-cushion research vehicle, 52
  TRANSPO '72, 183, 206, 222, 224
Treaty (see also International cooperation), 213, 214–215, 218, 282, 299, 311, 316, 331, 338, 357
  biological warfare, 134, 311
  environmental protection, 193–194, 217
  outer space use and exploration, 48–49, 134, 174, 203
  space object damage liability, 311, 341
Treaty on Principles Governing the Activities of States in the Exploration and Use of Outer Space, Including the Moon and Other Celestial Bodies, 195, 203
Treaty on the Limitation of Antiballistic Missile Systems, 201–202, 224–225, 338
Triad OI–IX (transit satellite), 306–307, 447
Trident (undersea long-range missile system) (ULMS), 31, 69, 185, 263
Triplet-triplet transfer of energy, 18
TriStar (jet transport), 307, 422
Tristate Regional Planning Commission, 383
Trubshaw, Brian, 350

Truman, President Harry S., 258
Truszynski, Gerald M., 73-74, 114, 130-131
TRW Inc., 107, 329
 Systems Group, 220, 243, 294-295, 370, 382
Tsander, F. A., 345
Tsiolkovsky Crater (moon), 144-146
Tu-2 (U.S.S.R. dive bomber), 442
Tu-95 (Bear) (U.S.S.R. reconnaissance aircraft), 321
Tu-104 (U.S.S.R. jet airliner), 442
Tu-134, 442
Tu-144 (U.S.S.R. supersonic transport), 157, 183, 263, 285, 325, 326, 368, 442
Tu-154 (U.S.S.R. jet airliner), 442
Tullahoma, Tenn., 321
Tupolev, Aleksey A., 263
Tupolev, Andrey N., 326, 368, 442
Turcat, André, 350
TV. See Television.
TVA. See Tennessee Valley Authority.
TWA. See Trans World Airlines, Inc..
Tycho's Star, 383
Tyrrhenian Sea, 228
Tyson, Robert C., 280
Tyuratam, U.S.S.R., 329, 445

U

U-2 (reconnaissance aircraft), 72, 215-216, 255-256, 280-281, 385
UAL. See United Air Lines.
U.A.R. See United Arab Republic.
UAW. See United Auto Workers.
UCLA. See California, Univ. of, Los Angeles.
UFO. See Unidentified flying object.
The UFO Experience: A Scientific Inquiry, 300-301
UHF: ultrahigh frequency.
Uhuru. See *Explorer 42*.
U.K. See United Kingdom.
ULMS (undersea long-range missile system). See Trident.
Ultraviolet (UV), 378
 Apollo 16 investigation, 102, 142-144, 179-180, 184, 306
 extreme, 430
 far, 102, 270-271, 306, 433
 satellite investigation, 208, 254, 386, 430
 solar, 210, 385, 427, 430, 443
 sounding rocket experiment, 374, 427, 449
 stellar, 254
U.N. See United Nations.
Undersea long-range missile system (ULMS). See Trident.
Underwater training facilities, 301
UNESCO. See United Nations, Educational, Scientific and Cultural Organization.
Unidentified flying object (UFO), 241, 300-301, 383

Unidentified satellite, 25, 77, 108, 117, 154, 190, 200, 252, 305, 346, 437, 439
Union of Soviet Socialist Republics. See U.S.S.R.
United Air Lines (UAL), 216-217, 382
United Aircraft Corp.
 Hamilton Standard Div., 132, 277, 293
 Pratt & Whitney Div., 20-21, 45, 90, 124, 176, 292
 Sikorsky Aircraft Div., 303, 362
 United Technology Center Div. (UTC), 35, 234
United Arab Republic (U.A.R.), 310, 329
United Auto Workers (UAW), 46
United Kingdom (U.K.), 167, 381-382
 aerospace industry, 60, 133-134, 329, 372-373
 Apollo 17 coverage, 444
 award, 202-203, 350
 Conference on the Human Environment, 218-219
 Environment Ministry, 307
 Farnborough Air Show, 307
 House of Commons, 288
 international cooperation, 200, 341, 372-373, 438
 international cooperation, space, 208, 251-252, 299-300, 378, 421, 422, 423
 meteorology, 105, 123
 National Space Authority (proposed), 373
 satellite, 208, 389, 446, 453
 Science Research Council, 202, 354
 Solar Energy Symposium, 330
 sounding rocket, 120
 space program, 125-126, 373
 supersonic transport, 157, 200, 248, 297, 301, 341
 treaty, 188
United Nations (U.N.), 173, 231-232, 311, 317, 384, 389
 Committee on the Peaceful Uses of Outer Space, 122
  Legal Subcommittee, 134, 174
  Scientific and Technical Subcommittee, 173
 Educational, Scientific and Cultural Organization (UNESCO), Communications Commission, 365
 First (political) Committee, 348, 356
 General Assembly, 49, 122, 218, 286, 311, 357, 380
 Governing Council for Environmental Programs (GCEP), 218-219
 Hall of Fame for Spacemen (proposed), 358
 International Space Log, 252
 Outer Space Committee, 122
 Voluntary Fund for Environment (proposed), 47
 World Meteorological Organization, 84, 105, 218-219, 453
United Nations Conference on the Human Environment, 217, 218-219, 223, 225, 226-227, 232, 236, 285
United Press International (UPI), 133

United States (U.S.)
  aerospace industry, 50, 62, 314, 319, 349, 425
  atomic energy, 55, 421
  Bicentennial, 45, 131, 454
  budget, 133, 410
  communications, 190–191, 208, 361, 410
  Convention on International Liability for Damage Caused by Space Objects, 230
  defense, 62, 72, 229–230, 316, 319–320, 328–329, 399, 445
  disarmament, 24–25, 203, 206, 207, 229–230, 236, 288–289, 335, 362, 390
    SALT, 90, 121, 214, 218, 252, 310, 316, 331, 356, 390
    test ban, 327
    treaty, 188, 190, 201–202, 224–225, 265, 299, 332, 338
  domestic policy and problems, 8, 55, 60, 98, 190–191, 446
  economy, 46–47, 60, 406–407
  employment, 64–65
  environmental program, 226–227, 285, 289, 298, 301–302
  exhibit, 307, 434
  Federal support, 291
  flag, 347
  foreign policy, 48–49, 55–56, 159–160, 228–229, 238, 240, 380
  international cooperation
    AIDJEX, 128
    Canada, 174
    Germany, West, 303–304
    P.R.C., 72
    U.K., 134
    U.N. participation, 47, 218–219, 311, 348
    U.S.S.R., 195, 201
      biological warfare treaty, 134
      congressional hearings, 203
      environmental protection, 177, 190, 191, 193–194, 195–196, 201, 285
      industry contacts, 201
      joint working groups. See Joint Working Groups.
      medicine and health, 190, 191, 194, 203, 207
      science and technology, 127, 195, 199
      scientist exchange, 136
      summit meeting, 189, 190, 191, 193–194, 195–197, 203, 206, 207, 303
  international cooperation, space. See International cooperation, space.
  metric system, 93, 182, 295, 297
  national goals and priorities, 167, 175–176, 314, 407, 443
  national security, 17, 46–47, 55–56, 64, 87, 232
    classification system, 187
    ecological, 154, 232
    Hot Line, 20, 377
    intelligence, 329, 445
    surveillance technology, 69, 154, 190–191, 321
  patent policy, 299
  research and development, 289
  science and technology, 37, 49, 131, 139, 151, 159–160, 175, 300–301, 326–327
  Solar Energy Symposium, 330
  space program. See Space program, national; Space race.
  trade, 49, 53–54, 86, 98, 126, 161, 425
  transportation. See Transportation; Supersonic transport.
  Vietnam war. See Vietnam war.
  world leadership, 29, 32, 50
United States and Soviet Progress in Space: Summary Data Through 1971 and a Forward Look, 28
United States Foreign Policy for the 1970's: The Emerging Structure for Peace, 48–49
United States Space Science Program: Report to COSPAR, 178
UNIVAC 1108 (computer), 131–132
Univac Systems Div. See Sperry Rand Corp.
Universal Atlas Div. See U.S. Steel Corp.
Universe, 176, 213, 298, 374
Universities, 267
  astronauts, temporary leave to, 205
  computer access, 71, 445, 452
  cooperation, 68, 74–75, 100, 234, 315, 347
  employment, 112
  foreign, 291
  funding, 31, 51, 258, 326
  graduate programs, 201, 267
  NAS membership, 95
  research, 43, 74, 94, 95–96, 118, 134, 169–170, 213–214
  satellite utilization, 114, 226–227, 325
  sounding rocket experiment, 44, 54, 61, 70, 288, 422
Sustaining University Program, 330
Universities Space Research Assn., 347
UPI. See United Press International.
Upper Atmosphere Observatory, 174
Ural Mountains, 421
Uranium, 14, 148, 203, 421
Uranus (planet), 441
Urban affairs (see also Manned Spacecraft Center, Urban Systems Project Office; Social problems), 52, 133, 206, 236–237, 383
Urban Systems Research and Engineering, Inc., 161
Urban Technology Conference and Technical Display, 270, 271
Urey, Dr. Harold C., 20–21, 162
Urokinase (enzyme), 247

U.S. Air Force (USAF) (see also individual bases, centers, and commands, such as Air Force Cambridge Research Laboratories, Air Force Materials Laboratory, Air Force Prototype Systems Program Office, Air Force Systems Command, Andrews AFB, Arnold Engineering Development Center, Space and Missile Systems Organization, Vandenberg AFB, etc.)
  accident, 222, 230, 233, 236
  Airborne Warning and Control System (AWACS), 42
  aircraft (see also individual aircraft, such as C-5A, F-III, etc.)
    accident, 222, 230, 233, 236
    air superiority, 263
    anniversary, 319, 350
    balloon, 319
    costs, 31, 93, 120, 123
    defects, 131, 209
    development, 50-51, 67, 77, 110, 154, 233, 236, 241, 242, 263, 311, 340, 393, 395
    AMST, 77, 382-383, 443
    disappearance, 355
    engine, 332
    flight test, 50-51, 67, 282, 292-293, 396
    future, 321
    prototype, 378, 397, 443
    reconnaissance, 162
    record, 61
  anniversary, 258, 319, 320, 350
  Armed Forces Unification Act, 258, 320
  award, 20-21, 29, 258, 263, 267, 313, 320
  budget, 31, 233-234, 294
  contract, 16, 51, 113, 122-123, 154, 209, 337, 340, 382-383, 443
  cooperation
    NACA, 350
    NASA, 7, 50, 67, 77, 79, 102, 154, 295, 347, 447
    NATO, 370
  launch
    balloon, 364
    missile, 174
    satellite, 25, 77, 100, 117, 154, 190, 200, 216, 252, 305, 336, 346, 379, 437, 439
  lifting body, 347, 447, 449
  missile program, 16, 65, 191
  personnel, 189, 192-193, 269, 303
    appointment, 110, 163, 308
    award, 20, 29, 303, 313, 353, 403
    death, 26, 222, 362
    promotion, 39, 60, 172, 404
    resignation, 221
    retirement, 18, 277, 303
  press comment, 96
  Project Blue Book, 300-301
  satellite, 77, 117, 216, 233-234, 252, 306, 329, 354, 379, 437, 439
    reconnaissance, 25, 96, 100, 154, 190, 200, 305, 346

  sounding rocket, 394
  space program (see also Defense, Dept. of), 210-211, 216, 241, 245, 286-287, 354, 428
  space shuttle, 8, 139-140, 181-182
  space tug, 390
U.S. Arms Control and Disarmament Agency, 93
U.S. Army (USA) (see also individual centers and commands, such as Army Air Mobility Research and Development Laboratory, Army Aviation Systems Command, Army Corps of Engineers, etc.)
  aircraft, 316, 350, 409
  Armed Forces Unification Act, 320
  Ballistic Missile Agency Treaty, 451
  contract, 72, 202-203, 288
  cooperation, 40
  personnel, 343
  satellite, 399
U.S. Bureau of Mines, 62, 67-68
U.S. Coast Guard, 242
U.S. Conference of Mayors, 304
U.S. District Court, Manhattan, N.Y., 122
U.S. District Court, Washington, D.C., 46, 255
U.S. Forest Service, 138
U.S. Geological Survey (USGS), 15, 39, 42-43, 62, 170-171, 225-226, 235, 396-397
U.S. Geological Survey Regional Geophysics Group, 259
U.S. Information Agency, 384
U.S. Marine Corps., 446
U.S. Navy (USN) (see also individual centers and stations, such as Naval Air Station, San Diego, Calif., Naval Missile Center, Naval Observatory, etc.)
  aircraft, 26, 123, 220, 271, 316, 322, 355, 425
  Apollo 17 postal covers, 312-313
  award, 265-266
  budget, 31
  contract, 151-152, 355, 425
  facility, 309
  launch, 306
  personnel, 130, 172, 236, 289, 335, 404, 429-430, 441
  satellite, 80, 192, 306, 369, 399, 447
  sounding rocket, 394
  Space Projects Office, 301
  surface effect ship, 252
  VAST, 215
U.S. Postal Service, 32, 73-74, 232
U.S. Public Health Service (USPHS), 65-66, 131
U.S. Signal Corps, 26
U.S. Steel Corp.
  Universal Atlas Div., 309
USA. See U.S. Army.
USAF. See U.S. Air Force.
USDA. See Agriculture, Dept. of.
USGS. See U.S. Geological Survey.

USN. See U.S. Navy.
USPHS. See U.S. Public Health Service.
U.S.-U.S.S.R. Cooperative Agreement on Environmental Protection, 285
U.S.-U.S.S.R. *Cooperative Agreements*, 303
U.S.-U.S.S.R. Interim Agreement With Respect to the Limitation of Strategic Offensive Arms, 316, 331
U.S.-U.S.S.R. Joint Working Groups. See Joint Working Groups.
U.S.-U.S.S.R. Treaty on the Limitation of Antiballistic Missile Systems, 331
U.S.S. *Forrestal*, 321
U.S.S. *Point Barrow*, 309
U.S.S. *Ticonderoga*, 148-149, 162-163, 367, 405, 419, 435, 437
U.S.S.R. (Union of Soviet Socialist Republics) (see also Soviet headings)
  agreement. See individual agreements and International cooperation; International cooperation, space.
  aircraft, 263, 326, 329, 365, 368, 425, 442
    accident, 349
    reconnaissance, 321
  anniversary, 14-15, 108, 338, 348, 377
  Apollo 16 mission reaction, 150, 151
  Apollo 17 mission reaction, 421, 424, 437
  astronomy, 123, 151, 187, 208-209, 255, 329, 375, 446
  award, 2, 326, 368
  boycott, 218-219
  communications, 188, 331, 348, 355, 361, 377, 380, 404, 426
  cosmonaut. See Cosmonaut.
  Council of Ministers, 18
  defense, 61-62
  disarmament (see also U.S.S.R., treaty), 24-25, 189, 190, 201-202, 203, 206, 207, 229-230, 232, 316, 331, 335
    SALT, 90, 121, 214, 218, 252, 310, 316, 331, 356, 390
  ekranoplane, 381-382
  Embassy, 255-256
  exhibit, 307
  International Biophysics Conference, 291
  international cooperation, 201, 203, 215-216, 236-237, 375, 377
    environment, 177, 190, 191, 193-194, 195-196, 201, 285, 323, 327
    exhibit, 256
    Hot Line, 20, 377
    industry, 62, 201
    medicine and health, 190, 191, 194, 203, 207
    meteorology, 105, 216, 403
    nuclear energy, 94, 259
    oceanography, 375
    power stations, 341
    science and technology, 127, 136, 195, 199, 213, 236-237, 352, 374-375
    space applications, 239-240
    summit talks, 189, 190, 191, 193-194, 195-197, 203, 206, 207, 264, 303
    visit to U.S., 374
  international cooperation, space (see also Apollo-Soyuz Test Project), 19, 208, 213, 214, 215-216, 219, 222, 231-232, 240-241, 357, 377, 436
    Canada, 174
    Czechoslovakia, 132, 243
    ESRO, 208, 397
    France, 128, 174, 208, 242
    Germany, East, 243
    Hungary, 132
    India, 75, 208, 223, 239
    lunar cartography, 168, 177, 229, 305-306
    lunar sample exchanges, 13-14, 100, 138, 299-300
    Mongolia, 132
    peaceful use, 214, 239-240, 323, 452
    planetary exploration, 28, 177, 178, 187, 189-190, 228, 273, 310-311, 376-377
    Poland, 132
    Romania, 132
    U.N. General Assembly, 285
    U.S., 49, 133, 151, 174, 178, 185, 190, 198-199, 201, 203, 206-207, 208, 229
      agreements, 134, 170, 195-196, 204-205, 303
      Apollo-Soyuz Test Project. See Apollo-Soyuz Test Project.
      joint mission (see also Apollo-Soyuz Test Project), 15, 27, 28, 100, 127, 128, 129, 130, 136-137, 151, 158, 182, 375, 407
      press comment, 117, 201, 204-205, 206-207, 276, 422
      "rendock," 182
      space shuttle, 10
      space station, 173
  launch
    failure, 277
    probe
      Luna 19, 46
      Luna 20, 52
      Venus 8, 119
    satellite
      Cosmos, 7, 9-10, 15, 33, 44, 52, 69, 77, 87, 102, 113, 119, 123, 127, 131, 135, 139, 155, 167-168, 174, 186, 200, 221, 236, 239, 241, 243, 250, 251, 253, 256, 257, 264, 265, 274, 280, 292, 294, 297, 301, 317, 319, 322, 330, 338, 340, 346, 355, 361, 366, 371, 373, 395, 428, 439, 443, 444, 445
      Intercosmos 6, 132
      Intercosmos 7, 243
      Intercosmos 8, 403
      Meteor 11, 123
      Meteor 12, 243
      Meteor 13, 363
      Molniya 1-20, 128
      Molniya 1-21, 350

Molniya 1-22, 404
Molniya II-2, 188
Molniya II-3, 331
Molniya II-4, 426
Prognoz 1, 139, 242
Prognoz 2, 242
Sret 1, 128
Salyut (space station), 174–175, 277
sounding rocket, 228
Soyuz 11, 167
meeting, 343, 360, 365
meteorological satellite system, 87
Ministry of Health, 170
missile and rocket program, 65, 87, 96, 177, 201–202, 229–230, 320, 329, 331, 352, 355, 397, 429
  ICBM, 54, 69, 89, 335
  MIRV, 221
  SS-4, 87
  SS-5, 87
  SS-9, 115, 252
  SS-11, 350, 355
  SSN-8, 210
  SSNX-8, 71,335, 397
  test, 71, 335, 350, 393, 445
missile threat to, 377
Nixon, President Richard M., visit to, 96, 185, 189, 190, 191, 197–198, 203, 207, 213, 232, 241
nuclear capability, 49, 55–56, 175, 289, 358, 421
nuclear energy facilities, 2, 10, 14–15, 55
press, 133
probe, 96
  Intercosmos 6, 235
  Luna 3, 237
  Luna 16, 13–14, 33, 52–53, 71, 179, 291, 299–300
  Luna 17, 52–53
  Luna 18, 52–53, 166
  Luna 19, 38, 46, 52–53, 94, 166, 337
  Luna 20, 52–53, 71, 179, 452
  Lunokhod 1 (U.S.S.R. lunar surface explorer), 7, 13–14, 53, 128, 166, 228, 256, 337
  Mars 2, 10–11, 28, 38, 60, 86, 90, 113, 166, 244, 299, 369
  Mars 3, 10–11, 28, 38, 60, 86, 90, 113, 166, 167, 244, 299, 369
  Venus 8, 140–141, 267, 273, 275, 284, 294, 312, 315–316
Radio Physics Institute, 331
satellite, 128, 132, 139, 167, 175–176, 188, 192, 321, 399
science and technology, 2, 10, 13–14, 15, 18, 123, 173, 176, 201, 213, 221–222, 228, 368, 375, 385–386
Science Institute of Space Research, 189–190
space program, 7, 18–19, 64, 68, 142, 166, 167–168, 175–176, 218, 353, 394, 399, 407
  achievements, 85–86, 108, 138, 243
  biology, 360
  budget, 167–168, 343

  competition with U.S., 9, 27, 137, 140–141, 167–168, 394
  launch record, 449
  lunar exploration (see also Luna probes), 13–14, 100, 128, 179, 426
  military applications, 64, 118, 352
  plans, 28, 86, 96, 100, 111, 248, 338
  press comment, 117
  press coverage, 133
  satellite inspection ability, 9–10
  secrecy, 167–168, 185
  Soviet Space Program, 1971 (report), 166
  training, 110–111, 201, 204, 205, 228, 229, 239–240
space station, 58, 61, 175–176, 206–207
spacecraft. See U.S.S.R., launch; and individual spacecraft such as *Luna 16, Mars 3, Molniya 1-20, Salyut 1,* and *Soyuz 11.*
Special Astrophysical Observatory, 208–209
State Committee for Science and Technology, 199
supersonic transport, 157, 183, 263, 285, 325, 326, 341, 368, 404
trade, 90, 133–134, 241, 435
treaty, 327, 331, 338, 380
  arms limitation (see also U.S.S.R., disarmament), 188, 201–202, 224–225, 229–230, 232, 283, 288, 299, 316, 331, 338
  biological warfare treaty, 134
  space, 174, 195
  weapons, 49, 120, 201–202, 229–230, 328
Usher, Scott, 220
Utah, Univ. of, 323
UTC. See United Aircraft Corp., United Technology Center Div.
Utility Tactical Transport Aircraft System (UTTAS), 303
UTTAS. See Utility Tactical Transport Aircraft System.
UV. See Ultraviolet.

V

V-2 (long-range ballistic missile), 202–203, 205, 321
VA. See Veterans Administration.
VAB. See Vehicle Assembly Building.
Vachon, Dr. Reginald I., 294
Valery, Nicholas, 251–252
Van Allen, Dr. James A., 115
Van Allen radiation belts, 115, 128, 227, 244
Van de Graaf Crater (moon), 152
Vandenberg AFB, Calif.
  anniversary, 328
  facilities modifications, 139–140
  launch
    missile, 174
    satellite, 216
      Radcat, 336

reconnaissance, 154, 200, 346
Stp 72-1, 336
Triad OI-IX, 306
unidentified, 25, 154, 190, 200, 346, 439
launch failure, 80
satellite launch vehicle, 108, 117, 252, 279-280, 305
space shuttle, 139-140, 181-182, 448
Vandenberg Tracking Station, 337
Van Hoeydonck, Paul, 153, 266, 317
Van Nuys, Calif, 173
Vanguard (satellite), 75, 395-396
Vanguard 1, 109
Vanguard-A History, 75
Vasilyev, V., 253
VAST. See Versatile avionics shop test.
Vega (star), 255
Vehicle Assembly Building (VAB) (KSC), 46, 139-140, 150, 298, 384
Vela (nuclear detection satellite), 332
Velikovsky, Dr. Immanuel, 291, 400-401
Venera (U.S.S.R. interplanetary probe), 108
Venus (planet) (see also Mariner Venus/Mercury mission and Venus probes), 64, 291, 294-295, 441, 444
atmosphere, 113, 176, 267, 312
probe, 56, 64, 65, 94, 140-141, 151, 312, 408
international cooperation, 178, 228, 311
Mariner 2, 429
Mariner 10, 307-308
U.S.S.R., 119, 167-168, 257-258, 267, 275
surface, 267, 312
Venus 4 (U.S.S.R. interplanetary probe), 119
Venus 5, 119
Venus 6, 119
Venus 7, 119, 267, 275, 337
Venus 8, 119, 140-141, 257-258, 267, 273, 275, 284, 294, 312, 315, 338
Versatile avionics shop test (VAST), 215
Vertical or short takeoff and landing aircraft. See V/STOL.
Vertical takeoff and landing aircraft. See VTOL.
Vertical-temperature-profile radiometer (VTPR), 352
Vertol Div. See Boeing Co.
Very-high-frequency landing system. See VHF landing system.
Very-high-resolution radiometer (VHRR), 352
Very-large-array (VLA) radiotelescope, 104
Veterans, 129-130, 187-188, 193, 209
Veterans Administration (VA), 73-74, 343
Veterans Administration Hospital, Buffalo, N.Y., 248
Veterans of Foreign Wars (VFW), 89
VFR. See Visual flight rules.
VFW. See Veterans of Foreign Wars.

VFW-16 (European commuter transport), 94
VHF: very high frequency.
VHF landing system, 67
Vick, L/C Dewain C. (USAF), 263, 350
Victorialand, Antarctica, 171
Vienna, Austria, 340, 343
Vietnam war, 151, 232, 247, 328, 329, 393, 407-408, 440
Viking (program), 28, 30, 407, 408, 428
Viking (rocket), 345, 395-396
Viking Mars mission, 56, 65, 85, 153, 313, 376-377, 384, 388
Viking Mars lander, 131, 290, 295, 364, 428
parachute system, 163, 256, 272-273, 290, 295
Viking Press, 211
VIMS. See Virginia Institute of Marine Science.
Vinogradov, Dr. Aleksander P., 179, 299-300
Virginia, 136, 157, 188, 206
Virginia Institute of Marine Science (VIMS), 216
Virginia Polytechnic Institute, 281
Visual flight rules (VFR), 93
VLA. See Very-large-array radiotelescope.
VLF: very low frequency.
Volcano, 77, 228, 250-251, 385-386
Volpe, Secretary of Transportation John A., 20, 49-50, 155, 161, 190, 203-204, 251, 272, 291, 323, 392, 411, 420, 425
Voluntary Fund for Environment (proposed), 47
von Beckh, Dr. Harald J., 373
von Braun, Dr. Wernher, 211, 352, 383, 440
anniversary, 117
awards and honors, 202-203, 220
documentary, 9
exhibit, 117, 185
press comment, 217
retirement, 202-203, 205, 451
space program, 98, 120-121, 151, 344, 399-400, 406, 409
von Dohnanyi, Aerospace Minister Klaus (West Germany), 262, 421
von Kármán, Theodore, 321
von Kármán, Theodore, Award, 320
von Kármán, Theodore, Memorial Lecture, 80
Vostok 1 mission, 138, 446
Vought Missiles & Space Co. See LTV Aerospace Corp.
"Voyage Beyond Apollo" (cruise), 409
Vreuls, Fred E., 129
V/STOL (vertical or short takeoff and landing) aircraft, 31, 67, 141-142, 161, 186, 187, 353, 355
VTOL (vertical takeoff and landing) aircraft, 30, 40, 56, 66-67
VTPR. See Vertical-temperature-profile radiometer.
Vulcan Materials Co., 309

W

Waddell Gallery, 266
Waddy, Judge Joseph C., 242
Wade, Gen. Horace M. (USAF), 110
Wake-turbulence research, 92
Wakelin, James H., Jr., 280
Wald, Dr. George, 387–388
Waldheim, U.N. Secretary General Kurt M., 286, 348, 358
Waldis, Alfred, 313, 345
Wallops Island, Va., 345
Wallops Station (NASA), 9–10, 56, 142, 287, 394
  airborne laser, 136
  contract, 293
  launch
    Explorer 46, 290
    Gravitational Redshift Space Probe Experiment, 194
    sounding rocket, 105
      Arcas, 173
        Super Arcas, 8, 39, 410, 460
      Astrobee F, 327
      Javelin, 51
      Nike, 73
      Nike-Apache, 39, 410, 460
      Nike-Javelin III, 120, 122, 127
      Nike-Tomahawk, 91
    Trailblazer II (reentry vehicle), 275
Walt, Gen. Lewis W. (USMC, Ret.), 317
War (see also Cold war; Vietnam war), 9–10, 39, 49, 134, 207
Ward, Frederick W., Jr., 454
Wark, Dr. David Q., 20–21
Warner, Secretary of the Navy John W., 265–266, 355
Warsaw, Poland, 20, 28, 183, 257
Warsaw, Univ. of, 20
Washburn, Bradford, Award, 368
Washington, 42, 166, 169, 252, 256, 264, 365
Washington, D. C. (see also District of Columbia), 86, 134, 188, 214, 277, 355, 429–430, 434
  awards presented in, 131–132, 210, 258, 263, 265–266, 350, 372–373, 380–381
  civic center (proposed), 322, 338
  clubs and associations, 83, 93, 263
  exhibit, 74
  Handler, Dr. Philip, interview, 127
  Hot Line, 377
  meetings, 20, 33, 50, 59, 98, 121, 153, 159–160, 162, 170, 177, 187, 193–194, 199, 203–204, 221, 223, 225, 226, 293, 298, 305, 319, 328, 352, 354, 443
  NASA personnel, 205, 222
  National Air and Space Museum, groundbreaking, 388
  press conference, 8, 39, 90, 128, 185, 197–198, 221, 340
  speeches, 89, 162, 226–227, 375
  U.S. District Court, 46, 242–243, 255
  visits to, 230, 384, 392

Weather Watch data, 263
Washington State Univ., 54
Washington, Univ. of, 241
Washington Univ., St. Louis, Mo., 264, 315
Washington, Mayor Walter E., 131–132
Wasielewski, Eugene W., 258–259
Wasserburg, Dr. Gerald J., 363, 380–381
Waste disposal, 293
Water pollution, 5, 63, 73–74, 101, 409
Water-recovery and waste-incinerator system, 194–195
Water resources management, 1, 2, 17, 77, 126, 302
Watson, James Craig, Medal of Science, 160
WB–57 (NASA research aircraft), 292–293
WBVTR. See Wide-band video tape recorders.
Wdowiak, Dr. Thomas, 181
Weakley, V/A Charles E. (USN, Ret.), 441
Weapon systems (see also Disarmament; Nuclear weapons), 50, 162
  biological, 134
  cost, 263
  electronic countermeaures (ECM), 61–62
  evasion technology, 65
  space shuttle, 157
  strategic, 24–25, 39
  U.S.–U.S.S.R. capability comparison, 49
Weather modification (see also Meteorology), 236–237, 247, 250, 408
Webb, James E., 89, 98, 205, 455
Webster, Grove, 269
Weightlessness effects, 15, 24, 41, 65–66, 73, 129, 136, 241, 279–280, 321–322, 337, 360, 419
Weinberg, Dr. Alvin M., 406
Weinberger, Caspar W., 380
Weinstein, Richard H., 126
Weiss, Howard M., 225
Weitz, Cdr. Paul J. (USN), 23, 219, 282
Weizenbaum, Joseph, 180
Wessels, Col. Robert R. (USA), 302
West Germany. See Germany, West.
West Indies, 184, 195
West Virginia, 18
Western Hemisphere, 54
Western Telegraph Co., 283
Western Test Range (WTR), 95, 115, 184, 269, 350, 389, 422, 430
Western Union International, Inc., 335, 386
Western Union Telegraph Co., 283, 286
Westinghouse Astronuclear Laboratory, 59
Westinghouse Corp., 337
Westinghouse Education Foundation, 429
Westminster School, 101
Whalley, Rep. J. Irving, 220
Whipple, Dr. Fred L., 372
Whitcomb, Dr. Richard T., 73
White, E. E., 356
White, L/C Edward H., II (USAF), 382
White House, 125, 229–230, 273, 446

announcement, 191, 197, 224–225, 230, 236–237, 253, 356
astronauts, 41, 162–163
ceremony, 59, 275, 338
Hot Line, 20, 377
meeting, 60, 236–237
press conference, 198–199, 201–202
space program, 185, 228–229
visits to, 350, 374
White House Council, 208
White Sands Missile Range (WSMR), N. Mex., 202–203, 301
  Aerobee launch anniversary, 394
  launch
    Aerobee 150, 156, 237
    Aerobee 170
      aeronomy experiment, 306
      astronomy experiment, 180, 259, 288, 319
      cosmic ray experiment, 189
      planetary atmosphere spectrum, 378
      soft x-ray experiment, 288, 422
      solar brightness measurement, 425, 427
      solar physics experiment, 158, 225, 242, 255, 274, 276, 366, 385
      x-ray astronomy, 61, 127, 189
    Aerobee 200, 388
    Black Brant VB, 360
    Black Brant VC, 210, 222
    Nike-Apache, 37, 222, 375
  test, 256, 272, 290, 295, 319, 428
White Sands Test Facility (WSTF), N. Mex., 80
White, Gen. Thomas D., Space Trophy, 267–268, 313
Whiteside, John J., 325
Whitney Museum, New York, 385
Wide-band video tape recorders (WBVTR), 235, 269–270
Wilford, John N., 110–111, 117, 127
Williams, William D., Jr., 220
Williamsburg, Va., 188
Wilson, George C., 9–10
Wilson, Robert R., 314
Wilson, T. A., 331
Wilson, Dr. William J., 365
Wind, 228, 250, 373
Wind tunnel, 67, 73, 110, 191–192, 283, 353
Wing, aircraft, 330, 378
  antisymmetrical, 191–192, 449
  augmentor, 169, 172, 324
  supercritical, 50–51, 73, 107, 214, 312, 449
    flight tests, 72–73, 94, 104, 109, 113, 123, 155, 163, 184, 186, 193, 274–275, 396, 449
  swing, 319, 329, 340, 355
Winifred (asteroid), 434
Winters, Jonathan, 134–135
Wisconsin, Univ. of, 61, 131–132, 152, 254, 288, 422
Wollenhaupt, Dr. Wilbur R., 433
Women, 15, 276

Wood, Dr. John A., 14
Woodall, Jerry M., 250
Woodbury, N.Y., 220
Woods, Robert J., 350
Woodward, William H., 85
Woomera, Australia, 120, 438
Worden, L/C Alfred M. (USAF), 1, 13–14, 20, 28, 35, 61, 194, 232, 256, 303, 314
  Apollo 15 mission, 41, 282–283
  awards and honors, 114, 210, 267–268
Working Group on Rendezvous and Docking (see also Joint Working Group on Compatibility of Docking Systems and Tunnels), 119, 127, 158, 197
Working Group III, 409
World Airways, Inc., 425
World Bank, 218
World Heritage Trust, 285
World Meteorological Organization (U.N.), 84, 218–219
  Commission for Instruments and Methods of Observation, 105, 453
World War I, 26
World War II, 441, 442
World Weather Program, 172–173
World Weather Program for Fiscal Year 1973, 172–173, 262
World Weather Watch, 262–263
Worlds Fair (1964–1965), 388
The Worlds of von Braun (TV documentary film), 9
Wrench, ultrasonic, 45
Wright Brothers Day, 422, 434
Wright Brothers Memorial Trophy, 326, 430
Wright, David B., 184–185
Wright Memorial Dinner, 430
Wright, Orville, 26, 422
WSMR. See White Sands Missile Range.
WSTF. See White Sands Test Facility.
WTR. See Western Test Range.
Wyld, James H., Propulsion Award, 399
Wyoming, Univ. of, 43

X

X–1 (rocket research aircraft), 233, 311, 350
X–5, 311
X–15, 79, 311
X–24A (lifting body), 312
X–24B, 347
XB–70 (supersonic aircraft), 311
Xerox Data Systems, 372
XLR–129 (rocket engine), 332
Xonics, Inc., 92
X-ray
  astronomy, 61, 90, 107, 127, 139, 189, 242, 243, 287, 288, 342, 360
  Crab Nebula study, 159
  lunar surface analysis, 14, 183–184, 411
  mapping, 107
  soft, 288, 422
  source, 56–57, 332, 386, 394

stellar, 332, 394, 438
telescope, 108

Y

Yak-40 (U.S.S.R. jet transport), 365, 425
Yale Univ., 287
Yankee Clipper (*Apollo 12* CM). See Command module.
Yeager, Capt. Charles E. (USAF), 350
"Year of the Skylab," 408
Yeliseyev, Dr. Aleksey S., 111
Yerkes Observatory, 302
Yevtushenko, Yevgeny, 142, 150, 387
YF-12 (research aircraft), 312, 327
YF-16 (swept-wing fighter), 378
YF-17, 378
YJ101 (turbojet engine), 378
Young, Capt. John W. (USN), 119–120, 162–163, 165, 345
  Apollo 16 mission
    commemorative medal, 264
    Congress, report to, 184
    debriefing, 180
    EVA, 148–149, 169
    launch, 142
    medical aspects, 120, 135, 150, 166, 420
    Nixon, President Richard M., 162–163, 228
    preparations for, 123
    press conference, 109, 172, 228–229
    tribute to, 163–164
  Apollo 17 mission, 194
  Apollo-Soyuz Test Project, 228–229
  awards and honors, 187
Youth Science Congress, 154–155
Yugoslav Academy of Sciences, 35
Yugoslavia, 28, 35, 41

Z

Zagreb, Yugoslavia, 35
Zeigler, Henri, 271
Zeta (star), 296
Zinn, Dr. Walter E., 341
Zirconium hydride reactor, 80
Zisk, Stanley H., 403
Zond (U.S.S.R. spacecraft), 108
Zraket, Charles, 396
Zumwalt, Adm. Elmo R., Jr. (USN), 151–152
Zvezdny Gorodok (Star City), U.S.S.R., 110–111, 117, 354

NASA HISTORICAL PUBLICATIONS

Histories

- Robert L. Rosholt, *An Administrative History of NASA, 1958–1963*, NASA SP-4101, 1966, GPO, $4.00.*
- Loyd S. Swenson, James M. Grimwood, and Charles C. Alexander, *This New Ocean: A History of Project Mercury*, NASA SP-4201, 1966, GPO, $5.50.
- Constance McL. Green and Milton Lomask, *Vanguard—A History*, NASA SP-4202, 1970; also Washington: Smithsonian Institution Press, 1971, $12.50.
- Alfred Rosenthal, *Venture into Space: Early Years of Goddard Space Flight Center*, NASA SP-4301, 1968, GPO, $2.50.
- Edwin P. Hartman, *Adventures in Research: A History of the Ames Research Center, 1940–1965*, NASA SP-4302, 1970, GPO, $4.75.

Historical Studies

- Eugene M. Emme (ed.), *History of Rocket Technology*, Detroit: Wayne State University, 1964, out of print.
- Mae Mills Link, *Space Medicine in Project Mercury*, NASA SP-4003, 1965, NTIS, $6.00.**
- *Historical Sketch of NASA*, NASA EP-29, 1965 and 1966, NTIS, $6.00.
- Katherine M. Dickson (Library of Congress), *History of Aeronautics and Astronautics: A Preliminary Bibliography*, NASA HHR-29, NTIS, $6.00.
- Eugene M. Emme (ed.), *Statements by the Presidents of the United States on International Cooperation in Space*, Senate Committee on Aeronautical and Space Sciences, Sen. Doc. 92-40, 1971, GPO, $0.55.
- William R. Corliss, *NASA Sounding Rockets, 1958–1968: A Historical Summary*, NASA SP-4401, 1971, GPO, $1.75.
- Helen T. Wells with Susan Whitely, *Origins of NASA Names*, NASA SP-4402 (1974).
- Jane Van Nimmen and Leonard C. Bruno with Robert L. Rosholt, *NASA Historical Data Book, 1958–1968*, NASA SP-4012 (1974).

CHRONOLOGIES

- *Aeronautics and Astronautics: An American Chronology of Science and Technology in the Exploration of Space, 1915–1960*, compiled by Eugene M. Emme, Washington: NASA, 1961, NTIS, $6.00.
- *Aeronautical and Astronautical Events of 1961*, published by the House Committee on Science and Astronautics, 1962, NTIS, $6.00.
- *Astronautical and Aeronautical Events of 1962*, published by the House Committee on Science and Astronautics, 1963, NTIS, $6.00.
- *Astronautics and Aeronautics: A Chronology on Science, Technology, and Policy, 1963*, NASA SP–4004, 1964, NTIS, $6.00.
- ————, *1964*, NASA SP–4005, 1965, NTIS, $6.00.
- ————, *1965*, NASA SP–4006, 1966, NTIS, $6.00.
- ————, *1966*, NASA SP–4007, 1967, NTIS, $6.00.
- ————, *1967*, NASA SP–4008, 1968, GPO, $2.25.
- ————, *1968*, NASA SP–4010, 1969, GPO, $2.00.
- ————, *1969*, NASA SP–4014, 1970, GPO, $2.25.
- ————, *1970*, NASA SP–4015, 1972, GPO, $3.10.
- ————, *1971*, NASA SP–4016, 1972, GPO, $3.80.
- James M. Grimwood, *Project Mercury: A Chronology*, NASA SP–4001, 1963, NTIS, $6.00.
- James M. Grimwood and Barton C. Hacker, with Peter J. Vorzimmer, *Project Gemini Technology and Operations: A Chronology*, NASA SP–4002, 1969, GPO, $2.75.
- Ivan D. Ertel and Mary Louise Morse, *The Apollo Spacecraft: A Chronology*, Vol. I, *Through November 7, 1962*, NASA SP–4009, 1969, GPO, $2.50.
- Mary Louise Morse and Jean Kernahan Bays, *The Apollo Spacecraft: A Chronology*, Vol. II, *November 8, 1962–September 30, 1964*, NASA SP–4009, 1972, GPO, $3.20.
- R. Cargill Hall, *Project Ranger: A Chronology*, JPL/HR–2, 1971, NTIS, $6.00.

---

* GPO: Titles may be ordered from the Superintendent of Documents, Government Printing Office, Washington, D.C. 20402.

** NTIS: Titles may be ordered from National Technical Information Service, Springfield, Va. 22151.

www.ingramcontent.com/pod-product-compliance
Lightning Source LLC
Chambersburg PA
CBHW081713170526
45167CB00009B/3565